THE ECONOMICS OF PRODUCING DEFENSE

Illustrated By The Israeli Case

THE ECONOMICS OF PRODUCING DEFENSE

Illustrated By The Israeli Case

Yaacov Lifshitz
Israel

(in cooperation with the Jerusalem Institute for Israel Studies)

SPRINGER SCIENCE+BUSINESS MEDIA, LLC

Lifshitz, Yaacov.
The economics of producing defense; illustrated by the Israeli case
p.cm.
Includes bibliographical references and index.
ISBN 978-1-4613-5062-0 ISBN 978-1-4615-0409-2 (eBook)
DOI 10.1007/978-1-4615-0409-2

Originally published by Kluwer Academic Publishers in 2003
Softcover reprint of the hardcover 1st edition 2003

"The study was supported by grants given by the Charles H. Revson and the Gass foundations."

Printed on acid-free paper.

CONTENTS

LIST OF TABLES

PREFACE

Defense economics is about the large variety and complex interrelations between national security and economics. Indeed, the reciprocal influences of military power and economic wealth concerned modern economic thought since its very inception, but nonetheless the emergence of defense economics as a recognized branch of study and as a special sub-discipline of the general theory of economics was relatively late. The breakthrough came in the early 1960s, under the influence of the Cold War, and research on the economics of national security issues in fact expanded significantly only during the 1980s. At the turn of the twenty-first century, defense economics is still a relatively young discipline, yet it offers a wide selection of theoretical analyses and empirical findings that may be helpful in contending effectively with problems arising when nations endeavor to enhance their national security.

This book was originally written in Hebrew for an Israeli audience. Security matters were a central theme in the life of the state of Israel from her first days. It was founded in the crucible of the War of Independence, and throughout the years had to cope, and still does, with severe national security problems. From an economic standpoint, it was manifested in constantly allocating an exceptionally high proportion of economic resources to defense, with far-reaching consequences for the economy. At the same time, evolving economic conditions affected the ability to mobilize resources for the military and the affordability of defense efforts. In writing on defense economics for the Israeli audience, the prime purpose was to introduce them to lessons and findings of the discipline as developed elsewhere. The hope was to convince policy-makers, other practitioners and as wide circles as possible of the public at large of the contribution defense economics could make to more effective management of practical policy and implementation problems in this important area.

The purpose in introducing the book to a broader, English reading audience is very much the same, and this aim influenced the book's structure and style of presentation. The discussion of each subject begins with some general background, presenting definitions, interpreting basic concepts and outlining relevant historical developments briefly. A theoretical analysis follows, illustrated and verified by empirical findings from research works carried out in different countries. On this basis, the same issues are examined for Israel, and the Hebrew version also included detailed, practical recommendations regarding current related questions. Since such policy prescriptions are specific in nature and would presumably be of limited interest and value for the present more general audience, they were omitted from the English version as a rule. However, the remaining discussion of the Israeli case can benefit readers outside Israel as well; it elucidates many issues of defense economics, and besides, as a small economy compelled to allocate a considerable share of its resources for defense, Israel's case carries a lesson in dealing efficiently with defense economic problems. The style of presentation is intended to make a new field more accessible, and thus avoids technical terms and complicated modeling as far as possible.

The scope of defense economics is too broad to be dealt with thoroughly, in full, in one book. The approach adopted here follows the distinction between demand-related and supply-side issues, with most chapters devoted to the latter, i.e. to questions about producing and supplying national security. This choice too derives from the main goal of the book; focusing on the economics of producing national security makes it possible to indicate more concrete conclusions, and thus to demonstrate convincingly the practical value of defense economics. Admittedly, however, the supply side of national security could not be covered fully either. For example, there is just a brief mention of the varied economic methods and quantitative analytical tools developed to improve defense management (budgeting techniques, cost-benefit and cost-effectiveness analyses of programs, costing models of complex weapons systems, incentive contracting methods, etc.).

The first chapter presents the main topics of defense economics, and the unique features that distinguish it from other branches of economics. It also elaborates on the factors that retarded its development in the earlier stages, and on those that later accelerated its formulation as a distinct sub-discipline. Over the years, focal points of interest shifted with changes in the nature of armed conflicts, in the international system and in the technological and economic environment, and the chapter traces those shifts of interest since World War I. The second chapter completes the presentation of defense economics' "visiting card" by providing a historical perspective of how national security became an economic issue. An enlightening economic-historical analysis of the subject was suggested by Adam Smith in *The Wealth of Nations*, published in 1776, and the economic aspects of defense became increasingly significant since then, concurrently with the changes in the nature of war, and especially with the totality of war. The chapter brings the main ideas of Adam Smith, and then discusses the consequences of the major changes that took place up to the mid twentieth century in mobilizing human and physical resources for defense. It proceeds to the views held by central schools of economic thought since the mercantilist doctrine on the interrelations between military power and economic wealth, and concludes with the macro-historical debate as to whether wars encouraged processes that advanced human social welfare or hindered them, and from the opposite direction – whether economic processes affected the role of wars in international relations.

While the first two chapters provide a general introduction to defense economics, the third and the fourth introduce defense economics in Israel. The third chapter examines economic fundamentals in Israel's national security doctrine, or more precisely, points to defense perceptions influenced by economic considerations, and suggests interpretations in line with economic ways of thinking and in economic terms for some of the principal elements in Israel's national security doctrine and in the defense policy derived from it. The fourth chapter surveys central issues of defense economics on Israel's agenda in the past, as well as current issues related to the possibility of a "peace dividend", if and when peace prevails in the Middle East.

Producing and supplying national security depend on the scope of defense expenditures, and on the production factors – military manpower and defense products – acquired with them. Hence the book proceeds with "pairs" of chapters as follows: the fifth chapter deals with the nature of defense expenditures and with measuring and comparing them, while the sixth illustrates the general discussion by analyzing the development of defense expenditures of Israel; the seventh chapter is devoted to economic issues of military manpower recruitment, to deciding its composition and to problems arising in employing it, once again illustrated, in the last section, through the economic aspects of recruiting and employing military manpower in Israel; the eighth chapter considers the defense products and the defense industries that develop and produce them, while the ninth focuses on Israel's defense-industrial base. Finally, since no country maintains autarky in supplying her defense requirements, and all nations, some to a greater and some to a lesser degree, import weapons systems and other defense equipment, the book concludes by analyzing the world arms market, the international arms trade and the consequences of foreign military aid in chapter ten, and discusses Israel's defense exports and imports and the special implications of the American military aid to Israel in chapter eleven.

The detailed discussion points at two broad and complementary conclusions: -

First, over the years, defense establishments became less autarkic in producing national security, i.e. less of the goods and services needed to meet national security requirements are produced within the armed forces. This trend is clearly consistent with developments towards a higher degree of specialization and division of labor typical of modern economies in general. Supplying national security lagged conspicuously behind the general trend, but it is now rapidly catching up. In the post-Cold War era, shrinking defense budgets apparently made it necessary to search for ways and means to produce an acceptable level of security, while economizing considerably on costs and operating more efficiently. As a matter of fact, the tendency to buy outside was greatly reinforced, i.e. buying was preferred to making. In some ways, the growing tendency to abolish compulsory military service and rely exclusively on all-volunteer forces recruited through market mechanisms, could also be regarded as part of the same trend, like the expanding employment of "substitutes" for regular military manpower, and especially the transfer of roles once considered purely military to civilians, either within the armed forces themselves or through outsourcing.

Secondly, The design, development and production of weapons systems and other defense products, clearly essential to and integral part in supplying national security, were carried out traditionally in a unique manner, with the defense-industrial base of countries separate and conducting business differently in many respects from other sectors in the economy. Things are changing in this area too. For a host of reasons, arms production turns out to be more and more a branch of general industrial production, and the defense industries become less unique and exceptional. Concurrently, economic globalization affects the defense production as well, and the trend towards internationalization spreading rapidly over defense sectors worldwide is a significant departure from the national nature that characterized them during the Cold War era. Instead of large national companies owned and controlled by nation states, or critically dependent on domestic demand from their governments, defense companies and other emerging organizational formations are less linked, having less affinity to any particular country. As a consequence, the nature of international arms trade changes, and it too loses something of its uniqueness in the interrelations among countries, becoming increasingly similar to other international economic activities and businesses.

The two conclusions are complementary in that they manifest the growing weight attached to economic considerations and motives in producing and supplying national security, as well as to their economic consequences. Hence also the deep conviction that defense economics may contribute to national security policy-making, and to the more effective management of the efforts to supply national security.

Defense economists believe openly or implicitly, deep in their hearts, that in-depth understanding of the interrelations between security and economics, and especially acknowledgement of the price nations pay for enhancing security, will stimulate efforts towards achieving a new world order, with less frictions and threats in the relations between countries. However as realists they understand too that power and military capabilities might be a necessary condition for achieving such a new order, and later an effective guarantee for preserving it, and therefore seek to demonstrate ways to obtain more security with fewer economic resources. Indeed, this, in brief, is what defense economics is all about.

This in the main is the state of affairs in the Israeli case as well. In the hope and to whatever extent that there will be real progress towards peace in the Middle East, Israel's military power will still continue to guarantee her existence, and indeed that of the peace itself, for many years to come. Therefore the need to allocate large resources for defense, and the need to cope with problems of defense economics is not likely to disappear soon.

In writing this book, I drew extensively upon the experience accumulated through my years in government service, as the chief economic adviser in the Ministry of Defense, the Director General of the Ministry of Finance and Chairman of the Board of Israel Military Industries (IMI) Ltd.

No less a contribution to the writing were the opportunities extended to me to teach defense economics in academic institutions in Israel, and to lecture on the subject to various audiences. All along the way, many individuals contributed a great deal to this book without knowing they were doing so, and thus assisted me. I benefited much from comments generously extended on parts of earlier drafts by Prof. Pinhas Zusman, Prof. Yehezkel Dror, Dr. Ze'ev Rotem, Prof. Asher Tishler, Mr. Nehamia Hasid and two anonymous referees selected by the Jerusalem Institute for Israeli Studies. I am grateful to them for helping to determine the structure of the book, to clarify the presentation and to avoid mistakes. The Jerusalem Institute supported the work and provided an essential accommodation for the writing, for which I am thankful to its management. The English version could not be brought to its successful completion without the dedication and professionalism of Mrs. Betty Sigler Rozen in translating the Hebrew manuscript. Valuable suggestions as to the English version also came from occasional exchanges with Dr. Uri Gluskinos.

The book is dedicated to Major General (Ret.) Zvi Zur, and to the memory of Prof. Pinhas Zusman. Their life stories, each in its own way, represents an outstanding integration of defense and economics. Zvi Zur was the sixth Chief of Staff of the Israeli Defense Forces, and then Deputy Minister of Defense. Later on, he underwent "conversion", and filled central roles in the civilian economy, heading some of the largest business companies in Israel. Pinhas Zusman was a senior commander in Israel's War of Independence who gave up a military career for assignments in agriculture, particularly in instructing immigrants who settled in distant parts of the country. Later on, he became a professor of agricultural economics in the Hebrew University, yet after the 1967 War returned to the Ministry of Defense, first as an economic adviser, and later as the Director General. I was privileged to work with both of them for many years. They were my mentors, I learned much from them, and especially was inspired by their unique approach to issues of defense economics.

Y. L.
March 2003

1
ECONOMIC ASPECTS OF NATIONAL SECURITY

National security is a broad concept, relating to the large variety of issues encountered by nations faced with threats to the physical safety of their inhabitants, to their territorial integrity or to other vital interests. Governments usually accord high priority to national security, and while contending with actual or potential threats are willing to dedicate large human and material resources to defense. Allocating resources for defense has, however, economic implications, and national security thereby becomes an economic issue. Defense economics addresses the many questions arising from the interaction between national security and economics.

1. THE SCOPE OF DEFENSE ECONOMICS

The science of economics focuses on the efficient allocation of scarce resources among competing ends. Defense economics, as a sub-discipline of economics, applies this notion of resource constraints to the domain of national security, and for that purpose adjusts concepts, theoretical methods and empirical tools commonly used in general economic analyses to the special structural and institutional features of the defense sector.

Defense economics is not based on any particular ideological premises: it does not favor militarism, nor does it preclude, on a normative basis, buildup of military power or high defense expenditures. Defense economics adopts a positive rather than a normative approach, seeking to explain the observable behavior of nations and its consequences. In some cases, the economic approach provided a clearer and a more complete formulation of the defense problem, while in others it was useful in designing solutions to problems already defined. In fact, this is another remarkable virtue of defense economics; it is a policy-oriented discipline that weighs advantages and disadvantages of potential means, making recommendations on preferable courses of action. On many issues, admittedly, economic analysis pointed out unexpected results.

During the second half of the twentieth century, the scope of defense economics broadened, and economists became interested in numerous defense-related issues (see Table 1.1). The large number of issues can be subdivided in various ways. A pioneering study in the field distinguished between the higher level of defense policy-making and the lower level of policy implementation (Hitch & McKean, 1971). The higher level relates to the allocation of resources between defense and civilian ends (the macroeconomic aspects), while at the lower level the focus is on efficient utilization of resources within the defense sector (the microeconomic aspects). Other studies added a third level: the international aspects of defense economics (Kapstein, 1992).

Table 1.1 Issues of Defense Economics (Partial List)

Deterrence, war avoidance, war initiation and termination
Strategic interactions, arms races, military alliances, arms control
Measuring defense expenditure, international comparisons of defense spending
Providing resources for defense in market and command economies
Determining factors of the demand for defense expenditure
Macroeconomic effects of defense spending, the effect of defense expenditure on economic growth, economic benefits and burden of defense
International arms trade
Economic warfare, foreign military aid
Optimization and efficiency in force level and composition, measurement of military capital, military capital-labor ratios
Budgeting, organizational structures and internal markets for efficient allocation of resources within the armed forces
Conscription versus volunteer military manpower
Acquiring arms and military equipment, procurement arrangements, management of research and development
Defense industries, defense-industrial policy, effects of the military-industrial complex
Economic mobilization for war, war recovery, consequences of disarmament and conversions
Extra-governmental threats: terrorism, insurrections, drugs, refugees, ethnic and religious fanaticism

Source: Hartley & Hooper, 1990; McGuire, 1995: 15

However, there were other points of departure. For certain purposes, it has been suggested to draw a line between demand-related and supply-side defense issues. In another case, a threefold classification was preferred. The first category includes standard topics of economic theory that require special treatment because of the structural and institutional peculiarity of the defense sector. The second category deals with problems of international relations possibly better understood by applying constructs from economic theory, though they are not purely economic in nature. The third category extends the scope even further, to economic aspects of a broader definition of national security, not limited to military affairs only. With that, subjects such as environmental protection, population movements and foreign labor minorities, preservation of depleted natural resources and trade barriers are brought into the domain of defense economics (Reppy, 1991: 269-270). The first two categories should certainly be included within defense economics. In the case of the third category, however, it has been argued that broadening the scope so much may obliterate the special identity of defense economics (Intriligator, 1991; Sandler & Hartley, 1995: 6).

Like any other field of positive economics, defense economics requires a basic behavioral theory as a core from which all other issues are derived, and this comes from explanations suggested by economists for the strategic behavior of nations in the international arena. Hence it might be methodologically important to classify subjects according to the relative dominance of defense versus economic elements in them. Subjects in which defense elements dominate constitute the core, while others in which economic elements dominate – though no less important – are derived from the former and examine their implications and significance (Intriligator, 1991: 273; McGuire, 1995: 24).

The group of subjects in which defense elements dominate is large and diversified. Economists apply the notion of optimal choice under constraints, models of duopolistic and oligopolistic markets, public goods theory, constructs of game theory, rent-seeking and other theoretical economic concepts to explain how nations determine strategic goals, to identify possible strategic responses, to define effective deterrence, to analyze arms races and military alliances, to suggest efficient arrangements for arms control, to better understand acts of terrorism and more.

No less diversified is the group of subjects where economic elements are dominant. One subgroup focuses on economic fundamentals of national security, stressing the two-way interrelation between economic strength and national security. The economic strength is the resource base that determines capability of countries to build and maintain military power. At the same time economic assets are often targets for the aggression of enemies, and should be protected. Specific questions include, for example, assessment of the national capability to sustain the economic burden of defense or comparisons of the relative efficiency of different economic regimes in providing resources for defense. A broader range of issues might include economic causes for conflicts among nations, and the promotion of security and foreign policy goals by economic means (economic statecraft).

A second subgroup consists of national security issues that can be formulated as problems of choice under constraints, and then analyzed by the usual economic methods of efficient allocation of resources. At one end stands the question of acquiring the highest level of security with available national resources, and answering it requires analysis of the macroeconomic consequences of defense expenditure in general, and of certain defense spending categories in particular. This involves issues such as the impact of defense expenditure on economic growth, the employment opportunities created by defense demand for skilled labor, positive spillover and negative crowding-out effects of defense research and development, economic consequences of disarmament processes and so forth. At the other end stand the requirement for defense establishments to utilize scarce resources efficiently, and the resulting need for microeconomic formats and mechanisms to assist toward that end. In this context, defense economics treats the government as a producer of outputs that uses inputs acquired at variable prices within a limited budget. This approach makes it possible to define conceptually the quantitative and utility dimensions of defense outputs, thereby allowing for cost-benefit and cost-effectiveness analyses to support choices between alternative defense programs. In addition, it is possible to calculate rates of substitution between various force components, thus selecting the most efficient mix for performing specific missions under budgetary constraints. Finally, giving monetary values to defense outputs and inputs is necessary for calculating their present values, making it possible to arrive at optimal multi-year decisions.

The large variety of topics involved in defense economics is both an advantage and a disadvantage. It leads to boundaries not strictly defined, and thus other branches of economics – e.g. public finance, international economics, industrial economics, game theory and regional economics – also deal occasionally with national security issues. At the same time, political science, the theory of international relations and strategic studies often penetrate the realm of defense economics, while on the other hand, defense economics mingles in all these fields, gaining from and contributing to them. In retrospect, the absence of clear borders may have slowed down the development of defense economics as a recognized sub-discipline of economics (see section 4 below). However, flexible borderlines have also been an advantage: because national security issues are inherently complex, applying a narrow disciplinary approach in this area is doomed to fail and to produce limited and unsatisfactory results.

2. MAIN THEMES SINCE WORLD WAR I

The second chapter provides a historical perspective of defense economic issues prior to World War I. Since the Great War, the turbulent course of twentieth century events brought in many new attention-demanding questions, and the issues in the forefront of defense economics changed rapidly. Following the War, economists concentrated mainly on questions of managing war economies, and particularly on making resources available for the military efforts during wars. One subject intensively discussed in those days compared taxes versus loans as means to finance wars (Pigou, 1921). Another controversial issue was the possible negative influences of the reparations paid by Germany on the economic stability of recipient countries, and on the international system as a whole (Keynes, 1919). Later, during the 1920s, most countries reduced their armed forces and defense budgets, and consequently general interest in defense economic issues waned too. But the interlude was short. The accelerated military buildup toward World War II soon renewed awareness of the economic aspects of military preparations before wars, and drew attention to the "economic war potential" of countries, i.e. their capability to mobilize maximum economic resources for military needs. In this context, discussions concentrated on measurements of economic war potential, on determinants of its scope and quality, and on structural characteristics that affect the ability to convert production from civilian to military uses within a relatively short time. In addition, economists evaluated alternative policy instruments for mobilizing the economic war potential, and commented on the role of governments in preparing the economy for war and in directing the economic system during war. In the mid 1930s, economists and others also obsessively debated the hypothesis that greedy arms producers and merchants actively encouraged war initiatives (Brodie, 1980: 249-253).

World War II introduced many practical economic problems. Economists had to understand the resource base that supports the national security of their countries, and assist in solving production bottlenecks and logistic complexities. Moreover, they had to study the structure of their enemies' economies too, since they were asked to recommend targets for strategic bombing and means of economic warfare that could effectively paralyze economic activities there. Meanwhile, their attention to the impact of war on economic stability rose. During the war, the main issue was how to manage a mobilized economy with a tolerable level of inflation, and later – how to avoid a recession while dismantling the huge war machine and sharply reducing government demand.

The Cold War years can be divided into two sub-periods (McGuire, 1995: 23-29). The first, approximately until 1970, was marked by adaptation to the idea that allocating large resources for defense is a permanent phenomenon. Modern military technologies, and notably nuclear weapons, have brought about major changes in the nature of wars, and the prevailing view suggested that the prospects for long and total wars, in the World War format, have considerably diminished. Short wars could be won only by using military power accumulated and trained in advance, and hence, from a strategic standpoint, economic capabilities count only as long as they are diverted for defense before wars break out, and are less important in the sense of economic war potential that was relevant for long wars (Hitch & McKean, 1971; Knorr, 1957, 1977). Indeed, as for the superpowers, their security increasingly relied at that time on deterrence through a balance of power continuously maintained, and consequently on permanently allocating large resources for defense. Yet the new approach extended to countries engaged in local conventional wars too. It was argued that in the new era conventional wars would be limited in scope and duration, or otherwise they could develop into an overall nuclear conflict. Thus not only superpowers, but also other countries that could be involved in a conventional conflict should prepare for short wars and build their military power in advance, allocating resources for defense on a continuous basis.

Defense economics had to deal then with the emerging consequences of permanent allocation of large economic resources for defense, and before long new problems appeared on its agenda. First, permanently living with high defense burden sharpened the dilemma of "guns versus butter", and expanded the theoretical and empirical work on measuring defense expenditure and opportunity costs (Benoit & Lubell, 1966; Benoit, 1968). It also raised interest in substitution patterns between defense spending and civilian uses in the short and longer terms (Russet, 1970). Secondly, attention was paid to the ability of socialist economies to sustain large allocations for defense, and to comparisons between capitalist and socialist economies in this respect. Thirdly, by contrast with wartime, peacetime allocation of resources for defense does not rely on the maximum mobilization principle, but rather reflects public choice between competing uses under resource constraints. Economists tried to find systematic and common patterns in the way defense expenditures are determined, and suggested for that purpose neo-classical equilibrium models, political economics theories including neo-Marxist interpretations and the military-industrial complex hypothesis, and behavioral explanations of bureaucratic organizations. Fourthly, the new circumstances emphasized the need to achieve any required level of national security with the lowest costs possible. This brought about a great deal of work on microeconomic aspects of defense economics, and led to innovative methods for planning, budgeting and cost analyses, and eventually to an overall new doctrine of "defense management" (Hitch & McKean, 1971; Enke, 1967). Simultaneously, research on the defense market and procurement also expanded, and novel contracting methods were devised for the management of research, development and production programs of weapon systems (Peck & Scherer, 1962; Scherer, 1964). In the early 1960s, in the United States, economists held central positions in the Department of Defense. They had a rare opportunity to apply defense economics to day-to-day decision-making, and to make use of the newly developed methods and tools (see section 4 below).

The year 1960 was an important milestone in the development of defense economics. That year, three books were published, each a pioneering study in its field and a cornerstone for further research. First, there was the previously mentioned book by Hitch and McKean, which showed how economic concepts of efficient allocation could be applied to defense issues. The second book was written by Louis Richardson (Richardson, 1960), and presented an arms race model incorporating economic constraints. In the third book, Thomas Schelling (Schelling, 1960), elaborated on deterrence and on the characteristics of credible threats and other strategic responses, using ideas and concepts such as expectations, incentives and equilibrium situations common in game theory. Soon afterwards Kenneth Boulding's book (Boulding, 1962) was published, describing conflicts within the framework of analytical models, and pointing out to the role of economic factors – ownership of properties, dominance in markets, availability of resources, etc. – in conflicts. In the mid 1960s, Olson and Zeckhauser contributed an innovative economic analysis of military alliances (Olson & Zeckhauser, 1966). By applying the theory of public goods, they proved that burden sharing among nations within a military alliance tends to be disproportionate to their respective economic capabilities, and that the level of security provided by a military alliance would be sub-optimal. These important works placed the strategic behavior of nations at the focus of defense economics, and made defense economics an important source of analyses and explanations for strategic interactions between nations.

The high economic price of permanently maintaining a deterrent balance of power stimulated a search for international arrangements of arms control and disarmament; another topic defense economics started to deal with in the 1960s. Economists made international comparisons of defense expenditure and defense burden in order to set up criteria for reducing military buildup. They examined behavioral patterns of nations within the framework of such potential international arrangements, using models of imperfect market competition, the prisoner's dilemma model and others taken from game theory. In addition, they analyzed the economic consequences of disarmament for industries and regions, and discussed the conversion of resources from defense to civilian uses.

During the first years of the second sub-period of the Cold War, defense economics was largely influenced by the public trauma that the Vietnam War caused in the United States. There was a general atmosphere of disappointment with the theoretical achievements and practical lessons of defense economics, and interest in the field almost faded away. Most senior economists who served in the American Department of Defense left their offices, and in academic institutions the collapse of defense economics was nearly complete (Leitzel, 1993: ix). One notable sign of the times was the declining role of microeconomic applications and of the advanced analytical devices offered for the management of defense (McGuire, 1987a: 761). However, in the course of time, new issues emerged, and they soon attracted economists' attention. In the United States, as a by-product of the Vietnam War, there was a public debate over the abolishment of conscription and the transition to an all-volunteer force. As militaries tend to be labor intensive, the costs of manpower mobilization for the military had occupied economists' interest in previous periods too. However, in the public debate of the early 1970s, defense economists went further than simply making comparisons of budgetary versus opportunity costs of conscription. As social justice considerations were important, they argued that a draft is a kind of an implicit head tax, and emphasized the income distribution consequences of military service. They also demonstrated, using econometric estimations of labor supply curves, that volunteer armed forces are economically feasible and suggested preferable compensation schemes.

Another focal point of interest was the issue of strategic overreach (Olson, 1989; Kennedy, 1992). United States became increasingly concerned by its shrinking economic competitiveness, especially in comparison with Japan, and suspicions arose that this might be an outcome of allocating large resources for defense over long periods. Economists started to examine the impact of defense expenditure on capital investments, technological progress, the supply of skilled labor for the civilian economy and other relevant processes. At the same time, interest in the relationships between defense and economic growth in the Third World also increased. Defense economics devoted much attention to these interrelationships due to the Third World countries' rising weight in the international arena. Other no less important reasons were proliferation of regional arms races, increasingly frequent local wars and the rapid growth of military sectors and defense industries in developing countries. In the early 1970s, Emil Benoit presented his econometric findings (Benoit, 1973). They suggested that economic growth is positively associated with defense expenditures, and with that started extensive theoretical and empirical debates that still continue.

A third central theme that increasingly gained the attention of defense economists in those years was the world arms trade and its consequences for national economies and for the international economic system at large. Modern military technology dramatically increased the costs of weapon systems, and no single country could maintain absolute autarky in providing for its defense needs. Indeed, all countries had to import arms and other military equipment to a greater or lesser extent, and arms transfers became an important element of international trade. At the same time, various forms of foreign military aid developed. They accorded the donor countries great influence in the international system, and became a significant economic factor for most recipient countries. During the 1970s, largely due to the petrodollars that flowed so generously into oil-producing countries, world arms trade expanded, and defense economists soon started to investigate the field intensively. Pioneering and comprehensive data collection was undertaken by the Stockholm International Peace Research Institute (SIPRI, 1971), and first attempts to analyze international arms transfers systematically were made by Robert Harkavy (Harkavy, 1975). On the export side, economists examined the possibility of economies of scale in production, and of lower unit costs of equipment for domestic customers, due to greater overseas sales. They also tried to estimate the macroeconomic and industrial effects of defense exports. On the import side, discussions concentrated on the relationship between defense imports and the large foreign debts accumulated by developing countries, but also on the apparent contributions of offset arrangements to the growth of indigenous defense industries and to the international flow of advanced technologies.

A fourth central topic relates to defense research, development and production, and to the defense-industrial base. This subject too has to do with the rapid technological advance in the military sphere, and its unavoidable effect on the costs of weapon systems. However, there were also other concerns. In the early 1980s in the United States, economists warned that the defense-industrial base was losing its ability to support the technological superiority on which American national security policy had been based since World War II (Gansler, 1980). On the other side of the Atlantic, economists criticized the redundancy and waste stemming from the national fragmentation of the West European defense industry and from the protectionism adopted by European governments in their defense markets.

The measures used to protect local defense industries sharply contradicted the attitude and progress toward a Single European Market in other sectors of the economy. Defense economists borrowed concepts and analytical methods from industrial economics, and analyzed the structural features, behavioral patterns and performance of defense industries. In the later 1980s, and even more so in the 1990s, the traditional aspects were supplemented by a rising interest in new developments of cross border industrial cooperation and the globalization of arms production.

In the early 1990s, following the end of the Cold War and the hopes for a new world order, the focus of defense economics shifted again. Disarmament, reduction of defense budgets and the search for ways to realize the "peace dividend" outlined two main groups of topics. First there were issues relating to conversion of defense plants from military to civilian uses and the resulting reduced demand for skilled labor, emphasizing questions about the role of governments in this transition. Secondly, there was the challenge of providing national security at an acceptable level despite decreasing budgets and increasing unit costs of weapons systems. This meant reexamining traditional approaches toward force levels and composition – i.e. tradeoffs between military manpower and equipment, regular and reserve forces, etc. – as well as toward defense procurement policies, research and development programs and the desirable size of the defense-industrial base. Given reduced demand, defense companies tended to merge. They strove for economies of scale and other synergetic benefits, thereby creating huge entities that inevitably led to higher concentration and diminishing competition in the defense sector. As noted above, for similar reasons companies from different countries also sought to cooperate, and the number of joint ventures across borders increased substantially. Indeed, the environment of defense businesses changed, and the new circumstances broadened the horizons of defense economics.

The conflict between the major superpowers ended, but regional security problems did not disappear. Defense economics, which provided reasonable explanations and models for the arms race between superpowers, now confronts arms races of a different nature. It must further develop its theory to accommodate regional arms races fed by an open world arms market, and which influence one another. Moreover, unlike the superpower arms race, regional races do not usually reach balancing deterrence; they lack the stability of the former and can easily deteriorate into an active war. Defense economics should also address the threats of proliferating weapons of mass destruction. For that purpose, it should expand the economic models developed for analyzing disarmament and arms control treaties between superpowers. The expanded versions should explicitly take into account possible reactions of the new "players" to arrangements unilaterally designed to limit their military actions. At present, the threat of weapons of mass destruction is amorphous, and far less direct and structured than it was in the past, within a much more complicated strategic environment of contrasting interests and overlapping coalitions. No doubt the theoretical and practical questions the new situation raises are more complicated and complex than any previous ones. The emergence of extragovernmental threats – terrorism, ethnic rebellion, etc. – also adds an important dimension to defense economics. In recent years, economists have begun to analyze terrorism, insurrections and other low-intensity conflicts by using equilibrium models of benefit-maximizers or rent-seekers. Since such conflicts involve negative externalities whose influence reaches beyond national and regional borders, national security problems become international in nature, a universal issue of collective risk management. Security has in fact become a global public good, calling for permanent military alliances, as well as for ad hoc coalitions and special international forces for peace enforcement and peacekeeping. Adopting such prescriptions for international security raises questions about financing, burden sharing and other problems that usually accompany the provision of public goods.

3. UNIQUE FEATURES OF DEFENSE ECONOMICS

Defense economics is simultaneously confronting two kinds of doubts. It should be convincing of its contributing potential to national security policy-making and management, and prove its special features that justify differentiating it from other branches of economics. The variety of subjects, with which defense economics has been dealing over the years, could be, by itself, an indication of its contribution and productivity. Otherwise, busy economists who are willing to be helpful would supposedly refrain from investing time and analytical and empirical efforts in the field. The uniqueness of defense economics can be attributed to the special features of the "product" with which it is involved, the special structural and institutional characteristics of the environment in which this product is usually "produced", and the special circumstances under which it is typically "consumed". Most probably, some of these features have their counterparts in other branches of economics. However, only in defense economics all of them seem to jointly take place, and it is their aggregated weight that makes defense economics a unique field of specialization.

3.1 Measuring Defense Output

The output of defense expenditure (defense output) is intended mainly to deter potential enemies, and should deterrence fail – to protect against offensive actions and minimize casualties and damages created thereby. Deterrence and the ability to protect have many elements, each of them may be measured in different terms, and it is impossible to consolidate them into one single aggregate unless some common denominator is provided. In other complex phenomena market prices are usually used for this purpose, and their dimensions are then measured in monetary terms. However, defense outputs are not sold on the market, they have no market prices and cannot be simply expressed in monetary values. In the absence of market prices in other public services, economists assume that the level of output is a monotonic function of the level of government expenditures. But in the case of defense expenditures, this method is not applicable, or at best is of limited value, due to three conceptual difficulties: -

i. *Flow versus stocks*: Conceptually, deterrence and the ability to protect are stocks rather than flows. At any given point in time they depend on the size of human and physical military capital stocks accumulated in the past (after deducting depreciation), and thus reflect the outcome of continuous spending throughout the years rather than the defense spending in any single period. Increased defense spending in one period over the previous one will not necessarily lead to a higher level of net military capital stocks, and the level of security available will not necessarily be higher either.

ii. *Externalities*: The level of output or utility that any particular country derives from defense expenditure – current or accumulated – depends not only on its own spending, but also on that of other countries, and on additional strategic factors. For example, when a country increases its defense expenditure, it is reasonable to assume that rival countries will not remain indifferent and will increase theirs too. Consequently the first country's level of security may not change – it may even decrease – despite the initial increase in its defense expenditure. Thus the monotonic relationship between output and expenditure holds only in a static environment without any further changes and interactions.

iii. *Joint products*: Besides external security per se, defense expenditure may provide other benefits, such as internal security, external and internal political advantages and economic benefits. These are joint products of defense spending, and the overall social utility that countries obtain may vary not only due to changes in the level of defense expenditure, but also through changes in its composition, resulting in different quantities of the various joint products (van Ypersele de Strihou, 1967).

However, despite these shortcomings, defense expenditure is considered conceptually less improper and technically more convenient than any other single measure of aggregate defense output. Hence besides its economic connotation, defense expenditure has also important strategic interpretations. It is a key factor in strategic assessments, and therefore has a major impact on the dynamics of arms races. Likewise, it is an important factor in shaping military alliances, and sometimes provides the basis for designing processes of arms control and disarmament.

The conceptual limitations involved in measuring and comparing defense expenditures are often more severe than similar measurements and comparisons in other economic branches. Moreover, from the technical standpoint, measuring and comparing defense expenditure data requires not only the proper handling of inherent statistical difficulties, to which economists are usually accustomed, but also overcoming secrecy and intentional disinformation (see chapter 5). This combination of problematic measurements and resultant important decisions is not equally typical of other economic areas.

3.2 "Producing" National Security

As "producers" of national security, governments do not follow common economic rules. They are not guided by profit maximization, and in acquiring the resources needed for "producing" national security, routines are substantially different from those usually applied in producing and providing civilian public services. Some resources, especially manpower, are not acquired in markets through price competition, but rather are recruited by law. Indeed, defense is the only area in which even advanced democratic countries still collect taxes in-kind: young individuals are obliged to serve their governments directly, i.e. to provide them with their time, and not only to pay taxes out of their money income. Other inputs, particularly weapons and specific military equipment, though acquired in markets, are often procured through uncommon processes. The military is the only customer for those products, and there are few manufacturers – or sometimes a sole producer – so that acquisitions take place under imperfect competition and prices are set through negotiations. Under these circumstances, there is no assurance that defense will be efficiently produced. Since prices of inputs do not reflect their values of marginal product, the mix of inputs will not necessarily minimize the opportunity costs of a given level of defense output.

The production of national security is also different from other public services because of the multi-period implications of its output. Current readiness should be prioritized vis-à-vis building up capabilities for peak demand periods, when military threats grow more acute or war actually breaks out. This often involves complex tradeoffs that require decisions under conditions of high strategic and technological uncertainty.

Problems of efficient production in a non-market environment, under high uncertainty, and within large systems that have to account for multi-period considerations are not common in other fields of economics.

3.3 National Security: Consumption or Investment?

Domestic uses of economic resources are divided between consumption and investment. Defense expenditures are different in this respect too, as they have features of consumption and investment products at one and the same time. They include pure consumption components that provide for the daily living of the military and the maintenance of current readiness. But except for periods of active armed conflict, the greater part of defense spending is usually devoted to future needs. Most of the defense products (weapons systems, other military equipment, military infrastructure and constructions, as well as the intangible outputs of defense research and development, military training, etc.) are durable in nature, and provide flows of defense services for many years. Also, the production factors and the typical manufacturing techniques employed in producing defense products have more in common with the production of investment goods, and in practice compete with them, while their relationships with the production of consumption goods are rather loose. Thus, although defense expenditures are recorded as current consumption in formal economic accounting (since they do not contribute to future production capacity of the economy in the usual sense), they clearly have qualities of investment goods.

Usually, countries demonstrate fairly stable patterns in allocating economic resources between the present and the future, and changes occur mainly within consumption and investment. However, as defense expenditures do not fall into any of these categories, sometimes they may crowd-out investments, and on other occasions may restrain consumption. Besides, because of their investment character, defense outputs have a special two-way relationship with economic growth. Increasing defense spending at present may stimulate or hinder economic growth, thereby determining the possible level of national security in the future. Defense economics must deal with these special aspects of seemingly routine macroeconomic issues often neglected elsewhere in the field.

3.4 National Security as a Public Good

It is common to assume that defense expenditure produces a pure public good, namely a product whose benefits may not be excluded from citizens whether they pay for it or not, while its consumption by any additional individual will not reduce the benefit derived by others. In fact, not all defense outputs maintain such features to the same extent. Defense outputs may be divided between those that provide pure deterrence (e.g. strategic forces of nuclear superpowers), on one hand, and those that provide pure protection (e.g. radar or other early warning systems), on the other hand. Between them there are defense outputs that provide combinations of deterrence and protection in varying degrees (e.g. most conventional forces and weapon systems, like tanks and fighting aircraft). Pure deterrence complies with the features of a pure public good. But as the ratio of protection to deterrence increases, the defense output becomes less of a pure public good. In such cases it is possible to deprive individuals in certain regions of defense benefits merely by moving troops away to other regions. Also, given the limited size of military forces, increasing their density in one region will unavoidably decrease the supply of security for individuals elsewhere (Sandler & Caully, 1975). Hence defense economics deals simultaneously with both pure and impure public goods.

Moreover, the economic theory of determining optimal levels of government expenditure on public goods has serious limitations when applied to defense. To arrive at the optimal level of such expenditures marginal propensity to pay should equal marginal cost of providing public goods, and for that purpose aggregate demand functions of all citizens for those goods must be defined. Furthermore, social demand for public goods can be significant only when consumers' preferences are unrelated, or exogenous, to the political process. These conditions are hardly met in determining defense expenditure. First, as noted before, it is difficult if not impossible to estimate the demand for defense spending due to the problems involved in defining the relationship between utility and varying expenditure levels. Even experts have such difficulties, the more so individuals with less information. Aggregating individual demands for defense cannot be interpreted in the same binding fashion as, for instance, the aggregate demand for public parks or highways. In most cases it will only reflect popular and not well-founded perceptions. Moreover, since the use of military power can be directed to reinforcing internal control as well, defense expenditures may be favorable for some and dangerous for others. Secondly, the demand for defense expenditure is to a large extent endogenous to the political process. The attitudes and actions of local political leaders influence public demand, and so do the defense expenditures and political attitudes of other relevant parties.

The public good with which defense economics deals is a peculiar one also because of its contrasting meaning for the individual country and for the world as a whole. While the expenditures of any individual country on domestic infrastructure or for improving its own human capital contribute in parallel fashion to world production capacity, and thus unambiguously increase global welfare, this is not the case with defense expenditure. The spending country will benefit from its defense expenditure, but other countries may feel insecure, and the consequences for global welfare are undefined. In this respect, in a world of two non-cooperating countries, defense expenditures will be too high. By reaching an appropriate agreement, they can reduce their respective defense expenditures and increase social welfare, while maintaining the balance of power between them intact. In somewhat more extreme words (Boulding, 1986: 12-13), while for any single country defense expenditures produce security, for the world as a whole they produce insecurity and potential destruction.

3.5 National security: exogenous or endogenous variable?

Economics treats defense expenditure, and in broader terms - national security, as a partially exogenous and partially endogenous variable. Since the desired level of security is derived in relation to external threats, the general tendency was to consider defense expenditure as an exogenous variable determined by governments outside the economic system. Later, observers pointed out that defense expenditure varied without significant changes in the strategic environment, or when such strategic changes alone could provide only partial explanations for increases or decreases in defense spending. Consequently, defense economists tried to identify economic and political factors that might also have an influence. In other words, they suggested that national security is not only a matter of "higher politics" in the realm of international relations, but also of internal social pressures and local political economy considerations (Mintz, 1992: 1). Recent works in defense economics acknowledge that defense expenditure could be endogenous to the economic system, and assert that the extent to which it is endogenous rather than exogenous inversely relates to the urgency and acuteness of external threats (Fontanel, 1994: 91-92). Again then, defense economics is confronted with an unusual ambiguity that contributes to its uniqueness.

4. ACCELERATING AND RETARDING FACTORS IN DEFENSE ECONOMICS DEVELOPMENT

Defense economics gained recognition as a sub-discipline and a special field within economics relatively late. From a broad perspective, this can be attributed to general reluctance to national security studies. Social scientists seem to have feared that interest in defense issues might be wrongly interpreted as favoring militarism, and the common approach was to leave such matters to military experts. A change began after World War II, particularly following the emergence of nuclear weapons and the new international realty created thereby, when war and strategy studies became an independent academic field (Harkabi, 1990: 26-27). But there were also some specific reasons related directly to economics and economists for the late development of defense economics.

In earlier stages, when economics made its first steps as an independent discipline, it asked for recognition in a relatively narrow range of subjects. In the nineteenth century and the beginning of the twentieth, economists focused mainly on value and distribution theories and on market mechanisms. Issues that could not be methodologically treated through competitive market analysis, national security included, were considered exogenous to the economic system. Moreover, mainstream economic theory revolved around the notion of equilibrium, and equilibrium was connected to concepts such as normality and harmonization. Within this conceptual world, analyzing defense issues would mean interpreting war as a normal event, and causing destruction as a reasonable phenomenon of human behavior. That approach, however, clashed with notions of enlightenment and rationality that dominated mainstream economic thought at least since Adam Smith

(Biddle & Samuels, 1991: 87). Finally, economists were reluctant to apply positive economics to the public sector, and the theory of public expenditure remained a neglected area of economics for many years.

There was another aspect too. At the beginning, defense economics made use of basic theories and the simplest analytical tools, while other fields of economics increasingly adopted complex mathematical models and sophisticated quantitative techniques. To quote a senior economist in the American Department of Defense in the 1960s: "The economic theory we are using is the one most of us learnt as second year students" (Enthoven, 1963: 422). Most likely then, researchers did not find economic problems of national security challenging enough (Leonard, 1991: 282). In addition, quantitative analyses require detailed information, and data availability is always an especially difficult problem in the domain of defense economics.

Since late 1950s, several developments in economics combined to accelerate the progress of defense economics. First, the growth of the public sector and its increasing share in advanced industrial countries required an explanation of the economic behavioral patterns of governments. Secondly, advances in macroeconomic theory, which treats the national economy as a single unit, laid the foundation for analyzing the actions of nations in the international arena in terms of a "rational actor". Thirdly, dividing lines between economics and other disciplines were removed to a large extent, interdisciplinary collaboration strengthened, and economists became interested in a variety of subjects that until then were considered outside the domain of the profession. Those developments gave rise to a general trend of specialization and subdivisions within economics.

Another accelerating factor, though of a different nature, was the rise of institutionalized economic advice on defense issues. Economists were given an opportunity to implement their thinking and apply economic analytical tools to real problems, and thus influence national security policy-making in practice. The first significant move took place in the Statistics Department formed within the British War Office at the end of 1939. The economists in that department collected and analyzed data concerning the war economy, predicted shortages, designed rationing plans to cope with them, and suggested methods for converting resources from civilian to military uses. No less important, during war as at any other time, they had to restrain the requirements of the military for additional resources (Leonard, 1991: 263). In the United States, economists recruited from academic institutions were employed in the Research and Analysis Branch of the Office of Strategic Services formed in 1942. Some of them were stationed in England to assist in selecting targets for air bombing. From an analytical perspective, the problems they dealt with were defined in optimization terms – how to maximize damages for the German war machine with a given campaign of air strikes. Forty years later, Prof. Kindleberger who served in that group explained that they were intuitively applying input-output and capital theory analyses. The former demonstrated how economic activity could be stopped by eliminating certain inputs - fuel for example - while the latter pointed out at the consequences of substituting labor for the capital assets damaged by air bombing (Kindleberger, 1980). The thoroughness and quantitative nature of the economists' analyses, and the concrete and immediate implications of the recommendations they made, attracted the attention of the highest military echelons to their work. The economists' authority was enhanced, and economics was granted priority over other humanistic disciplines. Moreover, there were important consequences for the economists themselves: they gained confidence in the applicability of economic paradigm to handling national security issues (Leonard, 1991: 262-268).

At the end of World War II, Gen. "Hap" Arnold the ambitious commander of the United States Air Force approved a budget of 10 million dollars for a research project called RAND. Gen. Arnold believed that by maintaining close relations with scientists, and particularly by integrating their analytical methods in decision-making processes, he was promoting Air Force superiority over other services. In the beginning, physicists and engineers were the dominant figures in RAND, but later on things changed. The economic department established in RAND played a crucial role in developing methods of systems analysis, and its work had a far-reaching influence on modern strategic thinking (Brodie, 1980: 394). In 1961, on the initiative of Robert McNamara, the Secretary of Defense in President Kennedy's administration, three senior economists from RAND joined the Pentagon, thus gaining an opportunity to apply the solutions of systems analysis at the highest levels of decision-making (Enthoven & Smith, 1974). Indeed, starting modestly from handling simple problems at lower levels in the early stages of World War II, by the 1960s economists had reached the highest echelons of national security policy-making in the United States. However, as mentioned before, in the early 1970s, a few years after McNamara and the team of economists that came from RAND left, the Golden Age of defense economics and systems analysis in the Pentagon came to an end.

By the 1980s, defense economics developed into an institutionalized academic profession. An international association of defense economists was established (*International Defense Economics Association – IDEA*), and three international scientific conferences were held under its auspices. In 1990, a new professional journal fully dedicated to defense economics' issues was published – *Defense Economics: The Political Economy of Defense, Disarmament and Peace* (in 1994 the name became *Defense and Peace Economics*).

However, even now, after gaining recognition as a sub-discipline of economics, fundamental questions regarding the essence and limitations of defense economics are still raised. Can economic analysis determine the optimal level of defense expenditure, i.e. how much is enough? Or is its contribution limited to presenting the economic factors that influence the decision, and to evaluating its economic consequences once made? Is there a basis for generalizing the interrelations between defense expenditure and economic variables, and are such formulations valid and sufficiently robust for a large number of countries and different periods? Or do the relationships tend to depend on the special circumstances of every case, not obeying any general rule that prevails in all places and times (Chan, 1985)? To what extent do economic considerations and incentives influence the behavior of the decision-makers on national security issues, and what are their relative weights in comparison with strategic, political, bureaucratic and social factors? Should defense economics develop an analytical framework that merely accounts for the special institutions of the defense sector? Or should it go further and also explain the development of such institutions (Reppy, 1991: 270)? Is it possible to achieve a comprehensive and coherent theoretical paradigm, or is defense economics doomed to remain a collection of separate answers given by economists to questions raised in the broad domain of national security (Hilton, 1987: 307, 312-313)?

Enthoven and Smith admitted that given the complexity of national security issues, economic analysis is, at best, incomplete. However, they argue that the economist can spare the analytical work from the decision-maker, leaving for him the more difficult questions that require his judgment (Enthoven & Smith, 1974: 90). Charles Hitch claimed that in order to handle national security issues successfully, economists should address modest targets. Optimal solutions are unachievable in a world of uncertainties and variables that are measured by incomparable units. But economic behavior does not necessarily require optimization, neither in the military nor in the economy. Simply, it requires finding better solutions than those obtained by other means, and this is often both feasible and important (Hitch, 1960: 442),

Setting modest practical targets does not mean giving up high hopes. Defense economists, or defense and peace economists, believe openly or implicitly that in-depth understanding of the interrelations between security and economics will lead not only to achieving more security with less economic resources. They hope that by presenting the price that nations pay for security they will encourage the political leadership to seek a different global order that reduces threats in the international system. If this is accomplished, it will create an opportunity to divert dear resources from warfare to economic development and the enhancement of human welfare.

2
A HISTORICAL PERSPECTIVE: HOW NATIONAL SECURITY BECAME AN ECONOMIC ISSUE

Everywhere and throughout the ages, the economic resources at the disposal of rival parties have influenced their power and their fighting ability. To paraphrase Thucydides, historian of the Peloponnesian War, the war was not so much a matter of arms as of expenditure, through which arms could be put to use. This chapter offers a historical view of the interrelations between military power (national security) and economic wealth.

1. NATIONAL SECURITY AS A PUBLIC GOOD IN ADAM SMITH

In *The Wealth of Nations*, (Part 5, Chapter 1) Adam Smith describes how security became an economic issue (Smith, 1776). In pre industrial societies, the costs of war were small, and borne privately by the actual participants. There was no need for special training as the routine of civilian life prepared young men for war, and wars, usually short and fought between agricultural high seasons, did not cause any substantial economic loss. Change came with specialization and the division of labor that accompanied industrial development. This created occupations where nature contributed nothing to income, and when industrial workers had to leave their regular economic activity to go to war, their income dried up completely and the population at large had to bear their living expenses. Thus in industrial society wars involved social costs, and question arose as to how they were to be met. Moreover, military activities became more complex, and like in every other field required specialization and division of labor. Hence it became the only or the principal occupation of a special group of citizens who served society as a whole, and society had to take care of their subsistence expenses, and meet their training expenditures on a continuous basis even in peacetime, not only during wars.

Historical and economic analysis led Smith to three practical conclusions about supplying security. The first referred to military organization: consistent with his perception regarding specialization and the division of labor, he declared that only a standing army – not a militia – could assure the continuous existence of an industrial society. The second conclusion related to army size: since the work of others supports the soldiers, their number is limited by output surplus after civilian subsistence needs are met. Economic limitations thus restrict warfare to a relatively small proportion of the populace (no more than one percent of the European population according to Smith). The third was that in the industrial age, responsibility for protecting society rests on the sovereign.

Smith was aware also of the special features of defense demand, and understood that the benefits of defense to individuals can neither be separated nor identified. Therefore, he argues, there is no guarantee that market mechanisms will assure sufficient resources for defense, or in modern terms there could be market failure in supplying national security. He adds another observation: economic growth and industrial advance increase not only the wealth of nations but also the costs of armaments and military activity, so that just when countries become a more attractive target for aggression, relative prices may lead them to neglect security (Goodwin, 1991: 24-25; 32-33). In Smith's view, then, fear of insufficient demand for defense, not only supply conditions, justified removing defense expenditures from the laissez-faire framework and placing the responsibility on the sovereign. Accordingly, he put forth two proposals for financing defense expenditures: defense serves the public, so society as a whole has to bear its costs, and because defense consumption is collective in nature, individuals should participate in defense financing on the basis of their ability to pay (Kennedy, 1975: 25-26).

In essence, Smith's analysis explains the process whereby security changed from a private to a public good. He also stresses the reciprocal influence between defense and economics, and how it has grown and became continuous over time. The trigger was socioeconomic change: the specialization and division of labor that went hand in hand with industrial development. These developments granted defense expenditures their social significance and led to their growth. But the influence was not unilateral. Since defense changed from a private to a public good and the economic system had to raise resources to finance it, there was a limit to army size, and in fact military power came to depend on the economic ability of society to support it. Finally, from an economic perspective the distinction between wartime and other times became less important as defense demanded economic resources in preparation for war too.

2. THE TOTALITY OF WAR

Until the two great revolutions at the end of the eighteenth century – the French Revolution and the Industrial Revolution – connections between the wars and fighting armies and the population at the rear were few and tenuous. The French Revolution introduced the "army of the masses" and brought the war to the doorstep of every citizen. At the same time, the Industrial Revolution paved the way for the mass production of armaments without which large armies were of no use, and made industrial capability the basis of military might. Influenced, though not solely, by these two developments, war became total: it touched every aspect of life, and required the utmost of national resources (Harkabi, 1990: 391).

The following discussion deals with one dimension of the growing reciprocal relationship between defense and economics: the mobilization of resources, human and material, for national security. At first, manpower was most important, since human strength and spirit were the main factors in waging war. Gradually, over time, inputs composition in producing national security changed. Industrial development and technological innovations offered a variety of armaments and auxiliary equipment, and fighting became capital intensive. In effect, war became a contest of materials, and materials became crucial in determining its outcome.

2.1 Manpower Mobilization

The economic basis for mobilizing manpower for security missions changed over the years. Eighth and ninth century fighting forces were based on knights on horseback. They needed heavy and costly defensive armor, and to gain mobility had to invest in training special horses. In addition, long campaigns required more than one such horse, as well as a groom, an armor bearer, a mounted scout and guards on foot. With staffs so expensive, the question arose as to how to properly reward knights and support their families, so they could devote themselves to military service at the royal command. The kings' incomes did not suffice, hence they began to grant the knights lands (the first to do so was Charles the Hammer, ruler of the Franks (685-741)), and the estate awarded in return for a loyalty oath and the obligation to give military service thereby became the cornerstone of feudal society. The search for a solution to a pressing defense problem played, then, an important role in creating the socioeconomic system that ruled Europe for hundreds of years.

However, the feudal mobilization system had many weaknesses. The service period was short (usually 40 days a year), and for long campaigns in remote places rulers had to pay in other ways. Call-up and deployment were lengthy and too slow for protection against surprise raids, so that standing forces became necessary. A unified command was impossible, as the vassal was only obliged to serve his direct overlord, and a few prominent, rebellious noblemen could damage the kingdom's military might. No less important was the diminishing land supply available to expand the circle of warrior knights. Hence, besides the forces the feudal system provided, kings began to look for mercenaries, and a new economic means of raising an army – money salaries - developed.

In fact, paid military service already existed in the early Middle Ages, but only in the fourteenth century did the French and English kings begin to base their armies mainly on it (Howard, 1985: 26-27). Since soldiers were prepared to serve as long as needed, the transition to a mercenary army made it possible to wage longer wars and of greater scope. However, the new economic base had its own weaknesses. Firstly, it depended on the rulers' financial capabilities, so that until the middle of the seventeenth century wars remained sporadic, occasionally stopping short because of royal bankruptcy. Secondly, armies based on mercenaries and on foreign contractors, who led them, had no loyalty to their employers. Control over such armies was weak, and particularly during the Thirty Years' War (1616-1648) warfare degenerated into depths of violent cruelty and futility.

At the end of the seventeenth century, with increased government ability to control economic resources, they retrieved control over the means to wage war too. Hence, in the coming century, professional armies who were in every sense servants of the state waged European wars, and they occupied soldiers devoted to their service in return for regular salaries, steady employment and chances for promotion. The basis for enlisting military manpower continued to be financial returns, but the quality of manpower was now quite different from that of previous centuries' mercenaries.

The next great change came in France, after the Revolution, in Napoleon Bonaparte's time. The Revolution granted sovereignty to the people, and the people who had taken power over the state were called on to defend it. The idea took root that every citizen had a duty to fight for his people. Thus the age of the army of the masses based on conscription began, and for the first time mobilizing resources for defense acquired dimensions of totality.

Ever since conscription and the army of the masses came into being, their nature, advantages and shortcomings have been heatedly debated. However, no new alternative ways to enlist manpower have been created. Most modern armies have been based on different combinations of a professional career army and soldiers called up by law. Compulsory service ended in the early 1960s in Britain and a decade later in the United States, both of whom now rely exclusively on volunteers recruited through market mechanisms. In the 1990s many other countries followed this path (See chapter 7, section 1).

2.2 Mobilizing the Industrial-Technological Potential

Until the end of the Napoleonic Wars, the adoption of industrial and technological innovations into the military sphere was rather slow. However, some innovations did have an important historical role. For example, long range, improved cannon were important in reinforcing central government and in establishing monarchies from which nation states later evolved. Only kings could afford artillery, and when they threatened the stone bastions of barons' castles, the barons had no choice but to surrender to the central power. The same was true of the free cities: they too had no shelter from the king's cannon.

The kings were aware of the military-industrial potential, and took care to develop it. Until the mid-seventeenth century, it was done within a loose framework, based principally on mutual interests of the kings and their contemporary "business circles". The rising bourgeoisie of merchants and economic entrepreneurs was no less interested than the kings in doing away with the feudal system, and at the same time perceived the business potential of producing weapons for the royal armies. On that account they placed the financial means for developing the army at the king's disposal and established arms factories, gaining royal protection and privileges in return. In the later seventeenth century, the central power was already stronger, and became more involved in the economy and in defense production.

State involvement was especially direct and deep in France. Jean-Baptiste Colbert (1619-1683), Louis XIV's influential minister, took a series of steps aimed towards military self-sufficiency, and established a local military-industrial base under state ownership (Kapstein, 1992, 21-22). In England too the state became more deeply and widely involved in industry, but its means were not so direct. Most defense production was in private hands, and the government encouraged and protected defense-related industries mainly by granting them monopolistic privileges. The differences in degree and means of state involvement were significant, and appeared to correspond, *inter alia*, with differences in industrial development, especially of the private sector (Kapstein, 1992, 21-22).

England led, however, in mercantilist integration of economic and defense objectives at sea. Her marine transportation and the fisheries were seen as important for both the economy and defense, since they trained sailors and employed them continuously, making it possible to shift them into military activities in wartime. This dual approach was clearly manifested in the Navigation Acts, first promulgated in 1660. It was asserted that maritime trade had to be carried out by English ships, that all ships over a certain size had to be built in England, and that the fleet had to be manned by English sailors. Other European nations followed suit.

In the nineteenth century, after the Napoleonic Wars ended, Europe had 30 years of peaceful international order – the Concert of Europe era – creating conditions for economic development that fundamentally changed the material environment in almost every sphere of life. The military was no exception, and when wars began again in the mid century they differed in almost every respect from the Napoleonic Wars. Firstly, the use of technology for military purposes accelerated, yielding cardinal innovations in weaponry. Innovations in small arms included, for example, the rifled gun, the breech-loading gun and the Gatling machine-gun; in artillery long range, breech-loading steel cannon with superior ballistics and improved recoil braking systems were introduced, and at sea steamships replaced the sailing boats (Howard, 1985: 115-117; Zook and Higham, 1981: 191, 205).

Secondly, wars were dramatically affected by developments in land transport and communications. Following the appearance of steam engines in the 1830s, railway lines were laid across Europe. Their first military use was in the war between France and the Austrian Empire in northern Italy in 1859: French forces of 120,000 soldiers reached the front in 11 days instead of the usual two-month foot march (Howard, 1985: 111). In subsequent wars, rail transport was extended: in the American Civil War it was a notable advantage of the North (Catton, 1979: 183), and in Europe it contributed to Prussia's victory over France in 1870 (Howard, 1985: 111). Railways changed war through a transformation in strategic mobility: it sped up mobilization and reduced the time needed to concentrate troops at the front. At the same time, it increased the staying power of armies, as soldiers reached the front in better physical condition, while the sick and wounded could be evacuated rapidly and replaced with fresh troops. Furthermore, the staying power has been enhanced through improved logistics: there was no longer a need to store supplies along the army's advance routes, and the soldiers' needs could be supplied directly and continuously to the front from sources in the home front. Finally, continuous supply by trains removed limits as to the size of an army (Howard, 1985: 112-113).

Electric communications began in the 1830s, and very soon became the means of military communication. Improved communication made it possible to control larger, more dispersed forces, while shorter time spans contributed to operational secrecy and greater possibilities of surprise. In the more general military context, the front could be commanded from posts at the rear, making for tighter connections between statesmen and military commanders. At the same time, war stopped being remote, something the civilian population heard about sporadically, after the fact. Military correspondents too used the communication lines, and their reports became a theme of intensive public debate.

The developments in military technology, in land transport and communication penetrated all aspects of warfare and of military organization. Central command and control strengthened, time and space dimensions expanded, offense-defense relationship changed, strategic and tactical surprises became feasible options and the importance of intelligence increased. In addition, special forces and auxiliary units proliferated, military bureaucracy grew and army life became bureaucratized.

The economic and industrial implications were no less significant. On land, new arms with their ever-increasing firepower required huge and growing supplies of ammunition, creating wider demand for factories using mass production techniques. At sea, battleships became a unique symbol of national might, demanding the state's most advanced technological achievements and the best of its industrial capabilities. The railways became supply lines between production centers and fighting armies, and together with improved communication systems made the industrial and technological potential an integral part of war efforts. Economic, industrial and technological strength became very important, and war, instead of being limited to armies, became in fact a struggle between the overall human and economic resources of countries involved.

War, then, took on another dimension of totality: not only mass human mobilization, but maximum mobilization of material resources too. The implications were far-reaching. First of all, in total war the nation's future was at stake and war objectives became broader, justifying greater investment in the war machine. A vicious circle was set in motion as broad objectives and huge investments accelerated one another. Secondly, war ceased to be only a struggle at the front, but spread throughout the country. Industrial production and supplies depend on the home front, which becomes a source of strength and hence a potential target. At the same time, means were developed to damage the rear, so that civilian defense became an integral part of war. Indeed, the separation between the front and the rear in wartime acquired new meaning. Thirdly, the broadening of war to the point where it included the full human and economic resources of the countries involved, produced more prolonged wars of attrition and fewer short, decisive ones. In long wars, economic factors become more important and results may depend on the total scope of economic resources available and the ability to manage "war economies", i.e. to mobilize larger shares of the economic potential for the war effort. Fourthly, in the new circumstances the results of war are determined by preparations for it in peacetime no less than on steps taken during the war itself, and battles become a kind of public contest over the superiority attained in intervals of peace. Hence economic capabilities are important not only during war, but also at all times.

Modern total war is generally acknowledged to have begun with the American Civil War (Amidror, 1979: 9-10). The agricultural society of the South based on slavery and for a generation resisting manifestations of the industrial revolution, struggled against the industrial North, whose superior economic resources determined the outcome (Catton, 1979: 178-188). Nonetheless, on the eve of World War I, the significance of total war does not seem to have been entirely clear. The rival parties misinterpreted the effect of full mobilization of resources on the course of the war and its possible length, and assumed that the growing interdependence of military ability and economic resources dictates short, decisive wars. Indeed, when the war broke out in August 1914, the general view was that the soldiers would be home for Christmas (Howard, 1985: 125-126). Consequently, at the early stages of the War, the countries involved had neither stocks of arms and ammunition nor the industrial capacity to supply large armies over a long period, and soon all combatants ran into severe shortages. In the meantime, the crucial role of the defense-industrial base was manifest in the clearest possible way: the front was at a standstill while the home front was "steaming" with activities to increase war production. All the participating nations carried out widespread industrial mobilization, allocating manpower to industries involved in the war effort and limiting production for other purposes. Military expenditures, which previously required some 4 percent of the GNP of the main combatants, grew substantially, and in 1916 accounted for one fourth to a third of it (Kapstein, 1992: 24).

World War I not only illustrated the link between industrial production and military ability, it elevated that link and applied it to new spheres. At the turn of the new century the Industrial Revolution was still largely a civilian phenomenon, and the unprecedented quantities required by the War led to full application of advanced production techniques in the military sphere. The War, then, was an important turning point in converting the fruits of the Industrial Revolution to military ends (Kapstein, 1992: 24). World War I was also the first major military conflict in which scientists played an important role. In the United States, Thomas Edison was asked to direct scientific programs in search of new ideas with military potential. In Europe, scientists developed the hydrophone and the geophone (to locate underwater and digging noises), depth charges, aerodynamic shells, the field telephone and other devices. The Germans appealed to scientists to break the trench warfare stalemate, and chlorine gas and gas masks were developed to this end, while the Allies developed smoke screens and flamethrowers. Moreover, scientists were employed not only to develop weapons but on the operational-tactical level too. In Britain, for example, they considered ways to cope with German submarines, and the mathematician F.W. Lancaster laid the foundation of what was later called operations research. The War also brought governments back into defense production. In the second half of the nineteenth century, large private firms had replaced government arsenals in defense production, but mobilizing industry into the war effort made it necessary to reverse the trend of privatization. In fact, governments' involvement in other economic areas grew as well, and most combatant states became "command economies" (Kapstein, 1992: 24). Finally, considerable attention was given to economic staying capabilities during wartime and to economic warfare. The Allies, for example, tried to bring down the German economy by the considerable means they invested in strangling Germany's foreign trade.

In the 1920s, large sectors of the defense industry, so greatly expanded during the War, reverted to civilian use, and the development of military technology slowed down considerably, particularly in the victorious nations. However, there remained at least two legacies of the War in the reciprocal relations between defense and economics. Firstly, the need for permanent state involvement in military research and development was recognized, and government presence in this area broadened and became institutionalized (Kapstein, 1992: 25). Actually, most new technologies utilized during World War II were developed in the 1920s and 1930s at government initiatives. Secondly, the lessons of economic warfare led Germany to prepare for a blockade it considered unavoidable if and when war broke out (Kapstein, 1992: 30-31). In a way, Hitler's blitz strategy also was, *inter alia*, an admission of the economic difficulty of waging a prolonged warfare.

Between the two wars another dimension of defense-economic reciprocity arose: arms producers were seen as warmongers (Brodie, 1980:

249-253). In the United States it was claimed that pressure from arms manufacturers forced the government into the War. They incurred heavy debts supplying arms to European governments, and apparently demanded that the United States enter the war on the side of the debtors so as to protect their own investments. This mythical claim led, among other things, to an official investigation (the Nye Committee) that defamed the firms engaged in defense production. When World War II broke out, many American firms refused to produce military equipment again.

World War II represents a peak level of economic mobilization. In Britain and Germany, harnessing economic resources to military ends was exploited to its very limits, while in the United States the share of GNP allocated to defense increased from 2.4 percent in 1938 to 41.6 percent in 1944. The link between the front and the civilian rear was also stronger than ever, and joining in defense production was considered not less important than joining the army. Actually, in Britain, manpower between 18 and 39 years of age was divided almost equally between the military and industry (Zook & Higham, 1981: 310-311). The extraordinary mobilization of industry produced unprecedented huge quantities of weapons, and led to substantial improvements in industrial efficiency. Finally, innovative technological developments (e.g. radar, computers, the jet engine, the atomic bomb) and reliance on scientists reached their peak too.

The end of World War II ushered in the nuclear age in military affairs, and it was again necessary to update perceptions about defense-economic interrelations. Within a very few years after nuclear weapons were invented, it became possible to use them to wipe out entire populations and economies. There was little time, then, for economic analyses to evaluate the significance of the new state of affairs.

3. NATIONAL SECURITY AND ECONOMIC WEALTH IN ECONOMIC THOUGHT

The history of nation states can be perceived as a long and continuous endeavor simultaneously to pursue military power and economic wealth as twin goals. No wondering then that the relationship between the two was an important axis in central schools of economic thought.

3.1 National Security and Economic Wealth: Complementary or Competing Goals?

During the 300 years from the sixteenth to the end of the eighteenth century, political-economic thought was dominated by the mercantilist doctrine. For the mercantilists national interest was of top priority, and they saw in power and wealth two national goals of equal rank, mutually supportive and each a means to attain the other (Baldwin, 1985: 73). Economic wealth was measured by the gold and other precious metals at the state's disposal, which would also serve to meet military expenditures when necessary. Thus states seeking military power and economic wealth strove to accumulate maximum reserves of these metals, and since metals were used as international means of payment, it became imperative for them to create a positive trade balance with other states. However, there was no way that all states simultaneously could enjoy export surpluses, and it was assumed that the chance of one state to gain economic advantage over others depended on its military power. In particular, it was asserted that states should maintain military power and strong naval fleets that would enable them to conquer and hold colonies, and rule the trade routes (Oser & Blanchfield, 1978: 25). The mercantilist perception thus saw the economy as static: a world with fixed economic resources where a zero-sum game was played among the nations, so that one of them could increase its wealth only at the expense of the others. Economic wealth was needed for achieving military power, which was in turn needed to increase economic wealth, and static mercantilist economics viewed the simultaneous pursuit of both ends as a viable policy without internal contradictions. In a world with fixed resources the dynamics set in motion by diverting resources to build military power was ignored, although it could eventually result in economic decline.

The economic liberalism that began to gain ground in the second half of the eighteenth century took a different approach. Adam Smith and his successors did not see the economy as a static system, and international trade was not for them a zero-sum game. They maintained that efficient use of resources through competitive market mechanisms would increase economic wealth, and so would do international trade when based on specialization according to the principle of comparative advantage. It follows that wars are not the way to increase a nation's wealth at the expense of other nations, but the opposite is true. Wars interfere with the "invisible hand" of the market and with the efficient international division of labor. By contrast, the freer the flow of goods and capital among nations and the more their economic wealth grows, the stronger become the economic incentives against going to war. When free trade prevails, war is irrational, and armed conflicts between advanced countries having economic ties with one another are unreasonable. According to the rules of classical economic liberalism, promoting national wealth through rational political economy increases national security as well.

The differences between the mercantilists and the liberals arose from their differing perceptions of the economic world, and therefore related to "how" rather than "what" (Baldwin, 1985: 79-87). Both schools thought in terms of national interest, and considered security and economic wealth as its key components. Likewise, neither school preferred power to wealth, nor subordinated economic to political interests. Their differences focused on appropriate ways to serve the national interest and to promote the two goals of statecraft effectively. Instead of the conflict approach of the mercantilists, the liberals suggested an alternative of laissez-faire.

Within the liberal view there was room for deviance, and national security considerations at times justified a departure from market economics. Smith, as previously noted, was aware that defense consumption was collective in nature, and that there might be market failure in supplying security. Hence he removed defense from the laissez-faire framework, and placed the burden of protecting society on the sovereign. Similarly, though in a limited way, liberals agreed to allow departure from free market principles for security reasons in foreign trade and marine transportation. Smith supported the protectionist Navigation Acts on the grounds that expanding international trade required a strong navy to protect the shipping lanes from hostile governments and from pirates (Whitaker, 1991: 39). Malthus favored protective tariffs to create autarky in grain production, and Ricardo agreed that immediately after a war agricultural imports should be limited, to allow for gradual reduction of excess production capacity accumulated during wartime (Kennedy, 1975: 28). The transition from war to peacetime economy worried the liberals in other contexts too, and to make adjustments possible they recognized that government might have to intervene.

The classical British economists were obviously influenced by their country's special conditions. On the Continent, under different circumstances, the German Friedrich List advocated an alternative approach – economic protectionism – that combined liberal and mercantilist ideas (Baldwin, 1985: 87-88). List shared the conflict perception of international relations, thought wars to be highly probable and preached a buildup of military power. As he assumed that industrial development was a decisive factor in military capability, he favored state intervention in this area, and supported protective tariffs on finished products to encourage undeveloped German industries. For List, however, protectionism was merely a way station on the road to free trade and peace among the nations.

During the twentieth century, certain aspects of mercantilism were renewed through the interpretation of international relations suggested by the school of political realism. That school rose as a reaction to the optimism that accompanied the founding of the League of Nations. It stressed that the international system was still anarchic, and that the League lacks the power to enforce international peace and prevent wars. The realists' theory asserts that nation states are rational, seeking to maximize their power. International relations, therefore, are based on conflict and on zero-sum games – one country grows stronger at the expense of others – so that for survival, a nation must rely only on its own power. It further argues that national economic resources and their efficient use are the basis of power, and particularly, since modern warfare depends heavily on industrial capacities, the industrial rank of a country determines its rank also in the international power system. The analysis leads to several conclusions. First, security considerations justify state intervention in the local economy. Second, the world economy is a zero-sum game just like international relations in general, and in their power struggle nations shape the world economy to serve their own interests. Third, international trade based on liberal principles of comparative advantage and division of labor does not take place by means of a global invisible hand. It requires a hegemonic power that is able and willing to provide the world with collective economic stability and international security. Political realism principles have been used to explain the rise and fall of the central world powers over the past 500 years, and the analysis shows a clear correspondence between military power and economic capability in the long term (Kennedy, 1992).

3.2 National Security and Economic Wealth: Reciprocal Influence

Political economy dealt with the reciprocity between national security and economic wealth on several levels. The "accountancy" approach presented the direct profits and losses of wars. The discussions at the macroeconomic level explained war's effects on overall economic performance, particularly wartime booms and the recession that follows. The approach one might call macro-history dealt with the accelerating or retarding effects of wars on major socioeconomic processes and on scientific and technological progress. It also examined the unifying or confrontational influences of major economic developments on the nature of wars, and on the role they play in international relations. On all levels, despite the negative consequences of wars, it was impossible to ignore some apparent positive effects that they had on the economy.

3.2.1 The Profit and Loss of Wars

As noted before, the mercantilists saw wars as a way to increase economic wealth, since wars led to control of colonies and thus made possible the exploitation of their cheap raw materials and labor. In the eighteenth century, French economists of the Physiocratic School and the liberal English economists disagreed with the mercantilist view of the so-called profits of war. They argued that the conduct of wars, and maintaining colonial conquests, required a disproportionately expensive military force, to the point that they consumed the entire profit – if there was any. Smith, for example, asserted that the interest on the debts that financed England's wars in his time was higher than the annual value of the goods that could be exported to the colonies, and certainly greater than any profit thereof (Brodie 1980: 228-229). He also criticized the mercantilists for presenting only direct money outlays as the costs of war, while the true cost should have included all output consumed directly or indirectly by wars (Goodwin, 1991: 27).

Ricardo, who was influenced by the Napoleonic Wars, pointed to the economic losses resulting from war disturbances to international trade and the distortions they impose on local production. According to his observations, during wars local production usually obeys principles contrary to those of comparative advantage. In addition, he was concerned with the costs of transition from peace to wartime conditions, and back. These costs include, *inter alia*, lost production and unemployment, and Ricardo asserts that they should be reckoned with the costs of wars (Goodwin, 1991: 30-31). His French contemporary, Jean Baptiste Say, added the human lives lost to the costs of war. In the terminology of our time, he accounted for the loss of human capital that includes not only accumulated costs of the past, but also future income that cannot be realized (Kennedy, 1975: 31-32).

Robert Giffen made a pioneering cost study of the Franco-Prussian War (1870-71). Direct costs included government expenditures on the war, the value of physical capital destroyed, the future value of soldiers' pensions, and on the French side also that of property the Germans confiscated. Indirect costs included temporary losses of civilian income due to mobilization and other disturbances to businesses, the future value of such permanent losses, and loss of human capital (estimated by the lost income flow). Since the war was fought on French territory, France's costs were four times as high as Germany's. Moreover, at the war's end Alsace-Lorraine was annexed to Germany and France paid reparations, so that the Germans earned a net economic profit equal to some 20 percent of their annual GNP from the war, while French losses grew to the equivalent of 16 months' GNP. Finally, Giffen calculates the change in national wealth, adding the consumption Germany saved in the course of the war to its economic profits and subtracting the consumption saved in France from the losses it underwent. In conclusion, Giffen asserts that the economic wealth of advanced societies renders war expenses bearable (Whitaker, 1991: 52-55). It is worth noting that in the postwar years victorious Germany fell into an economic recession, while in vanquished France a boom developed. There certainly were other reasons, however liberal economists found support in this chain of events for the view that war is economically unprofitable.

At the turn of the twentieth century, acquiring territory or other real assets through war became rare, and the discussion focus shifted to the economic implications of reparations payments. One common view at the time was that reparations in money lead to rising prices in the recipient countries, and thus to loss of competitive advantage in export markets and unemployment. The Anglo-American journalist Norman Angell used this and other economic reasoning to argue that so-called profits from war were a great illusion, and his 1910 book of that name was an internationally influential bestseller (Barber, 1991: 62). After World War I and following the Versailles Treaty, the question of reparations became especially controversial. Keynes recognized that reparations might adversely affect recipient states (Barber, 1991: 81), while Pigou, by contrast, rejected the idea that a victorious state could not benefit economically from war reparations. He explained that reparations were like an interest free foreign loan that need not be repaid, so that usually states profit from war reparations, and occasional damage can occur only when they choose to use them unwisely (Barber, 1991a: 132).

3.2.2. Marxism and the Economic Benefit of Wars

While the liberals considered wars as unreasonable and opposed to the spirit of the times, Marx and Engels saw them as deeply embedded in the structure of capitalism. They adopt a conflict approach, but do not preach military power like the mercantilists; rather they are resigned to it like the liberals. According to Marxist doctrine, wars are to disappear - and with them the need for military power and defense expenditures – with the withering away of states after capitalism is abolished.

Differently, however, from the liberal economists who accepted allocation of resources for defense as a necessary response to external threat, the Marxist explanation is based on internal motives and economic advantages that keep the ruling class in power. Capitalism creates two distinct social classes: the bourgeoisie and the workers. The ruling bourgeoisie pushes the capitalist state into war for an internal reason, as a refuge and distraction from its inherent contradictions, and for an external one, which is competition with capitalist groups in other countries. Marx and Engels themselves did not explain precisely how the capitalist class profits from wars. This was done later by John Hobson, a liberal British economist, by the German socialists Rosa Luxemburg and Karl Kautsky, and by Vladimir Lenin. Despite some differences all regarded wars and defense expenditures as means to deal with the crises of the capitalist system, arising from overproduction and under-consumption. Lenin's theory of competitive imperialism, for example, stressed falling profit rates as capital accumulated: additional profits and accumulation become critically dependent on foreign markets and investments, and governments compete, going to war at times, for control of new colonies. Colonies are a potential market for output surpluses, and having abundant cheap labor and raw materials they provide opportunities of particularly profitable investments (Oser & Blanchfield, 1978: 337-342; Brodie, 1980: 247).

There was an important innovation in the notion that defense expenditures do not result from external threat, but rather from the desire of a ruling class to maintain its position. Later on, the assumption that defense expenditures may be connected with internal political factors paved the way for new perceptions of political economy – the neo-Marxist views - holding that governments manipulated defense expenditures to improve the political atmosphere in times of economic slowdown. Another version of the same approach explained increased expenditures on defense as responses to the pressure of the military-industrial complex.

3.2.3 The Macroeconomic Effects of War

The classical economists saw war as a state of economic imbalance that interferes with current economic activity, and adversely affects opportunities for future growth. Thus they firmly denied that war and defense spending encouraged economic activity, and even more that this was a legitimate way to do so. John Stuart Mill called this "the fallacy of the glut", and regarded as ridiculous the wondrous solutions of governments that created demand by hiring more soldiers and sailors or buying military and naval equipment (Goodwin, 1991: 31-32). However their argument did not always fit the empirical facts. Not only did wars tend to cause economic booms, economic activity did not suffer substantially after they were over, and sometimes it was the defeated nations that grew faster. The need to accommodate theoretical principles to the facts led to supply-side and demand-related explanations. In the light of modern macroeconomic theory, however, they provided partial justifications at best.

Smith maintained that defense consumption grew mainly at the expense of private consumption, so that wars do not adversely affect investments. Hence even if production drops during wartime, postwar production capacity and output are unaffected. Indeed, imposing higher taxes to meet war expenditures may reduce savings and with it investments, but this is temporary and balances out after the war when taxes are reduced (Goodwin, 1991: 28). Another supply-side explanation comes from the Scottish emigrant to Canada John Ray. From his new home he perceived economic growth to be not exclusively a matter of capital accumulation, but also one of inventions and innovations, which may be enhanced by wars, in defeated nations in particular (Goodwin, 1991: 28-30).

Malthus explained the contradiction from the demand side. He worried about the ratio of population to available economic resources, and regarded war as both the result of pressures from population surpluses, and as an adjustment mechanism – due to wartime loss of human life - that holds back population growth. However, economic prosperity tends to raise natural increase, and since wars lead to economic prosperity, it is not clear whether they ultimately ease or exacerbate population pressures. Malthus replied that wartime economic booms are transient and short, and the greater they are, the severer the postwar recession. Indeed, it was only after the introduction of the Keynsian theory of aggregate demand in the mid 1930s that economists could clarify satisfactorily the economic cycles produced by war.

The way defense expenditures were financed, and especially the choice between taxes and loans, had important macroeconomic implications as well. Smith was aware of the irrational factors that impelled towards war and to superfluous military expenditures, and to put political difficulties in the way of financing them he favored taxes over loans (Whitaker, 1991: 39). However, he had an economic reason too: financing through taxation would come for the most part by reducing private consumption, while the use of loans would harm investment during the war, and because of the distributive effects of repaying the debt – also after the war (Goodwin, 1991: 31). Ricardo too recognized the need to make it politically difficult for governments to engage in expensive and unnecessary wars, and besides, argued that taxation would help stabilize prices during the war, thereby reducing the costs of postwar price adjustments ((Goodwin, 1991: 31; Kennedy, 1975: 29).

The unprecedented mobilization of resources in World War I, required new thought with respect to financing policy. Pigou, in a book dedicated exclusively to wartime political economy (Pigou, 1921), elaborated on the issue. He asserts that the first objective of economic activity in wartime is to maximize the war fund, placing the largest possible quantity of resources at the disposal of the war effort. Overtime work and expanding the labor force, primarily with youngsters and women, is insufficient, and there must be a reduction in private consumption and investments. Reallocation through the market may be too slow and to lead to rapid inflation, imposing a disproportionate share of war costs on people with low incomes. Hence the government has to intervene: it must control the supply of essential goods and services directly, and have priority in purchasing necessary quantities at agreed prices, then making the surplus available for private use. Under these conditions, where excess demand was certain, Pigou recommended to apply a rationing system and price control. The basis for rationing cannot be to equalize sacrifice, that is, it cannot reduce consumption uniformly, because there are many individuals whose consumption level is so low that they cannot possibly give anything up. Hence, by analogy with maximum aggregate welfare in peacetime, wartime rationing should be based on minimum aggregate sacrifice. The objective of maximizing the war fund leads Pigou to prefer loans to taxes as a means of meeting government expenses. He explains that high tax rates, particularly in a prolonged war, create negative economic incentives that will deplete the war fund. To reduce the potentially regressive influences on income distribution that may be caused by servicing the debt in the postwar years, he proposes repaying the debt right after the war in one installment financed by a one-time capital tax.

4. NATIONAL SECURITY AND ECONOMIC WEALTH: A MACRO-HISTORICAL PERSPECTIVE

Scholars of the history of civilization think that in ancient times peace seeking cultures attained greater economic prosperity than those that made war a central ideology and way of life (Clough, 1951: 47-48). For later periods, economic historians have no shared idea as to whether wars encouraged processes that advanced human social welfare, or hindered them. Between the two World Wars, Werner Sombert, a professor of sociology and economics in Berlin, and the American historian John U. Nef fiercely debated this issue. Sombert argued, contrary to the Marxists, that war does not result from social structures but creates them. Only wars could have produced the conditions needed for European capitalism to develop: they disciplined the masses, created incentives to set up large-scale enterprises, e.g. arms factories, and rationalized production and trade. Nef responded that though wars produced the need for large economic units to supply fighting materials, they also impeded the development of new economic units that could have served peacetime markets, making wars the revenge of industrialization, not the cause of its development (Winter, 1975: 4-5).

The facts and reasons economic historians advance for assigning war a positive role in developmental processes divide into two main groups: those that indicate accelerating influence on discoveries and technological innovations, and others that represent conditions that spurred industrialization and advanced production methods. The search for answers to military needs brings the finest minds into technology, and the intensity of the war effort, along with the pressure created by the sense of existential danger, encourages bold, unconventional ideas and speeds up the maturation of programs. Economic considerations are pushed aside, and much greater resources are devoted to technological innovation than would be otherwise available. Inventions and technological breakthroughs arising from or accelerated by wars might ordinarily have been delayed for decades or even generations, and once they came into the world for military reasons, they were found to have important uses in civilian life too. Supporters of this view take their examples from widely varied spheres and times: Roman roads, the bellows furnace for separating metals, drills to make cannon barrels, and the fruits of World War II in surgery, electronics and atomic energy.

No less convincing are the arguments for the contribution of wars to industrialization: modern warfare creates the demand for huge quantities, modern military organization must operate according to predefined specifications and uniform standards, all of which dictates mass production methods, standardized processes and large industrial plants. In addition, the modern army requires administrative skills and managerial knowledge; military service provides the experience necessary for operating large units, and these capabilities are eventually transferred to the civilian sphere.

Those who attack the historical interpretation placing war at the center of technological and industrial advancement do so on several grounds. First of all, they maintain, civilian needs have provided more challenges than war needs, and wars simply drew attention away from questions with broad implications and focused it on relatively narrow immediate military requirements. In fact, most technical innovations used by the military were diverted from earlier civilian uses: gunpowder was originally used for work in quarries; the casting methods for producing cannon were known many generations earlier and were used successfully to cast church bells, etc. There is also considerable doubt as to whether technologies used first in military applications actually found their way later into civilian use. Secondly, new developments in industrial organization too came from the civilian sector: the large enterprise originated around salt production and coal mining in response to normal market expansion, while military industry it was that remained split into small units for centuries. Thirdly, important industrial developments came from peaceful regions isolated from the principal theaters of war. In the fifteenth and sixteenth centuries industrial production grew more rapidly in England and in northern Europe, which were remote from war areas in the middle of the Continent, while in the next century in those very countries industrial development slowed down when they became involved in naval warfare. Likewise, the Industrial Revolution took place in England, which was relatively isolated from the Napoleonic Wars, while on the Continent it was delayed until those wars ended and peace prevailed (Clark, 1966: 31-14; Winter, 1975: 5). Fourthly, sometimes the military tended to be extremely conservative and backed away from the technological innovations offered them by civilian entrepreneurs. The British Navy, for example, refused to use steamships on the grounds that refueling would reduce sea time and cruising distance, and the change was forced on it only when it had to make its armored ships less cumbersome (Harkabi, 1990: 60).

To prove their views, those who denied the positive contributions of war also noted some retarding effects that wars had on economic progress. War gives massive encouragement to industries that it needs, they produce far in excess of ordinary demand, thus possibly leading to severe crisis when the war is over. This happened to American agriculture in the Civil War and World War I, and during the latter to the British steel industry and shipyards too. By contrast, industries low on the list of war priorities have trouble obtaining raw materials and other inputs, or transportation problems cut them off from their traditional markets, obliging them to downsize, sometimes to the point where they can no longer survive. In England in World War I, the textile industry was almost paralyzed for want of coal, and lack of transportation led to the loss of important markets subsequently taken over by others (Clark, 1966: 13). Wars retarded science too: The Thirty Years' War, for example, cut short a brilliant period of continuous research in the natural sciences, from Copernicus (1543-1573) to Kepler (1571-1630) (Clark, 1966: 13). The wars confined scientific talent to narrow areas, with preference for applied research at the expense of basic scientific inquiry, interfering as well with the spread of knowledge and reciprocal stimulation among scientists in different countries (Clark, 1966: 20).

Conflicting opinions also developed as to the effect in the contrary direction, i.e. the effect of economic processes on the role of wars in international relations. In the early nineteenth century Auguste Comte, the French sociologist and philosopher, a founder of the Positivist School, developed the view that industry along with scientific and technological progress would change the existing militaristic world society to a peace-seeking one. According to him, if material scarcity was the reason for international rivalry and wars, then the affluence ushered in by industrialization would eliminate the need for wars, and their very raison d'être would disappear. Moreover, since the source of material wealth lies in industrial production, conquering and holding foreign colonies become superfluous. Finally, with industrialization come science and technology, and since these have no national affinity they are likely to bring the homogeneity and solidarity that will eventually overcome the cultural differences and other sources of friction between states. Harmony will reign in international society, and material abundance will thus cause wars to disappear (Harkabi, 1990: 265-266). While Comte stressed the effect of industrialization on international relations, the liberal economists focused their arguments on international trade. They regarded trade as a unifying factor, since it fosters mutual interests and creates interdependence among nations, and maintained that global free trade would inevitably lead to peace, or at least reduce probabilities of war (Baldwin, 1985: 78-79). The perception of economic processes as a calming influence on international relations was popular at the turn of the twentieth century, and still set the tone on the eve of World War I. The chance that war would break out between nations bound together by highly developed economic links was seen as remote, even more so was the chance that if such a war broke out, it would be a long and total one.

But there were other views too. The Marxist-Leninist interpretation of international relations also focused on industrialization, but reached different conclusions: industrialization enriches the capitalists and impoverishes the workers, internal tensions arise, and to overcome them and for additional economic reasons, nations turn to economic imperialism and wars. Another school – the Historical School – arose in Germany in the 1840s. Its exponents were List and Wilhelm Roscher, whose arguments focused on differences between countries in their levels of economic development. They held that as long as there was economic imbalance concealed friction, not harmony, would prevail between nations. Hence rational considerations guide the conduct of international relations only after all countries reach the same level of economic development, while until then competition and reciprocal economic dependence will give rise to international conflicts and it will not be possible to establish peace (Barber, 1991: 67-69).

World War I proved conclusively that industrialization did not abolish wars but empowered them to wreak more destruction. Trade too created new points of friction and sometimes exacerbated international tension. Nonetheless, views that saw economic processes as a conciliatory influence, bringing nations closer together, did not disappear. They were even encouraged by the appearance of multinational corporations, which they regarded as increasing interdependence between nations, and creating a new dimension of integration between national economic systems. After World

War II, some people saw the introversion of national policies and the tendency to focus on raising the standard of living as a new basis for the role of economics and trade in ushering in peace and a new international order. They held out the examples of Germany and Japan, which seemed to show that economic rather than military power determines a nation's position in the global system (Harkabi, 1990: 65, 287).

The judgment of history on the reciprocal influence of wars and economics, then, is far from unequivocal. Wars have had at one and the same time positive and negative influences on economic development. Wars may have spurred the development of new technologies, encouraged the creation of large industrial organizations and helped spread advanced means of production and methods. At the same time, however, wars held back economic development by diverting resources from productive civilian use, and no less important – by distracting public attention from objectives directly concerned with enhancing the welfare of the country and its inhabitants. Economic progress increased the wealth of nations and expanded international trade, at the same time increasing the potential economic damage caused by wars. Apparently, using warfare as an instrument in international relations became less reasonable, but to the same extent economic progress gave wars greater resources and improved operational means, and economic interdependence led to friction and confrontation that often degenerated into actual fighting. Economic forces that might have blocked this trend appear not to have been strong enough to make wars disappear; the more so as it is not only economic factors that explain why wars break out. With the end of the Cold War and the disintegration of the Soviet Union, the view was heard once again that economics will now be more important than security problems, and that economic competition may replace military confrontation (Toffler, 1994: 29-30). Most regrettably, hopes for a new world order dissipated within a short time.

3
ECONOMIC ELEMENTS IN ISRAEL'S NATIONAL SECURITY DOCTRINE

The state of Israel was founded in war, and national security concerns played prominent role in shaping its development throughout the years. Security conditions have affected the social fabric, the political setting, individual life-styles, and in effect almost every aspect of life; they have certainly had substantial impact on Israel's economic development. The main defense economics issues that have been on Israel's national agenda over the years are dealt with in chapter 4. In this chapter, however, the focus is on influence in the opposite direction, i.e. influence of economic considerations on Israel's national security doctrine and defense thinking.

1. THE GROWING INFLUENCE OF ECONOMIC FACTORS

David Ben-Gurion, soon to become Israel's first Prime Minister, accorded defense the top national priority in 1947. He surmised that the Arabs would forcibly try to prevent a Jewish state from arising in the Land of Israel, and that an armed conflict would soon erupt between the small Jewish community and the entire Arab world surrounding it. By establishing defense as the central problem, he sought to assure mobilization of maximum resources for the military effort (Ostfeld, 1994: 23-26).

Shortly after the War of Independence ended in 1949, Israel hoped for a stable peace with her neighbors, and was willing to reduce defense expenditures (see chapter 6, section 1). However, before long it became clear that those hopes were remote. Israel realized that it would have to live with unique security realities for many years, and that defense should be granted a permanent and special place on her national agenda. The security conditions were perceived as unique because of the existential threat to the country, and the extraordinary length of time in which the Israeli-Arab conflict persisted as a "sleeping war" that roused itself every few years. Another major factor was the overwhelming quantitative disproportion between Israel and the Arab states in population, in area and in economic resources (Peres, 1970: Chapter 1; Horowitz, 1985: 57).

Perceiving the security situation as an existential danger and a long-term reality required the allocation of the greatest possible amount of resources to defense. It pushed aside the economic approach that regarded defense allocations as an ordinary problem of choosing between alternatives, and opposed to weighing defense on the same priority scale together with other categories of national expenditure. For almost two decades, the prevailing approach was, then, that defense must be financed according to its needs.

Open public debate about the size of defense expenditures and their price in social and economic terms first took place after the Six-Day War in 1967, and increased with the end of the War of Attrition at the Suez Canal in 1970. Defense expenditures continued to rise, and for the very first time, there was a wave of social protest demanding a larger share of national resources for social purposes at the expense of defense. Participating economists from academia (Gross, 1975) explained that defense could not be given a status separate from other categories of national expenditure. Decisions with respect to allocations affect all uses of resources simultaneously, and any attempt to avoid an explicit choice by dividing the process into two stages – to budget for defense first, and then to allocate the remaining resources among civilian uses – is simply impractical. Moreover, ignoring civilian uses and disregarding opportunity costs of defense contradicted the view that defense was an ongoing problem that had to have an economic and social basis in the years to come as well. However, the public debate and the economists' arguments had little effect on policy, especially since a short time later, in 1973, the Yom Kippur War broke out. After the war, the Israel Defense Forces (IDF) urgently needed to be rehabilitated and re-equipped, so defense expenditures continued to grow.

Defense expenditures and the economic burden they imposed apparently had to escalate further for awareness of their consequences to sink in deeply enough. The change occurred in the 1980s, when it was recognized that Israel had reached the limits of possible allocations for defense, and the conclusion emerged that henceforth defense policy should not be based on quantitative expansion (Rabin, 1979: 19; Inbar, 1983: 22; Dror, 1987: 10).

Over the years, economic impact on national security matters became progressively stronger due to increasing defense budgets and the concurrent growing economic burden, but no less significant was the decline of the hitherto general consensus about national security. Issues once the province of a small circle of decision-makers became themes of public debate, and instead of almost automatic and virtually assured support for the government's decisions on defense matters, there were disagreements and opposed positions. The decline of the consensus developed gradually, and reached the proportions of general shock in 1982. Then, for the first time in the history of Israel's wars, there was incisive and bitter dispute about whether the war in Lebanon was either just or necessary (Yariv, 1985). The barrier fell and the way opened up for the increased influence of non-military factors, economic factors included, in the sphere of national security.

Furthermore, in the early 1980s unprecedented hyperinflation developed in Israel and the need to balance the state budget demanded sharp cuts in government outlays, including the defense budget. In 1985 the government initiated an economic stabilization plan, and its successful implementation reinforced the restrictive attitude toward the state budget deficit.

From that time, then, due to the three factors above – the defense economic burden, the importance now attached to budget deficit size as the key to economic stability, and the controversy around defense issues - defense expenditures were no longer treated as a variable exogenous to the economic system. Economic considerations were now equal to those of purely security criteria in determining the size of defense expenditures.

2. THE BROADER CONCEPT OF NATIONAL SECURITY

Most of the time, the ruling perception in Israel was of national security as the weighted sum of military power and other factors: economic, social, demographic, technological and moral. Therefore, if national security is not a matter of military power alone, allocating excessive resources for defense should be avoided, as it may undermine the country's staying capabilities by weakening its economy or the social fabric. The broader concept of national security, then, implied limits on resources for defense in the interests of national security itself. Ben-Gurion, as Prime Minister and Minister of Defense, was a prominent exponent of this broader perception. In a session of the Knesset (Israel's parliament) in June 1950, he declared: "National security does not rest on the army alone, or on military equipment or the arms industry alone. Our security depends on the general ability of the nation in the economic sphere, in agriculture, in manufacturing, in technology and in the intellect. Even branches of the economy that do not seem to have a direct connection with the army are a decisive factor in war…" (Ben-Gurion, 1971: 141-142). On many occasions he asserted that overall national objectives had to be considered in determining defense expenditures.

Three main issues stood out in this context, demonstrating apparent tradeoffs between the level of defense expenditures and national security in the broad sense. One was the need to weigh present versus future security or, more specifically, to limit present defense expenditures and allocate more resources to developing the economy as a basis for greater defense expenditures in the future (Sadan, 1985). Another was the implication for social instability and society's inner strength of restraining living standards relative to income over a long period to make resources available for defense (Zusman, 1984). The third was "the sweetness and the sting" of reliance on foreign military aid: it gives the economy valuable resources, but also creates undesirable economic influences, and exposes the country to pressures that undermine its political independence (Razin, 1984). These themes are discussed further on.

Manifestations of the broader concept changed over time as military, international, economic and social circumstances changed. In the early 1990s, for example, following international and regional developments apparently leading toward a more peaceful horizon, there was a sense that the share of military power in the national security equation might be reduced

(Inbar, 1996: 48-51). On these grounds, the government indeed diverted the fruits of the rapid economic growth at that time to education and infrastructure development, while defense expenditures were not increased. By contrast, the collapse of the Middle East peace process and the worsening of the regional situation at the turn of the twenty-first century may well bring about new weighting of national security elements in the broad sense, and with that, also new priorities in public expenditures.

3. THE RESOURCE GAP BETWEEN ISRAEL AND THE ARAB STATES

There is a great difference in human and material resources between Israel and the Arab states. This absolute fact, and its far-reaching economic implications, have been and remain of first importance in formulating Israel's national security doctrine and policy. First, it has been recognized that war, specifically defeat in war, has an asymmetrical effect on the two sides. As Ben-Gurion noted in the 1950s, "Our trouble is that we cannot afford defeat, because then we are finished ...they can be defeated once, twice, if we defeat Egypt ten times it means nothing" (Dayan, 1976: 213). After the war in 1973, when the Arab states had huge oil revenues and the gap in resources between them and Israel widened, Sadat, the Egyptian president, remarked cynically that the Arabs could allow themselves another defeat, while Israel could hardly withstand another such victory (Horowitz, 1985: 96). This asymmetry translated into the clear knowledge that Israel would have no second chance and once defeated, would never rise again (Tal, 1996: 49).

Secondly, because of the vast differences in resources, not only defeat but in victory, however great, Israel could not anticipate that the security problem would be solved. The strategic results obtainable through wars were perceived as limited, so that even after victory it was necessary to prepare for another round of war. Ben-Gurion expressed that view in the 1950s (Ben-Gurion, 1971: 219), and it did not change substantially 30 years later when the then Minister of Defense Moshe Arens assessed the matter once again: "We can defend ourselves. We can make the Arabs suffer. We can destroy their armies for a time. But to solve the problem once and for all is beyond our power" (Yaniv, 1994: 18). In the early 1990s, Prime Minister Itzhak Rabin expressed a similar view: "Under the existing circumstances, Israel cannot devise a defense policy in which war or conquest of Arab countries will force [on them] the peace that we want" (Rabin, 1996: 2-3).

Knowing that there was no second chance, and particularly the recognition that war could not resolve the conflict, dictated to Israel a preventive approach and a defensive strategy. As Rabin put it: "If there is no significant goal in initiating a war, the best thing for Israel is to prevent war" (Rabin, 1996: 3). These considerations also underlie the warfare principles and structural concepts on which the Israeli armed forces were based, and particularly the clear priority given to offensive forces over staying capabilities. Strong offensive forces allow for military achievements in a short time – quickly carrying the war into enemy territory, destroying enemy forces, etc. – though not assuring long-term dominance. However, the resource gap does away with Israel's chance to contend successfully in staying capabilities. With that, there was the hope that after a few rounds of quick victory, the Arabs would conclude that Israel could not be conquered on the battlefield, and eventually they would give up the path of war.

4. AMBIVALENT APPROACH TOWARD FOREIGN AID

The resource gap between Israel and the Arab states gave rise to ambivalence toward foreign aid. Under Ben-Gurion's influence in the early 1950s, despite inferiority in manpower and resources, Israel's security doctrine adopted self-reliance as a fundamental principle (Horowitz, 1984: 111-114). It meant, first, "foreigners won't fight our battles", and second, that Israel should maintain autonomous control over military means in order to be able to cope with all potential threats. The self-reliance policy was clearly manifested later in establishing a large and advanced indigenous defense industry (see chapter 9).

Nonetheless, self-reliance was never seen as implying autarky. Policy makers were aware that autarky was impossible either as regards military supplies, or in the sense of renouncing financial help from world Jewry and from friendly states. Given the resource gap, there was a general agreement that in no instance can Israel meet all national defense requirements from her own resources. Accordingly, a distinction was made between military and economic dependence, and to maintain military independence, or freedom of military action, there was willingness to pay in economic dependence.

However, continuous economic dependence might eventually erode military independence as well. Indeed, it has been claimed that in the 1990s, facing new regional threats and admitting economic constraints, there was some erosion in adhering to the principle of self-reliance, and a growing readiness to consider international involvement in regional security arrangements (Inbar, 1996: 57-63).

5. THE ECONOMIC PRICE OF WARS

Apart from the War of Independence and the minor harm done by the Iraqi Scud missiles during the Gulf War in 1991, the wars between Israel and the Arab states did not damage the economic infrastructure or civilian property. There were, however, other economic prices to pay. Since Israel cannot afford a large regular army, it must call up the reserves each time additional forces are required in the battlefield. This disrupts and may even halt economic activity, and the resulting economic losses, some of them irreversible and with long-term consequences, grow in proportion to the length of the war and the mobilization periods. Similarly, Israel's geopolitical position requires it to maintain strategic stocks (fuel, basic foodstuffs, medicines, etc), since supplies from overseas might fail to arrive in wartime, and the longer the anticipated war, the greater their size and costs. Another price element relates to defense expenditures. During war, it is necessary to increase current expenditures for operating and maintaining considerably larger forces, while after the war the army must be rehabilitated and restored to its previous capabilities.

Like the resource gap, the economic price of war was also a key factor in promoting the preventive approach that dominated Israel's defense thinking. Moreover, the high price of war placed at the center of the defense policy a unique combination of deterrence and "decisive victory". Israel's military power was designed to deter the Arabs from going to war, but if deterrence failed, Israel could not afford a prolonged war of attrition because of its economic cost, so had to be able to defeat the aggressor quickly. A doctrine of pure defense – i.e. absorbing the first strike and being able to initiate a second one later – was also impossible, as Israel lacked strategic depth, and centers of population and of economic activity were too close to the front lines.

The combination of defensive strategy based on deterrence, with an offensive doctrine intended to achieve rapid decisive victory, produced a host of principles that together formed the core of security doctrine and policy. It justified preemptive war and escalating the state of "not war and not no war" to actual war. It established the need to conduct short wars and the imperative to carry the war to enemy territory. As noted earlier, it was reflected also in force structure, with clear preference for offensive forces over staying capabilities.

6. OFFENSIVE VERSUS DEFENSIVE DOCTRINES

Besides the offensive doctrine that dominated the force structure and its mode of operation, Israel's national security thinking adopted some fundamentals of a defensive doctrine too. Over time there were occasional shifts in emphasis, mainly as a result of changes in military conditions. However, resource limitations and relative cost considerations were also important in deciding the delicate balance between the two doctrines.

On the surface, an offensive doctrine appears to imply a more complex, more sophisticated military organization, with operational capabilities over longer ranges and substantial logistic support. Thus, compared with a defensive doctrine, it might require more national resources (Levite, 1988: 13-14). But in Israel's defense thinking the offensive doctrine is actually seen as cheaper. First, a preemptive war could eliminate the need to fight on two fronts simultaneously. Secondly, advance preparation and surprise provide tactical advantages, making it possible to use the force most effectively and economically. Thirdly, the war is likely to be shorter. Fourthly, rapidly transferring the war to enemy territory minimizes physical and economic damage to the civilian economy. Finally, an offensive doctrine saves resources by concentrating efforts in relatively limited zones, while a defensive campaign requires dividing forces and deploying them over the entire area (Levite, 1988: 37-38; Tal, 1996: 82).

In the early 1950s, military planning and organization in the main followed the defensive doctrine. A "territorial defense" system based on border settlements was established, which, with a small nucleus of conscript soldiers, had the task of blocking attack for the time needed to call up the reserves. In the planning of that time, then, transition from defense to offense was postponed to a second stage, until the reserves reported for duty. However, in the mid 1950s, after Egypt was massively equipped with armored vehicles, mobile artillery and aircraft (the Czech arms deal), the balance swung towards an offensive approach. It became clear that defense based on settlements was inadequate in the face of the growing force across the border, and that the Israeli army needed an air force, mobile infantry and auxiliary forces with heavy firepower, and a concept of offensive operation. This tendency was reinforced by the lessons of the Sinai War in 1956, and in the coming years emphasis had shifted to armored units and the air force. But the costs of the new weapons systems required heavy investment, so there was no choice but to thin out the "territorial defense" system and civilian defense (Yaniv, 1994: 45-46; Levite, 1988: 42-43).

The approach changed once again as a result of the Six-Day War. The War broke out after a long waiting period and a high state of preparedness, thus demonstrating the limitations of the offensive option, and especially those of a preemptive attack. Besides, Israel now held large conquered areas, with new borders far from civilian population centers and the right, once seemingly self-evident, to launch a preventive war or even a preemptive strike, was now far less obvious. Moreover, given Israel's new strategic depth, military planners could now consider the absorption of an enemy strike without trying to preempt it for economic reasons also; the defensive option became less expensive. Thus while the Israeli Army continued to acquire offensive weapons, the center of political and military strategy shifted to defense (Yaniv, 1994: 226; Horowitz, 1984: 139-140).

The renewed preference for defense did not last long. The Yom Kippur War demonstrated all too clearly the unbearable cost of absorbing the enemy's first strike, even given strategic depth, and with the army deployed along supposedly defensible borders. Nonetheless, the pendulum did not swing back fully to the offensive doctrine, and apparently, at least until the early 1980s both doctrines were promoted in parallel and equally (Inbar, 1983: 17-20). From an economic perspective, it seemed as if resource limitations had vanished, and with them the need to choose between doctrines. The reason appears to be the increase in American military aid that could be used only through purchases of major weapons systems and other military equipment in the United States. Decision-makers perceived the foreign military aid constraint as separate from and independent of the domestic resources constraint, and vice versa, and by referring to each separately they did not consider the various uses of resources as competitive. Indeed, offensive capabilities were developed within the constraints of foreign military aid, while defensive capabilities were promoted within the limits of domestic resources. Later on, however, the interrelations became clear: even when financed by foreign aid, the larger forces have to be trained and maintained, mainly through local resources.

In the early 1980s, realizing the reciprocity between increased forces and increased maintenance expenses, and admitting that it was impossible to allocate greater resources for defense, Ariel Sharon, then Minister of Defense, declared an all-out preference for an offensive approach. Sharon announced that Israel would no longer increase the quantity of armaments at her disposal, but would develop or purchase only replacements for what was outdated. But since the Arab states were arming at an accelerating rate, and doubts arose as to whether Israel could successfully ward off an attack and mount a counterattack, Sharon felt that preemptive military initiative from Israel was justified (Inbar, 1983: 22). He even spelled out explicitly and publicly causes that might lead to such an initiative (Sharon, 1985: 161-162).

The debate over doctrine continued in the 1990s. It concentrated mainly on the implications of technological changes in the battlefield of the future, and more than ever, economic considerations, particularly cost comparisons of alternative options, played an important role. Experts argued that advanced military technologies not only offer enhanced military capabilities, but might also save costs, thus alleviating the economic burden of defense.

7. QUALITY VERSUS QUANTITY

Israel's inferiority in resources made it necessary to devise a national security doctrine, and in particular military operational and organizational principles, that will produce maximum security at any given level of resources. Interpreted in practice, this meant a preference for quality over quantity, or in economic terms: preference was given to the quality of inputs in producing defense and to efficient production technologies that would achieve maximum utilization of available resources.

Optimal utilization of human resources was achieved through the three-tier structure of military manpower. A small permanent career army serves as the backbone of command and of professional skills. Conscript soldiers, after basic and advanced training in the various military professions, carry out most current defense tasks. The reserve army makes up the main force in war times, and to preserve its high military capability reservists are called up at intervals for refresher exercises. Besides, special call-up and emergency mobilization procedures have been devised to assure arrival of reserve units at the front in the shortest possible time. This system enables Israel to make do with a relatively small regular force and at the same time to enlist a large proportion of the population in time of need, thus reducing the impact of its demographic disadvantage on military force ratios. In addition, the system is designed to minimize losses to the civilian economy.

Several questions, however, have arisen over the years. Due to the increasing complexity of the battlefield and of military tasks, as of weapons and support systems, numerous military functions require long training and experience. Hence there have been doubts as to whether it is possible to maintain high, up-to-date abilities through a relatively short annual spell of reserve service. Besides, acknowledging that in "ordinary" years only a small proportion of the reserves are called up, created concerns that the resulting unequal burden sharing could provoke disturbing social consequences.

Raising productivity in defense production by improving the quality of inputs meant promoting the professional skills of manpower, and upgrading the technology of weapons and support systems. The IDF has continuously attached great importance to its human resources and invested in improving their skills, seeing that as the key to advantages in quality. To this end, the army developed a sophisticated and rigorous screening system, making it possible to choose "the finest for flight," or to select the right candidates for officer training and for elite units. In fact, the screening system made it possible to achieve a high degree of compatibility between the supply of qualities and skills among the soldiers and the demands of various military tasks. Furthermore, an advanced system of instruction and training was developed, which combines elements of individual and group competition, thereby stimulating, as a basic norm, endeavors toward high professional standards in every field and at all levels. The professional skills of combat units are directed toward optimal operation of weapons systems: shooting down enemy planes in air battles, accurate and effective firing in confrontations of armored forces, etc. At the same time, soldiers in support units are trained to attain maximum utilization of military hardware: more sorties per day with a given fleet of aircraft, more continuous and longer mobility of armored vehicles, speedy repairs to damaged equipment that returns it to use as soon as possible, etc. All are examples of improved input-output ratios or of higher productivity of manpower in producing defense.

High quality manpower is also a precondition for absorbing modern technologies. Over the years, the IDF maintained a policy of technological superiority that would balance out its quantitative inferiority. This technological superiority showed itself notably in the air, but there were many examples in other arenas too. On land, the IDF was ahead of the Arab armies in introducing mobile artillery, and Israeli tanks had improved firepower control systems and operated advanced ammunitions that the Arabs did not have. At sea, the Navy made use of sea-to-sea missiles indigenously developed that outperformed any Arab navy similar equipment. Stated in economic terms, the IDF aimed not only at higher productivity of manpower, but also – through giving priority to technologically advanced weapons systems – at higher productivity of the military capital.

The military capital productivity issue became more acute as the prices of advanced weapons systems rose drastically, while increases in defense budgets slowed down. In the later 1980s, there were arguments for a new formula of force structure combined with a more effective warfare doctrine that would make it possible to reduce both army and budget size (Wald, 1987; Yaniv, 1994: 386-390; Levite, 1988, 120-121; Pedatzur, 1990; Dror, 1989: 80, 85-86). The idea was to use precise, guided munitions and other sophisticated weapons extensively, on the assumption that they were more cost-effective than statistical weapons. Proponents of this view also recommended that the new policy should incorporate fruits of indigenous research and development as much as possible. They asserted that unique indigenous developments that cannot be acquired in the international arms market, and do not reach Arab hands, create possibilities of tactical surprise, since their capabilities remain unknown until used in battle, and that could add up to strategic advantages. Finally, it was in line with the traditional approach to prefer quality to quantity, as it made possible to broaden further the technological gap between IDF and Arab weapons. It was also seen to rest on Israel's comparative advantage: the Israeli research community's intimate knowledge of new developments in microelectronics, computers and sensors, and the high quality manpower that could absorb and make optimum use of weapons systems based on the newest technologies.

Efficient production technologies are an outcome of a wide range of factors, many of them formulated in rather vague terms, and are thus hard to define and even harder to quantify. They include operational flexibility, adherence to tasks, quality of command and control, initiative and improvisation, communications and reporting credibility, fighting spirit and the like. All find expression on the battlefield and, with some adjustments, in logistic activities performed in the rear. In these Israel could obtain an advantage over her opponents by combining social and military-doctrinal factors, and this qualitative advantage was perceived as an essential and integral part of the overall effort to compensate for quantitative disadvantage. As an example, the principle of operational flexibility determines that manpower and other resources need not be assigned in advance within rigid formations. Instead, a more economical course of action would be to assemble forces and use them as developing conditions indicate. Another example prefers adhering to a task over adhering to a plan. It asserts that in face of changing circumstances every plan can be regarded as a basis for adjustments. Moreover, with credible reporting, judgment and decision making about such adjustments can be delegated to lower echelons closer to the field, thus allowing for more rapid reaction and for savings of resources (Horowitz, 1985: 65-67).

4
DEFENSE ECONOMICS ISSUES IN ISRAEL

In the 1950s and 1960s, prior to the 1967 War, open arguments about national security were virtually unknown in Israel. National security enjoyed a broad consensus, and besides, there was concern lest public argument on defense-related subjects be construed as a sign of national weakness. As for defense economics issues, there was another reason too: the economy grew very rapidly, and growth in defense expenditures was not exceptional compared to other areas. Defense expenditures were not thought to harm economic stability or to hinder growth. After the 1967 War, professional economic interest in defense arose as defense expenditures were growing rapidly, as was their relative weight in the economy, and national consensus on defense issues began to fade.

1. MEASURING DEFENSE EXPENDITURES

Defense expenditures as reported in the government budget were incomplete: some were kept secret, others were recorded under "civilian" budget headings, and reported data included expenditures on tasks performed by the military but unrelated to defense. However, it was not merely a matter of confused categories. There were also price distortions: an unrealistic exchange rate undervalued the cost of imports, and low, administratively-determined compensations to conscripts and reservists undervalued the real economic cost of military manpower (Patinkin, 1965: 52-54). Without an improved measurement of expenditures, the practical worth of discussing "How much is enough for defense?" was limited.

Several attempts were made to reach more comprehensive evaluations. For the years 1955-1969, price adjustments of conscripts' and reservists' compensation and of defense imports added some 15 to 25 percent to reported expenditures (Kohav & Lifshitz, 1973: 4; Lifshitz, 1974: 88-90). A later evaluation, to the end of the 1970s (Berglas, 1983: 17-23), augmented price adjustments of labor and imports with expenses outside the defense budget, such as costs of civil defense, stocks for civilian use in wartime and veterans' pensions. In these years, while distortions in foreign currency items dwindled to negligible size, underestimation of local currency expenditures had swollen to 50 percent or more of the amounts reported. Another study extended to the 1980s, showed similar results (Ben-Zvi, 1993: 23-26). The Central Bureau of Statistics (CBS) published its own evaluation for 1993-1995 (CBS, 1996). It added to accepted definitions an estimate of the economic value of regular and reserve compulsory service, imputed insurance costs to account for the higher personal risk of military service, expenses for building shelters, for maintaining stocks of essential products and defense-related expenses of civilian ministries. The estimates indicated that costs not recorded in the Ministry of Defense budget amount to over 30 percent of domestic defense consumption.

Relative measures of defense expenditures were controversial too. Usually, relative weight of defense in the economy is measured by the ratio of defense expenditures to GNP, but in Israel it might be overstated. Defense expenditures in the numerator include a relatively large element of direct and indirect imports, while the denominator, relating only to GNP, ignores imports and import surpluses that always contributed substantially in the Israeli case to available economic resources. From the 1970s onward, when most defense imports were financed with American military aid, calculations became even more complicated. Since the military aid could not be used for anything except financing defense imports from the donor, it was argued that such imports should be neutralized when calculating the economic burden of defense. Eventually, it was accepted that for estimating the relative weight of defense in the Israeli economy several indices should be considered in parallel. A study that brought together definitions of the defense burden from various publications and articles found no fewer than 15 alternatives, and they varied, for 1981, from 8.8 to 27.9 percent (Gilshon & Beenstock, 1989).

2. DEFENSE BURDENS IN ISRAEL AND IN THE ARAB STATES

Systematic comparisons between the defense burden of Israel and the Arab states were first made only in the late 1960s (Safran, 1969; Shefer, undated). These showed that Israel succeeded in improving the ratio between her defense expenditures and those of her principal adversaries without significantly increasing the relative economic burden, chiefly because Israel's economy grew more rapidly at that time (Safran, 1969: 170-171). Closing the defense expenditure gap continued until 1972, but in the next ten years the trend was reversed. Moreover, as growth rates in Israel slowed down, and as Egypt, Syria and Jordan enjoyed relative prosperity, the share of defense expenditure in Israel's GNP rose, while in the neighboring countries it fell somewhat (Shefer, 1985; 1987).

Another comparison of the defense burden over time was attempted by using estimates of military capital stocks. It covered 1973-1984, and showed the combined military capital stocks in Egypt, Syria and Jordan to be more or less equal to Israel's. However, the defense burden – this time measured by the military capital to GDP ratio – rose much more steeply in Israel than in the confrontation states (Halperin, 1986).

A third method examined changes in the balance of power between Israel and her neighbors over time by employing the economic theory of production functions. The idea was to place quantitative estimates of military manpower and capital within production functions, and thus to derive weighted measures of relative military power for the rival parties. A comparison made with a Cobb-Douglas production function concluded that from the early 1950s till mid-1980s, the balance of power did not change significantly (Rotem, 1988; 1990).

However, there were serious reservations about using economic data to evaluate security threats and military force ratios, and even more so about referring to the Arab states' defense budgets in determining Israel's defense spending. First of all, there were doubts as to data quality: Arab states do not report their defense expenditures fully, and particularly do not include most of their defense imports, a major cost item, in official publications. Moreover, expenses are stated in local currencies, and for comparisons it is necessary to choose an appropriate exchange rate, an exceedingly complex problem in economies with a multiple exchange rates system (Shefer, 1985; 1987). Secondly, it was argued that economic analyses are based on data from the past, and as regards the future, they can at most indicate long-term economic potential, while security threats may emerge in the short term even when they contradict long-term economic trends (Nevo, 1975). Thirdly, the reservations called attention to the narrow economic base of both Israel and the Arab states, which makes it impossible to maintain the arms race without massive foreign aid. Thus analyzing local economic potential was said to be of minor importance at most (Allon, 1988: 169). In view of all the reservations above, and perhaps because of a general attitude that underestimated economic influences on defense issues, Israeli defense policy makers did not usually pay sufficient attention to the economic constraints of the Arab countries. This changed after Sadat's peace initiative in the late 1970s, attributed at least in part to economic motives, and even more after Syria's economic decline in the mid 1980s, clearly manifest in the military area as well. From then on, there is an increased tendency to incorporate economic data on Arab countries in military evaluations.

3. DEFENSE – AT THE EXPENSE OF WHAT?

The continuous growth of defense expenditures after 1967 put the guns versus butter issue on the public agenda, and with it, the need to consider the opportunity costs of defense (Bruno, 1975; Michaely, 1975). This time the social and economic costs of defense expenditures – the forgone alternative civilian uses, the larger balance of payments deficit and the resulting foreign debt – were presented explicitly, not only the implications for national security in the broader sense (see chapter 3, section 2).

At first, the economists' approach was not generally accepted, and was often criticized emotionally for slaughtering cows properly considered sacred. One argument, however, was to the point: it claimed that since defense expenditures generated employment and promoted industrial development, technology and exports, not only opportunity costs but indirect benefits too must be weighed when deciding on defense budgets. Economists responded to this argument, and while not denying the positive side effects, asserted that there are direct and more cost-effective ways to increase employment, encourage economic growth and improve the balance of payments.

Explicit discussions of the opportunity costs of defense led to empirical attempts to quantify them, answering the question "defense – at the expense of what?" An analysis relating to 1955-1969 estimated marginal substitution rates of alternative uses of resources, and showed that when defense expenditures had to be increased, the preference was to cut investment and to a lesser degree – to restrain consumption. However, increased economic dependence on other countries through import surplus was avoided (Lifshitz, 1974: 97-102). Another work distinguished between standard of living and quality of life, as expressed in the extent and level of public services, and concluded that defense needs harmed the quality of life more than the standard of living (Zusman, 1984: 24). In the later 1970s and the 1980s, when the rate of GDP growth decreased, attention focused on the effect of defense expenditures on economic growth (see section 5 below).

4. DEFENSE – AT WHOSE EXPENSE?

Defense opportunity costs may be influenced by the financing methods of defense expenditures, whether taxes or loans. In the first years, the perception was that defense expenditures secure the future of the state, and like any other investment for the future can be passed on to the next generations through loans. But financing investments through loans assumes that when the loan is due, the return on those investments repay it. By analogy, financing defense spending through loans, assumes that future security conditions will allow for lower expenditures in the coming years. In the Israeli reality, however, that assumption is virtually incorrect. Due to the continuous nature of the Israeli-Arab conflict, the burden of repayment adds on to the current one, and together they create intolerable conditions for a properly functioning system of public finance. Moreover, loans, in particular compulsory loans as was the practice at that time, were seen to contain a tax element, so that in any case the lender generation, not the next ones, bears the main burden (Ronen, 1968). That being so, the view that defense should be financed through taxes grew stronger in time (Zandberg, 1970).

Financing defense expenditures through taxes has difficulties too. It means relatively high taxation, and an inevitably increased "excess burden" that raises the overall cost of defense. Furthermore, since defense expenditures tend to fluctuate, sometimes rather steeply, adjusting taxes accordingly may disrupt overall economic activity, adding once again to defense costs (Berglas, 1983: 5, 39). Moral points were made too: supporters of taxation and belt tightening thought this would demonstrate a self-help effort and increase the chances for foreign aid (Frumkin, 1970). By contrast, concern was expressed that taxation and a lower living standard especially in difficult security conditions, would harm public morale and undermine its staying power (Berglas, 1970).

Financing methods determine who bears the defense burden, since they affect the distribution of income. Loans have a regressive effect: obviously for people with high incomes, government loans are an alternative saving to a far greater extent than they are for low-income people. As for the taxation, it has been claimed that were it not for high defense expenditures, Israel could do with a quite different tax system: income tax, which is progressive, would play a greater part, and there could be fewer indirect, and by definition regressive, taxes (Avnimelech, 1978: 113).

An interesting point was raised as to possible regressive effects on income distribution of defense expenditures. It argues that since defense is a public good, all individuals consume it in equal quantities. However, the utility – or imputed income – derived from an equal quantity of defense services is not the same for all individuals: it varies according to differences in their marginal substitution rates between defense and income, which are inversely related to their marginal utility of income. Since the marginal utility of income is less for those with higher incomes, they derive higher marginal utility from the same quantity of defense than do people with lower incomes. According to 1972 data, while the Lorenz index of disposable income inequality for salaried employees was 0.266, the inequality index of imputed income from defense was more than twice as high – 0.522 (Avnimelech, 1978: 114-116).

Naturally, the regressive influence of defense expenditures, and of financing them, on income distribution became severer as defense spending grew and assumed greater weight in the economy.

5. DEFENSE EXPENDITURES AND ECONOMIC GROWTH

Like the defense economics literature in general, Israeli opinions too disagree about the effect of defense on economic growth. Those who claim it has positive effects, point to the technological and scientific advance of Israel, and to the emerging sophisticated modern defense and civilian industries that substituted for imports and accomplished outstanding export records. For them, all these beneficial developments were clearly associated directly and indirectly with defense expenditures. Their opponents mention three types of possible retarding influences. The first is on domestic factor employment. On the demand side, due to the larger direct and indirect import component in defense, increasing defense spending at the expense of civilian uses may reduce overall demand for domestic output. At the same time, on the supply side, since production of defense is skilled-labor intensive and requires considerable amounts of foreign currency, increasing defense spending might create structural problems: limiting the availability of imported raw materials for civilian use, or crowding out scientists and engineers of civilian industries. The second influence is on the growth rates of labor, capital and overall productivity that eventually determine the economic growth potential. Allocating resources for defense at the expense of alternative uses may, *inter alia*, reduce domestic investment and thus slow down the growth of physical capital stocks. Security conditions may negatively affect immigration or even lead people to emigrate, thus influencing the size of the population and of the labor force. The growth rate of the civilian labor force is directly affected also by the rate of military enlistment. Finally, the relative size of the defense sector influences the competitiveness of the economy, the scope of research and development in civilian areas etc., all of which find expression in a lower rate of growth of overall factor productivity than otherwise would be possible. The third influence is on economic policy: since rising defense expenditures make for budgetary deficits and inflationary pressures, thus enlarging the balance of payments deficit and the external debt, the government must sooner or later adopt a policy of economic restraint that may hinder growth.

The possibility that defense expenditures hinder growth suggests that by reducing them temporarily, to allow for increased investment and output, it might be possible to expand the economic base for greater defense expenditures in the future. This dilemma became more acute when for over ten years, until the mid 1980s, the economic growth rate continued to be slower than anything Israel had previously known. However, empirical examinations did not support that course of action. For 1955-1969, according to econometrically estimated rates of substitution between defense and other uses of economic resources, when defense consumption increased, investments decreased just a little more than private and civilian public consumption combined did (Lifshitz, 1974). For a longer period, until the early 1980s, once again no clear association could be found to show that a lower level of defense expenditures would have brought about positive changes in investments and growth (Gilshon, 1986: 16-20).

The relationship between defense expenditures and economic growth was also estimated using simulations. At first, defense consumption share of GDP was reduced by one percentage point, assuming that released resources would increase investment and accelerate GDP growth. In the next stage, calculations were made to figure out how long would it take, given the faster growth rate, to return to the initial absolute level of defense consumption without imposing a greater defense burden (i.e. without raising the defense consumption share of GDP). With data of the early 1980s (Zusman, 1985) and of the mid 1980s (Gilshon, 1986: 27-29) it was verified that the required period would be very long, or alternatively, that the required "defense deficit" would be of impractical magnitude, and in view of the prevailing security risks it was not a realistic option. Moreover, there was no guarantee that resources released by defense will find their way to investments, and not, say, to higher standards of living. The conclusion could not be avoided, then, that reducing defense expenditures cannot be relied upon as a strategy for renewing growth in the Israeli economy (Hasid & Lesser, 1981: 44; Katzir & Shadmi, 1984; Tropp, 1989).

6. DEFENSE EXPENDITURES AND ECONOMIC STABILITY

Discussing the influence of defense expenditures on internal and external economic stability in Israel calls for the analysis of three subjects: the role that changes in defense spending may play in a stabilizing, anti-cyclical economic policy; direct and indirect effects on employment, and direct and indirect effects on the balance of payments.

6.1 Defense Expenditures as a Means of Stabilizing Economic Policy

In 1952, the government required a new economic plan to cope with a rapid inflation and a severe lack of foreign exchange. The government undertook to restrict its regular budget expenditures to current tax revenues, and Ben-Gurion, as Minister of Defense, decided to cut the defense budget by 20 percent, mainly through reducing military manpower. Yigael Yadin, then Chief of Staff of the Israel Defense Forces (IDF), assumed that with such budget and personnel cuts, the IDF could not carry out its duties, and he resigned. This event, however, was exceptional in Israel's defense economic history, and defense expenditures have not generally been used as a means of fiscal restraint. Nonetheless, it provided two important lessons, as it later appeared. Firstly, the state budget structure with the relatively large share of defense expenditures, and the political balance of power among the various government ministries, make it impossible to reduce government expenditures without a large, effective cut in those related to defense (Barkai, 1990: 63). Secondly, to attain a large, effective cut in defense spending, its objects must be precisely specified (Greenberg, 1991: 20-21).

Once again, 30 years later, a large cut in the defense budget was needed for fiscal reasons, then too because of an economic crisis and galloping inflation. Research on Israel's inflation in the 1970s and 1980s finds its root in the huge budget deficit, beginning in the late 1960s, principally due to the rise in defense expenditures (Barkai, 1980, 21-24; Bruno, 1989: 93-94, 98). Likewise, the successful stabilization policy in 1985 is credited to the fiscal balance attained, with general agreement that cutting defense expenditures contributed substantially to the outcome (Bruno, 1993: 103-104, 128-130). It might be argued correctly that the inflation could have been stopped had the deficit been reduced by cutting back other expenses. But in both the 1950s and the 1980s, and in most of the years between and afterwards, the way to fiscal balance in the prevailing budget structure went through defense expenditures. Their relatively great weight in Israel's public spending gives them a special status with respect to inflationary processes and other aspects of economic stability, even if their effect per se does not differ qualitatively from any other category of public expenditures.

Just as reducing defense expenditures serves as fiscal restraint, so expanding them can stimulate economic activity. The hypothesis that defense expenditures have been raised to improve economic conditions was examined for 1960-1984 (Mintz & Ward, 1989). The findings indicated that while security considerations decisively determined defense expenditures over the years, election campaigns and deterioration of defense companies' profitability also had some minor influence. Specific examples of defense expenditures used to ease unemployment can be found in the 1966-1967 recession (Greenberg, 1997: 70), and again in the early 1990s. On the latter occasion, special funds were appropriated for creating jobs in defense industries to help absorb the huge wave of immigrants from the former Soviet Union (Ministry of Finance, 1991: 37).

6.2 Effects on Employment

Domestic defense consumption creates a demand for labor directly, and through domestic purchases – also indirectly. Besides, the overall demand for labor in the economy might be affected by fluctuations in reserve service. As reserve service implies temporary absence from work, employers seeking to assure uninterrupted operation may demand more workers than they actually require (Klinov, 1993: 8). During the years when defense employment was expanding, there were arguments that it created a manpower shortage, pushed wages up and exacerbated inflation (Neubach, 1984: 180).

For the period from the mid 1950s to the late 1960s, no statistically significant links were found between changes in domestic defense consumption and unemployment rates, despite considerable growth in the regular army and in the domestic defense consumption share of GDP (Lifshitz, 1974: 94-95). The main explanation lies apparently in the rapid growth of labor supply during those years; expanding defense employment helped absorb the new labor force participants that otherwise would have been unemployed. A study of the 1970s reached a similar conclusion under different circumstances. It was found that the growth of the regular army and of reserve service were central in preventing increased unemployment, despite slower economic growth at the time (Klinov, 1993). An analysis based on 1980s data verified the assumption that defense demand was relatively labor intensive: calculations showed that diverting resources from defense to improving living standards, without increasing foreign debt, might have raised the level of unemployment (Katzir & Shadmi, 1984: 121-122).

6.3 Defense Expenditures and the Balance of Payments

The main effect of defense expenditures on the balance of payments is through direct and indirect defense imports, the latter an estimate of the import component in domestic defense purchases. In fact, since indirect defense imports are an approximate figure, the exact scope and influence of overall defense imports are not well defined. It should be also noted that as import components differ from one industry to the other, overall defense imports may vary not only with changes in the volume of domestic purchases, but also with changes in their composition.

Neither is the export side sufficiently clear. Domestic defense demand led to the establishment of an indigenous defense industry that exports a considerable part of its output. Usually defense exports require ministerial permits, but there may be additional exports of unknown volume that do not require permits, yet come directly or indirectly from the same industrial infrastructure. Moreover, in many cases defense exports were pioneers, opening foreign markets to civilian goods. Defense contribution to overall exports might then be considerably larger than implied by recorded defense exports. By contrast, security conditions also impeded Israeli exports, mainly because of customer' doubts whether, under certain circumstances, orders could be delivered steadily and on time. That was evident, for example, in the months preceding the Gulf War of 1991 (Lifshitz, 1991: 317), and again later, when the security situation deteriorated due to Palestinian terrorism.

Israel's foreign trade was also affected by the Arab boycott. While the exact quantitative influence is unknown, undoubtedly Israeli exports were kept out of large potential markets, and restrictions on import sources forced Israel to pay more for imports like oil (Elizur, 1997).

The security situation had its direct and indirect effects on unilateral transfers and on the capital account in the balance of payments too. The most conspicuous direct effect appeared in military aid from the United States, first as special credit and later, for many years, as grants (see chapter 11). However, civilian aid from the United States, usually a smaller volume, was also related in part to defense realities. Indirect effects appeared in donations and loans from world Jewry: especially during the first decades, such transfers grew whenever the security situation worsened and the economic burden became heavier. On the other hand, security-induced uncertainties were without a doubt a key factor in retarding substantial foreign investment. Only in the 1990s, following the Middle East peace process, when conditions for relative stability emerged, did foreign investments to Israel grow.

7. DETERMINANTS OF DEFENSE EXPENDITURE TRENDS

Econometric research on factors affecting the development of Israel's defense expenditures is relatively limited. For the 1960s, one study (McGuire, 1982; 1987) found that the propensity to spend on defense out of available income (GDP plus private transfers from abroad) was positive though relatively small, and that it gradually diminished over the years under review. It also found that defense expenditures of the Arab states influenced Israel's indirectly more than directly, by increasing the United States willingness to extend military aid to Israel. As regards American aid to Israel, there were two additional important side effects: it enabled Israel to maintain relatively high investments, thus contributing to its economic growth and making it possible to devote resources to defense; and it affected relative prices, in particular reducing the relative price of defense consumption, thereby encouraging it to expand.

Another study (Rotem, 1985; 1988) examined data for a longer period, 1950-1987, and pointed out to strong, statistically significant association between Israel's defense expenditures and those of her immediate neighbors. By contrast, the peripheral Arab states (Iraq, Saudi Arabia and Libya) had an asymmetrical influence: lower defense expenditures there had little effect, because the main threat continued, while increased defense expenditures on their part was seen as adding to the existing threat, spurring Israel to further increase her defense expenditures. The findings showed another asymmetry too: when defense expenditures of the confrontation states increased, indicating a higher level of threat, Israel's GNP had no statistically significant impact on her defense expenditures. However, when the enemy's defense expenditures decreased and less threat was perceived, the influence of GNP became significant. A consideration of economic constraints emerged, then, mainly at times when defense tensions relaxed.

Other studies (Ward & Mintz, 1987; Mintz & Ward, 1989; Mintz, Ward & Bichler, 1990) distinguished between domestic expenditures and defense imports, and looked for separate explanatory factors. However, since defense imports in one period imply increased domestic expenditures later, imports were taken as an explanatory variable in the equation of domestic expenditures. Surprisingly, these studies do not consider GNP or GNP per capita as factors that might explain the course of defense expenditures. It was found that from the early 1960s to the mid 1980s, Israel's domestic defense expenditures were influenced mainly by those of the confrontation states and by the consequences of wars. Elections too increased domestic defense expenditures (apparently an evidence of internal political advantages), as did a downward trend in the profitability of defense companies (apparently an evidence of the military-industrial complex influence). Contrary to expectations, however, no statistically significant association was found between domestic expenditures and defense imports, or any significant influence of current economic conditions (measured by the rate of inflation). As for defense imports, they were almost exclusively determined by American military aid, particularly since 1967. The researchers noted the very slow pace at which defense expenditures adjusted to demand as determined by the model, and concluded that current security conditions are of less weight than long accumulated experience in determining defense expenditures. This conclusion, however, has been criticized on grounds of econometric technicalities (Beenstock, 1998: 197)).

A later study (Beenstock, 1998: 199-207) for 1960-1994, showed, firstly, that the long-term elasticity of defense spending with respect to GNP was positive and greater than unity, and that the elasticity with respect to population size was negative and even larger. It follows, then, that there are substantial economies of scale in defense expenditures, meaning that protecting a larger population requires a smaller per capita expenditure, and that only if GNP grows faster than population, that is, when GNP per capita grows, the defense expenditure share of GNP might grow too. Secondly, the long-term elasticity of Israel's defense expenditures with respect to those of the confrontation states was positive and less than unity, suggesting that the Israeli response to Arab defense spending did not lead to escalation and destabilization in the regional arms race. Thirdly, the elasticity of defense expenditures with respect to their relative price was negative and greater than unity; i.e. there was a tendency to economize on defense expenditures when it became expensive relative to other uses of resources. Fourthly, over longer time periods, American military aid was not an important determinant of Israel's defense expenditures, and with adding the civilian aid the overall influence was even negative, though small. That surprising result does not indicate that the American aid was not important, but rather shows that Israel accords high priority to national security and would have allocated economic resources on a similar scale in any case. However, if that was so, the higher level of defense expenditures would have set Israel on a completely different economic path, eventually reducing its ability to carry the defense burden, and leading to a different equilibrium mix of resource uses, with a lower level of defense expenditures.

Despite the different findings in the studies above, two general conclusions emerge. Israel's strategic circumstances forced her to give defense a very high priority, but not only the security threat determined the level of resources allocated – economic considerations also played an important role. American aid in itself did not determine the demand for defense expenditures, but nonetheless, its indirect effects on overall economic developments, on the ability to bear the defense burden and on relative prices were very important indeed.

8. MANAGING THE DEFENSE BUDGET

Effective management of the defense budget is essential for efficient defense production. This was affected by the reciprocal relations between the Ministry of Finance and the Defense Establishment (the Ministry of Defense, its civilian side, and the IDF, its military side), as well as by the different views within the Defense Establishment itself.

8.1 The Ministry of Finance and the Defense Establishment

Throughout the years, what was really at issue between the Finance Ministry and the Defense Establishment is the size of the defense budget, and even when seeming to discuss particular questions, both sides always weighed them as to their implications for its overall scope. One such issue was comprehensiveness. The Finance Ministry consistently sought to include in the defense budget as many defense-related items as possible for two main reasons. On the macro level, it assumed that presenting the full extent of resources devoted to defense would convince decision-makers and the public that reductions could be made. The micro level idea was that only if all expenditures were included under one budgetary constraint would the Defense Establishment consider tradeoffs by the same, single scale of priorities, thus attaining an internal optimization that would make it possible to economize on defense expenditures, or alternatively, to produce more security at any given expenditure level, thereby avoiding pressures to increase the budget. On the face of it, it seemed right for the Defense Establishment to agree to the Finance Ministry's approach, which gave it control over larger sums. In fact, however, there was opposition in most cases: there were concerns that in future years when items transferred from other budgets had to be increased, the Finance Ministry would see it as an internal affair of the defense budget, unavoidably impairing essential military matters. In the course of time, many changes were indeed implemented in line with the Finance Ministry approach. Nonetheless, some deviations from the principle of comprehensiveness remained, often interfering with efficient management of the defense budget.

Another still unresolved issue relates to the multi-year nature of defense spending, given the length of time needed to building a military force. From time to time the Defense Establishment presented multi-year programs that included budgetary implications, but the Finance Ministry would not commit to them, fearing to lose flexibility in managing its economic policy. The result for the Defense Establishment was great uncertainty, and in many cases it had to alter plans based on financing assumptions that did not materialize. The frequent changes of priorities and plans inevitably thwarted efforts to derive the optimal level of defense from limited fiscal resources.

The numerous issues in the defense budget that required professional expertise prevented the Finance Ministry from being effectively involved in its details. Hence the preference was to maintain a distance, and with it to gain some immunity to pressures for new budgetary requests. Most of the time, then, the Finance Ministry was content to preserve the general framework of the budget, and its division between local and foreign currency. Yet there were exceptions. The Finance Ministry would intervene in internal decisions of the Defense Establishment whenever convinced that they might affect other segments of the public sector or interfere with central economic policy issues, most notably wage policy (Gadish, 1984: 187).

The lack of detailed involvement on the part of the Finance Ministry, and in fact, the non-involvement of any outside factor at all, made the government exclusively dependent on the Defense Establishment in determining the particulars of defense expenditures. No alternatives to the plans proposed by the Defense Establishment were presented, and usually the government did not carry out independent analysis before making decisions. Without external administrative checks and balances, it could be assumed that the Defense Establishment, which operates in a non-market environment, would become inefficient.

8.2 Inside the Defense Establishment

If, despite all this, defense resources are managed with acceptable efficiency, credit is due to the internal checks and balances. As mentioned before, the Defense Establishment is a dual structure: on one side is the IDF headed by the General Staff, and on the other – the "civilian" Defense Ministry. Not surprisingly, the two have often been unable to see eye to eye on budgetary issues. In the early 1950s, the Defense Ministry demanded responsibility for the budget on grounds that a civilian authority is better positioned to manage it in line with the broader perception of national security. By contrast, the IDF maintained that only the General Staff could properly evaluate security threats, take calculated risks and propose appropriate solutions, and therefore it and only it could be held accountable for an effective budget (Greenberg, 1993: 53-54).

In 1953, in his capacity as Minister of Defense, Ben-Gurion intervened in the controversy and determined the authority and responsibility of the civilian wing as regards budgeting and financial matters (Ben-Gurion, 1981: 8). A few years later, the pragmatic attitude of the then Chief of Staff, Moshe Dayan, coincided with the principles outlined by Ben-Gurion. Dayan wanted the army released from tasks outside its main function of preparing for war, and agreed to a working formula that handed over to the Defense Ministry hegemony in financial matters (Greenberg, 1993: 67). The way opened, then, for a greater involvement of the Defense Ministry in the budget, and soon a clear change in policy appeared: allocation of resources to domestic defense industry increased, either by direct financing of investments, or indirectly, by preferring domestic purchases to imports. Ministry heads justified that approach, *inter alia*, by asserting that it promoted and strengthened the economy. On the other side, however, the General Staff thought that diverting resources from off-the-shelf purchases abroad to domestic developments and production, meant higher uncertainty as to quality and delivery time, and thus undermined preparations for war (Greenberg, 1993: 69). What was at stake, then, is the relative weight of factors beyond immediate military needs, and particularly economic factors, in managing the defense budget.

During the 1970s two trends gained momentum: the expansion of each Service's budget at the expense of the central budget hitherto controlled by the Defense Ministry, and the transition to a "consumer" rather than a "supplier" approach in budgeting. Both trends had a common rationale: in a large, complex, widely dispersed organization, allocation of resources according to a detailed central planning cannot possibly be efficient, and solutions through internal quasi-market mechanisms are preferable. Consequently, "final consumers" were allowed to decide what and how much they need, while the central budget gradually became limited to spending categories that demonstrate externalities or clear differences in time preferences between individual units and the military at large, basic research and development for example. A complementary step was to set up "closed economic entity" rules for units within the IDF that supply goods and services to final consumers, e.g. maintenance centers.

At that time, economists – usually enthusiastic proponents of market mechanisms – were gaining influence in the Ministry. They felt that transferring the center of gravity in internal budget decisions to final consumers, and introducing quasi-market mechanisms, would make for more efficient use of resources, so seemed willing to give up some of the Ministry's positions of power. However, there were at least two other reasons for the successful incorporation of economic principles into internal allocation processes. Firstly, the 1970s rapid expansion of the Defense Establishment, and the increased professional specialization and technological sophistication that accompanied it, reinforced the Services' and professional corps' demand for greater autonomy in budgetary decisions. Secondly, emerging acquisition possibilities in the United States on one hand, and growing export opportunities for the domestic defense industries on the other hand, somewhat reduced the weight of the argument traditionally held by the Defense Ministry, namely that the budget had to support domestic development and production capabilities.

The search for organizational structure, division of authority and work procedures that would assure efficient use of defense resources continued in later periods too. For example, in the late 1970s, the Director General of the Defense Ministry proposed to eliminate redundancies by a division of labor based on maturity time of projects. He suggested that responsibility for procurement of short term projects, up to three years, would lie with the IDF, while in longer projects the Defense Ministry would be responsible (Ma'ayan, 1985: 302-303). Another influential factor was the desire to keep military personnel from direct contact with suppliers. This trend accelerated due to an unprecedented instance of corruption and fraud (the Dotan case) that came to light in the late 1980s. As a result, civilian control over procurement became broader and closer, and some say that the shock and overreaction so upset the balance between the military and civilian wings of the Defense Establishment that procurement processes became cumbersome and less efficient. In the mid 1990s, however, stronger voices were heard to the contrary, asserting that it was the IDF that accumulated undue influence over the budget. Behind the arguments was the Defense Ministry's inability to extend financial support to domestic defense companies, which after the Cold War went through a severe business crisis due to declining international arms markets. The critics held that the greater influence of the IDF reinforces short-term considerations, while neglecting those for the longer term, and in particular underestimating the need to preserve the technological infrastructure of the domestic defense-industrial base (Bonen, 1995; Tal, 1996: 106).

9. MICROECONOMIC ISSUES IN DEFENSE MANAGEMENT

Decentralizing budget processes made it necessary not only to transfer responsibility to final consumers, but also to provide them with price lists for goods and services. Budget costs do not always reflect opportunity costs – costs of conscripts and reservists are clear examples – and it became necessary to set special costing principles in these cases. In this context, a particular question arose with respect to the pricing of American military aid. Since American grants could be used exclusively for financing acquisitions of military equipment from the United States, this equipment appears to bear no opportunity costs. Economists argued that this might be partially correct for the national economy as a whole, but the case is different for the Defense Establishment. At any given time, as aid grants are limited to a certain sum and do have alternative uses within the Defense Establishment they carry "shadow prices". Hence it was recommended to use the domestic cost of export dollars' added value for internal calculations (which expresses the opportunity cost of foreign currency to the economy).

Correct costing of expenditures is a necessary but not a sufficient condition for efficient allocation of resources. Costing principles had to be complemented by a set of economic rules for program analysis. Thus, for major weapon systems or possible combinations of them, life cycle costing models were introduced, which account not only for acquisition costs, but also for discounted future costs of absorbing, training personnel, operating and maintaining such equipment.

Economic principles were applied to procurement too, especially when carried out in non-competitive markets and based on negotiations. The rules were the same for all suppliers and made no distinction between government and privately owned companies. Besides, as the rules were common knowledge, they limited the room for bargaining, and buyer-supplier relations became more stable. The economic procurement principles proved their viability in changing economic circumstances, and were particularly helpful during the hyperinflation of the early 1980s, allowing defense purchasing to proceed satisfactorily when other sectors had difficulty continuing economic activities based on long-term contracts.

Through the years, western military establishments have tended to prefer "buy" to "make" and expanded outsourcing. In Israel's first years, the IDF had to establish internal capabilities for manufacturing, upgrading and maintaining equipment, as well as for the supply of numerous services not found in the civilian market. But with time, the civilian sector developed, defense and other industries were able to meet military needs, and an increasing number of such activities had no further place within the military. The emerging new division of labor between the IDF and outside suppliers of goods and services offers a great potential for savings. Outsourcing makes it possible to exploit advantages of specialization, economies of scale and competition, all of which may reduce costs. By contrast, manufacturing and service units that operate within the IDF are not exposed to competition, and since their costing is incomplete, administrative control over them is also ineffective. Actually, there are hardly any mechanisms to prevent excessive expenses, or to create incentives to efficiency and savings. Furthermore, internal units lack the flexibility necessary to adapt manpower and production capacity – initially built up to meet demand-peak requirements – to changing levels of demand, or to sell output surpluses in the open market. Proponents of internal production systems and of self-sufficiency objected to outsourcing initiatives on the grounds of an apparent conflict with the need for readiness. They also added that even if making is less efficient than buying and outsourcing per se, there are important indirect contributions, mainly in training personnel for a large variety of military tasks. In fact, however, the tendency has been to expand purchase and outsourcing, the military doing less by direct means, although considerable room still remains for further steps that might promote efficiency and yield substantial savings.

10. THE PEACE DIVIDEND

In the 1990s, the great global changes, and especially the strategic changes in the Middle East, seemed to have created, for the first time, chances for political arrangements and peace agreements between Israel and the Arab states (besides Egypt, with whom a peace agreement was signed in the early 1980s). Here was a new item for the heavy agenda of defense economics in Israel: how to maximize the peace dividend?

The literature of defense economics calculates the peace dividend as the difference in civilian GDP over time, between peace and "not peace" scenarios. The assumption, explicit or implicit, is that the first will be larger for a combination of reasons: output losses and damage to capital stock due to recurrent wars are avoided; resources can be diverted from defense to civilian use, especially to investment; a peaceful atmosphere encourages foreign investments since perceived risks are lower, and the same might apply also to foreign trade and tourism.

Different estimates were made as to the anticipated peace dividend for Israel (Gottlieb, 1994; Kanovsky, 1994). All analyses distinguished between direct and indirect influences, as well as between short and long-term effects or between the transition period and the steady state to follow. Direct effects include avoiding war damages, possible reduction in the defense budget, possible shortening of the compulsory service and alleviating the reserve service burden. As to the economic significance of avoiding wars, it is agreed that estimates based on previous wars are inadequate, since in evaluating future war potential damages, unlike in the past, one cannot disregard the consequences of possible long range missiles attack on the home front. Opinions differ, however, as to reducing defense expenditures and transferring resources to civilian use. Since the mid 1980s, the defense consumption share of GDP has declined steeply, and in the mid and later 1990s, the economic burden of defense reached the level of the early 1960s (see chapter 6). Some think that the decline was too steep, producing an intolerable defense deficit, especially in face of the need for the IDF to guarantee Israel's security in the interim period of political arrangements. Others maintain that Israel's defense burden is still high by international standards, and that given the new strategic circumstances, it can and should be further reduced. In any case, no one disputes that in the interim period defense expenditures will have to rise in order to pay for new deployment of forces, for warning systems, intelligence and other items.

Indirect effects also involve several components. First, a stable peace is likely to encourage immigration to Israel, and to reduce emigration, leading to population and economic growth. Secondly, an influx of foreign investment is likely, both because of new opportunities in the region and lowered risks. Third, new markets in neighboring states may open up to Israel, and because of the relatively short distance they may provide opportunities for exporting goods and services that are usually considered non-tradable in nature. At the same time, in the opposite direction, neighboring countries may offer Israel sources for cheaper imports. However, most experts agree that trade potential between Israel and its neighbors is rather limited, and point out, *inter alia*, to the discouraging twenty years of experience in developing trade between Israel and Egypt.

Fourth, peace between Israel and the Arab states, and revoking the Arab boycott may open up markets in third party countries that previously avoided economic ties with Israel. Fifth, if the Middle East becomes a peaceful, stable area, tourism to Israel is likely to grow. On that, however, doubters hold that security tensions are not the only reason for regional instability, and more changes are needed to establish an image of the Middle East as a safe tourist destination (Kanovsky, 1994: 5).

Just as peace arrangements are likely to produce economic advantages, so economic advantages may reinforce peace. Nonetheless, one cannot ignore the great economic disparities between Israel and her neighbors, and that the road to overcoming the Arab states' economic distress – which has nothing to do with their relations with Israel – is long and hard. World history is filled with wars and armed conflicts rooted in economic distress. Thus peace, however great its contribution may be, will not free Israel in the foreseeable future from the need to invest in defense and to confront the problems of defense economics.

5
MEASURING DEFENSE EXPENDITURES

Defense expenditures are an aggregate measure of inputs; they express payments to military manpower and to suppliers of goods and services in a given period. But as in other cases where the output of government expenditures is difficult to evaluate, the common practice is to see military expenditures as an aggregate measure of output too. Defense expenditures are considered as an output index in strategic contexts mainly, while in economic contexts they serve as inputs index (Wiberg, 1983: 162).

In both contexts, defense expenditures are frequently used in comparisons. Comparisons are made between the defense expenditures of different states and inter-temporally for the same state. Besides, defense expenditures are compared with various economic variables to express their economic significance, or with population, area, length of borders and the like, to indicate strategic implications. All these uses involve problems of definition and evaluation, not to mention technical difficulties of collecting data widely regarded as state secrets, and often manipulated by considerations of internal and external politics.

1. OBJECTIVES, LIMITATIONS AND BIASES

The following anecdotes illustrate possible objectives and some of the difficulties involved in measuring defense expenditures: -

i. In the early 1960s, based on intelligence information regarding numbers of divisions, NATO commanders assessed that the Warsaw Pact conventional forces were far larger than those of the West, and tended to give up the conventional option in advance. But economists in the American Department of Defense thought differently. They converted the intelligence data into budget terms, and showed that the Soviets would have needed a quite illogical level of defense expenditures to maintain such a force. After further calculations, they concluded that the capabilities of a Soviet division were about a third of one in the United States Army, and their findings led NATO to rethink the conventional option (Enthoven & Smith, 1974: 102-109).

ii. In 1981, President Reagan justified the largest ever peacetime increase in defense spending on grounds of the expenditure gap between the Soviet Union and the United States (Deger & Sen, 1990:3). The argument over what this gap signified for the balance of power between the two countries began after the Vietnam War, and continued for a decade (Holzman, 1980). Throughout that time the Soviets officially reported a constant, implausibly low level of defense expenditure, 25 billion dollars a year. American experts

used various calculation methods to correct the Soviet data, and obtained figures ranging from 75 to 140 billion dollars (Boston Research Group, 1981: 63). When in 1989 for the first time the Soviet Union published "true" data, they indicated that western estimates have been exaggerated by some 30-40 percent (Deger & Sen, 1990: 62-69).
iii. In 1973 the United Nations General Assembly called on its members to reduce defense expenditures by ten percent and to devote some of the money saved to international development programs. But since there was no prior agreement on how to define and measure defense expenditures, it was impossible to give this initiative any real content (Ball, 1988: 97, 104). Similarly, in the late 1970s, NATO countries agreed to increase their defense spending by three percent a year. But here too no agreed measurement accompanied the decision, so that members could apply it as they saw fit, and increases appear in fact to have been lower (Deger & Sen, 1990:8).

Following these and other examples, it is possible to outline the principal uses of defense expenditure measures at the higher level of foreign and defense policy-making: -
i. Changes in defense expenditures of rival states, often interpreted as changes in the level of military threat, may accelerate or moderate arms races, bringing wars nearer or putting them off.
ii. Adversary states may rely on changes of each other's defense spending in supporting foreign policy steps such as requests for aid, demands for international sanctions, etc. Likewise, governments may use information on opponents' defense expenditures for political purposes at home.
iii. Defense expenditure data of allied states are important too, as they may deter a common enemy. Besides, when states cooperate on military affairs, defense expenditure data may serve as a basis for just burden sharing.
iv. Information on defense expenditures is an essential precondition for multilateral arms control and disarmament agreements, especially those based on reduction of military spending.

Defense expenditure measures may also be important on the lower level of military planning. By reference to main expense categories, it is possible to define tradeoffs between alternative defense policies, to set up priorities for developing military capabilities, to allocate resources for current readiness as against future buildup of forces, etc. Without aggregate measurement in monetary terms, all of these are either difficult or impossible, since the items under consideration cannot usually be reduced to a common denominator in other terms.

In the economic context, defense expenditure measures are required for analyzing the macroeconomic and industrial implications of defense, and particularly for calculating the opportunity costs and the economic burden of defense. Governments are certainly interested in such information for their own country, but often when relating to other countries as well.

Military capabilities are the result of numerous interacting quantitative and qualitative elements. It is, therefore, doubtful whether any single measure, including measures of defense expenditure, can express them in a meaningful manner. However, defense expenditure measures have clear advantages over alternatives in this respect. Assuming that military establishments act rationally, i.e. seek maximum military strength at every level of defense expenditures, military power (or at least its purchasable elements) is a monotonic function of accumulated defense expenditures. Besides, monetary values make it possible to sum up partial elements into aggregates without any special assumptions, while aggregate calculations in other terms, if at all possible, require abstractions that frequently produce results having no practical content. Finally, other measures are not routinely available, while defense expenditure measures of one quality or another are available for most of the countries. Thus defense expenditures may serve as a raw, primary indicator of military power, one that must be complemented by more information before reaching practical conclusions.

Reported defense expenditures are not a perfect measure of defense opportunity costs either, as they do not usually reflect the full value of forgone civilian uses. Differences arise mainly because certain expense elements are reported in values below their market prices. However, in the economic context as in the strategic one, there is no better point of departure than defense expenditure for expressing the opportunity costs of defense.

Reporting defense expenditures is not a purely statistical matter for governments. Usually they consider it as a sensitive topic for both external security and internal politics reasons, and often hide information and even misinform deliberately: some figures are biased downward while others may be inflated (Wiberg, 1983: 166-167; Brzoska, 1981: 264). World defense expenditure statistics, then, have to contend with a total absence of data, with partial data of uncertain quality, with data not based on uniform definitions and sometimes even with data deliberately falsified.

2. STANDARD DEFINITIONS

Broadly defined, national expenditure for defense should express the total value of economic resources that contribute to national security in a given period. Compared to a hypothetical situation in which no threats exist, it is also the value of the economic resources that could be saved and diverted to alternative uses. Standard definitions, however, are narrower in scope. First, they relate only to public expenditures, ignoring defense costs borne by households and business firms. Secondly, they recognize expenditures that support the nation's military capabilities directly and concretely, while overlooking those that contribute to the quality, internal strength and cohesion of society (e.g. expenditures for education). Similarly, standard definitions do not include investments in civilian infrastructure (roads, sea and airports, communications systems and the like) or expenditures on basic research and for developing technologies of a general nature, though they too serve national security in the broad sense. Thirdly, extra costs of civilian infrastructures and construction due to specifications dictated by defense considerations are also excluded. A special case of excluded indirect costs is that of conscript soldiers (see below). Thus in following standard definitions, most countries and international institutes, and eventually most of the empirical research on defense economics, all make do with the narrower approach. The desires to maintain operative content, as well as practical considerations of data availability and consistency, have often overcome more qualitative arguments.

The budget of the government department in charge of defense, whatever it is called, is the starting point in measuring defense expenditures in all countries. This budget typically includes expenditures related to the daily living of the military – expenses for personnel, current operations and maintenance and training; and expenditures for the buildup of military forces – weapons acquisition, military construction and research and development. These expenditures account for the primary inputs in producing defense, and hence are the hard core of any definition of defense expenditures. However, the defense budget may include expense items of a civilian nature that should be omitted, while defense-related expenditures that are budgeted to civilian ministries and other public agencies should be added. As a first step, then, measurement of defense expenditures requires the re-sorting of items in the public budget. In Britain in the 1960s, that process increased the Defense Ministry's budget by ten percent (SIPRI, 1973: 18), and in the United States in the 1970s – by 45 percent (Boston Research Group, 1981: 74-75). In some developing countries additions may be much larger (Ball, 1988: 116-117; Brzoska, 1995: 49).

2.1 Functional Criteria

Reclassification of public budgets as mentioned above is usually based on functional criteria, i.e. public expenses that do not appear on the defense budget are included in defense expenditures if they are clearly linked to the core of military activity, ignoring organizational boundaries. The following borderline cases illustrate the functional approach: -

Paramilitary forces: The police, espionage and counter espionage services, border and coast guards, civil defense units and the like. It has been suggested to include such expenses in defense expenditures only when the forces involved are organized, equipped and behave like the military, and thus are able to carry out operations similar to those of the military in terms of objectives and results (Wiberg, 1983: 164). Accordingly, expenditures of civil defense forces are not included, and the tendency is also to exclude expenditures of forces dealing mainly with internal security, like the police.

Units in civilian ministries and public agencies dealing with defense or "mixed" matters: Atomic energy authority, space research agency, etc. These are units outside the defense ministry, frequently as a result of political considerations or administrative convenience, which could equally well be within it from a functional point of view. If the civilian and military budgetary elements can be separated, the latter are included in defense expenditures. In other cases, the tendency is to include the budgets of these units in defense expenditures if the expenditure for military uses, actual or potential, is thought to be greater than for civilian uses.

Services by military units to the civilian population: These may be given as a matter of course (e.g. civilian air-traffic control, meteorological services) or on occasion (e.g. rescue operations in case of natural disasters). Since these serve clearly civilian purposes, the expenses involved are excluded from defense expenditures.

Foreign military aid: Military aid extended by countries in money, equipment or actual military forces, contributions to the finance of permanent military alliances or ad hoc coalitions, etc. The NATO definition includes the costs of military aid in the defense expenditure of the donor countries, although it is generally intended to increase their political influence, not to contribute to their military capabilities. The Stockholm International Peace Research Institute (SIPRI) deducts military aid from the defense expenditures of the recipient countries so as to arrive at a more accurate measure of the economic burden of defense.

Strategic stocks: Fuel, metals and other raw materials for the production of arms, medicines; basic foodstuffs and the like. These stocks are designated to enable continuous production and consumption of certain goods and services, despite the likely interruption of supplies in an emergency. As a rule, costs of accumulating strategic stocks are not included in defense expenditures, though in specific cases some finished goods with a definite link to military activity, like jet engine fuel, might be included.

Aid to defense industries: This may be open and explicit, by means of grants, loans or direct investment in special purpose production equipment, or indirect, by means of tax exemptions, debts write-off and the like. Indigenous production capacity for defense clearly contributes to national security. It reduces dependence on foreign sources for arms, thereby expanding the freedom of strategic action. It also makes it possible to adapt equipment to local theaters and warfare doctrines, thus improving military capabilities. Besides, financial support to defense industries sometimes substitutes for higher prices of finished goods when sold to the defense ministry. All this, then, seems to justify including aid to defense industries in defense expenditures. However, government aid to defense industries often supports export activities too, and there is no practical way to distinguish that part which is relevant only for domestic uses. Furthermore, in some instances such aid is extended as part of a government policy to encourage employment or capital investments in general, so once again separating the aid given exclusively for defense is not a simple matter.

Basic technological research: Public financing of research programs at universities, civilian research institutes or in industry. One approach maintains that if the results of such research are likely to be applied to military ends, its costs should be included in defense expenditures, especially when the financing is provided by the defense ministry (Benoit & Lubell, 1966: 102). Another approach objects to including expenditures for basic research, even if financed by the defense ministry, since its military contribution is not yet clear, and if it does materialize, it is likely to become operational only after years of heavy additional expenditure.

2.2 Timing Criteria

Public budgets are run on a cash basis. Therefore, to express defense benefits accrued at any specified period, or forgone alternative uses of resources at that period, recorded defense expenditures should be adjusted. In the strategic context, payments that yield defense benefits at another time should be subtracted, while costs of defense benefits previously paid for, or to be paid for in the future, should be added. Similar adjustments are required for economic analysis. The following examples of time-based distinctions may further clarify this issue: -

Veterans' pensions: When a budgetary pension system applies, pension payments to veterans are delayed expenditures for past services, contributing nothing to present or future military capabilities, so they should be subtracted from current defense expenditures as an index of defense output. Likewise, at the time pensions are paid, they are a kind of transfer payments from the government to individuals, and do not involve diversion of real resources from alternative civilian uses. They too should be omitted, then, from defense expenditures as a measure of defense opportunity costs. By contrast, when an accrual pension system applies, money is continuously deposited in funds from which future pension payments will be made to persons now serving. Such deposits, when made, are in every way labor expenses, and should be included in defense expenditures as measures of both output and of input. In fact, under the budgetary pension system, current costs of labor – and therefore also defense output and its opportunity costs – are undervalued. This issue is discussed later in the context of valuation problems.

Compensation to war invalids and bereaved families: These payments do not contribute to present or future military capabilities either. From the economic standpoint, they are transfer payments from the government to individuals, and at the time they are paid, they do not involve allocation of real resources to defense at the expense of alternative civilian uses.

Payments for procurement: These include advance and progress payments before delivery, payments on delivery and repayments of suppliers' credit after delivery. As contributions to military capabilities, the relevant time to report procurement is the delivery date. As for economic implications, real resources are withheld from alternative civilian uses for the entire production period, and for defense imports on credit – on the date the debt is created (generally the delivery date), since the same credit could have financed – directly or indirectly – other imports. That being so, when payments do not coincide with production and delivery, using actual payments for the measurement of defense expenditures may be misleading, both as regards military capabilities and the economic burden. Nevertheless, for most local purchases it is unusual to take time differences between payments and delivery into account, whether due to technical difficulties or to the implicit assumption of "corrective offsets" between old and new transactions. For defense imports, a single sum is usually reported on arrival, in a way that makes it independent of financing sources and payment schedules.

Another issue of timing may result from deviations between actual spending and the budget. To reflect strategic and economic implications correctly, actual spending should be preferred to budgets as authorized. But reports on actual spending are published late, which may make the measurement of defense expenditures a worthless exercise, especially as regards evaluating the capabilities of rival states. Besides, budgets are usually more detailed and can be categorized and analyzed in a way that is impossible for data on actual spending. At times the practicalities override qualitative considerations in this matter too (Benoit & Lubell, 1966: 100).

3. SPECIAL DEFINITIONS

Standard definitions, even if adjusted according to functional and timing criteria, still might be insufficiently accurate either for evaluating military capabilities or for estimating the opportunity costs of defense.

3.1 Measuring Military Capabilities

There is no obvious relationship between defense expenditures in a given period, which are a flow, and military capabilities, which are a stock. First, only expenditures for the buildup of forces contribute directly to military power. Second, the stock of military power at a point in time is the result of accumulated past investments, not only of current buildup spending. Thirdly, current expenditure for the buildup of forces does not represent additional military capabilities for that year only, but the anticipated value of services from the new assets throughout their operational life. In view of all these distinctions, annual changes in defense expenditure are an inadequate basis for evaluating the evolution of military power in those years; just as differences in defense expenditures of rival countries cannot be considered an accurate representation of the balance of power between them. Hence attempts were made to evaluate military power through certain components of defense expenditures rather than through their totals. Another alternative is to estimate the value of military capital stocks directly from physical data.

One method uses operations and maintenance (O&M) expenditures, assuming that they are proportional to military power at any given time (SIPRI, 1973: 16-17, 29-30). The greater the forces as to manpower and materiel, the greater the expenses for salaries, food, spare parts, repairs and the like. Moreover, it is reasonable to assume that the more technologically sophisticated and superior in military performance the equipment is, the dearer it is to maintain. Thus greater expenditures for O&M represent a higher level of military power in quantity and quality. But these are only ordinal conclusions, and additional assumptions are required to quantify them. Particularly, it should be assumed that O&M expenditures and military power are not only positively associated, but that the ratio between them is constant over time and identical in all countries. Indeed the assumption of a constant ratio over time has some support. For one thing, military organization is based on rigid definitions of units and sub-units and on fixed quantitative standards of manpower for each unit, and military logistic planning too has its fixed "keys". These organizational features are translated into fixed factor ratios, possibly leading to constant relationship between expenditures (in real terms) and output too. For another, rapid technological progress has shorten cycles for renewing the arsenal, thereby limiting the opportunities for learning and productivity increase in operations and maintenance that otherwise may affect the O&M expenditures to military power ratio over time. By contrast, it is more far-fetched to assume that ratios are the same in the armed forces of different countries.

A second method is based on expenditures for the buildup of forces (Zusman, 1985: 189-193). It departs from the assumption that purchase prices reflect fully differences in initial military effectiveness between items. On such grounds, then, it measures the military power at any given time as the sum of all past expenditures for the buildup of forces, less those that relate to items no longer in use, and after allowing for depreciation. However, calculating depreciation and appropriately expressing obsolescence are not simple tasks (Hildebrandt, 1990: 161-166). These questions are common to the third approach.

The third approach evaluates the military capital stock directly. This includes weapons systems, other equipment, ammunition inventories, military infrastructures and buildings and additional fixed assets. The evaluation begins by collecting physical data. At the next stage, physical data measured in different units has to be given a common denominator, recognizing relative weights. If the purpose is to find out military capabilities, relative weights must reflect the relative military effectiveness of every component. However, as no appropriate units of effectiveness are available, monetary purchase prices are used instead. Certainly, such evaluations require enormous, detailed data, and since they are constructed bottom-up, there is often a risk of partial coverage and of underestimation. Furthermore, beyond technical problems lie some conceptual questions: -

i. Using purchase prices instead of units of effectiveness limits the value of the results. First, purchase prices may not accurately differentiate between identical items of different ages, or in other words – may underestimate the possible effect of physical depreciation on operational performance (Halperin, 1989: 3). Similarly, with this method technological obsolescence – the emergence of technologies that offer more effective military equipment – and operational obsolescence resulting from acquisitions of superior means by rival states, are also overlooked (Hildebrandt, 1990: 161-166). Secondly, in this method greater military capital stock means greater military power only if prices of military assets are highly correlated with their relative military effectiveness. However, military effectiveness is affected by varied factors, like theater conditions and military doctrines, and these are hardly reflected in purchase prices. When weapons are imported, it is reasonable to assume that their prices reflect relative effectiveness under circumstances prevailing in the producer country, which may be quite different from those of the importer (Halperin, 1989: 6-8).

ii. The value of military capital stock expresses the expected flow of services over the life of the assets included. For military operations, however, interest lies in the immediate period, thus current and future services must be separated through discounting the value of the latter at a rate that expresses time preference. For instance, if the remaining life is ten years, and assuming an annual discount rate of ten percent, the relevant share of the military capital stock for the current year will be about 16 percent, compared with only one tenth when time preference is ignored (Hildebrandt, 1990: 161-163). The method requires, then, an estimation of the life spans of military assets, and choosing a discount rate that expresses strategic time preference, inevitably involving somewhat arbitrary assumptions.

iii. The productivity of military capital depends not only on its size, but also on features of the production function, economies of scale for example (Halperin, 1989: 15-18). Moreover, it is influenced by the size and quality of the manpower operating the military capital. Thus if defense production functions should be specified anyway, it might be advantageous to evaluate military power by explicitly factoring data about manpower and the stock of military capital into the equation. This approach was applied, as noted before, to evaluate the relative military capabilities of Israel and the Arab states (see chapter 4, section 2).

3.2 Measuring Opportunity Costs of Defense

To measure opportunity costs of defense, expenditures that represent resources with no alternative civilian uses must be subtracted from standard definitions. Resources may lack alternative uses for a number of reasons.

There could be restrictions on the use of certain resources, as with foreign military aid earmarked for arms acquisitions from the donor only. Another reason might be widespread unemployment; it can be assumed that if soldiers were discharged, some would not find jobs in the civilian economy in any case. Unique specialization in defense production could also make certain resources of no use elsewhere. Similarly, expenditures that substitute for civilian uses should be omitted completely or in part. They represent resources that even if not needed for defense, would have been consumed in any case. Military manpower living expenses such as food, clothing and medical services are a case in point; if not included in defense expenditures they would have a similar value in private consumption. Another example is infrastructure expenses – roads, communications systems, etc. – serving both the military and civilians; they would be included in civilian investments.

Opportunity costs of defense should also express the effect of present defense expenditures on the future level of alternative civilian uses. On one hand, some expenditure for training soldiers may contribute to the quality of the civilian labor force and raise productivity, while some expenditure for defense research and development may contribute to overall technological progress, thus expanding future production capacity for all uses. On the other hand, however, defense expenditures may come at the expense of capital investments in other sectors of the economy, and of social investments in education, health and the like, or may deplete the civilian economy of high quality human resources, thus hindering future economic growth. Contribution to future production capacity reduces the opportunity costs of defense, while impeding its growth raises them.

Current opportunity costs of defense are augmented from time to time by one-time opportunity costs caused by wars, notably output losses during wars, erosion of human capital and damage to fixed capital assets.

4. CORRECTIONS AND ADJUSTMENTS OF MONETARY VALUES

A relatively small share of defense expenditures is spent on goods and services acquired in competitive markets. The major part of the expenditures includes economic resources for which the prices paid often deviate substantially from their market value, and at times, the military even uses some economic resources (e.g. lands and other real estates) free of charge.

4.1 Manpower Costs

The most obvious example of value distortion relates to the labor cost of conscripted soldiers. It is measured by low administrative prices that do not reflect the market value. As a first approximation, it could be assumed that the deviation equals the difference between the alternative civilian income of the soldiers and their military income in money and in-kind. Thus in order to correct reported expenditures the alternative civilian income should be estimated, which is a complex matter. Conscripts are young, usually without work seniority, and their education is, on the average, different from that of civilian labor force participants. Since these qualities clearly influence income levels, it is impossible to rely on average civilian income as given. Besides, without conscription different supply and demand conditions might prevail and the equilibrium level of income in the civilian labor market might change, particularly in the age group of which the conscripts are a part.

Furthermore, when defense expenditures are meant to reflect opportunity costs of defense, not only the forgone civilian income at the time of military service is important, but also income differences likely to arise in the future. Because learning and work experience have a positive effect on income, the actual income of a soldier in his first year of civilian work, say after three years of military service (the Israeli case), is likely to be less than his hypothetical income had he not had to serve and had this been his fourth year of work. Moreover, the negative difference is likely to persist, though gradually diminishing, and to disappear only after several years. But on the other hand, military experience and training, and particularly training in skills transferable to the civilian sector, may raise the income of veterans to a point higher than the hypothetical level in their first year of civilian work, accelerating the rate at which the income gap is closed in subsequent years, and even make the differential positive instead of negative. In terms of the economic theory of human capital, the equilibrium value of human capital is equal to the present value of the flow of income it is likely to earn in future years. Corresponding to the possible income differences in the years after military service as outlined above, their present value may be negative and add on to the loss of income during the service itself, reduce the value of human capital and thus increase the opportunity costs of military manpower or, by contrast, be positive and reduce the loss during the service period and the respective opportunity costs (Knapp, 1973).

Alternative civilian income represents the cost of military service to the individual, while for the economy as a whole, under full employment, the value of the output that could be produced in the absence of conscription, not just labor income, should be considered. The average output per worker is higher than the average labor income as it includes returns on complementary production factors too (SIPRI, 1973: 10).

Two deviations unconnected with the legal right to enlist manpower should also be considered. First, as previously noted, under the budgetary pension system the conventional reporting of defense expenditures does not reflect the full cost of voluntary manpower either, because part of its compensation is delayed to the period after active service. In this matter, apparently, career military manpower is not different from other public sector employees. However, the military service retirement age is earlier, and the ratio between "pension time" and active service time is higher, thus – in relative terms at least – the underestimated costs are larger. Secondly, reported labor costs do not express the special dangers of military service. As the casualty rate in military service is relatively high, it might be appropriate to add an imputed risk premium to current labor costs. However, such an adjustment would be applicable only when defense expenditures are taken for a measure of defense output. In the economic context, the higher risk of military service does not imply a potentially greater civilian GDP loss.

4.2 Procurement Costs

The reported value of defense purchases may need correction too: -
i. Sometimes purchase prices are determined arbitrarily by law, by ownership relations or other special arrangements unconnected with market conditions. Typical arrangements are exemptions from indirect taxes on goods and services, importing defense equipment or inputs for domestic defense production at an official rate lower than the effective exchange rate, etc.
ii. Most defense procurement is carried out under imperfect competition, through negotiations. Prices are affected, then, by the degree of competitiveness, by the parties' bargaining power and by controls over development and production costs. Comparison between countries may yield misleading results as to relative military power simply because the oligopolistic profits that suppliers of military equipment in one country rake in are greater than those in the other (Wiberg, 1983: 170).
iii. When military equipment is purchased through foreign aid, its monetary value sometimes goes completely unreported since no payment was required, or only token prices are listed. In other instances, where the purchases may be financed by loans on preferred terms, opportunity costs should be adjusted by deducting the value of this benefit. Finally, defense imports are sometimes by barter; i.e. are paid for in raw materials or other civilian goods, without any mention of money value.

4.3 Evaluating Defense Expenditures of One Country in the Prices of Another

Data on defense expenditures may be misleading if prices are substantially divorced from market values. Assuming that this was so with respect to Soviet defense expenditure statistics, the American Central Intelligence Agency (CIA) developed a measurement method that multiplied detailed physical data for the Soviet armed forces and their mode of operation by American prices and cost coefficients. The result expressed the expenditures that would have been necessary to support the size and patterns of activity of the Soviet order of battle, and to develop and acquire new equipment, in the economic environment of the United States.

Critics pointed out the arbitrary nature of the method. For example, each time American soldiers' pay was raised, so was the estimate of Soviet defense expenditures, although no real change occurred in their military capabilities (Fallows, 1981: 179). Another argument alleged that the system ignored differences in military efficiency. The price of an American main battle tank, for instance, might be higher than that of a Soviet tank due to more advanced technology and better military performance, thus multiplying the number of Soviet tanks by the price of an American tank exaggerates Soviet expenditures (Holzman, 1980: 5-7). A third contention asserted that the effect of relative prices was ignored. If relative prices in the Soviet Union resembled those in the United States, this would obviously affect the structure of military forces, the capital-labor ratio, the tasks carried out by the armed forces and their mode of operation. With rational behavior that strives to minimize costs by responding to relative prices, Soviet defense expenditures would have been lower. Evaluating one nation's defense expenditures in another nation's prices can be valid, then, only when basic economic conditions are similar, where one can reasonably assume that differences in relative prices are insignificant (ACDA, 1979: 13).

5. INTER-TEMPORAL AND INTERNATIONAL COMPARISONS

The absolute size of defense expenditures is of limited practical value, and measurements are made principally for the sake of comparison between different periods, between states or groups of states, and the like. Comparisons other than those based on defense expenditures suffer, as noted earlier, from the inherent difficulty of assigning appropriate relative weights so that the military power or defense opportunity costs can be expressed in a single aggregate measure. Expenditure based comparisons appear to provide an answer in that they rely on monetary values as relative weights. With that, however, comparison of defense expenditures raises complex questions too.

5.1 Inter-temporal Comparisons

Comparisons of defense expenditures between two periods are designed to present the real (constant price) change in military power or in opportunity cost of defense over time. The two following examples illustrate the questions arising in the calculation of real changes in military power and in defense opportunity costs: -

i. In the early 1970s, conscription was abolished in the United States, whose armed forces are now made up entirely of volunteers. The change was accompanied with increases of reported defense expenditures, though there was not necessarily an increase in military power or in opportunity costs, and if there were – it would almost certainly have been less than the one in expenditures. The increased expenditures had three components: some hitherto hidden economic costs now appeared in the budget; costs of the All-Volunteer Force are likely to be higher because its military efficiency is apparently greater, and over time wages regularly tend to rise, while this is not usually the case with the administrative pay of conscripts.
ii. Suppose a squadron of obsolescent aircraft is replaced by a new generation, without increasing the total number of planes. Comparing military power on the basis of numbers, no change is evident. If, however, the defense expenditures for the two years in which the acquisition took place are compared, there might be an increase that reflects the higher price of the new aircraft. In part, the higher price reflects qualitative improvements or superior performance, and in part - regular price increases over time in development and production costs.

An analysis of real changes in defense spending must distinguish between changes in military organization or activity patterns, labor productivity increase and quality upgrade of equipment, and general inflationary price changes. The first type, if they only turn concealed economic costs into open budgetary items, have to be rectified by correctly measuring defense expenditures in general, as discussed in earlier sections. The measurement of real changes in defense expenditures over time focuses, then, on measuring overall price increases, and on separating inflation from qualitative changes.

Ideally one should define a fixed "basket" of defense output and follow changes over time in the money needed to purchase it. This would yield a price index of military power or national security, like the one for the cost of living. However, due to difficulties in defining a basket of defense output, the routine is to make do with a fixed basket of inputs required for producing defense. The resulting price index, then, may apply to the measurement of defense opportunity costs in constant prices, but is not straightforwardly applicable for indicating changes in military power. For the latter purpose, it should be assumed that budget allocations are optimal, i.e. strive to attain maximum power within the given budget, and that some specific input-output relationships are maintained (Skons, 1983: 196-200).

Constructing a defense price index based on a basked of inputs begins by dividing defense expenditures in a base year into categories. Next, price changes in each category are measured directly or by available price indices calculated routinely for other purposes. Obviously these "borrowed" civilian sector indices do not always adequately describe the behavior of defense expenditure prices, and it should be considered in evaluating the quality of the results. The last stage sums up the price changes measured for the various expenditure categories, taking their relative weights in total defense expenditures in the base year. It follows, then, that fluctuations in defense expenditure's composition over time also affects the quality of the results.

Few countries construct a special defense price index. In 1980, only 12 countries reported data on defense expenditures to the United Nations while using special price indices (Skons, 1983: 202). Most states calculate real changes in defense expenditures by discounting with one price index – the GNP or GDP deflator or the consumer price index. These indices are chosen because they represent a wide variety of goods and services, but mainly for want of something better. In countries that do calculate special defense price indices, they can be compared with the GDP deflator or with the consumer price index. For example, during the 1970s, for advanced industrial countries having available data, on the average, the accumulated increase of defense expenditure prices was 10 percent higher than both GDP and private consumption prices (Skons, 1983: 202). This difference is not insignificant. Possibly the greater increase in defense prices is because of insufficient attention to improved input quality, but it might also be that defense prices are indeed rising at a more rapid pace (Brzoska, 1995: 53). A relatively large share of the inputs for defense production comes from economic sectors where inflation rates are above average – sectors that employ high salaried personnel, high tech industries and the like. Moreover, defense production is often criticized for being insensitive to costs ("gold-plating"), and as already noted, most defense procurement takes place in monopolistic and oligopolistic markets. Not surprisingly, then, several studies found a price increase at an average annual rate of some 10 percent for succeeding generations of major weapon systems (see chapter 8, section 4.7).

Economic logic implies that to calculate real changes in defense opportunity costs, using civilian price indices is justified. For instance, if defense expenditures come wholly or largely at the expense of private consumption, to reflect correctly the real change in their opportunity costs, price increases should be deflated according to the consumer price index. Similarly, assuming that the distribution of forgone alternative civilian uses resembles their shares of the GDP, discounting by the GDP deflator may be appropriate. In other words, to measure opportunity costs of defense in constant prices there is no need for special price indices, and civilian price indices may well be more suitable, if they are weighted according to the assumed distribution of alternative uses (Brzoska, 1995: 4). This gives rise, however, to two further questions: is it possible to surmise the distribution of alternative civilian uses, and is it reasonable to assume that it will remain constant over time?

When defense expenditures are calculated in constant prices to indicate quantitative changes in military power, qualitative improvements should be separated from the inflationary element of price increases, since the former reflect additional military power, and only the latter should be discounted. Qualitative improvements represent increased productivity of military manpower and higher performance standards of military equipment. The productivity of military manpower rises over time due to training and experience, to the growing capital-labor ratio in the military sector, and as a result of technological advances embodied in the equipment in use. However, since there is no way to measure manpower output, it is also impossible to measure directly changes of labor productivity in the military sector. Instead, the common practice is to assign to military manpower the same rates of productivity growth as in the civilian sector. Hence if the pace of technological improvements in military equipment is faster or the increase in the capital-labor ratio in the military sector is greater than in the civilian sector, the growth in military productivity is underestimated (Skons, 1983: 200). It follows, then, that constant price calculations using civilian labor productivity data might underestimate military power growth too. As to improvement in equipment performance over time, a distinction is made between those due to advancing from one generation of weapons systems to the next, and current improvements during the life cycle of any given system. To derive the rate of improvement of successive generations, development and production costs, in constant prices, of each generation are compared. To measure current improvements, annual changes in purchase prices of a particular model are compared with annual changes in the price index of the entire "family" of that equipment (e.g. purchase price of a particular aircraft model as against the general price index of military aircraft). The ratios so derived for each category of improvements, when summed up are considered to represent the overall rate of improvement in equipment performance over time. Using this and similar methods to isolate qualitative improvements, it has been shown that rates of increase in defense prices over time do not differ significantly from parallel civilian price indices (Skons, 1983: 209-210).

5.2 International Comparisons

Assuming full information about the quantities in the defense expenditure baskets of states A and B, and further assuming that the two baskets contain identical items, to compare them so as to derive meaningful strategic conclusions the different quantities of each state should be multiplied by a common set of prices. A question arises then: which prices to choose? The ratio of A's expenditures to B's will be higher if we choose B's prices, and lower if we choose A's. To illustrate: in 1976, American CIA officials estimated that the Soviet Union's defense expenditures were 44 percent higher than those of the United States when calculated in American prices, and only 29 percent when Soviet prices were used (Holzman, 1980: 4-5). Since calculations with index numbers do not tell which ratio is "right", the range of ratios or their geometric average should be considered. Moreover, the difference, in relative terms, between the two ratios (about 11 percent in this case) depends on the extent to which the defense expenditures are detailed – with more detailed data, not only does the difference widen, but opposite results might emerge.

In addition, multiplying quantities of one state by prices in the other ignores, as noted, possible quality and efficiency differences. Theoretically, when assigning soldiers' pay in one state to soldiers in the other, efficiency differences may be accounted for by relating them to only a part, say half, of average wages. This is true too of equipment prices. However, the difficulties involved in estimating relative efficiency may lead to arbitrary corrections, doing away with any practical content in the estimates.

Due to the difficulties inherent in international comparisons of baskets, alternative methods have been devised. These are based on converting one single measure – reported or adjusted defense expenditures – stated in each country's local currency to a uniform currency, usually United States dollars. In this method the main issue is choosing an appropriate exchange rate.

Many countries, in particular developing countries, employ multiple exchange rates, and if there is a single official rate, it does not always reflect the market value of foreign currency because of controls, import taxes, export subsidies and the like. In other countries, exchange rates may be influenced by speculative currency flows so that here too, rates do not always accurately reflect the relative prices of goods and services. A technical difficulty may arise also: when exchange rates fluctuate widely and defense expenditures are unequally distributed over the year, selecting a representative rate is no straightforward matter, and the results obtained might vary substantially, depending on the choice made.

Nor do exchange rates seemingly representing competitive equilibrium in foreign currency markets express the relative prices of all goods and services. They reflect the ratio between local and foreign prices for tradables only, while for nontradables it may be different. Defense expenditures usually include a relatively large non-tradable component, so that converting them from local to foreign currency according to ordinary exchange rates may be especially misleading.

To convert baskets of both tradable and non-tradable goods and services, the common practice makes use of purchasing power parity (PPP) exchange rates. With this method, exchange rates are determined by the ratio between local currency expenditures required in different countries to buy identical baskets of goods and services. However, the possibility of applying the PPP method to comparisons of defense expenditures depends on ability to define identical defense baskets for different countries, and this is no trivial task. But the problem relates mainly to comparisons between military capabilities. For comparing opportunity costs, PPP exchange rates in any case should reflect the relative prices for alternative civilian uses, and those obviously can be calculated more readily.

A pioneering effort to calculate world defense expenditures using PPP exchange rates was made in the mid 1960s. In the absence of exchange rates based on military baskets, it used those calculated for consumer baskets, and the resulting estimate for worldwide defense spending was 13 percent higher than the one derived from official exchange rates (Benoit & Lubell, 1966: 110-111). At a later period, the difference was estimated to be even higher, some 25 percent (Brzoska, 1995: 55). Certainly, for individual countries the differences could be far greater (Brzoska, 1981: 271).

PPP estimates usually increase the value of goods and services of poor developing countries more than that of other countries; thus they may suggest a different global defense expenditure distribution. For example, in 1980, the developing countries' share was 23 percent when calculated with official exchange rates, and 38 percent when based on PPP rates (Hewitt, 1991: 32). By contrast, in the mid 1980s, a group of experts calculated the defense expenditures of OECD countries – where the price structure is similar – by using PPP rates especially adjusted for that purpose, and then compared them with estimates based on ordinary exchange rates; the differences were indeed insignificant (Brzoska, 1995: 55).

International comparisons are often made over time, not for one single year, thereby raising a question of sequencing. Since the point of departure is local currency data in current prices, should one first calculate the defense expenditures of each country in constant prices of a common base year by using domestic price indices, and only afterwards convert them into dollars according to that year's exchange rate? Or, alternatively, should one first convert annual defense expenditures in local currency for the entire period into dollars at the relevant annual exchange rate, and only at the second stage discount price changes, this time however, by applying a uniform index, say an American price index? Both methods apparently yield data in constant prices for a particular year and in a single currency, but nonetheless, the results are not identical.

6. THE RELATIVE WEIGHT OF DEFENSE EXPENDIURES IN THE NATIONAL ECONOMY

Defense expenditures are compared with economic aggregates to express their relative weight in the national economy. It may indicate, among other things, the economic burden of defense, or from a different angle – the positive contributions of defense to overall domestic economic activity and employment. It is worth noting that defense expenditures and economic aggregates are both measured in monetary values, thus calculated ratios have neutral dimensions that help to overcome some of the difficulties attendant on inter-temporal and international comparisons.

The most frequently used economic relative measure is the ratio of defense expenditures to the gross national product (GNP) or to the gross domestic product (GDP) of the same period. To interpret this measure as an expression of the defense burden, the numerator should include all the opportunity costs of defense. At the same time, certain expenses not representing consumption of the defense sector per se must be deducted. This includes primarily transfer payments to other sectors, like compensation to the disabled and to bereaved families, which are anyway recorded in the national accounts, when spent, as private consumption, or investment grants to defense industries, which are eventually reflected in the national accounts as part of domestic investments. Defense expenditures, after subtracting transfer payments, are defined as defense consumption. As for the denominator – the GNP or the GDP – it only partially reflects the resources of the economy. It includes only the output of local production factors or, more precisely, the value added by those production factors, and does not account for imported resources. Since defense consumption generally includes direct and indirect import components, to compare it only with the GNP may exaggerate the relative share of defense. The smaller the GNP is in relation to total resources on one hand, and the greater the component of direct and indirect imports in defense consumption on the other hand, the greater the discrepancy will be. In view of such a possible discrepancy, the ratio of defense consumption to total resources available might be preferred. However, with this method, foreign military aid grants earmarked for defense imports, if available, should be subtracted from both defense consumption and total resources. Finally, if reported defense expenditures are augmented by an estimate of military manpower opportunity costs, meaning that greater added value is credited to domestic production factors, GNP data too should be corrected by an equal sum.

It was argued that the ratio of defense expenditures to GDP or to overall resources might be misleading and irrelevant in poor countries, which unavoidably allocate a larger proportion of their economic resources to basic existential needs. Hence it was proposed to calculate the ratio of defense expenditures to "hypothetical maximum surplus", defined as GDP less a minimum per capita consumption (usually measured by the absolute poverty level), multiplied by population size. According to calculations made for African states in 1989, whereas the ratio of defense expenditures to GDP was only 4.4 percent, the defense burden in relation to "hypothetical maximum surplus" reached a level of 50 percent (Brzoska, 1995: 57). Following a similar line of thought, sometimes a measure of the marginal defense burden may be relevant. This relative measure compares the supplements to defense consumption with the supplements to total resources in a given period. If in a certain year defense consumption decreased in relation to the previous year, the marginal defense burden will be negative although the overall burden is still positive.

Defense consumption subdivides into domestic defense consumption and direct defense imports. The former includes services of domestic production factors directly employed by the defense establishment, and goods and services purchased in the local market, which further divide into services of local production factors and imported inputs. It is thus possible to combine the respective values of direct and indirect services of local production factors, on one hand, and the direct and indirect imports, on the other hand, thereby obtaining a final division of defense consumption into two main components: domestic added value and total imports. Domestic added value in defense consumption is a component of GDP, and the ratio between the two reflects what proportion of the total services of domestic production factors is allocated for defense. It is thus both a component of the defense burden and a measure of the defense sector contribution to the employment of domestic production factors. Defense imports constitute part of the total imports of goods and services to the economy, and its relative share expresses the implications of defense for the balance of payments.

The GDP or total resources share allocated for defense does not reflect political priorities of governments. While governments produce and provide defense, they do not control all the resources at the disposal of the economy at a particular time. Thus the accepted practice is to measure the relative weight of defense expenses in the public or government sector. For that purpose, statistics from national accounts can be used, comparing defense consumption to overall public consumption. Alternatively, it is possible to refer to public revenues and expenses data, and express the budgetary or public finance burden of defense. Partial calculations have also been made, i.e. defense expenditures were compared to other categories of public expenditure, to education, health and welfare in particular. Such comparisons were perceived to reflect national priorities for defense vis-à-vis social development (Brzoska, 1995: 58-59).

In this context, the functional structure of the public sector may be important. Where the central government supplies most public services, the defense expenditure share may be lower than it is in countries where local authorities are charged with providing services like education and health. Similarly, differences in tax and transfer policy are significant too. Where governments collect taxes on a larger scale and then "recycle" them by way of transfer payments to the private sector, the share of defense expenses in the budget may be lower than it is in governments with different policies (Brzoska, 1995: 58).

Completing the picture, to express the full weight of the defense sector in the national economy defense exports should be considered too. Defense exports are often a byproduct of domestic defense consumption, and could hardly have existed without the basis of a domestic demand. At the same time, export orders increase the volume of production, thus providing economies of scale to the advantage of the domestic customer, and besides, producing for export often constitutes a kind of reserve stock that can be diverted to internal use in an emergency.

6
ISRAEL'S DEFENSE EXPENDITURES

At the turn of the twenty-first century, Israel's defense expenditures were about twenty times as high in real terms as they had been in the early 1950s. During these years the trend and rate of change of defense expenditures went through different sub periods, as did their composition and relative weight in the national economy. This chapter surveys these developments, showing practical applications of the definitions and measures discussed in the previous chapter.

1. OVERALL DEFENSE CONSUMPTION

After the War of Independence, expenditure for defense consumption, in constant prices, decreased for five consecutive years, and in 1953 it was less than half what it had been in 1948, and the lowest it was to be at any time thereafter. Israel came out of the War anticipating peace with her neighbors, and prepared to reduce defense expenditures considerably. However, there were other reasons too. First, the major powers, striving to maintain the status quo in the Middle East, limited arms supplies to the region, which obviously kept Israel from acquiring significant quantities of foreign weapons. Secondly, at that time Israel faced severe economic constraints, absorbing huge waves of immigrants and making its utmost to meet their urgent needs.

The next half-century divides into two main periods: from 1954 to 1975, the defense consumption grew rapidly and almost continuously, while since 1976, the pattern changed to a moderate decrease with occasional fluctuations. In both, changes were affected by three groups of factors: security conditions, including active wars from time to time, current military threats and the regional arms race; the intermittent options of acquiring weapons from abroad, and economic constraints.

1.1 The Rapid Growth Period, 1954-1975

Towards the mid 1950s, the armistice agreements made after the War of Independence weakened, and raids by hostile Arab infiltrators (*fedayoun*) abounded. In 1954 Israel realized that she had to prepare for a second round of military confrontation with her neighbors, and therefore decided to increase its defense expenditures and accelerate the buildup of the Israel Defense Forces (IDF). At that time too the regional arms race, to continue for decades, began: Egypt turned to the Soviet Union, the Czech arms deal was concluded and the status quo on arms supplies to the Middle East fell to pieces. Israel, on her part, applied to France, and began to acquire arms and military equipment in a quantity and of a quality previously unknown. In October 1956 the Sinai Campaign broke out. All these tripled defense consumption in constant prices within three years (see Table 6.1).

After the Sinai Campaign, growth in defense consumption stopped, but the respite was short. While military victory sharply reduced current security incidents, the arms race continued. Egypt led it on the Arab side, but Israel had to relate not only to the Egyptian threat, and strove to maintain a military balance vis-à-vis a potential coalition of several Arab states. Moreover, as sophisticated weapons systems appeared in the region, Israel grew more doubtful as to whether her qualitative advantage in manpower, a decisive factor in previous wars, would now be adequate. Consequently, the demand for advanced weapons systems increased, and a large-scale program was carried with the purpose of building a strong air force and mobile armored forces (Luttwak & Horowitz, 1975: 186-201). On the supply side, massive armament was possible first of all because France continued to sell Israel state-of-the-art equipment, and more so as new sources opened up in West Germany and in Britain. However, no less important in making the substantial increase of defense consumption possible was the rapid growth of the Israeli economy during the late 1950s and the early 1960s: expenditures could be increased without creating an unbearable economic burden. In 1966, on the eve of the third round, the Six-Day War, defense consumption was already some 70 percent higher in constant prices than in 1956.

After the war in 1967, four principal factors led to continued rapid growth of defense consumption. The first was a variety of expensive and unfamiliar new tasks: constructing fortifications in the Sinai peninsula; developing land and air transportation infrastructure and communications systems to assure control over broad additional spaces, and coping with manifestations of low-intensity warfare in the populated occupied territories. Secondly, a harsh war of attrition against the Egyptian army continued along the Suez Canal. Thirdly, the regional arms race accelerated. The Soviet Union became more deeply involved, and expanded the supply of armaments to the Arab states who urgently needed to rehabilitate their destroyed armies, while Israel, having lost the traditional French source of supply, now obtained long sought after sources in the United States. Finally, after the trauma of the French embargo, Israel adopted a defense-industrial policy of self-sufficiency, and invested large amounts in establishing independent capabilities of developing and producing major weapons systems.

In those years, economic conditions for increasing defense consumption were comparatively favorable. Before the war the country was in recession, so that rising defense consumption employed hitherto idle production factors almost without crowding out alternative uses. At the same time, foreign currency resources increased: first of all, the military victory stirred a wave of enthusiasm and identification with Israel in western Jewry, and financial contributions from abroad poured in. Secondly, the United States began to provide special financial aid for defense needs. In the early 1970s however, defense expenditures stopped growing briefly. Sharply increased defense consumption made it impossible to promote vital economic and social objectives, and when the Suez Canal war of attrition ended and relative calm prevailed, evolving social protest demanded a change in national priorities.

Table 6.1 Development of Defense Consumption, 1954-2001

Years	Average Annual Rate of Change*	Accumulated Change for Sub-periods*	Index** 1953=100
1954-55	26.0	58.7	159
1956	100.0	100.0	317
1957	-31.6	-31.6	217
1958-60	4.4	13.8	247
1961-66	13.8	116.9	536
1967	76.5	76.5	945
1968-70	23.8	89.7	1792
1971-72	-3.0	-5.9	1686
1973	63.9	63.9	2764
1974-75	7.1	14.7	3170
1976-77	-16.4	-30.1	2216
1978-85	1.5	12.8	2499
1986	-13.7	-13.7	2157
1987-91	1.8	9.1	2353
1992-95	-3.4	-12.8	2052
1996-01	2.8	18.0	2421

* Constant prices, in percentages.
** Constant prices, the last year in each sub-period compared to 1953.
Source: to 1955 – Kohav & Lifshitz, 1973: 5; from 1956 – CBS, 1996: Table 4, 38; Bank of Israel 2002: Table A-App.-3.6 (1).

The Yom Kippur War that broke out in October 1973 once again reshuffled all the cards. In the war year itself, defense consumption rose by about 64 percent and in the two following years by another 15 percent. On the demand side was the urgent need to rehabilitate the army, make equipment functional again and replace depleted inventories, while on the supply side unprecedented opportunities to acquire state-of-the-art weapons systems from the United States opened up. In 1975, Israeli defense consumption reached an all-time record.

1.2 Moderate Decrease with Occasional Fluctuations, 1976-2001

The next period, from the mid 1970s to the beginning of the 2000s, constitutes a new and different chapter in defense consumption (see Table 6.1). While in the first period there was an upward trend, and a constant price decrease in defense consumption occurred only in three of the 22 years, in this second period the general trend was downward, and of the 26 years there were 15 of constant price increase and 11 of constant price decrease. In 2001 defense consumption was 23 percent less than at its peak in 1975.

Since the mid 1970s, the path of defense consumption followed somewhat erratic changes taking place in other areas. In the late 1970s a substantial change for the better occurred in Israel's strategic situation when a peace treaty with Egypt was signed. In a short time, however, the effect was offset by Syria's growing power designed to attain strategic balance with Israel, and by deteriorating current security conditions on the northern border with Lebanon. In 1982 the Lebanese War broke out, after which the IDF remained deployed there in large numbers for a long time. Later came another change for the better: chances of all-out war on the eastern front diminished as an economic crisis forced Syria to reduce her defense expenditures and render part of her army inoperative, and Iraq sunk into prolonged war with Iran. But in the late 1980s and early 1990s, current security problems on the Lebanese border continued, and a violent uprising in Judea, Samaria and Gaza (the first *Intifada*) broke out. Moreover, distant threats from weapons of mass destruction aimed at Israel emerged from Iraq and Iran, and became a great concern. During the Gulf War in 1991, dozens of Iraqi Scud missiles were indeed fired, and for the first time in Israel's wars, the rear became a battlefront. In the first half of the 1990s, global changes and diplomatic processes in the Middle East exerted a positive effect, and the threat of all-out wars between Israel and her immediate neighbors subsided further. However, the more distant threats remained, as did, with varying intensity, security problems on the Lebanese border and in the relations with the Palestinians. At the end of 2000, following the breakdown of peace talks with the Palestinians, a wave of terror broke over Israel (the second *Intifada*), and security deteriorated to new depths.

Security conditions were obviously a key factor, but economic conditions too had their effect on defense consumption fluctuations. After the Yom Kippur War, the Israeli economy went into prolonged stagnation, rapid inflation, deficits and balance of payments instability, and in the mid 1980s, as part of a comprehensive effort to stabilize the economy the defense budget was substantially reduced. In the early 1990s, when economic growth resumed, the economy had to cope with a huge influx of immigration, so once again economic and social goals took precedence over defense needs.

2. THE COMPOSITION OF DEFENSE CONSUMPTION

Defense consumption is made up of domestic and imported components, the former divided between labor costs and domestic purchases of goods and services. Over the years, the composition of defense consumption changed (see Table 6.2), partly because of different quantitative growth rates of the various components, and partly due to changes in their relative prices. The most significant change relates to the share of labor costs. In the early 1950s, it was about half of total defense consumption, while in the mid 1970s, after continuously decreasing, it was less than 20 percent. In the following years the trend reversed, and the labor cost share grew steadily to more than 40 percent during most of the 1990s. In the course of these changes, two phenomena stood out. First, when the labor cost share declined, it was mainly as a result of quantitative growth rates lower than those recorded for other components of defense consumption. By contrast, when the share increased, particularly in the last decade, it was due to relative prices; the price of labor grew faster than prices of other components. Secondly, labor cost share decline corresponded with the rapid growth of overall defense consumption, while its increase occurred when overall defense consumption decreased moderately. The inverse relationship between changes in overall defense consumption and labor costs presumably indicate the relative rigidity of expenditures related to military manpower.

Table 6.2 The Composition of Defense Consumption, 1955-2001 (Selected Years)*

Years	Labor Costs	Domestic Purchases	Defense Imports
1955	49		
1960	35	40	25
1966	29	38	33
1970	18	37	45
1975	15	35	50
1980	23	37	40
1984	26	39	35
1990	37	37	26
1995	47	33	20
2001	43	32	25

* Percentages of total defense consumption, in current price.
Source: CBS, 1996: Table 1, 33; Bank of Israel, 2002: Table A-App.-3.6 (1).

The changes in the defense import share form a kind of mirror image of changes in the labor cost share. There was a continuous rise, reaching 50 percent in the mid 1970s during the massive acquisitions of American advanced weapons systems after the Yom Kippur War, with a drop in the years that followed. It seems as if the major part of the increase in defense expenditures was allocated for acquisitions of new weapons systems and equipment from abroad, and that those items were the first to suffer from budget cuts thereafter. In fact, however, expansion of imports was not so much the result of planning, but rather a reflection of the changes in military aid from the United States. Since the mid 1980s, military aid stabilized at a fixed nominal sum, and import options have consequently eroded in real terms considerably (see also chapter 11).

3. DEFENSE CONSUMPTION AND OVERALL OPPORTUNITY COSTS OF DEFENSE

Reported defense consumption does not reflect fully the overall opportunity costs of defense. Current estimates of the missing components are not available, yet partial estimates for specific points in time may indicate the magnitude of the gap and whether, over time, it has widened or has become less significant: -

i. *Expenditures of civilian ministries and local authorities*: defense-related public consumption not commonly included in reported defense consumption added 1 to 2 percent in the 1970s (CBS, 1983:67) and 2.5 percent in the 1990s (CBS, 1996: 39). Increased terrorism in the 1990s seems to have increased defense consumption not included in the defense budget, particularly of the police.

ii. *Expenditures for shelters and strategic stocks*: For the early 1980s, the investment in shelters for both residential and public buildings, and the costs of maintaining stocks of essential goods, mainly fuel, were estimated to be some 4-5 percent of reported defense consumption (Berglas, 1983: 21-22). In the mid 1990s, these expenditures added only 3 percent (CBS, 1996: 39).

iii. *Labor costs of conscripts and reservists*: In the early 1980s, the potential contribution of conscript soldiers to GNP was estimated at 5 percent (Berglas, 1983: 18-19), about 20-25 percent of defense consumption for that period. For reservists, the gap between reported costs and opportunity costs narrowed over the years, as the rules determining their compensation were adapted so as to cover most of their civilian earnings. Moreover, since 1986 there has been a substantial quantitative decline in reserve service (Ministry of Finance, 1994: 67), and the related expenditure and its share of defense consumption decreased considerably too. In the mid 1990s, additional labor costs due to conscript and reserve service were estimated at 17-20 percent of reported defense consumption (CBS, 1996: 39). It appears, then, that by comparison with estimates for the early 1980s, gaps arising from labor costs have narrowed.

iv. *Expenditures on defense imports*: These were underestimated in the 1950s and most of the 1960s, because official exchange rates used in statistical reports deviated widely from effective exchange rates. Over the years, however, the differences diminished, and from the early 1970s, there have been no gaps in evaluating the defense imports.

v. *Interest payments on the government's defense-related foreign debts*: Foreign loans, even if designated for defense, release scarce foreign currency for other uses. Thus any attempt to assign part of the government debt or the respective interest payments to defense, is quite arbitrary. Interest payments identified with the government's defense-related foreign debts added some 6 percent to defense consumption in 1977, 14 percent in the 1980s and 11 percent in 1992 (Ministry of Finance, 1992: 68).

vi. *Aid to defense industries*: In the late 1960s and early 1970s, when domestic defense industries were growing rapidly, government aid, if extended, was recorded in the defense budget. However, in the early 1990s, large sums were allocated to government owned defense companies outside the defense budget. These expenditures added up to 2-3 percent of annual defense consumption (Ministry of Finance, 1994: 72).

Developments in contrary directions make it difficult to determine whether, all in all, the gaps between overall opportunity costs and the measured defense consumption actually changed. Since labor costs have always been a main component of that gap, in view of the above comparisons it may be assumed that there has been no increase with respect to the early 1980s. In comparison with earlier periods, since defense imports are no longer undervalued, it would seem that the gaps have even narrowed.

4. THE RELATIVE WEIGHT OF DEFENSE EXPENDITURES IN THE ECONOMY

Analysis of the relative weight of defense expenditures in the Israeli economy would be incomplete if only statistical measures regularly applied in defense economics literature are employed. In what follows, a relatively large number of alternative measures are considered simultaneously.

4.1 Overall Defense Consumption and Total Economic Resources

The discussion below relates to three relative measures: overall defense consumption to GDP, overall defense consumption to total resources, and defense consumption to total resources, after subtracting direct defense imports from both numerator and denominator. The second relative measure is required because throughout the years available resources have significantly exceeded GDP, while the third accounts for the foreign capital inflows that were available in most years, whether related directly (American foreign aid) or indirectly (donations and loans from world Jewry) to the financing of defense imports. All three measures outline a similar course over time: after a decrease in the early 1950s, the defense shares remained more or less the same, and relatively low, until the 1967 War. Between then and the Yom Kippur War in 1973 there was an upward leap, and the relative weight of defense in the economy doubled. The upward trend continued for the next three years, while from 1977 there was a continuous downward trend. In the mid 1990s the relative weight of defense returned to the level typical of the years before the Six-Day War (see Table 6.3).

Table 6.3 Defense Consumption Shares of GDP and Total Resources, 1950-2001 (Selected Years)*

Years	Defense Consumption to GDP	Defense Consumption to Total Resources	Defense Consumption to Total Resources, without Direct Defense Imports
1950	8.9	7.1	
1955	7.7	5.6	
1960	7.9	6.0	4.6
1966	10.2	7.6	5.2
1970	25.1	16.7	9.8
1975	32.1	19.8	11.0
1980	23,0	14.4	9.1
1986	15.8	10.3	7.4
1990	13.8	9.5	7.2
1995	9.7	6.6	5.3
2001	9.4	6.5	5.0

* Percentages, in current prices.
Source: CBS, 1996: Table 5, 7; Bank of Israel, 2002: Table A-App.-1.1, A-App.-3.6 (1).

The defense share rose and fell with parallel changes in defense consumption, yet changes in the rate of economic growth were very important too. Despite growing defense consumption, until the Six-Day War its share remained steady and relatively low, since those were years of rapid economic growth. By contrast, in the mid 1970s, the relative weight of defense in the national economy peaked, as rising defense consumption combined with a declining rate of economic growth. During the last half of the 1970s and the first half of the 1980s, while the general trend was a decrease in defense consumption, its share decreased very little, because economic growth was slow. In fact, only when growth accelerated in the 1990s, did the relative weight of defense drop to the low 1960s levels.

Admittedly, throughout the years, the relative weight of defense in the Israeli economy was very high. Had Israel had to meet all defense needs from her own resources, in most years she would have had to allocate more than 10 percent of GDP to defense, in a third of those years 20 percent, and in the peak period, some 30 percent. Even taking total resources into account, the fraction allocated to defense exceeded 10 percent more than half the time, some years approaching a fifth. The comparison between the second and third measures (see Table 6.3) indicates to what extent foreign military aid alleviated the economic burden of defense.

4.2 Domestic Defense Consumption and GDP

It may prove useful to analyze separately the share of domestic defense consumption in GDP, and the effect of defense imports and their financing on the balance of payments. Domestic defense consumption that demands locally produced goods and services, affects the employment of production factors, price trends and various structural aspects of the economic activity. When measured by the domestic defense consumption to GDP ratio, it appears that through the 1960s and until the mid 1970s the influence of domestic defense demand over economic activity grew, but after that it declined (see Table 6.4).

According to accepted national accounting practices, the relationship between returns on labor in the defense sector and GDP is a measure of the direct contribution of defense to GDP. From the early 1950s to the early 1970s, this relative contribution doubled and since then it remained more or less constant at about 4-5 percent (see Table 6.4). Changes in relative contribution may come as a result of change in the size and composition of military manpower or from differential rates of increase in soldiers' wages and in GDP prices, and these factors may operate in the same direction or may offset each other. Until the mid 1970s, real wages in the defense sector (in terms of GDP prices) rose, while in subsequent years they fell. Changes in real wages alone, however, do not fully explain the change in the defense sector's contribution to the GDP. Apparently, then, there were changes in the size and composition of military manpower too: in the first period they increased the contribution of defense to the GDP, while in the second one they offset the influence of the decline in real wages. However, reported data underestimate the contribution of defense to the GDP, since expenditures on conscript soldiers do not reflect their economic cost. As noted above, according to an estimate for the early 1980s, the costs of conscripts was undervalued by 5 percent of the GDP, so that this correction alone would almost double the direct contribution of the defense sector.

Table 6.4 Shares of Domestic Defense Consumption and Its Components in the GDP, 1952-2001 (Selected Years)*

Years	Overall Domestic Defense Consumption	Returns on Labor	Domestic Purchases
1952		2.4	
1955		3.7	
1960	6.0	2.8	3.2
1966	6.8	2.9	3.9
1971	14.2	4.4	9.8
1974	17.2	5.1	12.1
1980	13.8	5.3	8.5
1986	11.0	4.9	6.1
1990	10.2	5.1	5.1
1995	7.7	4.5	3.2
2001	7.1	4.1	3.0

* Percentages of GDP, in current prices.
Source: CBS, 1966: Table 1, 33 and Table 5, 37; Bank of Israel, 2002: Table A-App.-1.1, A-App.-3.6 (1).

The share of domestic purchases in GDP reflects the impact of defense demand on the various sectors in the economy, particularly on industry. In the early 1960s the impact was rather minor, but within less than a decade the ratio of domestic defense purchases to GDP more than tripled. In 1974 it peaked, reaching 12 percent of the GDP, but later it declined (see Table 6.4).

4.3 Defense Expenditures and the Balance of Payments

Defense needs have affected the balance of payment in three principal ways. First, in most years, direct defense imports made up a relatively high proportion of the goods and services account, and it would have been even higher had imported inputs for domestic defense production been added. Secondly, until the late 1980s, American military aid was extended in part as long-term loans, and the subsequent debt service payments became a large expense item in the balance of payments. Thirdly, American military aid and defense-related capital imports were in most years important source items on the unilateral transfers and international capital flow accounts.

Before 1967, direct defense imports made up less than 10 percent of total imported goods and services, almost doubling after the Six-Day War, and raising another step after the Yom Kippur War. Between 1968 and 1981, direct defense imports made up some 45 percent of the overall import surplus (see Table 6.5). However, in the following years the trend changed: the average annual level of direct defense imports rose slowly, and its share of total imports decreased. During that period, as noted, defense imports were tightly linked to the amount of military aid, which since the mid 1980s remained nominally constant. Moreover, it was agreed that increasing sums out of the annual aid could be converted for financing domestic expenditures (250 million dollars in the early 1980s and 475 million dollars in the 1990s). Finally, direct defense imports financed from Israel's own resources also diminished (about 350 million dollars in the mid 1970s and only 150 million dollars from the mid 1980s). In the mid 1990s, the direct defense import share of total imports of goods and services was, on the average, lower than in the 1960s, and apart from exceptional supply years, defense imports became almost insignificant in Israel's balance of payments.

Table 6.5 Defense Expenditures in Foreign Currency and the Balance of Payments, 1958-2001

Years	Direct Defense Imports (Annual Average, Million Dollars)	Direct Defense Import as % of -		Own Foreign Currency Defense Expenditure* as % of Total Exports of Goods and Services	Net Unilateral Transfers to the Public Sector as % of Foreign Currency Defense Expenditure*
		Total Import of Goods and Services	Overall Import Surplus		
1958-59	42	7.3	12.6		
1960-66	95	9.2	20.5		
1968-72	451	17.3	44.0		
1974-76	1,482	20.1	44.1	11.8**	79**
1977-81	1,539	13.2	46.0	8.8	78
1982-85	1,465	9.6	33.1	10.3	108
1986-90	1,692	8.2	35.4	7.7	127
1991-95	1,673	5.2	19.7	6.0***	135***
1996-2001	1,913	4.2	20.0		

* Foreign currency defense expenditure: direct defense imports plus principal and interest payments on the defense debt to the United States. Own foreign currency defense expenditure: foreign currency defense expenditure less direct defense imports financed by American military aid funds.

** For 1975-76.

*** For 1991-93.

Sources: to 1972 – Kohav & Lifshitz, 1973: Table 3, 9; from 1974 – Bank of Israel, 1996: Table A-14, 348-349; Bank of Israel, 1997: Table F-App.-11, 295; Bank of Israel, 2002: Tables D-App.-1.1 and 1.2; CBS, Statistical Abstracts of Israel, 1998: 4-7; Ministry of Finance, 1984; Ministry of Finance, 1992: 68.

While the influence of direct defense imports on the balance of payments was decreasing, increasing amounts were required for servicing the defense debt (principal and interest payments). Those sums, along with direct defense imports not financed by aid funds, represent the defense foreign currency burden, as far as Israel's own such resources are concerned, and to derive meaningful economic conclusions they might be compared with current income from exports of goods and services. From the mid 1970s to the mid 1980s, defense required about a tenth of the revenues from exports. However, the rapid rise in exports, particularly in the early 1990s, reduced the burden to only 6 percent (see Table 6.5).

Defense needs created demand for foreign currency resources, but at the same time they were an important factor on the supply side. Apart from American military aid, sums raised specifically to meet the country's defense needs could hardly be separated from other unilateral transfers and foreign capital inflows. Indirectly, by considering the ratio between net unilateral transfers to the public sector and defense expenditures in foreign currency (see Table 6.5), it seems that since the mid 1970s Israel has had convenient foreign currency sources for financing her defense needs.

4.4 Defense Expenditures and the Public Sector

The high level of expenditure that followed from defense needs is no doubt a key factor in the relatively high weight assumed by the government in Israel's economy.

Table 6.6 Defense Expenditure Shares of Total Public Expenditures, 1960-2001 (Selected Years)

Years	Percentages*
1960	20.8
1965	19.5
1970	42.6
1975	40.6
1980	29.4
1985	28.7
1990	22.1
1995	16.9
2001	15.8

* In current prices.
Source: Bank of Israel, 1998: Table E-App.-1a, 300-301;
Bank of Israel, 2002: Table A-App.-3.2 (2)

Until the Six-Day War, defense accounted for a fifth of all public expenditures, rising to 40 percent in the first half of the 1970s. Subsequent years saw a decrease, and at the end of the 1990s the defense share was lower than it was in the early 1960s (See Table 6.6).

The large share of defense within public expenditures severely reduced flexibility in managing the government budget and other related issues of public finance policy. During the last decades, the crux of the financing issue concerned local currency expenditures, since most defense expenditures in foreign currency were covered, as noted, by external sources. When it became necessary to increase domestic defense expenditures, the government faced three basic options: reducing other public expenditures, increasing taxes revenues or deficit financing.

Until 1966, tax revenues fully financed domestic public expenditures. In the latter half of the 1970s, however, when domestic defense consumption rose by nearly 10 percent of GNP (compared to early 1960s), the government did not attempt to cut back civilian public consumption, and all additional tax revenues (about 20 percent of GNP) were allocated to, though did not suffice for, increased transfer payments (some additional 26 percent of GNP) (see Table 6.7). In the later 1960s, and even more so in the 1970s, increased defense expenditures relied, then, on deficit financing. New classification of public expenditures and revenues makes it impossible to compare data from 1980 onward with those of earlier periods. Nonetheless, in the 1980-2001 period itself, the decrease in domestic defense consumption by 6.7 percent of GDP was clearly an important factor in reducing the domestic public deficit. A more detailed analysis for sub-periods shows, however, remarkable differences in tradeoff patterns. In the latter half of the 1980s and the first half of the 1990s there was a clear tradeoff between domestic defense consumption and the domestic civilian demand of the public sector. In the first half of the 1980s, the share of both components was reduced in an effort to restrain the public deficit and stabilize the economy. By contrast, in the later 1990s, though again both shares declined in parallel, this time it was designed to allow increased transfer payments without increasing the overall tax burden and the public deficit. In any case, because of their high relative weight in public spending, even if adjustments in domestic defense expenditure are not perceived as an explicit means of macroeconomic policy, it is impossible to ignore them in this context either.

Table 6.7 Financing the Domestic Defense Consumption, 1957-2001*

Years	Domestic Defense Consumption	Domestic Civilian Demand**	Net Transfer Payments	Taxes	Other Domestic Expenses	Public Deficit
1957-61	6.0	10.9	9.8	29.3		
1962-66	6.1	11.1	11.3	31.1		
1968-73	13.0	10.8	17.2	37.1		
1974-80	15.5	11.9	36.3	50.1		
1980	13.5	19.9	19.7	41.4	-0.2	11.5
1985	12.0	17.3	15.9	42.4	3.7	6.5
1990	10.1	19.3	13.5	37.6	2.3	7.6
1995	7.5	23.6	12.8	38.9	0.6	5.6
2001	6.8	22.9	14.3	40.3	1.3	1.5

* Until 1980 - percentages of GNP; from 1980 – percentages of GDP.

** Until 1980 – includes civilian public consumption only; from 1980 – also includes government investments.

Source: For 1957-1980 – Berglas, 1983: Table 4, 38.
From 1980 – Bank of Israel, 2002: Tables A-App. 3.1 (2) and 3.3.

7
MILITARY MANPOWER

Producing and supplying national security are labor intensive and require different types of manpower: military manpower; civilians, whether performing jobs within the armed forces or in other branches of the defense establishment, and employees in defense industries and elsewhere in the civilian economy, who produce goods and services that defense requires. The military manpower, namely officers and soldiers in uniform, divides into active duty personnel, or the regular forces, and reserve forces. In many countries the regular forces further divide between conscript soldiers drafted by law for a given period of compulsory service, and volunteers recruited in the labor market for relatively longer periods. The different types of manpower serve as substitutes for one another to some extent, and economic analysis may help choose the optimal mix, i.e. the one that produces the desired level of security at the lowest costs. With this end in view, defense economics analyzes theoretical and empirical questions of manpower demands, supply and employment in producing national security.

1. UNIVERSAL CONSCRIPTION VERSUS SELECTIVE VOLUNTEER FORCE: MILITARY, POLITICAL AND SOCIAL ASPECTS

Labor services not acquired through markets, but that rather become available by special enlistment laws, is unique to the production of national security. The historical process that brought about the "army of the masses" based on conscription has already been discussed (see chapter 2, section 2.1). In brief, the revolutions that stimulated the rise of nation states at the end of the eighteenth century, promoted the idea that good citizenship combines the rights it grants with the duty to defend the state. At the same time, industrial and technological developments made it possible to have military organizations based on large numbers of soldiers, and the two trends conjointly provided the political-social justification and the practical conditions for mass armies and universal military service. However, to attain universal military service it was necessary to deprive individuals of their freedom of choice, and to refrain from using market mechanisms and economic incentives in recruiting military manpower.

Adam Smith, who wrote before the emergence of the army of the masses, estimated that a standing army would be more expensive than militias. Nonetheless, faithful to the principle of specialization and division of labor, he asserted that only an army of regular forces could attain the professional level needed to operate ever more sophisticated weapons, and assure the continuous existence of an advanced industrial society (Smith, 1776: Part 5, Chapter 1). In fact, however, matters developed differently. A national army based on universal conscription could place huge numbers of soldiers in the field in a short time and a small standing army was perceived as insufficient, so most countries soon adopted the new framework, if only as a means of self-defense. There were, however, doubts about the efficiency of an army exclusively based on compulsory service, hence standing armies of professionals were also maintained, and in the nineteenth century most European national armies combined conscripts and career service personnel. With that, there were some exceptions. Britain, then the great power, had only permanent career forces, possibly due to the nature of military missions required to maintain control over her colonies, or because she was an island state difficult to attack (Wallach, 1976: 477). The United States, by contrast, both for ideological and geopolitical reasons, made do with militias only (Amidror, 1979: 10-21).

During the second half of the twentieth century, an increasing number of states ended conscription. Britain decided to abolish conscription in the late 1950s, and since 1963 her armed forces have been based entirely on volunteers. The United States introduced an All-Volunteer Force (AVF) in 1973. At the same time, in other western countries that still maintained a mixed structure, the armed forces came more and more to resemble those based exclusively on volunteers. In the 1970s and 1980s, most of those countries cut down their regular forces, and in particular reduced the ratio of conscripts to volunteers. In fact, the difference between overall manpower requirements and the numbers of volunteers determined the extent of compulsory enlistment, and conscription rates eventually declined. In Western Europe in the 1970s, for example, it averaged only 57 percent of the relevant age groups (Mellors & McKean, 1984: 29). That being so, compulsory military service became selective rather then universal, and to avoid making it even more selective, service time was reduced. In the early 1990s, in most NATO states compulsory military service lasted no more than a year (Tuohy, 1991). The end of the Cold War changed threat perceptions in many countries, while the 1991 Gulf War highlighted the efficiency of the American and British volunteer forces. Consequently, more countries were encouraged to rethink their attitudes to conscription. In 1995 Belgium did away with it, as did Holland in 1997. France, the country in which the army of the masses was born, decided to phase out conscription gradually, and it was to end, after more than two centuries, in 2002.

1.1 The Military Aspect

From a military standpoint, the attitude toward the army of the masses, and the compulsory service it was based on, was marked by the ambivalence inherent in issues of quantity versus quality. Those who supported conscription maintained that it gives the armies numbers of soldiers unobtainable by any other means, and as long as quantitative force ratios are decisive in wars, it is impossible to relinquish mass compulsory enlistment. Others differed, having doubts about the possibility of transforming large numbers of conscripts into a military advantage. At first, reservations mainly related to logistic difficulties and limitations of command and control. Later, as military activities became more capital intensive, and equipment more technologically advanced, concerns focused on possible quantitative and qualitative imbalances between manpower and equipment. Eventually, doubts arose with respect to the importance of quantitative force ratios per se.

Napoleon said, "God is on the side of the big battalions". Clausewitz, the great commentator on the Napoleonic Wars too held the view that in future wars nations with larger armies would have an advantage, and firmly supported universal conscription (Harkabi, 1990: 460; Wallach, 1980: 54). However, when wars broke out again in the mid nineteenth century, the effectiveness of mass armies was called into question. Ardant du Picq, a French officer, maintained that in any case only a relatively small portion of the conscripts participate actively in actual fighting (Wallach, 1980: 136-137). Helmuth von Moltke "the Elder", the Prussian army Chief of Staff, concluded from the Franco-Prussian War in 1870-71 that the mass army did not change the course of the war, and pointed out specifically the lack of planning and control that marked its fighting (Wallach, 1980: 80). Friedrich Engels, Marx's collaborator, who was no less brilliant as a military commentator than he was as a social philosopher, put forward a different argument. He was deeply impressed by the new arms that appeared on the battlefield, notably in the American Civil War, and criticized the conscript army because its short service time did not allow for the training needed to operate the new arms effectively (Wallach, 1980: 252-253).

The phenomenon of gigantic armies peaked in World War I, although in its advanced stages the supposedly inherent advantage of numbers was observably undermined. At the end of World War II, there was no more room for doubt: novel weapons gave relatively small groups the ability to decide individual battles and change the course of entire campaigns, considerably reducing the importance of quantitative relationships in manpower (van Creveld, 1985: 165-167). Thus technological developments that made possible the rise of the army of the masses in the eighteenth and nineteenth centuries were also crucial factors in accelerating its decline in the twentieth.

At the same time that relative manpower quantities declined in importance, the importance of capital-labor relationships in military activity grew. The German General Hans von Seeckt, who between the two world wars worked to establish the new Reichswehr, realized that the rising cost of weapons makes it impossible to give them to many soldiers. Therefore, he maintained, large numbers of redundant soldiers may actually disrupt effective proportions of service personnel to equipment (Wallach, 1980: 159-

162). The British J.F.C. Fuller took a similar position; he preached enthusiastically in favor of a small, professional army relying on quality rather than quantity, especially on mobility and armor, and understood that the costs of establishing a mobile, armored combat force made a mass army impossible (Wallach, 1980: 192-193). The British B.H. Liddell Hart, among the most important military thinkers of the twentieth century, argued to the same effect. He asserted at the end of World War II that compulsory military service had seen its day, as it adheres to quantitative norms in an age where quality has become decisive (Wallach, 1980: 231).

With the emergence of nuclear weapons, the mass army era was thought to have come to an end (Janowitz, 1973). In retrospect, however, nuclear weapons fulfilled an important function in strategic deterrence, but had no real effect on manning the armies. Meanwhile, military manpower issues were extensively influenced by the rapid technological progress in conventional weapons. Operating and maintaining more sophisticated and complex weapons systems required higher quality manpower, and it became evident that the optimal composition of military forces is not merely a quantitative issue of capital-labor ratios, but rather – and perhaps mainly – a problem of quality. From a military standpoint, the comparison between universal conscription and recruitment through market mechanisms shifted, then, from the quantitative to the qualitative sphere, and the main question was which system could bring higher quality soldiers into the armed forces.

Before the transition to AVF, there was much doubt in the United States as to the quality of volunteers that market forces could bring in. While the draft had made it possible to take a relatively small proportion of the relevant age groups, and by applying various screening methods to obtain soldiers with higher education and psychometric scores, there was concern lest the volunteer force turn into an employment shelter for those without high school education, unable to find a civilian job. Indeed, in the first years the quality of recruits decreased, at least in terms of formal education and psychometric scores. In the 1980s, however, the trend reversed. In the 1960s and the early 1970s, when the draft was still in force, 64 percent of soldiers had high school education while in the volunteer force of 1984, 89 percent did; the share of volunteers with the lowest psychometric score decreased from 20 to only 12 percent, respectively (Gansler, 1989: 290). In addition, the transition to a volunteer force raised average seniority, and with it – accumulated professional experience. In the Army, the share of those serving four years and more rose from 33 percent in 1974 to 50 percent in 1990, and similar trends were noted in other Services (Warner & Asch, 1995: 351-352).

1.2 Political and Social Aspects

From political and social points of view, there are two main issues in the dilemma of choosing between enlistment systems: their potential impact on political institutions and on the social order, and the inherent conflict between individual rights in democratic societies and the forcible intervention in individual choice inevitably implied by compulsory military service. On the first issue, the United States Congress clearly asserted its views in 1784. It maintained that in peacetime standing armies are contrary to the principles of governance of a republic; they may serve to establish tyranny, and are thus a danger to the liberty of the people (Amidror, 1979: 11). After the War of Independence Washington's army was disbanded, and until the Civil War the prevailing perception opposed a professional standing army on a full scale, or even as a nucleus, and regarded the militia as the only acceptable form of military organization (Amidror, 1979: 16-21).

A permanent career army might alienate the military establishment from civil society. Moreover, when a military ethos and strong military elite develop, the delicate balance at the very foundation of free democracy may be upset. The civilian political leadership could lose control over military power, and the army, out of seemingly patriotic motives, might drag the nation into adventures in foreign and defense policy, and even intervene in domestic politics. By contrast, when the whole nation serves in the army, the army cannot be alienated from the people, and a natural buffer system is created against undesirable outcomes. The opposition to this argument did not deny the dangers, but maintained that enlistment methods, and the military organization they give rise to, are less important than the political culture and the democratic tradition; if these are strong enough, there will be a barrier against undesirable developments.

Another aspect of the debate touched on the relationship of citizens to the state and its institutions. Supporters of universal compulsory service stressed its contribution to civic responsibility, patriotism and loyalty, and the bond it creates among different social strata. Those against it thought that the very compulsion in military service undermines respect for the state and its institutions, and since conscription affects every young person, it could become a dangerous focus of mass protest against the nation's authority.

Military conscription is based on a consensus that certain circumstances authorize governments to oblige citizens to lay down their lives for their country. This authority is absolute, justified by the responsibility governments undertake for national security, namely the safety of the citizens, the territorial integrity of the state and its internal stability. But authority given to the government limits the role of individuals in defining their duties to the state, thereby creating tension and contradiction with the democratic perceptions of individual free choice as the highest of all values. In the past, only a negligible minority, mainly people of religious and pacifist views who rejected any type of organized violence denied the state's authority to force citizens to serve in the army. In recent decades, however, criticism has gone beyond traditional conscientious objection and has developed, in the western industrial democracies mainly, into politically based opposition (Burk, 1989; Mellors & McKean, 1984: 27). Since there have long been no active wars, in Europe at least, external threats no longer seem real, and it is harder to "sell" the public a national security policy justifying interference with individual free choice. Then too, when wars occurred, as they did in Korea, Algeria and Vietnam, governments could hardly convince their citizens of any link between those wars and the national duty to participate in them. Moreover, in the absence of military tasks, soldiers were sent to perform civilian jobs, some of which, unlike military service, are outside the public consensus, while others, e.g. operating public services struck by labor disputes, in particular aroused controversy. There were economic factors too, which are discussed further on.

More than other reasons, the loss of its representative character brought about political opposition to compulsory service. A broad consensus favored conscription when it was seen as a just mechanism, distributing the national security burden fairly among all citizens. But in the mid 1960s, the children of the baby boom that followed World War II created an excess supply of potential recruits, and the actual share of draftees within the relevant age groups dropped. Moreover, there were no clear criteria for exemptions and deferments, and the inevitable result was outright discrimination. When the draftee army was no longer representative, and the defense burden became unfairly distributed, it lost an essential element needed for long-term survival in a free democratic society. Apparently, the army's representative nature is in danger too when it recruits by means of market mechanisms. Volunteer forces, based primarily on economic incentives, are likely to attract a disproportionate quantity of individuals from the weaker social strata, with no better employment opportunities in the civilian economy. If a volunteer army is not representative, however, it is because of free individual choices and not because of selective compulsion by the law. In fact, the American volunteer forces are no less representative than the draft army they replaced.

2. CONSCRIPTION VERSUS RECRUITMENT THROUGH MARKETS: THE ECONOMIC ASPECT

From an economic perspective, the choice between conscription and volunteer forces has been analyzed in two complementary directions. The first estimated and compared the cost of the two military organizations. The second studied their allocation and distribution effects.

As early as the second half of the nineteenth century, when many thought that cost considerations, not those of strategy and politics, made conscript forces inevitable in a world of large armies, economists rejected that simplistic conclusion because it failed to account for the full costs of conscription. Robert Giffen maintained that military pay, unrelated as it was to market conditions, did not reflect the full economic cost of manpower. John E. Cairnes argued that conscription is like a tax levied on those forced to serve, and said that this special tax should be factored into the costs of defense. Cliff Leslie stressed that the full cost of soldiers goes beyond alternative civilian earnings during their service period, and should include the implications for subsequent earnings too (Whitaker, 1991: 47-49).

2.1 Compulsory Military Service as an Implicit Tax

Most individuals compelled to serve in the army see it as a change for the worse in their personal welfare. They have to devote time to military service despite the lower money and money-equivalent returns that they receive, as compared to their alternative civilian income, and despite the fact that most of them prefer civilian to military life-style. Individual welfare loss due to military service may be considered an implicit tax (Amacher, Miller, Pauly, Tollison & Willett, 1982: Part I): instead of giving the government part of the money income from their work, conscripts are required to give that work itself. This is undoubtedly a deviant tax; a tax in-kind, which, except for compulsory military service, is virtually unknown in modern fiscal systems. Moreover, only a defined part of the population, the young in particular, bear that implicit tax. Others are exempt, and by comparison with alternatives to the draft – which require open budget increments and thus higher monetary taxes on the public as a whole – they are likely to pay lower taxes. At the same time, the entire population shares the national security the conscript soldiers supply. In such conditions, then, conscription is likely to increase inequality in income distribution and social welfare.

The implicit tax has two components: the difference between alternative civilian income and military income, and the comfort differential between civilian and military life-styles. Suppose that an individual could estimate the additional income that would induce him to choose to serve in the army. In the first stage he would estimate the income difference between civilian and military employment, but since he would prefer the former even if the two were equal, in the second stage he would demand additional compensation for giving up the comfort of civilian life. In other words, he would demand military pay (Wm) that is greater by a given rate (α) than his alternative civilian income (Wc). The addition to military income that would induce individuals to choose military service is therefore a possible measure of the conscription tax (t), which can be expressed as $t = Wc + \alpha Wc - Wm$.

Three observations may be made here. Firstly, the size of the conscription tax depends on the preference of each individual, and therefore, although α is usually assumed to be positive, for individuals who prefer the military life-style it might be a negative quantity. It also depends on the ability to estimate the positive and negative benefits of military service, thus for lack of information or for other reasons, it is possible that individuals will underestimate the benefits they may derive from military service, and overestimate the conscription tax. Secondly, military earnings (Wm) include cash compensation and income in-kind such as food, housing and medical services. These components are generally evaluated by their costs to the army, not by their value to their recipients, and a bias may arise. On one hand, soldiers may genuinely value certain components of the income in-kind at less than their actual cost, while on the other hand, because of economies of scale the army may purchase or produce certain goods and services at a particularly low price, and their value, as soldiers see them, will be higher than their costs. When soldiers undervalue their income in-kind, the conscription tax as estimated by actual costs will be biased downward. Thirdly, alternative civilian income (Wc) is not always higher than military income. Under certain market conditions, when unemployment is high for example, some individuals may be in the opposite situation. The conscription tax may then be negative, and military service may increase individual welfare. In this context, it should be noted that the conscription tax relates to the alternative civilian income and not to the current income of prospective draftees, if any. For example, students usually do not have current income, as they voluntarily gave it up when deciding to devote their time to studies. Nonetheless, they too pay a conscription tax – the tax equal to their forgone income – if military service keeps them away from their studies.

The conscription tax can be defined in another way as well: instead of the income increment the individual would require to make him prefer military service, it could be thought of as the amount he would be willing to pay for exemption. These are two different points of departure, each of which implies a different amount of conscription tax. Which measurement, then, is correct? The answer lies in basic perceptions of the interrelations between the individual and the state, and hence is outside the realm of purely economic analysis. If the prevailing perception recognizes the state's right to impose military service on individuals, the relevant measurement will be according to the willingness to pay for exemption. However, if the individual's right to avoid military service is recognized, then the relevant measurement is the income increment that the government has to pay. The compensation system in armies of volunteers is based on the second approach.

When compulsory and voluntary services exist concurrently, volunteers too pay an implicit tax. If there were no supply of conscripts, more volunteers would be needed, and higher wages would be required to obtain them. Thus the difference between that higher hypothetical wages and what the volunteers actually receive is an implicit tax borne by them. It does not, however, affect the volunteers' behavior, since their actual wages represent the minimum amount for which they are willing to choose military employment in any case, and hence it has no effect on the allocation of resources. On the other hand, this implicit tax affect the income distribution amongst individuals, as it represents the forgone economic rent that volunteers would have earned in the absence of compulsory service.

Two methods have been used in attempts to measure the implicit tax imposed by compulsory service. One method derived the wage necessary to secure the required quantity of service personnel from an estimated supply equation of volunteers, and then subtracted from that amount conscripts' compensation in money and in-kind (Oi, 1967). This method, however, is applicable only when prospective draftees can choose to volunteer, as was the case in the United States before 1973, and even then it is likely to underestimate the implicit tax. Prospective draftees may volunteer as a way to minimize the expected loss associated with compulsory service (Bradford, 1968), so that supply equations of volunteers (based on actual behavior) underestimate the "true" reservation wages and the implicit tax. On the other hand, when volunteering is an option available only to those who complete compulsory service, as it is in Israel, the supply equation of volunteers does not express the preferences and employment opportunities of prospective draftees, and is practically irrelevant to calculations of the conscription tax.

The second method evaluates the alternative civilian income of draftees from data on income in the economy in general, and not from wages demanded by volunteers. If data are available, an attempt is made to consider the average income of civilian employees as similar as possible to the draftees as to age, education and the like (Davis & Palomba, 1968). With this method too, however, estimates of the implicit tax may be biased. First, it ignores the compensation increment required for military service, implying that for an equal wage, the individual will have no preference for civilian employment over compulsory military service. Secondly, the average civilian income is likely to be less than the alternative income of conscripts if, following medical and psychometric screening tests, the army succeeds in enlisting the most capable, who are also those with better chances for relatively high civilian income. Alternatively, the average civilian income is likely to be higher than the relevant income of conscripts if the "brightest and the best" are more successful than others in avoiding compulsory service (Schwartz, 1986: 565).

Compulsory service involves a loss of income during the service years themselves, but may also affect the civilian income of conscripts for better or for worse in subsequent years, and even during their entire work life (see also section 6). Having joined the civilian labor force late, the conscript's income may lag behind that of his peers who did not have to serve. Alternatively, skills and experience acquired during the service may give him a higher entry level income and more rapid promotion, so that after several years his income will not only equal but even exceed that of individuals who did not serve (Miller & Tollison, 1971; Fredland & Little, 1980: 50-52). Taking later influences into consideration, the implicit tax is not limited to the actual service years, and the present value of income differentials in future years has to be accounted for too. This latter element may be either positive or negative, resulting in an implicit tax greater or smaller than that during actual service only, and veterans may even enjoy an income premium.

Empirical studies made in the United States on data from the 1960s found that conscription tax rates based on alternative civilian income for service years only, ranged from 60 percent for those with low level education to a little over 80 percent for those at higher levels (Davis & Palomba, 1968). When the tax was calculated in relation to the present value of the income flow over the entire working life, rates of between 0.7 and 3.6 percent were obtained, respectively (Miller & Tollison, 1971). Conscription tax remained, then, a positive quantity even after including income differences for the years after military service. An estimate for Belgian conscripts in the early 1980s, relating only to years of service and to military cash compensation, found a tax rate higher than 90 percent, and when payments in-kind were added – more than 60 percent. Here too tax levels were found to rise with education levels (Kerstens & Meyermans, 1993: 282-283).

Given the influence of education level on conscription tax, for tax considerations alone it is not worthwhile to defer military service in order to acquire higher, say academic, education. During actual service years, the more highly educated lose more alternative civilian income, and as regards future income flow – it is reasonable to assume that abilities acquired in the service will be of less value to them than to lower educated conscripts. Similarly, deferring enlistment to a later age also raises the implicit tax, as civilian income generally rises with age (Knapp, 1973).

2.2 Distributive Effects

Theoretical analysis and empirical findings verify that the implicit tax imposed on the conscripts themselves is progressive: tax rates are higher for those with higher alternative civilian incomes. If so, the conscription tax meets criteria of vertical justice. But the tax rates imposed on conscripts are much higher than those borne by civilians at the same income level. In the United States, when it was found that draftees with higher education "paid" an average tax of more than 80 percent, no group of civilians paid nearly as high an average rate. The same was true of Belgium: the tax rate on conscripts was some eight percentage points higher than what the highest civilian percentile paid (Kerstens & Meyermans, 1993: 282-283). Thus the conscription tax does not meet criteria of horizontal justice, and though progressive, it enlarges income distribution inequality among individuals, The conscription tax may also discriminate between generations: since military manpower requirements change over time, conscription rates change from one age group to another, and with them the implicit tax that different age groups in the population have to pay (Amacher et el., 1982: Part I).

When conscription is abolished, and the armed forces rely wholly on volunteers, more of these are required and it obviously becomes necessary to raise military pay to obtain the required numbers. As a result, if all other expenditure items remain equal, the defense budget will grow, and new explicit taxes will apparently be required. A comparison is thus called for between the implicit taxes embodied in conscription and the additional explicit taxation needed to finance larger volunteer forces. If the budget increase is smaller than the implicit taxes implied by compulsory service, then abolishing conscription will increase social welfare. Moreover, even if the sums are equal, and an explicit tax merely replaces an implicit one, since explicit taxes are imposed on the entire population through the regular taxation system, a more equitable income distribution is likely to ensue, and social welfare will increase in this case too. However, another influence must be considered: an increase in explicit taxation may distort efficient allocation of economic resources. It can be shown that under certain circumstances – when there are no significant productivity differences between conscripts and employees in civilian production, and a large enlistment becomes necessary along with a particularly large increase in defense expenditures (as in wartime) – the implicit conscription tax as a substitute for explicit taxes may increase overall social welfare (Garfinkel, 1990). (See also section 2.4 below for cost comparison of conscription and volunteer forces).

There is still another approach to the role of conscription within a general fiscal policy framework. Models of optimal taxation have shown that in a world of "second bests", given the differences in abilities and preferences among individuals, it is possible to improve equilibrium social welfare by imposing quantitative constraints that correct tax-induced deviations between social values and market prices. Thus when the existing tax system encourages individuals to consume more leisure and offer less of their labor, appropriately designed compulsory service may be an optimal policy instrument for redistribution in-kind, since it forces draftees to decrease their leisure consumption (Kersens & Meyermans, 1993: 274).

2.3 Allocation Effects

Conscription interferes with individual allocation of time, and thus limits labor mobility between productive activities. Consequently, the allocation of economic resources between defense and civilian uses is not Pareto-optimal, or in other words: a different factor allocation would raise the defense output obtainable at any given level of civilian uses, or the level of civilian uses without harm to national security, or would raise both the defense output and the level of civilian uses simultaneously.

Soldiers recruited either by law or through the market mechanism are at all times no more than a fraction of the population fit to perform military tasks, so that under both systems recruiting involves screening and selection. Suppose that two candidates equal as to their military abilities but different as to their alternative civilian income report for enlistment. Economic efficiency criteria dictate that the one with lower alternative civilian income should be inducted. Or, in general terms, among all those fit for military service, induction should begin with those with lower alternative civilian income, and proceed in ascending order of alternative income until the required number of soldiers is obtained. Other criteria will increase the opportunity costs of defense and impose deadweight losses on the economy (Hansen & Wiesbrod, 1967; Miller, Tollison & Willett, 1968; Amacher et al., 1982: Part II). Conscription, however, is not intended to minimize opportunity costs to the economy, and accepted selection processes are designed to obtain the most suitable manpower for military tasks. Only rarely and exceptionally, then, will the good of the military coincide with that of the economy, while in most cases those with qualities and abilities that fit them for military service are also those with the greatest potential in civilian employment. By contrast, recruitment through the market considers alternative civilian income directly and deliberately: it shows itself in candidates' supply patterns, namely in their reservation wage, and to recruit those making a greater contribution to the economy they must be offered higher military pay. Thus recruiting through the market is likely to produce a more efficient allocation of economic resources, and from this standpoint armed forces of volunteers are preferable to those of conscripts. Moreover, in relative terms the greater the number of candidates than the number of soldiers required, the greater will be the advantage of the volunteer force over the army of conscripts. The deadweight loss imposed by conscription is inversely associated with the proportion of recruits in the relevant population.

However, even when recruitment takes place through the market mechanism, efficient allocation of economic resources may suffer. The army is a relatively large employer and its demand for manpower may affect wage levels in the economy, so that the marginal cost of employing volunteers could be higher than their reservation wage. Under such circumstances, if the recruiting authorities act rationally, they will behave as a monopsonist: the equilibrium quantity of recruits will be determined by equalizing their value of marginal product to their marginal cost, and not to their wage, and thus it will be also higher than their marginal opportunity costs to the economy (Borcherding, 1971). Here too, then, the equilibrium reached is not Pareto-optimal, and allocation efficiency could be improved by transferring workers from civilian employment, in which the value of their marginal product is lower, to military employment, where it would have a higher value.

In both conscription and a volunteer force factor allocation may not be efficient, and may impose deadweight losses. What is required, then, is a comparison of loss size in both situations, and it is generally impossible to determine in advance in which of them the losses will be greater. The size of the loss depends on the additional quantity of conscripts required compared to the Pareto-optimal quantity (when conscription prevails), and on the extent to which the quantity of volunteers has decreased compared to the Pareto-optimal quantity due to monopsonistic behavior (when there is only a volunteer force), or on the relationship between the two. The quantity differences between each of the situations above and Pareto-optimal equilibrium are determined by demand and supply elasticities, and in the case of conscription also by the administratively determined military income of conscripts. If the additional quantity of conscripts is greater than the decrease in the number of volunteers due to their increasing marginal cost, then conscription will impose a greater deadweight loss, and vice versa.

The recruitment system may also adversely affect the efficient allocation of resources within the defense establishment. If conscripts' pay is administratively determined, unconnected to prevailing wage levels in the economy, the relatively low price of manpower may lead to inefficient factor combinations, i.e. combinations of labor and capital yielding less defense output than could have been obtained from the same total quantity of resources. The central argument is that conscription results in excessive use of manpower in producing defense, as compared to the alternative of recruitment exclusively through the market, since the relative price of labor decreases. This occurs first because the administrative price of conscripts is always less than the demand and supply equilibrium price of volunteers, and secondly, because conscription on one hand reduces demand for volunteers, and on the other hand raises the supply of volunteers trying to avoid the draft, thus leading to a lower equilibrium price of volunteers. In a mixed structure of conscripts and volunteers, then, the price of manpower recruited through the market will be lower than in one made up entirely of volunteers. *Ceteris paribus*, the lower price of both military manpower components encourages labor-intensive production, and may retard the transition to alternative military technologies requiring a larger capital-labor ratio.

Besides the relative factor prices effect, conscription expands the use of manpower for other reasons. The relatively short compulsory service term leads to rapid personnel turnover, and therefore, at all times a greater quantity of soldiers is engaged in training and exercises, and hence unavailable for current military missions. Likewise, short service prevents conscripts from making progress along the learning curve to the same extent as career soldiers do, their military productivity is lower and more of them are needed for given tasks. Finally, conscription increases the control of the military planner over the timing and quantity of labor inputs, and planners prefer to rely on inputs in assured supply (DeBoer & Blackley, 1990: 90).

The recruiting system affects not only the capital-labor mix, but also the manpower mix and its allocation among different activities. In mixed structures, conscripts will usually assume a larger part of the overall service personnel than do career soldiers or civilians. Since their opportunity cost to the economy is ignored, tasks are carried out by conscripts even when they could be done as well or better by career military personnel or civilians, i.e. even when other combinations could produce more defense output with the same quantity of manpower. Moreover, because of the seemingly low price of conscripts, the army tends to prefer direct to indirect labor employment: to do more work in-house rather than to buy and outsource to commercial firms, even though true economic considerations indicate that the latter course is more cost-effective. When Britain and the United States adopted an AVF system, the relative proportion of civilians in the armed forces rose, particularly in support units, as did outsourcing (Gansler, 1989: 287-289).

2.4 Cost-Effectiveness Analysis

Obviously, despite the allocation and distributive effects and their economic and social implications, the choice between draft and volunteer forces is dictated first of all by direct comparison of the two in cost-effectiveness terms. To develop the cost of either system, three components are to be considered: the opportunity cost of recruits, training costs, and the deadweight loss from taxation imposed to finance the defense budget. If the population of candidates for military service is greater than the required quantity of soldiers, for an equal quantity of manpower it can be reasonably assumed that the first component will be larger under the draft. The draft may include individuals whose opportunity costs are higher than equilibrium military pay (i.e. than the pay determined when recruitment is effected exclusively through market mechanisms) at the required quantity of manpower. Likewise, because of rapid turnover, the second component too may well be larger under the draft. By contrast, the budgetary cost of draftees is usually less than for an equal quantity of volunteers, so that the third component is likely to be lower under the draft. Thus for an equal quantity of soldiers, it is impossible to predetermine which of the two types of forces will be more expensive. But cost considerations are not the only ones.

Common perceptions regard a volunteer force as more effective and as producing more security than a draft force of the same size for four main reasons. First, research shows that soldiers' productivity grows with experience, and volunteers have more experience than draftees due to their longer service time. Second, due to the more rapid turnover among them, a relatively larger proportion of draftees are engaged in training at any given time, meaning that with a volunteer force of the same size more soldiers are available for operational missions. Third, the military productivity of volunteers is assumed to be higher because of greater motivation. Fourth, influenced by relative prices of manpower and equipment, the volunteer army will be more capital intensive (Warner & Asch, 1995: 377; Warner & Asch, 1996: 304). Hence differences in effectiveness make it possible that a volunteer force is preferable, even when its cost is higher, to a draft force.

A complete cost-effectiveness examination must consider "the social surplus of defense" that derives from each recruiting system, i.e. the difference between the social value of defense output and the total cost of military manpower. Without specific data, here too it is impossible to predetermine which system will yield the greater social surplus. However, analysis points to several possible conclusions (Warner & Asch, 1996: 307-310). First, when demand for defense decreases, a volunteer force is preferable. However, when the demand for defense expands, it could reach a level high enough to give preference to a draft force. Secondly, the "surplus" derived from each system depends not only on demand level, but also on the elasticity of demand for defense, and the greater it is, the greater the likelihood that a draft force will produce a larger social surplus than a volunteer force. Changes in demand elasticity are related to threat perceptions, and *ceteris paribus*, it will be larger as perceived threats grow more acute. Thirdly, when the military productivity difference between the two types of forces grows, as it does with technological improvements in military equipment for example, a volunteer force will be preferred. Fourthly, technological advance in military equipment not only increases the productivity gap between the more experienced volunteers and the conscripts, but also requires greater investment in training and practice. The faster the turnover of personnel, the greater the investment required, so that the marginal cost of conscripts will rise relatively faster than that of volunteers. Hence, the two influences of technological advance increase the social surplus derived from a volunteer force as compared to a draft force.

To complete the picture, it should be noted that each recruiting system involves administrative costs. Conscription means the high costs of keeping a roster of the population eligible for service, a call-up and reporting system, and means to enforce the duty to serve, which include dealing with evaders. Voluntary enlistment involves high advertising costs and incentives to recruiters. Usually costs of conscription depend on the size of the relevant population, while in voluntary enlistment they depend on numbers required. Hence the lower the ratio between manpower required and the relevant population, the more likely it is that operating costs of the conscription system will be higher than those of voluntary enlistment.

3. THE SUPPLY OF VOLUNTEERS FOR MILITARY SERVICE

Analyzing the supply of volunteers is important when voluntary and compulsory service exist side by side, and even more so when abolition of the draft and total reliance on a volunteer force is under consideration. There is a most understandable doubt as to whether the market mechanism can supply the necessary quantity and quality of manpower, and in particular, whether it can support the rapid expansion of manpower needed as threats become more acute or active wars break out (King, 1977).

3.1 The Income Effect

An individual who can choose between civilian and military jobs compares the income flow from each choice, and the economic analysis assumes indifference if military income (Wm) is larger than civilian income (Wc) at a rate (α) that expresses compensation for the special conditions of military service. The rule for deciding between the employment options is therefore $Wm \geq Wc + \alpha Wc$, and a supply curve for volunteers can then be derived as a function of military and civilian incomes, or of the ratio between them (Wm/Wc) (Fisher, 1969; Withers. 1972; Ash, Udis & McNown, 1983).

Naturally, prospective recruits differ in their productivity in civilian employment, meaning that there will be differences in alternative civilian income among individuals. The higher the military income, the more individuals will have alternative civilian incomes that are lower, and hence, other things being equal, the higher the military income or the higher the ratio between military and civilian incomes, the greater the supply of volunteers. Prospective volunteers differ too as to preferences between civilian and military life-styles. Most prefer the former, but the higher the military income, the more individuals will find in the difference between military and alternative civilian incomes sufficient compensation for the choice of military employment. For this reason too, then, the volunteer supply will increase as military pay rises.

The supply-quantity of volunteers at every level of military income, or the elasticity of supply with respect to income, depends on the variance in preferences and in characteristics that determine the civilian income of individuals in relevant populations (Warner & Asch, 1995: 352-353). If individuals are identical as to preferences and as to determinants of their civilian income, there is some level of military income at which all are indifferent as to military service versus civilian employment, and the elasticity of volunteer supply will be infinite. Conversely, the more heterogeneous the population of potential recruits, the less elastic the supply with respect to income.

Empirical studies in Britain (Ridge & Smith, 1991) and in the United States (Dale & Gilroy, 1984: 203; Berner & Daula, 1993: 333; Warner & Asch, 1995: 358) generally found a significant positive relationship between supply-quantity of volunteers and military income. The studies also calculated supply elasticity, most of them as regards ratios between military and civilian incomes, and estimates centered on values close to 1. Hence a certain rate of increase in the ratio between military and civilian incomes increased the number of volunteers by approximately that rate. Some studies, however, reported higher or lower supply elasticity values with respect to income, indicating to a need to explain the diverse results.

One factor that affected findings was the measure selected to represent alternative civilian income: civilian income of young people in the age group of potential volunteers (Berner & Daula, 1993: 324-325) or a more general measure that goes beyond immediate earning opportunities (Horne, 1985: 36). The second approach, when the average income of all workers in manufacturing industries was selected, showed higher supply elasticity with respect to the ratio between military and civilian incomes.

Other explanations focused on the relative importance of allowances and benefits in-kind. Certain benefits may be effective for some groups, and not for others. Scholarships for academic studies, for example, may influence high school graduates to enlist, and be less influential for the less educated

(Dale & Gilroy, 1984: 196-197, 201-202; Brown, 1985: 231). Benefits in-kind such as military housing, medical services and the like and price discounts (e.g. purchases in special stores, sports club memberships, etc.) will be more effective for family heads than for single people. In general, fringe benefits were found to influence retention decisions more than the primary decision to enlist, because those already serving recognize their value better than those without service experience (DeBoer & Brorsen, 1989: 856). Thus changes in composition of military pay between basic salary and benefits may influence supply elasticity and increase or decrease supply-quantities of volunteers even when total income remains constant.

Similarly, the income share received on retirement may also affect supply elasticity. Pension payments usually start as soon as soldiers are discharged, so that their share within total military income flow is relatively large. Deferment of income that does not detract from the present value of total income flow makes military service more attractive for individuals with a high marginal propensity to save, while for those with a low marginal propensity the influence of deferment is in the opposite direction.

Empirical research on the relationship between the supply of volunteers and income revealed two other findings that required explanation. One is that the estimated coefficients for the income effect were more statistically significant as regards enlistees of an above median psychometric score, and among them, especially those with high school education (Brown, 1985: 230-232). To interpret this finding, a distinction between two stages in the recruitment process was suggested. Recruiting authorities first enlist the maximum quantity of high quality candidates, but since at existing wage levels demand always exceeds supply, the difference is made up with candidates of lower quality. Thus while supply determines the quantity of high quality recruits, the demand of the military authorities determines the quantity of lower quality recruits, and with it the overall quantity of recruits. In other words, in general supply equations that do not relate to specific populations demand-constrained observations too are included (Brown, 1985: 230-232; DeBoer & Brorsen, 1989: 854-855). On this assumption, it is possible, then, that the elasticity of supply with respect to income as estimated from all observations is actually overstated, and if estimates were based only on supply-constrained observations they would be lower.

The second finding came from later studies that focused on the institutional features of the recruitment process. It appeared that in regions where the ratio of military to civilian incomes was known to be lower, fewer volunteers were expected, and goals for recruiting units were predetermined accordingly. It follows, then, that the military to civilian incomes ratio affects recruitment both through the individuals' choice and by determining recruiters' goals. Estimates of supply elasticity with respect to income do not distinguish between influences, thus are biased upward from this reason too, and the bias will be greater, the greater the impact of actual income differentials on determining goals is (Berner & Daula, 1993: 331-334).

3.2 The Effect of Civilian Employment Opportunities

Since the decision to volunteer for military service is a choice between alternative employment options, employment opportunities in the civilian labor market are another important factor affecting the supply of volunteers. Obviously, the higher the unemployment rate in the civilian economy, the longer will be the line of job seekers at the recruiting offices. They compare the expected income flows in military and civilian employment, and in so doing take into account the reduced expected civilian income due to limited civilian employment opportunities (Fisher, 1969). However, Early empirical studies were not sufficiently clear as regards the effect of unemployment rates on the supply of volunteers. The values commonly estimated for the elasticity of supply with respect to unemployment rates ranged between 0.2 and 0.5 (Dale & Gilroy, 1984: 203). But there were also some findings indicating that unemployment had no effect whatsoever on volunteering for the army (Ash, Udis & McNown, 1983).

Several hypotheses were put forth to explain the inconclusive results. First, it was suggested that the number of contracts candidates signed in each period, not the number that actually enlisted, should measure reactions to economic conditions; signing the contract reflects the decision to enlist, while actually entering the service may take place later, under different economic conditions. Moreover, since reactions lag behind economic changes, the unemployment rates that prevailed a few months before the contracts were signed should be considered (Dale & Gilroy, 1984: 195; Dale & Gilroy, 1985; Horne, 1985: 35, 37). Secondly, it was argued that rising unemployment enables the recruiting authorities to obtain the full desired complement even of high quality manpower, so that observations of actual numbers of enlistees are in fact demand-constrained (Ash, Udis & McNown, 1983). Thirdly, there may be a problem of simultaneity in resultant estimates: enlistment itself may reduce unemployment rates among young people, particularly as relatively few from those age groups participate in the labor force (DeBoer & Brorsen, 1989: 855-856). Later studies that took some of these arguments into consideration generally produced significant results, and elasticity estimates were found to be of higher values, though less than unity (Berner & Daula, 1993: 333).

3.3 Conscription as an Incentive to Volunteer

When conscription and volunteer forces exist side by side and volunteering is an option for those fit for compulsory service, the "threat" of conscription affects the supply of volunteers. Surveys made in the United States in the first half of the 1960s showed that 38 percent of the volunteers would not have taken this step had they not anticipated conscription (Bradford, 1968: 622). A significant relationship between the probability of being drafted and the supply of volunteers was also found in econometric research. In Britain where conscription existed until 1963, a supply equation of volunteers was estimated for the period between 1953 and 1987 using a dummy variable to distinguish between years when conscription was in force and those when the military was made up entirely of volunteers. The coefficient of the dummy variable was positive and statistically significant (Ridge & Smith, 1991: 288-289).

Since in any case the result will be the same and individuals are going to serve as soldiers for a given period of time, why does the "threat" of conscription affect voluntary enlistment? Economic analysis assumes that the answer lies in the effect of probable conscription on expected civilian income (DeBoer & Brorsen, 1989: 855), and considers voluntary enlistment of those subject to conscription as a strategy to minimize expected losses from compulsory service (Bradford, 1968). Potential draftees do not know for certain if and when they will be called up, and each date may have a different opportunity cost in terms of civilian income loss. Their main chance of minimizing the opportunity cost of conscription is to volunteer for military service on the date or at the age that suits them best. Theoretical analysis shows that volunteers under this circumstances are not necessarily those for whom the absolute opportunity cost of military service is the lowest, but rather individuals whose opportunity cost rises earlier and faster with age (Bradford, 1968: 635).

If the military authorities see conscription as a complementary way to obtain quantities of soldiers unavailable through the market mechanism, the probability of being drafted becomes endogenous to the volunteering process, i.e. it not only influences it but is influenced by it. The more volunteers there are, the lower is the probability of being drafted, and the effect of conscription on the supply of volunteers will weaken (DeBoer & Brorsen, 1989: 853-855).

3.4 The Compensation Increment for Special Conditions of Military Service

As noted before, voluntary enlistment requires military pay in excess of alternative civilian income, and the determinants of that compensation increment too affect the volunteer supply. Economic analyses explained this in different ways. One approach followed the theory of labor economics, which relates wage differences among jobs to differences in working conditions. Military service often involves working long hours, prolonged separation from one's family and social milieu, extraordinary physical and disciplinary demands, and an unparalleled high risk of injury or even death. All this suggests that the compensation increment varies with the features of military duties, place of service and the like.

An interesting attempt to examine the relationship between income demanded by volunteers and service conditions assumed that differences in quit rates from similar occupations in the American Navy and Air Force reflected differences in service conditions. Furthermore, it was assumed that a differential wage policy could equalize the quit (or the retention) rates. The research found that in five out of seven groups of similar occupations the Navy quit rates were greater than in the Air Force, and calculations showed that the Navy would have to pay premiums of between 4 and 24 percent in order to equalize them (Solnick, Henderson & Kroeschel, 1991). Although the study did not compare military to civilian occupations, it supports the assumption that the supply of volunteers depends on an income increment that compensates for differences in service conditions.

The pre-volunteering level of welfare also may affect the compensation increment demanded by prospective enlistees. The analysis of welfare economics shows that the greater the initial level of welfare, the higher the compensation increment demanded, meaning also that this increment is monotonously associated with the alternative civilian income of potential enlistees. The military wage policy adopted in Britain in 1970 followed this line of thinking. Wages for military occupations were made to accord with prevailing levels for similar civilian occupations, and in addition to special allowances (e.g. flight pay), it was found necessary to add a more or less uniform compensation of 5 percent. Since the rate was uniform, the higher the individual's alternative civilian income, the greater the absolute sum he received (Tarr, 1981).

Another approach relates to the effect of security threats on personal welfare. At any level of GNP, security threats reduce personal welfare, and at the same time, at any given level of security threats, the greater the military force facing those threats, the less the welfare decrease. However, the influence of security threats is not equal for all individuals; it affects soldiers' welfare more than that of civilians, and among soldiers – according to what their military duties are. In this framework, the compensation increment is designated to balance out differences in marginal welfare arising from changes in security threats, and hence is linked to changes in the threat level and military force size (or quantity of volunteers). With a given quantity of volunteers, a rise in the threat level implies a greater compensation increment, while at a given level of threat, enlarging the military force makes it possible to make do with a lower compensation increment. In extreme situations, because of contrary influences of threat and military force size, it can be shown (Harford & Marcus, 1988) that the supply of volunteers will decrease with increased income, or that the special income increment will reduce the quantity of volunteers, rather than increase it.

The compensation increment may be affected also by societal attitudes toward the military as an occupation; it tends to be higher in societies that accord the military lower status (Fisher, 1969; Ash, Udis & McNown, 1983; DeBoer & Brorsen, 1989: 862).

3.5 The Effect of Population Size

The supply of volunteers naturally depends on the size of the relevant age group. In a study for the years after the American draft was abolished, the population of 18-year-olds was explicitly introduced as an explanatory variable in the supply equation. Supply elasticity with respect to population was found to be 1.56, actually larger than for any other variable in the supply equation of volunteers (DeBoer & Brorsen, 1989: 861-862). By contrast, in Britain for 1953-1987, no statistically significant relationship was found between quantities of volunteers and size of the 15-19-year age group. Possibly the army correctly foresaw the demographic trends and adapted its recruitment policy accordingly or, alternatively, population size change may be too raw a datum, and using it without any other distinctions – quality level for example – makes its influence on volunteer supply truly insignificant (Ridge & Smith, 1991: 289-290).

3.6 Influence of Recruiters and of Recruiting Methods

Recruitment policy and methods used by recruiting units also affect supply. The second generation of empirical studies, after the mid 1980s, used models that separated the effects of recruitment resources and recruiters' efforts on the supply of volunteers from those of other variables (Warner & Asch, 1995: 354-360). The point of departure was that differential efforts had to be invested in recruiting different candidates, and particularly, that the greater the candidate's alternative civilian income, the greater the efforts required. It was also assumed that recruiters' efforts could be adjusted to level and targets through quotas and incentives. These studies found a statistically significant association between supply of high quality volunteers and numbers of recruiters, quotas and advertising expenditure. The number of recruiters was the most significant (Warner & Asch, 1995: 357).

Those later studies made it possible to evaluate the cost-effectiveness of alternative recruiting methods, thereby recommending an optimal mix of resources for varying recruitment goals. For instance, supply is relatively elastic with respect to military income, and hence raising basic military pay may be considered an effective recruiting technique. But this is not only a more effective means; it is also a costly one. It is impossible to raise basic pay to attract new enlistees without applying that raise, in parallel, to those already serving. When success was defined in terms of higher quality recruits, the number of recruiters, advertising and education benefits were found to be preferable in cost-effectiveness terms to raises in basic pay and enlistment bonuses (Warner & Asch, 1995: 359-360).

3.7 Retention

At any given period, volunteer supply includes those who have never served and those who reenlist after completing one or more service contracts. When there is conscription, those who reenlist may also come from conscript ranks. Economic analysis holds that retention decision-making is based on income comparison between two alternatives: serving n additional periods in the military or immediate retirement from the service (Warner & Asch, 1995: 360-363). In the first, the present value of military income for n additional terms of service has to be summed up along with pension payments and other retirement benefits and civilian income flow from period $n+1$ on. In the second alternative, civilian income must be considered as a replacement for military income in the first n periods too. The individual will decide to reenlist if there is at least one span of time (n) for which the difference between the first and second income flows exceeds the compensation increment he demands for the special conditions of military service.

Many factors influence the relationship between the two future income flows. First, it depends on the chance of military promotion, which affects military income. Secondly, it depends on the influence of military training and experience on after-service civilian income, which in turn may be associated with the additional service time n (see below, section 6). Thirdly, pension payments and other retirement benefits too vary with n. Retirement vesting usually requires a minimum service time, so that the income flow in the first alternative above may be marked by discontinuity at a time point known in advance. Fourthly, there may be a one-time reenlistment bonus, which also varies with n.

In empirical research studies, most of which were made in the United States after the transition to AVF (Warner & Asch, 1995: 363-366), retention elasticity with respect to income was for the most part found to be greater than unity. Reenlistment bonuses had a positive, statistically significant though lesser influence, with one-time bonuses more effective than those paid over time. In both cases, as expected, retention elasticity was less for hardship postings and those that demanded greater physical effort. Retention elasticity with respect to income was also less when promotion possibilities were specifically considered: when reenlistment is under consideration, the possibility of promotion seems to substitute partially for current income.

There were some other interesting findings. Married people tended to reenlist more than singles, apparently due to the value they attach to benefits in-kind and to the non-monetary advantages of military service. Volunteers with higher education tended to reenlist less than others with lower qualifications, apparently because of their higher alternative civilian income, although this effect decreased with time-in-service. The chance to acquire a higher education encouraged retention for a limited period. With that, quit rates increased once studies were completed, seemingly because of the wish to realize the acquired human capital. Finally, high unemployment rates in the civilian economy led to higher reenlistment rates.

3.8 Volunteer Supply for the Reserve Force

In most countries regular service personnel, conscripts or volunteers, are the nucleus of the armed forces, augmented by reserve forces in times of emergency. Experience after the two world wars shows that at least in countries with large populations, even when war breaks out the army does not need all qualified citizens, and volunteer reserve forces may be adequate. Hence the implications of using market mechanisms to enlist volunteers for the reserve forces should also be examined.

In the beginning, the economic discussion of volunteer supply for reserve service was perceived as a kind of extension of the analysis of volunteer supply for regular service. Like joining the regular military service, which is a choice between full-time military and full-time civilian employment, the decision to join a reserve unit was regarded as a choice between part-time military service and a secondary civilian job, thereby subject to explanations derived from the theory of moonlighting labor supply (Mehay, 1991). Indeed, the fact was that most reserve volunteers had full-time civilian jobs. Later on, however, some notable differences were observed between the two situations. For instance, since in ordinary times reserve service is close to home and combined with civilian routine, deciding to join a reserve unit is influenced chiefly by local economic conditions, while volunteering for regular service is more influenced by countrywide economic conditions. Another difference is that volunteers for regular service are usually without previous military experience and join at the bottom, while many volunteer reservists – about half of them in the American instance (Warner & Asch, 1995: 388) – served previously as regulars, and from the start were posted to relatively senior positions. Finally, there are significant differences in the institutional characteristics of reserve service and of secondary civilian work. For one example, reserve volunteers undertake in advance to serve for relatively long periods, while the outstanding advantage of secondary civilian work is its adaptability to changes in the individual's economic and family situation. Given that, and the special conditions of reserve service (e.g. life in a military setting, compliance with military discipline, and long separations from family and from one's primary civilian work in emergency situations), the very possibility that individuals might see the two types of work as alternative, competing choices within the secondary job labor market is questionable.

Supply models of secondary work analyze the choice between work and leisure, and assume that the additional work hours the individual offers are a function of the difference between desired working hours, determined by income level and leisure preference, and actual working hours in the main job. Generally, the greater the working hours and income from the main job, and the greater the family's income from other sources, the less the supply of extra working hours. The hypothesis that joining the reserves is like choosing an additional civilian job implies that full-time employees in the civilian economy who volunteer for reserve units, and others in the same capacity who take on a secondary civilian job are responding identically to the same factors. An American study for the late 1970s showed that increased income and more working hours on the main job indeed reduced the supply of overtime hours in civilian occupations, but the opposite was true for the effects on volunteering for the reserves. The effect of additional family income on joining the reserves was as expected, but was not statistically significant as regards additional civilian work (Mehay, 1991: 330-332). On the basis of these findings, the hypothesis that volunteering for reserve service resembled an ordinary decision about additional work was rejected.

However, empirical research did confirm that economic factors carry some weight in the decision to join reserve units. The research above found that a rise in the ratio of reservists' military income to the income from additional civilian work tended to increase the quantity of volunteers for the reserves, and decreased overtime hours in the civilian sector (Mehay, 1991: 332). Other studies too found statistically significant income effects (Warner & Asch, 1995: 388-389). Furthermore, it has been found that local economic conditions, for example unemployment rates, also have an effect on the supply of volunteer reservists (Mehay, 1990: 358-359; 1991: 333; Warner & Asch, 1995: 388-389). The significant effect of economic factors indicates that market mechanisms can be relied on to recruit reserve forces as well.

4. THE DEMAND FOR MILITARY MANPOWER

Economic analysis of the demand for military manpower is in essence a discussion of the efficient production of national security. Defense economics sees the armed forces as a gigantic firm – or as a collection of firms: tank divisions, air force squadrons, navy fleets, etc. – using various factors to produce security. Accordingly, military commanders face an identical problem to that of business profit-maximizing firms in competitive markets: they have to choose the optimal combination of inputs necessary to obtain maximum security at every level of costs. Likewise, military demand for manpower is no different from the demand for production factors of business firms: it is determined by the desired level of output (or security threat level), by relative factor prices and technical opportunities for factor substitution, and the quantity of soldiers demanded will be the one for which their value of marginal product equals their marginal cost.

In practice, however, the analogy between producing security and business firms is far from perfect. First, there are no clear measures for defense output; hence deriving military production functions and making productivity calculations are of limited practical value. Obviously, if the value of marginal product of soldiers is not known, nor the way quantitative or qualitative changes (more or fewer soldiers, changes in education level or in military experience, etc.) alter it, it is impossible to determine the desirable quantity of manpower in the same straightforward manner as business firms do. Secondly, the army has a different status in both the product and factor markets than do business firms operating under competitive conditions. In the product market of each state, the army is the sole supplier of national security. In the labor market, as a major employer, the army maintains monopsonistic power, which is particularly relevant in market segments for specific skills with limited use outside the production of security, and as the sovereign, when conscription is in force, the army faces a manpower supply of infinite elasticity (within the relevant range of quantities) at a cost entirely unrelated to market conditions. Thirdly, in producing security, profitability cannot be a measure of performance, and besides, it is by no means even certain that military organizations try to minimize costs. It is not groundless to assume that the behavior of military commanders is more plausibly described through economic models of bureaucracy, in which other objectives, sometimes contrary to the dictates of economic efficiency, are promoted (Sandler & Hartley, 1995: 160-161).

4.1 The Effect of Relative Factor Prices

Military technologies in their broad sense – technical features of weapons systems, warfare doctrines and the operational principles derived from them, and features of military organization – along with administrative reasons, all combine to limit labor-capital substitution or possible tradeoffs between different types of manpower. Hence even if the army tries to minimize costs, it frequently lacks the flexibility needed to respond to relative factor price changes. For instance, an advanced, sophisticated weapons system requires specially trained teams to operate it and specific technicians to maintain it, thereby creating constant proportions between soldiers and equipment, as well as fixed combinations of various types of manpower, and both can often be altered only by giving up entire systems or by introducing completely new generations of weapons. Since processes of developing, producing and absorbing of weapons systems are relatively long, changes in input composition and in the quantities of manpower demanded also take time. Factors substitution in supplying national security, if they occurred, could have been observed, then, only over long periods.

In Britain between 1948 and 1973, approximately the same level of defense expenditures was maintained, while service personnel decreased. As a result, average defense expenditures per soldier in constant prices increased about 3.8 times (Hartley & Mclean, 1981). In the United States, in 1982 constant prices, stocks of military equipment per employee in the armed forces rose from 31,000 dollars between 1929 and 1945, to 56,000 dollars between 1950 and 1963, and to 83,000 dollars between 1964 and 1987 (DeBoer & Blackley, 1990: 88). It appears, then, that in both countries the capital-labor ratio in producing defense increased. During those years both countries went over to all-volunteer forces, and since the transition was accompanied by an increase in the relative price of labor, the growing capital intensity in defense production evidently reflects adaptation to changes in relative factor prices.

Several studies attempted to evaluate the effect of relative factor prices on input composition in defense production. One study (Kelly, 1977) tried to explain the capital-labor ratio variance in 12 NATO states in 1973 by differences in military pay level. It was assumed that military technologies were similar in all states, and that capital costs were close enough so that differences in military pay adequately accounted for differences in the relative prices of capital and labor. The number of tanks per 1000 soldiers in combat units represented the capital-labor ratio, averaging 7.5 for all 12 states, albeit with a wide range, from 2.5 in France to12.5 in Holland. Indeed, that range in itself indicated that even with a given technology there were opportunities for factor substitution. The estimated equation of the relationship between the capital-labor ratio and military pay provided a positive, statistically significant coefficient for military pay, and differences in military pay among states explained a considerable part of the differences in capital intensity (R^2=0.71). In another study supply and demand equations for military manpower in the United States between 1955 and 1986 were estimated, and it was found that manpower demand responded positively to changes in prices of military equipment: manpower demand elasticity with respect to military equipment prices was 0.15 for 1955-1973, and more than doubled, to 0.36, for 1974-1986 (DeBoer & Brorsen, 1989: 862). Positive demand elasticity indicates that manpower and equipment were mutually interchangeable, but the relatively low values show that they were at best weak substitutes. Besides, the differences between the periods may be interpreted as indicating that abolishing conscription in 1973 led to greater consideration of equipment prices in determining the desired quantity of manpower. A third study compared the ratio of military manpower to front-line weapons systems in 13 NATO countries, in Sweden and in Israel for 1989 and 1993. Great differences were found, but unlike the findings in the studies above, they could not be explained by differences in relative prices of manpower and equipment among the countries involved (Owen, 1994: 280).

Some studies also estimated substitution elasticity between military manpower and capital (i.e. the rate of change in the ratio of military manpower to capital, divided by the rate of change in their relative prices). In the United States Navy, for 1956-1972, substitution elasticity between capital and labor was estimated at 1.13 (Clark, 1978). In Britain, according to data for 1952-1987, substitution elasticity between manpower and all other inputs in the production of defense (defined as capital, though including civilian employees in the armed forces) was also almost unity (Ridge & Smith, 1991: 291-292). In the United States, in the years 1929-1945 the substitution elasticity between manpower and military equipment was substantially higher than 1, while after World War II it was lower than 1. Between the two periods differences were also found, and in the same direction, in the demand elasticity for each production factor with respect to its price, thereby suggesting that changes in military technology in the last decades, principally with the appearance of long range missiles, reduced the scope of choice between inputs in general, and limited the possibilities of labor-capital substitution in particular (DeBoer & Blackley, 1990: 88).

The last two studies also evaluated the effect of conscription on use of labor and capital in producing defense. The British research found that conscription, when was in force, contributed significantly to increased demand for military manpower, while the American study concluded that abolishing conscription led to replacement of manpower with equipment beyond the dictates of changes in relative prices (DeBoer & Blackley, 1990: 93). A study from Belgium used simulation to examine the probable effect of equalizing the price of conscripts to their full economic price, and alternatively, the effect of abolishing conscription and relying completely on volunteers. It was found that in both cases capital intensity would be likely to increase. In the first case, however, demand for capital and for volunteers would increase, while in the second, capital demand would grow at a higher rate, while demand for volunteers would drop because of the sharp rise anticipated in their wages (Kerstens & Meyermans, 1993: 278-280). By contrast, the study comparing the ratio of military manpower to front-line weapons systems, cited above, arrived at different findings regarding the downsizing effects of ending conscription on manpower use. In Britain and Canada, the only two countries in the sample with all-volunteer force, the ratio of soldiers to equipment was no lower than in countries with conscription (Owen, 1994: 280).

In conclusion, most empirical research confirms that changes in relative prices of labor and capital over the years have affected the composition of inputs in producing defense. In addition, the empirical findings suggest that capital-labor substitution elasticity was approximately unity, but it possibly diminished over time due to new military technologies that limited substitution opportunities. Finally, abolishing conscription has usually been a driving force for greater capital intensity, and its effect seems to be beyond the implications of relative price changes themselves.

4.2 The Effect of Changes in Defense Output

Empirical studies have shown that changes in defense expenditures, expressing as they do the desired level of defense output, influenced manpower demand positively and significantly. While their effects were felt throughout the armed services, they stood out particularly in the ground forces, which are more labor intensive than the navy or the air force. A British study for 1952-1978 found that adaptation of manpower to changes in the desired level of defense output was usually very slow, and attributed this to contractual constraints, namely to the fact that volunteers enlist for relatively long terms. However, an adaptation time of seven years, found for the Army, appears to indicate that manpower was being hoarded as well (Sandler & Hartley, 1995: 161).

Changes in the level of defense output influenced the tradeoff between labor and capital too. In the United States it was found (DeBoer & Blackley, 1990) that whatever were the changes in defense output, the respective shares of labor and equipment changed in opposite direction. However, until the mid 1960s labor share in defense output changed parallel to changes in output size, while later, as defense output increased, the labor share decreased. A possible explanation derives, as noted before, from technological developments, especially from the central role missiles now have in military arsenals. At the same time, possibly there is no symmetry, and factor substitution patterns in the course of military buildup may not be identical with those that exist in periods of downsizing.

An interesting pattern appeared in the above research on the immediate response of 13 NATO states to the end of the Cold War. Every state reported a decrease in military manpower, but there were significant differences in the rate of decrease in different countries, and sometimes even between services in the same country. Analysis showed that the higher the ratio of manpower to front-line weapons system was in 1989, the greater the rate of accumulated manpower decrease until 1993, and the conclusion was that military authorities took advantage of the opportunity to repair imbalances that evolved over the years (Owen, 1994: 283-284). Possibly, then, economic considerations grow stronger in selecting inputs for producing defense when fundamental changes in the international environment create the opportunity for overall reforms in the structure of the armed forces.

4.3 The Effect of Manpower Productivity

Measuring the productivity of military manpower, as noted earlier, encounters difficulties because it is impossible to quantify defense output. Empirical studies in this field, most of them from the United States, produced the following main findings (Warner & Asch, 1995: 368-370): -

i. At a given level of capital, military manpower marginal productivity tends to be positive. For example, the United States Navy found that marginal increase in the manning level reduced ships' maintenance downtime, and made the ships fully mission-capable a larger proportion of the time.

ii. Service personnel productivity increases with experience. It was found, for instance, that pilots' performance is influenced by career flying time more

than by recent flying time. Experience influence on productivity differences was greater in military assignments that required higher professional skills.
iii. Seniority in a position also has a favorable effect on productivity, while frequent rotation decreases the performance of units. For instance, target shooting by tank crews improved when the average seniority of the tank commander and the gunner was greater.
iv. There is a positive relationship between productivity and the quality of enlistees in terms of education and psychometric scores. For example, the greater the proportion of high school graduates in the crew and the higher their psychometric scores, the shorter was the maintenance downtime of ships. In simulated battles with Patriot missiles, performance improved substantially when operators' psychometric scores were higher.

A higher quality, more experienced force is likely to perform military tasks better, but will apparently be more expensive. Optimizing force mix thus requires the weighing of increased manpower productivity against additional costs.

4.4 The Effect of Substitutes for Regular Service Personnel

Regular service personnel can be substituted, albeit with limitations, by reserve forces, by widening the scope of military tasks that civilians perform, and more indirectly, through outsourcing. These new-old issues of planning and managing human resources that produce national security gained increased attention after the end of the Cold War, in view of the substantial reductions in defense budgets.

Reserve forces definitely offer cost advantages, but matters are much less clear as to their relative efficiency, and thus as to the actual substitutability between regulars and reservists. In medical units, for example, there is perfect substitution. In the Navy and Air Force, by contrast, the accepted view is that the nature of operational tasks, as well as the maintenance of complex, technologically advanced systems, require greater skill and more experience than could be acquired through reserve service. In the ground forces it may be possible to use more reservists, but here too the price may be a lower level of preparedness, with a long period of refresher exercises before the reserves are actually ready for operational activity. In this connection it has been stressed that the larger the proportion of reservists who have gone through regular service, the greater the substitution possible. But when regular forces are reduced, the supply of veterans available for the reserves eventually declines too.

In countries that have gone over to a volunteer force, civilians now man more posts. Economic analysis suggests that increased employment of civilians, as well as the expanding of outsourcing (see below), is a consequence of certain features of the military compensation system. With conscription, the administrative cost of soldiers is not only low but also uniform, and does not express their individual differences in productivity. To a somewhat lesser degree, because of the rigid military pay structure, the price mechanism is also less effective in allocating manpower in a volunteer force. Usually, uniform wage grades are applied to large groups, and thus the military pay inevitably embodies an economic rent that could be saved had a more flexible and differential system was in use. However, a differential military pay system may contradict military organizational principles (see below, section 5), so that other methods are sought. Employing civilians does not disrupt the military hierarchy, and compensating them at accepted civilian labor market rates does not undermine prevailing relativities among ranks and duties. Furthermore, civilians compensation system can be more easily designed to strengthen the link between performance and rewards, which is no simple matter in the military pay system.

Outsourcing is another alternative. Instead of military personnel performing certain activities in-house, a large variety of goods and services may be purchased commercially from civilian firms, indirectly replacing military manpower with civilian workers. In many areas it may be more cost-effective, since it introduces explicit market tests and competition into supply processes, and thus no wondering that the long-term trend was clearly to expand outsourcing. Before World War II, the United States Army and Navy manufactured much of their equipment in government-owned facilities and shipyards. Later on, the greatest part of military materiel was purchased from commercial firms, while those plants still owned by the government were handed over to private contractors for management and operation. Over time, outsourcing was extended to varied services: catering, cleaning, facilities management, transportation, vehicle and equipment maintenance, and even certain aspects of military training. As a matter of fact, not only cost economies encourage outsourcing; the more complicated military equipment becomes, the harder it is to recruit people skilled enough to maintain it.

Defense economics literature discusses outsourcing in the broader context of establishing institutional frameworks to improve allocation efficiency within the defense establishment, and in fact as a substitute for such arrangements (Sandler & Hartley, 1995: 175-176; Hitch & McKean, 1971: Chapter 10). Unlike business firms, where efficient production and optimal factor combinations are the outcome of the profit motive and of competition, the "military firm" operates in a different environment, which lacks these incentives and does not guarantee efficient use of resources. To tackle this problem, economists proposed, in different versions, to create quasi-markets within the military sector, and to set up operational guidelines that simulate competitive behavior. A rather extreme proposal to this end was presented as early as World War II (by Prof. Abba Lerner). It suggested that the entire defense establishment be organized as a network of markets, and that allocation decisions be delegated to numerous bodies: commanders would receive budgets for acquiring equipment (from procurement bodies) and personnel (from recruiting authorities), using some form of tenders

(Hitch & McKean, 1971: 211-214). The proposal was not implemented, but its basic idea was adopted in less far-reaching arrangements. For example, certain activities in the defense establishment were organized as "closed economic entities" that sell goods and services to other, budget-holder units, and to that end acquire personnel and other necessary inputs, making their own judgments in order to minimize costs. Such arrangements sometimes contradict military organizational principles, and when difficulties arise in applying them, it is preferable to outsource activities to commercial firms. With that, outsourcing has to assure true competition among suppliers. But specifically in this area, close connections and daily contacts with suppliers make for a tendency to become a captive client, and thus to miss out on the advantages of market competition.

5. MILITARY COMPENSATION SYSTEMS

Most Western armies base their compensation system on similar principles (Warner & Asch, 1995: 380-381). The current compensation consists of basic pay that varies with rank and time-in-service; allowances and benefits in-kind that also depend on rank, and sometimes on marital status, and many special payments like those linked to military duties (e.g. flight and sea pay), enlistment and reenlistment bonuses and so forth. Countries resemble one another as regards the basic pay tables by ranks, and differences mainly relate to the method by which time-in-service, marital status and other personal data are factored into the calculations. Countries may also differ as to the impact of civilian wages on the military compensation system. Most Western armed forces also have similar retirement systems. Retirement vesting is delayed until a relatively long period of service is completed, and for those who become vested, pension payments begin right after discharge. However, those who leave without completing the minimum period are not entitled to any pension rights. In most cases, pension is a significant item in total military compensation.

The military compensation system is designed to achieve two simultaneous objectives: to enable the army to recruit the required quantity and quality of manpower through the labor market, and to stimulate officers and soldiers to perform their duties and carry out military tasks, often under extraordinary circumstances, as well as possible. The first objective stresses the need to set up a compensation system that weighs alternative civilian incomes, and the two objectives together have to consider the exceptional conditions of military service and the unique features of military organization. However, contradictions occasionally arise, and some compensation components that serve recruitment well may be hindrances to an effective performance-oriented incentive system, and vice versa. Therefore, and since the compensation system has to obey budgetary constraints, its various components must be most carefully selected.

The dilemma of differentials versus uniformity is a major concern in this context. Uniformity stems from the need to avoid contradictions between the task hierarchy and the wage structure that otherwise could impair military organizational functioning. Compensation is thus usually uniform within each rank except for seniority differences. But because individuals have different preferences, compensation uniformity apparently implies that the army is paying some people too much and they enjoy an economic rent, while others are paid too little. Uniformity, then, interferes with minimizing costs. No less important, because of uniformity the compensation system fails to express the relative scarcity of talents and skills in demand, and hence could be attractive to people of little importance to the army, while falling short in recruiting or reenlisting those more important to it (Sandler & Hartley, 1995: 168-171). Finally, uniformity also mars the incentive system and reduces the possibility of encouraging outstanding performance.

Uniformity and the resultant economic rent arise, *inter alia*, because entry is always at the bottom of the military organization (Warner & Asch, 1995: 384). This compels the army to recruit a sufficient reservoir of quality personnel who over time, and after multi-stage screening, will be eventual candidates for the higher positions. At the time of enlistment, there is no way to identify true ability with any certainty and in fact there is no basis for differential compensation. Yet raising the pay of all new recruits improves chances to attract a greater proportion of potential promotion candidates. Thus even if most recruits do not serve more than one or two periods and do not advance above the lower rungs of the ladder, equal compensation for all is unavoidable, even if it means an economic rent for most.

Differential compensation is closely linked to the notion of performance-oriented incentives. At every rank, efforts that individuals of a given capability level invest in performing their duties depend on chances of promotion and the returns anticipated from it. Anticipated returns are defined by the basic pay scale, and include the non-monetary compensations for each rank. The probability of promotion depends on performance evaluations made by direct commanders who, in turn, are influenced by effort invested and by the number of promotion candidates. It is thus possible to formulate an equilibrium model in which the level of effort that individuals invest in performance is a function of compensation differentials between ranks, the value attached to relevant non-monetary benefits and the number of candidates competing for promotion at each rank (Warner & Asch, 1995: 382-384). According to the model, the higher one gets on the ladder, the higher the differentials should be to offset the diminishing probability of further promotion. Otherwise the tendency will be to reduce effort in those same higher positions that are particularly important to overall military performance. Another result of the model shows that seniority increments for years of service render promotion a less effective performance incentive, making it preferable to link such increments to rank.

Retirement arrangements too raise several dilemmas. Is such a long vesting period justified? Is there not some injustice in not compensating those who leave earlier? And from a different perspective, is it possible to reduce the pension costs that increasingly burden the defense budgets (Gansler, 1989: 297-299)? In the military organization, pensions have a role different than in the civilian sector, because of the "joining at the bottom" limitation and the need for a self-sorting system that encourages the best to stay and those unsuitable for promotion to leave voluntarily (Warner & Asch, 1995: 384-386). The relatively long period required for pension vesting causes those see their promotion chances as low, and chances to complete the minimum time for pension rights as slight, to quit. At the same time, it encourages those with promotion potential to invest efforts in their jobs so they will not be forced to leave before they become vested with retirement benefits. Additionally, generous retirement schemes for those who become vested comes from the desire to rejuvenate the ranks and encourage even "the best" to leave. Without the incentive to retire, senior staff whose military income could well be higher than alternative civilian income, might stay on and reduce promotion opportunities for younger people. This is especially likely for those trained in the military-specific skills.

6. CIVILIAN RETURNS ON MILITARY TRAINING AND EXPERIENCE

Military service may affect the income of veterans in subsequent civilian employment, and this later influence may be important for both individual enlistment and retention decisions, and for military compensation policy. There is also a macroeconomic aspect, namely the extent to which military training and experience contributes to productivity and economic growth in the civilian economy. Empirical research has sought to find out whether veterans earn a positive or a negative premium in return for their military service, and to identify possible factors that may increase or decrease that premium. The theoretical explanations for a possible income premium came from the economic theory of human capital and from the labor economics model of signaling.

The human capital theory attributes income differentials over time to differences in initial stocks of human capital. Thus if veterans earn a positive premium, it must come from their larger human capital relative to their peers who did not serve, when the former begin work in the civilian sector. On one hand, while the soldier devoted his time to the army, the civilian of his age was accumulating training and experience in civilian employment, thereby increasing his human capital. On the other hand, experience accumulated during military service, and particularly training in certain specific skills, may contribute to veterans' human capital. Thus the question focuses on the relative size of the different influences.

The contribution of military experience and specific training to veterans' human capital is no foregone conclusion. The theory makes distinctions between a "general" and a "specific" contribution to human capital. A specific contribution to human capital yields an income premium only if the worker uses it directly, i.e. occupies a position where skills acquired through that specific contribution are explicitly applied. By contrast, a general contribution to human capital – whether from general experience or specific training – may yield an income premium in a variety of occupations. Firstly, then, the contributions of military training and experience must be evaluated as to whether they add to human capital only specifically, or are they a general contribution as well. Secondly, not all military training and experience, either as a specific contribution to human capital or as a general contribution, are equally transferable to civilian occupations, and moreover, there could be changes in their transferability over time. Apparently, recent changes in the nature of military tasks due to developments in military technology, in weapons systems and other equipment, have broadened the overlap between required skills in the two sectors. Finally, the analysis is concerned with the relative development of human capital, meaning that the contribution of military service to acquired human capital must be compared with the parallel human capital acquisition of those who do not serve, including, in particular, comparisons of the contribution of military courses with that of civilian vocational training frameworks .

Having isolated effects of other variables, most empirical research found that military experience and training did contribute to the human capital of veterans. Some findings indicated that contributions were not only specific, but general: military service contributed more to the civilian income of those with less education, apparent evidence that military service substituted at least partially for formal schooling and acquiring a general education (Knapp, 1973; Berger & Hirsch, 1983: 460). By contrast, the contribution of military vocational training, as distinct from military service in general, tends to be specific only: those who received such training and used it in their civilian work earned a substantial income premium, while those who did not use it enjoyed no benefit beyond the general influence of military service on veterans' income (Fredland & Little, 1980: 50-52). It was also found that the return on military vocational training tended to be lower than on civilian training. Obviously, military training is determined by the army's needs, not those of the civilian market, while in the civilian sector vocational training matches the occupations in demand, with their promise of returns. However, a low absolute return does not necessarily imply that the rate of return is lower: military training often costs the individual nothing, or at least far less than the costs borne by his civilian counterpart. Finally, research shows, as expected, that the contribution of military experience and training to civilian income varies between different occupations (Fredland & Little, 1980; Goldberg & Warner, 1987), appearing to reflect differences in the transferability of military know-how and experience to civilian occupations.

Military training and experience, then, contribute to veterans' human capital, but that contribution must still be compared to forgone investments in human capital due to delayed entry into the civilian labor force. In the United States in the early 1980s, the income of white male veterans ten years after discharge was still found to be some 15 percent lower than that of men who did not serve. In the terms of that time, the income difference equaled two years' seniority in civilian work. Because the average term of military service was longer, although the positive contribution of military service and training did not entirely cancel out the result of delayed entry into the civilian labor force, it reduced that effect considerably (Angrist, 1990). Similar evaluations were derived from research on conscripts in Holland, although with opposite conclusions. In 1989, the income of veterans nine years after their service was found to be 8 percent lower than that of individuals of same age who did not serve. Here too the difference was calculated as equal to the returns for two years' seniority. The typical Dutch conscript, however, served only 14 months. Thus military service increased the income differential caused by delayed entry into the civilian labor force, that is, it eroded the human capital that conscripts brought with them to the army (Imbens & Klaauw, 1993). These seemingly contradictory findings may indicate that length of service is an important factor: the longer military service is, the greater the human capital that may be acquired. Furthermore, it is reasonable to assume that there may be a minimum service time for acquiring any meaningful human capital that will contribute to future civilian income.

Many non-measurable personal characteristics affect labor productivity, and employers cannot identify most of them in advance. An army veteran passed aptitude tests when he enlisted, and met conduct and performance criteria during service; hence it is reasonable o assume that his non-measurable ability is higher than that of others. A military discharge certificate is thus a valuable screening tool for civilian employers (De Tray, 1982): they tend to relate differently to veterans and to individuals who did not serve. They assume that the potential productivity of the former is greater, so are prepared to offer better entry-level pay and faster promotion tracks. This, then, is another possible explanation for the difference between veterans' incomes and the incomes of those who did not serve.

Empirical studies designed to support this explanation tried to show that under certain conditions, veteran status might serve as a more effective signal than under other conditions, and these specific conditions, when present, yielded a relatively higher income premium to veterans. For this purpose, cohorts with different rates of enlistment were compared. The argument was that low enlistment rates leave a large group of high quality individuals out of military service, people who could have completed their service successfully, and therefore in this situation the veteran status losses its effectiveness as a signaling device. By contrast, when recruiting rates are high, it is reasonable to assume that those rejected were turned down as unsuitable, so that the greater the recruitment rate, the more effective veteran status is as a signaling device. Statistical findings supported the theory: veterans who belonged to the higher rates of enlistment cohorts indeed enjoyed relatively higher income premiums. Further evidence for the signaling role of military service was derived from a comparison of veterans' income premiums at different education levels, assuming that, *ceteris paribus*, the relative contribution of unmeasured abilities to productivity decreases with years of schooling. Statistical analysis showed that veteran status yielded a larger income premium to men with less than 12 years of schooling as compared to others with higher education (De Tray, 1982).

In conclusion, empirical research bore out the assumption that military service affects civilian incomes favorably after discharge. However, according to most findings, this contribution was not more than what is obtained from accumulated experience in civilian employment. The relationship between the income premium gained through military service and the contribution of civilian experience varied according to length of service and nature of military occupation.

7. THE ECONOMICS OF MILITARY MANPOWER IN ISRAEL

Military manpower in Israel is based on a permanent career service nucleus, conscripted soldiers and reserve forces. This system, established in the early 1950s, met two conditions: it made it possible to have a relatively small regular army whose apparently negative effect on the civilian labor supply was tolerable; and the system, effectively used, attained high rates of enlistment in emergencies, thus reducing the military implications of the population gap between Israel and her neighbors. But there were shortcomings. From the military standpoint, it required adequate early warning to call up reserve units, and dictated short war scenarios to avoid extended periods of large-scale mobilization. Socially and economically, military service became a permanent and integral factor for decades of most men's lives, imposing a heavy burden and arousing serious questions of sharing and social justice. No less disturbing were the macroeconomic consequences. The discussion that follows suggests economic considerations for evaluating the recruitment methods, organizational structure and employment conditions of military manpower in Israel.

7.1 The Need for Reform

Demographic developments led to a surplus of candidates for compulsory service, raising the question of whether the Israel Defense Forces (IDF) size should be determined by demographic supply, or recruitment policy adapt to the desired dimensions of the army. The government has not laid down an explicit policy, and the IDF has responded to the increased supply by more selective conscription on one hand (Cohen, 1995: 238-240; 1997: 93-96), and by a more liberal policy of early discharge from compulsory service on the other hand. But this does not seem to solve the problem of conscript surplus. At the same time, it became possible to decrease the reserves. Since 1990, soldiers in combat units serve in the reserves until the age of 45 and others until 51, instead of 54 as they did before. Moreover, in total, fewer days per year are actually served now.

From a purely economic point of view, reducing the number of days served by reservists, and replacing reservists with conscripts are positive developments. Since income rises with age, and reservists are older than conscripts, it is virtually certain that substitution between the two types of manpower decreases the overall opportunity cost of defense. Furthermore, factor interdependence makes it likely that the absence of reservists from their regular work reduces the civilian GDP by a greater amount than their incomes indicate. With that, it is impossible to consider the other changes as positive. Selective conscription and early discharge do not account for alternative civilian income. In fact, the IDF gives up the compulsory service of people with a relatively low alternative civilian income, so that at best, inefficient allocation of manpower between defense and civilian economic activity is no worse than it was. Most serious, however, are the implications for internal inefficiencies within the IDF. As the steps so far adopted do not resolve the basic problem of surplus conscripts, their availability at a uniform low budget price is a barrier to efficiency. It delays transition to more capital-intensive methods, encourages using conscripts in tasks better done by other types of labor, and hinders outsourcing initiatives and acquisition of goods and services from commercial firms, though by true economic criteria substantial resources could have been saved by these means.

On the demand side, technological advance and the changing nature of military tasks have increased demands for skilled, professional manpower with experience acquired during a long period of service. There are now more and more tasks that only permanent service personnel or the IDF's civilian employees can fill with acceptable efficiency. The IDF responded in several ways, the main one being to establish elite units specialized in operating and maintaining sophisticated weapons systems. This required special training and promotion tracks, the latter combining compulsory service and an advance undertaking to serve for a stated period of time in the permanent service corps, and sometimes pre-military training too. Thus, seemingly, the best of both worlds has been accomplished: low cost due to the conscript element and high productivity because of prolonged service (see below). Meanwhile, the decreased importance of physical abilities and increased advantages of skill and practical experience enlarged the circle of military duties in which age limitations no longer apply, making it necessary to rethink retirement and pension policies as well.

Another factor that makes change necessary is the increasing relative share of labor expenditures in the defense budget. That trend began in the 1970s, while in recent years labor expenditures reached 40 percent of overall defense consumption, and more than half of domestic defense consumption (see chapter 6). Increased expenditures on manpower resulted from quantitative growth, but also from the rapid rise of wages and fringe benefits. In recent decades in Israel, there has been a general tendency for the share of labor in national income to rise, and admittedly the IDF is not isolated from the environment in which it operates. Moreover, military activity is labor intensive, in labor-intensive industries productivity growth lags behind the economy's average, and since these industries compete for workers with other less labor-intensive industries, where the rise in productivity is faster, their labor income-to-output ratio grows more rapidly than the average for the economy as a whole (Barkai, 1996). Civilian labor-intensive industries responded by importing low-cost foreign workers and by moving their plants to countries with relatively cheap labor. These options are obviously not possible for the military, and the IDF had to confront the upward trend of wages by adapting the compensation of service personnel to the developments in the civilian economy. It is undesirable, and ultimately impossible, to keep the military pay apart from compensation standards prevailing in the rest of the economy, so other ways must be considered for stopping the rise in the labor expenditure share of the defense budget.

Finally, the need to reexamine military manpower policy arises from changes in the Israeli public's attitudes to the military in general, and in the attitudes of service personnel to their duty and their profession in particular. The military service lost some of its earlier glorification, and service personnel too no longer see it as a national mission only.

7.2 Possible Reforms in Recruitment Policy and Military Service Tracks

Abolishing conscription and basing the military solely on volunteers recruited through the market mechanism is foreign to the security perception and the conceptual world of Israeli society. Nonetheless, it should be considered a long-term goal. Undoubtedly such a change has many implications and requires most thorough planning. Furthermore, it calls for a long transition period. In the interim, however, some major reforms that both confront the current problems discussed above, and make headway toward that long-term goal, should be considered: shortening compulsory service; expanding service tracks that combine in advance compulsory and volunteer terms, and increasing enlistment and retention of volunteer personnel.

A partial answer to the manpower imbalance may be to continue the universal conscription of relevant cohorts, and at the same time cut the compulsory service period substantially across the board. Shorter service would reduce waste, and the open and disguised unemployment, especially in the rear echelons and in general service roles, so defense would presumably be produced more efficiently. The macroeconomic advantages are clear too: the implicit conscription tax would be much reduced, and with it – the undesirable influence on welfare distribution and on the efficient allocation of resources. As indicated by the analysis above (see section 2.4), at certain levels of demand for defense, net "social surplus of defense" is inversely associated with the number of conscripts. Shorter compulsory service may also promote voluntary enlistment, or in other words increase chances of successful recruitment through the market mechanism. In terms of the supply model for volunteers, under the new conditions potential volunteers would be younger with lower alternative civilian income. It is also reasonable to assume an inverse relationship between length of compulsory service and the reservations it creates about additional service time, so that potential volunteers may be content with a smaller income increment to compensate for their natural preference for civilian life-style.

Opponents of shorter compulsory service base their main contention on the level of military training and the professional skill attainable in a limited time. It can also be argued that there will not be enough manpower for current military tasks, and that the conscript army's role, as a corridor to the reserves will suffer. But shortened compulsory service is not an end in itself here, and should be examined in the context of an overall reform that will, *inter alia*, redefine the duties of the conscript army and the reserves.

Integrated service tracks combining compulsory and voluntary service are not a new idea. An "old" example are the combat pilots: since their course takes up most of the compulsory service period, candidates must undertake in advance to serve for a stated length of time beyond what is required by law. That way the Air Force obtained a return on its costly investment in pilot training, and most importantly, due to the accumulated flight experience, pilots greatly improved their performance capabilities. The innovation of recent years is in applying the system to a variety of roles having two main characteristics: relatively expensive basic training, and potential productivity growth with time-in-service and experience. High cost may be the result of extended training, of using expensive training materials or of both together. Time-in-service and experience positively affect productivity when there is a potential for on-the-job learning. Such conditions obtain for most jobs in the operation and maintenance of modern weapons systems, and grow more important with their increasing sophistication and complexity. Thus in view of military technology developments and the "battlefield of the future" outlook, the spectrum of military jobs suitable for integrated service tracks will only grow broader.

Defense economics analysis makes it possible to evaluate the significance of integrated service tracks from the microeconomic standpoint, i.e. as to their effect on individuals subject to conscription, and from the macroeconomic aspect of efficient labor allocation in the economy and the opportunity cost of defense. From the first, choosing an integrated service track option gives conscripts a chance to influence the nature of the military service forced on them, and thus, *ceteris paribus*, it improves their welfare. Moreover soldiers on integrated service tracks are likely to accumulate more human capital than in other options. In many cases that choice transforms conscript service, in terms of accumulating human capital, from "lost time" to highly productive years. In addition, in some positions, the human capital accumulated is transferable to civilian uses without substantial transition costs. Thus, beside other considerations, readiness to enlist on an integrated service track may be regarded as recruits' decisions to invest in human capital. They agree to extend their military service beyond what the law demands, since they anticipate a civilian income premium at the end of it.

The macroeconomic point of departure is that the IDF demands a defined quantity of trained, experienced people. Hence it compares the opportunity cost of soldiers on integrated service tracks with the explicit cost of the same quantity of career service personnel, of equal training and experience, on the conventional track. Because of the soldiers' age and education, the cost of the integrated service tracks will presumably be lower. But other influences may come into play. On one hand, since the IDF does not reckon with opportunity costs but only with budget prices, it sees integrated service tracks as a relatively cheap means to obtain manpower for complex tasks. Thus it will tend to increase the number of soldiers beyond the optimal quantity, meaning that from the point of view of the economy as a whole, an ineffective allocation of manpower between defense and civilian activities will arise. On the other hand, graduates of the integrated service track, with their greater accumulated human capital, will possibly contribute more to the productivity of civilian sectors, once they complete their service, than other veterans will. Indeed, economic developments in Israel show that service personnel who acquired advanced technological skills in the military contributed a great deal to the growth of flourishing, high tech industries.

Finally, it is well to recall that since more soldiers on integrated service tracks renders compulsory service more effective, due to higher average labor productivity, it also makes more soldiers redundant, so that it is highly compatible with a shortened compulsory service term.

Along with shortening compulsory service and reducing the reserve force, the permanent service corps should be enlarged, which means increased reliance on market mechanism recruiting. Defense economics analysis shows that recruitment through the market mechanism, which takes into account alternative civilian income, leads to a more efficient allocation of manpower between defense and civilian economic activities. For the same reason, it assures the choice of efficient combinations of labor and capital in producing defense, and more efficient internal allocation of manpower within the military. With that, increased reliance on the market mechanism raises concerns that it may not be possible to recruit the desired quantity and quality of soldiers. Elsewhere the supply of volunteers has been shown to be elastic with respect to income, and that difficulties in enlisting certain types of manpower may be overcome by offering specific benefits. Israeli experience is no different. In several instances, when there was an increased demand for career service personnel after the Yom Kippur War, or in the later 1980s when resignations from the permanent service corps increased, the required quantities were successfully recruited by means of wage increases and other suitable benefits. As a matter of fact, in some military sectors supply even exceeded demand. Eventually, then, successful recruiting through the market depends to a great extent on the military compensation policy.

7.3 Adaptations in Military Compensation Policy

There have to be changes in compensation policy in order to recruit the necessary quantity and quality of volunteers for the permanent service corps, and to combat the rising proportion of labor costs in the defense budget. Changes in the compensation system must therefore meet three main criteria: making military service attractive as compared with civilian employment alternatives, promoting productivity and reducing the IDF's labor costs. Some possible changes may include: -

i. *Reexamining the variety of compensation elements.* As a general rule, a compensation system that has developed over the years will have components that are no longer relevant to changed circumstances. The IDF compensation system is no different. Reexamining the basket of cash and in-kind benefits would presumably bring about changes that increase both productivity and the relative attractiveness of military service. This refers in particular to benefits eroded by changing circumstances in relation to their costs. Hence economies can be realized by replacing such benefits with cash payments, with no harm done to the welfare of career service personnel.

ii. *Wider differentials and increased variety of compensation options.* As shown in the general discussion, in time economic rents, even substantial ones, unavoidably become part of the compensation system. This stems from the organizational principle of joining at the bottom, from uniform wages for relatively large groups and from the tendency to "level upwards" when it comes to wages and benefits. To reduce economic rents, greater differentials should be established between groups of service personnel, based on service conditions and job requirements on one hand, and on civilian employment opportunities and alternative civilian income on the other hand. In addition, to achieve greater compensation differentials, and at same time avoid conflict with traditions of uniformity and hierarchical relations, there have to be numerous service tracks, differentiated by their requirements, and accordingly by the compensation components attached to them. The modern work environment, military or civilian, is becoming less homogeneous and more individualistic, and the military has to follow suit in adopting a larger variety of mandatory periods of service, promotion tracks, voluntary retirement arrangements and so forth. Furthermore, in certain service tracks the "sacrosanct" principle of joining only at the bottom can be relinquished. There are more and more professional jobs with parallels in the civilian economy, where training and experience accumulated in the civilian sector equal or even surpass what is acquired during military service. This not only increases the reservoir of potential candidates and not only makes it possible to recruit more productive people at lower cost, but it also saves the high costs of training and seniority.

iii. *Replacing fixed by variable payments and strengthening the compensation to performance linkage.* Despite difficulties in measuring output, formulas that relate compensation to contribution can be devised in many military tasks. Substituting variable for fixed components may increase productivity without increasing costs. Besides, if formulas are adapted to express relative scarcities, it may improve the IDF power to attract people in key professions.

iv. *Employing more civilians.* Individuals who have had outside vocational training can be hired, thus saving expensive training costs. Furthermore, civilian employees are not as tied to rigid promotion tracks as the forces in uniform typically are, which generally makes greater compatibility between abilities and the job demands possible. Finally, compensation to civilians can be less uniform and more flexible, allowing for performance-oriented incentives even where similar arrangements cannot be applied to military personnel.

v. Changes in retirement policy. A view widely held is that the military promotion rate is too fast, and that the retirement vesting age is too low. Appropriate changes in both areas will lead to a longer average service period, and due to more professional skills and experience will raise productivity. No less important, it is reasonable to assume that those whose promotion chances are vague, and thus the probability that they will complete the term required for pension benefits is lower, will retire voluntarily at an earlier stage, while others, with better chances to move up, will invest more effort in their duties so as to enjoy the incentives that go with higher ranks. Furthermore, there will be a change in the division of income flow between the years of active service and those that follow. The present value of retirement bonuses and pension payments to career service personnel is some 57 percent of their income during active service, as against 21 percent only among other retirees from government service (Ministry of Finance, 1996: 82-83). Such large differences suggest that the flow of military income over time is not optimally divided, and that replacing deferred with current income may save expenses, and at the same time increase the welfare of career service personnel.

8
DEFENSE INDUSTRIES

Producing and supplying national security require along with manpower a wide variety of hardware products – weapons systems and other military equipment – generally termed as defense products. This chapter examines the supply of defense products and elaborates on the defense industries that design, develop and produce them.

1. THE BOUNDARIES OF DEFENSE INDUSTRY

The traditional approach of industrial economics defines industry boundaries by the degree of substitution in demand between products. Defense products, however, usually perform different functions in military activity, and in most cases cannot be an operational replacement for one another. Even companies commonly thought to belong to the same segment of defense industry often do not produce interchangeable products. The aerospace industry, for example, includes producers of fighter aircraft, military transport aircraft, surveillance satellites, helicopters and guided missiles, each of which has a different military function. Thus defense industry is not an "industry" in the usual sense (Beard, 1993: 30). Moreover, for statistical and other purposes, defense companies are assigned as a rule to "ordinary" industrial categories (metal products, electronics, etc.), and do not count as a separate sector.

Defining the boundaries of defense industry for analytical purposes requires, then, a somewhat unconventional approach that combines the differences between it and other industrial activities on one hand, and the common characteristics of its various components on the other hand. One possible point of departure may still be defense products (Dunne, 1995: 402-403). From a functional standpoint they divide into three main groups: weapons systems, i.e. means designed to kill, destroy or damage various targets; special purpose auxiliary equipment such as command and control systems, intelligence and early warning systems, communications systems and the like; and general purpose products (e.g. vehicles, fuel, medical equipment). For definition purposes, the borderline could be drawn between the first two groups and the third, but the close interdependence among all three in modern warfare removes any significance from this traditional functional distinction. Similarly, it is impossible to draw a clear line between defense and other industries by referring to the technological features of defense products. The accepted taxonomy in this respect describes a hierarchy of complexity: at the top are integrated systems (fighter aircraft, main battle tanks, battleships), below them are subsystems (engines, warheads) and sub-assemblies (sights, sensors), down to components (integrated circuits) and materials (semiconductors). Descending the scale, the ratio between specific and generic qualities varies, and differences between defense and civilian production become blurred.

Another possible basis for definition could be the defense companies, but often the differences between them are more pronounced than the similarities. In defense industry there exist concurrently large and small companies, government-owned and private companies, companies with specialized defense-related capabilities only and others that can produce for broad civilian uses too, companies whose sales depend largely on defense demand and others where that dependence is small. Defense companies also differ in the kind of work they do: research and development (R&D), serial production in large quantities, integration of systems in small quantities, or maintenance and logistic support. Some companies specialize in one particular activity, while others combine different types of work. Other differences among companies relate to their position in the supply chain: the first tier is made up of prime contractors, the second of subcontractors, and the third of suppliers of components and materials. Belonging to one tier or the other is not unchangeable, and prime contractors of one weapons system are sometimes subcontractors for another one. Suppliers at lower levels, particularly of multi-purpose products, may not be aware of the final destination of their products, and often do not consider themselves as part of the defense-industrial base.

A more promising approach suggests the customer as a key for defining the defense industry. According to this approach in its broader version, the defense-industrial base consists of all suppliers of goods and services to the ministry of defense and to the armed forces. Furthermore, not only actual but potential suppliers are included, i.e. companies that might enter defense production in times of peak demand, and especially those that could be converted more easily than others to defense production in emergencies. The definition thus obtained is admittedly very clear, yet its boundaries become so wide that their value is doubtful. A narrower version is therefore more practical, one that considers the differences between suppliers as to their degree of dependence on defense demand and their contribution and importance to the defense-industrial base and to national security. Since there is no reason to assume that dependence and contribution are positively associated, a two-dimensional matrix derives, in which a descending order of suppliers may be presented (Dunne, 1995: 401-402): suppliers with high dependence and high contribution belong to the core of the defense-industrial base, low dependence and low contribution are characteristics of low affinity, while between the two are suppliers with low dependence and a high contribution, or high dependence and low contribution.

The defense-industrial base may also be defined from the perspective of public policy, namely the total of organizations, property and industrial activities towards which governments maintain an explicit policy, intervening to assure their existence for national security reasons. In the past these were linked mainly to industries related to the processing of strategic materials (e.g. petroleum, steel, special metals) while today the emphasis is on industries committed to high technology (Kapstein, 1992: 92).

Since the boundaries are not clear, data regarding defense industry activities are not usually presented separately in national and international economic statistics. As for the empirical research on defense economics, it distinguishes between defense and non-defense companies in two principal ways (Ratner & Thomas, 1990: 58-60): from the standpoint of the companies, according to their degree of specialization in defense supplies as measured by the proportion of their output sold directly and indirectly to the ministry of defense; and from the government's point of view, according to the importance of suppliers to national security as measured by their defense sales in absolute terms. The first method includes in the defense-industrial base companies who sold more than a certain percentage of their output to the ministry of defense, and the second – companies that in a given period sold the ministry goods and services above a certain absolute sum.

2. EXPLAINING THE GROWTH OF DEFENSE INDUSTRIES

In the past, defense-related research, development and production were almost entirely the domain of advanced industrial states. In recent decades, however, defense industries have grown rapidly in developing countries, and arms production has become a worldwide phenomenon. Proliferation has taken place primarily for strategic and political reasons: nations invested in developing their own capabilities of arms production so as to be independent of supplies from other countries, and the extent of investment was determined largely by their real and perceived strategic needs. But these do not fully explain either the rapid growth in industrialized countries or the spread of arms production to the Third World. To further explain these trends, two main lines of argument were suggested. The first adopts the neo-classical approach, seeing the size and composition of the defense-industrial base as a rational choice that balances strategic needs and economic considerations. The second rejects the idea that the dimensions of the defense-industrial base are determined to produce an optimal level of security efficiently, attributing them to a coalition of interested parties that derives various profits from a larger scope of defense production.

According to the first approach, although defense industries are established initially for strategic and political reasons, technological and economic incentives accrue over time: the desire to improve profitability through economies of scale, to assure employment, to enhance technological progress, as well as balance of payments advantages. There could be contributions to economic growth too, especially in the Third World: defense industries may be a focus for industrial development, a vocational training framework and a source of technological innovations for other sectors. Economic and technological aspects that once seemed secondary gradually acquire independent status, demanding their place among national priorities, and defense production is no longer perceived to serve only strategic and political goals, but also as a means to promote economic and social objectives.

But these advantages often create tensions with, and at times operate against, the dictates of economic efficiency. In most countries, autarky in meeting defense needs is not a realistic economic option, nor does local defense industry allow for real self-sufficiency. At the same time, nations can benefit from specializing and trading with each other. Importing arms may be economically preferable to developing an indigenous defense-industrial base and relying on local supplies, while protecting and subsidizing domestic defense production may lead to distortions in allocating economic resources. Thus strategic and political motives and economic and technological considerations sometimes complement one another and sometimes are contradictory, and the course along which the defense industry develops represents a balanced compromise between national security demands and efforts towards self-sufficiency, and their economic implications.

The second approach maintains that within the realm of national security a complex, dialectic process gives rise to reciprocal demand-supply influence. In many countries there are evidences of coalitions of interests between the military, the defense bureaucracy, politicians, arms manufacturers and defense industry workers. Such coalitions accumulate power and influence over defense expenditures and the size of the defense-industrial base. President Eisenhower, in his farewell address in 1961, coined for that alliance of interests the term "military-industrial complex", and expressed deep concern that it might lead to decisions that do not benefit national defense. According to this perception too, the external threat provides the first justification, but mutual interests of power groups may impel the country to increase defense expenditures and to continuously initiate programs for developing and producing new weapons systems, thereby expanding the dimensions of the defense industry. Moreover, a vicious circle is set in motion: the companies that make up the military-industrial complex develop a culture of inefficiency and waste, and the less competitive they become, the more they dependent on defense contracts, and the greater their pressure for new government orders.

The military-industrial complex approach drew its theoretical basis from behavioral theories of elite and power groups. In time, other versions were added (Dunne, 1995: 409-411). For one, the neo-Marxist argument asserted that the military-industrial complex does significant service to capitalism; its inherent inefficiency is an important factor in avoiding accumulation crises. Others differ from the military-industrial complex approach, seeing its explanations as having descriptive rather than analytical value. They note correctly that pressure groups influence political decisions on other issues too (Dunne, 1995: 411), and consider the military-industrial complex as no more than a transitory structure – one of many in the long history of producing means of violence – which developed in the twentieth century, notably in the Cold War years, gradually declining in its aftermath (Lovering, 1993: 136).

3. DEVELOPMENT OF THE MODERN DEFENSE INDUSTRY

The foundations of modern defense industry lie in the mechanization of war in the nineteenth century. At first, arms production was entirely an activity of the state and a purely national matter, but with the growth of industrial capitalism the greater part was privatized, and the privately owned defense companies adopted an international orientation. Hence at the turn of the twentieth century, a few large private companies dominated world arms production; they divided the market and coordinated their prices, and later on took advantage of the surging demand during World War I, with no little cynicism, to amass enormous profits. After the War the demand dropped, and with it capacity to produce arms shrank as well. Moreover, in those years public attitude towards defense industries grew extremely hostile, arms manufacturers were perceived as "merchants of death" (see also in chapter 10, section 2.1), and many companies distanced themselves from defense business. As a result, when military buildup resumed before World War II, the infrastructure for arms manufacture was rather limited. Governments had to support the rebuilding of arms production capabilities, and the pendulum swung once again towards public involvement in defense industries. Meanwhile, the international orientation was disappearing too: the defense industries built in Europe, Japan, the United States and the Soviet Union concentrated on responses to local demands, and the volume of international transactions considerably decreased. During World War II, when national economies became "militarized" to an exceptionally high degree, the defense-industrial base took the framework of a national entity, and has remained so for decades.

In the United States after the War, defense-related plants were converted en masse to civilian use. With that, technological opportunities, the results of prewar and wartime innovations, soon led to the founding of new defense industries. Yet differently from the 1930s, when government arsenals produced almost all of Army ordnance items and much of Navy ordnance and ships (Peck & Scherer, 1962: 98), this time the government did not intervene directly. As a matter of fact, it refrained from applying a protectionist policy of any sort with regard to the defense-industrial base. In Britain, the government defined defense-industrial potential as a leading priority in the postwar reconstruction of the economy, and continued its traditional policy of protecting and supporting a core of defense companies, especially in aerospace and related industries (Lovering, 1990: 455). In both countries, defense industries resumed growth concurrently with the escalation in the Cold War. In other Western European countries defense industries were almost totally destroyed, and reconstruction took about ten years in France and even longer in Germany and elsewhere. An accelerated growth started in the 1960s, when NATO shifted its emphasis from nuclear deterrence to conventional weapons systems.

3.1 Defense Industry in Industrial Countries During the Cold War

There were several trends common to the development of defense industries in all industrialized Western countries during the four decades of the Cold War. First, frequent changes in expenditure levels and in the scope of defense purchases led to corresponding fluctuations in industrial activity, and defense enterprises inevitably went through cycles of expansion and contraction that made efficient operation and planning forward difficult. Structural changes usually followed each such cycle, increasing the degree of concentration and reducing possible competition (Gansler, 1980: 9-11).

Secondly, developments were highly influenced by the mode and degree of government intervention. In the United States, where most defense production was transferred to private hands, it was assumed that market forces could assure desirable structure and efficient operation of the defense-industrial base, and the government avoided an explicit industrial policy on defense matters (Gansler, 1980: 9). In Europe, by contrast, governments did not rely on market mechanisms and competitive procurement, but rather actively intervened to assure defense-industrial capabilities. Moreover, in line with Clause 223 of the Rome Treaty of 1957, they adopted an approach of preference for local manufacturers, and took measures to limit access of foreign producers to their national defense markets. Motives were in the main political, reflecting the desire to maintain freedom of action on defense issues, but there were economic considerations too. In particular, the common view was that developing and producing defense products could enhance technological progress and create employment opportunities for skilled labor. British policy changed in the 1980s: defense production was privatized, competitive procurement became the rule, and the market opened to foreign competitors. On the other hand, in France, leading defense companies remained under government ownership, protected from foreign competition through most of the 1990s.

Thirdly, public opinion in Western democracies is sensitive to the use of taxpayers' money in general, and has been particularly acute as regards defense spending and procurement. It demanded equal opportunities, public fairness and transparency, thus leading to over bureaucratization of defense procurement. Procurement processes became cumbersome, anchored in thousands of pages of instructions, and the companies had to allocate substantial financial and administrative resources to develop special skills for working with procurement authorities. Moreover, public criticism enflamed by occasional scandals urged governments to initiate frequent reforms in procurement policy, thereby forcing suppliers to adapt to new rules and augmenting the already high business uncertainty in which they operate (Gansler, 1989: 242-243; Lovering, 1990: 456; Hooper, 1995: 63-64).

The fourth trend is linked to the increasing technological intensity of defense products. American strategy emphasized technological superiority and accorded high priority to perpetual improvements in the performance of weapons systems, spurring the industry to continuous technological progress and innovation. Western Europe followed essentially the same track. The ongoing technological race had several major implications. As for the governments, they had to assist in developing advanced technological capabilities, and hence, besides their role as customers of finished products, they became the central financial source of defense R&D. The defense industries, on their part, grew into an R&D intensive sector, sometimes up to a point of creating imbalances between production and R&D abilities (Gansler, 1980: 10-11). Furthermore, increased technological complexity brought about continuous increase in unit cost of defense products, and with it reduced quantities of each product were produced. Consequently, adaptations in production facilities and processes became necessary.

Fifthly, rapid technological advance, bureaucratization and the unique ways of doing business separated and distanced the defense industries from other industrial sectors, and in most economies defense companies became an isolated, standalone sector. Furthermore, companies that engaged in civilian activities as well tended to concentrate defense production in independent divisions run on special principles. At the same time, within the group of defense companies a high degree of continuity prevailed, and leading companies of the 1940s and 1950s, sometimes with different names, made up the nucleus of the defense-industrial base of the 1980s. In Europe this isolation affected, *inter alia*, organized labor: unions were highly influential, creating special labor relations and employment patterns.

Finally, from the 1970s on, Western defense industries grew more international in nature. They became increasingly dependent on foreign sales on one hand, and on foreign supplies of critical components and materials on the other hand. In Europe, due to relatively small national markets, the international orientation was more deeply rooted, and urged on by competition with American producers led to special formats of joint projects in which companies from several countries took part (see section 7.4).

In the 1980s serious problems appeared in American and Western European defense industries. In the United States, at the prime contractor level, observers pointed out particularly the aging in plants and equipment, large excess production capacities, difficulties in engaging young scientists and engineers, and an unsound financial structure of companies. At lower tiers of the industry, the mass withdrawal from defense businesses was noted, meaning fewer alternative sources of subsystems and components, bottlenecks in production, and eventually longer response times and supply periods (Gansler, 1989: 242-243; Kapstein, 1992: 96-97). In Western Europe, market fragmentation along national lines and absence of competition impaired efficiency, led to technological backwardness vis-à-vis American producers, encouraged labor hoarding and reduced investments in equipment renewal. Since civilian sectors were progressing towards a Single European Market at that time, the defense exception was even more conspicuous. Yet the Cold War had to end and procurement budgets decrease by tens of percents, for an in-depth restructuring of the defense-industrial base to occur.

A brief reference to the Soviet Union and Japan is in place here. Beginning in the 1930s, the Soviet defense industry continuously received priority in the allocation of scarce resources, and before the Soviet Union collapsed in 1991, it was the world's largest defense-industrial complex, employing six million workers. Defense plants, like the rest of Soviet economy, operated according to five-year plans, thus having a high degree of stability, in sharp contrast with the fluctuations typical in the West (Kapstein, 1992: 108). In the 1970s, following a comprehensive modernization plan and huge investments, there was a technological and industrial leap forward. The Soviet defense industry evidently acquired new competencies, and could successfully incorporate advanced electronics and other novel military technologies in its weapons systems (Kapstein, 1992: 109). Comparing it to the defense-industrial base of the West, there were three distinctive features (Sandler & Hartley, 1995: 193-194): separation of research and design centers from production facilities; allowing for competition in research and design only, and to this end maintaining two parallel agencies for each type of equipment (e.g. Mikoyan and Sukhoi for fighter aircraft); standardization of defense equipment across the Warsaw Pact states, making it possible to obtain scale economies through long production runs of each type of equipment. As for the defense production in Japan, its renewal was marked by an exceptional policy of developing and applying dual-purpose technologies. Commercial specifications were used extensively in manufacturing defense products, and defense and civilian production were often integrated within the same manufacturing plants (Sandler & Hartley, 1995: 193). In following this course, Japan actually preceded the Americans and Western Europeans by almost two decades.

3.2 Defense Industries in Developing Countries

Even before World War II, non-industrialized countries made attempts to embark on defense production. However, in 1945 only four Third World countries – Argentina, Brazil, South Africa and India – were capable of producing major weapons systems (Moodie, 1979: 295). From the 1960s on, the number of developing countries with defense industries increased steadily, and by the 1980s more than 50 of them were producing a relatively large variety of military equipment (Rosh, 1990: 57).

Specific motives for establishing a defense-industrial base varied from case to case, yet they have been always a combination of military, political and economic factors. Military factors included real or potential threats from neighboring states, and in some cases ambition for regional hegemony. There was also the idea that weapons systems designed indigenously, and thus specifically adapted to the needs and conditions of the local theater, might improve military capabilities. Political reasons combined considerations of internal and external prestige with the endeavor to end dependence on developed countries. Local weapons production was seen as an expression of sovereignty, and as a tangible evidence of progress into modernity. At the same time, release from dependency on foreign supply sources made it possible to reduce the political leverage of defense producers in developed countries who often took advantage of their special position, and interfered blatantly in their customers' foreign and domestic affairs. Moreover, eager to shake off the influence of developed countries, developing countries lacking the means to develop arms industries of their own also took care to diversify their supply sources, and when they diverted their demand to the new producers an additional incentive for growth in the Third World defense industries evolved. From the economic standpoint, indigenous defense industry was regarded as an important means to stimulate industrialization and to save foreign currency. Obviously, defense industries did not develop at the same pace everywhere, and empirical research found that growth rate differences were associated with such factors as previous experience with embargos, level of industrial development in general and the dimensions of the overall economic base (Rosh, 1990: 60) (see also in section 5.3).

Progress generally took place by stages, infrastructure and experience obtained at one stage serving as a foundation for the next one. A typical course of development was: building maintenance and repair facilities for imported weapons; final assembly of systems under license, using imported sub-assemblies and components; import substitution of simple components, produced domestically under license, while continuing to import the more sophisticated components (engines, electronic devices, etc.); at this point, exports of domestically produced components to the licenser or to other destinations may commence, as well as foreign sales of the locally assembled systems; local design of certain parts, importing fewer components and manufacture of complete systems under license; domestic production of locally designed systems, still combining sophisticated imported elements; production of locally designed systems, using components most of which are locally produced. Moving from one stage to the next generally requires foreign know-how. In fact, without technology transfer from the industrialized world, real defense industry could not come into existence in the Third World (Ball, 1988: 355-356). Industrialized countries, on their part, transferred military technologies to the Third World as a means of competition for markets, and to gain the benefits of cheap labor. Since the 1970s, technology transfer has become an integral part of the world arms trade, particularly through different offset arrangements (see chapter 10, sections 4.2 and 5.3).

In evaluating the achievements of the defense industry in developing countries and its implications for the international system, several issues were emphasized. First, even after decades of investment, not one country in the Third World has freed itself from significant dependence on outside supply sources. All continue to import new major weapons systems, and countries that cut back on importing complete systems still depend on technology transfers and critical component imports. Furthermore, in most developing countries, the existence and continued development of defense industry depends on capital imports for investment and on external assistance in training manpower. Possibly there is less concern about supplies being cut off in times of emergency and short-term political pressures. Likewise, it is possible that the bargaining power of developing countries vis-à-vis the developed world as regards arms acquisitions has improved. However, maintaining a domestic defense-industrial base that can adequately respond to challenges of new threats is itself conditional on external factors. While its content has changed, dependence itself has not disappeared.

Secondly, it has not been proved that a domestic defense industry makes it possible to adapt military equipment to specifically local military needs (Moodie, 1979: 303-305). In most developing countries R&D capabilities lagged behind production capacities, and in license-based production, the know-how transferred was only partial and insufficiently detailed for implementing changes and adaptations. In other instances, prestige considerations pushed developing countries beyond their ability, resulting in poor quality systems and inferior performance. Finally, local adaptation became less important with time, since the traditional suppliers, particularly in Europe, were ready to offer their own adapted versions in the effort to obtain new export orders.

Third, domestic arms production in developing countries seems to have flattened the world power pyramid somewhat, but its contribution to international stability is doubtful. The larger arms producers in the Third World are located in areas considered unstable, and most of them are traditional adversaries. Changes in regional security conditions may cause them to further develop their defense production capabilities, thus turning the advancement of defense industries into a new dimension of an arms race, eventually undermining regional security and exacerbating contention. At the same time, Third World arms production ability interferes with developed countries' control over regional conflicts: Iraq and Iran could wage war for most of the 1980s, importing most armaments from the Third World.

3.3 Defense Industry in the post-Cold War Era

When the Cold War ended, the main threat behind four decades' race to develop and acquire defense products disappeared with it. Defense establishments, particularly those of countries directly involved in the East-West conflict, had to undergo drastic changes. They had to examine new potential threats, to reformulate national defense perceptions and adapt their armed force size, structure and equipment to the new global order. Meanwhile defense budgets and acquisitions dropped sharply. In the NATO states, there was no further need for current production of large quantities of arms and ammunition, and doubts arose as well with respect to ongoing development programs basically designed to meet Cold War operational requirements. Western defense industry, then, faced steadily decreasing demand, and as a consequence – large excess production capacities and labor surpluses, implying inevitably a painful process of downsizing. Moreover, defense companies had to cope with great uncertainty, waiting for new threats to be defined and for strategic perceptions as to the roles and missions of the armed forces in the post-Cold War era to evolve. In the Soviet bloc, the Cold War's end was intertwined with an overall systemic collapse of the Soviet Union, and with far-reaching changes in the regimes and economies of her former allies. Hence the huge military-industrial complex there was affected not only by changing threats or strategic and military considerations, but also by economic chaos and dysfunctional law enforcement system. On top of that, since defense production in the Soviet bloc rested on reciprocity and interdependence among plants now in politically separated states, production was disrupted by shortages of subsystems and components previously supplied from other states. Elsewhere in the world, outside NATO and the former Warsaw Pact, change was less uniform, and some countries even expanded defense production.

In 1996, global defense production was estimated at 195-205 billion dollars, as against 260-295 billion dollars in the mid 1980s (SIPRI, 1999: 407-411), a decline of some 30 percent. However, the global average rate conceals great variation in the scale of reduction between the main centers of production, as well as some differences in timing (SIPRI 2000: 314-317). The sharpest decrease took place in Russia. It started immediately after the Soviet Union collapsed, and continued steadily from 1992 to 1997, amounting to a total decline of 90 percent. In 1997 then, Russia's defense production was only 10 percent of the Soviet Union's on the eve of its collapse. In the United States, continuous decrease was reported from 1987 to 1996, reaching a total of 47 percent. In Britain too, the decrease began in the late 1980s and continued steadily until the mid 1990s, adding up to a total of 30 percent. By contrast, in France the decrease began only in 1991, continuing until 1995, for a total of 37 percent. For Germany, partial information indicates that in the mid 1990s the value of defense production was about half what it was at the beginning of the decade. Finally, Japan reported the least decrease (some 28 percent), and in the shortest period (1992-1995). In the mid 1990s the decline came to a halt; in all prime production centers increased production was reported, although not one returned at the end of the decade to the production level of ten years earlier.

A similar downsizing picture emerges from employment data. In the United States, the Department of Defense estimated that during 1989-1997 total defense-related employment declined by 39 percent (Weidenbaum, 1997: 595), or by about 1.5 million workers. In the three main defense production centers in Western Europe, declines of 27 percent in Britain, 29 percent in France and 64 percent in Germany were reported for 1990-1997 (SIPRI 1999: 399), implying that a total of about half a million employees were laid off. Other Western European countries too saw considerable declines in employment (*Defense News*, 14.6.1999). Estimates for Russia point to a drop from 6.4 million in 1991 to fewer than 3 million employees in 1996 (SIPRI, 1997:255), and according to an estimate for 1999 numbers further declined to only 2 million employees (*Defense News*, 18.10.1999).

Notwithstanding the drastic downsizing, in most countries adaptation of production capacity lagged behind the decline in defense demand. In the United States critics noted that even after a decade of defense budget cuts employment remained 23 percent higher than at the budgetary low point of the Cold War (Gholz & Sapolsky, 1999: 6). Another illustration of the same point: in the mid 1990s, defense factories were operating at about 60 percent capacity, compared to 83 percent for American industry as a whole (Weidenbaum, 1997: 598). Besides, adapting production capacities and employment to demand was not uniform across industries. Especially in Europe, while the aerospace industry went through large-scale consolidation and became more efficient, in industries producing land systems structural changes were slow, and large excess capacities remained.

From the industry standpoint, not only demand level is important, but also possible changes in its composition due to new military tasks. At first, the missions that the Western armed forces would have to undertake in the post-Cold War era were not clearly defined, and there was great uncertainty as to what equipment would be needed. With time, however, strategic perceptions adapted to the new situation have emerged, acknowledging, in essence, the responsibility of developed countries for the world order, and asserting their willingness to fill an active role in peace enforcement and peacekeeping, even including military intervention in remote parts of the world. These new perceptions draw, *inter alia*, on lessons derived from experience in more than 20 armed conflicts during the 1990s in the Persian Gulf, the Balkans, Africa, East Timor, Chechniya and more.

Some elements of the new strategic perceptions affect defense industry directly: -

First, the new approach implies that military institutions should prepare for a wide variety of scenarios involving symmetrical and asymmetrical threats, military and nonmilitary in nature, from low-intensity conflicts, terrorism and global crime by non-state actors to conventional wars between national armed forces. In addition, capabilities should be developed to contend with future adversaries trying to remain below the threshold of clear aggression, as well as with rescue campaigns and humanitarian missions.

Secondly, to confront different types of crises effectively, armed forces must implement organizational and doctrinal changes, including far-reaching adaptations of weapons systems and equipment. Two illustrations may demonstrate the extent of change in this respect. First is the need for rapid deployment of forces in distant regions. It requires equipment light enough to transport, yet sufficiently robust and of a high degree of survivability, so that it can fight on arrival. In addition, rapid and flexible operations in changing theaters and against a wide variety of enemies, make it necessary to provide commanders with advanced command and control systems, implying that relatively small units must be equipped with the advanced means once available to large formations only. Moreover, rapid deployment of forces and supporting them over long periods requires special logistic means specifically adapted to complex environments. A second illustration is precision engagement capabilities and maximum avoidance of casualties and damages. Western public opinion is sensitive to casualties among its forces, and political considerations and the unwavering glare of the international media make it imperative to avoid collateral damage as well. In the 1999 war in Kosovo, although it was one of the most carefully controlled operations in history, NATO commanders did not escape criticism for the few errors committed, which caused heavy losses. What is needed, then, is a combination of advanced systems providing high quality targeting information (satellites, airborne surveillance systems, unmanned air vehicles with remote sensing capabilities, etc.) with sophisticated, precision standoff weapons. Besides precision, in order to obtain controlled effects man-in-the-loop systems are required, i.e. systems allowing for human intervention after firing. For the same reason, demand for non-lethal weapons is growing too.

Thirdly, since accepted perceptions see threats to collective international security, the prevailing approach is that responses too must be collective, reinforcing the requirement for international coalitions of forces. Coalition warfare requires a high degree of interoperability in equipment, which, in turn, depends, *inter alia*, on the technological capabilities of national defense industries. Lack of interoperability and significant technological gaps among coalition partners were evident already in the 1991 Gulf War, and became even more conspicuous later, in the Kosovo War. While the American forces used the most modern munitions and could maintain continuous operations, day and night and in all weather, their European allies could barely play a secondary role in a confrontation in their own backyard.

Finally, the "revolution in military affairs" is worth mentioning. It has been argued that technological innovations in weapons, and the accompanying changes in military organization and doctrines, add up to a revolution in conventional warfare, presumably to the extent of eventually superseding the current paradigm of high intensity conflicts dominated by tanks, manned aircraft and the like. In particular, it was suggested that the new technologies of long distance precision targeting, communications and sensors could be combined and a "system of systems" could be built, providing full control over thousands of square kilometers anywhere on earth.

From the viewpoint of defense industries, the variety of threat scenarios translates itself into a rich array of requirements, challenging the companies to innovate and to harness advanced technologies to nontraditional tasks. Yet at the same time it creates greater uncertainty, since difficulties in balancing demands, mainly in budgetary terms, often delay decisions or lead to frequent changes in ongoing projects, including rescheduling and sometimes outright cancellation. In a similar manner, adapting weapons systems and equipment to new operational demands is also at one and the same time good news and a source of worry. It creates demands for the development and production of new products, but on the other hand, leading companies may find out that capabilities advantageous to them in the past are no longer relevant, and valuable assets, tangible and intangible, have become worthless. Established industrial activities may decline and even vanish completely. The "revolution in military affairs" may have a similar effect. Adopting the new perception, even gradually, might imply calling off major projects already under way, either because they become less essential or because budgets must be diverted elsewhere, thus too creating an existential threat to top-of-the-line defense contractors. Regarding coalition warfare, it makes it necessary for European defense industries to narrow the technological gaps between themselves and the American defense industry. It creates a powerful incentive to accelerate consolidation and to strengthen cross-border cooperation within Europe, and at the same time stresses the importance of transatlantic defense-industrial ties.

Changes having no direct connection with strategic developments also affect defense industry. While procurement budgets declined, the performance and costs race in military equipment escalated further. Rapid technological progress makes available weapons systems with superior performance, but the more advanced generations are also more expensive. According to several estimations, the unit cost of successive generations of battleships, submarines, fighter aircraft, guided missiles and the like doubles about every seven years (see section 4.7 below). The combination of decreasing budgets and increasing unit costs has far-reaching implications for the defense industry. It is possible to acquire fewer types of equipment and smaller quantities of each, which leaves room for fewer producers. Fewer platform manufacturers are needed, and at the same time, manufacturers of the advanced subsystems that give weapons their special qualities become more important. Prime contractors tend to increase the scope of their own production, and for that take over specialized companies on one hand, and reduce orders from subcontractors on the other hand. Different production facilities are needed: smaller areas, different types of machinery, fewer unskilled production and assembly line workers, and so forth. There are also increased incentives to national and international cooperation, aiming at higher degree of specialization and economies of scale that will hopefully slow down rapidly rising costs. Moreover, in these circumstances development programs are fewer and farther between, so that losing a competition for developing a major weapons system may determine the fate of a large defense company. Indeed, the McDonnell Douglas failure in the early stages of the competition over development of the American fighter aircraft of the future, the Joint Strike Fighter, was a decisive factor in ending its existence as an independent entity, and the world largest defense contractor of the early 1990s was acquired by its archrival Boeing in 1997.

Another factor that impacted on defense industry was development in the interrelations between military and civilian technologies. For years defense industry had led in developing innovative, product specific technologies, but invested little in modernizing process technologies that could economize on costs. Moreover, the isolation of defense industry prevented the transfer of such new efficient production methods, currently being developed in the civilian sector. Change began in the late 1970s and the 1980s, first in the effort to mitigate the rising unit costs of weapons systems, and later on also due to the quality advantages of civilian technologies, which meanwhile had progressed rapidly. In fact, advance in the civilian sphere reversed the old relationship between military and civilian technologies, and in many areas civilian development now took the lead. This change in relationship had several notable ramifications. First, it shed new light on the longstanding claim supporting the national defense-industrial base: apparently technological benefits from defense production were not as large as they had been (Brzoska, 1998: 78). Secondly, with shrinking defense markets and continuing growth of markets for high tech civilian goods, the share of defense production in high tech industries diminished, thereby increasing in relative terms the costs of maintaining technological autonomy in defense production. Third, as it grew more dependent on civilian technologies for all these reasons, defense industry drew closer to civilian industry. The institutional and functional separation between the two became blurred, producing arms became to a large extent a branch of general industrial production, and defense industry became less exceptional. Finally, the tendency to focus on dual-purpose technologies was reinforced. These technologies allow defense producers to diversify their industrial activities and enter the civilian market, thus reducing their dependence on defense demand, and yet enabling them when the need arises to return to producing for defense within a relatively short time.

Globalization, an important feature of today world economy, affects defense industry too. Numerous transnational defense companies were created either by cross-border mergers between existing companies or by establishing new joint companies. An even more common format was multinational joint ventures that have proliferated in almost all sectors of defense production. The trend started in electronics and aerospace, where it is still most prominent, but in later years other sectors – armored vehicles, battleships – followed suit. Globalization in defense production is a significant departure from the national character typical of the Cold War defense-industrial base, and in particular it changes the traditional reciprocal relations between governments and defense companies fundamentally. Instead of large national companies owned and controlled by nation states, or critically dependent on domestic demand from their governments, defense companies and other emerging organizational formations are less linked to a particular country. In addition, it became impractical or even harmful for governments to hold on to the traditional preference of domestic producers, whether for national defense reasons or in an attempt to derive economic benefits. Globalization also clashes with essential national defense interests. Vital defense production comes to depend on foreign hardware and software, and there is growing concern about unauthorized transfer of sensitive technologies, loss of military advantages and uncontrollable arms proliferation worldwide. Lastly, globalization increases competition in national and international markets, imposing greater efficiency, thus leading to layoffs and unemployment. It also exposes defense companies to economic fluctuations and financial crises in different world regions. The financial crisis in Asia in the late 1990s, for example, was harmful to many Western defense companies, especially in Europe.

Important information about the development of defense industry comes from data on the world's 100 largest defense companies (for technical reasons, companies in Russia, Eastern Europe, China, South Korea and Taiwan are excluded). The data was gathered and processed separately by the Stockholm International Peace Research Institute (SIPRI) and by the weekly *Defense News* since 1990, both of which present more or less a similar picture. According to SIPRI data (*SIPRI Yearbook*, various years), between 1990 and 1999 the defense sales of the Big 100 in constant prices decreased by 30 percent, mostly until 1995. However, the sales of the top companies behaved differently: they decreased mainly in the first two years of the decade, by 20 percent only, steadily increasing since 1993. The largest companies, then, apparently reacted rapidly to changing market conditions, and hence were successful in increasing their sales despite general decline in demand. The other implications of this development were the eventually growing market shares of the largest companies and the higher degree of concentration in the defense market. Market share of the Big Five (in respect to total defense sales of the 100 largest companies) reached 43 percent in 1999 as compared to 22 percent in 1991, and of the Big 25 – 75 percent compared with 65 percent, respectively. The data also highlights changes in dependence on defense-related sales. After some adjustments allowing for changes in company identity due to mergers and acquisitions, it appears that from the beginning to the end of the decade, dependence of the Big Ten on the defense market grew, but there was no real change for the Big 25.

Concurrently, export became an important strategy for survival in the defense sector: many companies increased exports substantially, and the share of export sales to overall sales grew. Finally, the list of the largest companies demonstrates the fundamental structural changes that defense industry underwent. Indeed, following an intensive process of acquisitions, mergers and divestments, far-reaching changes occurred in the identity and internal structure of companies. In fact, not a single company among the Big Ten of 1991 preserved its original format, and each appears on the 1999 list either with a new structure or as a subdivision of another company. The same applies to many companies at the lower levels as well. Out of the Big 100 at the beginning of the decade, 36 companies left the defense business or were sold and merged with other companies during the 1990s (SIPRI, 2000: 318).

Before summing up, it might be appropriate to add some details about the Russian defense industry, where the greatest changes took place. Both demand and supply factors combined to bring about a drastic decline in defense production. On the demand side, budget constraints cut sharply procurement for the Russian armed forces. For instance, in 1991, 414 new military aircraft were supplied to the air force, while in 1995 only seven aircraft (SIPRI, 1997: 255). Exports also decreased greatly; official figures indicate that sales in current prices dropped from 7.1 billion dollars in 1991 to only 1.7 billion dollars in 1994 (SIPRI, 1997: 257). On the supply side, the systemic collapse of the Russian economy left defense plants without financial means, disrupted the supply of raw materials and interfered with maintenance of equipment, forcing the defense industry into long shutdown periods. In the first years, the government neglected the defense industries, and except for establishing an organization to coordinate exports in 1993 (see chapter 10, section 2.5), no real steps were taken towards rehabilitating the huge military-industrial complex of the past. Change came at the end of 1995, with the issuing of presidential decrees outlining comprehensive structural reform, notably privatization and consolidation. Government policy also called for converting defense plants to civilian production, though actual progress in this direction was rather limited. In 1998-2000, there was apparently a turnaround, and the value of defense production almost doubled (SIPRI, 2001). With that, growth resumption and industrial renewal seemed to depend chiefly on whether the Russian defense industry could regain its central place in the international arms trade (see chapter 10, section 2.5).

Analyzing the impact of the trends and processes outlined above leads to several main conclusions: (a) The end of the Cold War brought about substantial decrease in the absolute dimensions of defense industry, and in its share in the GDP and employment of the main defense production centers. During the 1990s they were smaller than in any previous period since the end of World War II, and this situation will most probably continue at the beginning of the twenty-first century. Possibly, then, the macroeconomic implications of defense production too will be relatively limited; (b) Changes in the composition of defense demand dictated by new threats and different military tasks may give rise to fundamental changes in the industry: long established activities may lose their value completely, and leading defense companies may disappear; (c) Several developments, especially the growing reliance on civilian technologies, made defense industries less unique, and they are increasingly a branch of general industrial production; (d) Reciprocal relations of governments and defense industries changed fundamentally: government control over defense companies is diminishing, and so is the affinity of companies for any particular country. As a consequence, possible effective control over global proliferation of arms and sensitive military technologies also becomes limited.

4. MICROECONOMIC ISSUES

Since supplying national defense is subject to economic constraints, defense industry not only has to develop and manufacture products that meet national defense needs, but it must as well be economically efficient. The microeconomic discussion deals with aspects of economic efficiency in defense industry.

4.1 Criteria of Economic Efficiency

Economic efficiency in defense industry has long been criticized. There were complaints, *inter alia*, about gold plating, costs overrun, delayed deliveries, excessive profits, low labor productivity and hoarding redundant manpower (Sandler & Hartley, 1995: 177). These were not always justified, and certainly did not adequately consider the various limits imposed on the defense industry, which often interfere with efficient operation.

Since defense production takes place in a regulated sector that does not act according to competitive market rules, standard efficiency measurements based on output-input ratios or on cost ratios do not usually apply. But in this defense is apparently no different from any other regulated industry. Three specific factors, however, make it particularly difficult to evaluate economic efficiency in defense industry. First, defense products provide different outputs simultaneously – tanks, for example, provide firepower, protection and maneuverability – and therefore only by separating them and relating costs to similar outputs can the economic efficiency of manufacturing companies be measured meaningfully through costs comparisons (Beard, 1993: 35). Secondly, with defense products it is not enough to compare production costs, or even development and production costs. American studies showed that operation and maintenance usually comprise no less than half the life-cycle costs of a typical weapons system, and that they are closely associated with design, development and production costs (Gansler, 1989: 157). Hence comparisons should account for the overall life-cycle costs of defense products. Third, tradeoffs between performance, costs and time should also be accounted for. In developing and producing a weapons system, the aim is to achieve the best performance in the shortest time for the lowest costs. Yet giving up certain performance requirements might save costs or shorten schedules, so that economic efficiency have to be examined in each case in relation to the overall balance between them.

It should be noted also that companies' survival, which is usually taken as an indirect indication of their efficiency, does not apply here either. Due to the special, complex relations between governments and major defense contractors, governments tended to intervene in times of difficulty, and extended financial assistance to companies that otherwise would have ceased to exist (Beard, 1993: 34).

One approach in studying defense industry efficiency focuses on individual programs, comparing their actual results in terms of performance, costs and time with pre-assigned goals, with results obtained in previous programs, with parallel measures for defense industries in other countries or, when relevant, with civilian programs. Such comparisons, however, often encounter technical difficulties that lead to ambiguous conclusions. Another approach was based on the structure-conduct-performance model of industrial economics. It examined possible relationships between industrial performance – e.g. technical efficiency, price-marginal costs ratios, product diversification and profitability – and structural features such as concentration, barriers to entry and exit, contestability and economies and diseconomies of scale.

4.2 Unique Ways of Doing Business in the Defense Market

On the face of it, defense markets seem like any other market where buyers, generally defense ministries, meet sellers. Indeed, most transactions involve standard, price sensitive items, purchased in large quantities at low unit cost. However, the greater part of procurement budget funds is appropriated for a small number of major weapons systems and other complex, technologically advanced military equipment. Here small quantities are purchased at high unit cost, and procurement decisions are performance-sensitive. This second category of transactions is carried out in a special way that makes the conduct of business in the defense market unique.

4.2.1 The Defense Market Versus the Civilian Market

The uniqueness of the market for defense products (the defense market) is often demonstrated by comparing it with commercial markets for civilian goods (the civilian market) (Fox, 1974: 39; Gansler, 1989: 159-160): –

i. In the civilian market there are usually many buyers and sellers for each product, while in the defense market there are a few, or possibly even a single, monopsonistic customer. The number of sellers for each product is also small, and at times a single buyer confronts a single seller.

ii. Buyers and sellers in the civilian market act independently, at arm's length. In the defense market, however, there is strong interdependence between seller and buyer, and they maintain constant contact throughout the procurement process. Thus in the defense market there is usually no way to hold one side or the other responsible for deviations, and disputes are resolved through adjustments that preserve the contractual relations.

iii. In the civilian market sellers initiate new products, usually on the basis of market research that provides only general information about sales prospects, while in the defense market the buyer establishes the requirements for a product, and development and production are carried out on the basis of contractual orders and approved budgets.

iv. Many decisions, especially those regarding product specifications and quality, that are commonly taken in the civilian market within companies, are determined in the defense market outside them, namely by the defense ministry and the armed forces. Besides, the military customer reserves the right to introduce changes in specifications, quantities and schedules as a routine procedure, thus subjecting programs to frequent modifications. The content of most civilian transactions is defined in advance and remains intact, and significant changes may lead to cancellation.

v. In civilian markets producers finance the development-production effort. In the defense market the buyer bears most of the development cost, at times provides equipment and facilities for the use of the producer, and as a standard practice extends advance and progress payments to finance current production.

vi. Civilian products are traded in large amounts, while most defense products are purchased in small quantities.

vii. The civilian market offers the buyer a wide range of choice in every product category, while only rarely does the defense industry produce different products simultaneously for the same mission.

viii. Since adequate substitutes are available for every product, market price is a dominant factor in civilian buyer's choice. In the defense market, however, price is only one of the determining factors, possibly less important than product performance or date of operational availability.

ix. In the civilian market prices are primarily determined by competition, while in the defense market they are based on cost evaluations. Consequently, while reduced demand for civilian products brings down their prices, in the defense market a drop in demand may raise prices, since producing smaller quantities implies higher unit costs.

x. Profits in the civilian market reflect risk and efficiency. By contrast, in the defense market profits are controlled: they are determined as a percentage of costs, and should costs decrease, thereby reducing the price for subsequent orders, the supplier's profit decreases too. In the defense market, then, profits lose their significance as incentives to efficiency and risk-taking.

xi. In the civilian market sellers compete for market share, while in the defense market the winner often takes all, so in many cases competition is "all-or-nothing".

xii. In general, demand for civilian products changes slowly and in relatively small increments. With defense products, however, new threats and technological opportunities, as well as budgetary decisions, can suddenly alter demand, subjecting it to sharp fluctuations.

xiii. In civilian markets, supply-side adjustments to demand-related changes are relatively quick. Supply-side adjustments in the defense market, on the other hand, require a long period. First, development and production periods for defense systems are exceptionally long. Secondly, the more complex and technologically sophisticated the products of companies become, the narrower their specialization, and the harder it is for them to adapt equipment, workers' skills, production processes and organization to changes in demand. Finally, technological, marketing and administrative factors created barriers to entry and exit, making movement into and out of the defense sector difficult or even nonexistent.

xiv. In most civilian sectors, government intervention as a sovereign is limited, and usually does not extend beyond setting standards, enforcing anti-trust regulations and the like. In the defense market, however, government intervention in that capacity is far broader and deeper. Since most activities in the defense sector are financed with taxpayers' money, governments maintain close control over suppliers, set up priorities in granting orders to small businesses or to areas of high unemployment, etc.

There may be many reasons for the unique ways of doing business in the defense sector, and for differences between defense and civilian markets. In their pioneering study on American weapons acquisition process, Peck and Scherer suggested two fundamental explanations (Peck & Scherer, 1962): an extraordinary combination of uncertainties, and the non-market environment in which the defense industry functions.

4.2.2 High Level of Uncertainty and Its Numerous Sources

Uncertainty means relative unpredictability of the outcome of a contemplated action, and thus the greater the deviation between original planning and actual results, the greater uncertainty is thought to be. In a sample of 12 programs, Peck and Scherer found that actual development costs were 3.2 times higher, and the time needed to complete development was 36 percent longer than what was originally predicted. Examinations made then by other researchers generally showed even higher deviations in time (of 50 to 60 percent). Deviations in performance quality when relating specifically to features like airspeed, maximum range, accuracy and the like were in most cases positive, varying from 0.8 (i.e. 20 percent below predicted level) to 2.0 (Peck & Scherer, 1962: 19-25). Subsequent periods also provide evidence of significant deviations. In the late 1960s and early 1970s, it was reported that 90 percent of the major weapons systems acquired by the United States Department of Defense cost twice as much or more than the first estimates, and that of 35 major development and production programs only two met, or were ahead of their schedules (Fox, 1974: 3). Projects completed in the United States and in Britain in the 1980s and 1990s showed actual costs that deviated by 40 to 70 percent from original plans (Sandler & Hartley, 1995: 122).

Some of the numerous sources of uncertainty are external to the defense companies and the acquisition process, while others are embedded in them. An important external factor could be new intelligence assessments that dictate changes in operational requirements. Another external source is political in nature: changes in government may bring about changes in strategic perceptions, and budgets for defense procurement programs may be expanded or cut off accordingly. Macroeconomic factors may also alter budgets. From the standpoint of the companies involved, possible changes in demand imply a higher degree of uncertainty in receiving new orders, and at times concern lest programs be terminated or rescheduled over longer periods. Since defense demand is made up of "bulks" (reflecting the large size of individual projects), demand changes might have an all-or-nothing effect, and thus for the individual company, or for certain plants within it, they may be a truly existential question.

Demand fluctuations and the uncertainties created thereby may have other, less far-reaching, though not insignificant implications. Studies in industrial economics show that differences among companies as regards investments in fixed assets can be explained by the degree of fluctuations in demand for their products. Thus for two companies with an equal average demand, the greater the fluctuations are, the less the investment in fixed assets (Beard, 1993: 37-38). Indeed, American surveys indicated that defense contractors invested less – almost half only, on average – than comparable companies in the civilian economy in new machinery. Furthermore, the higher the share of defense in total sales was, the lower the investments to sales ratio (Gansler, 1980: 248; 1989: 251). Large, frequent demand changes may lead also to inefficient use of variable inputs. Regarding labor, this influence was manifest in two different ways. On one hand, particularly in the United States, demand fluctuations led to unsteady employment and high labor turnover, so that defense companies, considered unstable workplaces, had to pay about 20 percent more than other companies to workers with the same skills (Gansler, 1989: 248). On the other hand, mainly in Europe, companies responded by labor hoarding, especially of technical personnel, thereby decreasing labor productivity and increasing unit product costs (Sandler & Hartley, 1995: 196-197).

Yet another source of uncertainty lay in the progress of military technology since World War II. Military innovations led to rapid technological obsolescence, bringing about frequent cancellations and far-reaching changes in programs, and often undermining companies' competitive capabilities. The advanced technology intensity of weapons systems is also an endogenous cause of uncertainty, since R&D thus entails greater risks, and more numerous, more complex problems may arise in defense projects.

Technological uncertainty impelled defense companies into the forefront of technological development and into heavy investment in R&D. It showed as well in the types of manpower and other assets they employed, in work organization and in their general corporate culture. Besides, it imposed narrow specialization: companies specialized not only in a certain category of weapons systems (e.g. missiles), but rather in special types within each category (e.g. air-to-air radar-guided missiles). Thus workers' skills, equipment and production organization became more narrowly specific which, in addition to other implications, intensified the differences between defense and civilian industrial activities. Moreover, the prevailing opinion held that ability to deal with uncertainty and technological innovation depended on combining separate, generic technologies, and that large, vertically integrated companies could perform multi-disciplinary R&D and make use of potential synergies better. Hence governments diverted the greater part of their procurement budgets to larger contractors (Gansler, 1980: 221), concentration grew and competition in the defense sector diminished (In fact, contrary evidence was found as regards the relationship between company size and technological innovation (see below, section 4.3.1)). Finally, with time, as rapid technological advance increased costs sharply, not even the largest companies could finance the necessary R&D alone, and various forms of strategic alliances emerged in which independent companies collaborated in producing knowledge and integrating their resources, while sharing costs and risks.

4.2.3 The Special Status of Governments in Defense Markets

The non-market character of the environment in which defense industries operate is mainly an outcome of the special status that governments maintain in defense businesses. Defense ministries are sole customers of the defense industry in their countries, and often take advantage of their monopsonistic position vis-à-vis local suppliers. Governments also regulate the foreign demand for indigenously produced defense goods through export controls, and by assisting local companies that compete in the world arms market. Furthermore, due to their special position governments may influence the supply side of the defense market too (Dunne, 1995: 406). To this end they sometimes adopt a protectionist policy or extend financial aid, make investments in general infrastructure or in specific production facilities, establish regulations to prevent foreign takeovers, to encourage mergers or discourage them and the like. Through such measures governments may determine the dimensions and structure of defense industry, competition level, entry and exit of companies, prices and profits, technological level and nature of ownership. Government involvement that penetrates into every aspect of defense business leaves, in fact, little room for market mechanisms, and thus limits substantially their role.

Defense ministries, like all large bureaucracies, are not homogeneous, and assuming that one single objective guides their attitude to defense companies oversimplifies matters. In fact, they "wear different hats" simultaneously, and their activity in the defense market combines several behavioral models. Ideally, the defense ministry should act as an agent of the public, making use of its monopsonistic status to attain the highest level of national security at minimum costs, and to this end seek to obtain defense products at the lowest possible prices. Wearing this hat the considerations of the defense ministry are mainly short term, and its relations with the companies are conflictive in nature, based on bargaining, designed to broaden and sharpen competition, confronting potential suppliers with each other in order to prevent collusion. Usually, however, defense ministries are not content with their role as cost-minimizing agents, and act to assure the long term development and production capabilities of the national defense-industrial base as well. In this context, defense ministries could be described as cartel managers for the defense industries, also including private companies (Beard, 1993: 36). Several observations support this view: defense ministries divide work among prime contractors to assure employment, they intervene in selecting subcontractors, they transfer technological and financial risks from producers to government by financing R&D and other investments, they support marketing efforts of defense exporters, etc. Governments are ready as well to extend aid to defense companies facing financial difficulties, sometimes sparing them bankruptcy and closure. Such a policy, however, whether explicit or implicit, even if justifiable from the strategic or political-social standpoint, negates important competitive forces that adjust industry size to efficient dimensions, thereby enhancing economic productivity. In particular, companies artificially made immune from consequences of bad decisions will lack incentives to avoid them in the future, and this effect of moral hazard can be of extraordinary importance in some cases (Beard, 1993: 38).

Defense ministry behavior is also affected by imperatives of sound public conduct. Exposure to public criticism has made lack of confidence and the fear of possibly abusing administrative loopholes the point of departure to governments' attitude toward defense suppliers. As a consequence, despite the close contacts, contractual relations are basically conflictive, and defense businesses are subject to rules and regulations exceptional in quantity and in detail. There is a tendency to uniformity too, thereby neglecting possible differences in the economic conditions under which companies function, and ignoring their differing, and sometimes-contradictory needs and interests (Fergusson, 1995: 32-24). Matters are further complicated when procurement regulations seek to assure that public funds will simultaneously promote national security and other goals (e.g. encouraging small businesses, helping regions with high unemployment, etc.).

Finally, ownership of defense companies – government or private – also affects defense ministries' behavior (Dunne, 1995: 420-421). All defense companies, regardless of ownership, seek to promote their own interests through government purchases. Direct government ownership would seem to provide more effective means, usually unavailable when dealing with private companies, to cope with possible manipulations. But governments have a greater commitment to assure profits, continuous production and stable employment in enterprises they own, and therefore are more exposed to pressures and political leverage of communities, labor unions and managements, especially in placing orders. Orders may thus be placed with government companies not on a competitive basis and without considering the comparative efficiency of other suppliers, the more so since transfer of rents within the government sector is not always transparent and true costs are easily concealed.

4.2.4 Barriers to Entry and Exit

The non-market character of the defense business environment is a result too of barriers faced by companies wishing to enter into or exit from the defense market. These barriers stem from technological, marketing and procedural factors (Dowdall & Braddon, 1995: 111-118; Dunne, 1995: 408-409). Defense companies employ highly specialized physical and human assets that cannot easily be converted to civilian uses without their value depreciating significantly. For example, machines and equipment used in defense factories have special features, designed to meet extraordinary specifications not in regular demand in civilian production, and the same is true of personnel specialized in design techniques and in production and testing methods rarely if ever used elsewhere. Converting specific equipment and labor skills to civilian use is very difficult, and it is even more difficult to change managerial culture. The culture of defense companies focuses primarily on product performance and quality, with costs taking second place, while civilian companies stress large quantities at low costs.

The same technological barriers observed from the opposite direction, make it very difficult if not impossible for civilian companies to enter the defense market. In addition, potential entrants find it difficult to demonstrate their technical capabilities in advance, and the military customer who assigns great importance to proven capabilities and wishes to avoid seemingly unnecessary risks is likely to prefer experienced, familiar producers.

Personnel specialization stands out too in marketing and contracting departments. Defense and commercial marketing are entirely different. In the first, personal contacts developed over time are most important, and marketing staff tends to consist largely of retired senior officers and former defense ministry officials. The other face of the coin is obviously the absence of marketing experience and sales promotion skills needed in competitive commercial markets. In particular, since defense demand is relatively insensitive to price, and in addition, defense procurement regulations allow for renegotiation and for price adjustments even after contracts have been signed and orders are being carried out, marketing departments in defense companies almost totally lost their competence for the price competition so central in commercial marketing. This is, then, another major barrier confronting companies who wish to leave the defense businesses and enter commercial markets.

Barriers to entry were also created by the administrative complexity of the procurement process. Procedures multiplied and became complicated not only due to the technical sophistication and high cost of defense products, but also for lack of competition to assure that taxpayers' money would be used effectively and fairly. Thus the cause (lack of competition) paradoxically led to the result (barriers to entry) that eventually reinforced and perpetuated this lack of competition. Particularly important were the accounting, costing and reporting procedures that suppliers were required to adopt. Differing from accepted business practices, on one hand they imposed an administrative burden on the companies concerned, but on the other hand, those who learned to thread the bureaucratic maze could increase income and profits. In fact, expertise in defense procurement regulations became a necessary precondition for doing business in the defense market, and a specific asset at the disposal of defense companies. However, for companies seeking entry, not knowing the rules of the game was a hard barrier to surmount. Finally, the administrative complexity and the special organization needed for dealing with complicated procedures increased defense company overheads, making them still less competitive, and creating another barrier to exit from the defense market.

4.2.5 Special Features of Defense Contracting

Defense procurement requires special contracting formats, and a voluminous literature developed to analyze the pros and cons of various contract forms that might be suitable for the defense business. Analysis was based on information economics and game theory models, and particularly on the analogy with the principal-agent framework. The point of departure is the limited information that "actors" in the procurement process possess, and the possible effects of incentives on their behavior (Rogerson, 1995; Sandler & Hartley, 1995: Chapter 5).

In any given contract, the government seeks to minimize procurement costs. In many cases, however, this cannot be accomplished through price competition, and to determine prices the government must rely on cost estimates generally based on partial and uncertain information. Eventually, ex-post costs may indicate that the return determined in advance was too high, yielding the supplier excessive profits, or that it was too low and caused him losses. Neither outcome is desirable for the government. The first case is incompatible with minimizing costs, while too low a return and losses may prevent completing orders satisfactorily. Moreover, the financial results of one program affect suppliers' willingness to participate in others, and losses may reduce the government's options in selecting suppliers in the future or, in broader terms they may erode the defense-industrial base.

Suppliers, on their part, seek to achieve maximum profit in every contract. Uncertainty prevents them too from predicting costs accurately, yet the situation is not entirely symmetrical. Companies generally have "private" information that the government lacks, so they are in a better position to predict costs, and in particular, they can evaluate the relationship between the efforts they invest in carrying out the contract and actual costs. Companies can take, then, steps to economize on costs, thereby increasing their anticipated profits. However, such steps may involve higher risk and induce a level of disutility in them (e.g. assigning the best engineers to defense programs may interfere with performance in other areas), so they will decide on their "effort level" by comparing these two opposite effects. The government, although aware of the level of efforts-costs relationship, and despite the various administrative methods applied, cannot monitor all company steps effectively, and thus cannot control their effort investment and assure cost minimization. Similarly, the suppliers too may eventually confront undesirable results: notwithstanding their efforts to save, the uncertainty inherent in defense activities may lead to ex-post cost above that estimated in earlier stages, and they might even encounter losses.

So much for contracts at predetermined, fixed prices. By contrast, if the government offers a cost-reimbursement contract, i.e. undertakes to pay back all actual costs to the suppliers, the latter become immune from the danger of encountering losses, but incentives to save are eliminated too. Suppliers cannot benefit from becoming more efficient and lowering costs, and as for the government – nothing assures that procurement costs will be minimized. Thus neither the fixed-price nor the cost-reimbursement contract makes it possible simultaneously to minimize costs to the government and to insure suppliers against losses. Hence special contracting methods had to be developed in the attempt to find an optimal balance between the two competing goals through a combination of risk-sharing and incentive fees. Risk is divided between government and suppliers by reimbursing actual costs, so that the supplier's risk is limited to the anticipated profits. With that, to encourage the supplier to lower costs the government entitles him to an agreed part of the difference between some pre-calculated target costs and actual costs. Since the government retains the remainder of the difference, both sides actually enjoy the savings.

The special character of contracting methods adds another dimension of uniqueness to defense businesses. First, negotiations determining the main parameters of the contract – target costs, profit rate, division of ex-post savings and the like – are long, wearisome and adversarial, demanding great skill on both sides. Second, decisions on contractual parameters and supervising actual costs later are both based on suppliers' "open books" that must obey defined rules, different from those ordinarily accepted in accountancy. Third, where contracts are based on target costs, the government might enjoy cost savings in yet another way: target costs in future contracts are based on actual costs in earlier, completed orders. Suppliers' behavior will be affected, then, by considerations relating to future profits as well, and thus profits lose some of their significance as incentives to economy. Fourth, in complex long-term contracts, renegotiations of central parameters are usually allowed. While both sides have that option, the suppliers usually invoke it. For this reason and because of asymmetrical information there is some concern lest incentive contracts degenerate into mere cost plus agreements, with no real incentives to save costs.

4.3 Structural Features of the Defense Industry

The special way of doing business in the defense market naturally affected the structural features of the industry. In general, the industry maintains a high degree of concentration, defense companies are averagely larger than those in other industrial sectors, and the internal structure of the companies and of the entire supply network differs from other sectors in several important aspects.

4.3.1 Company Size, Degree of Concentration, Takeovers and Mergers

A British study for the early 1980s showed that defense companies were on average 5.3 times larger by their fixed assets, and 4.4 times larger by number of employees, than other companies in the sample (Dunne, 1993: 98-99). Defense companies grew steadily because of the tendency to concentrate contracts in a few large companies. In the United States in 1990, the ten largest companies supplied 29 percent of the Department of Defense acquisitions, and their respective share grew to some 38 percent in 1995 (*Economist*, 13.1.1996). In Britain, according to 1987/88 data, 20 percent of Defense Ministry orders were placed with five suppliers, and about 40 percent with 15 suppliers (Martin, White & Hartley, 1996: 332).

In economic terms, the relatively large size of defense companies and the tendency to concentrate contracts in the hands of a few large suppliers can be explained by possible economies of scale in defense development and production. If present, these encourage internal growth, but also impel companies towards takeovers and mergers, thus reducing the number of actors, leading to a higher degree of concentration and to a less competitive environment. Hence for governments interested in more efficient defense production and at the same time wishing to maintain competitive procurement, economies of scale, if they exist, present a dilemma.

The early study of Peck and Scherer did not provide a definitive answer as to the possibly positive relationship between size and efficiency in defense companies. The authors pointed out that shifting emphasis from relatively simple products to complex systems, and the increasing share of R&D in defense production, alter the traditional approach to the size-efficiency relationship. Modern weapons systems contain sophisticated components based on a variety of technologies, and developing and producing them require various kinds of expertise. It is not always possible or desirable to maintain all expertise within a single large company, and it is almost certainly possible to find specialized small and medium-size companies capable of producing, and particularly of developing, some subsystems better and more efficiently (Peck & Scherer, 1962: 181-189). Later studies confirmed these intuitive arguments. It has been shown that large, multi-divisional, vertically integrated companies are not structurally adapted to the dynamics of innovation, while small and medium-size companies were more efficient, more aware of costs (Deutsch & Schopp, 1982: 2), and yielded more innovations per development dollar (Latham & Hooper, 1995: 13).

Defense companies are frequently too large relative to their actual output, and maintain excess production capacity. One possible explanation lies in the fact that most major weapons systems are produced in quantities smaller than originally planned. Under these circumstances, assuming that companies adjust their size to produce formally planned quantities with maximum efficiency, they inevitably build too large plants. Another explanation regards expansion to inefficient dimensions as deliberately manipulative, designed to influence procurement decisions by presenting lower marginal costs, and so presumably encourages larger orders (Rogerson, 1991: 235-236). Possibly too, excess capacity could result from introducing new production technologies – robots, computerized numerical control, etc. – without complementary adjustments, for example in the number of workers.

In companies too large for what they produce, i.e. their output is sub-optimal, average production costs decrease, but this in itself does not necessarily prove the presence of economies of scale. To examine empirically whether decreasing average costs reflect inefficient size or economies of scale, cross-sectional data for programs differing in the quantities produced were studied. The assumption was that with economies of scale the proportion of excess production capacity and the rate at which average costs decrease would be less in programs of larger quantities. Findings indicated, however, that the two variables were independent of quantities produced, thus validating the hypothesis that company size was indeed inefficient (Rogerson, 1991: 242-247).

Another aspect of the relatively large size of defense companies is the high degree of concentration typical of defense industry, manifested mainly in two ways. One is that a small number of large domestic producers supply most national defense ministries' acquisitions, and the second is that only a few potential producers are available in each product area. While the former is *inter alia* an outcome of the typical "purchase unit", i.e. the very large money value of individual contracts, the latter reflects the high degree of specialization required of defense producers. For either reason, over time the number of defense producers decreased in most industrialized countries. Some left the industry, some prime contractors became subsidiaries of larger companies through acquisitions, and others merged into huge conglomerates. A high degree of concentration limits competition and increases mutual dependence between the military customer and the defense companies.

From time to time the trend toward concentration was accompanied by a wave of takeovers, acquisitions and mergers. As a matter of fact, the rapid advance of military technology was an important motivating factor here too: to maintain competitiveness and profitability in an environment of rapid innovation, defense companies have to be at the cutting edge of technological developments, which in turn requires resources normally beyond the reach of individual companies. Besides, technological progress substantially changed the composition of activities in many areas. For instance, aerospace, the largest sector in the defense industry, produces fewer aircraft every year, airframes per se has become less important, at least in terms of their share in total costs, while avionics and electronic systems have grown in importance. Adapting to these changes, aircraft producers expanded into the electronics business, often through acquisitions or mergers. Governments, on their part, usually encouraged consolidation, seeing in the emerging industrial setup an assurance for the companies' survival, as well as a means toward greater efficiency that eventually would reduce costs.

Industrial economics distinguishes three types of mergers: horizontal, between competitors in the product market; vertical, between buyer and seller; and conglomerate mergers among unrelated companies. Horizontal mergers in defense industries were often considered irrational: there was no real consolidation, the plants were still treated as separate entities with their independent engineering, marketing and managerial staffs, and only legal frameworks were eliminated. Competition was thus reduced but not costs, so the overall balance of horizontal mergers in terms of economic efficiency could have been negative (Gansler, 1989: 255). As for conglomerate mergers in defense industry, it has been suggested that there were special incentives, unknown in other industrial mergers. In particular, to prepare for competitions over large, complex development contracts companies had to hoard experienced staff and other technical resources in idle periods, and conglomerate mergers provided opportunities to use them in unrelated areas in which there was demand (Beard, 1993: 36-37).

4.3.2 Structural Aspects of Companies and the Supply Network

Most of the large defense companies adopted organizational principles of vertical integration. The theory of industrial economics maintains that companies have chosen that format mainly for costs saving reasons. However, in the case of defense companies, where saving costs is not a high priority, this explanation is at best incomplete (Beard, 1993: 33). Vertical integration appealed to defense companies because of the complex nature of military systems, or more concretely, since prime contractors who are held responsible for systems as a whole, felt obliged to organize in a way that assured full control over the development and production of subsystems and components. They adopted vertical integration, then, out of conviction that effective coordination and integration of development and production activities – necessary conditions for succeeding in complex projects – could be more conveniently implemented within the boundaries of one company than across several companies. At the same time, priority of performance over costs, absence of competition, cost-reimbursement contracts, and the possibility that profits would be cut down as efficiency increases, all combined to dull sensitivities to the negative consequences of the cumbersome structure of vertical integration. With time, however, the drawbacks became more conspicuous. Especially, the size and the centralized structure made rapid response impossible, prolonging the laboratory-to-field time span, and organizational rigidities deterred creative young employees, making it difficult to bring in the varied types of technological expertise needed. Finally, even vertically integrated companies, although grown very large, could not afford to undertake the huge investments required for developing and producing the next generation of weapons systems alone.

Defense supplies thus came to rely on a network of companies made up of a few prime contractors and many subcontractors and suppliers of components. In fact, the second and third tiers account for much of the value of the contracts: for major weapons systems their share is typically some 40 to 60 percent (Gansler, 1989: 258). However, describing the supply network by noting the respective share of its various tiers does not sufficiently stress their interdependence (Dowdall & Braddon, 1995: 107). Some suppliers operate simultaneously within several supply chains headed by different prime contractors, sometimes as their immediate supplier and sometimes as suppliers of lower levels, and at the same time they are customers of other companies in the same or in other supply chains. The defense supply network, then, is indeed complex and ramified.

Relations between prime and subcontractors were based on competition and conflict. The former encouraged keen price competition among potential subcontractors, and whenever production was possible either within or outside the company, between internal units and outside suppliers as well. Often they were very strict about technical specifications as regard the work of subcontractors so as to assure themselves safety margins, and in time of lower demand they tried to pass down the burden by retaining a larger work share, reducing orders from subcontractors, narrowing their profit margins and demanding more convenient terms of payment. There were also significant differences in government attitude toward prime contractors and suppliers in the lower tiers. For instance, governments invested in production facilities and equipment for the first but not usually for the others. They also helped prime contractors in financial difficulties, but very rarely subcontractors and component manufacturers (Gansler, 1989: 260-261). In time, subcontracting in the defense sector became unattractive, and the number of producers dwindled. In the United States between 1967 and 1980 the number of suppliers for the aerospace industry alone decreased from 6,000 to 3,000, and in many areas only one or two remained (Gansler, 1989: 258-259). Hence when demands surged, critical bottlenecks appeared, supply times grew longer and prices rose.

Since the 1980s, civilian production has changed fundamentally, as new production technologies and processes and new corporate organizations replace old-established ones (Latham, 1995: 179-184). The new paradigm, sometimes called postfordism, rests on lean production and flexible specialization that simplify and expedite the changeover from one product to another, making it possible to produce smaller quantities with high quality and at reasonable cost. In addition, postfordism applies concurrent engineering techniques, which from initial product design stages factor in manufacturing and life-cycle maintenance considerations, in contrast with the serial approach of the past. New forms of production organization appeared too, giving preference to self-managing teams of multi-skilled workers over structured groups of workers specializing in one single operation. Other novel ideas include zero-defect quality control and total quality management, which focus on on-line prevention instead of post-production detection of defects, and inventory management according to just-in-time principles. No less important were the emerging new concepts of corporate organization; they replaced vertical integration by formats such as "solar complexes" and strategic alliances based on longer-term, cooperative relationships with suppliers. All these developments in the civilian economy were chiefly the result of increasing competition and of high price sensitivity. In the defense sector, however, such motives are much less influential, and hence the diffusion of the new paradigm has been delayed.

Furthermore, it has been argued that should the new paradigm expand into defense industry, it will differ from the one in the civilian sector (Latham, 1995: 184-189). The new industrial approach centers on flexibility and on willingness to take risks, while the defense industry – due to high uncertainties, rigid performance demands and the severe reactions to deviations, the military customer's priority for proven production processes and the like – is naturally reluctant to adopt such ideas. Besides, government supervision designed to prevent fraud and collusion requires, for instance, that components and subsystems be purchased through competitive tenders, thus preventing longer-term relations between prime and subcontractors, as well as limiting the options for total quality management and just-in-time supply. Likewise, the procedure of dual sourcing, i.e. the requirement for at least two potential suppliers of each subsystem or even of each important component, is also a hindrance. In essence, dual sourcing is intended to assure continuous supply and to create incentives to lower costs through competition, but it puts product developers at risk of not receiving production orders, and sometimes of having to transfer technical information to competitors. Fear of losing competitive advantage if required to do so, prevents producers from integrating new technologies in their designs in the first place. Thus defense and civilian sector production are likely to remain different even in the post-fordist era.

4.4 Defense Companies and the Labor and Capital Markets

The special nature of defense businesses and of defense companies also imposed special behavior patterns in the markets for production factors.

4.4.1 Manpower Policy

In some senses the defense industry of the Cold War period resembled a huge program of relief works. It created a host of employment opportunities under conditions divorced from the general labor market, though most of the output was never used. At the same time, given their demand for a technically skilled, experienced labor force, large defense companies were prepared to invest more in training their employees than were parallel civilian companies, giving defense industry a prominent role in vocational training. Defense industry was also an important catalyst in promoting higher studies in science and technology: the more technologically intensive defense production became, the more scientists and engineers it needed.

The relatively high labor demand maintained for several decades, combined with the special nature of defense business and the special structural features of defense companies, gave rise to some typical phenomena. Due to differences in business culture, these were more prominent in Britain and Western Europe than in the United States:-

i. With limited competition and cost-reimbursement contracts, company managements could more easily agree to institutionalization of organized labor and the increased power of labor unions. Thus special collective bargaining patterns developed, and internal labor markets were created, with their own rules of conduct and codes regarding employee status, promotion tracks, compensation systems, etc. Generally, promotion chances and working conditions were better than in other industries.

ii. The workforce size in defense companies gave political power to the workers and their representatives. Governments were sensitive to the political implications of defense companies' labor policy, and often opposed layoffs and other labor-related cost saving measures designed to accommodate to changing market conditions.

iii. When demand decreased, there were also nonpolitical reasons not to hasten to lay off workers, which led to labor hoarding. Cost-reimbursement contracts made it possible to hoard workers at public expense, and their peculiar logic, especially in purely cost plus contracts, induced companies to do just that as a means to increase profits. But there were, however, sounder economic reasons as well. During the Cold War years, companies usually considered demand decreases as temporary, hence had to be prepared to compete for new development and production contracts as they appeared. Since the greater part of the workforce consisted of scientists, engineers and highly skilled technicians, the cost of laying off and rehiring them, and particularly costs of training and of assembling compatible new work teams, could be extraordinarily high, thus providing an economic rationale for retaining redundant workers in idle interim periods.

In the 1990s, employment patterns changed, as did other significant features of the defense industries (Lovering, 1993: 126-130). First, the number of workers was sharply reduced. Second, the proportion of scientists, engineers, technologists and other skilled workers, always high relative to other manufacturing industries, became even higher, as the rates of decline in all other occupational categories were greater. Noticeably, entire layers of middle management were dismissed. Third, labor union influence decreased, and hiring, promoting and compensating workers came to depend far more on external than on internal factors. Fourth, changes were more prominent in large companies than among subcontractors and small suppliers who always depended less on defense markets, were less apt to hoard labor, and did not follow institutionalized work procedures as much. Consequently the ratio between direct and indirect employment in defense development and production declined. In the post-Cold War era, then, the defense labor market is becoming more like the civilian labor market.

4.4.2 Financial Requirements and Sources

The financial requirements of defense industries are unlike those of other industries in several respects. Obviously, the large scale and extended time span of defense projects are important factors in this area. Besides, while most industries require financial means to invest in physical assets (buildings and production equipment), in defense industry a large proportion goes into intangible assets (R&D, know-how acquisition, and preparing competitive proposals for developing new weapons systems).

Differing requirements led to the emergence of special sources, and as in many other areas, they relate mainly to the deep involvement of the customer, i.e. the government, in defense procurement. Because of the large amounts involved, and particularly due to the high specialization of assets, making it impossible to use them outside their original purpose, governments agreed to finance much of the investment in fixed assets. Production facilities, machines, testing equipment and the like are thus set up on the premises of defense producers and operated by them, even though they are state property, and in many cases users' fees are low, in effect a government subsidy to the companies involved. Sometimes entire plants may be state property, managed by private companies on a long-term contractual basis (GOCO – government owned, contractor operated). Likewise, there are also differences in sources of working capital. Because of the long time spans needed to develop and produce weapons systems, only rarely is full payment made after delivery. In fact, in most cases government pays producers much of the total return throughout the course of the contract by means of advances and progress payments upon meeting predetermined milestones, thereby reducing the need for current financing from banking or other sources.

Even so, defense companies still require significant financial resources. They have to invest in developing and preserving technological and technical capabilities on a continuous basis so as to be suitable candidates for future government contracts. They need large sums to prepare detailed responses to newly issued requests for proposals. Very large amounts may also be needed for bridge financing of work-in-progress between milestones, while delays in development and production, by no means rare in defense projects, postpone revenues further, possibly exacerbating the financial burden. The financial community has been very cautious as regards defense companies' financial requirements. Due to technological uncertainty and highly fluctuating demand, financing defense businesses was considered a high risk, and companies often had difficulties finding the financial backing they need.

Studies that compared the financial strength of defense and non-defense companies found no justification for these reservations of the capital and money markets. For example, an American study for the years 1978-1992 (Bowlin, 1995), compared changes in the Altman Z' score (a weighted aggregate of liquidity, profitability and debt load used to predict bankruptcy) for defense and non-defense companies. While the financial condition of both deteriorated, mainly after 1983, deterioration in defense companies for the entire period was found to be less severe than the general situation, and for each individual year the financial condition of the two types of companies was equal. The financial deterioration of defense companies could not, then, be associated with specific reasons, but rather stemmed from market conditions affecting other companies as well.

4.5 Profitability of Defense Businesses

The profitability obtained by defense companies reflects not only their technical and economic performance, but is also the outcome of government policies, often a peculiar mix of tight supervision, protection and support. Important too are the security conditions (war, accelerated buildup) and economic circumstances (degree of competition) in which defense companies operate (Sandler & Hartley, 1995: 198-199).

Studies that examined the profitability of American defense industries reached several main conclusions. One related to differences between profit rates as determined in contracts and actual profit earned, which was often substantially less. For example, during and after the Vietnam War, until the beginning of the 1980s, average contractual profit rates were about 6.5 percent, while in fact they were 4.9 percent during the war and about 4.7 afterwards (Gansler, 1989: 252). Another conclusion related to profit differences between defense and civilian sales. When profitability was measured as a percentage of sales, it was lower in defense sales. By contrast, when measured by the rate of return on equity there were no significant differences, and for large defense companies it was even higher. This latter finding represents the lower equity-to-sales ratio required in defense companies, particularly in the larger ones, due to government facilities and equipment placed at their disposal, as well as due to the progress payments and other financial arrangements common in defense work (Fox, 1974: 309-313). Profitability was also examined by computing the total market rate of return to investors owning stock in companies (i.e. annual dividends paid to shareholders plus stock price appreciation during the year, the sum divided by the value of the stock at the beginning of the year). A study for 1948-1968 (Stigler & Friedland, 1971), and another for 1970-1989 (Trevino & Higgs, 1992), found that the total market rate of return for defense companies exceeded the comparable return for shares in general, and there was a positive, statistically significant association between that rate and the share of defense sales in total sales of the companies concerned.

British studies show that the transition to a competitive procurement policy in the early 1980s reduced the relative profitability of the defense industry: it was above that of industry in general until 1985, and below it afterwards (Dunne, 1995: 99). It was found further that following that policy change, the profitability of companies selling a relatively larger proportion of their output to the Defense Ministry was lower and fluctuated more over time than did the profitability of other companies that sold the Ministry a smaller proportion of their output (Martin, White & Hartley, 1996: 333-335).

In the early 1990s, despite the steep demand decline, some defense companies increased their profitability considerably, and the improvement in business results showed in the market value of the companies, which appreciated by more than the average rate. For instance, in 1993 the market value of leading American, British and French defense companies grew by almost 50 percent, while for all companies listed in those countries the rate of market value appreciation was 7 to 23 percent (SIPRI, 1995: 458). For the seven leading American defense companies, the profit rate rose from 3.1 percent of sales in 1992 to 5.4 percent in 1995 (*Financial Times*, 9.1.1996). It looks, indeed, as if there was an inverse relationship between defense expenditure trend and that of the profitability and market value of defense companies. Between 1991 and 1995, authorizations in the American defense budget decreased by some 12 percent, while the weighted index of defense companies' stock prices rose by about 340 percent (The stock price index of Standard & Poor 500 rose about 90 percent in that period) (*Interavia*, April 1997). One explanation lies in the restructuring and other efficiency-enhancing steps taken by the companies concerned, which were especially far-reaching in view of shrinking markets. However, changes in the nature of defense business in the post-Cold War era appear to have had some effect too. It has been noted, for example, that emphasis shifted from platform production to sophisticated electronic systems ("black boxes"), which are as a rule more profitable. Another change was observed as regards investments in fixed assets and R&D. While during the Cold War profits obtained in previously completed programs were invested in new ones, with longer intervals between generations of weapons systems and thus with fewer new programs en route, investments declined. In the early 1990s, then, companies could accumulate cash surpluses that enabled them to reduce debts and improve their financial structure, leading eventually to increased profits (*Financial Times*, 9.1.1996; *Interavia*, April 1997).

4.6 Technological Intensity of Defense Products and Defense R&D Efficiency

Defense products are technology-intensive, and this quality – particularly notable since World War II – has had a wide-ranging influence on the industry, as mentioned in other contexts. Here our focus narrows to the efficiency of defense R&D and to the management of R&D programs.

The absolute preference given to performance, together with the transfer of risks from the companies to their customer, the government, through cost-reimbursement contracts, turned defense R&D into a "culture of innovation" that is remarkably different from civilian R&D (Molas-Gallart, 1995: 143-144). Firstly, in the civilian economy the impetus to innovate comes from market forces, and demand elasticity with respect to price is an important consideration. In the defense sphere innovations are dictated by military operational necessities, but often also by technological opportunities, i.e. technology itself becomes a driving force towards obtaining novel weapons systems, and whatever the trigger, there is no willingness to compromise on performance to reduce prices. Secondly, in the civilian sector time is critical in realizing competitive advantages, and therefore preference is given to gradual and incremental improvements that reduce the time needed for introducing new products into the market. In the defense sector, by contrast, most innovations come in great leaps, technological gaps between successive generations of weapons systems are significant, and the development period is substantially longer. Thirdly, civilian R&D is designed for the production of large quantities, and thus, besides improving product performance, stresses process technologies in order to reduce costs. Military products, on the other hand, are produced in small quantities, so that defense R&D focuses on product technologies, clearly prioritizing performance over costs.

Defense R&D involves both applied and basic research, and the tendency in allocating resources between the two generally favored applied research for two reasons. First, because of decision-makers' time-preference: they consider actual results in shorter time spans as much more important than capabilities that might mature into products only in the distant future. Second, because of free rider behavior: the multipurpose results of basic research and of developing technological infrastructure, or their positive externalities, prevent potential users from investing in them, though they hope to enjoy their fruits. But neglecting the technological base may limit the potential for innovation in the future, and, over time, detract from applied military developments. To give long-term interests adequate representation, some countries set up central organizations to promote basic research, especially in selected key areas not affiliated with missions of particular users. In the United States, for instance, at the end of the 1950s, DARPA (Defense Advanced Research Projects Agency) and in the 1980s SDIO (Strategic Defense Initiatives Office) were established for this purpose (Gansler, 1989: 237-238). Independent central organizations, however, have their problems; they may evoke alienation, and traditional organizations may reject the fruits of their efforts.

Defense R&D management may adopt an approach of technological breakthroughs, aiming at introducing radical innovations, or that of incremental technology changes, each approach leading to differences in performance, schedules and costs. The incremental approach option depends on the degree of modular flexibility of weapons systems, i.e. the possible interchangeability of major conforming subsystems of a final product for similar subsystems with relative ease, without risking unpredictable system behavior. However, the growing complexity of weapons systems, and particularly the growth in the numbers of interdependent components, has raised doubts as to whether modular flexibility is a feasible option in the defense area. It was suggested, then, that military technology must be dominated by radical changes, with no place for gradual development processes and incremental improvements. The truth apparently lies somewhere in the middle: modular flexibility is a matter of degree and varies from one defense system to another. Also, in certain defense systems it might be more important than in others. For example, it is especially important for defense systems subject to an endless action-reaction spiral: electronic warfare measures and electronic countermeasures, tank protection and anti-tank ammunition, etc. Active action-reaction processes unleash rapid technological change and fast obsolescence, and introducing continuous improvements by upgrading subsystems and components, enhances chances of contending successfully in the performance race. The importance of modular flexibility and of incremental R&D and improvements was demonstrated convincingly in the development of missiles against large weapons platforms such as aircraft, tanks and battleships. While the incremental improvements through module replacement and adaptation were not the only path taken, that approach was decisive in according missiles their central role in the modern military environment (Molas-Gallart, 1995).

Most defense R&D is carried out by business companies within contractual frameworks in which the government assumes full risks, reimbursing all actual costs plus a profit. From time to time the question arises of whether the government should leave risks with the companies by adopting a total package procurement policy. In this method, companies compete for both development and production, committing themselves to a single fixed-price contract for the two phases of the acquisition cycle. Moreover, sometimes the package may include spare parts and maintenance for several years thereafter. The assumption is that the contract as a whole would be profitable enough to induce companies to take the R&D risks upon themselves, even if that phase by itself might imply losses. The "package contract" approach was used in the United States in the 1960s, but rejected after bitter experience with enormous losses that forced the government to bail out the contractors (Sandler & Hartley, 1995: 152). If total package procurement is at all useful, it would be only for small programs or those with a relatively high degree of certainty.

As R&D costs surged and development time grew unreasonably long, managers of defense R&D had to find ways to halt or at least slow down the two trends. Both arose from the same causes inherent in the way R&D programs were carried out: e.g. the desire to thrust in as many technological innovations as possible, even ones that had not yet matured; or unrealistic budgeting due to financial constraints that failed to allow for unpredictable developments so normal in the course of R&D work. Cost increases lead to extended schedules, and vice versa. Moreover, it appeared that when the two combined there was a special effect. Empirically, it was found that increased costs arising from extended schedules reduced the quantities purchased at a rate twice as high as costs that increased for other reasons (Lichtenberg, 1995: 453). Arguably, when the program is completed after long delays, the weapons system might be considered technologically obsolete, not up-to-date operationally, or inferior to what potential rivals had. Efforts to cut costs and shorten schedules encouraged the development of varied, at times highly sophisticated methods for planning, managing and controlling R&D projects.

Another central planning and management problem is the optimal pattern of outlays in pursuing a given R&D program over time (Lichtenberg,

1995: 453-455). On one hand, because of inherent technological risks there appear to be good reasons for proceeding cautiously, and to increase outlays as technical problems are solved. In this way, by transferring expenditures from earlier to later stages, the discounted value of overall program costs is reduced. On the other hand, the level of expenditure may affect progress, and more spending in earlier phases – trying alternative approaches in parallel, for instance – may shorten the process and eventually reduce costs. Such a relationship, if present, is not known with any certainty either to customer or developer. Empirical studies indicated that the first pattern was usually preferred, and investment increased as the project approached completion. However, there was no evidence that increasing outlays as programs progressed was linked to greater uncertainty about difficulties of completion.

Identifying the common features of successful projects provided important insights as to managing R&D programs (Gansler, 1989: 232-233). In successfully completed projects the work was usually assigned to a small, experienced team operating in a decentralized fashion and isolated from the company's other activities. Close contact with potential users was also important, as were tests, including large-scale operational tests, in earlier stages of the program. Production of prototypes for both the whole system and for subsystems was useful in demonstrating their technical feasibility and operational effectiveness, as well as for the examination of their production and support costs. Successfully completed projects made maximum use of proven technologies, components and subsystems, including commercial components and specifications. They also stressed development of technological alternatives at the system level, and particularly at the level of subsystems and components, thereby reducing risks and making it possible to consider tradeoffs between performance and costs. As far as costs were concerned, determining targets for unit and life cycle costs as a primary requirement in designing systems was shown to be most important, as was a realistic budgetary approach. Finally, avoiding frequent changes in schedules, planning for a smooth transition from development to production, and freezing improvements at a certain point during the development process yet creating options to add them at a later stage, all helped to shorten development time.

4.7 The Rising Unit Cost of Defense Equipment

Researchers found that the continuous rise in unit costs, especially since World War II, was a robust phenomenon that withstood changes in political and military climate, in economic conditions and in techniques for selecting and procuring military equipment. Moreover, unit costs rose at an almost uniform annual rate across a wide range of apparently disparate items, as if they were obeying some central dictate. Implications are far-reaching: with constrained defense budgets, smaller quantities of advanced weapons can be acquired, which affects force structure, the nature of military missions and military planning in general. Norman Augustine, later chairman of Lockheed Martin, the world largest defense contractor for most of the 1990s, put it dramatically: calculating the rate at which unit costs were rising he anticipated that in 2054 the entire American defense budget could buy only one tactical fighter aircraft (Augustine, 1983: 55). Certainly, rising unit costs carried significant implications for defense development and production and for defense companies too, as noted already in various contexts.

The rising costs of weapons systems have several expressions that though interrelated, should be considered separately. The first noteworthy phenomenon is the more rapid pace at which prices of defense products rose as compared with general price indexes. For industrialized countries it was estimated that prices of defense products rose annually by 3-5 percent more than the general inflation rate (Fergusson, 1995: 35). The second is costs overrun, or the tendency for cost underestimation manifested in deviations of actual from planned costs that repeatedly occur in defense programs (see section 4.2.2 above). The third is the rising unit costs of successive generations of weapons systems. Studies made in the United States and in Britain indicated, after adjusting for general inflation, the following annual rates of cost escalation in this respect: destroyers and submarines – 9 percent, helicopters – 9.5 percent, frigates – 10.5 percent, guided missiles and fighter aircraft – 11 percent (Pugh, 1993: 181; Kirkpatrick, 1995: 265). Cost escalation rates, then, were almost uniform for all equipment, despite differences in technology or military purpose, and no less important – annual increase of some 10 percent implies doubling unit costs every 7.25 years.

At first, explanations for rising unit costs focused on the rapid pace of technological progress in the military sphere and on the increasing technical complexity of most defense systems. Adherence to this view exclusively, however, ignores contrary trends that characterized the price development of technology-intensive products in the civilian economy (Gansler, 1980: 17-18). Comparison with consumer electronic products or civilian computers is relevant here, since these systems make use basically of the same technology that contributes significantly to every modern weapons system. In both spheres technological progress was rapid and brought far-reaching improvements in performance, but in military products unit costs rose, while in civilian products they decreased substantially. Additional explanations, beyond technological advance, were required, then, for the steady escalation of unit costs of weapons systems, and they focused on the competitive nature of arms races between rival nations, and on the need for each side to avoid inferiority vis-à-vis the other. The security derived from military equipment depends not only on its performance, but also on its superiority (or inferiority) to what opponents have, so that every state initiates improvements in its own weapons systems when becoming aware of improvements made by its adversaries. Thus a vicious circle is set in motion, pushing costs upward, or creating a performance-costs race.

The explanation of military competition does not deny the role of technology in the rise of costs; in fact, technological advance is the process through which military competition is carried out. The increased threat triggering military competition stems from technological advances on one side, obliging the other to invest in technological advance to balance them. Military competition then, is like climbing a technological ladder where each step involves increased unit costs. But technology spreads across borders, and sooner or later the pace of technological progress becomes the same on both sides. Therefore despite rising unit costs, neither side attains superiority for more than a short time (Kirkpatrick, 1995: 276).

Under conditions of military competition, rising unit costs are an outcome of rational behavior that seeks maximum military effectiveness within a given budget. However, other interests may affect and reinforce the process. For example, defense companies wishing to obtain new orders must at all times maintain a full range of technological capabilities, implying, *inter alia*, preserving jobs. To cover the costs involved, they must plan for increases in unit costs at least matching the rate of increase in unit labor cost, and it can be shown – on the basis of a typical defense producers' cost structure – that this may lead to a unit cost rise of about 10 percent a year (Pugh, 1993: 190). The similarity of the rate of escalation in unit costs that appears to arise from industrial interests, and the rate of increase observed in unit costs of successive generations of weapons systems as noted above, may raise suspicions as to a possible link between rising unit costs and the machinations of the military-industrial complex. It is plausible, however, to argue to the contrary, namely that defense industry is a victim not a beneficiary of increased unit costs. Over time, rising unit costs forces the industry to downsize and reduce its workforce (Kirkpatrick, 1995: 277).

Prospects of halting or slowing down substantially the rise in unit costs do not seem promising. Improvements in production technologies or organization increase productivity, but even if it doubles, say from 2 to 4 percent annually, it will have little significance: unit cost will merely double every nine instead of every 7.25 years (Pugh, 1993: 182). For possible innovations in procurement policy the situation is similar. It was argued, for example, that stimulating competition in defense procurement might bring considerable cost savings. Proponents of this course pointed to estimates made in the United States showing that competition could save 20-25 percent in many defense programs (Fox, 1974: 25, Sandler & Hartley, 1995: 145), and to the savings brought about by the 1983 reform in Britain, where the stated goal was a 10 percent saving, while actual savings range from 10 to 70 percent (Sandler & Hartley, 1995: 197). However, this is merely a one-time effect; assuming, for example, a cost reduction of 20 percent, the result boils down to a 23-month respite in the present annual unit cost escalation of some 10 percent (Pugh, 1993: 182). Another direction recommended is cooperation among defense industries in different countries. Sharing development costs and reducing production costs by means of longer production series appear to offer a potential for economy. In fact, however, differences in detail in national demands, duplicated R&D, and distortions arising when work is assigned out of chauvinism rather than making use of comparative advantages, all combine to cancel out much of the saving, so that cooperation does not seem likely to reduce unit cost rise substantially either (Kirkpatrick, 1995: 279) (see also below, section 7.4). Neither did ordinary defense exports generally yield significant savings in unit costs (see chapter 10, section 4.3). Limited chances to slow down the costs race need not mean that efforts to make industry more efficient, or to expand competition in the defense market, are superfluous. At the same time, the difficulty in changing that robust phenomenon makes it imperative to factor its implications into both military planning and defense-industrial policy.

The various implications of rising unit costs in the defense industry were discussed previously. In addition to what was already noted, there is also the effect on the choice between domestic production and importing, which eventually influences the dimensions of the domestic defense-industrial base and course of its development. Cost escalation usually induces governments to consider off-the-shelf offers of foreign manufacturers. However, the influence is not uniform: it is likely to be more significant in equipment where fixed costs, including development costs, count for a larger part of total costs. In ships, for example, development and other fixed costs constitute a relatively small proportion, so that domestic development and production remain an attractive option for most countries, while for guided missiles the reverse is true, and purchase in another country may be preferable (Kirkpatrick, 1995: 282).

5. MACROECONOMIC ISSUES

As stated before, governments first divert economic resources from alternative uses and invest them in a domestic defense industry to obtain strategic-military advantages, but later realize its potential macroeconomic benefits as well. The macroeconomic analysis attempts to evaluate the cost and the benefit of investments in the defense-industrial base from the perspective of national economies.

5.1 The Strategic Criterion of Self-Sufficiency

Investing in a domestic defense-industrial base should be considered vis-à-vis the alternative of importing weapons and military equipment. The import option has its advantages: cost may be lower; delivery is probably faster since countries usually import defense products already in use in the producer countries; for the same reason, imports can provide equipment of proven qualities, or even battle-tested arms; and imports make for wider choice and thus allow for more flexible military planning. With all their potential advantages, however, importing armaments may lead the buyer into political and military dependence on the seller. Large investments in a defense-industrial base are therefore, among other things, the price states are ready to pay for self-sufficiency.

Self-sufficiency can be interpreted in various ways. For pariah and isolated countries, there are very restricted import options or none at all, so they must establish their own independent defense-industrial base. For other countries, import possibilities may be limited to certain weapons systems, or the arms obtainable abroad may not suit theater conditions or accepted military doctrines, so domestic capabilities are required to complement the arsenal with items unattainable by other means. Indeed, the growing world arms trade in recent decades, the increased number of suppliers and keener international competition broadened import possibilities (see chapter 10). Moreover, these recent developments induced producers in exporting countries to develop and offer foreign clients weapons systems in versions specially adapted to their needs. Yet while self-sufficiency possibly lost its primary, original meaning, another problem became more acute: increasingly open international arms markets make it harder to conceal acquisitions from abroad, and remove the advantage of surprise on the battlefield. With widespread, uncontrolled proliferation of state-of-the-art military technologies, domestic development and production became an irreplaceable source of unique solutions, unavailable in the world market, and thus a source of force multipliers. Furthermore, self-sufficiency is often linked to concerns over interrupted military supplies, and paradoxically interruptions become more likely when the supplies are more needed, i.e. when war breaks out. States are ready to sell arms under the assumption that they are assisting toward building a regional balance of power to deter wars, so may cut off supply to avoid active involvement in armed conflicts between third parties when they break out.

Absolute self-sufficiency is not on any country's agenda. Autarky in military affairs is neither a practical goal nor even desirable, and in every country some level of dependence has to be accepted. The self-sufficiency issue must therefore be examined in relative terms of adequacy. What size is adequate for the domestic defense-industrial base? What capabilities are essential, and in particular, are production capabilities enough or should there also be indigenous R&D capacities? Should such capabilities be maintained over a wide range of defense items, or just in some selected areas? Every level of self-sufficiency involves different opportunity costs, and the decision is ultimately a question of priorities and political choice.

Assuring any desired level of self-sufficiency involves two main issues: dependence on critical components that domestic industry does not manufacture, and ability to expand domestic defense production extensively and rapidly in an emergency. In the first case, it is often claimed that relying on critical foreign components deprives self-sufficiency apparently achieved through indigenous assembly and production capabilities of any real significance. In the short run, production and supply continuity may suffer, while in the longer term, giving up production could lead to loss of engineering capacities to design essential components, or to prevent such local capacity from being developed. In most cases, however, motives for importing components are economic, and as in every other context they are dictated by comparative advantages affecting international division of labor and trade. Defense systems producers in industrialized countries use components from foreign sources since production costs, and labor costs in particular, are lower. Likewise, companies in industrialized countries that once produced these components gave them up for more profitable civilian products. As a rule, the market test that determines whether domestic production is worthwhile factors in the uncertainty of relying on imports, and in case of a market failure, specific government intervention could cope with it. It is sometimes suggested that a distinction be made between economic and technological reasons for importing components, the latter implying the absence of indigenous technical capability to develop and produce them. But this ignores the crucial role of technical capabilities in determining the comparative advantages of economies, hence bringing no new real content to the self-sufficiency issue. As to possible rapid expansion in times of need, it became less important because in any case the complexity of military equipment demands long organization and production periods, while long conventional wars are considered unlikely. For example, a close look at the option of accelerating production of F-16 fighting aircraft indicated that it would take more than three years to attain any real increase (Gansler, 1989: 265). In such conditions, ability to expand production does not seem to have any significant effect on immediate military capabilities. Nonetheless, the post-Cold War period may rouse some new interest in the issue. In the new international circumstances, there may be longer warning periods regarding threats and conflicts on one hand, while on the other hand automation and novel production methods change the dimensions of possible industrial response. To illustrate: on the eve of the Gulf War in 1991, the United States expanded production of Patriot missiles, and indeed all those fired during the war were produced after the Iraqi invasion of Kuwait; in Britain, industry assisted in adapting military equipment to desert conditions, rapidly produced protective means for armored vehicles, and accelerated development and production of new missiles (Taylor, 1993: 117-118).

5.2 Economic Criteria: Positive and Negative Macroeconomic Effects

Some macroeconomic effects attributed to the defense-industrial base are positive, others negative. On the positive side, the defense sector was said to be a promoter of industrial development and of economic growth in the economy as a whole. It created employment opportunities, especially for scientists and engineers, and provided a source of technological know-how and innovations that in time spilled over into the civilian economy. Critics noted, however, that because of the esoteric nature of military technology, its spin-offs were few and of little significance, while spin-ins from civilian to defense industries increased in recent years. They claim further that the defense-industrial base, protected from competition as it is, grew into an inefficient industrial structure. Its companies have no real incentives to economize on costs, and developed bad habits of wasting resources that eventually extend to other sectors. The growth of defense industry, then, adversely affected industry as a whole, and in fact undermined the competitiveness of national economies. Moreover, critics continue, the defense-industrial base crowded out physical and human resources from civilian activities, thus possibly slowing down growth and technological progress there. Since influences are both positive and negative and seem to operate concurrently, the ultimate test appears to be quantitative-empirical. In the best case, however, evidence provided by empirical studies (Dunne, 1995: 424; Sandler & Hartley, 1995:196-199) is mixed, and there is no clear conclusion as to whether economic benefits are greater than the direct and opportunity costs of developing and maintaining the defense-industrial base over time. Moreover, if positive effects on employment, growth and balance of payments do exist, they have to be compared with results obtainable by alternative policy means.

Technological spin-offs attracted special attention in the macroeconomic context (as well as in specific industries and individual companies), and their implications were studied from several perspectives. The impact on labor productivity was examined for six countries – Canada, France, Germany, Japan, Holland and the United States – for the period from the mid 1950s to the end of the 1970s (Deutsch & Schopp, 1982: 10-15). The estimated equations presented productivity growth as dependent on changes in real wages and in government funded R&D for military and civilian purposes, and the resultant calculations pointed to an inverse relationship between labor productivity and the ratio of military to civilian government R&D expenditures. Given the relatively larger size of military R&D projects, the results above were interpreted as possibly indicating diseconomies of scale in R&D programs. Another examination as to the effect of R&D programs on productivity was made in the United States, distinguishing between R&D projects privately financed by companies, most of them in civilian fields, and those financed by the government, most of which were defense-related. It was found that company financed R&D had a positive, statistically significant effect on productivity, while government financed programs did not (Lichtenberg, 1995: 447). With a fixed level of non-defense R&D, it appears, then, that government financed R&D does not affect productivity. The findings do not rule out, however, the possibility that defense R&D affect the scope of civilian R&D itself either positively or negatively, and thus indirectly influence productivity. Responding to a survey, executives of industrial companies indeed indicated that changes in government financial support for R&D led to changes in the same direction in company funded R&D programs, yet econometric analyses of this issue produced ambiguous results (Lichtenberg, 1995: 448-450).

Defense R&D expenditure may influence civilian R&D in two opposite directions. On one hand, technological knowledge and innovations accrued in the defense R&D may spill over into the civilian economy, thus making R&D resources employed there more efficient and laying the foundations for novel civilian industries (spillovers effect). On the other hand, the demand for specialized production factors, in particular demand for scientists and engineers arising from defense R&D raises their price, thereby diverting them from civilian to defense activities (crowding-out effect). The crowding-out effect is likely to be particularly evident under full employment, due to factor supply inelasticity in the short term, but it presumably becomes weaker in the longer term, the more so in competitive markets where factor supply responds more flexibly to relative income changes.

The crowding-out effect hypothesis was examined empirically in Britain (Buck. Hartley & Hooper, 1993: 167-171). Wages in the defense sector were compared to those in the civilian sectors, assuming that higher wages in the former draws manpower away from civilian R&D. Data for 1989 showed that R&D employees in defense industry were paid less than their peers in the civilian industry in both absolute and relative terms. The defense sector seems not necessarily to have attracted suitable R&D workers with higher wages, while civilian industry had to compensate them for non-monetary benefits like status, employment security, more advanced technology and the greater degree of freedom in the scientific research typical of defense work.

Since ideas and technological knowledge move from one sector to another with people, the spillover effect is supposedly connected too with manpower mobility from the defense to the civilian sector. Surveys carried out in the United States in 1982, and then again in 1986 (Lerner, 1992: 233-240), found among respondents who were active in defense industrial work in 1982 a relatively extensive mobility out of those occupations during the four years under consideration, despite the accelerated military buildup in that period. The least mobility was from defense-dominated industries, e.g. ordnance and aerospace, while most of it came from industries with lower defense to total sales ratios, or where defense projects were occasional. It was also found that those who moved from defense to civilian activities were significantly younger, more poorly paid and less experienced, and as to occupation – engineers switched less frequently than workers in other occupations. Hence most workers who switched from the defense sector to the civilian economy had less job-related knowledge or human capital, so there was little, if any, potential for spillover effect. Among respondents active in defense in 1986 extensive mobility was found in both directions in the four-year interval, and there too the young, the poorly paid and the less experienced were more mobile than others. Based on the data assembled, an attempt was made to explain mobility by applying statistical methods, and some findings seemed to support the presence of a positive spillover from the defense to the civilian sector. Younger workers were more likely to switch in either direction, but once age was controlled for, those with higher pay and more company-specific experience were more likely to make transitions. Likewise, engineers were more likely and production personnel less likely to switch from defense to civilian activities. Finally, in dual-use industries where defense and civilian activities were more or less balanced, transition of workers from defense to civilian employment was fairly common, less so transfers in the reverse direction.

The benefits of military technology to specific industries were also examined (Chakrabarti, Glisman & Horn, 1992). Reference was made to economic performance measures (growth rate, competitiveness, profitability) and to indicators of technological achievements (employees' scientific publications or numbers of registered patents), assuming that the quantity of defense contracts would create differences between industries. Specifically, the influence of defense contracts on the degree of competitiveness of industries was examined by considering their respective export-to-import ratios, and on innovation or creation of knowledge by comparing patent output per employee. The findings did not point to any systematic relationship. For example, the transportation equipment industry, with the largest proportion of defense contracts, had a patent productivity well below the average for all manufacturing industries, and was near the average regarding international competitiveness, while chemicals, with the lowest proportion of defense contracts, performed very well regarding both knowledge production and international competitiveness. Defense contracts, then, did not seem to contribute to a higher level of economic or technological performance. A similar conclusion was reached by comparing technological achievements in 1970-1975 and in 1980-1985: during the accelerated military buildup of the early 1980s the performance in terms of registered patents did not improve, and the number of scientific publications even decreased. Since analysis at the industry level might ignore substantial disparities among companies, the examination was carried out for individual companies as well. Here too, however, the share of defense contracts in sales could not clearly explain economic and technological performance.

5.3 Macroeconomic Consequences of Defense Production in Developing Countries

Statistical research on defense production in Third World countries began in the 1980s. Attempts were made to explain the scope of defense production in terms of motives (involvement in armed conflicts, previous experience with arms embargos) and capabilities (demographic and economic size, level of economic development and technological advance). For instance, a study of defense production in 72 developing countries (Rosh, 1990), found that GNP and industrial development level (measured by the degree of processing of exports) had a significant influence on domestic defense production, while the defense burden and defense imports did not. Previous experience with embargos also had a positive influence on the value of defense production, and states that experienced embargos tended from the start, at every level of the other variables, to produce more defense equipment than others. The lack of influence of defense imports, as if domestic production was not meant to reduce dependence on foreign acquisitions, is a surprising result. It possibly recognizes that even with an established domestic defense industry, problematic dependency on imported critical components and advanced major systems remains.

In establishing domestic defense industry, developing countries had three principle economic objectives: to lower prices for defense products, to enhance economic and industrial development, and to reduce defense acquisition-related foreign currency expenditures.

5.3.1 Promoting Industrialization

The potential influence of domestic defense industry on the industrialization of developing countries remains controversial. On one hand, it may encourage industrialization by creating demands for industrial inputs that stimulate the establishment of new manufacturing enterprises, introduction of advanced production processes and technologies, and the training of engineering and technical personnel. On the other hand, establishing defense industry may crowd out physical and human resources from civilian sectors, thereby retarding development and industrialization in the rest of the economy. Furthermore, in countries devoting great resources to arms production, defense industry requirements often dominated the entire industrial structure, and when new industries did emerge, they were capital-intensive and highly specialized, contributed little to employment, and produced very few of the normal, day-to-day necessities of the population. Indeed, even proponents of indigenous defense industry rarely deny that industrialization by direct investment in the civilian economy could be more advantageous. The right question, then, is not whether defense industry has to be established to promote industrialization, but rather about the influence on industrialization when prevailing political and military circumstances in any case require the establishment of a domestic defense-industrial base.

To produce arms, industrialization must definitely be given priority, but possibly the causal sequence is the other way around, namely first industrialized growth occurs, creating the industrial potential for arms production that then stimulates the expansion of domestic defense production. One study (Wulf, 1982: 11-15), defined industrial arms production potential of an economy by the output and number of employees in relevant industries, and by the number of scientists, engineers and technicians involved in R&D in the economy as a whole. It found a high degree of correspondence between the actual rating of countries on the defense production scale and their arms production potential, and exceptions – i.e. countries with a well developed industrial base and appropriate labor force that did not count among the major arms producers – could be explained by specific constraints (lack of foreign currency, licensing difficulties and the like) or by explicit policy decisions. Another study (Brauer, 1991) divided developing countries into four groups according to defense production level: non-arms producers, occasional producers of arms, those that consistently produce arms although on a limited scale and of little sophistication, and those that produce arms at an advanced level, both in quantity and quality. Further on, three measures for industrial arms production potential were defined: the relative size of the relevant industrial base (measured by the value share of arms production-relevant industries in the total value of the output of manufacturing industries), industrial diversification (measured by the number of active sub-categories of arms production-relevant industries) and creation of human capital (measured by rate of school enrollment). The three measures above were then compared for the various groups of countries. From the mid 1970s to the mid 1980s, there was a clear correspondence between arms production potential and actual arms production. For example, the relative weight of the arms production-relevant industries in total industrial output was less than 20 percent in the first group, and between 45 and 48 percent in the countries that produced advanced arms in large quantities; in the group of countries that only occasionally produced arms, there were an average of 55 active sub-categories of arms production-relevant industries, while in the fourth group there were 127; and the higher countries were on the scale of arms production, the higher the rate of school enrollment at all learning levels, with greater gaps at the higher ones. Thus apparent arms production potential was indeed translated to actual production. This study also found some countries whose capabilities were greater than actual production, and others that expanded production beyond their capabilities. Industrial capability and human capital are therefore necessary but not sufficient conditions for defense production, and there must be political and military incentives for countries to invest in developing a defense-industrial base.

Industrial strategy was also an important factor in determining the scope and diversification of defense production in developing countries. General industrialization in developing countries followed either an import substitution strategy or an export-oriented strategy. Accordingly, in the defense-industrial sphere it might be relevant to distinguish between a strategy of self-sufficiency and one of internationally integrated arms production (Wulf, 1982: 16-24). An obvious parallel between import substitution industrialization and domestic arms production was evident, for example, in Argentina, in Peron's time, and in India since the 1960s. By contrast, in Brazil, Israel, Taiwan, Korea and Singapore, for example, the explicit desire to export was a major driving force in developing the defense industry. Success in achieving self-sufficiency may be measured by the size of the remaining import component, which generally did not fall below 30 percent (Wulf, 1982: 17; Ball, 1988: 376). Moreover, often the import residue did not reflect the full extent of technological and industrial dependence on foreign sources: as the domestic component grew, so did dependence on foreign experts. Countries adopting the export-oriented strategy advanced by stages: first they produced components and subsystems under license, and later moved on to producing and exporting domestically developed weapons systems, using imported subsystems. In these countries importing technologies and components is a normal and integral part of the process, and reducing the import component is not a goal in itself.

An important test of the influence of defense production on industrial progress lies in the ties that develop between defense and civilian sectors. In some instances, defense demand accelerated the establishment of new industries that otherwise would not have emerged, and initially producing for military purposes, they later expanded into civilian production. In general, in defense production the added value of prime contractors is relatively small, that is, they buy a large proportion of inputs from other sectors, and the resulting backward linkage effects are likely to encourage industrialization in the economy as a whole. However, there were many cases in which no close interdependency developed (Wulf, 1982: 25-26). Since defense products are usually manufactured in small quantities, the derived demand for inputs may be too small to justify investment in special equipment and plants. Besides, the advanced technology of modern defense production is of little use in manufacturing products typical of consumer demand in developing countries.

Defense development and production programs have at times been justified in that they increase employment. It is doubtful, however, whether defense production is an efficient way to create employment in developing countries. Capital and technology-intensive processes, common in defense industry, employ relatively few workers and at a high average cost per employee. On the other hand, demand for unskilled or semi-skilled labor – constituting most of the labor supply in developing countries – is very limited, since most defense products never reach the stages of standardization, maturity and mass production. Similarly, defense industry is seen as promoting the professional skills of its employees, but here too, such a positive influence, if present, affects a relatively small number of workers only. According to data from the early 1980s, in most arms producing Third World countries the share of those employed in defense industry was no more than two percent of all industrial employees (Ball, 1988: 367). Neither could defense industry workers always use their training in the civilian sector, and if they did, it might deflect industrial development towards laborsaving techniques and the production of goods the economy does not need. Nonetheless, there were cases where defense industry helped stem the brain drain of scientific and engineering talent to the industrialized countries.

In evaluating the impact on industrialization, technology transfer from the defense industry to the civilian economy should also be considered. In fact, most developing countries could not create their own technological capabilities and R&D infrastructure, and their defense production was based largely on imported knowledge. In the early 1980s, only ten developing countries produced any type of weapons systems of their own design, even then domestic development required help from foreign staff, and the more sophisticated components were acquired abroad (Wulf, 1982: 6-7). In these circumstances, technological spin-offs were limited, and any hopes of technological spillovers to the civilian sector were barely realized.

5.3.2 Economizing on Foreign Currency Expenditures

Foreign currency savings attainable through domestic defense production were not great either. Firstly, in the short and middle term, large foreign currency expenditures were needed to finance imports of machinery and production equipment, royalty payments for know-how and licenses, salaries for foreign personnel, and the like, so direct and indirect import components still remained high. Based mainly on India's experience, it was estimated that developing countries could not expect significant foreign currency savings for the first 20 years of domestic defense production (Ball, 1988: 378). Secondly, attempts to increase the domestic added value may involve sharp price increases, so that even at a later stage, when domestic production is based on local development, the import component is unlikely to be less than 30 percent. The value of imported inputs may remain high also because of growing ambition to produce more sophisticated equipment based on expensive components and materials that can only be purchased abroad. Thirdly, domestic production never meets all needs, and equipment not domestically manufactured is usually the more complex and more advanced, and therefore the more expensive. Attempts to reduce the foreign currency defense burden were made not only through expanding domestic production and substituting direct defense imports, but also by means of exports and offset agreements (see chapter 10, sections 4.2 and 5.3).

6. GOVERNMENT POLICY TOWARDS THE DEFENSE INDUSTRY

The need for governments to adopt an active, explicit policy towards the defense industry, beyond the usual customer involvement, derives not only from the high priority accorded to national security, but also from the degree of imperfection of the defense market. It is impossible to rely only on the invisible hand of the market, and to assume that business decisions based on the narrow self-interest of companies, though possibly correct from their own point of view, will of themselves develop an industrial structure capable of meeting essential, long-term national interests in any adequate fashion. However, to no less an extent there is also doubt as to whether governments can successfully develop and implement an appropriate industrial policy in general, and an effective defense-industrial policy in particular. Government bureaucracies lack entrepreneurial qualities, and may fail to recognize technologies worthy of support, or to spot "champions" that should be encouraged. Moreover, governments are subject to political pressures and respond to them; they seek support and popularity among their voters, the industrial policy may become captive to private interests, inevitably leading to a governmental failure no less harmful than the market failure it is trying to overcome. Opponents of government intervention in the defense sector note cynically that defense acquisitions are intended primarily to protect the citizens, not the defense companies.

The American point of departure was that government intervention in the operation of the defense-industrial base should be limited, and effected as far as possible through market mechanisms. By contrast, in Western Europe there were no reservations as to government involvement in the defense industry, and the central dilemma, especially in recent decades, was whether to preserve the national character of defense industries, or to open local markets to foreign competition and encourage international cooperation. Policy issues also changed with demand trends: questions of industrial mobilization, typical of the Cold War era, gave way to problems of preserving development and production capabilities in a shrinking market, and of converting excess production capacity from defense to civilian needs in the years thereafter. Some regard defense industry as a special sector, eligible for government assistance in the restructuring forced on it by the declining market; others think that government should treat defense like any other declining industry.

6.1 Defense Procurement Policy

In establishing their defense procurement policy, governments face the dilemma of choosing between competition and regulation. This dilemma is rooted in the non-market nature of the defense business environment, and in the contradictory goals pursued simultaneously: superior military performance, economic efficiency and cost economies, reasonable risk-sharing between customer and supplier, control over profits generated with taxpayers' money, and sometimes additional, more specific aims. For both reasons, it is impossible to rely exclusively on competitive mechanisms, and regulation is necessary. Regulation mechanisms, however, have their own severe deficiencies, and particularly, companies learn quickly how to live with them, or more bluntly – how to circumvent the constraints they impose. For example, company managers prevented from increasing profits will pursue other objectives, like enlarging personnel under their direct control (presumably increasing thereby their internal organizational power and prestige). In a broader sense, regulation may repress initiative and creativity, and turn managerial focus from outputs to inputs. Regulation also involves substantial costs; according to one estimate it may add some 5-10 percent to the cost of programs (Sandler & Hartley, 1995: 141). Often a vicious circle develops: public criticism leads the authorities to enforce regulation, regulation induces various negative consequences, sometimes leading to outright failures, thus reinforcing criticism and pressure to tighten regulation controls even further. Finally, regulation initially designed to remedy failures arising from the non-market nature of the defense business environment, are likely to exacerbate the lack of competition, since they make companies reluctant to do business with the government.

The impracticability of relying on competition only on one hand, and the drawbacks of regulation on the other hand, led to a search for a middle way, and beside regulation, which remained a major element, attempts were made to incorporate some competitive mechanisms as well in the defense procurement process. From the point of view of economic theory, taking corrective measures to partially improve market competitiveness relies on the "second best" approach. On that principle, reinforcing the influence of market forces does not require many producers, and two or three strong competitors in each segment of the market could be sufficient. Nor is there any need for more than one to engage in actual production, as competitive influence comes from the very existence of an alternative, or merely from the threat of competition (Gansler, 1989: 160-161).

In the United States in the early 1970s, the then Deputy Secretary of Defense David Packard (better known as the founder and owner of the Hewlett-Packard computer company), introduced competition among prime contractors into the development phase. Prior to that, total package procurement was the rule, awarding in advance one contract for both development and production to a single company, implying in effect a monopolistic situation in the transition from the development to the production phase. Under the new initiative, the government financed the development and production of prototypes by competing companies in parallel, thereby increasing the number of candidates before finally choosing the producer. A study that examined the consequences of the change showed that competition effected considerable savings in costs, and moreover brought about improvements in performance and schedules (Sandler & Hartley, 1995: 153). To the same end, in the 1980s the principle of dual sourcing was introduced. Indeed, the policy of maintaining alternative sources is not new; it was applied on a large scale during World War II and the Korean War, yet at that time the principal motive was to assure steady supplies, not competition and saving costs. This time, however, the principle was adopted with the clear intent of making competition a central strategy at every stage of the procurement cycle, with exceptions allowed only in extraordinary cases (Sandler & Hartley, 1995: 146).

In the 1990s the competition issue was addressed again in face of an unprecedented wave of mergers that flooded the American defense sector. At first the government encouraged the mergers, since it regarded them as a way to reduce the excess production capacity created in the defense industry with the end of the Cold War. Later, concerns grew that competition in the defense market would dwindle, and the policy became more selective. The full force of the dilemma emerged in the debate over the mega merger of Lockheed Martin with Northrop Grumman, announced in July 1997 and cancelled after a harsh struggle in July 1998. The Defense Department objected to the merger, basing its arguments mainly on the possibly negative consequences for competition. In particular, officials stressed that when one and the same company maintains both the capabilities of integrating a complex system and of producing its subsystems, other producers in the second and third tiers will not be given a fair chance to compete for the supply of subsystems, and with time they will disappear. Then, in the absence of competition, the prime contractor will have no incentive to keep costs low and to invest in innovation (*Jane's Defence Weekly*, 25.3.1998).

Introducing competition into defense procurement as well was the central theme of British reforms in the 1980s. Thatcher's Conservative government coming to power in 1979, adopted a large-scale privatization policy, and most of the government-owned defense companies were transferred to private hands. Later, fundamental changes in the traditional relationship between government and defense industry were introduced: *inter alia*, the share of competitive biddings rose, risks were transferred to prime contractors by adopting fixed-price contracts more frequently, overall life-cycle costs of weapons systems became a major consideration in the selection and contracting processes, and progress payments according to milestones became a more common practice. Consistent with the new approach, military requirements were spelled out in broad terms, instead of the detailed specifications used previously, leaving the suppliers to select the best response, while clearly presenting marginal costs for additional performance capabilities to the customer. The results were impressive: in 1982-1983, 36 percent of all acquisitions (by money value of contracts) were competitive and 16 percent on a cost-reimbursement basis, while for 1989-1990 the proportions changed to 49 and 4 percent respectively. Moreover, the changes implemented were reflected in price savings varying from 10 to 70 percent (Hooper, 1995: 63-64). However, although the cost saving advantages of the new policy were clear, some of its other implications were controversial. The number of companies active in the defense sector more than doubled between the beginning and the end of the 1980s, but growth concentrated in small companies, and competition increased principally in the market segment of middle-sized and small contracts. By contrast, the number of companies in the upper group of suppliers (selling to the Defense Ministry for more than 250 million pounds annually) decreased from nine to five (Hooper, 1995: 65), concentration grew, and competitive procurement options in the market segment of big contracts diminished. In those conditions, concerns developed lest the government be unable to withstand the pressures, and eventually would revert to the old relationships. In the 1990s, a National Audit Office report pointed out schedule overrun and other significant deviations in carrying out defense contracts, and the competitive process was held to be the chief factor in the apparent deterioration. It was claimed that the competitive policy distorts incentives and impels suppliers to excessive risks ultimately expressed in the results of programs (Bell, 2000). Indeed, the Labour Government that took power in 1997 brought in a new policy of "smart procurement". While competition remains central, the new policy is based on closer cooperation between government and industry.

Competition does not always pay. For instance, cost economies through competition were shown to depend on industry capacity utilization, and only when that is below the normal 80 percent can significant price savings reasonably be expected (Sandler & Hartley, 1995: 154). Moreover, since competition requires several strong participants in each sector, it implies more investments in R&D and production facilities, and high costs of technology transfers between companies. It may also increase costs by preventing economies of scale and learning effects. According to one estimate, maintaining alternative sources may add up to 40 percent to procurement costs (Sandler & Hartley, 1995: 154). A balance between competition and regulation, then, is no simple matter and clearly the search for new formats will continue.

6.2 The National Defense-Industrial Base in an Era of Globalization

During the Cold War, governments regarded the defense-industrial base as a strategic asset and preserved its national character, sometimes too zealously, especially in Western Europe. The Treaty of Rome in 1957 enabled members of the European Community to limit access of foreign producers to domestic defense markets, and protectionism emerged, isolating producers in the domestic defense market from external competition, and preventing foreign takeovers or mergers between local and foreign companies. As a result, excess production capacity and relative inefficiency developed. But governments accepted such "distortions" understandingly, and saw them a worthwhile price for independent capabilities to develop and manufacture arms. There were, however, other views, advocating the establishment of a free trade zone or a common arms market on a European or even a transatlantic basis, regarding them as a natural extension of the military treaty, and pointing out the considerable benefits that would result from opening markets. An early estimate in the mid 1970s evaluated the annual extra costs due to the isolationist national procurement policy within the NATO framework at some 11 billion dollars in 1975 prices (Hartley, 1995: 462). Another estimate from the early 1980s, asserted that equipment standardization within the NATO armed forces and opening the markets would make economies of scale possible, effecting a reduction of unit costs of major weapons systems by at least 20 percent not to mention savings by eliminating R&D duplications and the like (Anderton, 1995: 540).

As technological progress dictated inevitably more expensive weapons systems, it became increasingly clear that the demand in the fragmented Western European defense market could not support large, complex defense projects. At the same time, American competition increased, and in a number of competitions in the 1970s and 1980s the European industry suffered embarrassing defeats. In the late 1980s and early 1990s additional influences emerged, some previously noted: the steep decrease in defense budgets exacerbated difficulties in financing expensive projects; advances toward the European Single Market highlighted the deviance of national "guardianship" over defense industry; rapid technological progress in civilian areas undermined the image of military projects as promoters of technology-intensive domestic industry; and the increasing importance of multinational military missions and international military coalitions implied more emphasis on equipment interoperability. In the mid 1990s, in view of the defense mega-mergers in the United States, European governments and companies grew increasingly concerned over possible American hegemony in the defense industry worldwide, and pressures to consolidate European defense industry mounted. There appears to have been general agreement that European companies could compete with the American giants, or alternatively – cooperate with them, only if they attained similar size and efficiency. Opinions were divided, however, as to the desirable format and the steps needed to achieve it.

Cooperation on a limited scale between companies from different Western European countries, usually within specific projects, began as early as the 1960s. The initiative generally came from governments who regarded collaboration as way of sharing the financial burden, expanding markets and saving costs through larger production runs. Nor did they ignore indirect economic advantages: participating in an international project could lead to inflows of advanced technologies that otherwise could be acquired only through longer periods of substantial investments, and when used could favorably affect the domestic industry as a whole: the larger output as compared with national projects might reduce unit costs, thus making export sales potentially more competitive; chances to export to third parties could also improve due to the marketing power of large international industrial groups, an important factor in the competition with the American defense industry. All these considerations were weighed against disadvantages of dependence and uncertain supply. In fact, however, the potential advantages were realized only partially, if at all. Joint programs were based on the principle of "*juste retour*", namely on division of labor among participating countries according to their share in financing, or to the number of systems they intended to acquire, ignoring their comparative advantages and specialization. Moreover, political and local considerations often influenced governments in deciding which domestic companies were to take part, and they did not always choose the ones most suitable from the technological and economic standpoint. In many cases cost economies were offset by increased development costs because of different, special equipment characteristics demanded by each country (see section 7.4 below). Additionally, potential economies from longer production runs were not realized because governments insisted that final assembly take place in each country separately. In the 1990s it was understood that the traditional framework of joint programs had had its day, and should give way to mergers creating truly multinational companies; independent companies that would engage with subcontractors in different countries on the basis of cost considerations and technical abilities, free from concerns of "*juste retour*".

However, the road to cross-border mergers was paved with dilemmas. Firstly, a widespread view in the late 1980s held that national consolidation should precede such mergers, and in some countries – German and Britain and to a lesser extent Italy, though not France – there was in fact a wave of mergers among the leading defense companies. On the other hand, two arguments were advanced against the nationally-based consolidation concept: one was that "national champions" would try to conquer the entire European market, and it would be hard to merge them or even induce them to cooperate; another was that national consolidation would prevent the future European defense industry from being organized along lines of industrial specialization – aircraft manufacturers separately, missile producers separately and the like – but instead would create multi-disciplinary companies, lacking critical mass and inferior in the face of American competitors. Secondly, the companies to be merged were unequal as to profits and financial strength, but nevertheless, with government backing, they demanded equal control, so it was difficult to reach agreements. Thirdly, in 1998 when the most promising basis for consolidation in the European aerospace industry seemed to be to convert the consortium of Airbus Industries into a single company, British and German partners conditioned such a step on privatizing the French partner, which was government-owned. They were concerned that government involvement in ownership of the merged company might lead to conflict between commercial and national-economic interests (e.g. preserving jobs). Fourthly, views differed as to the desirable size for the merged companies, both from the standpoint of competing in the world market and of efficient management. Some felt that American mergers need not be the models, since European defense industry was used anyway to smaller production series. Others argued that the European base is not large enough, and strategic alliances or even outright mergers between American and European companies were preferable. Naturally there were no few political conditions and obstacles to progress either. Towards the end of the 1990s, the establishment of cross-border joint ventures in specific fields accelerated somewhat. Some of them could be considered more comprehensive than previous cases, since they included whole sectors of parent company activity, not merely individual product categories or predefined projects. Only in 1999, however, was there a real leap forward in internationalizing the European defense industry, when the first large transnational company in aerospace and defense – European Aeronautic Defense and Space (EADS) – was established under joint French-German-Spanish ownership.

The debate over the future format of the European defense market dealt with the demand side as well, and in particular with opening national markets to foreign suppliers and with various alternatives for joint procurement. In 1986 the Independent European Policy Group (IEPG) proposed a plan (in the Verdling Report named for the working group's chairman, a former defense minister in the Netherlands) with a view to abandoning the prevailing protectionist policy. The cautiously worked out plan was based on "balanced access" to markets, that is, the voluntary implementation on a reciprocal basis of competitive tenders open to producers from all European NATO states, but without necessarily guarantying equal chance to foreign offers. In fact, the plan preserved the distribution of defense-industrial capabilities of countries, without redistributing work on the basis of competitive advantages, and thus was no longer practical in 1990s conditions (Walker & Willett, 1993: 146-151). Beyond that there was no real progress in applying the approach that governments should not show preference to national suppliers and allow open competition, buying at the best price (Brzoska, 1998: 84). At the same time, some progress was made on the joint procurement track. The most significant step was the establishment of a joint procurement framework for France, Germany, Italy and Britain – *Organisation Conjoint de Cooperation en Matière d'Armament (OCCAR)* – at the end of 1996. Several joint procurement programs were managed within this framework, and some considered it a precursor for a wider procurement agency to include most Western European countries (SIPRI, 1997: 245; *Interavia*, January/February 1997).

Questions about the future of European defense industry are closely interrelated with its bi-lateral relations to the American defense industry, and here too the European approach is rather ambivalent. Asymmetry has been present for years, and the large defense import surplus from the United States has troubled Europe all along. Europeans who oppose a free transatlantic defense market are convinced that the big beneficiaries would be the American manufacturers, and that only a few European defense companies could survive the competition likely to evolve. The result may well be unbearable dependence of European countries on the United States. In a less pessimistic scenario, in the resultant division of labor the United States might focus on the development and production of state-of-the-art equipment, while European companies will be left to deal with the lower technology products (Hartley, 1995: 466), or as another argument states, since there are doubts as to whether European companies have any chances of being equal partners, they might be doomed to remain subcontractors and component suppliers, thereby over time losing their ability to produce integrated military systems (*Economist*, 13.1.1996). In addition, opponents of the transatlantic approach point out the differences in management culture and work habits that may interfere with true partnership (*Interavia,* April 1996). Indeed, even some who do not oppose transatlantic relations per se maintain that European defense industry should first merge horizontally so that it will be able to compete with or become a real partner of the American giants. By contrast, supporters of the open door between European and American defense industries point out that since substantial technological gaps exist in essential areas, the absence of transatlantic ties will cut off European defense companies' access to advanced military technology, unavoidably leading to technological and operational inferiority of the armed forces as well (Walker & Willett, 1993: 154). Besides, they see no contradiction between closer industrial interrelations with the United States and industrial consolidation within Europe, but rather the contrary – complementary steps (*Interavia,* August/September 1996). Finally, they stress the importance of transatlantic industrial relations for the future of NATO.

Transatlantic industrial relations are an important policy issue for the United States too. Western Europe is a significant target market for the American defense industry, and American government and industry both followed closely the developments on the other side of the Atlantic, trying to evaluate their implications. One opinion held that the consolidation process, by strengthening the European companies, creates better partners, and thus provides a broader basis for possible collaboration. However, the more prevalent opinion was that the industrial giants emerging in Europe would be strong enough to do without American partners, and therefore consolidation would shut the doors and create a "Fortress Europe". Those concerned with "Fortress Europe" assert that loss of European markets would harm not only the companies involved, but the Department of Defense as well, since exports help reduce prices of defense products. By contrast, should transatlantic partnerships arise, they would increase competition in the American market, highly concentrated following the mergers of the 1990s, and help prevent price increases (*Economist*, 22.11.1997). A further argument expressed concerns lest the lack of real links between American and European companies bring about keener competition in the world arms market, harming sales to third parties too (*Defense News*, 5.7.1999). Those suspicious of "Fortress Europe" agree that there is no place for a "Fortress America" either, indeed demanding that the American government make its arms exports and technology transfers regime more flexible in order to facilitate partnerships with European companies. Admittedly, the profit and loss balance of the American defense industry as regards open transatlantic ties is by no means a foregone conclusion. The American defense market will remain the largest in the world, and it is far from clear whether the benefits derived from additional business in relatively smaller foreign markets may compensate for the loss of orders to foreign producers, who would have eventually greater access to the home market.

The relations between the American and Western European defense industries, and especially the requirement for relaxing controls over defense exports and technological transfers, is part of a far larger issue occupying the attention of American policy-makers since the late 1990s. It has been recognized that the shrinking domestic market makes defense exports more important, and that doing business internationally often requires American companies to enter into partnerships with counterparts in target markets, or in general to cooperate more closely with foreign companies. This, however, raises the danger that sensitive technologies will leak out, and policy-makers are confronted with the dilemma of how to enable American defense companies to participate in the globalization process that increasingly dominates the industry, and still protect secret technological know-how. The issue turns to be even more complicated as defense production becomes more dependent on civilian or dual-purpose technologies, with no clear distinction between civilian and military technologies, while civilian high tech industries are already borderless. Several possible courses of action have been proposed; some focus on adjusting the Munitions Control List to include only the most essential capabilities, while others distinguish between target countries, recommending a preferential attitude towards countries with export controls resembling those of the United States. In any case, globalization is a challenge for governmental defense-industrial policy, particularly in the United States, with the world's largest and most advanced defense industry, and is likely to change fundamentally the clearly national nature of the defense-industrial base, as it existed throughout the Cold War.

6.3 Preserving the Defense-Industrial Base in a Shrinking Market

When the Cold War ended and defense demands decreased, world defense industry confronted substantial excess production capacity, and had to go through considerable downsizing. For governments it meant coping with new issues, mainly in two areas: preserving an adequate defense-industrial base in a shrinking market; and assisting defense companies in downsizing while deriving as much "peace dividend" as possible from the demilitarization that became possible in the new era.

American policy of the early 1990s, at the time of Secretary of Defense Richard Cheney, was based on the perception that notwithstanding the decline in demands, basic critical industrial capabilities should be preserved, since reconstructing them is an expensive, long process. Hence priority was given to continuity in planning, development, building and testing of defense system prototypes, and to creating conditions that would allow rapid transition to full-scale production, if and when concrete needs arose (Taylor, 1993: 118-119; *Business Week*, 16.3.1992). To implement the new procurement strategy diminishing budgets were distributed over numerous development programs, but covering the early phases only, up to the production of experimental prototypes. The new policy relied on economic and industrial considerations, but to no less an extent on the strategic assessment that the accelerated military buildup of the 1980s provided the American armed forces with superior capabilities, sufficient to cope with short term security needs, and that under prevailing circumstances, the United States has the time to examine new technologies and systems thoroughly before making procurement decisions (Fergusson, 1995: 30). But not everyone agreed with the new policy, and particularly not with its underlying industrial assumptions (Taylor, 1993: 118-119). It was argued that completing the development work and building experimental models was not enough, and that production technologies and systems integration had to be tested in at least limited serial production. There were also doubts as to the feasibility of producing from prototypes that had been on the shelf for a long time. Finally, there were concerns that the new policy was not providing sufficient incentives to companies to keep at their defense activity. The argument was that with so few new programs, and without new production orders in sight, losing in competition for a development contract could motivate companies to exit the defense market, and once converted to civilian business would no longer be able to turn back to defense production. The defense-industrial base might, then, further shrink, reaching undesirable, risky small dimensions (Fergusson, 1995: 31).

President Clinton's administration changed the policy: defense budgets were further decreased, but instead of canceling major programs, the new policy sought to obtain a lower expenditure level by economizing on a long list of items. In line with this approach, Secretaries of Defense Les Aspin and William Perry favored reducing the number of prime contractors in the various segments of the defense industry through mergers and acquisitions. They believed that this would reduce excess production capacity and increase efficiency, thus making it possible to reduce prices of a large variety of military equipment. The message was conveyed at a meeting with defense industry executives in 1993, a gathering subsequently called "The Last Supper", since it ushered in the end of independent existence for many companies. To encourage consolidation, and particularly in ways that eliminate excess capacity, the government was ready to reimburse companies for expenses connected with closing plants, integrating infrastructures, dislocating equipment and personnel, layoffs and the like (SIPRI, 1997: 242-243; *Economist*, 13.1.1996). As already noted, when defense industry later became highly concentrated, opposition to mergers and acquisitions grew for fear that competition would be severely restricted.

More problems arose at the end of the 1990s. The "indigestion" suffered in some of the larger mergers, together with huge debts accumulated in the course of acquisitions, led to deteriorating financial and technical performance among leading companies. At the same time, civilian high tech industries were flourishing, and the defense industry faced difficulties in recruiting talented young workers: a third of the technical personnel in defense companies was approaching retirement age, and only a negligible minority were 30 years old or less (*Defense News*, 29.5.2000). The government, then, was urged to assess the financial strength and the technological and managerial capabilities of the defense industrial giants created by the consolidation process. The task was assigned to the Defense Science Board (DSB), a consulting body made up of senior executives of defense companies, whose report included a long list of recommendations (*Defense News*, 3.7.2000; *Jane's' Defence Weekly*, 6.12.2000). Central to the recommendations were several steps intended to increase profitability and improve the cash flow of companies: to allow higher profit rates for exceptional performance in cost, quality and schedules; to adjust profit principles so as to increase returns on cost economies; to leave a larger share of efficiency-related savings with the companies; to increase progress payments and so forth. Raising the wage ceiling for managers and other key workers recognized by the government was proposed in connection with the personnel problems. The report also related to necessary reforms in defense export control, to needs for increased government R&D budgets and for encouraging company-financed R&D, for expanding the use of multi-year contracts, for creating new business opportunities by outsourcing industrial work, and citing as well the need for removing barriers that prevent the use of commercial technologies in defense products and the development of closer relations between defense and commercial companies.

Furthermore, when the new picture of American defense industries clarified, the wave of mergers appeared to have been chiefly financial in nature, and hopes for significantly reducing production capacity were not realized. A report from the Department of Defense early in 2001 pointed to idle facilities and to underutilized plants, whose maintenance adds hundreds of millions of dollars annually to the costs of budgeted projects (*Defense News*, 29.1.2001). A sharply critical article pointed out excess production capacity not only in relation to the declining demand, but also in view of the vast stocks of equipment at the disposal of the armed forces (Gholz & Sapolsky, 1999: 9-12). The economic explanation was that production capacity did not decrease more because the Department of Defense reserved most of the fruits of companies' increased efficiency for itself; indeed, the prevailing rules routed into the Federal Treasury 80 percent of such savings (*Defense News*, 29.1.2001). There was, however, a political-economic explanation too, attributing the course of events to a fundamental change in the interrelations among the various parts of the military-industrial complex (Gholz & Sapolsky, 1999: 16-22). By that argument, after the Cold War, when threats diminished, politicians' readiness to weigh purely military considerations diminished as well, especially when these conflicted with employment and other economic interests of their voting districts. Governments thus became more receptive to defense lobbies, and pork barrel politics came to dominate the defense procurement more than ever in the past. In these conditions, when prospects of defense contracts attainable through aggressive lobbying improved, it was reasonable for companies to avoid closures of idle facilities. Proposals made to encourage companies to rationalize at factory level included, for example, postponing the sharing of efficiency-related savings with the government for several years, accumulating savings in a special fund that would make investments for the mutual benefit of the Department of Defense and the companies, etc. (*Defense News*, 29.1.2001). However, some considered such proposals as inadequate, and recommended direct subsidies to companies that would write off sunk costs and direct compensation for workers left without jobs. For them, the indirect subsidy by means of superfluous contracts was by far more expensive (Gholz & Sapolsky, 1999: 43-45).

In the discussions about preserving an adequate defense-industrial base, closer integration between defense and civilian industries played a central role. The idea was that a larger portion of goods and services needed by the military should come from companies that were not unique in the previously accepted sense. In this way defense companies could operate in civilian markets and civilian companies could do business with defense ministries (Gansler, 1989: 273-282; 1993; 1995). It was not only a matter of acquiring civilian items wherever these met military requirements, but of giving up esoteric military specifications, of designing military systems in advance around commercially available parts and components, and of producing military and civilian products in the same plant, with the same equipment and the same engineering and production staff.

Technological developments encourage this train of thought (Gansler, 1995: 91). In the defense sector, technologies for producing modern weapons systems changed, replacing traditional heavy industry technologies with growing use of electronics, information technologies and new materials. In the civilian economy, at the same time, product technologies that have to withstand intense commercial competition offer products of high quality performance, adaptable to changing environmental conditions, and, most importantly, make it possible to produce at lower costs. Production technologies changed too. Companies have moved from mass to flexible production, and can reach high efficiency in small-scale production of multiple products. Finally, advanced technologies frequently develop in the civilian sector, and only later spill over into military uses. Due to all these developments, producing civilian and military products in the same plant became more feasible than in the past, and considering possible savings of expensive investments in special facilities and machinery, it is a more worthwhile option economically as well.

Supporters of integration between defense and civilian industry present it as an effective way to cope with both the increasing costs of weapons systems and the need of preserving an adequate defense-industrial base while eliminating excess production capacity. Defense-civilian integration would spare the military customer the need to bear the cost of a special industrial base alone, making it possible for him to gain leverage from civilian sector investments. Relying on proven civilian technologies will shorten development processes, another significant saving. At the same time, using equipment and components that meet the quality and price tests of commercial markets, a more extensive use of automation made possible by combining military and civilian demand-quantities, and the more competitive environment in which defense procurement would take place, all may help cut current costs. As for the defense-industrial base, it might benefit first of all from expanding and diversifying the circle of suppliers. Besides, defense companies would have the option to divert hitherto burdensome excess capacity to civilian markets, maintaining the engineering and production infrastructure that when needed can be redirected to defense production. There will be no reason, then, to preserve idle capacity for the sole purpose of emergency expansion. Another advantage lies in recruiting manpower: integrating defense and civilian activities would make the defense companies a more attractive workplace, thus assuring, though indirectly, the military customer too of access to superior technical abilities.

To encourage such integration, unique military specifications should be kept to a minimum and the design approach changed. No less important is the removal of administrative barriers. Indeed, there has to be an overall reform in defense procurement, adjusting the way of doing business to accepted commercial practices as much as possible. The traditional approach to defense R&D should change as well. Dual-purpose applications and improving production processes should have priority in allocating resources, and R&D efforts at the lower levels of the supply chain should be encouraged. When the sole application is military, it is logical to finance R&D work at the prime contractors' level, since they are in any case the main users of the fruits of R&D efforts. In the case of dual-use applications, however, greater benefit may come from financing the R&D of component manufacturers and suppliers of materials.

The possibility of integrating defense and civilian activities in the same companies arouses no small measure of doubt (Fergusson, 1995: 32-33). Despite the impressive technological progress in civilian areas, and the increasing relevance of civilian technologies to military uses, the need for uniquely military technologies has not disappeared, and neither has the need for government finance and for the deep involvement of the military customer in R&D processes. Similarly, while civilian specifications can meet many military requirements, it is still impossible to do entirely without uniquely military specifications. Nor should one assume unreservedly that new production processes make it easy to move from defense to civilian production and vice versa, especially at the higher levels of systems integration. There are subjective barriers and hindrances too. For one thing, major defense companies oppose changes that undermine their unique status. For years they invested great resources in developing core competences that give them a comparative advantage in doing business with defense ministries, and naturally they are not interested in taking part in reforms that may erode these assets. They also see their potential in the civilian market as limited, have doubts as to their ability to compete successfully with established civilian companies, and prefer to invest in political lobbying for orders from their traditional customers in a familiar business environment. Secondly, among policy makers and procurement authorities too, attitudes are ambivalent. They agree that expanding the integration between defense and civilian industrial activities, and relying more on commercial solutions, carry a promise that defense systems will keep up with rapid technological progress, and at affordable costs, but are nonetheless concerned that blurring boundaries may erode government control over defense production and over the defense-industrial base. In particular, there are growing concerns that defense-civilian integration may accelerate the leakage of sensitive information, and thus undermine national security interests. As noted, globalization makes it more difficult to monitor and control the flow of military capabilities based on civilian technology. It remains an open question, then, whether defense will become a business like any other, and whether limiting the special features of the defense-industrial base might be an effective way to preserve it in an adequate manner.

In Western Europe, unlike the United States, when declining defense demands raised concerns over the possible loss of essential technological and industrial capabilities, governments did not hesitate to support defense industries, and especially government-owned defense companies, directly and for a long period. The French government extended substantial financial aid to Giat Industries, a government-owned major producer of land systems and ordnance (*Jane's Defence Weekly*, 3.5.2000) and to SNPE, a government-owned group that manufactures chemicals and explosives (*Jane's Defence Weekly*, 17.1.2001). The Spanish government did the same for Santa Barbara, producer of armored vehicles, and for the Bazan shipyards, both owned by a government holding company (*Jane's Defence Weekly*, 4.11.1998; 31.5.2000). In Italy, the government invested considerable sums in the equity of Finmeccanica, a government-owned conglomerate that incorporate much of the country's defense industry. The British government supported Royal Ordnance, an ammunition producer. The company was privatized in the mid 1980s and went through a comprehensive restructuring (the company workforce was reduced from some 20,000 to 4,000 by the late 1990s). Nevertheless, in 1998 its survival came under threat due to the post-Cold War general decline of domestic demand, and especially because of losing several large orders to foreign suppliers. To preserve domestic strategic capabilities in ammunition production, the British government was prepared then to place a ten-year order with the company, paying a premium above world prices (*Financial Times,* 26.8.1998; *Jane's Defence Weekly,* 28.6.2000).

In the early 1990s, governments still owned much of Western European defense industry. The two exceptions were Germany, with a long tradition of private ownership, and Britain, where most defense companies were privatized during the 1980s. By contrast, French, Italian and Spanish governments retained ownership of the main defense companies, and only in the later 1990s adopted a privatization policy. While direct motives were not the same and there were differences in methods, it appears that in all countries, preserving an adequate defense-industrial base was one of the major considerations in adopting privatization.

6.4 Converting Defense Resources to Civilian Uses

Conversion is defined as a process of transferring resources rendered unnecessary in defense activity due to the decline in defense demand, to activities in other economic sectors. A distinction is made between direct (or microeconomic) conversion in which efforts are made to utilize existing defense plants and their workforce to produce civilian rather than military goods and services, and indirect (or macroeconomic) conversion, where resources released from the defense sector are reallocated to the civilian economy (Hartley, 1995: 484-487; Willett, 1990). In both cases there is an attempt to find alternative employment for unutilized resources, and at the same time to preserve as much essential infrastructure and technology as possible for rapid response to future defense needs. In addition, while reallocating released resources attempts are also made to contribute maximally to economic growth and productivity. These goals are often contradictory, so it is no surprise that successful conversions are not frequent. Another barrier is the relatively narrow specialization of resources used in defense production. Finally, conversion often encounters political and social opposition since in the interim period, until resources released from defense are reemployed, unemployment may arise.

Conversion as defined above may be regarded as a specific case of two broader phenomena, and examined as one of their subsidiary issues. One is the general transfer of resources from defense to other economic sectors, not necessarily in the context of defense demand decline (e.g. the context of technology spillover, as discussed earlier, see section 5.2). The second, even broader phenomenon is adaptation to change in any, not necessarily defense companies. Some companies manage to adapt and survive, while some fail and cease operation. The questions then arise as to what leads to success, and whether failures have some common denominator.

Conversion is not a new phenomenon. It took place on a large scale after both World Wars, but then – after a long period of shortages – there was a great wave of demand for consumer and investment goods and for services that made it easier to absorb excess production capacity, even providing opportunities to apply technologies developed for military needs in the civilian sphere. Also, for many companies, mainly in the United States and Britain, conversion meant going back to their familiar pre-war occupations and civilian market business culture. However, not every conversion succeeded. For a well-known example, in the early 1950s the British company Vickers failed to convert part of its tank production to tractor manufacture (Sandler & Hartley, 1995: 286). During the long Cold War period there was, as a contemporary American expression put it, a "permanent war economy" (Melman, 1974), and save for short spells following the Korean and Vietnam wars, conversions were of limited scope, and only a few succeeded. After the Vietnam War, for instance, major American defense contractors tried to join the response to social and environmental needs at the center of public attention – controlling ecological pollution, building mass transportation systems, exploiting alternative energy sources and developing and manufacturing artificial organs. However, defense R&D and the production tradition and the business culture of the companies involved produced complex, expensive and usually inapplicable solutions (Willett, 1990: 476; Sandler & Hartley, 1995: 286-287).

Direct or microeconomic conversion is discussed below as one of the business strategies of defense companies (see section 7.3). Indirect conversion, by contrast, is a public policy issue. Defense economics deals with indirect conversion within the framework of disarmament processes, and considers their economic dimension in terms of investment (UNDIR, 1993; Intrilgator, 1998): in the beginning there are conversion costs and later, after resources were diverted to alternative uses, there is a return in the form of a "peace dividend". As for costs, there are direct costs of adapting production factors and of developing capabilities to produce non-military goods and services, as well as direct and alternative costs associated with factor unemployment in the transition period. The peace dividend, on the other side of the equation, is not usually a one-time event, and may spread out over a long period. The economic worth of indirect conversion thus depends on the ratio of future benefits to initial costs, and is affected by the assumed discount rate and by the length of transition and returns periods. But a positive ratio is not sufficient; according to welfare economics theory, the change is socially desirable if income is transferred from potential "beneficiaries" to potential "losers".

Inasmuch as indirect conversion involves reallocating economic resources and income redistribution and may generate positive social returns, governments seem called upon to take an active role in such processes. Not everyone agrees, however. One argument relates to defense sector uniqueness. Those in favor of active government involvement in conversion point to the unique strategic role of the defense sector, and assert that governments therefore have a particular responsibility to determine the scope and composition of the defense-industrial base in normal times, and even more so when demand decline threatens its continued existence. Opponents are less impressed with the uniqueness of defense industry, claiming that it is no different from other economic sectors that at one time or another have to deal with demand changes, and thus should be treated by governments like any other declining industry (Hartley, 1995: 487). The opposing views stem too from different evaluations of the nature of defense industry. Those who oppose special treatment argue that government support will usually be wasted on an obsolete industrial paradigm, while those in favor say it is anachronistic to think of defense industry in those terms (Lovering, 1993: 133-137). Proponents also justify government involvement that compensates for losses incurred in disarmament processes by pointing to the need to lower barriers that otherwise might hinder the development of a peaceful international environment. Naturally there is a temptation to retain active production lines, and when domestic demand falls off, pressure to export military equipment arises. Exports are usually to countries embroiled in regional disputes, and as a result arms races accelerate and instability increases worldwide. Finally, there is even a more far-reaching view, claiming that governments have to take an active part in macroeconomic conversion, since it provides an opportunity for fundamental and comprehensive changes in the economic and social order (Willett, 1990: 475-481). This approach, inspired by the Marxist analysis of historical materialism, regards conversion as far more than a provider of alternative employment and a preserver of technological capabilities, seeing it indeed as a lever in creating a new industrial system, where social and environmental values maintains a central position.

Opinions also differ on the practical level as to policy measures: is it enough to create general economic conditions that assist resources released from the defense sector to be absorbed elsewhere, or is there a need for a particular intervention focused on vulnerable regions, specific occupations or companies? On the face of it, given the possibility of market failures, government intervention seems justified in specific, individual cases. But analyses of public choice show that severe failures may accompany government intervention too. Furthermore, just as government intervention in building the defense-industrial base created a powerful lobby that influenced public expenditure priorities, so government support to converting individual companies might unintentionally create institutionalized power groups that would gain substantial economic rents. Hence one opinion maintains that assistance on a case-by-case basis should be avoided, and that governments should make do with implementing the right macroeconomic policy, i.e. a policy that assures stable demand level in the economy as a whole, and with changing the composition of public expenditures by diverting resources directly to preferred civilian goals. Opponents of that view hold that it exaggerates the efficacy of market mechanisms and of government expenditure. In particular, it relies on an untested assumption that the economic system is sufficiently flexible to cope with the required adjustment in the defense sector, and ignores the special institutional and behavioral features of defense industry. They also stress that transferring production factors between sectors may involve frictional unemployment, not only demand-related unemployment. As they see it, proper macroeconomic conditions are not enough, and microeconomic measures should also be applied: retraining personnel, improving information on employment opportunities, granting investment incentives, extending support for development of dual-purpose technologies, and the like.

In former socialist states conversion is taking place simultaneously with the transition from a planned to a market economy. One such undertaking alone is difficult enough, and to cope with both at the same time, the usual means cannot be adequate. Problems are particularly severe in Russia and the Ukraine, with their once huge military sectors, and some opinions hold that for those countries to meet their high conversion costs they must have financial and technological aid from Western states and international financial institutions (Intriligator, 1998: 42). The Russian authorities encourage the diversion of resources from defense to civilian production, and in the mid 1990s it was reported that in about half the companies under governmental control, civilian sales had reached 75 percent of the total, and only in 16 percent of all companies were military products still the major part of all production (SIPRI, 1995: 472). With that, the choice of civilian products was based on purely technological criteria, not relating to the needs of the economy or market opportunities, so commercial successes were relatively limited (Hartley, 1995: 486).

7. INDUSTRIAL AND BUSINESS STRATEGIES OF DEFENSE COMPANIES

Government defense-industrial policy was a decisive factor in determining the development of defense industry, but no less important were the strategies adopted by the companies themselves. Companies responded to changes in market conditions, in military, political, economic and technological circumstances, and to changing government attitudes. Their responses divided along two principle axes: the first includes specialization in defense business at one end, and making exit from the defense spheres at the other; the second axis distinguishes between national and international strategies. Companies that decided to specialize in defense business downsized by adjusting their manpower (laying off workers) and physical capital (closing production and other facilities), and withdrew into their traditional core business by divesting non-defense activities or, on the contrary, tried to establish their position and enlarge market share through acquisitions and mergers with defense companies in complementary fields. Sometimes they strengthened their grip on the defense market by joint ventures and other forms of cooperation, which amounted to less than creating new legal entities. Companies that decided to exit the defense business sold off their defense activities, and entered new civilian areas through direct conversion or diversification. On the whole, diversification was a more common strategy than conversion. National strategies focused on domestic markets, and included, *inter alia*, joint ventures, and mergers and acquisitions within the country. International strategies expanded business beyond the domestic market, creating interrelations and varied collaborative structures with companies in other countries. Many factors influenced the choice of a specific strategy, and later, its successful implementation. Indeed, choosing a strategy was sometimes accompanied by long deliberations and involved zigzags, with companies unexpectedly changing positions, and though consistently stating their intent to continue in the defense business, were then sold overnight and swallowed up within other companies. Apparently, in an unstable environment the ability to change strategies and adjust rapidly to new conditions is essential for survival (Augustine, 1997).

7.1 The Strategy of Downsizing

A strategy of downsizing is designed to achieve higher rates of utilization and increase factor productivity. Defense companies adopted a strategy of downsizing in previous transition periods too – after World War II and the Korean and Vietnam Wars – but in the 1990s, downsizing measures common in the past were inadequate. Defense companies seeking survival through downsizing had to cut back manpower and close production facilities and offices at rates far greater than the rate of decrease in their income (Lundquist, 1992: 80-81). The explanation lies in the different initial financial situation: in previous instances, when downsizing after a war, the companies were financially strong, whereas the end of the Cold War found most of them after years of low profitability, without financial reserves and overburdened with debts. Company size had to be adapted to a lower demand level, and at the same time enable the accumulation of surpluses so as to strengthen the corporate financial structure.

Successful downsizing should combine laying off employees with closing and integrating production facilities. GM Hughes, for instance, reduced staff to less than half in a few years, and following the acquisition of General Dynamics' missiles division, integrated the work previously performed in five production plants in one site. As a result, production capacity utilization rose from 30 to 85 percent, making it possible to reduce missile production costs by 30 percent (SIPRI, 1995: 458). In another case, the operational plan devised to accommodate the Lockheed-Martin Marietta merger called for reducing manpower by 12,000 (out of 170,000), closing 12 production facilities and laboratories and 26 offices, with estimated savings of 1.8 billion dollars annually (*Electronic News*, 3.7.1995).

The strategy of downsizing can be applied in another way. Most defense companies are conglomerates of individual projects that derive little if any benefit from sister programs or headquarters (Lundquist, 1992: 81). When examined on a stand-alone basis, some are not worth keeping, while others might be valuable if separated from unprofitable divisions and released from paying for corporate overhead and the inefficiencies of parallel, wasteful programs. Some of these seemingly mini-companies could become the nucleus of the company itself, but could also be suitable candidates for selling off, for partial equity spin-offs, for creating strategic partnerships or for mergers with other companies operating in the same sphere.

Applying downsizing strategy has its difficulties. Restructuring often involves several years of instability and uncertainty, and requires a clear commitment of all management ranks to persist in change despite the obstacles. Clear-eyed, almost cynical evaluation is needed as to the true economic value of each business unit in the company, which frequently means abandoning accounting procedures adapted to government procurement. Moreover, as in any change process, there may be unexpected or undesirable developments. In particular, there is the risk that not only unneeded workers will leave, but some of the company-specific human capital as well will be lost. Finally, substantial manpower reduction may raise strenuous objections from labor unions, community and politicians, leading to compromises that make for deviations from the original plans. In most such cases, if basic conditions have not changed, there is no choice but to complete the plan at a later date, after additional losses and financial erosion, with greater cutbacks, and at greater cost.

7.2 The Strategy of Mergers and Acquisitions

Mergers and acquisitions are not new in the defense industries of industrialized countries. The present wave began in the United States among aircraft manufacturers; they acquired companies specializing in military electronics while withdrawing from certain traditional airframe construction work, thereby becoming technically stronger, with a wider range of capacities for the production of major integrated systems. Beginnings were modest – half dozen acquisitions in 1986 – but in the next ten years more than 100 such transactions were completed (*Interavia*, November 1996).

Most mergers and acquisitions were of the horizontal and vertical type, designed to strengthen the absorbing companies' position in the defense market. Yet there were additional anticipated advantages: canceling duplications and reducing excess production capacity; increasing efficiency through economies of scale; expanding capabilities of complex system integration; and spreading inherent risks of large R&D and production programs. In a broader sense, mergers and acquisitions were perceived as providing an unusual opportunity to introduce fundamental change in organizational structure, in decision-making processes and in work procedures (Augustine, 1997: 89).

On the other hand, there were no few difficulties. Mergers may produce inflexible giants that cannot respond rapidly to market developments. Often the merging companies had different business cultures; preferring one meant negating whatever the other did, while to create a new culture with the best of both worlds is a long and complex matter (Augustine, 1997: 92). It would naturally arouse disquiet and concerns among workers, and key employees, those with unique skills, may leave. Managements of merging companies, then, should take special steps to assure workers' rights, as well as to look for ways to reinforce workers' affinity with the new entity. For instance, in the merger between Lockheed and Martin Marietta, 1,400 managers at all ranks were required to buy and hold shares of the merged company in amounts proportional to their salaries, in the hope that this would strengthen their commitment to the new organization (*Jane's Defence Weekly*, 19.6.1996). Another difficulty may arise when mergers and acquisitions are financed with credit. They burden the new company with debts, and servicing them in the coming years may severely impede current activities. From a different standpoint, critics of the large mergers warned against the lobbying power of such industrial giants and their stronger influence over the government.

In the United States in the mid 1990s, when most opportunities to acquire middle-sized companies or divisions of companies exiting the defense business had been realized, companies of more or less equal size merged and created giant groups: in 1994 Lockheed and Martin Marietta merged, and in 1996 the merged Lockheed Martin acquired most of Loral's defense business; Boeing at the end of 1996 acquired the defense business of Rockwell International, and then in 1997, its archrival McDonnell Douglas; in 1995 Raytheon acquired E-Systems, and in the following two years added to its portfolio the defense business of GM Hughes and Texas Instruments; Northrop took over Grumman in 1994, and a year later the merged company acquired the defense business of Westinghouse Electric.

The merger of Lockheed and Martin Marietta, and the acquisition of McDonnell Douglas by Boeing, aroused special interest, and are thought to be a major milestone in the restructuring of the world defense industry in the post-Cold War era. In essence, the two mega-deals were of a different nature. Lockheed Martin, after acquiring Loral in 1996, incorporated some 20 companies that only ten years earlier were independent. Since Lockheed produced aircraft while Martin Marietta and Loral specialized in military electronics, the outcome resembled a conglomerate merger, and many held that it could not provide synergic benefits, but merely additional size. The new partners naturally had a different view, and stressed two main advantages: the complementarities of various business areas that give the merged company exceptional capabilities in system integration; and considerable potential for savings through reduced overheads and eliminated duplications. Augustine, first chairman of the merged company, explained that when capabilities have to be combined, a "virtual company" is formed, which draws what it needs from the different units within the giant organization, thus preserving the flexibility of a "small business" within the framework of a large company (*Economist*, 17.5.1997). The acquisition of McDonnell Douglas by Boeing, by contrast, led to a typical horizontal merger. It combined Boeing's expertise in commercial aircraft with McDonnell Douglas's in the military sphere, creating thereby a company with two complementary focuses of activity, each in a different field. As both fields are subject to cyclical fluctuations, surging demands in one could compensate for a recession in the other. Besides, Boeing's commercial culture and cost control tradition could benefit the defense business that McDonnell Douglas brought to the partnership, while byproducts of military R&D could contribute to the production of commercial aircraft.

In Europe, until the late 1990s, mergers and acquisitions were few, and on a small scale. Exceptional in its scope was the Daimler-Benz takeover of four companies – Messerschmidt-Bolkow-Blohm, Motoren-und Turbinen-Union, Telefunken System-Technik and Dornier – resulting in a German aerospace industry integrated within one concern, Deutsche Aerospace (DASA). It took, however, seven years for the new group to derive any meaningful economic advantages from the mergers, and in the early 1990s it incurred heavy losses (Bullens, 1995: 168-172; *Flight International*, 28.1.1998). Large mergers took place in 1999 only: British Aerospace acquired GEC-Marconi and in continental Europe Aerospatiale Matra, of France, DaimlerCrysler Aerospace of Germany and the Spanish Construcciones Aeronauticas merged into European Aeronautic, Defense and Space (EADS).

7.3 Strategies of Diversification and Direct Conversion

Defense companies that adopted a strategy of diversification altered the internal composition of their activity, seeking to increase the relative weight of civilian business by adding new ventures, often through acquisitions. In direct conversion, manpower and equipment previously employed in defense production are utilized to produce civilian goods in the same plants, usually taking advantage of skills and technologies developed for military purposes. Direct conversion is considered a particularly attractive strategy, since it avoids the costs entailed by closing plants and dislocating production factors (Sandler & Hartley, 1995: 284). In addition, although some companies adopted direct conversion as a means to exit from defense business completely, most saw it as a way to preserve technological capabilities until the next cyclical reversal in defense demand. In the 1990s, however, responding to the decrease in defense demand that followed the end of the

Cold War, defense companies preferred the strategy of diversification to contending with the difficulties involved in direct conversion (Hartley, 1996: 177; Dowdall & Braddon, 1995: 114). Attempts at conversion, if made, were at the lower levels of the supply chain (*Economist*, 2.10.1993).

Large defense companies who adopted diversification carried it out in three principal ways. One was based on their technologies; the British GEC-Marconi, for example, used microwave technologies developed for military applications to produce direct broadcasting by satellite disk electronics for the household market (Gruneberg, 1995: 98). A second strategy transferred R&D resources to civilian areas; the American Westinghouse Electric, for example, expanded its R&D into the crime prevention and security areas (*Business Week*, 5.7.1993). A third strategy acquired enterprises with proven experience in the civilian market; the American Rockwell International, for example, acquired industrial automation companies. Whatever was the diversification strategy chosen, companies deliberated between avoiding the potentially negative effects of different business cultures on one hand, and deriving synergic benefits from them on the other hand. Hence in some cases new civilian activities were separated from traditional defense business, while in others organizational frameworks that interrelates the two areas were created. Sometimes only separate marketing systems were set up, and civilian products were sold under different brand names. In other instances, once civilian applications were developed, production and marketing licenses were sold to companies experienced in civilian markets.

Selecting direction and products is no less important than organizational solutions, and in this respect both diversification and direct conversion have a common denominator. The two strategies involve a two-dimensional change: one is the market and customers dimension, and the other is the technology and products dimension. In the first, it is possible to proceed from traditional military to paramilitary customers, then to industrial customers and ultimately to households. In the second dimension, the first step can be adapting existing military products to their civilian equivalents (hunting guns, commercial satellites), then proceeding to dual-use applications of technologies initially developed for military uses, and ultimately to purely civilian products, making use of the companies' accumulated know-how rather than of any specific technology. The farther out along each axis, the greater the risk, and chances of success increase when major changes on both axes simultaneously are avoided. In both strategies, results will be better as a rule when companies keep to areas of proven expertise and to activities like their previous ones. In making their selection of direction and products, companies must also consider demand forecasts and assess the commercial potential of new business, and should not be led by innovative ideas based on available technological and industrial capabilities only. In practice, however, defense companies unused to market uncertainties and competition had difficulty evaluating the profit potential in their new areas of operation, and frequently failed in their attempts to translate technological advantage into commercial success.

Success in diversification and conversion depends also on the ability of companies to change. Changes are required in technological and engineering processes, e.g. reducing development times, adopting design to cost methods and so forth, as well as in marketing, distribution and service. To accomplish such changes it is necessary to retrain workers and hire new ones knowledgeable in the relevant fields or, alternatively, to collaborate with companies having the needed capabilities. Yet this is not enough; rigid thinking, a consequence of working in assured markets, has to be overcome, otherwise the companies are doomed to fail in contending with rapidly changing market conditions. Indeed, there must be basic changes in organizational culture, in managerial approaches and in the willingness of owners, managers and workers to take risks, all of which develops slowly. Clearly the ability to change is affected by the previous combination of defense and civilian business, and companies with civilian market experience find it less difficult to cope with diversification and conversion. Similarly, ability to change depends on degree of specialization, and presumably for producers at the second and third tiers diversification and conversion will be less difficult than for prime contractors. Finally, both strategies are more likely to succeed if they are adopted in time. Only rarely do defense companies prepare in advance for decreased demand, and moreover, when it happens, the first preferred response is to lobby strenuously against reductions in orders, and only later – if lobbying does not help – do they lay off workers and decrease activity in the defense market (Brauer & Marlin, 1992: 158-162). The move towards the civilian market is, then, a crisis step, taken without previous planning, and with depleted financial resources.

Forces external to the companies themselves should not be overlooked (Brzoska, 1998: 89-91; Dowdall & Braddon, 1995: 114). It is understandably easier to find civilian business and enter it successfully when the market is growing, while in recession such endeavor is particularly hard. Besides, when the defense market shrinks, many defense companies try simultaneously to get into similar types of civilian business, and the crowding created thereby makes it hard to discern opportunities. There are differences too between industries: in some the barriers to entry are lower than in others; culture gaps between defense and civilian businesses are not uniform either, and this is true too of interrelations among R&D, production and marketing. In the aerospace industry, for example, it is easier to balance civilian and military activities than in the car or the electronics industries, since the last two are based on mass production and consumption. Government-owned companies have their specific difficulties; they are not free to choose new civilian business on the basis of solely economic considerations, and administrative directives often impose rules of conduct unsuitable for coping with competitive markets. Finally, in certain cases such government assistance as R&D subsidies, grants and loans on a regional basis (where unemployment was severe) or to certain industries (shipyards) proved an important factor in success.

In no few cases, particularly in the United States, expanding civilian business successfully led companies to withdraw from the defense market entirely. That happened with Westinghouse Electric and Rockwell International, whose diversification moves were mentioned earlier, and with GM Hughes who successfully developed a new business of commercial satellites. In the latter case the managerial attention devoted to diversification was said to have brought about neglect of defense work, leading eventually to the loss of important contracts to the point where the military division had to be sold to Raytheon (*Economist*, 18.1.1997).

In conclusion, attention should be drawn to a somewhat dialectical relationship: the more competitive defense markets become, and the more governments emphasize value for money, the greater the pressure on defense companies to acquire a civilian-type business culture, and the more likely they are to be prepared for entry into new, non-military spheres. Indeed, when defense establishments are perceived as only another customer seeking high quality, high performance goods at low cost, defense companies lose their uniqueness to a great extent, and conditions arise for applying strategies of diversification and conversion more easily (Gansler, 1995: 94).

7.4 The Strategy of International Collaboration

Strategic industrial alliances between two or more defense companies from different countries were the most common form of international collaboration. A strategic alliance is an organizational form in which companies contribute labor and capital resources to a common goal (e.g. share the work and the overall development and production expenses of a weapons system), while maintaining their individuality and independence, often remaining competitors in fields outside the alliance (Dussauge & Garette, 1993: 45-46). Strategic alliances, especially among European companies, proliferated in recent decades, since they were thought of as a means to increase economic efficiency, reduce development risks and broaden potential markets. Over the years three additional trends emerged: the average number of participant countries per project rose; the proportion of collaborative to total defense projects was higher in the 1980s than it was in the 1960s, apparently indicating a movement away from national projects to international collaboration (Hartley, 1995: 473-474); and as well, collaborative projects grew more varied as to equipment and organizational-management formats. There was another important development: as long as the main motive to cooperate was to reduce development and production costs due to budget constraints, most projects proceeded within ad hoc agreements between governments. Since the mid 1980s, however, there have been more company initiated international partnerships, usually on a business basis, which sometimes led beyond the frameworks of strategic alliances (Latham & Hooper, 1995: 18).

The accepted assumptions anticipate that collaboration will increase R&D costs by a factor equal to the square root of the number of participants. For instance, for two partners costs will increase by some 40 percent, and assuming equal sharing each party will bear 70 percent of the costs of an independent national program, thus saving 30 percent. Economies of scale in production are anticipated too, the usual assumption being that doubling production quantity saves some 10 percent of unit costs (Hooper, 1995: 66). On the other hand, collaborative projects involve special costs (Hartley & Martin, 1993: 197-198; Dussauge & Garrette, 1995: 121). One source of such special costs is the need for coordination, clearly associated with the number of participant countries or companies involved. Another is the duplication of R&D tasks and production work, since each partner seeks maximum adaptation to his requirements, or wishes to derive as much economic and technological advantage as possible from domestic activity. In addition, analyses of public choice anticipate increased costs and inefficiency in collaborative projects due to the involvement of various interest groups. Finally, critics add that consensus-based management that leads to cumbersome decision-making processes may cause prolonged delays, which increase costs too.

The various influences on costs were examined empirically. In a project of two countries collaborating in the development and production of an aircraft, with a total output of 300 units, savings were estimated at 10-18 percent compared to a national project, after allowing for the special costs of collaboration. Doubling the number of partners, with each new partner acquiring additional 150 units, increased savings rates to 20-30 percent. Similar savings of some 20 percent were obtained in official British estimates for the Eurofighter project of four partners (Hartley & Martin, 1993: 199). Another study (Hartley, 1995: 477-478) found that R&D costs in joint projects were higher than in similar national projects by 5 to 50 percent, most commonly 10 to 25 percent; unit production costs were 10 to 25 percent lower with a median of 20 percent; and development times were from 10 percent longer to twice as long, the median 20 percent, meaning an additional two to nine years of development work. The assumption that joint projects' output is greater than that of national projects was also validated (Hartley & Martin, 1993: 201-204): in military aircraft projects in Europe since 1960 the average output per national project was 226 units (22 projects) and 355 per joint project (8 projects). Contrary to expectations, however, the proportion of exports to total output, in aircraft quantities, was greater for national than for joint projects.

As noted above, development times in joint projects are longer than in national projects. Since development time may be subject to many influences, not only to whether the project is carried out nationally or collaboratively, a multiple regression model of aircraft development times was estimated (Hartley & Martin, 1993: 207-210). It considered the number of partners in projects, the companies' previous experience in developing similar aircraft, the "source" (e.g. a militarized version of civilian models), and the extent to which the aircraft in development embodies new technologies in terms of maximum speed, takeoff weight and range. A distinction was made also between American and non-American aircraft. The "American factor" influence may come from the larger size of the typical American aerospace company, allowing for more R&D resources and so shortening project schedules, or from technological change between successive generations of aircraft, usually less for American designs than elsewhere. The equations estimated could explain only 40 percent of the variance in development times, indicating that additional explanatory factors were involved. Besides, most of the coefficients obtained were statistically significant and in the directions anticipated. However, joint projects were not found to have a significant effect on total development time, and when special companies rather than loose and less binding organization were established, development was even more rapid.

Other studies examined the effect of organizational and management formats in collaborative projects. Two features were compared: degree of structural complexity and nature of governance structure. A low degree of structural complexity is evident in tasks performed separately, without creating an independent legal entity. Governance structure may be balanced or unbalanced, i.e. that one partner is dominant over the others. On the basis of these features, three main categories were distinguished (Dussauge & Garrette, 1993: 53-58; 1995: 122-124): (a) Unstructured projects where no new legal entity is created, development and production tasks are defined as modules, each assigned to one partner, who then carries it out separately. Partners carry out marketing and sales separately as well, usually following some geographical division of the market; (b) Semi-structured projects with a balanced governance structure, in which development and production tasks are assigned as before, and carried out by the partners separately. Marketing and sales, however, are assigned to a separate entity, in most cases equally owned by all partners. This marketing and sales organization is the only interface with the market and customers, thus playing a central part in defining the product, and in effect operates as a prime contractor vis-à-vis partner companies; (c) Joint ventures within a separate legal entity and with unbalanced governance structure, incorporating existing facilities of partner companies in the new framework, and frequently exchanging staff between facilities according to technical and administrative considerations. The first two categories are typically project-based and therefore are limited in time, while the third is business-based, and have no pre-determined time limit.

In an empirical attempt to statistically associate organizational formats with technical and commercial success in 30 joint projects, two principal findings were derived (Dussauge & Garrette, 1995: 124-129). First, among equally successful projects from a technical point of view, the semi-structured were thought to be more successful commercially than unstructured. Secondly, all projects that were ranked high technically and commercially were semi-structurally organized, while those ranking lower in both dimensions were for the most part unstructured, and a few semi-structured. Apparently then, a separate marketing and sales entity is a necessary condition for attaining technical and commercial success simultaneously. However, the presence of semi-structured projects among the unsuccessful ones too, suggests that it is not a sufficient condition for commercial success. Since there are no grounds to assume an association between separate sales organization and technical success, it appears that technically successful projects encourage common marketing, which then, on its part, contributes to the project's commercial success.

Another empirical study examined a sample of 70 strategic alliances between companies in the same or in different countries during 1950-1970 (Dussauge & Garrette, 1993). All companies, most of them European, competed in the defense market, and in 30 percent of the cases they cooperated in civilian areas. Comparison between strategic industrial alliances in the defense and civilian spheres pointed to three notable differences. First, most defense alliances were unstructured, while most civilian ones were semi-structured. Competition in civilian markets seems to justify a joint marketing effort through a separate organization, while in defense, particularly as regards domestic markets, marketing aspects are secondary and do not get that much attention. Secondly, most joint projects where several partners carried on similar activities simultaneously were in the defense area. Duplication, a source of inefficiency and increased costs, was almost always present in production, especially in the final assembly of weapons systems, and arose from considerations of political and military independence. By contrast, civilian projects operating under stronger competitive-commercial pressures, cannot endure the inefficiency that duplication entails. Thirdly, the governance structure of strategic alliances in defense was less balanced than of those in the civilian sphere. Possibly, government involvement in military programs compels potential partners to accept unbalanced partnership for political reasons.

Another comparison was made between multilateral and bilateral alliances (Dussauge & Garrette, 1993). Multilateral alliances led to the formation of separate legal entities in more cases than bilateral alliances did. Besides, in multilateral alliances ownership of the new legal entity was almost always divided unequally. It may be assumed that the greater the number of partners and the more complex the relations within an alliance, the higher the transaction costs involved, and the specific legal, unbalanced structure adopted is designed to limit them. However, unbalanced ownership may result also from the relative competitive positions of the partners, which are nonetheless rivals. A statistical examination rejected the latter hypothesis, thus reinforcing the transaction costs interpretation,

Finally, since the study covered a long period, it shed light on developments over time: earlier strategic alliances were unstructured, with tasks assigned among the partners; later semi-structured alliances appeared, with separate legal entities for marketing and sales, while business-based joint ventures were formed only in the late 1980s and the early 1990s. Collaboration among companies competing in the defense market may be considered a substitute for international industrial concentration, which is not politically feasible. The expansion of strategic alliances in size and scope over time suggests apparently weakened political barriers, at least relative to pressures for economic efficiency, and a trend towards the more traditional industrial concentration (Dussauge & Garrette, 1993: 60-61).

Public choice analyses too were applied to international defense projects. They describe the dynamics created among partners, and present the features and results of those projects as an outcome of the efforts made by the many participants in each country to promote their own interests (Hartley, 1995: 475-477). In conclusion they suggest that all joint projects inevitably move farther away from the ideal model and into inefficiency. To reduce negative consequences, three guidelines were suggested: allocating development and production work competitively, based on the comparative advantages of each country; assigning a prime contractor subject to contractual incentives and penalties, and hence to risks; and compensating losers outside the framework of the projects themselves.

In the past, strategic alliances and joint projects were common mainly in Europe. Towards the end of the 1990s, interest in transatlantic collaboration increased. On the American side, interest in collaborative projects with Western European companies assumes that most opportunities for significant mergers among American companies have been exhausted, and that future consolidation will take place mainly in the international arena (*Jane's Defense Weekly*, 9.7.1997). In Europe, collaboration with American companies is seen as an access route to advanced military technologies. Both sides attach great importance to the market potential of joint projects. Conspicuous in this context is the American Joint Strike Fighter (JSF) program. The aircraft is designed to replace most fighter aircraft in the armed services of the United States, and many other countries are expected to acquire it as well. The United States allows friendly countries to participate in the program in different capacities – as full partners, as associate partners or as observers – and some Western European countries have already joined. In particular, Britain is a full partner from the very first stages of the program, and BAE Systems is a member of the group headed by Lockheed Martin, the winner in the competition against the group headed by Boeing. Indeed, some see the emerging partnership within the JSF program as the nucleus for a future transatlantic aerospace industry.

In the post-Cold War era, a new direction for cooperation is opening up between Western defense companies and Russian producers. The former estimate that in the coming years the market segment dominated in the past by Russian weapons systems can provide considerable growth potential, and that penetrating into it requires cooperation with Russian producers in adaptations, upgrading and new developments. On the Russian side, the incentive lies in the need to adapt military equipment to Western standards and upgrading them with advanced Western technology. In the 1990s, French companies led in this area, though American, British, Italian and other companies also showed interest (*Interavia*, March 1997).

At times, companies adopting an international strategy advanced beyond the creation of strategic alliances for a particular program or defined business area, and acquired full or part ownership of defense companies in other countries. There were several notable transactions of this type in the 1990s. The British engine manufacturer Rolls Royce acquired the American engine company Allison in 1995 to gain a foothold in the market and access to government R&D funds in the United States (*Interavia*, August/September 1996). In 1998 British Aerospace acquired 35 percent ownership of the military aircraft division of the Swedish SAAB. Before it was sold to British Aerospace, GEC-Marconi acquired ownership of several American military electronics companies, and in 1998 leaped forward by acquiring Tracor, fifteenth on the list of the American Department of Defense suppliers, thus climbing up to sixth place among military electronics suppliers in the United States (*Jane's Defence Weekly*, 29.4.1998). The merged company BAE Systems at the end of 2000 acquired the Aerospace Electronics Systems division of Lockheed Martin, and its North American arm became a dominant player in the sensitive area of electronic warfare. Following that acquisition, BAE Systems employs some 25,000 workers in the United States, thus achieving its strategic goal of becoming the European defense company with the greatest American presence, and the American company with the greatest presence in Europe. In the opposite transatlantic direction, the most conspicuous transactions were the acquisition of the Swedish Bofors Weapons Systems by the American armored vehicles producer UDLP, and of the Spanish Santa Barbara by General Dynamics. The French military electronics giant Thales (formerly Thomson-CSF) adopted an extraordinary international strategy. Described by company executives as "multi-domestic growth and development strategy", it seeks to establish a diversified and ramified worldwide network by acquiring medium-size companies in different countries, though with a strong position in their domestic markets, thus getting closer to customers and attaining a domestic identity. Within a few years, Thales acquired holdings in defense companies in the United States, Britain, Australia, South Korea, Singapore, South Africa, Brazil and other countries. In 2000, 75 percent of the sales and half the personnel of the company were reported to be outside France (*Flight International*, 12.12.2000).

9
THE ISRAELI DEFENSE INDUSTRY

The growth of Israel's defense industry corresponds in many respects to that of emerging defense industries in other developing countries in the latter half of the twentieth century (see in chapter 8, sections 3.2 and 5.3). However, there were some noteworthy differences as well, in time driving the Israeli defense industry closer to the defense industries of developed countries. It might be claimed, then, that Israel's defense industry occupies a middle position between the defense industry of advanced industrial states and that of developing countries.

1. FROM STRATEGIC SELF-SUFFICIENCY TO TECHNOLOGICAL FORCE MULTIPLIERS

Political barriers encountered in the 1950s and 1960s in attempting to acquire arms and military equipment abroad evoked permanent uncertainty as to whether the needs of the Israel Defense Forces (IDF) could be met on the basis of imports alone. A dual policy was therefore adopted: no effort was spared to exploit opportunities for foreign purchases, and at the same time substantial resources were invested in establishing a domestic defense industry. Domestic defense production was thus perceived at the time first and foremost as a release from absolute dependence on foreign supply sources. Yet as long as supplies from France – the only state then openly willing to sell arms to Israel – were regular, domestic defense production remained relatively limited in scope, consisting mainly of small arms and ammunition, conversion and upgrading of obsolescent weapons systems, assembling light training aircraft under a French license and the like. But when France imposed an embargo in 1967, industrial goals broadened. The defense industry was called upon to provide strategic self-sufficiency for Israel, and in particular to release her from absolute dependence on foreign suppliers even for major weapons systems. Development and production of fighter aircraft, main battle tanks, missile boats and various types of missiles soon commenced, and the defense-industrial base was largely expanded. The self-sufficiency doctrine thus established dominated defense-industrial policy for the next 15 years, and only in the 1980s, due to stable, ongoing supplies from the United States, as doubts about the import option began to dissipate, its priority gradually declined.

Opportunities for procurement abroad were an important, perhaps the most important factor, but not the only one that determined the approach to the role of domestic defense industry. Even before the French embargo trauma, two opinions were in debate (Klieman & Pedatzur, 1991: 73-74). One asserted that it was unrealistic to expect that Israel could produce all the armaments the IDF required, imports would always be needed, and thus the domestic defense industry should concentrate primarily on improving imported systems and adapting them to the special theater conditions and to the IDF warfare doctrine. The other view held that even in optimistic political scenarios there are no grounds to assume that all requirements for military equipment could be met through foreign acquisitions, and therefore as large a proportion as possible of the necessary equipment should be produced locally. The first view was held mainly by military people, and reflected concerns that vast investments in domestic industry, bearing fruit only in the course of time, would impede force buildup and war preparedness of the army in the short term. It was rooted too in doubts about the technological capabilities of the domestic industry, and warned that with local products the IDF would be using inferior equipment, technologically backward as compared to what was obtainable elsewhere. The other opinion, held mainly by civilian officials who were particularly active in developing procurement ties with France and in continually seeking out other sources abroad, seemed more aware of problems created by dependence on foreign suppliers and more confident of the abilities of Israel's scientific and engineering community. Economic considerations as well influenced both groups. While the first one stressed the limitations of a small economy, the second emphasized that developing an advanced, technology-intensive defense industry fitted in with the country's general economic goals and with its comparative advantages. Events of the late 1960s, especially the French embargo and the British unwillingness to sell Israel advanced tanks, made a formal decision between the two views superfluous.

Throughout the years in which the self-sufficiency concept was dominant, it clearly did not mean absolute independence, and no strategic or economic justification was claimed for complete autarky in arms production. In the early 1980s, however, several reasons combined to justify a reassessment of the self-sufficiency policy even in its less binding form. Available procurement sources and American grants were, as noted earlier, very important. Yet even more important were economic constraints. In Israel as elsewhere, the race for superior performance raised the cost of developing and producing novel armaments to the point where it was doubtful whether a small economy could afford the economic price demanded by a self-sufficiency-oriented defense production policy. At the same time, massive supplies of American arms made it necessary to allocate increasing local currency budgets for absorbing and maintaining them, so reducing constantly the amounts left for domestic development and production of major weapons systems. Some maintained that in these circumstances adherence to the self-sufficiency policy in defense could lead to economic dependence with severe political, and eventually military, implications (Peri & Neubach, 1984: 34-37).

Specific claims were raised as well. First, despite large investments, only partial independence was attained: in the mid 1980s, the Israeli content of overall military capital, including the value of local improvements in foreign systems, was estimated at only 25-30 percent (Halperin, 1987: 1003). Secondly, in a small country, the relatively large resources tied up in development and production programs of major weapons systems imply great macroeconomic risks; even if absolute costs are lower than in large countries, in percentages of GNP they are far higher (Halperin, 1987: 1001). Third, dependence on foreign supplies has been replaced by dependence of the defense industry itself, particularly on critical imported components, and in some cases on foreign technologies. Finally, developing the defense industry retarded economic growth in other sectors (see below). Not all the claims were equally valid. In particular, evidence about the rate of self-sufficiency achieved, and about apparent replacement of one form of dependence by another, do not indicate an erroneous policy or a incompetent defense industry, but rather a conscious public choice of how much economic resources should be committed towards the self-sufficiency goal.

The 1980s marked the beginning of a gradual change in domestic defense production policy, reaching a turning point with the decision to cancel the Lavie program for developing a new fighter aircraft. The emerging perception focused on the idea that the industry has to supply the IDF with force multipliers by means of original, unique technological solutions, while self-sufficiency was relegated to second place. The importance of technological force multipliers grew at that time due to the accelerated technological advance of the battlefield, and original solutions became imperative because of ever more open international arms-markets and the resulting unprecedented proliferation of new military technologies. Israel was specifically challenged as well: with the collapse of the Soviet Union and the Russian arms industry crisis, the Arab countries turned increasingly to Western arms sources, and concerns deepened regarding a future Middle East arms race between Western weapons systems. These new conditions reinforced the idea that only indigenous unique development, unobtainable in the market and concealed until used in battle, could assure Israel of qualitative advantages and the ability to surprise.

The change in priorities had important implications for the defense industry. If self-sufficiency is not a principal goal, domestic products and imports may be considered as complementary rather than substitutes, and it is possible to acquire expensive platforms abroad, and then add on to them original, surprising features developed and produced locally. Such an approach reduces, however, the need for plants to produce and assemble platforms, and increases demand for plants specializing in systems development and production instead, implying a fundamental restructuring of the industry. Besides, emphasis on original and unique technological solutions requires increased R&D expenditures, and justifies government support in maintaining several parallel know-how centers in every principal technological area; competition among them is more likely than any other means to stimulate originality and innovation. There is another implication too: while self-sufficiency does not run contrary to an aggressive defense export policy, prioritizing force multipliers imposes severe restrictions on exports because of secrecy concerns. This latter issue indeed posed a serious dilemma. Over the years Israel's defense industry came to depend considerably on foreign markets, its comparative advantage was based chiefly on technological innovation, and limiting exports to less advanced or degraded models might well interfere with the chances of selling overseas.

2. MAIN DEVELOPMENTS

The foundations of arms production were laid in the 1920s and 1930s, and during the War of Independence in 1948-1949, existing workshops were capable of producing explosives, mortars and mortar shells, submachine guns, hand grenades, small arms ammunition and similar items that contributed significantly to the military effort. In the 1950s and early 1960s, the defense industry developed mainly within the government sector, and the number of workers almost tripled (from 5000 in the mid 1950s to 14,000 ten years later (Barkai, 1987: 36, 42)). In 1967, on the eve of the Six-Day War, however, defense industry relative weight in the economy was still low: employment accounted for some 6 percent of the workforce in manufacturing industries, and sales reached 3-4 percent of GDP. At that time, then, the defense industry had no real macroeconomic significance. The great leap forward occurred at the end of the 1960s, and accelerated growth continued for nearly 20 years, until the mid 1980s. With that, there were two distinctive sub-periods: in the first, until about 1976, rapid growth was led by domestic defense demand, and in the second, by rapidly increasing exports. During both sub-periods growth was faster than in other sectors, the relative weight of defense industry in the economy rose, and the defense industry became an influential factor, affecting macroeconomic and structural processes in the entire economy. The impact was especially prominent in manufacturing industries: in the mid 1980s defense industry made up some 27 percent of all industrial output, and employed about 20 percent of the workforce and 30 percent of the industrial capital stock (Barkai, 1987: 37).

A turning point came in the later 1980s: defense exports still continued to grow, but Defense Ministry orders decreased considerably. Following years of rapid inflation, the government adhered to a restrained fiscal policy, defense expenditures were cut, and moreover, within the defense budget there were growing needs in other, less flexible expenditure categories. According to one estimate, between 1985 and 1989, total defense industry sales decreased by some 5 percent (Tov, 1992: 651). The downward trend in domestic procurement continued in the early 1990s, but this time without compensatory export growth, and defense companies, in particular the large government-owned companies, were caught up in a crisis (see section 6).

Defense exports are discussed in detail later (see chapter 11, section 1). Here, however, it is worthwhile to note that from the mid 1970s to the late 1980s, exports increased approximately tenfold in current dollar prices, and their share in defense industry activities grew substantially. In effect, the industry came to depend on export sales: in the mid 1970s they constituted just 15-20 percent of sales, while in the second half of the 1980s they ranged from 60 to 80 percent of the larger companies' sales (see Table 9.1 for 2001). This is without any doubt an extraordinary rate, singling out the Israeli defense industry from others worldwide. Export growth counterbalanced domestic demand fluctuations, thus saving the companies high adjustment costs in manpower and production facilities. Yet it also required expanding the development and production infrastructure to meet specific needs of foreign customers, and when exports stopped growing, the industry was left with excess manpower and equipment.

In the 1990s, defense industry activity decreased while the economy as a whole grew rapidly, hence its relative weight declined by all criteria. At the end of the decade exports grew once again and domestic sales were more or less steady, but the relative weight of defense industries remained low, and they were no longer an influential macroeconomic factor.

From a technological standpoint, the development of Israel's defense industry resembled the typical patterns found in other developing countries (see chapter 8, section 3.2). Until the mid 1950s, it concentrated on producing small arms and ammunition, and on improving weapons systems purchased from surplus, some of which was truly scrap. In the second stage, local production under license commenced, nourished by fruitful cooperation with French companies. At that stage, for example, the Israel Aircraft Industry (IAI) built the Fuga Magister training jet. The third stage was marked by the implementation for the first time of locally designed changes and adaptations in products acquired abroad or produced in Israel under license, in response to new regional threats. Examples included replacing the original Sherman tank guns with ones of larger caliber, a response to the T-54/55 Soviet tanks; adding machine guns and rocket launchers to the Fuga Magister training jets, providing them with combat capabilities; and installing an electronic package developed in Israel in the French Mirage 3. The know-how and experience acquired in these projects made it possible to proceed to the fourth stage of developing and producing subsystems and sub-assemblies subsequently integrated in imported weapons systems and in platforms assembled in Israel. The central projects at this stage were associated with fighter aircraft, producing the Nesher, an Israeli version of the Mirage 5, and with upgrading the British Centurion tank. The transition to developing and producing major weapons systems in the late 1960s and early 1970s was thus a natural offspring of the earlier advances. From the

1970s on, Israeli defense industry developed and produced fighter aircraft, unmanned air vehicles, tanks, missile boats and varied types of missiles.

Another noteworthy characteristic of the development pattern was the close association between the emerging operational needs of the IDF and the themes of development and production in the defense industry. Changing threats and the lessons learned from wars shifted centers of activity from time to time, and dictated the industrial agenda. Thus for example (Bonen, 1994: 56), in the 1973 War the Gabriel sea-to-sea missile and certain local electronics systems contributed decisively to victory in the naval theater; lessons learned from the War of Attrition along the Suez Canal in the late 1960s focused attention on developing standoff weapons and varied supporting systems, and a complex combination of those means made it possible to destroy the Syrian surface-to-air missile batteries in the Lebanese War in 1982; the vulnerability of Israel's tanks in 1973 made it urgent to shift resources to programs designed to improve tanks protection; and the threat of long range surface-to-surface missile attacks, and the actual experience during the Gulf War of 1991, led to concentrated efforts in developing anti-ballistic missiles systems and surveillance satellites.

The Israeli defense industry participated in several American programs of developing advanced weapons systems. In the 1980s it was among the foreign industries invited to take part in the program of developing the strategic weapons of the future (Star Wars), and in the 1990s it contributed to the American efforts in the field of anti-ballistic missiles through its own development of the Arrow system. As noted previously, international collaboration became an important part of world arms industry activities, and participation in international development and production projects may well be the next stage in the development of Israel's defense industry.

3. CHARACTERISTICS OF ISRAEL'S DEFENSE-INDUSTRIAL BASE

The core of Israel's defense-industrial base consists of a few large defense companies that rank among the largest industrial enterprises in the country. In 2001, four of the 20 largest industrial companies in Israel were, by volume of sales, defense companies (see Table 9.1). Moreover, Israel's leading defense companies rank high, by their size, in the hierarchy of defense industry of developing countries, and indeed in the hierarchy of world defense industry as a whole. At the end of the 1990s, SIPRI's Big 100 included 11 companies from non-OECD countries, six of which were Israeli (SIPRI, 2000: 328-330).

The small number of large-sized companies resulted in a high degree of concentration: of 150 companies defining themselves as engaged in defense production at the beginning of the 1990s, the Big Ten employed 78 percent of the workers, and accounted for 82 percent of the sales and for 87 percent of exports (Tishler & Rotem, 1995: 475-476). Company size inevitably led as well to accumulation of power in the hands of managements and labor organizations, in turn influencing the government's defense-industrial policy, and especially the labor relations in the defense sector.

Table 9.1 Israel's Largest Defense Companies, 2001

	Rank	Total Sales ($ Million)	Exports as % of Sales	Number of Employees
Israel Aircraft Industry	3	2,089.0	76.7	14,500
Elbit Systems	14	764.5	70.4	5,000
Rafael	19	500.0	. .	4,300
Israel Military Industries	20	437.0	41.4	4,100
Elisra	26	306.1	62.0	1,590
Tadiran Communications	46	153.6	71.7	960
Urdan Industries	111	74.2	12.5	380
Soltam Systems	112	73.1	90.6	325

Source: Dun & Bradstreet International, 2002, 90-96.

Israel's defense industry is remarkably technology-intensive, resembling in this respect the defense industries of advanced industrial nations (Halperin, 1987: 1009). A demand-related explanation for the outstanding technological level derives from the basic quantitative imbalance between Israel and its Arab adversaries that impelled Israel to develop qualitative advantages (see chapter 3). On the supply side, technological intensity was obtainable due to the relative abundance of skilled labor, particularly of scientists, engineers and technicians.

The technological intensity of the defense industry stands out in comparison with other domestic industries: in the mid 1980s the ratio of defense to civilian industrial R&D was estimated at five to one, though in output terms civilian industry was three times as large and employed four times as many people. Similarly, though defense industry employed one fifth of all workers, it employed about half the scientists and engineers in manufacturing industries (Halperin, 1987: 1009). Over the years the differences diminished, however. One study indicated, for example, that between 1978 and 1987 the proportion of engineers in civilian industry workforce grew faster than in defense industry, especially in civilian industries employing a higher proportion of skilled labor (Berman & Halperin, 1990: 156). Differences in relative technological intensity between the two industrial sectors suggest that defense industry may have crowded out skilled labor from civilian industry, retarding its growth. By contrast, on grounds of the diminishing differences, particularly since the 1980s, it is plausible to assume that crowding-out effects gave way to spillover effects in time, thereby contributing to an accelerated development of technologically advanced civilian industry (see section 4).

Over the years there were changes in ownership of defense companies, however in no few cases in the opposite direction to that typical of other countries. As noted before, in the 1950s the industry developed principally through government bodies. Later on the circle widened, the number of non-governmental defense companies grew, and the government even sold its share in companies it helped to establish. Conspicuous in the 1960s and 1970s was the proliferation of companies owned jointly by Israelis and foreign partners. In the 1980s, however, most of the latter withdrew, and the companies went over to total Israeli ownership. Foreign investors had individual reasons to sell their holdings, and no common denominator could be found. Yet the trend, as it developed, was the reverse of the internationalization prevailing in recent decades in the defense industry of other countries. At the turn of the twenty-first century there are in effect almost no foreign investors in the Israeli defense industry.

As for government ownership of defense companies, despite some changes over the years, its share remained high both in comparison to other domestic industries and to defense industries in most Western countries. Among the eight largest defense companies (see Table 9.1), three are government-owned, their sales in 2001 comprising some 70 percent of all sales of the group. However, government ownership did not interfere with competition in the domestic defense market. As a rule, there was no noticeable preference for government-owned companies in awarding contracts. Also, the government did not usually stand in the way of private defense companies that established competing development centers or production facilities to those in government-owned companies, nor did it intervene in investment decisions of companies under its ownership that might have led to duplications and to competition between one government-owned company and another. Notwithstanding its relative small size, high degree of concentration and government involvement in ownership, the domestic defense market in Israel is, in fact, highly competitive, and in most segments, especially in advanced electronics, several companies are active and no single large one is clearly dominant.

If government ownership had any clear influence, it was on company culture and management patterns. Formal limitations in several areas deprive government-owned companies of the managerial flexibility necessary for operating in a businesslike fashion. For instance, they are not free to adopt employee compensation practices similar to those of civilian high tech companies with whom they compete over skilled manpower, and at same time they are committed to public sector wage policy, although it is often incompatible with their business situation. As another example, government-owned companies must obtain special approvals for moves such as setting up subsidiaries, entering partnerships, mergers and acquisitions, losing opportunities at times due to a slow-moving bureaucracy. Then too, the need arises on occasion to consider government political and public sensitivity and restrict or postpone unpopular steps (e.g. closing unprofitable plants). Sometimes government-owned company executives assumed, not without grounds, that the government would not let them down, and would bail out companies in distress. Such a moral hazard approach reduces sensitivity to losses, and eventually creates a corporate culture not typical of business companies. Indeed, through the years, government defense companies demonstrated clear preference for rapid growth over profitability, and even when demand conditions deteriorated, strove to preserve their size. The development patterns of government-owned companies produced similar effects. They all started as Defense Ministry departments, and only at a later stage – IAI in 1968, Israel Military Industries (IMI) in 1990 and Rafael in 2001 – became genuine business companies. The transition required organizational adjustments, setting up accounting and costing systems hitherto nonexistent, acquiring advanced management and control methods, and most of all, learning to think differently. Government ownership hardly encouraged such processes, and they moved forward only in the 1990s, under crisis conditions (see section 6).

Government influenced the defense companies not only as owner but also, as elsewhere, in its other capacities: as the policy-maker on national security and macroeconomic matters, as the companies' principal customer, as granter of export licenses and more. At the same time, as noted earlier, the size of the leading defense companies brought power to their management and labor unions, which they often did not hesitate to use in promoting seemingly "private" interests. Thus in Israel too there were extensive and complicated interrelations within the military-industrial complex. However, some unique features developed as well. Certain differences in comparison with military-industrial complexes in other western countries, the United States for example, are noteworthy (Mintz, 1985). Firstly, due to the relatively high government share in ownership, the defense industry does not serve the narrow interests of private capital, and private shareholders have only marginal influence on defense production policy. Secondly, government procurement decisions were not as a rule affected by local politics considerations, and pork barrel politics as such was rarely if ever evident. Decisions not taken on purely strategic grounds aimed at enhancing general economic and social goals, like creating employment and supporting development regions. Third, the vital importance of domestic defense production enjoys a broad consensus that cuts across political camps. Fourth, top-level officials in the Defense Establishment and industries come from all hues of the political spectrum, and are not specifically identified with a "hawkish" attitude. Fifthly, there is no unanimity within the Defense Establishment as to the degree of self-sufficiency that should be maintained.

In view of these and other differences, the Israeli model does not fit accepted perceptions of the mode of operations and influence of the military-industrial complex. In the Israeli reality there was never place for a radical theory of conspiracy among the partners in the military-industrial complex. If ideological, institutional and economic interests converged, it never resulted in a coordinated, uncontrollable joint action that deviated from the national interest. However, the military-industrial complex was indeed a powerful pressure group at most times. It advocated a broad, highly developed defense-industrial base, and due to the prominence of national security issues in Israel's reality often won priority over other lobbies and other public policy issues. Furthermore, for secrecy reasons and because of the special character of defense business, until recent years the military-industrial complex was exposed to relatively little public criticism. At the same time, the special close relationship within it had important practical advantages. The small dimensions of the country and the economy, a common background of military service and of studying at the same engineering schools, and mobility between the various segments of the military-industrial complex made for particularly effective communication (Peri & Neubach, 1984: 40-41). Thus direct contacts and continuous interaction between users, developers and producers made it possible to achieve superior operational performance while adapting varied defense products specifically to theater conditions and to the IDF's warfare doctrine, and no less important, to shorten processes and save valuable resources. These special features – short development periods and hence relatively low development costs – are among the Israeli defense industry's outstanding advantages, and contribute to its competitiveness in the world arms market.

The balance of forces that affects defense industry has changed since the mid 1980s. At the government level, the Finance Ministry became more involved in defense industry affairs; at first in the context of the controversial program of developing a new fighter aircraft, the Lavie, and later due to the financial crisis in government defense companies that called for a special budgetary assistance. The Finance Ministry usually adopts positions in line with its annual fiscal policy and other short-term, purely economic considerations. Within the Defense Establishment, following the expansion of Service's budgets at the expense of the central budget (see chapter 4, section 8.2), decisions as to full-scale development and production were transferred from the Defense Ministry to the IDF, and the latter naturally prioritizes preparedness and current requirements. Here too, then, a bias was created in favor of specific, short-term programs at the expense of developing basic capabilities. At the same time, the government defense companies' financial situation deteriorated, and they could no longer allocate adequate means for developing new technological infrastructures. Under these circumstances, worry was expressed that the defense-industrial base might gradually lose the ability to fulfill its strategic roles.

4. MACROECONOMIC AND STRUCTURAL EFFECTS

Until the latter half of the 1960s defense industry exerted no real macroeconomic influence, but this changed after the war in 1967. In the mid 1960s the Israeli economy experienced a recession, and the expanded domestic production for defense that followed the War and the French embargo imposed then, was a crucial factor in resuming economic activity. In particular, it created vast employment opportunities, and offered attractive jobs for scientists and engineers recently immigrated. It led as well to an extraordinary expansion of investments, and since defense demand concentrated in technology-intensive industries, industrial growth as a whole took that direction. Between 1968 and 1984 the share of industries such as electronics, aerospace and optical instruments rose from 6 to 24 percent of total industrial output, with most growth at the expense of traditional industries (Berman & Halperin, 1990: 150). The accelerated development had qualitative facets, not just quantitative ones. The level of mechanization rose, high quality standards were established, advanced quality control systems were introduced and so forth, and these improvements did not remain in the domain of the defense industry only, but spread through subcontracting into many other plants that now had to comply too with the higher standards. Defense industry was thus an important agent of modernization, contributing to productivity growth in the entire economy (Berglas, 1983: 36; Blumenthal, 1985: 135). In addition, defense production helped in the geographical distribution of population and of industry, which has always been an important policy goal in Israel. Finally, as noted before, in the later 1970s and in the 1980s defense industry grew through increased exports, thus contributing significantly to the balance of payments in those years (see chapter 11, section 1).

The adoption of the self-sufficiency doctrine, primarily for strategic and military reasons, and the resulting increased defense demand stimulated defense industry growth, but the remarkable rates of growth would have not been achieved without the preliminary economic conditions prevailing at the time. Since the late 1950s, with government aid, large resources were devoted to industrialization, laying the foundations for a diversified industrial infrastructure and providing managerial and technical experience (Kleiman, 1970: 56). Higher education and technical training systems expanded as well, and their ability to teach technical skills improved considerably. In fact, between 1968 and 1984, the number of scientists and engineers in manufacturing industries grew by 10 percent annually (Berman & Halperin, 1990: 151). The Israeli experience, then, does not help either to determine whether strategic motives that stimulate large investments in defense industry accelerate industrialization and promote quality of manpower, or vice versa, namely defense industries arise where there already is a developed industrial base with skilled manpower (see chapter 8, section 5.3.1).

The rapid growth of defense industry, however, raised criticism from time to time, pointing to at least three apparent undesirable macroeconomic effects. First, it was claimed that defense industry is by nature an unstable economic activity, subject to unpredictable demand fluctuations, and given its size may induce instability in the entire industrial sector, and even in the economy as a whole. Secondly, possible crowding-out effects were noted, supposedly retarding growth in the civilian sector. In particular, it was argued that competition over a limited quantity of scientists and engineers caused steep wage increases, making it hard for civilian companies, who sell their products in competitive markets, to recruit and employ highly compensated skilled manpower, thus eventually hindering their growth. It was further argued that the negative crowding-out effect was more significant than the positive spillovers effects of technology transfers from defense to civilian industries (Halperin, 1987: 1002-1003). The third criticism overlaps the crowding-out effect argument somewhat, but stresses that even if defense industry contributed positively to other sectors, the same resources allocated to defense production, had they gone elsewhere in the first place, would have yielded greater macroeconomic returns in terms of increased GDP, exports and the like. Indeed, comparisons between measures from defense and civilian companies as regards value of sales to R&D expenditures, average exports per skilled worker, etc., showed that returns in civilian industry were often higher (Berman & Halperin, 1990: 162-164).

In fact, however, concerns about instability did not materialize, and indeed no other economic sector in Israel enjoyed continuous growth or maintained a stable aggregate level of activity for so long as did the defense industry. When defense production eventually declined in the 1990s, its weight in the economy was already relatively low, and the macroeconomic effect was not great. As to arguments about the crowding-out effect, they may have been valid for a short time. Civilian high tech industry grew rapidly from the late 1970s, and within only five years its accumulated growth was similar to what defense industry achieved in about 15 years (Berman & Halperin, 1990: 156). Such development could hardly have taken place if civilian industry had not followed the path already opened for it by a sophisticated defense industry, and most important of all, without defense industry's outstanding contribution to the supply of skilled manpower. Employment opportunities offered by defense industry stimulated demand for education in the natural sciences and engineering, encouraged Israelis who studied technology abroad to return, prevented brain drain and helped absorb immigrants with academic education. Therefore even if defense industry did cause some delay in the development of certain industries, the technological and human infrastructure it put in place enabled those industries to close the gap rapidly and reach impressive achievements. Spillover effects in terms of direct applications of military technologies in civilian uses were indeed limited, but other more indirect effects led to outstanding results. For instance, some companies began operations in the military sphere, moved into civilian business based on their accumulated technological know-how and experience, and gradually became leaders in their new field. Other companies were set up by former defense industry employees, or had such people as their technological backbone, and joined the civilian market from the start, making use of know-how and experience acquired in defense development and production. In the 1990s, such companies numbered in the thousands. Most of them were still immature at the turn of the century, and the high tech crisis worldwide negatively influenced their prospects. Some, however, won international recognition as technological leaders, and not less important – were highly valued by the international capital market. The third claim ignored the non-economic reasons for establishing a defense-industrial base. Furthermore, it does not seem to have considered the full economic return yielded by indigenous defense production, particularly some microeconomic advantages.

5. THE MICROECONOMICS OF DOMESTIC DEFENSE PRODUCTION

Domestic defense production has to meet economic criteria, yet a simple comparison between domestic and import budgetary price is usually an inadequate measurement to prove its economic advantage. From the national economy standpoint, the cost of dollar saved through defense import substitution has to be compared with the costs of obtaining foreign currency in other ways (Kohav & Lifshitz, 1973: 13-14). However, following this line of thinking some maintained that in Israel's special circumstances, where most defense imports are financed with targeted American aid grants, domestic defense production, by definition, is not economical. As against dollars saved by domestic production there are dollars whose apparent opportunity cost is zero, and however efficient domestic production may be and however low the cost of dollar it saves, that cost can never be less than zero. But this is simply a flawed argument. When aid grants are limited and fully utilized, domestic production of some items releases grant funds for the procurement of others, so aid dollars too have an opportunity cost greater than zero. While aid grants certainly affect the price of foreign currency – when they increase, the exchange rate is likely to decrease and vice versa – they are not different in this respect from other inflows recorded in the balance of payments. The standard economic criterion for import substitution is therefore valid as well for domestic defense production in an economy provided with targeted foreign aid. At the same time, it does not imply that the dollar saved by domestic production must be below the effective exchange rate in every case. The criterion sets a reference price, and every deviation from it has to be justified by means of the non-economic advantages of domestic production, e.g. self-sufficiency, operational advantages and the like.

The main doubts in Israel regarding the economic rationale of defense production arose from the apparent impossibility of realizing economies of scale. The large fixed costs involved in defense projects – large investments in R&D, production infrastructure and special equipment – require large-scale, continuous production to break even, while in Israel production runs are relatively small, and often fluctuate over time for budgetary reasons. The truth is, however, that arms markets are oligopolistic in nature, and prices are not so closely associated with production costs in which advantages of scale lie (Berglas, 1983: 34). Moreover, weapons systems and military equipment are produced in small runs everywhere, and for many items Israeli production runs are no smaller, especially when local demands are augmented by export sales. At the same time, certain factors make defense production in Israel cheaper than abroad. In technologically intensive defense production, a country endowed with advanced technological capabilities may have a comparative advantage, and Israel maintains a highly developed scientific and technological infrastructure, as it has relatively abundant skilled labor. Indeed, the cost of skilled labor in Israel was lower than among producers of arms and military equipment in the major industrial states. Furthermore, battlefield experience gave Israeli users a unique ability to define and specify operational requirements, and close contacts between users and developers made it possible to translate requirements accurately and effectively into development and production programs. Such conditions led to economies in R&D resources, and shortened processes of malfunction detection and correction. Finally, the superpowers, and in particular the United States, tend to develop weapons systems and equipment that can be used anywhere in the world, which are thus more expensive. Domestic Israeli production was free to focus on a narrower range of performance capabilities, more suitable to the local climate and the size of the arena, and more compatible with IDF warfare doctrines. This too brought cost economies, and improved the cost-effectiveness of indigenously produced military equipment.

Possibly the most convincing proof of domestic defense production efficiency is the success of Israeli defense exports. Defense exports competed successfully in the world arms market, though in no instance were they granted direct government subsidies. Besides, at the time exports were growing rapidly, sales expanded mainly to Western markets where transactions are especially price-sensitive. Exports were generally profitable to the companies involved, and as regards the national economy, a study of two-thirds of the defense exports in the later 1970s showed that the cost of a value-added dollar in defense export was on average only 0.85 of the effective exchange rate for imports (Dvir & Ben Zackay, 1985: 145).

6. THE 1990s CRISIS IN GOVERNMENT DEFENSE COMPANIES

In the early 1990s, the three large government defense companies – IAI, IMI and Rafael – suffered large losses and severe cash flow problems. A comprehensive restructuring was unavoidable, including closing plants and laying off thousands of workers, and the government had to extend extensive financial aid to allow current operations to go on. Analysis of sales data rules out the possibility that the crisis was solely the result of demand decrease.

There were evidently other explanations, some common to all three companies and some specific. Distinctions must be made also between causes of the crisis and factors that influenced its course.

One common explanation lies in the large share of exports in total sales. In the late 1980s and early 1990s, exports were 80 per cent of all IAI sales, IMI exported 50 percent of its output and Rafael about a third, so all were negatively affected by changes in the world arms market. During the 1970s the world arms market prospered, but in the late 1980s and in the 1990s it turned into a clearly buyers' market (see chapter 10, sections 2.5 and 2.6), and the keen competition eroded producer profits. Another factor affecting all three companies was the relatively high degree of autarky they maintained. It developed in earlier periods, when Israeli industries were still less developed, but continued with minor changes only when the industrial base expanded, and potential subcontractors' capabilities had substantially improved. The companies preferred "make" to "buy" in production and in supporting services, thus incurring diseconomies of scale in some of their activities, with resultant higher costs and inefficiency. The avoidance of subcontracting and outsourcing stemmed to a large extent from rigid labor relations that prevented significant workforce reduction, even when engaging with outside sources was more economical. Among the company-specific causes, a change in IAI's export composition was a crucial factor: civilian exports, traditionally with lower profitability and occasionally incurring losses, grew, while relatively profitable defense exports declined. In fact, the two activities cross-subsidized one another, so as long as defense exports grew, less profitable or unprofitable civilian exports were bearable. When military orders fell off, however, the company ran into heavy losses, and within a short time, into severe cash flow problems as well. Another company-specific factor related to IMI: incorporated as a company in 1990, it did not manage to reorganize properly, and was virtually unprepared to cope with the drastically changing business environment of the post-Cold War era.

Company and government responses to the emerging crisis conditions were too slow and hesitant. In the beginning, both thought the problems were temporary and transitory in nature, therefore not justifying any significant steps to reduce production capacity or adjust costs in any other manner. Political and public concerns then arose, mainly in the government, which further delayed unpopular steps. But delays merely increased losses, exacerbated the crisis and made recovery steps more costly. In addition, company managements – especially in IMI, as a new company, and in Rafael, still a government department – did not have full, accurate accounting and costing information, hence could not easily single out focuses of losses and inefficiency, and even less so devise the necessary adjustment measures. Managements that for years prioritized technological and technical challenges and neglected business aspects also lacked the skills and experience to make proper business assessments, underestimated the resources and the time needed for an effective recovery plan, and all initial plans deviated and fell short of their assigned goals. A second round of recovery steps was then inevitable, requiring additional financial aid.

The main difficulty arose with adjusting manpower size, composition and costs to realistic demands and to the companies' financial capacity. The labor unions strenuously opposed recovery plans that included large-scale layoffs or significant cutbacks in wages and fringe benefits. When, after prolonged negotiations downsizing plans were finally launched, they involved extremely high severance payments that used up most of the funds the government allotted for the companies' recovery. Furthermore, though the absolute number of layoffs was large, their distribution by activities and types of workers did not always comply with actual requirements and efficiency-related considerations. The plans were based on "voluntary retirement", sometimes at a uniform rate for the entire organization, and did not allow for shutdown of losing business units and for reinforcing the more competitive, profitable ones. In addition, with "voluntary retirement" it was impossible to assure that the more essential workers would remain, and in some cases indispensable skills and know-how were lost.

Practical day-to-day consequences were felt in their full force in cash flows, stressing the companies' problematical financial structure, especially their lack of adequate equity. Both IAI and IMI, when transformed from the status of government departments to that of independent companies, were not provided with adequate initial equity. In the case of IAI (incorporated, as noted, in 1968), the government as owner did not invest in the equity of the company in the later stages either, despite the company's accelerated development. At the same time, fully owned by the government, the companies had no access to the capital market, and could not raise equity from outside sources to accommodate their growth. Expansion of production facilities, investments in company R&D and in establishing an overseas marketing network were all financed by accumulated profits, and on top of that, the companies were often required to provide interim financing for the Defense Ministry's orders too, mainly for R&D work, in view of current budgetary constraints. As a result, when expensive adjustment processes had to be financed, the industries lacked sufficient capital reserves, and their equity was eroded completely. With negative equity and current losses, the companies were cut off from their regular banking sources as well, and their continued existence came to depend totally on Ministry of Finance funds.

Although the government became deeply involved in defense industry recovery, at no stage did it develop an overall policy with long-term goals. Indeed, there was no small measure of inconsistency: demands for financial aid rejected at first were subsequently approved, with additions, and so it was with other processes like closing production lines and laying off workers. In addition, since the acute questions were financial, the Finance Ministry took the lead, and its officials were naturally more interested in the costs of programs and with short-term problems, neglecting other no less important aspects of the restructuring process. The plans, then, did not assure stability and profitability over time, nor did they provide the defense industries with a sound basis for long-term technological advantages.

Significantly, at one stage or another of the crisis each of the three companies tried to diversify by expand into civilian areas, and in all instances the results were unsatisfactory, at least in the short run. Besides the inherent cultural, organizational, technological and economic differences between defense and civilian businesses, there were specific difficulties due to the limitations binding on government-owned companies. In other countries, defense companies usually expanded into civilian markets by acquiring civilian companies or by establishing business partnerships with those of proven experience in the civilian sector, but such moves were doomed to fail in view of the clumsy, lengthy bureaucratic approval processes required of government companies. Furthermore, entering new, and especially different and unfamiliar business spheres requires management attention and vast financial resources practically unaffordable by companies struggling for survival, so it is doubtful whether such endeavors had any chances to succeed in the first place. A few years later, however, after reaching reasonable financial stability, IAI did have some impressive successes in expanding civilian activities. It certainly proved the importance of a company's general condition for successful diversification, yet it also indicated, as evident in other places too, that balancing military and civilian activities is more achievable in aerospace than in other sectors of the defense industry.

In conclusion and as a background for discussing a desirable defense-industrial policy, two important lessons clearly emerge. Firstly, managements of government-owned companies lack the flexibility needed to cope with a business crisis. The management of IAI, an independent company since the late 1960s, did not in effect have more flexibility than those of IMI or Rafael, because of labor union power, government sensitivity as a owner to political and public implications of recovery measures, bureaucratic ineptitude in making decisions, and lack of equity. Secondly, the government encounters great difficulties in functioning as owners in times of business crisis, exposed as it is to political pressures from interested parties. Inevitably, its decisions are and cannot be free of considerations alien to the issue itself. Nor is the government a homogeneous organ, and its ministries reveal different if not conflicting emphases. For all these reasons government responses are slow and hesitant, conspicuously unsuitable for coping with business crises.

7. ON A DESIRABLE DEFENSE-INDUSTRIAL POLICY

In Israel there is a broad consensus as to the need for a domestic defense-industrial base, but within that consensus are differing views as to the desired size, sales composition, consolidation level, ownership, the role of economic considerations and the like. There are also several inherent dilemmas, accompanying domestic defense production issues throughout the years: attaining a reasonable degree of self-sufficiency versus the economic and social price involved; the indispensability of unique, surprising force multipliers versus the need to export sophisticated systems to assure the continued existence of defense industry; contributions to short-term military preparedness versus developments of technological options, laying foundation for innovative warfare doctrines and force structure in the longer term; resource consolidation versus duplications that stimulate competing ideas. The difficulty in formulating a coherent defense-industrial policy is without any doubt real, but equally beyond doubt is the overriding importance of such a policy.

Defense-industrial policy must aim at a new industrial structure that will meet Israel's security needs and correspond to potential foreign demands, and at the same time complies with economic criteria that assure its business stability in the long run. The principles that follow may be considered as a preliminary framework for a desirable defense-industrial policy in the years to come: -

i. As long as the core companies of Israel's defense-industrial base remain government owned, special arrangements must be made to assure them managerial and operational flexibility. These arrangements are in a sense a necessary completion of recovery steps already taken, and a remedy against sliding back into depression. Circumstances changed frequently in the past, and will apparently do so in the future, hence an ongoing ability to accommodate change is of the utmost importance. There are four fundamental elements within the issue of managerial and operational flexibility: establishing labor relations that allow for rapid organizational and structural adaptations; creating a new compensation system for the industry's employees, independent of public sector wage policy, one in which compensation, performance and the company's business conditions are more closely associated; simplifying government procedures to enable companies to enter strategic partnerships, establish industrial alliances, sell, acquire and merge with companies in Israel and abroad; and broadening the companies' equity bases, making them financially sound, capable of raising investment and working capital by the regular means of the business world.

ii. A distinction must be made between three types of activity in all defense companies, irrespective of ownership, according to a combined test of strategic value and possible conformity with rules of business conduct. The first category includes the most strategically essential activities that on one hand have to continue in all circumstances, and on the other hand cannot be run on business principles since anticipated demands are too small to cover fixed costs. Those activities are therefore unlikely to break even. Technological infrastructures and special R&D centers, as well as specific production facilities, belong to this category, and they should be maintained as budgeted units of the Ministry of Defense, their full costs covered from those sources, though current management and operations may be contracted out to business companies on a multi-year basis. The second category, at the other extreme, consists of activities important to national security, in most respects resembling other business activities, which may well be managed according to usual profit and loss considerations. Most defense R&D and production activities belong here. The third category lies between the two, and includes activities essential for national security that cannot be given up, yet involve a higher than normal degree of uncertainty as regards anticipated demands, and thus may fail to break even if run solely on the basis of accepted business rules. This category includes principally strategic production lines within companies, and the Defense Ministry should undertake to reimburse the fixed costs necessary to maintain essential minimum production capacity at those facilities. In fact, the Ministry has to consider this essential minimum capacity as any other emergency stock. Yet reimbursement should be conditional on the extent of actual activity, and the government has to offer companies incentives to obtain as many orders as possible for these same production lines, thereby reducing its share in covering the fixed costs directly. Without this threefold distinction and the differing rules for each category, it may be impossible to assure a proper response to strategic needs in the long term. Even if in the short term companies manage to maintain production lines operating at a loss through cross subsidies, they will refrain from renovating them, and in the long run those lines will lose their technological and production capabilities.

iii. Government-owned defense companies should be privatized as soon as possible. This is the central conclusion drawn from analyzing the recent past, i.e. the crisis that befell those companies and its outcomes, but it is also an inevitable move in considering future prospects. Ownership change will help the companies acquire managerial and operational flexibility, and facilitate the implementation of the new structure of distinctive three categories outlined above. No less important, as private companies it will be easier for them to proceed along the path of international collaboration, since Western companies are reluctant to enter partnerships or engage in any other form of ownership association with a government. Implementing privatization plans raises questions that are far from simple. Should the government renounce ownership of defense companies completely, or would it suffice to reduce its share holdings and transfer control to private hands? Should a company be sold as a single unit, or is it better to sell coherent business units within it separately? Should the companies be privatized by issuing stocks to the public, or by selling to private owners on the basis of potential strategic synergism? Should recovery precede a change of ownership, assuring that the sale would arouse serious interest and bring a better price? How can essential security and economic interests be preserved after the change of ownership? Every alternative brings advantages and drawbacks, and besides, considerations have to be practical, accounting for the feasibility of each option. For instance, insisting on selling ownership in full may come to nothing, among other reasons because of strenuous objections from labor unions that feel securer if they retain association with the government, at least in the transition period. Similarly, selling business units separately increases chances of carrying out the plan, since there would be apparently more potential buyers for coherent businesses than for large, diversified companies. Possibly, then, instead of adherence to one single privatization strategy, it might be advantageous to apply a variety of methods, on a case-by-case basis. Some may require multi stage plans, with government gradually exiting from ownership.

iv. A desirable defense-industrial policy should weigh the increasing importance of dual-purpose technology properly, with the advantages of integrating defense and civilian industrial activities. That is important in the defense sector, and no less so outside it. In the relatively small Israeli economy, poor in natural resources, knowledge accumulated in decades of huge investments in defense R&D is one of the most, if not the most, important resource for economic growth and industrial development, and seeking ways to expand its contribution to the civilian economy must be central to any industrial policy. In this context, the successful experience of the early 1990s, when the government created instruments to encourage venture capital investments in the civilian sector, is instructive. Possibly a special government venture capital fund should be established for investing, together with private partners from the business sector, in potential civilian applications of technological ideas developed by defense companies. To overcome incompatibility between the organizational culture of established defense companies and the managerial style suitable for innovative startups, the new entities should be separated from defense industries. At the same time, in order to encourage essential cooperation from the defense companies, they should be entitled to royalties and stock options of the new firms.

Responsibility for formulating and implementing the defense-industrial policy rests with the Ministry of Defense. No government agency is more capable of weighing the complex considerations involved, and it also has the institutional power and necessary instruments for effectively applying the desirable policy. The Ministry of Defense should accept budgetary responsibility for issues related to the defense-industrial base, just as it takes budgetary responsibility for other aspects of national security, and should reorganize in a way that will assure balanced allocation of the defense budget among the many needs for which it is responsible.

10
THE ECONOMICS OF ARMS TRADE

Producing national security requires not only domestic economic resources, but goods and services acquired from other countries as well. Consequently, a wide-ranging international trade in arms and military equipment has developed, with effects on national and international security, bilateral and multilateral international relations, the economies of buyers and sellers and the world economy at large. Since world arms trade is not limited to purchases and sales in the accepted sense, it became common to use the terms arms transfers, suppliers and recipients in this context.

Though aware of the strategic and political influences involved, and without making light of them, defense economics elaborates on the following aspects of arms transfers: demand-related and supply-side causes of arms trade, structural features of the world arms market, behavior patterns of suppliers and recipients, and the impact of arms transfers on national economies. Moreover, since arms transfers often involve foreign aid, special financial terms and varied modes of offsets between suppliers and recipients, the economic discussion has been broadened to examine, as well, the economic implications of all adjuncts to arms transfers, not just those of the transactions themselves.

1. REPORTING INTERNATIONAL ARMS TRANSFERS

Until the 1970s, virtually all research on international arms transfers was descriptive, yet definitions and measurement questions arose even then. Sometimes the use of import items as declared is misleading, since certain civilian equipment can be used, as they are or after rapid conversion, for military purposes too. In particular, clear-cut distinctions may be complicated regarding subsystems and components: an inertial navigation system can serve civilian or military aircraft, transmission units may be good for heavy tractors or tanks and so forth. It is equally problematic to make distinctions according to the intended user. For instance, paramilitary forces often import equipment identical to what the army uses, and they too may engage in armed conflicts with an external adversary. A definition based on the declared user may be too narrow, while definitions reckoning with potential military applications of dual-purpose equipment may be too broad, and the resulting differences in measurements can be indeed vast.

Arms transactions may include, besides weapons and equipment, services like training, technical instruction, maintenance, construction of facilities, etc. These services are admittedly essential complements, sometimes without which the equipment cannot be used effectively, and ignoring them underestimates the military value of imports. Nonetheless, due to technical difficulties not all statistical reports on arms transfers include such services. Likewise, reporting is not uniform as regards trade in equipment, technologies and inputs for domestic defense production.

Frequently arms transfers are reported in quantitative terms, without a common denominator to express their aggregate volume. Reports in monetary terms on the other hand allow for aggregations, however they too present problems. Firstly, sometimes prices are unrelated to actual costs and are arbitrarily determined, so officially declared monetary sums do not really reflect the economic value of transactions. For years this was the case with Soviet arms exports, for example. Secondly, the monetary value of transactions depends on their financial terms. Payments may be in cash, through grants, by means of loans on ordinary terms or "soft" loans, or by all methods combined, and converting them to a common base, say to a cash payment equivalent, requires making assumptions, with resultant valuations possibly varying within a wide range. Thirdly, related arrangements may be even more complicated: barter transactions in which arms are exchanged for goods (e.g. oil), or transactions incorporating counter trade, co-production and other forms of offset. Their monetary value should reckon all accompanying conditions and arrangements, but available information does not generally provide that. Fourth, the volume of arms transfers can be measured upon signing the contracts, at time of delivery or when they are paid for. However, not all signed contracts are carried out, sometimes no payments at all are made, and delivery dates do not necessarily match the timing on which weapons systems become operationally usable, or on which the economic implications come into play. Finally, monetary measurement raises problems in making inter-temporal and international comparisons, similar to those discussed in detail as regards defense expenditures (see chapter 5, section 5).

Since official arms transfer agreements are usually classified, and besides, there is no small amount of covert, illegal arms trade, the gathering of arms transactions-related quantitative and financial details has to overcome many obstacles. Some progress was made in the 1990s, following the United Nations initiative in establishing the UN Register of Conventional Arms, intended to alert the international community to large-scale armament that might endanger world stability. All member nations are asked to report exports and imports of conventional weapons annually, thus allowing for cross checking of information from suppliers and recipient countries, and for highlighting suspicious gaps. Results, however, were mixed: about half the United Nations member states provide data on a regular basis more or less, but several others, indeed very active in the world arms trade – Middle East countries in particular – do not; other states report only on certain groups of weapons. In addition, to minimize reluctance, data are published as reported by member states without any further verification. There were discrepancies between suppliers' reports, which tended to be less reserved, and those of recipients, yet in some cases gaps arose merely from differing interpretations and other technical causes (SIPRI, 1995: 556-558). Finally, without undermining their importance in other respects, since the reports contain only quantities, and do not provide monetary values or financial information, their contribution to understanding the economic aspects of arms trade is limited.

Economic research on world arms trade tends to rely mainly on publications of the former American Arms Control and Disarmament Agency (ACDA) and of the Stockholm International Peace Research Institute (SIPRI). Both institutions maintained a consistent methodology over the years, and reported relatively long time series of actual arms deliveries in monetary terms. There are, however, significant differences between the two (Catrina, 1988: Appendix 1). SIPRI estimates begin in the late 1940s, but until the latter half of the 1970s related only to arms transfers to the Third World. ACDA estimates, by contrast, always included most countries, but began only in the early 1960s. SIPRI relates only to major weapons systems, while the ACDA definition is much broader. The two also use different valuation methods when arms are supplied free of charge, or for transactions where prices are either unknown or apparently much lower than actual production costs. SIPRI multiplies quantities by prices known from prior similar transactions, while ACDA is more sensitive to "political" pricing (i.e. different prices to different customers), attempting to adhere to the actual financial aspects of each transaction (Brzoska, 1994: 69-70). Different definitions and valuation methods resulted in different estimates for overall world arms transfers and for arms transfers of individual suppliers and recipients. For example, the two sources agree that world arms transfers reached an all time record in 1987, yet according to ACDA revised estimates they amounted to about 72 billion dollars, in 1990 prices (ACDA, 1997: 100), while by SIPRI to only 41 billion dollars (SIPRI, 1997: 267). ACDA estimates were thus higher by 75 percent. Likewise there were differences in estimating changes in volumes of trade over time: while according to both sources arms transfers decreased continuously after 1987, reaching their lowest level in 1994, accumulated decrease, in constant prices, was about 68 percent according to ACDA and only about 50 percent according to SIPRI.

SIPRI stresses that aggregation of data on physical arms transfers through applying monetary values is designed merely as a trend-measuring device, to permit the measurement of changes in the total flow of major weapons and to illustrate its geographical patterns (SIPRI, 1997: 267). Both institutions also have reservations about comparing their estimates of arms transfers with official economic statistics of GDP, public expenditures, export and import, external debt and the like; their methods of estimation differ remarkably in nature from accepted practices of national accounting, and hence using figures of arms transfers to calculate, say, the relative weight of arms imports in total imports in fact compares data that does not have a common base (ACDA, 1995: 170). For economic analysis purposes, ACDA and SIPRI estimates are lacking too in that they present only data on deliveries, ignoring the financial conditions of arms transfers. An alternative measurement (Brzoska, 1983; 1994), adopting the opportunity costs approach, showed that financial arrangements may shift the economic effects from the time of deliveries to later periods (see below, section 5.4).

2. THE DEVELOPMENT OF WORLD ARMS TRADE

Robert Harkavy, a pioneer in arms transfer research, pointed at four characteristics in which the arms trade since World War II differed from that of the earlier inter-war period (Harkavy, 1975): the structure of supply in the world arms market; supplier-recipient interrelations; financing modes, and the general diminishing dependency on arms transfers due to expanded capabilities of indigenous defense production. Another survey of modern arms trade (Gerner, 1983) elaborated on six themes: volume of arms transfers; private versus governmental trading; suppliers' identity; recipients' identity; purposes of arms transfers, and the general political, economic and military climate of the time. Other studies emphasized the change in weapons quality and the increasing complexity of transactions (Catrina, 1988: 26-41), and the growing importance of sub-national, trans-national and international actors relative to national actors (Kolodziej, 1979). A historical survey of arms transfers must, then, relate to many different facets.

2.1 From Unregulated Private Trade to Government Involvement

Before World War I, arms transfers were of relatively limited volume. Trade was carried out entirely by private merchants, without government intervention or public supervision, within an unconcerned environment that hardly paid attention to arms transfers. Arms, particularly small arms, were sold to anyone who wanted and could afford buying them, on pure profit grounds. After World War I, however, public attitudes to arms transfers changed drastically, spurring governments to assume an active role. In the early 1920s, a special committee formed by the League of Nations examined the apparent influence of arms dealers ("merchants of death" as they were called) on war accelerating factors, and its conclusions blamed the merchants for fomenting war scares and disseminating false reports on military and naval programs to stimulate armament expenditure, and for collusions and organizing international monopolies to raise armament prices and rake in huge profits (Kapstein, 1992:142-143). In the 1920s and 1930s, the various peace movements of the time held similar views; at the height of their powers their extreme suspicion contributed much to the emerging hostile public opinion against arms trade. By contrast, serious scholars who examined the issue in depth tended to reject the criticism that associated outbreaks of wars with supposed influence of private businessmen on governments (Brodie, 1980: 248-253). Nonetheless, in the United States and Britain in the mid 1930s, on-going public protest led to official inquiries on arms trade, their conclusions paving the way for greater government involvement in arms distribution. In the United States and in most arms-producing countries in Europe, governments acquired the authority to prohibit arms export in certain cases, and the foundations were laid for licensing and controls that still exist at present, though with modifications (Gerner, 1983: 8-9). Practically, however, there were no great changes: restrictions imposed on arms trade were generally of limited scope, and attempts to apply international control on arms transfers failed.

Toward the outbreak of World War II, and even more so during the war, arms transfers grew immensely. The United States, hitherto the world's third arms supplier, after Britain and France, and selling mainly to South American countries, became the first supplier, transferring military equipment worth tens of billions of dollars to her European allies between 1941 and 1945 (Kapstein, 1992:146-147). The war brought about closer government supervision of defense production, and with it, of the arms trade. The era of freedom of action for private producers and merchants came to an end, and from that point on arms transfers were controlled and effectively supervised by central governments. From another angle, arms transfers became an integral part and a central instrument of foreign policy. During the Spanish Civil War (1936-1939), German and Italy made use of arms supply to the Fascist forces to influence the war's outcome, and in World War II it was the United States that sent unprecedented quantities of military equipment to her European allies to swing the war in their favor. To this end, the Neutrality Act of 1935, which prevented arms transfers to belligerents during wartime, was revised, demonstrating publicly and explicitly, in a legal manner, the link between foreign policy and arms transfers.

2.2 The 1950s and 1960s: The Growing Role of Arms Transfers in International Politics

In the first years after World War II, there was a general decrease in arms demand, and the warring nations, notably the United States and Britain, were left with large surpluses of usable equipment. The two powers became the world's main arms suppliers, and willingly transferred equipment from their stocks to traditional allies, generally free of charges or for very low prices. Before long, however, there were far-reaching changes in the world arms market (Sutton & Kemp, 1966). In 1955 the de facto Anglo-American monopoly in arms supply was broken, and the Soviet Union too started to provide arms to countries outside its direct sphere of influence (the turning point came with the Czech arms deal, selling arms to Egypt). In the later 1950s, France, its industry rehabilitated, regained its important position in arms sales, and alongside the four largest powers, additional suppliers – Italy, Germany, Canada and some others – gradually emerged, competing effectively for market share, and selling to countries who preferred not to identify with the major powers. In parallel, the circle of recipients widened too. Arms transfers were no longer unique to the industrialized world, and included for the first time and on a scale that could no longer be ignored, sales to developing countries. Between 1945 and 1965 the number of independent states grew from 50 to 120, and the vacuum left behind by the de-colonization gave rise to regional ambitions and frictions among the new states that fed the demand for arms. At the same time, weapons quality also improved, with new equipment from production lines replacing obsolete surplus items in the trade.

The truly qualitative change, however, was that arms transfers became a central factor in the Superpowers' struggle. The United States began to supply weapons not only to her European allies, but also to other nations perceived to represent strategic interests (Greece, Turkey, Iran, Nationalist China, Japan, South Korea and more). The Soviet Union, on its part, became the main arms supplier for the Arab states and for several African countries. Military aid became an important political lever in contending for political influence in key regions outside Europe, competitive arms distribution to Third World countries proliferated, especially to countries embroiled in regional conflicts, and since each Superpower supported another side of the conflict, it fueled regional arms races and escalated them.

A comprehensive pioneering study by SIPRI (SIPRI, 1971) revealed the increasing role of the developing nations in arms transfers. It indicated that deliveries of major weapons systems to Third World countries grew fourfold, in real terms, between the early 1950s and the late 1960s. The main sources were the United States and the Soviet Union, which together supplied some 67 percent of all arms transfers between 1950 and 1969, followed by Britain and France, supplying about 20 percent (SIPRI, 1971: 11). Markets in the developing countries were especially important for Britain (about half her defense exports in the 1960s), less so for the Soviet Union and for France (about 40 percent) and even less for the United States (some 35 percent) (SIPRI, 1971: 5). The study further indicated that the Superpowers were guided mainly by ambition to attain control and influence over the recipients, while British and French motives were industrial, relying on exports to support their domestic defense industries, and hence were not too selective as to clients. Other suppliers – notably Switzerland, Sweden, West Germany and Japan – maintained a selective policy, and avoided selling arms that were likely to take part in internal or external conflicts. As for the recipients, several principal motives were identified: acquiring improved capabilities to withstand internal and external conflicts; raising national pride and reinforcing national identity, and assuring support for the government from the military elites. The research also pointed to the changing directions of arms transfer flows. In the first half of the 1950s the Far East accounted for 34 percent of all transfers of major weapons systems and for only 12 percent, when Vietnam was omitted, in the second half of the 1960s (with Vietnam included, the proportion of the Far East remained unchanged). Decreases were conspicuous too in the relative proportions of Greece and Turkey (from about 20 to 6 percent) and of Latin America (from some 18 to 5 percent). By contrast, the relative proportion of the Middle East rose from about 11 to about 31 percent between the two periods (SIPRI, 1971: 16).

2.3 The 1970s and 1980s: Quantitative and Qualitative Growth of Arms Transfers

In the mid 1960s, mainly under the influence of events in Southeast Asia, world arms transfers surged (see Table 10.1), and then again in the early 1970s, after several years of relatively moderate growth, this time influenced by events in the Middle East. Rapid growth characterized the years between 1976 and 1982 also, followed by leveling off, while from 1988, for the first time since World War II, seven years of steady decline in arms transfers were recorded. Throughout that period until the mid 1990s, there was a close correspondence between the course of total arms transfers and arms transfers to Third World countries.

Table 10.1 Arms Transfers Worldwide and to the Third World, 1963-1999*

	Average Annual Rate of Change		The Share of Third World Imports in World Arms Transfers
	World	Third World Imports	
1963-66	11.5	24.2	52
1967-71	1.7	2.0	62
1972-73	38.9	45.0	71
1974-75	-9.5	-12.5	67
1976-82	12.0	15.3	79
1983-87	0.8	0.1	78
1988-94	-14.8	-16.7	64
1994-99	2.6	-6.4	51

* Constant prices, in percentages.
Source: Catrina, 1988:21 and ACDA, various years.

Third World countries' defense imports increased not only quantitatively but also qualitatively (ACDA, 1979a: 16), reflecting the progress in military technology along with other, more specific factors. On the supply side, the competitive arms distribution that in the 1960s stemmed mainly from the struggle for political influence turned into an economic contest in the next two decades. Arms transfers were increasingly perceived as a way to obtain foreign currency and maintain employment, and especially as a means to preserve an active and economically sound domestic defense-industrial base. Therefore, when arms transfers were no longer a matter of foreign policy only, there seemed no reason to give up economic opportunities for political considerations, and suppliers were increasingly willing to reduce the technological gap between arms for export and arms for their own country's use. At the same time, the number of suppliers in the world arms market proliferated, and although for most of the period in question the United States and the Soviet Union supplied more than half the defense imports of the Third World, the remainder came to be divided among a growing number of competitors (more than 40 supplier nations in the early 1980s). In the keen competition that developed, the quality of performance and technological sophistication of weapons became key factors.

On the demand side, changing financing methods influenced equipment quality. With a reinforced commercial orientation foreign aid grants decreased, and sales were increasingly for cash or on commercial credit conditions. For instance, whereas until the 1970s in the United States there were no "regular" commercial sales, i.e. without implications in the federal budget, in the later 1980s such sales made up more than 30 percent of all arms export transactions (Catrina, 1988: 33; Kapstein, 1992: 151). In the later 1970s the process accelerated too because of some developing countries, in particular oil-exporting countries, that could now afford cash payments for arms acquisitions. Customers buying on commercial terms demanded dominance in selecting military equipment, which meant in practice that they demanded and received state-of-the-art equipment, identical with what their suppliers themselves used. The trend came to a halt in the 1980s, *inter alia* because of sharp increases in prices of sophisticated weapons systems, and concurrently with the leveling off in overall arms transfers, the demand for sophisticated equipment became also more moderate. In many instances, rather than replacing aging systems with the newest ones, customers preferred to acquire kits for upgrading and modernizing the weapons systems at their disposal (Catrina, 1988: 29-30).

2.4 Structural Developments in the Cold War Era

Within the world arms market, in time, four sub-markets developed, differing in volume of arms transfers, in technological level of equipment and in the nature of transactions (Catrina, 1988: 42-45). The first sub-market included arms transfers among industrialized nations, most of them in complex transactions involving offsets. While this sub-market maintained first place in technological sophistication, in the mid 1960s it receded to second place in volume of arms transfers. The first place according to volume came to be occupied by the second sub-market, consisting of arms transfers from industrialized to developing countries. Offset arrangements increased there too, although licensed production or programs of co-production remained beyond the technical capabilities of most developing countries, as was doubtful their economic rationale. The technological level of equipment varied widely, with the most up-to-date fighter aircraft traded alongside old generation, secondhand systems. The third sub-market was composed of arms transfers among Third World Countries, and the principal change there was the appearance of new producers of major weapons systems. In general, the new producers offered less sophisticated weapons systems than were available in the industrialized countries, yet the relative simplicity in operations and maintenance and price advantages appealed to potential customers. Arms transfers from developing to industrialized countries, a relatively new phenomenon, created a fourth sub-market. Insofar as complete systems and finished products were concerned, sales volume was small and most transactions were a one-time deal. Steadily growing, however, were component production and subcontracting undertaken by developing countries for defense industries in industrialized countries.

Conspicuous on the supply side of the arms market throughout the years was the growing number of supplier nations. As a consequence, firstly, an unprecedented abundance of conventional arms developed (Pierre, 1982: 123). Secondly, the technological gap between arms supply of industrial countries and the demand of developing countries was bridged to some extent, since the new suppliers offered, as noted, less technologically intensive weapons, more suitable for the less sophisticated needs of Third World customers (Catrina, 1988: 58). Thirdly, competition increased, spurring traditional suppliers to offer newer and better quality weapons. Fourthly, numerous sources and keener competition made recipients less dependent on suppliers, which, combining with other factors, changed the clearly sellers' market of previous years into a buyers' market in the 1980s. Finally, numerous sources made it virtually impossible to impose effective international restrictions on arms sales, and controlling arms transfers became much more complicated. This, however, did not reach the point of substantially undermining the dominance of established suppliers from the industrialized world on the supply side of the market. In fact, no developing country became the main supplier for any other (Pierre, 1982: 126). The arms industries of developing countries remained in the main suppliers of equipment of middle to low level of sophistication, and moreover, could not offer attractive financing packages and offset programs as their competitors in industrialized countries did (Catrina, 1988: 58-59).

Notwithstanding the growing number of suppliers, the supply side of the world arms market remained highly concentrated throughout the years. According to ACDA data (ACDA, various years), degree of concentration peaked in the 1960s and early 1970s, when the two leading exporters, the United States and the Soviet Union, supplied 75 percent of all arms transfers, and the Big Ten 97 percent. In the following years the share of the two leaders decreased, to about 60 percent in the 1980s, but the Big Ten continued to provide nearly 90 percent. In the early1990s, corresponding to the overall decline in trade volume, concentration grew once again (see below). The degree of concentration was high indeed in overall arms transfers, but it was particularly prominent in certain segments of the market. Not all suppliers offer all types of military equipment, and in every category the number of suppliers decreases with increase in technological level. In addition, increased concentration paralleled a remarkable consistency in the list of major suppliers. Over three decades, only 16 countries were ever among the Big Ten, and of the Big Ten of the early 1960s, only three were absent from that group at the beginning of the 1990s.

On the demand side, in the post World War II years, more countries wanted more arms. The number of arms importing countries grew with the increasing number of independent states, following the disintegration of colonial empires. New independent states accorded high priority to defense establishments, seeing them as expressions of sovereignty, and arms imports helped to strengthen the military, and at the same time reinforced its loyalty to and support of the political leadership. Yet demand for weapons expanded because of actual defense needs also. Many Third World countries, that as a group led the increase in world demand for arms, were actively involved in armed conflicts: of 120 armed conflicts worldwide, in 1950 to 1979, only six did not involve Third World countries (Pierre, 1982: 132). Finally, it was argued that Third World countries' increasing demand for arms is part of a general redistribution of political, military and economic power between developed and developing countries (Pierre, 1982: 3-4).

The country distribution of arms imports was naturally less concentrated than that of exports. Most of the time the Big Ten among the importing countries did not take more than half of world arms transfers, and the share of the top importer was only some 10 percent. Somewhat surprisingly, however, the group of leading importers too was relatively stable: though more than 150 countries import arms on a continuous basis, in three decades only 24 of them appeared in one year at least on the list of the Big Ten. Import concentration had still another facet: states tended to concentrate their imports in just a few sources. Data on the import distribution of 81 developing countries during 1978-1987, (ACDA, 1990: 25-29) indicated that a third relied exclusively on a single supply source, and another third acquired 50 to 85 percent of their defense imports from a single source. Import concentration was particularly great as regards major weapons systems (except for battleships): two-thirds of the countries imported fighting aircraft and main battle tanks from a single supplier, and 75 percent of the countries relied on a single source for surface-to-air missiles. Logistic considerations of uniformity and prompt supply, along with financial constraints, appear to have overcome desires for source diversification that could reduce strategic dependence and exposure to political and economic influence, or could bring economic benefits through competition.

Over the years, arms flows changed directions, resulting in a substantially different arms import distribution by geographical regions (ACDA, various years). Most conspicuous was the increase in the Middle East share, which from the early 1970s became the principal target for arms transfers. At first the increase in weapons acquisition was fed by the Arab-Israeli conflict, afterwards by the long Iran-Iraq War, and then by the Gulf War of 1991. At the same time, the proportions of Europe and East Asia in world arms imports declined considerably, those of African and Latin American states rose until the early 1980s and then declined, while South Asia increased its share, mainly during the 1980s. Interestingly, since the mid 1980s the United States share in world defense imports has also increased, though it is the world's largest manufacture and exporter of defense products.

2.5 Arms Transfers in the 1990s

The end of the Cold War brought about a general decline in arms demand, and as a result international arms transfers decreased considerably. As a matter of fact, the growth of arms transfers ceased in the early 1980s, mainly for economic reasons, the decrease in oil prices and the large debt accumulation in the developing countries among others. However, from 1983 to 1987 arms transfers remained stable at a relatively high level, real decrease began only towards the end of the decade, and accelerated significantly in the early 1990s. According to ACDA, by 1990 the world arms market had shrunk by a third, and from 1990 to 1994, by additional 50 percent (ACDA, 1997: 100). According to SIPRI, in reference to major weapons systems, the rates of decrease in the two periods were about 28 and 31 percent, respectively (SIPRI, 1998: 318). In 1995-1997 the decline stopped, and world arms transfers grew by some 30 percent (ACDA, 2001: 1; SIPRI, 2002: 409). Later on, however, the trend changed once again; at least as regards major weapons systems: for 1998-2001 there was an accumulated decline of some 35 percent, and the 2001 volume of international sales was even lower than that of 1994 (SIPRI, 2002: 409).

Shrinking sales volume was accompanied by far-reaching changes in the structure of the market. In the post-Cold War era, as the traditional division into Eastern and Western spheres of influence was eliminated, the dual structure of the market disappeared. The United States became the world's most dominant arms supplier, while Russia, replacing the Soviet Union, was relegated to a lower place. In the peak year of 1987, the Soviet Union's share of major weapons systems transfers reached 40 percent, while at the 1994 low point, Russia's share was only 4 percent (SIPRI, 1996: 479-483). The American share in supplying weapons rose from about 30 percent in the later

1980s to more than 50 percent in the first half of the 1990s, and in 1999 the United States provided 64 percent of all arms transfers, an all-time record (ACDA, 2001: 4). The reverse side of the coin was the recipients' transition from dependence on the Soviet Union to dependence on the United States. In 1997-1999, among the 43 countries whose defense imports exceeded 500 million dollars, 23 acquired two-thirds or more of their defense imports from the United States (ACDA, 2001: 6). Another notable structural change relates to the group of developed countries as a whole. In fact, at the end of the 1990s they became more dominant on both sides of the world arms market. Their relative share in supplying arms rose above 96 percent, compared with 92 percent ten years earlier, and for the first time in decades their relative share also exceeded that of the developing countries on the demand side (57 and 43 percent at the end of the 1990s for the developed and developing countries respectively, as against 35 and 65 percent ten years earlier) (ACDA, 2001: 1).

The motives of arms suppliers also changed in the post-Cold War era. Over the years, emphases in government attitudes toward defense export have shifted. At first, between the two world wars, governments became involved in arms transfers because of concerns as to their potential adverse effects on international relations, and thus emphasized control and supervision. Next they discovered the value of arms transfers as a foreign policy instrument, and later their economic benefits, so governments soon came to the aid of arms exporters in varied ways. Involvement veered, then, from control and restriction to assistance, seeking a balance between political, strategic and economic considerations. In the 1990s, when geopolitical competition between the Superpowers waned, the time seemed ripe for "de-politicizing" the arms trade. Moreover, there was no further concern of regional conflicts escalating into a Superpower confrontation, and trade in conventional arms, as long as it was not connected directly with active crises, did not seem to be an acute international security problem. Finally, the new international environment made cooperation in imposing arms embargos possible if and when necessary. All the above reasons reduced the weight of political and strategic factors in arms suppliers' calculations, and strengthened the economic ones.

Economic considerations grew stronger, too, because of the decline in domestic defense demand and the resulting excess production capacity of major manufacturers. In fact, many major weapons systems disappeared from local purchase lists in the United States and Western Europe, and were produced only for foreign customers (*Arms Sales Monitor*, 15.2.1995). To a considerable extent, then, export sales became an existential condition for many Western defense industries. In addition, excess production capacity and dependence on exports reinforced competition in the world arms market, and companies demanded more freedom of action in struggling for market share, pressuring governments to relax control over arms transfers and abolish cumbersome bureaucratic procedures that interfered with their efforts.

The United States seniority in world arms trade, beside its position as the only world superpower, renders its defense export policy especially important in determining the patterns of international arms transfers. Early in 1995, the American administration formulated a new policy, which, for the first time related explicitly to the relationship between arms transfers and the defense-industrial base. One of the five principal goals declared that arms transfers are intended to enhance the ability of the defense-industrial base to meet national defense requirements and maintain long-term military technological superiority at lower cost (SIPRI, 1995: 497). On a practical level, there were two approaches: the minimalist one made do with changes in the way of doing business, stressing the need for reducing bureaucratic cumbersomeness and adopting more businesslike procedures in Foreign Military Sales (FMS) programs, the framework for most American arms transfers (Layne & Metzger, 1995); the more far-reaching one demanded federal budget support for defense exports, including financing of expensive marketing campaigns and financial risk-sharing by means of long-term credit guarantees. Its proponents demanded in addition that American embassies worldwide help promote defense exports, as they indeed do with respect to other economic interests, and that senior government officials take part in convincing foreign governments to acquire American rather than other military equipment. Both groups were against the charges imposed on arms sales to foreign customers, and particularly against R&D recoupment charges that could amount to 5 to 25 percent. It was argued that such charges made prices too high and uncompetitive, and besides, purely budgetary considerations justified waiving them: exports allow producers to gain economics of scale, and thus make it possible for the American Department of Defense to acquire arms at a lower price; the loss of defense export opportunities undermines the industrial base, and expenditures of rehabilitating could be many times greater than revenues from charges. In the late 1990s, defense export policy questions focused on how to prevent the leakage of sensitive technologies, while at the same time allowing American defense companies to participate in the globalization that increasingly came to dominate defense production (see chapter 8, section 6.2).

Export policies of the principal Western European supplier nations – mainly Britain, France, Germany and Italy – were traditionally free of political inhibitions, and were indeed guided by economic interests during the Cold War as well. Besides, most of the time European governments supported defense exports massively. In the early 1990s in Britain, subsidies for defense exports were estimated at 20 percent of their value, and defense export share in total export credit was over 30 percent. France during that period extended buyers' credit to 30 percent of defense export transactions, and credit guarantees – to some 65 percent of them (SIPRI, 1997: 252-253).

A real revolution occurred in the priority of economic considerations in Russian defense export policy. In the Soviet Union, strategic, geopolitical and ideological considerations guided defense exports, while economic considerations, if any, were secondary. Weapons systems were frequently offered to foreign customers at arbitrarily determined low prices, with no relation to costs. Recipients' ability to pay was often ignored, and huge transactions were carried out free of charge or on long-term, low-interest credit, some of which was never paid off. In the new international reality and in view of deteriorating economic conditions, priorities changed, and Russian arms transfer policy came to be dominated almost solely by economic considerations. Russia now regards export of military equipment as a principal source of foreign currency, and a way to relieve the ailing defense industry without recourse to public funds. There is no further interest in subsidized arms sales, or in sales on credit, and traditional customers lacking purchasing power ceased to be attractive sales targets. At the same time, it was understood that the competitive world arms market requires entrepreneurial, dynamic behavior incompatible with cumbersome control mechanisms, and that the success of Russian defense exports depends on an efficiently organized commercial system, as on maintaining the right balance between governmental supervision and the rules of competitive markets. In 1992, the old organization of defense exports was reformed, granting some individual manufacturers considerable freedom of action in their export sales. The outcome, however, was disappointing, and a guiding hand conspicuously absent (Sorokin, 1993: 8-9). Consequently, a government trading company, Rosvooruzheniye, was established at the end of 1993 to coordinate arms exports, while some major producers were still allowed to approach potential customers directly. On the whole, the new structure seemed to promote foreign arms sales, as towards the end of the 1990s, indeed, Russia's defense exports increased (see below).

Russia's position in the world arms market, as a replacement of the Soviet Union, changed dramatically in the 1990s. In the mid 1980s the Soviets exported 20-23 billion dollars worth of armaments annually, while in the first half of the 1990s, according to Russian sources, sometimes providing contradictory data, exports did not exceed 2-3 billion dollars a year (SIPRI, 1997: 256-257). The immediate cause of the drastic decline lay in the systemic chaos into which the defense-industrial complex plunged. There were nonetheless other reasons, on both demand and supply sides. As noted before, Russia had no further interest in selling arms to traditional clients who could not pay. At the same time, some former clients, notably the Eastern and Central European states once part of the Warsaw Pact, had no interest in renewing procurement ties with Russia beyond essential acquisitions of spare parts, preferring for political and economic reasons to cooperate in military affairs with other countries, particularly with NATO states. Other countries preferred western suppliers who could offer offset agreements, while Russia could not, or refrained in principle like the Soviet Union in the past, from combining technology transfers with arms transaction. In addition, Russia adopted certain restrictions on defense exports for foreign policy reasons, and limited arms sales to internationally legitimate markets. Difficulties were evident, too, on the supply side, in particular with respect to the types of military equipment and support services that Russia could and would sell. In the market segment of advanced military technology, Russian products were inferior, and furthermore, there was no real separation from the Soviet tradition of maintaining technological gaps between the weapons currently in use by the Russian army and those offered for export (Sorokin, 1993: 11; SIPRI, 1995: 505). As long as weapons were transferred to traditional and to a great extent captive clients, either free of charge or at subsidized prices, technological gaps were no hindrance, but things changed in competing with Western suppliers and in carrying out transactions based on realistic cost-effectiveness calculations. Another limitation lay in the incompatibility of Russian equipment with Western standards. Finally, potential customers hesitated to acquire arms from Russia because of uncertainty as to whether she could continue producing certain types of equipment and keep up a steady flow of spare parts and technical support.

In the 1990s Russia made great efforts to expand defense exports and, *inter alia*, approached traditional customers of the West, particularly those seeking to diversify their procurement sources. Thus, for instance, Russia succeeded in signing contracts with Malaysia, the United Arab Emirates and Kuwait (Grimmett, 1995: 1). She also invested great marketing efforts in several countries in Southeast Asia, the Persian Gulf and Latin America, which were not Soviet clients (SIPRI, 1995: 501-502). The approach was on the whole pragmatic, unwilling to give up without special cause on opportunities to sell arms. According to Russian reports, in the second half of the 1990s average annual defense exports stood at 3.5 billion dollars (SIPRI, 2002: 392). In SIPRI data for 1997-2001, Russia occupied the second place among arms suppliers, and her share of world arms transfers was 17 percent (SIPRI, 2002: 375). However, the greater part was sold to only two countries – China and India – and the small number of major clients indicates a basic weakness that threatens Russian defense export prospects in the future.

2.6 Some Concluding Remarks

Globalization and the growing dominance of economic factors may deprive the arms trade of some of its unique features, making it increasingly like other international economic activities and businesses. Governments of major arms-producing countries will approach international arms trade in the same manner they treat any other bilateral or multilateral economic issue, and the economic consequences of arms transfers will be for them no less important than their strategic and geopolitical implications.

Globalization in economic matters brings into the defense domain developments similar in nature to those that occurred previously in the civilian economy, possibly creating thereby a new equilibrium in supplying defense products, between domestic production and reliance on international trade, with implications both for national defense-industrial bases and for international arms transfers. Two influences should be noted in particular. First, trade in defense products will not be limited to flows of finished products, but will include also an increasingly wide range of components, subsystems and parts of weapons systems, thus carrying with it advantages and disadvantages. On one hand, specialization and international division of labor is likely to make the world defense industry more efficient, and hence save costs, while on the other hand, defense production becomes exposed to possibly interrupted supplies of inputs from distant sources, and comes to depend on internal events in many different countries. Second, because of economic incentives globalization in defense production may expand beyond the circle of industrialized nations, and lead towards increased cooperation between them and Third World states. In the beginning, arms producers in the developed countries turned to co-production and agreed to technology transfers to Third World countries as part of their efforts to penetrate markets and compete for market share, while in the 1990s and in the future motives may be chiefly associated with reducing costs and obtaining other economic advantages. Especially, since some developing countries have highly competent scientific and engineering manpower, available at lower cost than in industrialized countries, reciprocal relations may no longer be limited to production, but rather deepen and broaden into joint R&D projects too (Gupta, 1993: 63-65). In time, then, sales from developing to developed countries are likely to grow. Moreover, in the new circumstances defense industries in developing countries may advance to a higher technological level, a goal they failed to achieve through their own independent R&D programs. From another standpoint, it will become less possible to control the proliferation of advanced military technologies.

The types of arms traded between nations are closely associated with the nature of military threats. Emerging threats to international stability in the post-Cold War era stem mainly from regional conflicts among Third World states, from sub-national armed conflicts initiated by insurrectional groups and from terrorist attacks, all of which increase the demand for less technologically advanced armaments. Third World countries are not usually expected to confront military superpowers, but rather their traditional rivals who also operate less sophisticated weapons. Besides, selling prospects are focused in new states splitting off from existing ones, or in insurrectionist groups, most of whom do not have the military training needed to operate advanced equipment, and look for simpler weaponry. There are also economic reasons; on one hand, prices for sophisticated military equipment are rising constantly, while on the other hand, the end of rivalry between the Superpowers substantially reduced foreign aid and subsidized arms transfers to developing countries, their effective purchasing power thus diminished.

Technological and industrial developments also affect the composition of arms trade. As noted (see chapter 8, section 3.3), civilian and dual-purpose components are used more and more for military purposes, thus increasing their relative weight in the arms trade (Anderton, 1995: 534). The trend may lead to a broader circle of direct and indirect participants in arms transfers, and as well makes effective supervision and control over global arms proliferation extremely difficult. In addition, combined with other changes, it may further erode the unique features of arms transfers, augmenting their similarities to other international commercial activities.

As regards specific sub-markets, special factors may be at work, bringing in their typical trade patterns. In the developing countries sub-market, special influence may derive from intentions to acquire weapons of mass destruction and the means to deliver it. International non-proliferation agreements make it difficult to acquire such complete systems. Technology transfers, trade in components and mobility of experts between nations, however, cannot be effectively controlled, and thus increasingly become the principal channel through which Third World states in search of such military capabilities operate (Gupta, 1993: 65). In the sub-market of industrialized nations, arms transfer patterns may be affected by the growing tendency to manage crises and carry out peacekeeping and peace enforcement missions through multi-national coalition forces, which on their part require interoperable weapons systems. Besides, Western Europe's defense industry consolidation and the course of development of transatlantic defense-industrial ties may also be important here (see chapter 8, section 6.2). Finally, if Russia is regarded as part of the developed country sub-market, industrial collaboration between its military-industrial complex and Western defense industries may also affect patterns of world arms trade.

Lastly, as mentioned above possibilities of supervising arms transfers effectively and of controlling military equipment proliferation are becoming more doubtful. Over the years, there were conflicting opinions as to the prospects of international arms control arrangements (Neuman & Harkavy, 1979: 320-321). "Reformists" advocated international limitations on arms transfers, presenting them as an effective means to attain quantitative reductions and as a way to halt the advancing sophistication and lethality of weapons systems distributed throughout the world. The "cynics" on the other hand cast doubt on the likelihood of any practical outcomes of such international agreements; seeing arms transfers as one aspect of general political and strategic rivalry, they maintained that their quantitative and qualitative dimensions could be downsized only through reducing international rivalries themselves. A third group shifted the crux of the matter from the potential effectiveness of arms control to its discriminatory implications. For them, limitations imposed by suppliers in the developed countries on arms transfers to the Third World were an arrogant manifestation of a patronizing policy, and an attempt to preserve the unbalanced global distribution of power. By the 1990s these arguments lost much of their original significance, and new contradictions emerged. The political and military environment seemed riper now than at any previous time since World War II for arrangements that would restrict arms transfers, but economic and industrial incentives operated in the opposite direction. In a world buyers' market, with shrinking domestic markets and the need to assure existence of national defense-industrial bases, international agreements could well prove ineffective. Indeed, while cooperation in peacekeeping created appropriate conditions for imposing embargos on arms transfers in extreme cases, opportunities to sell arms under other circumstances were not easily given up. There are objective difficulties too in enforcing control due to the influence of globalization and the increasing reliance on dual-purpose technologies and commercial components for military purposes. Finally, covert arms trade that circumvents embargo decisions and export laws should not be ignored either. Covert arms trade always existed, occasionally flourishing and reaching large proportions. Since covert trade develops in response to restrictions on arms transfers, the more restrictions there are and the more strictly enforced, the greater the chances that illegal deals will proliferate. From the point of view of international control too, then, arms trade is becoming "normalized", and prospects of restricting it are indeed rather limited.

3. STRATEGIC AND GEOPOLITICAL FUNDAMENTALS OF ARMS TRANSFERS

Arms transfers are not simply part of the regular circulation of goods among countries. They involve strategic, political and economic motives, and affect all these areas. Hence the strategic and political fundamentals of arms transfers should be noted briefly before elaborating on their economic aspects. For several main reasons (Pierre, 1982: 275-276), arms transfers became central to international relations in the second half of the twentieth century. First, their geopolitical implications are multidimensional; a given arms transaction affects not only the parties directly involved, but also the relations between each of them and other potential recipients in the region, and between them and other, even more remote potential arms suppliers. Secondly, arms transfers are perceived as expression of political support, and as such became more important as traditional instruments of political support, e.g. permanent positioning of troops, faded. Thirdly, arms transfers proved effective in redistributing power from the industrial to the developing nations, with greater and more immediate effect than other means.

3.1 Arms Transfers and Armed Conflicts

Arms transfers are designed to improve recipients' military capabilities, and ideally, at the same time to grant strategic and geopolitical benefits to suppliers. However, the possibility that arms transfers will be a positive-sum game depends on whether they contribute to a deterrent balance of power that holds back potential aggression or, on the contrary, feed the arms race, undermine stability and accelerate the outbreak of war. The issue extends further to the effect of arms transfers on the course and severity of wars.

Advocates of arms transfer control maintain that countries become involved in armed conflicts when they possess the capacity and means to do so, and the more they have, the fiercer the military confrontations are. The argument had a dynamic version also: adversary states fear that the additional military power created on the other side by arms transfers is meant for offensive purposes, they may react by procuring arms themselves, and thus accelerate the regional arms race and raise tensions, or go even further and launch a preemptive strike. By contrast, those who see arms transfers as a normal international relations phenomenon stress that the arms themselves do not cause international conflicts nor do they lead to wars. Causes of wars are deeper, fueled by political, economic, territorial and ideological rivalries (Pierre, 1982: 3-5). According to this approach, arms transfers indeed create a stabilizing dynamic: military weakness may invite aggression from local rivals, or perhaps even domination by a state seeking regional hegemony, and arms transfers that improve the recipient's military capacity may help maintain relationships of mutual deterrence, and thereby reduce the danger of war. In global terms, since arms transfers are usually from militarily strong nations to weaker ones, they result in a more equal distribution of military capabilities worldwide, thus eventually increasing stability. Between the two extremes there is obviously room for a middle way. Its point of departure is the acknowledgement that a whole host of factors determine the influence of arms transfers on military conflicts: characteristics of the conflict (disputed issues, intensity of antagonism), existing weapons stocks, quantity and quality of weapons transferred, leaders' goals and political perceptions, commitments of third parties, probability of external intervention, and the like. Therefore, it holds, the influence of arms transfers vary from one case to the other depending on specific circumstances, and should be evaluated empirically (Kinsella, 1994: 19-20).

Theoretical models that attempted to identify stabilizing or destabilizing effects of arms transfers concluded that the degree of asymmetry created thereby in the balance of power, or in what the parties could achieve by actively using the arms transferred is a significant factor (Anderton, 1995: 548-550). Empirical research presented several noteworthy findings. One study analyzed armed conflicts of the 1960s in the Third World that used imported weapons from developed countries, and showed that the influence of arms transfers varied with the effectiveness of the weapons received, effectiveness being not only a matter of technical features, but also the result of the specifics of the military and political environment (Anderton, 1995: 553). In the opposite direction, another study examined whether conflicts feed the demand for arms and lead to a greater volume of trade, and showed that the monetary value of imported weapons, especially from second rank suppliers, did indeed grow in the course of military confrontations (Anderton, 1995: 553-554). A later study analyzed the influence of arms transfers on ten wars between states, and reached the following conclusions: arms transfers were a factor in decisions to wage war since they affected perceptions of changes in the balance of power; supplier states had little leverage in conditioning or determining the outcomes of wars; arms transfers generally prolonged and escalated wars, increasing the suffering and destruction they caused; arms embargos helped to contain actual fighting, but did not suffice to compel warring parties to stop wars (Anderton, 1995: 554).

Another group of studies used econometric techniques to identify action-reaction behavior patterns in supplying and acquiring arms. The assumption was that if such patterns prevailed, arms transfers would most probably accelerate regional arms races rather than contain them. In the 1960s and 1970s action-reaction behavior was found on the supply side at the level of major powers and allied blocs of states, while on the demand side, although states generally responded to arms imports of their rivals, there were other factors – previous arms imports, for example – that explained imports better (Grener, 1983: 21). Another study, limited to recipient countries only, found that the action-reaction pattern in arms imports was common in the Middle East but not in Asia. Confrontations in the Middle East were apparently more bilateral in nature, while in Asia the parties directly involved in a given conflict were affected by the behavior of other regional states (Mintz, 1986). Other studies presented a broader question, namely do arms transfers stimulate conflictual or cooperative behavior of nations in the international arena? While there were more cases of accelerating conflicts, there were enough exceptions to rule out a decisive conclusion (Anderton, 1995: 550-552). However, it was suggested that ambiguity appears to arise from relating to whole regions, while if any links at all exist, they should be sought in relation to specific rivalries (Kinsella, 1994: 21-22).

3.2 Strategic-Military Cost-Benefit of Arms Transfers

From the recipient standpoint, the principal strategic-military benefit of arms transfers lies in enhancing military capabilities. These depend on several factors: correspondence of equipment with operational requirements, theater conditions and warfare doctrines; operational absorption capacity and the technical ability to maintain the equipment; possible restrictions imposed on the use of equipment and the like. Recipients cannot usually choose items freely from suppliers' arsenals, and even less can they demand special development, or major adaptations of equipment to specific needs. Therefore there may be large gaps between what the recipients need and what suppliers are willing to sell, and the military contribution of actual arms transfers is likely to be partial and limited. Moreover, the actual military benefit was sometimes less than anticipated even, paradoxically, when suppliers agreed to sell technically advanced equipment; recipients could not keep it technically operational or use it effectively. Finally, when military and strategic interests of suppliers and recipients do not fully overlap, suppliers may restrict the use of the arms, e.g. limiting it to defense only or barring its use against some specific opponents, and military benefit to the recipients is again impaired (Kolodziej, 1979: 10-11).

Recipients pay for the option of enhancing their military capability by becoming militarily dependent on suppliers (Catrina, 1988: Chapter 9). Armed forces based on weapons systems and equipment from a certain supplier, having accommodated to the hardware with appropriate warfare doctrines and logistic organization, find it extremely difficult to replace suppliers. It entails usually comprehensive, far-reaching changes which themselves require huge resources, and no less important, imply prolonged transitory period of reduced preparedness (as demonstrated, for example, when Egypt cut itself off from Soviet sources and applied to the West). Besides, dependence on a steady flow of spare parts and other disposable items for the daily routine develops, and if interrupted, may, too, impair military preparedness and deterrence. In war, re-supply becomes necessary, and suppliers willing to sell arms in peacetime in the belief that they contribute to deterrence and stability, are not usually eager to continue supplying while battles rage.

From the suppliers' standpoint, arms transfers to allies or to other friendly countries are designed to deter a common enemy or to preserve regional stability, and so to promote their own strategic goals, sometimes indirectly. There may also be more direct and immediate military benefits: the Nixon Doctrine that evolved in the United States following the Vietnam War regarded increased arms transfers as an effective substitute for military presence and for direct American involvement in wars; arms transfers can be the "means of payments" for rights to set up military bases in the recipient's country, to fly in his airspace, or to conduct intelligence operations out of his territory; arms transfers may also provide an opportunity to test new weapons in combat conditions, thereby gaining valuable technical and operational feedback (Pierre, 1982: 21-22); arms transfers to military allies may facilitate the adoption of equipment standards maintained in the supplier's state throughout the alliance, etc.

But for suppliers too there is another side to the coin. There is often a risk that arms transfers will escalate arms races and war will break out. Then, defeat of allies may harm the supplier country's strategic interests and international status; he may have to continue deliveries in the course of the war, in contravention of his own policy, in quantities and of qualities unthinkable under other conditions; massive arms transfers within a short time could deplete inventories and reduce his own military preparedness, and in extreme circumstances the supplier forces could be drawn into active participation in the war. Moreover, since control over arms transferred to others is limited, they may be used for undesirable purposes or fall into unfriendly hands, eventually causing direct military damage to the suppliers. For instance, the supplier may incur high costs due to an earlier, unintended disclosure of operational and technological secrets, and worse still, may find himself threatened, or even hurt by his very own weapons. Nor can suppliers guarantee effective operation of weapons systems in other hands, and their products' reputation, as their own prestige in the world arms market may suffer unjustifiably from the poor combat performance of weapons due to the recipients' inferior abilities.

The strategic-military cost-benefit balance is indeed complex for both suppliers and recipients, separately and jointly. The net result depends on specific circumstances, but it is also affected by the degree of correspondence in the parties' strategic-military interests. The greater the correspondence, the greater the benefit to each side, and arms transfers are likely to be a positive-sum game. By contrast, in the absence of converging interests, a dynamic of bargaining will develop, each side striving to enhance its own interests by means of the other. As each side relates to its own goals and ignores reciprocal implications, chances for an optimal response to the strategic-military needs of both sides diminish, and arms transfers may even turn into a negative-sum game (Kolodziej, 1979: 9).

3.3 Political Cost-Benefit of Arms Transfers

Suppliers expect to gain political influence over recipients, and in particular to acquire leverage in specific policy issues. Leverage means in this context manipulating supply relations so as to coerce or stimulate the recipient to adapt its policy, or certain activities, to the supplier's wishes. Leverage through coercion may include, in ascending order of severity, refusal to supply particularly sophisticated products, refusal to increase military aid, suspension of selected or of all deliveries, and reduction or ultimately complete cessation of military aid. Leverage through incentives may be manifest in promises to increase aid or to enhance its technological level (Wheelock, 1978: 123).

Arms transfers, however, also create political commitment on the part of suppliers. Refusing requests for new acquisitions may be interpreted as shirking political commitment, and other customers as well may regard suppliers not adhering to their commitments as untrustworthy. Thus searching for political influence, suppliers may lose their political flexibility. Furthermore, suppliers' political flexibility may be limited by a kind of reverse leverage: recipients may threaten to apply to alternative sources, or to transfer their political affinity to another camp, thus pressuring suppliers to meet their requests. Recipients are usually aware of the political leverage gained by procuring arms, and in many cases preferred government-to-government transactions to purely commercial ones (Gansler, 1980: 206-207). The political dimension of arms transfers, then, is not unidirectional, and the interrelations created thereby frequently limit the political freedom of action of both parties.

Attempts to gain political influence may sometimes boomerang. While arms transfers strengthen the ruling group and gain its support, they may become thorns in the flesh of its opponents, and a unifying symbol for efforts to overthrow the government. In the case of a revolution, suppliers will almost certainly lose their political influence, and weapons previously supplied may be directed against their interests and cause them political damage. That happened to the United States' relations with Iran after the Shah was overthrown, for example (Kapstein, 1992: 145). But even in lest drastic circumstances, arms transfers and military aid may arouse antagonism and strong protest against their suppliers (Gerner, 1983: 29-32). In times of internal dispute, arms transfers give the rival parties the means either to expand it and make animosity more violent, or to stamp it out brutally. In addition, arms transfers represent inflow of advanced technology to undeveloped countries, thereby creating rapid structural change in societies not mature enough to absorb it, which in turn may cause an economic and social divide that exacerbates internal violence. Finally, arms acquisitions and defense expenditures in general, come at the expense of responses to economic and social expectations of the people, and neglect in these areas may stir up frustration and unrest. In all such cases, hostility may eventually be directed against the arms suppliers.

Thus the political cost-benefit balance of arms transfers has its ambiguities too. While suppliers anticipate political influence over recipients, the recipients see arms transfers as a political commitment of the suppliers, and each side has some leverage over the other. Suppliers and recipients become politically captive to one another, and the inherent reciprocity in their relationship urges both to restraint and mutual consideration. Hence the political influence that suppliers acquired remained in fact latent, an option executed moderately, if at all, for fear of rupturing ties and boomerang effects. Recipients on their part preferred in most cases to substitute security risks for the risk of being exposed to political influence, and were usually ready to take the political path marked out for them.

An empirical study attempted to identify the factors that helped the United States gain political influence over arms recipients between 1950 and 1992. Of a total of 191 cases there were 80 successful ones, and they were examined in depth to identify factors that proved effective in one or more cases. Quantitative analysis pointed at six core determinants of success: type of sanction applied, type of policy over which influence is sought, recipient regime, the status of the United States in the international system, degree of recipient's dependence on the supplier's weapons and availability of alternative sources for arms. More specifically, influence attempts were more likely to succeed through positive rather than negative sanctions; when focused on altering the recipient's foreign rather than domestic policy; when the recipient was a civilian regime, not a military one; when the United States was more powerful in the international system; when the proportion of recipient's imports from the United States was greater, and when the recipient had less alternative supply sources. However, some findings were rather surprising: for example, a wider scale of indigenous defense production in recipients' countries did not seem to diminish chances of gaining political influence, while on the other hand, active involvement of recipients in external conflicts did not seem to enhance them. The main conclusion appears to be, then, that gaining political influence depends on a set of variables, not on one or two principal ones (Anderton, 1995: 556-557).

3.4 Arms Transfers as an Interaction of Interest Groups

Nation states, or national actors, are not the only participants in the arms transfer process. Systemic examination of arms transfers may identify at least four groups of actors (Kolodziej, 1979): -

i. *National actors*: governmental authorities that formally determine nation states' foreign and defense policies, for whom arms transfers are instruments to obtain security, political influence or macroeconomic advantages.

ii. *Sub-national actors*: large, national public and private bureaucracies, such as military organizations seeking security and military capability through arms transfers; industrial organizations looking for economic benefits (e.g. profits); techno-scientific centers – universities, research institutes and individual researchers – interested in new knowledge, technological progress and product development; and government agencies that supervise the other sub-national actors.

iii. *Transnational actors*: multinational companies for whom producing and marketing arms is part of their regular business activity, their behavior guided by economic interests; and revolutionary movements having relations with third parties outside the country, for whom arms transfers advance the efforts to replace the ruling government and effect internal political change. Transnational actors differ from sub-national actors in having relations and interests that cross national borders, and from national actors in lacking legal standing and legitimacy both in their own countries and internationally.

vi. *International actors*: collective bodies composed of nation states, with recognized international status in arms transfers (e.g. military alliances), and organizations or other collaborative forms for arms control.

Some actors belong to more than one group. For example, defense companies engaged in collaborative programs with companies in other countries may be considered as sub-national and transnational actors at one and the same time. Moreover, over time the goals driving the behavior of various actors, or their relative importance, may change.

From this viewpoint, all explicit and implicit influences, as well as horizontal and vertical interrelationships within and across the various groups involved in arms transfers should be considered in order to understand the occurrences in the world arms market. While the determinants of arms transfers are primarily political disputes and armed conflicts between states, the defense-industrial, bureaucratic-administrative and techno-scientific structures that emerged in every country in response to national security exigencies, have themselves become interested parties in arms transfers, usually favoring their expansion, and demanding recognition and influence on the security issues and conflicts that foster them. At the same time, institutionalization of worldwide division of labor to save costs and of cooperation among technological talents within multinational companies or other cross-border frameworks, have bound separate national centers into a complex transnational superstructure, with its own interests in expanding arms transfers, also demanding influence over security issues. Finally, international military alliances legitimize arms transfers on grounds of national and collective security and, moreover, generate need for additional arms trade to attain standardization and interoperability.

This approach gives primacy, or at least a most important role in arms transfer processes to pressure group politics. All the groups favor arms transfers, but each seeks to promote different goals through them, and harmonizing all the goals is difficult or impossible. Thus negotiations and bargaining develop among local elites, in formats that vary from one political or economic regime to the other, and these processes eventually determine the course of the world arms trade. The outcome, then, reflects a compromise in internal power struggles, not maximization of economic and strategic benefits on the national level. For example, as regards suppliers, while there is a positive supply of arms only if sub-national actors derive benefits from arms exports, there is no way to assure that the overall balance of arms exports at the national level will be positive. The advantages of arms exports are direct, explicit and concentrated among certain interested parties, while costs are less clear, dependent on alternative resource uses and spread over the entire community (Smith, Humm & Fontanel, 1985: 243). At times not only do sub-national actors convince governments to agree to arms transactions that are doubtful from strategic, political and economic standpoints at the national level, but even pressure them to extend costly aid to such transactions. Aid may take different forms, but the common denominator in most cases is that gaps emerge between social and private benefits and costs. The private cost-benefit balance of exports may be positive, while the social balance may entail loss (Sandler & Hartley, 1995: 247-248). Similar conditions may prevail among recipients.

4. THE ECONOMICS OF ARMS EXPORTS

While strategic and foreign policy considerations motivated countries to commence defense exports, economic considerations encouraged them to continue and expand it. Sooner or later, the maintaining of an indigenous defense industry brought about problems of efficient resource use, and exports provided an attractive solution. Somewhat paradoxically, since defense exports helped maintain economic activity and avoid unemployment of labor and capital, it could serve as a bridge between domestic economic-social and external strategic-political goals. On the macroeconomic level, arms exports may stimulate GDP growth, employment and public sector revenues, and contribute to foreign trade and the balance of payments. On the microeconomic level, since they make larger runs and continuous production possible, defense exports may save costs to the domestic customer.

4.1 Macroeconomic Implications

Obviously, producing military products for export, like any other production, contributes to employment and to GDP. The question therefore is not if but how much, in relative rather than absolute terms, in comparison with other demands, do defense exports contribute to economic welfare.

In the United States at the end of the 1970s, employment resulting from exports within the FMS programs (at an average level of 8 billion dollars a year), was estimated at 0.3 percent of the labor force, and other federal programs, spending the equal amounts, was shown to create more employment (Cahn, 1979: 179). Neither did an analysis by industries reveal significant employment implications. Calculations on the basis of the Michigan Model of World Production and Trade showed that a unilateral American embargo on arms trade in 1980 would have reduced employment in three industries only – transport vehicles, electric equipment and metal products – and in no case would the estimated reduction in employment have been more than 2 percent (Grobar, Stern & Deardorff, 1990: 112-113). In Western Europe, from the same analysis, a multilateral embargo on arms trade would have led to a relatively greater movement of workers than in the United States: in 1980, 0.28 percent of French workers would have changed employment, 0.16 percent in all countries of the European Common Market, and only 0.12 percent in the United States (Grobar, Stern & Deardorff, 1990: 114-115). Possibly the arms trade was more significant for employment in Western Europe than in the United States, although the calculations were affected by overall unemployment, the flexibility of the labor market and other structural features of the economies concerned. In any case, even among the main Western arms supplier nations, defense exports were not a significant element in employment. A similar observation emerges with respect to the relative contribution of defense exports to GNP. ACDA data indicate that defense exports in supplier nations contributed to GNP about 0.2 percent in the 1960s, 0.3 percent in the 1970s and 0.4 percent in the mid 1980s. For more than two decades, from the mid 1960s to the mid 1980s, the proportion of defense exports in GNP in any year exceeded 1 percent in only four countries (Catrina, 1988: 242-243)

The macroeconomic significance attributed to defense exports is not therefore a result of its quantitative contribution to employment or to GNP, but arises mainly from three other causes. First, defense exports affect outputs and employment in advanced technology industries, and thus encourage knowledge-based growth in the entire economy. Secondly, defense exports create employment opportunities for scientific, engineering and technical manpower, contributing to the economy's human capital. Thirdly, defense exports balance out fluctuations in domestic defense demand, which tend to be larger than fluctuations in other demand components, thus helping toward economic stabilization. Opinions are divided as regards the first two causes, just as there is no agreement as to the effect of defense production and defense expenditures as a whole on economic growth and on technological progress. The third cause derives from a seemingly inverse relationship between domestic defense demand and arms foreign sales, an assumption that had to be verified empirically.

The economic situation in Western Europe worsened in the 1970s and the early 1980s, mainly due to surging oil prices, and unemployment increased. A study of the four major European arms suppliers found that arms exports rose in those years, and explained that in times of economic difficulties, when governments allocate more resources to goals unrelated to national security and the defense share in the state budget decreases, defense industries are impelled to make up domestic demand shortage by increasing exports. However, it was also found that the proportion of defense exports to overall exports did not change, implying that other exports also responded to the deteriorating economic conditions, fulfilling a stabilizing role no less than defense exports. Apparently, then, there was nothing unique about the course of defense exports under economic recession circumstances (Catrina, 1988: 242). Another study examined the assumption of an inverse relationship between domestic demand for airplanes and aerospace industry export in Britain between 1948 and 1967, and a significant negative relationship was found. The relationship was also negative when examining the effect of domestic demand for airplanes on the relative share of the British aerospace industry in world aircraft trade (Sandler & Hartley, 1995: 254-255). A third study examined how the levels of defense expenditures and their annual changes during 1958-1983 affected exports of major weapons systems from the United States and from the four main exporters in Western Europe to the Third World. It was assumed that a long-term positive trend in overall defense expenditures would influence exports favorably, since it represents a growth in domestic acquisitions that brings cost advantages, making the country's defense industry more competitive on the world market. On the other hand, however, in the short run, domestic acquisitions compete with exports for production capacity, so that a negative association between defense exports and annual changes in defense expenditures may be expected. A short-term negative association was found in the United States, France and Italy, indicating that exports may have a stabilizing effect. By contrast, no short-term negative association was found in Britain and West Germany, and moreover, in those two countries and in the United States the long-term effect of defense expenditures on defense exports contradicted the earlier assumption. The explanation put forth held that France and Italy, as a matter of policy, adapted weapons systems in advance to potential foreign customers requirements, while other countries' defense-industrial policy adhered to local priorities (Smith, Humm & Fontanel, 1985: 244-247). Hence the chances for arms exports to balance out fluctuations in domestic demand and fill a stabilizing role in the economy depend on the extent to which defense production policy considers world market requirements in advance.

Defense exports may also benefit the state budget, and within it the defense budget. For example, if defense exports reduce the cost of military products, they make it possible to economize on defense expenditures. In the United States in the later 1970s, FMS programs were estimated to have saved 0.5 percent of defense expenditures (Cahn, 1979: 180), and a later estimate for Britain, France and West Germany indicated annual savings of up to one billion dollars for each country as a result of their defense exports (Anderton, 1995: 540). In addition, exports of surpluses or obsolete generations of weapons systems contribute to state budget revenues, as do royalties on R&D investments and charges for administrative expenses. Finally, when defense industries are government-owned, and accrued profits or losses find their way to the state budget, exports also exert indirect influence as a determinant of companies' business results.

Defense exports relative share of all exports usually measures their contribution to the balance of payments, despite possible inaccuracies (see section 1). According to ACDA data, the ratio of defense exports to total exports worldwide, peaked at 2.9 percent in 1984, decreased continuously to 0.8 percent in 1996, and stood at 1 percent in 1999 (ACDA, various years). Arms exports, then, did not have great relative weight in world trade as a whole. There were, however, considerable differences among countries, and in some there were significant changes over the years. In the Soviet Union the relative contribution of defense exports was the greatest, growing from 13 percent in the later 1960s to 20 percent and more in the 1980s. Later there was a steep decline, and the estimate for Russia in 1994 was about 2.5 percent and in 1999 about 4 percent. In the United States, the relative shares were lower and indicated a downward trend: about 8 percent in the early 1970s, 5-6 percent in the mid 1980s and 4-5 percent in the later 1990s. An example of relatively wide fluctuations was provided by France: the relative contribution of defense to total exports rose continuously from about 1 percent in the late 1960s to about 4 percent in the mid 1980s, then dropped to less than 1 percent in the beginning of the 1990s, and rose again to more than 2 percent in the latter half of the decade.

The discussion of the special influences of defense exports on the balance of payments comprises several issues. One issue examines whether exporting arms helps or hinders civilian exports. Defense exports may open doors to civilian exports, but it is also possible that countries acquiring arms may in the short run cut down civilian imports for lack of foreign currency, so that defense exports come at the expense of civilian exports (Cahn, 1979: 175-176). The second issue is about the role of arms exports in assuring import sources of strategic raw materials, especially energy supplies. This became very important in the 1970s, following the surge in oil prices, and also included the question as to whether arms exports could be instrumental in recycling petrodollars and mitigating the effect of oil prices on the balance of payments of industrialized countries. One study found evidence linking oil flow to arms flow – big consumers of oil sold more arms, and big oil-producing countries acquired more arms – and concluded that in Western industrialized states arms exports did indeed performed such a role (Catrina, 1988: 241-242). With that, however, as noted earlier, the non-defense exports of Western European countries grew as well, so it is doubtful whether arms exports fulfilled an exceptional role here. An inverse causality relation was also postulated, not specifically with respect to oil, i.e. that developing countries raise prices of their strategic raw materials to obtain finances for their arms acquisitions (Catrina, 1988: 72). The third issue relates to the net foreign currency balance of defense exports. The import component in defense production and exports may be high, especially for new suppliers. The true value of exports, moreover, may be lower than the nominal value due to the financial conditions of arms transactions. In addition, credit sales are not always paid for. The Soviet experience provides abundant examples, but in the 1990s the British government too had to make good on one fourth of its credit guarantees for arms exports (SIPRI, 1997: 253).

In sum, the overall macroeconomic balance of defense exports is not conclusive. Nevertheless, countries continue to promote foreign arms sales, and governments tend to assist and to participate actively in marketing. Along with non-economic explanations, possibly short-term economic advantages or alternatively, the attempt to overcome immediate difficulties, swing the balance and cover up undesirable economic implications that may emerge only over longer periods.

4.2 The Influence of Offset Arrangements on Arms Exporters

As already noted, the world arms trade in recent decades tended to move from "simple" arms transfers to more complicated transactions, *inter alia* applying various kinds of offsets. In the early 1970s, 15 countries adopted offset arrangements, while in the beginning of the 1990s the number of countries that did so grew to 130 (Hall & Markowski, 1994: 173). Notwithstanding the many variants, all such arrangements had a common purpose, namely alleviating the financial burden of importing arms by creating economic, industrial or technical advantages for the buyer. The various arrangements fall into one of two categories: creating for the importing country foreign currency income opportunities to offset the foreign currency expenditures of importing arms, or adding elements to the military equipment included in the transaction, which increase its value to the importing country at no extra cost. Offsets may be further divided into direct and indirect arrangements. The first involve offsets through goods, services and know-how related directly to the arms transaction, while the others extend to elements totally unrelated to the transaction itself. A given arms transaction may provide a basis for direct and indirect offsets simultaneously. The principal forms of offsets include: -

Counter trade: a supplier's commitment to buy goods and services from the recipient country, often defined as a percentage of the transaction's value.

Capital investment: a supplier's commitment to build infrastructures, production plants and service centers in the buyer's country either alone or with local partners.

Co-production: an intergovernmental agreement making available to the recipient country technical information and know-how for the domestic production of complete systems or subsystems developed and routinely produced in the supplier country.

Joint production: a division of labor between seller and buyer in producing a weapons system designed and developed by the seller. Sometimes, the buyer may also take part in development.

Production under license: a commercial agreement between the original producer of the equipment in the seller country and a local producer in the buyer country, in which technical information and production know-how are transferred so the equipment can be produced in the buyer country.

Subcontracting production: a commercial agreement between producers in the two countries, in which the suppliers purchase subsystems and parts produced in the buyer country. The commitment to buy from subcontractors may be limited to subsystems and parts for systems designated for the buyer, or broadened to include systems intended for third parties.

Technology transfers: assisting R&D centers in the buyer country, participation in joint R&D programs there, technical assistance to subsidiaries established in the buyer country, and the like.

Marketing and support service rights: the buyer obtains rights to market the equipment to third parties on the supplier's behalf, or is authorized by him to extend support services to other clients having his equipment.

Some forms of offset are of purely economic significance (counter trade), others carry techno-economic implications (co-production, production under license), but certain arrangements may benefit the buyer economically, enhance his technological ability, and no less important – reduce his strategic-military dependence on the supplier at one and the same time. Recipients are naturally interested in reaping maximum benefit from every arms import transaction, and the competition among suppliers in the arms market since the 1970s worked in their favor. Offsets, then, became, a critical consideration for buyers, and a central sales promotion strategy for sellers. With that, on the face of it most offsets are a zero-sum game, and when buyers profit – sellers lose. Moreover, there are grounds for the claim that offset agreements are in essence a kind of restriction on the free trade among nations, thereby causing trade diversion, distorting efficient allocation of economic resources and eventually reducing social welfare.

From the standpoint of the arms exporters' balance of payments, in the short run offsets reduce the net foreign currency returns, since they involve foreign currency outlays for the acquisition of goods and services in general (counter trade), for procuring subsystems and components of weapons systems in particular (production under license, subcontracting), and for investments in the buyer country. Buyer countries sometimes obtained accumulated offsets of more than 100 percent, presumably implying a negative net contribution to the exporter country's balance of payments. An examination by the American government found that in export contracts with offset arrangements, the average offset rate for 1980-1984 was 55 percent (as not all contracts had an offset clause, the average rate in overall defense exports was substantially lower) (Catrina, 1988: 36). In the longer term, if offset arrangements strengthen the industrial and technological capabilities of the recipient, they may impair prospects for future exports as well; not only is indigenous production likely to reduce demand from the buyer himself, but he may also compete in sales to third parties with the supplier. It seems, then, that suppliers have no interest in entering offset arrangements, and their readiness to undertake such commitments calls for further explanations.

The principal explanation presents offsets as another dimension of complex transactions and an additional bargaining point (Martin & Hartley, 1995: 127). In complex transactions, negotiations are conducted on quantity, quality, prices, delivery dates and the like. There are usually trade-offs between those elements, and the contracts, when signed, represent a balanced compromise between seller and buyer. The offset becomes part of the equation, leading to another balanced compromise and to a new equilibrium between the elements of the transaction. Offset may be regarded as an alternative to price discounts and often, in anticipation of demand for offsets, suppliers request in advance a higher price that compensates for them (Hall & Markowski, 1994: 187). Negotiating possible offset arrangements may also draw attention away from other details of the transaction and from making comparisons with competing alternatives (Hall & Markowski, 1994: 186). There are complementary explanations too: competition in defense markets is imperfect, producers enjoy economic rents, and offsets reflect readiness on their part to share that rent with the buyer (Udis & Maskus, 1991: 154-156; Hall & Markowski, 1994: 185); since in most cases prime contractors, who subcontract work, are parties to offset arrangements, by agreeing to offsets they merely give up work done by others (Udis & Maskus, 1991: 153); in this context, too, offset commitments spur prime contractors to seek out suitable subcontractors in the buyer country, and they may indeed find more efficient producers and thus save costs (Martin & Hartley, 1995: 127); know-how and technology transfers to developing countries may turn out to be a calculated risk that carries with it additional profits – the recipient may have difficulty implementing the transferred technology, and continue to buy components, production equipment and technical support from the supplier.

Once offset agreements are reached, suppliers try to reduce their impact as far as they can. They will naturally strive to prolong implementation time as much as possible, and in counter trade commitments, for example, they will try to buy goods that in any case would have been traded between the two countries. When there is already large scale bilateral trade, counter trade arrangements can hardly assure any net additions to the normal level of trade carried on even without them.

4.3 Microeconomic Implications

Defense exports make it possible to save on costs due to continuous production and larger runs. Larger runs decrease average unit costs because fixed costs are applied to more units, and as a result of economies of scale and learning effects. Continuous production saves the costs of closing and reopening production lines.

By definition, producing larger quantities does not save fixed costs, but allows for apportioning them among more customers so that each carries a smaller share. Technically, reduced unit costs for the local customer may be expressed in one of two ways: applying fixed costs proportionally to the units sold abroad and to those sold domestically, thus lowering the prices for the local customer in advance; or alternatively, charging the local customer with the full fixed costs, and later, following export sales, crediting him with recoupments from the units exported. The second method is usually used for R&D expenses. Sometimes the decision to develop and produce a new weapons system is based on the anticipated average unit cost, and export potential may well be a decisive factor.

Economies of scale and learning make real cost savings possible. Economies of scale are a function of production volume at a given time: a larger volume per unit of time makes it possible to save costs since it pays to employ specialized machines and apply a more carefully planned division of labor. Learning is a function of cumulative output over time: workers acquire skills and discover more efficient methods while producing in the first period, and thus save costs in the second one. In both cases overall production quantities are important, but their distribution over time exerts contrary influences; while concentrating production within a shorter period requires more workers and may result in larger economies of scale, spreading the same production quantity over a longer period requires fewer workers and may yield greater learning effects.

Export orders are likely to balance out fluctuations in the budget and in local demand, and thus help maintain a steady production rate. A steady production rate, known in advance for long periods, in turn, allows for production cost savings, since it enables companies to organize labor and machinery more efficiently, to schedule material purchases more accurately and so forth. In addition, export orders may prolong production periods, keeping production lines open ("warm") after local orders have been delivered. Due to time preference, once weapons development is successfully completed, the local customer is eager to acquire the new system as soon as possible, and production facilities may suffer idle periods. Business companies cannot wait idly until the need arises for a new generation of weapons, and will either convert production lines to other uses or close them down. Defense exports can bridge gaps between local procurement cycles, thereby saving costs of closing and reactivating production lines.

Microeconomic advantages are expected to be greater in exporting technologically advanced weapons systems, where the ratio of fixed R&D costs to variable production costs is higher. Thus the composition of exports too may determine the savings for local customers (Catrina, 1988: 263-264). Besides, savings in local procurement depends on the timing of producing for the two markets: if export units are manufactured at the end of the series, economies of scale and learning effects will not affect the costs of local procurement, which would now be completed. Finally, manufacturers may add features the domestic consumer does not need, designed only to improve the export chances of equipment. In these cases, exports may raise costs and make local procurement dearer rather than cheaper (Gansler, 1980: 205).

Competition in the world arms market in fact placed obstacles in the way of reaping potential microeconomic advantages. In many cases arms suppliers could not afford to charge foreign clients for fixed costs, and had to determine export prices according to marginal costs. To the extent that economies of scale and learning effects reduced marginal costs, it was export clients who benefited while local customers covered all fixed costs, and paid eventually higher prices.

4.4 Arms Supply in the World Market

Over the years, the commercial orientation in international arms sales was reinforced. Producers regarded military equipment as any other product with a profit potential, and the impact of profit considerations grew stronger (Kolodziej, 1979: 15). In essence, when greater profits were to be made from export than from local sales, profit-maximizing producers preferred to sell weapons abroad. In the 1970s, foreign sales were often more profitable than local sales: a study of the American defense industry in 1976 showed that export profitability was 2.5 times higher than that of domestic sales, and similar evidence was obtained as regards Britain (Gansler, 1980: 205). Since that time, two major trends influenced the relative profitability in opposite directions: on one hand, increased competition in the world arms market eroded export profits, while on the other hand defense budget cuts in the late 1980s and early 1990s tightened already strict controls over profits in local sales (Sandler & Hartley, 1995: 243).

Traditional theory explains international trade in terms of comparative advantage. Comparative advantage may arise from differences in production technologies or in the relative abundance of production factors in different countries, as well as from special local market conditions for the finished products. The distinctive features of the international arms trade, however, sometimes raise doubts as to the adequacy of the accepted economic rules in explaining suppliers' conduct. For example, the neoclassical model of international trade anticipates that a country will trade with countries different from itself, a suitable explanation for the arms trade between developed countries and the developing Third World, but not for the wide-ranging trade among the industrialized countries, alike in their relative abundance of production factors, in technology and in tastes. One answer is that more detailed distinctions should be made between countries. Following this line of thought, arms trade within the industrialized countries sub-market is an intra-industry trade that obeys the comparative advantage principle too, yet identifying the causes requires consideration of more specific variables (Anderton, 1995: 539). For instance, for 67 states in 1991, it was found that defense to total exports ratios could be explained econometrically by the country's intensity in human capital of specific types and by a high local demand for defense products (Tishler & Rotem, 1995: 490-492) (see also chapter 11, section 1.1).

It is not entirely accepted, however, that more detailed examination of differences between countries is all that is necessary to explain the arms trade, and there is even disagreement with the underlying assumption that relative cost advantages are a major determinant of arms flow worldwide. The arguments divide into three groups: those that stress structural reasons and other characteristics of arms transactions; arguments about government involvement in arms transfers; and those that suggest adding explicitly non-economic factors to the model.

Structural reasons and other inherent characteristics of the arms trade make for supply prices of weapons systems that are not always determined by production costs and maximizing economic profits, as in conditions of perfect competition, and so they do not always scrupulously reflect comparative advantages. First, the high degree of concentration on the supply side of the market leads to oligopolistic competition among suppliers, and in oligopolistic competition supply prices tend to deviate from costs. Secondly, in seeking self-sufficiency countries created excess global production capacity, and rational, profit-maximizing producers are urged in the short run to sell in the world market at marginal costs, lower than their average production costs, although incurring long-term losses (Kolodziej, 1979: 15). Third, since arms transactions are long-lasting in nature, and the acquisition of major weapons systems is usually followed by many years of support and spare parts supply, suppliers interested in maximum profits over the entire life cycle of any given transaction are prepared to offer the major weapons systems themselves at low prices, seemingly unrelated to costs, taking advantage of the buyer's consequent dependence to charge highly profitable prices in subsequent phases (Sandler & Hartley, 1995: 248).

Government involvement in arms trade derives not only from purely economic considerations, and sometimes supports transactions lacking economic rationale. If the sale is not worthwhile for the company, but is in the government's strategic and political interests, it may subsidize the transaction in different ways. By contrast, if it is worthwhile for the company but undesirable for other reasons, the government can prevent the sale by administrative means. Government policy and involvement thus affects arms supply no less than do the fundamentals underlying comparative advantages.

Acknowledging non-economic factors, a political economy analysis of the arms trade developed. The arms market is considered a special case of international trade in conditions of imperfect competition, where suppliers decide on sales not only according to economic profits, but also in view of their security repercussions (Levine, Sen & Smith, 1994). In technical terms, the objective function of the supplier depends on the profit from the sale and on the security repercussions derived from changes in the stock of arms held by the recipient. Arms sales increase the recipient's stock of arms and add to his military capability, thus also altering the security of the supplier. He will be ready to increase (decrease) sales at any given profit level when the recipient's additional military capability contributes to (impairs) his security.

Arms transfers that have a positive short-term effect on the security of suppliers may turn into negative effect in the long run, and vice versa. For example, arms transfers originally intended for legitimate self-defense may accumulate to the point where the recipient acquires offensive capabilities, threatening thereby the supplier's interests. In the reverse case, arms transfers to a regime with aggressive intentions may be potentially dangerous for the supplier in the short run, but may also give him leverage over the recipient that can be used to restrain illegitimate behavior, leading eventually to a positive long-term effect on the supplier's security. Thus arms supply of myopic exporters differs from that of forward-looking exporters who weigh possible developments after the supply period as well. The first are interested merely in economic profits and seek to maximize them, while the others seek to maximize the aggregate present value of economic profits and of the security repercussions over time. Time is significant for the recipients too; they see suppliers' willingness to maintain longer-term relationships as crucial, particularly regarding the supply of spare parts and re-supply in times of conflict. Suppliers' credibility is therefore an important factor, and a distinction must be made between situations where recipients believe that suppliers intend to supply them in the future and situations where they do not. Recipient behavior will be different if he considers the supplier credible or not, and in either case, depending whether he himself is myopic or forward-looking. Accordingly, in these various cases the security repercussions for the supplier and hence on his supply, will be different. .

Defense companies and other advocates of arms exports argue that suppliers must ignore the externality from the sale, or the security repercussions, because of competition. It is often maintained that 'in a competitive market, if we do not sell the arms someone else will, and the country will get neither the money nor the security'. In other words, if security repercussions are inevitable, arms export decisions have to be made on a purely economic basis. Hence competitive conditions in the international arms market are important too, and the model accounts for them by introducing a measure of the number of competing suppliers.

Assuming that suppliers know the demand function of recipients, and are aware of the impact that changes in recipients' military capacity have on their own security, it is possible to apply an equilibrium analysis and derive several propositions as to features of the arms supply: -

i. Positive security repercussions increase supply, while negative ones decrease it. Arms supply will be positive as long as economic profits exceed negative security repercussions.

ii. Myopic suppliers who ignore security repercussions will offer allies less arms and rivals more arms than will forward-looking suppliers.

iii. Recipients' behavior does not affect the decisions of myopic suppliers. With myopic suppliers, then, the arms supply will be the same, irrespective of whether the recipients are myopic or forward-looking.

iv. Should both suppliers and recipients be forward-looking, when credible suppliers pre-commit themselves to a continuous relationship, arms supply to allies will be greater, while without such commitment it will be less. In sales to adversaries, there will be opposite outcomes. The explanation is as follows: when recipients trust suppliers' commitments they assume a higher present value for future flows of supplies, or consider their potential military capability greater, and react accordingly. As a result, the security repercussions for the suppliers will be also greater: in supplying allies they are positive, and thus increase arms supply, while in supplying adversaries they are negative, reducing arms supply. With myopic recipients, there is no significance to a commitment or an absence of it.

v. Whatever the security repercussions, increased competition, i.e. a greater number of non-cooperating suppliers, increases the supply from each individual supplier, assuming it is positive.

Integrating economic and political-strategic factors makes it possible to expand the analysis in other directions: to examine the chances of cooperation among suppliers and its possible influence on supply; to consider differences in suppliers' strategic position in the international system (e.g. to distinguish between the United States with its global security outlook and consequent leadership in the world arms market, and second rank suppliers like France and Britain who follow the United States and are motivated mainly by economic considerations); to include several recipients in the model, and to differentiate between those capable of producing arms domestically and those that are not, etc.

Explanations of supply behavior in the world arms market should relate its qualitative aspects as well. All suppliers abandoned the idea of limiting arms transfers to surpluses and obsolete generations, and influenced by competition, are prepared to sell state-of-the-art arms systems right off the production line. What is left to examine, then, are under what conditions, and to what degree suppliers are prepared to develop and produce military equipment especially for export. In the international marketing of civilian products the question as such is virtually irrelevant; products or versions of products are routinely developed and produced specifically for export. By contrast, in the international arms market some major suppliers were more attentive to special requirements of recipients and some were less attentive, yet only rarely were weapons systems developed and produced exclusively for export. In the United States, for example, in President Carter's time, government policy attempted to reduce the world arms trade, and forbade companies to produce or to adapt weapons especially for export (Kapstein, 1992: 148). In other countries, avoiding production or adaptations specifically for export was the outcome of pressure from the military establishment who regarded over attention to export clients' demands as coming at the expense of coping with national security problems (Sandler & Hartley, 1995: 251-253). The main reason, however, was economic: from the producers' standpoint, expected returns on development and production of weapons systems for export only did not justify the investment. Besides the usual risks of military development programs, exports introduce special problems: there are seldom any assured customers in the development stage, and potential orders may not materialize because of changing needs during the long development cycle, the appearance of better solutions from competing sources, or cost overrun that make the product non-competitive. Furthermore, sales may not take place due to political developments involving either suppliers or recipients, and at times due to third party intervention (Smith, Humm & Fontanel, 1985: 242).

5. THE ECONOMICS OF ARMS IMPORTS

Countries import arms primarily to improve their military capability in coping with external threats, but at times also, *inter alia*, to strengthen the government against rivals from within, to satisfy military ambitions to modernize and so to reinforce army support for the political leadership, and to gain prestige in the international arena. Economic advantages from arms imports, if any, are seen as by-products. With that, the demand for arms imports becomes effective only when there are the means to pay for it, so economic constraints determine the possibility of importing arms, and at the same time defense imports have immediate and long-term important economic implications.

5.1 Economic Determinants of the Demand for Arms Imports

Deciding the scope of arms imports is in essence choosing between alternatives in building the country's military capabilities: as against self-sufficiency in meeting military needs, there is the option of importing arms from abroad, which may be advantageous in some cases and less so in others. The decision impinges on many expense items; besides direct outlays imports entail indirect costs of building new infrastructure, training and current maintenance over the entire life cycle of the weapons systems. Furthermore, defense import decisions have a special dimension since they have to reckon not only with economic constraints in general, but with foreign currency constraints in particular. Then there are also multi-period considerations: keeping imported equipment operational and using it effectively depends on spare parts being available over time, themselves imported in most cases. The scope of defense imports, therefore, is determined in a multi-period optimization process, subject to general and specific resource constraints, and taking relative prices into account.

Inasmuch as the decision on defense imports is an economic one, their volume has to correspond with the level of overall economic resources and of foreign currency resources, and to indicate tradeoffs with other expense categories at every given level of resources. Empirical findings can hardly elucidate the issue; they all come from research on related questions, and generally offer only partial and indirect evidence. A cross-section study (Pearson, 1989), examined the relative importance of various factors in explaining the variance in defense imports among countries. The regression equations included many variables: area, population, type of regime, security-related features (defense expenditures, number of soldiers, nuclear capability), economic characteristics (GNP, balance of payments), involvement in a foreign or domestic conflict, and foreign policy alignment. The analysis explained about a third of the variance, with economic variables having significant influence yet secondary to that of security-related features and foreign policy alignment. A second study (Mintz, 1986), examined action-reaction patterns in defense imports in six pairs of countries in the Middle East and Asia, all involved in armed conflicts between 1960 and 1980. The annual change in each country's defense imports was estimated as a function of the defense imports of its rival in the previous year, of the defense burden (defense expenditures as percentage of GNP) in the previous year, and of a dummy variable representing involvement in war in the current year. The estimated coefficient for the defense economic burden was negative in 10 out of the 12 countries, and in nine of them it was statistically significant. The economic burden of defense, then, constrained defense imports regardless of regional differences or differences in arms sources.

Another study (Looney, 1989), while focusing on the role of defense expenditures and defense imports in the debt accumulation of 61 Third World countries, also pointed to factors affecting arms imports. The countries were divided into two groups by their relative foreign currency positions, and the underlying assumptions were that defense imports are associated with the effective demand for arms (indicated by overall defense expenditures), with foreign currency constraints (by the external debt), with possible substitution between imports and domestic supply, with the composition of military forces as between manpower and equipment and the ability to substitute one for the other (by the size of the armed forces personnel), and with the external or internal threat level. The statistical estimates referred to 1981, and the results supported the hypothesis that foreign currency constraints induced an economic choice in deciding the scope of defense imports. First, for the sample as a whole and for the more foreign currency-constrained group of countries, there was a statistically significant, inverse relationship between defense imports and military manpower, while for the second group no such relationship was found. It appears, then, that while relative scarcity of foreign currency led to substitution between defense imports and manpower, the relative abundance of foreign currency sources made it possible to increase both defense imports and military manpower. Secondly, in the sample as a whole and in the two sub-samples there was evidence for substitution between defense imports and indigenous defense production, but the scope of substitution was unequal: the group of countries with relatively abundant foreign currency could substitute more imports for domestic production. Apparently, larger foreign currency sources enabled those countries to establish a larger-scale defense-industrial base, thereby creating more opportunities for arms import substitution.

An additional study (Hess, 1989) also found significant substitution between defense imports and domestic defense production. The study examined a sample of 76 developing countries between 1978 and 1984, and used the volume of defense exports as an indicator of domestic defense production capacity. It showed that every dollar of defense exports reduced defense imports by 1.24 dollars (when Israel, whose defense imports are financed almost entirely by American aid, was omitted from the sample, defense imports decreased by some 3 dollars for every defense export dollar). By contrast, a positive, statistically significant relationship was obtained between defense imports and the ratio of military manpower to population, and it was taken to indicate that manpower and weapons are complementary factors in producing national security. However, six Middle East countries that imported most of the military equipment and received most of the foreign aid in the sample influenced the results considerably. When they were omitted from the estimates, the positive association between defense imports and military manpower became much weaker, thus further verifying that the relationship between defense imports and military manpower depends on the relative scarcity or abundance of foreign currency sources.

5.2 Arms Imports and Economic Growth

Analyzing the possible effects of arms imports on economic growth is part of a broader issue, namely the interrelations between defense expenditures as a whole and economic growth. However, it has also some unique aspects. The critics contend that arms imports divert scarce foreign exchange resources from uses that contribute to social and economic development directly. They emphasize in particular the follow-on import requirements of defense import transactions, i.e. the import component of local expenses related to the absorption and current operation of imported weapons systems, as well as the ongoing need for imported spare parts, concluding that the retarding effects on growth are continuous as well. Moreover, the negative effect may extend in scope and time as a result of the regional arms race dynamics: arms imports by one country may induce rivals to acquire arms abroad, thus spurring both sides to continue importing arms over a long period. Besides the direct effects, they point to three indirect factors: transferring complementary domestic production factors, skilled manpower in particular, from the civilian to the military sector; accumulating an external debt that imposes principal and interest payments on the economy in future years, and increased dependence on arms suppliers from advanced industrial nations (see below).

There are, however, opposing opinions too (Neuman, 1979). Rejecting the above criticism, they note that there is no assurance that reducing defense imports will release economic resources precisely in ways that contribute to social and economic development, while on the other hand arms imports, and especially the transfer of advanced technology that often comes with them, may well accelerate modernization and technological development in recipient states. They stimulate development of modern infrastructures and training of technical personnel, as well as acquisition of managerial and organizational skills, and these positive changes are not long confined to the military. Economic development, the proponents assert, need not begin in the civilian economy, and accelerating factors originating in the military sector are likely to spread out in time, and to bring growth and development to the economy as a whole. The first one affected is the building industry: new bases, improved airfields, roads and communications systems are needed. Soon the industrial sector becomes involved too, supplying not only larger quantities of its products, but also products of higher quality. The industrial sector is likely to benefit from the transfer of new technologies and from the flow of advanced production and managerial know-how, *inter alia* by recruiting army veterans trained by the arms suppliers or who gained technical experience during their service years. Later on the influence diffuses through to the service industries as well; the education system, for example, has to offer new, more advanced vocational training options. In the long run, then, the circles widen to a point where the processes set in motion by arms imports affect most sectors of the economy. The resultant improved capabilities may in turn affect arms acquisition policies: on one hand it is possible now to meet a greater part of the defense requirements indigenously, and on the other hand requests for foreign arms too will advance to a higher technological level. Thus a new cycle of reciprocal influences begins, accelerating economic growth even more, and the process keeps repeating itself. A military-nucleus strategy of development, then, may achieve military power through arms imports and economic growth simultaneously, making them complementary, and not competing policy goals.

According to this view, the positive effects of arms imports on economic growth are not limited to specific transaction forms. It is assumed, however, that arms imports combined with offsets will benefit industrial development, and technological advance in particular. Incorporating offsets into arms transfers commenced in the industrialized countries as a means of creating employment and of acquiring state-of-the-art technologies, and soon became an integral part of arms transfers from industrialized to developing countries. This reflects the improved bargaining position acquired by arms importers in the Third World, but to no less an extent the importance they attached to the apparent economic advantages of these arrangements. Even if most offsets were originally designed to ease the financing of foreign procurement, or to reduce future dependence on foreign suppliers, they also contributed to economic development. Counter trade, for example, was designed to increase the importer's foreign currency income, and thus help finance arms procurement per se. However, when applied skillfully, turning the initial penetration into a continuous flow of sales in future years, it may become a permanent factor, enhancing industrial development and exports over the longer run. Likewise, co-production, joint production, production under license or subcontracting – all broaden the possibility of military self-sufficiency, and at the same time support the domestic defense-industrial base and contribute towards its growth. Finally, foreign investments in infrastructures, in production and maintenance facilities, as well as technology transfers through collaborative R&D programs, support for research centers and the like, may all be assumed from the outset to serve economic development and technological progress more than any other purpose. Incorporating offsets into arms imports, then, increases the chances that they will contribute to economic development. With that, however, basic conditions in developing countries did not always make it possible to realize the economic advantages apparently inherent in arms imports and the offsets that went with them (see also the next section).

Empirical studies that considered the group of developing countries as one entity concluded that arms imports impeded economic growth (Sandler & Hartley, 1995: 256). Similarly, a study that related to a whole region, the Middle East, showed too that under certain circumstances a considerable decrease in defense imports might over time accelerate significantly the GDP growth in the countries involved (Sandler & Hartley, 1995: 258). Less unequivocal results, however, were obtained from analyzing the association between defense imports and economic growth in different groups of developing countries (Looney & Frederiksen, 1993). The study related to a sample of 62 developing countries from 1972 to 1989, dividing them into two groups according to their actual rate of economic growth in the 1980s. The division corresponded to a great extent with the distinction between resource constrained and unconstrained countries (relative resource availability being defined by a large variety of economic variables). Data showed that in countries that grew at a relatively low rate in the 1980s, the proportion of defense imports to total imports was, on average, higher than in countries that grew relatively faster: it was twice as high in the 1970s, and about 50 percent higher in the 1980s. Nonetheless, the proportion of defense imports to total imports was found to be only fifth in importance among the variables explaining countries' rating according to growth rate in the 1980s. Later on, a regression equation was estimated, defining the 1980s growth rate as a function of investments, external debt, defense expenditures and defense imports. As regards the sample as a whole, defense imports had no effect on growth. When low-growth (resource constrained) countries were gradually eliminated from the sample set, so that the relative weight of high-growth countries increased, defense imports tended to impact negatively on growth, although the effect was fairly weak and statistically significant for only part of the regressions. By contrast, when high-growth countries were gradually eliminated, arms imports did not appear to have any statistically significant impact on the economic growth in the 1980s. Hence on one hand defense imports retarded the economic growth of relatively rich countries, although a much smaller share of their imports was defense-related. On the other hand, defense imports did not affect the growth in the more resource-constrained countries. Possibly part of the explanation lies in the means of financing defense imports: resource-constrained countries tended, more than others, to finance arms imports through increases in their external debt, i.e. through foreign currency otherwise unavailable and consequently of seemingly low opportunity cost (Looney & Frederiksen, 1993: 250).

5.3 The Influence of Offsets on Arms Importers

Direct offset arrangements (licensed production of components, subcontracting, co-production or joint production) are a middle way between domestic development and production of military equipment on one hand, and off-the-shelf purchases of foreign equipment on the other hand (Martin & Hartley, 1995: 123). All three are alternative methods of procurement, yet they carry significantly different implications for costs, technological and financial risks, control over equipment specifications and wider industrial and economic benefits for the purchaser. Thus import transactions that include direct offsets may result in more expensive weapons systems compared with off-the-shelf purchases, but involve fewer technological risks than domestic production. They may also reduce control over the operational and technical features of weapons systems compared with domestic development and production, but are commonly perceived as offering more industrial and economic benefits than "simple" imports. In addition, direct offsets provide political convenience; governments may present arms import transactions that include such arrangements as contributing to domestic employment and enhancing the industry technologically. However, since governments are usually unwilling to pay a high premium for involving domestic producers in procurement programs, indirect offset arrangements have developed as well.

Central to offset arrangements is the notion of the buyer recouping part of the import costs, or gaining some other benefit (see section 4.2). But since arms import transactions are not coerced, the buyer is almost certain to make some sort of sacrifice too (Hall & Markowski, 1994: 178-179). Indeed, evaluating the true consequences of an offset arrangement for importing countries is not simple, as both sides have incentives to exaggerate the apparent benefits. For instance, when Britain acquired the AWACS airborne early warning systems from Boeing in 1986, the American supplier estimated that the offsets created 40,000 new jobs in Britain, the British government's estimate was 37,5000, while other calculations showed that at most it was a matter of 4,800 man years (Martin & Harley, 1995: 136).

Offset arrangements imply trade diversion since they transfer economic activity from one country to another, and therefore may ultimately decrease welfare in importing countries, and in fact of both parties to the transaction. In fact, offsets present some kind of a prisoners' dilemma: suppliers anticipate offset demands so they raise prices in advance, while purchasers, who assume prices to include offset compensation, insist on incorporating such arrangements for which 'they have paid anyway'. Following this approach, both sides may benefit if they agree in advance not to include offsets (Hall & Markowski, 1994: 187). However offsets, persisting and proliferating for decades, suggest that they do not necessarily harm welfare, and perhaps on the contrary even enhance welfare. Several explanations were put forward to settle the seeming ambiguity. Microeconomic explanations emphasized the oligopolistic nature of defense markets and the economic rent derived by suppliers in such circumstances. According to this view, purchasing governments insist on offsets not only as means of protecting and encouraging domestic industry, but also as a strategy that forces suppliers to transfer some of their economic rent to the clients (Udis & Maskus, 1991: 154-156). Another microeconomic explanation points to the possibility of saving costs through joint transactions due to economies of scope. Suppliers and recipients may prefer transactions involving packages of products, rather than an equivalent number of separate transactions, one for each product, when the larger scope obtained through joint production, procurement and distribution may result in cost economies. Joint transactions may contain elements that have no functional connection with the primary system, and offsets could be such elements (Hall & Markowski, 1994: 180-183).

There are macroeconomic explanations too. First, offset arrangements sometimes help to overcome non-tariff barriers, thereby encouraging international specialization and division of labor which otherwise would not be possible, expanding international trade, and eventually enhancing welfare. Secondly, governments insist on offsets that include as many technologically intensive and new products as possible (Martin & Hartley, 1995: 128). As a result, manpower will move to sectors having above average productivity, and the GDP will increase (Udis & Maskus, 1991: 156). Thirdly, advanced technology transfers through offsets are likely to be more successful than under other conditions, since responsibility rests with the seller, and concerned about his reputation, he will make sure to carry out the transaction in the best possible way (Udis & Maskus, 1991: 156-159).

The chances that offsets will enhance welfare vary with categories of products. They are, for example, relatively high in complex air systems (e.g. fighter aircraft): those transactions are usually large in scope, involving a great deal of industrial activity; they contain many advanced technologies that are likely to spill over into other sectors, inducing significantly positive effects there; their underlying production programs are characterized by economies of scale, and they are carried out in oligopolistic conditions that give rise to economic rents, which could be shared between the parties (Udis & Maskus 1991: 153). Indeed a large proportion of offset agreements are related to transactions involving the aerospace industry.

A survey of British companies performing work for foreign producers, most of whom were American, attempted to evaluate the actual results of offset agreements (Martin & Hartley, 1995: 130-134). Three principal conclusions appeared: the profitability of offset works was similar to that from other sales, i.e. no advantage was taken of the enforced contractual obligation to increase profitability, nor were prices reduced to penetrate the American market; the contributions to the technological progress of companies, or to their competitiveness, through new products and modern production techniques were negligible; only few companies, mostly the smaller ones, reported ongoing sales after contractual obligations were met.

Also in Britain, in a retrospective analysis of a specific case – the largest ever offset obligation of the contract for AWACS systems – it was found that some 60 percent of the offset work involved existing production, including the Rolls Royce engines, which would apparently have been purchased in any case. Moreover, a substantial portion of offset work had nothing to do with advanced technologies (Martin & Hartley, 1995: 134-137).

5.4 Arms Imports and External Debt

The economic analysis of arms imports refers broadly to the financial aspects of transactions. Countries financed imports with grants from supplier or third party governments, with loans, by cash payments or through barter agreements. In time, as noted earlier, suppliers became more commercially oriented, which made financial conditions more commercial as well. In particular, there was a clear trend for a smaller proportion of grants. Table 10.2 presents the changing financial conditions of arms transfers from the United States. Western European conditions were similar until the mid 1980s, but differently from the United States, European suppliers continued to provide extensive long-term credit to finance defense exports into the 1990s (Brzoska, 1994: 74). Arms transfers from the Soviet Union were mostly within military aid packages until the 1970s. Although they were presented officially as sales on credit, there was often an implicit understanding that the debt would never be repaid. In the 1970s, arms transfers began to be viewed as an important source of foreign currency income, and sales with no actual return decreased to some 25 percent (according to other estimates, only to 60 percent) ((Brzoska, 1994: 70-71). After the Soviet Union collapsed, all arms transfers from Russia are carried out for cash or barter. Arms transfers from other suppliers without a real economic return were usually very limited. Finally, some arms transfers were financed by third party grants, notably from oil-producing countries. This began in 1973, following the surge in oil prices, and the recipients were mostly Middle East states. In the 1990s, common estimates hold that half to two-thirds of arms transfers worldwide are paid for in cash or in credit.

Credit financing of defense imports postpones the financial burden from the years when the actual transfers are made to the years when principal and interest have to be paid off. Thus inconvenient budgetary tradeoffs can be postponed, and in real economic terms, there is no immediate need to reduce living standards or investments. Over time, however, debt accumulation affects economic performance directly and indirectly: foreign currency resources must be allocated for servicing the debt at the expense of other uses, and difficulties may rise in obtaining new foreign currency resources, because of foreign investors' and lenders' doubts about the country's creditworthiness.

Table 10.2 American Arms Transfers by Financing Sources

	Grants	Cash Payments	Loans
1950s and 1960s	From 80-90 percent in the 1950s to 50-60 percent in the 1960s	20 percent	Negligible
1970s	Slightly over 50 percent in first half; Negligible afterwards	80 percent in mid decade; 60 percent at its end	14 percent in the beginning; 6 percent in mid decade; 33 percent in second half
1980s	Negligible in the beginning, and some 50 percent in second half	25 percent in mid decade; Increase thereafter	33 percent in first half; Negligible thereafter
1990s	25 percent	Above 70 percent	Negligible

Source: Catrina, 1988: 33; Brzoska, 1983: 272; Brzoska, 1994: 70, 72.

Empirical attempts to estimate the economic effects of arms imports-related foreign debts were fraught with identification problems. Countries may pay in cash for defense imports, yet at the same time borrow abroad to finance their civilian imports. In such conditions external debt appears to accumulate from the latter, but it would be lower were it not for the defense imports. To overcome identification problems, an opportunity cost approach was adopted (Brzoska, 1983; 1994). First, foreign military aid grants were subtracted from defense imports, leaving the defense imports paid for in cash or on credit. At the next stage, it was assumed that countries whose net external debt did not rise in a given year paid, in that year, for the rest of their defense imports in cash, while countries whose net external debt rose, paid for them on credit, in the amount of the full addition to net external debt (if defense imports less grants are greater than the net addition to external debt, the additional defense imports are considered to have been paid for in cash). This method, then, assumes full fungibility of foreign currency credit, making no distinction as to the source or explicit purpose of loans, and thus provides a maximum estimate of the credit financing of defense imports. In line with the logic of the opportunity cost notion, however, that estimate does not necessarily represent credit-financed defense imports, but rather indicates the extent to which the country's external debt could have been lower were it not for defense imports. The first study (Brzoska, 1983) related to arms importers in the Third World in the 1970s, pointing out the clear increase in credit financing of defense imports: from one billion dollars in 1970, about one fourth of defense imports, to some nine billion dollars in 1979, somewhat over half. To investigate the economic implications further, assumptions had to be made as to interest rates and repayment periods, and then, debt burden estimates had to be compared with balance of payments statistics. Several interesting results were obtained. First, principal and interest payments on accumulated defense debt in countries that financed defense imports on credit in the 1970s were greater in 1979 than the cost of new defense imports that year. Secondly, payments on defense debt in 1979 made up more than 75 percent of total value of defense imports in all the developing countries. Thirdly, interest payments on the defense debt were a fifth to one fourth of all interest payments by Third World states, and the outstanding defense debt was about one fourth of their total foreign debt. Fourthly, if not for defense imports, about one fourth of all credit inflows to developing countries could be avoided at the end of the 1970s.

A continuing study (Brzoska, 1994) extended the scope to 1987. Arms transfers to Third World countries reached a record of about 46 billion dollars in 1982, and by 1987 they decreased to some 36 billion dollars. In the years 1980-1983 the development countries required an average credit of 20 billion dollars a year to finance defense imports, more than double the amount in 1979, while for 1984-1987 the average annual sum was about 14.5 billion dollars. The relative proportion of credit financing decreased as well, and stood at some 40 percent in 1987. Here as in the first study, to evaluate the economic implications involved, calculations were also based on various assumptions. The principal conclusion was that despite the decrease in defense imports between 1982 and 1987, there was no decrease in the defense import-related financial burden, defined as cash payments for new imports plus principal and interest payments on the defense debt. Indeed, in the late 1980s it was higher than at any previous time.

The relationship between defense expenditures, defense imports and external debt accumulation in developing countries was examined in another study (Looney, 1989; other findings of this study were discussed in relation to determinants of defense imports, see section 5.1). In the sample as a whole, and in the foreign currency unconstrained countries, defense expenditures did not significantly influence the external debt, but in countries whose foreign currency situation was relatively poor, defense expenditures contributed significantly to external debt accumulation. Similarly, in developing countries as a whole, defense imports were not significant in accumulating external debts, but countries with particularly severe foreign currency constraints relied to a great extent on foreign loans for defense imports, and much of their external debt seems to arise from that source.

5.5 Arms Imports and Recipient Dependence on Suppliers

Opinions emerging during the 1970s held that arms transfers give suppliers varying degrees of control over recipients, and may become a mechanism through which a small group of industrialized countries perpetuate their own superiority and the international power hierarchy. In the economic arena, they claimed, it could be manifest in an international division of labor that impedes growth and development in Third World countries (Pierre, 1982: 307-308; Catrina, 1988: 15-17). The argument was tested by examining possible links between arms transfers and three dimensions of foreign economic influence: concentration of foreign trade (the relative weight of arms supplier states in the general exports and imports of recipients), extent and status of foreign investments (e.g. whether investments from supplier states have some sort of a monopoly status), and exploitation of raw material by foreigners. In general, economic dominance and dominance in the arms trade were found to go hand-in-hand, although there were differences in this respect between suppliers. For example, in the case of the Soviet Union, arms transfers and other economic interrelations were tightly linked, but not at all in the case of France (Gerner, 1983: 26).

Dependency is the coin in which recipients repay suppliers for what they cannot pay for in any other way. Hence free-of-charge arms transfers, along with their apparent economic advantages, make recipients more dependent on suppliers than fully paid transactions do. Dependence is usually measured as the relative share of defense imports in total military acquisitions, or from the opposite angle, as the level of self-sufficiency. So simple a measurement, however, does not give the full picture. First, the distribution of imports among suppliers has to be considered, and the broader that distribution, the less dependence there is. Secondly, options for expanding self-sufficiency or for replacing suppliers should be taken into account. Finally, dependence varies with the severity of security threats to recipients (Catrina, 1988: 167-172).

Taken as a whole, since the 1970s there seems to have been a decline in recipients' dependence on suppliers: arms transfers within foreign military aid programs decreased, while the relative proportion of regular commercial transactions increased; new arms suppliers emerged, and available procurement sources became abundant; more countries established their own defense industries, thus gaining ability to produce certain types of weapons and to support and maintain imported weapons systems independently. Moreover, in the 1980s the world arms market changed from a sellers' to a buyers' market, and arms sales became no less important for the suppliers than acquisitions were for the recipients. However, there is an inherent asymmetry in supplier-recipient interrelations: while suppliers are interested principally in economic benefits, for recipients, arms procurement may be a question of survival. Hence not only the chances to obtain arms, which seem to have increased, must be weighed, but also the price of failing to do so.

A new type of dependence emerged in the late 1970s, namely that of principal arms manufactures on countries supplying subsystems and components for major weapons systems. Indeed, economic globalization made autarky not worthwhile even for the main industrial powers, including the United States (Kapstein, 1992: Chapter 8). In the short run, dependence on imported components exposes prime contractors to the risk of interrupted supply as a result of events in their country of origin. In the long run, however, implications can be more severe: having given up the capacity to produce them, the ability to design and develop components for a new generation of systems may also be lost. In the regular course of business there is readiness to accept the drawbacks of commercial dependence, given the clear advantages of international division of labor; and in ordinary circumstances importing components for weapons systems is not different, resting on the same economic reasoning. However, in emergency situations relying on imported components may have far-reaching strategic significance. This, then, is another dilemma arising from the clash between national perceptions of security and global perceptions of economics, and the solution is a matter of costs, priorities and calculated risks.

6. FOREIGN MILITARY AID

Arms transfers were frequently linked to foreign military aid programs. In these instances, supplier governments extended the financial means to import goods and services intended to improve recipients' military capability. The means usually included grants and loans under preferred conditions, even though regular commercial loans could have been regarded as foreign aid. Obviously, arms transfers or the services of military advisers for which no return whatever was required also fall within the boundaries of foreign aid. As noted earlier, third parties were sometimes sources of foreign aid, and as far as economic implications for recipients are concerned, there is no difference between such cases and "ordinary" ones. From the economic standpoint, the essence of foreign military aid is that it eliminates, reduces or postpones the economic burden of defense imports.

Foreign military aid may be targeted, implying that it can be used for acquiring arms from the donor only. There are, however, other possible forms of aid, with varying degrees of restriction on their use. Economic analysis shows that even targeted aid is fungible to some extent, allowing the recipient to use it as he sees fit. Thus in discussing the economic implications of foreign military aid, its effects have to be analyzed first in the same terms as any foreign aid, then proceeding to the special influences arising from the relationship between military aid and defense imports and expenditures.

6.1 Economic Implications of Foreign Aid

Foreign aid adds external economic resources to the economy. Discussions about the possible impact of resource transfers from outside on the economic developments in recipient countries began with the reparations imposed on Germany after World War I (see chapter 2, section 3.2.1). Famous economists like Keynes, Pigou and Ohlin took part in that debate, and though reaching different conclusions as to the overall consequences, all agreed that foreign unilateral transfers generate not only income effects, but also relative price effects, thus altering production structure, income distribution and the balance of trade. After World War II, the United States initiated the Marshall Plan, within which it transferred huge economic resources to help rehabilitate Europe, and the impressive success of that plan shaped to a great extent the perception of the relationship between foreign aid and economic growth. Later on, a similar approach was applied to encourage economic development in Third World countries (Baldwin, 1985: 290). The underlying assumption was that the main obstacle to their economic development was shortage of capital resources, since poor economies cannot save enough of their national income to support a desirable investment level ("the savings gap"), and at the same time they lack export revenues to pay for the imports of machinery, equipment and other inputs required for industrial development ("the foreign currency gap"). Foreign aid seemed to be an effective way to overcome both constraints, that of resources in general and that of foreign currency in particular.

In the 1970s, however, doubts arose as to the role of foreign aid and its effectiveness in promoting the economies of developing countries (Mahdavi, 1990: 111-112). The early criticism focused mainly on the following issues: -

Donor self-interest: Foreign aid is not a one-sided gift, but rather a bilateral exchange of benefits. Countries extend aid to enhance their own strategic, geopolitical and economic interests, and while this does not prevent the aid from contributing to the economic development of the recipients, contradictions may arise (Ball, 1988: 257-263). First, aid donors supported projects designed to bring them economic benefit, even if the value to the recipient nation was doubtful. Second, in many cases aid was linked to the purchase of goods and services from the donor, thus undermining its value, since recipients were prevented from choosing the cheapest source, and moreover sometimes had to pay inflated prices. Third, where the interests of aid donors matched those of ruling governments in recipient nations, it often led to selecting projects on the basis of political prestige, disregarding economic returns on the investments.

Influence on recipients' economic policy: With foreign aid recipient governments tended to put off reforms essential for enhancing economic development (e.g. food aid made it possible to refrain from agrarian reforms), and often adopted popular economic policies that retarded economic growth in the longer term. Besides, since the foreign aid was usually channeled through governments, it expanded the public sector and reinforced its economic influence at the expense of the business sector.

Influence on domestic savings and capital formation: Since there was no way to assure that foreign aid would be invested, not diverted to increasing consumption, nor that it would not decrease domestic savings, critics asserted that foreign aid may end up in replacing domestic capital resources with external ones, without actually adding to them. In classical economic analysis, there is no reason to assume that foreign aid will be used entirely for investments. On the contrary, *ceteris paribus*, it is likely that the income effect resulting from additional resources will increase both investments and consumption. The view that conditioning aid on specific uses would assure its flow into investment was shown, as well, to have but limited validity because of fungibility. Finally, to the same extent and for the same reason, the view that loan aid is preferable to grant aid since it dictates productive uses that will make it possible to repay principal and interest when due, turned out also to lack a factual basis. Indeed, foreign aid may decrease domestic savings because of displacement effects (Mahdavi, 1990: 113-117). In particular, aid enables recipient governments to reduce public savings by increasing public consumption or by always-popular tax cuts; in the last case causing increased private consumption as well. When foreign aid replaces domestic savings, not only is it impossible to achieve a higher growth rate, but also sustaining a given growth rate becomes dependent almost entirely on aid continuity. One way or another, the effect of foreign aid on capital formation in recipient states is far from clear, and consequently its contribution to accelerated growth is doubtful as well.

In later years, economic analysis concentrated on the influence of foreign aid on resource allocation between private and public uses, on the composition of government revenues and expenditures and on the industrial structure of recipient economies. Besides, economists pointed out the transitory nature of foreign aid, comparing its consequences with those of the so-called "Dutch disease", i.e. the economic phenomena arising from a temporary "gift of nature", that affect the economy positively for a limited period, and at the same time induce structural changes that might undermine its competitiveness in the longer term.

Central to the influence of foreign aid on resource allocation, and on appropriating public revenues for various expenditure categories, is the notion of fungibility. Aid resources, even if originally designated for a particular target, i.e. targeted aid, may in fact serve other purposes: directly, by circumventing donor-imposed restrictions, or indirectly, by replacing and releasing domestic resources formerly assigned to the original target. Allocation effects of foreign aid thus depend, *inter alia*, on the portion of resource supplement that can be used in any way the recipient chooses, i.e. on its fungible component. Theoretical and empirical studies showed that the fungible component of targeted aid could be statistically estimated from recipient demand equations (McGuire, 1987: 849, 869). For example, a study that examined the influence of American aid on the eight largest recipient countries in the years 1972-1987 found that this fraction was not significantly different from 1 as regards military aid, meaning that military aid was perfectly fungible, while with respect to civilian aid the fraction was lower, about 0.645 (Khilji & Zampelli, 1994).

Foreign aid, then, affects resource allocation through both its targeted and fungible components. The second induces a purely income effect that may increase demands for all normal resource uses (i.e. those whose elasticity of demand with respect to income is positive), including the demand for that category of expenditure for which aid was originally extended (e.g. defense expenditures in the case of foreign military aid). The first component includes, in addition, a substitution effect as a result of relative price changes. Targeted aid reduces the relative price of designated uses and thus, *ceteris paribus*, increases the demand-quantity of those uses, while at the same time, through cross-elasticities of demand with respect to price, may increase or decrease the demand for other uses. The ultimate result, then, may well be an overall change in the composition of resource uses, determined by the division of foreign aid between targeted and fungible components, the marginal propensity to spend on various uses from the purely income supplement, and by the "regular" and cross-elasticities of demand with respect to prices of the various uses. The ultimate scope of the uses for which the aid was originally given can indeed hardly be estimated in advance, unless all the relevant parameters are known.

If most foreign military aid is fungible, and the propensity to spend on defense from what may be considered as a purely income supplement is positive but relatively low, the net addition to defense expenditures will be small. Defense expenditures will increase by the amount of the aid on one hand, yet on the other hand local defense expenditures will decrease, and the difference, although likely to be positive, will be small. In addition, the price, or the opportunity cost of defense will decrease as a result of the influence of the targeted component of military aid, thereby increasing demand-quantity, but local defense expenditures may shrink further, if demand elasticity with respect to price is less than 1. Hence with a given amount of foreign military aid, the smaller the fungible component and the greater the propensity to spend on defense and the demand elasticity of defense with respect to price, the greater the increase in overall defense spending.

From a different point of departure, although foreign aid is usually extended to governments, because of fungibility part of it may leak out to the private sector through some tax relief mechanisms, thus limiting the maximum increase in defense expenditures to the portion still at the public sector's disposal. The fungible component of foreign aid, then, is of primary impact here too. However, another consideration comes to the fore. It was empirically observed that government spending from additional external fungible resources is different than that from additional internal resources, the so-called flypaper effect. The differences show in the marginal propensities to spend on defense and on civilian uses from each type of resources, defining a measure, π, which, when equal to 1, indicates the absence of a flypaper effect, or that there is no difference between expenditure patterns as regards the two income sources, while when equal to 0, all the additional foreign income is channeled to the private sector. In a study on Israel, an estimate of π did not differ significantly from 0 (McGuire, 1987: 859), while, by contrast, in the study on the eight largest recipients of

American aid, π was found to equal 0.85 and was not significantly different from 1 (Khilji & Zampelli, 1994: 359).

As a source of budgetary revenues, foreign aid may not only increase public expenditure and reduce taxes, but may also substitute for loans, either from the central bank, from the public or from abroad, meaning in fact less inflationary financing. Empirical finding as regards this issue indicated that a significant part of foreign aid to governments indeed replaced domestic loans (Cashel-Cordo & Craig, 1990).

The influence of foreign aid on the industrial structure of the economy is a direct result of changes in relative prices. The income effect of foreign aid leads to a general increase in demand for goods and services, and in full employment conditions will eventually bring about an overall increase in prices. However, in an open economy, prices of tradable goods exposed to foreign competition tend to rise less than prices of non-tradable goods and services, and new relative prices are established. Furthermore, in a regime of flexible exchange rates, foreign aid that increases the foreign currency supply may result in revaluation of the exchange rate, thus making tradable goods relatively cheaper on the domestic market, while non-tradable goods and services become relatively dearer. In time, the accumulated impact of relative price changes increases wages and profits in the non-tradable sector, thereby attracting labor and capital from the tradable sector. The resultant structural change undermines the economy's competitiveness in international markets, the balance of payments deteriorates, and eventually the rate of economic growth rate declines too. These processes resemble in essence the phenomena associated with the "Dutch disease".

Interesting explanations for the economic effects of foreign aid were put forward by analyses of public choice (Landau, 1990). In the public choice approach, governments (or powerful individuals within them) that wish to stay in power seek to strengthen support from voters in general, and from selected groups (e.g. the military) in particular. For that purpose, they can adopt policy measures that enhance the general welfare, and others that produce transfers or create rents for selected groups. However, in view of constraint – budgetary limitations, administrative capabilities, public criticism and the like – each type of policy involves opportunity costs in terms of the other. Besides, it is plausible to assume a diminishing marginal benefit to the government from either of the two types of policies, and moreover, that transfer-creating activities produce negative reactions from the general public at an increasing rate as the transfers increase. Maximizing the benefit to the government requires allocating scarce resources between the two policies in a way that equalizes their net marginal benefit. When equilibrium in this respect prevails initially, the inflow of foreign aid alters the equilibrium levels of the two policy activities. If aid resources are used to enhance the general welfare – to accelerate economic growth, for example – to restore equilibrium the government has to reallocate resources and expand rent-creating activities for selected groups. These, in turn, may slow down economic growth, reducing or perhaps even canceling out any growth-enhancing effects of the foreign aid. Empirical findings for 60 developing countries in the years 1960-1985, supported the hypothesis that as a result of foreign aid recipient governments reallocate resources from activities that enhance general welfare to those that increase rents for selected groups.

6.2 Special Economic Implications of Foreign Military Aid

Although a large part of intergovernmental aid was designated for military purposes, economic literature has dealt little with foreign military aid as a separate subject. When extended unconditionally, military aid was considered, from an economic standpoint, as any other foreign aid, and when targeted, as directly and solely financing defense imports, without further implications on the economy. Theoretical reasons and empirical findings, however, suggest that foreign military aid may have some unique economic aspects.

One such aspect arises from the special defense imports-military aid linkage. Defense imports may follow the financial possibilities extended by military aid, or alternatively, aid donors may consider recipients' actual defense import requirements and provide them with foreign aid accordingly. As frequently assumed, if foreign aid results from aid-donors' interests, the first causal relationship may be more common, and in the absence of foreign aid defense imports adapt to the lower level of available financing. But if aid-donors' decisions accommodate the defense needs and financing possibilities of recipients, then defense imports determine the extent of foreign military aid, and in the absence of aid, foreign currency almost certainly will be diverted from civilian to defense uses. Research on American aid to Israel (see in chapter 4, section 7) found that changes in the aid level between 1960 and 1979 could be explained to a great extent by changes in defense expenditures of the Arab states, seen as a measure of threat or an expression of Israel's defense needs, on one hand, and by changes in Israel's own economic resources (the GDP and unilateral transfers from abroad, excluding American aid), on the other hand (McGuire, 1987: 863, 867). It appears, then, that at least in the case of the United States and Israel, defense imports, or Israel's defense needs, have influenced aid level, not vice versa.

Another study related to a sample of 76 developing countries in the years 1978-1984, and simultaneously estimated demand equations for defense imports and foreign aid, i.e. the equation for defense import demand included foreign aid as one of the explanatory variables, while the equation for foreign aid demand included defense imports as an explanatory variable (Hess, 1989). Three of the resulting conclusions are noteworthy here. First, foreign aid was second in importance among the variables explaining defense imports, and an added dollar of foreign aid increased defense imports, on average, by about half a dollar. At the same time, defense imports were the most important explanatory variable of foreign aid, and the relationship between them, at the margins, was close to 1. The findings are important primarily in that they support the assumption of a close reciprocal relationship, however they do not provide conclusive evidence as to the causality direction. With that, the degree of association and magnitude of the coefficients give the impression that defense imports more often determined the scope of foreign aid than vice versa. Secondly, when the Middle East states, which are defense import intensive and enjoy considerable foreign aid, were eliminated from the sample set, the symbiotic relationship between defense imports and foreign aid disappeared. It seems, then, that in countries having relatively few defense imports and little foreign aid, there is no significant reciprocity between them. Finally, the equation for defense import demand included, as a measure of defense needs, the military manpower to population ratio, and the equation for foreign aid demand included, as an expression of economic needs and possibilities, a weighted index of socioeconomic variables. These variables were found to be important explanatory factors in the two equations, indicating that defense imports were greater when security needs required it, and foreign aid was greater when domestic economic resources were inadequate to meet them.

A second unique aspect relates to military aid fungibility. Although foreign military aid was usually targeted, empirical findings (see above) indicated that its fungible component was fairly large, and moreover, higher than that of civilian aid. It turns out, then, that military aid recipients directly, or more often indirectly circumvented the restrictions imposed on its use, reducing domestic defense expenditures or expanding the roles of the defense establishment within the economy. In countries with developed defense industries, for example, foreign military aid made it possible to divert domestic industrial capacity to produce for defense exports (McGuire, 1987: 869). Explaining the somewhat surprising results, it was suggested that in countries under acute external and internal threats, governments continuously devote a substantial proportion of their own resources for defense purposes, thus living under constant pressures to allow higher levels of civilian resource uses. Military aid, when extended, provides the opportunity to address such pressures, enabling governments to release some resources that would have been otherwise allocated to defense for non-military uses (Khilji & Zampelli, 1994: 358).

A third issue deals with the influence of principal and interest repayments for past military aid. Since not all aid is extended through grants, and part of it creates an external debt to be repaid with interest in subsequent years, the question arises as to whether servicing the defense debt affects future defense expenditures. In most countries such principal and interest payments are not included in the defense budget, and are usually excluded as well from accepted measures of defense economic burden. Nonetheless, indirect and implicit effects on allocation processes cannot be ruled out. In Israel, for example, it was found that principle and interest repayments were indeed taken into account in government's budgetary decisions (McGuire, 1987: 855, 858). In any case, whether the debt service is accounted for explicitly or not, in the course of time, as long as there are no qualitative changes in security threats and needs, foreign aid through loans in the present undermines the ability of the country to assign adequate resources to security and cope successfully with military problems in the future.

Present military aid also affects the future level and composition of defense expenditures in other ways. First, aid makes it possible to enlarge the defense establishment, and allows the armed forces to become accustomed to living beyond their means. When the aid stops, and since downsizing is not a simple matter, governments might be forced, in the short term at least, to transfer resources from other uses to defense beyond what would have been the "normal" level without the aid in previous years (Ball, 1988: 294). Indeed, the influence of military aid was sometimes paralleled to "an addiction", or in technical terms, thought to produce some sort of a ratchet effect, reflected in a marginal propensity to spend on defense when military aid increases, which is higher than the marginal propensity to economize on defense spending when military aid diminishes (Intriligator & Brito, 1989: 120). Secondly, when military aid is used mainly for acquiring major weapons systems, as it often is, it broadens the force structure base, requiring increasing allocations from the defense budget for training, maintenance and other current expenses in subsequent years. Some of the additional needs make it necessary to increase local defense expenditures, while others, to divert future foreign aid to importing subsystems and spare parts. In any case, at a given level of defense expenditure, it might become impossible to invest further in force buildup, or even to maintain a satisfactory level of operational capabilities within existing forces. In other words, over time, military aid may cause an imbalance in defense expenditures, and as a result impairs the country's military capabilities. Thirdly, the composition of defense expenditures is affected as well by the relative price changes caused by foreign military aid. Foreign aid makes imports relatively cheaper, thus discouraging the development of a domestic defense-industrial base. Furthermore, foreign military aid makes military capital cheaper too, thereby changing the relative prices of labor and capital, and as a consequence may increase the capital intensity involved in producing national security. The relative factor price effect is not limited to the military sector, and in other sectors it may take an opposite direction, thus retarding the transition from labor-intensive to more technologically advanced production. Such influences should be seriously considered, since there is always a high degree of uncertainty as regards the long-term continuity of military aid.

11
DEFENSE EXPORTS AND IMPORTS OF ISRAEL

The development of Israel's defense exports has followed the growth patterns typical of other new arms suppliers who joined the international arms trade since the 1960s and 1970s. However, there were also marked differences. It is plausible to assume that Israel's defense exports occupied a middle position in the world's arms market between traditional major suppliers in advanced industrial countries and new, smaller suppliers in developing countries. In like manner, the development of defense imports to Israel has also had some special features. At first, import patterns were affected mainly by lack of available sources, while later, since the 1970s, they were determined almost exclusively by American military aid. Foreign aid became a major factor in defense planning, and carried important economic implications as well. This chapter discusses the economics of Israel's defense exports and imports, and elaborates on the special role of foreign military aid.

1. THE DEFENSE EXPORTS OF ISRAEL

Like the exports of other new arms suppliers who emerged in the Third World, Israel's defense exports first arose from considerations of foreign and defense policies, but within a few years, the influence of economic factors gradually increased. There were other similarities as well. In the beginning, export included outdated weapons systems withdrawn from the Israel Defense Forces (IDF), along with relatively simple small arms and ammunition produced locally, and later expanded to more sophisticated systems and equipment. In their early phases, defense exports concentrated on the sub-market of developing nations, but then successfully penetrated advanced industrial countries, with a constantly growing proportion of goods sold to these countries. In many instances export activity involved industrial cooperation with domestic companies in target countries, and particularly in recent decades took advantage of the growing demand for upgrading customers' military equipment through such collaborative frameworks. However, some special features differentiated Israel from other new arms suppliers.

For one, Israel's defense export grew at an exceptionally rapid pace, almost continuously for four decades. In the 1960s, exports were limited, amounting in the middle of that decade to 12-15 million dollars a year (Klieman, 1992: 43). At the beginning of the 1970s they had grown to 70 million dollars, and by 1975 amounted to 170 million dollars. Since that time the growth of defense exports accelerated further, reaching about 480 million dollars a year toward the end of the 1970s, and about 800 million dollars annually at the beginning of the 1980s. In 1985 they exceeded the 1 billion dollars mark for the first time, reaching 1.5 billion dollars by the end of the decade. That annual level was maintained for several years, but in the second half of the 1990s growth resumed, and at the turn of the century defense exports exceeded 2 billion dollars.

Another special feature was the exceptionally high share of export sales in the total output of Israel's defense industry. As already noted (see chapter 9, section 2), the domestic defense industry began to grow rapidly after the 1967 War, and until the mid 1970s it supplied mainly the domestic demand. For about a decade thereafter, acquisitions by the Israeli Ministry of Defense remained at approximately the same level in real terms, but after 1985, they began decreasing. In the meantime, however, most major defense companies succeeded in compensating for domestic demand changes by increasing sales abroad, and the increased export sales along with a shrinking domestic market unavoidably led to a heavy dependence on international markets. In fact, while in the mid 1970s exports accounted for merely 15-20 percent of total sales, since the late 1980s they provide more than 70 percent of sales of the major companies (see Table 9.1). Such high export shares are doubtlessly unique, unparalleled either among the new arms suppliers or among the traditional ones.

1.1 Success Factors of Defense Exports

It is usually claimed that countries developed indigenous defense industries to decrease their dependence on imports, and thus reduce their political and military vulnerability. Nonetheless, they soon needed to turn their production capacity to export to ensure employment, to attain economies of scale and reach profitability. This development was in principle valid for Israel also; yet there were some specific arguments that justified defense exports in their own right.

Several measures may indicate competitive advantages of a country's industries in the international market: their proportion in the country's export, their market share in the world trade (Porter, 1990: 744), and the direction and rate of change of their exports compared with global changes in the relevant markets. All these indicated that Israel's defense industries had competitive advantages in the world arms market. In 1975, defense exports contributed about 17 percent of Israel's industrial exports, and in the following years their rapid growth led to a relative share of one fifth to one fourth of the total exports. All of Israel's exports amounted to only a few thousandths of total world trade, yet its defense exports were about 2% of the peak of world arms trade in 1987, and reached 6-7% in the 1990s. Moreover, in certain relevant market segments, after excluding types of major weapon systems not produced in Israel, Israel's share was much higher. Finally, the continued rapid growth of its defense exports in the 1980s and most of the 1990s, contrasted with the general stagnation followed by a steep decrease in the international arms trade, also bore witness to Israel's competitive position.

Israel's success in exporting military products and her defense industries' competitive advantages in the world arms market may be attributed to several factors. One was the proven performance of Israeli equipment on the battlefield. Another was the pragmatic political approach that did not take exception to Israel's involvement in world arms trade per se, and decided about military sales on the merit of each case. In the early stages, this approach was a result of Israel's enforced diplomatic isolation. Defense export was justified as an important instrument in competing for international influence, gaining political support and cementing bilateral relations. In addition, defense export was seen as a signal to both friends and foes that Israel is strong and can defend its vital interests, and as a means of tightening relations with the Western world in general and with the United States in particular (Kleiman, 1992: 61-84). Significant results were achieved, and defense exports paved communication routes where no other relations existed: in the 1960s – with African states; in the 1970s – with the Shah's Iran; later – with China, and others. Political interests gained, and they in turn contributed to the growth of defense exports.

Some countries preferred to acquire military products from Israel due to political considerations of their own; others did so for lack of choice, because of restrictions other suppliers imposed on them; and still others refrained from buying Israeli defense products for political reasons. Changes occurred from time to time, and while some markets closed, others, previously closed, opened. Developments in the 1990s provided several such examples. After the Cold War, countries that were formerly in the Soviet sphere of influence and refrained from any commerce with Israel, found a suitable partner in the Israeli defense industries for upgrading and adjusting Soviet arms systems to Western standards. Likewise, the growing openness vis-à-vis Israel, accelerated through the Middle East peace process, dispelled the inhibitions of Western and Northern European countries that hitherto barred Israeli companies from acquisition tenders for weapons systems and military equipment. In both cases, these new developments produced significant transactions.

Other success factors may be attributed to the market characteristics and the business nature of arms transfers. Here three main factors enhanced Israel's defense exports in the early stages (the first two actually apply to any small defense producer): -

i. The oligopolistic circumstances typical of the international arms market make it possible to set prices above production costs, so that diseconomies of scale are less significant, and are not necessarily translated into low profitability (Berglas, 1983:34). Besides, production series of military

equipment of other arms producers tend to be small too; and in many items, Israel's production series were large enough to achieve economies of scale.
ii. Due to the secrecy enveloping the arms trade, it usually avoided public criticism in the customer's country, and transactions were decided according to prestige considerations and other interests of a small inner circle. Thus prices were not the main factor considered by most of Israel's Third World clients in choosing their supply sources.
iii. Israel could offer know-how, and often was ready to cooperate with countries wishing to establish their own defense-industrial base. The demands of developing countries for offset arrangements, then, found pragmatic open-mindedness in Israel. At the same time, Israel was considered to be a source of advanced technology and to have up-to-date and proven experience. Moreover, Third World countries perceived Israel as closer to their own economic and social problems, and often preferred the Israeli source and interaction with Israeli experts to cooperation with Western industrial countries.

Technological factors were also important in the relative success of Israel's defense exports. As noted (see chapter 10), the technological over-sophistication of Western weapons systems was often a deterrent in transferring them to the Third World. On the other hand, most developing countries that established their own defense industries were unable to penetrate markets in advanced industrial countries, because their technological level remained low. Israel offered a variety of systems that met the needs of both developing and advanced industrial countries. Since the 1980s, with the expanded demand for upgrading weapons systems, the rich experience of Israel's defense industries in this area also came to the fore.

As to purely economic explanations for the relative success of Israel's defense exports, they follow mainly from the classical factor proportion theory of international trade, attributing the comparative advantage of countries to their different factor endowments. Israel is relatively abundant in high quality human capital, and particularly in scientists, engineers and other technologically oriented workers extensively employed in defense industries, usually in higher proportions than in other industries. In addition, Israel is endowed with an abundance of knowledgeable and experienced manpower in the military domain. The knowledge and experience accumulated during the military service are later translated into specific expertise in development, production and marketing of military systems when former service personnel join the civilian labor force. The economic explanation assumes, then, that high quality human capital and military know-how and expertise have endowed Israel's defense industries with competitive advantages, and have enabled it to succeed in the world arms market.

This assumption was examined empirically (Tishler & Rotem, 1995: 480-481). The study was based on defense exports data for 1969-1991, and included, besides measures of the relative abundance of high quality human capital and military expertise, three variables: the size of the domestic defense market, worldwide demand for arms and the effect of wars. An attempt was also made to find out whether the scopes of Israel's civilian and defense exports were mutually associated in any manner. In all the estimated equations, the relative abundance of high quality human capital was found to be the most important factor, and its influence on the scope of defense exports was positive and statistically significant. The effect of military know-how and expertise was also positive, although not statistically significant (the problem may possibly lie in choosing suitable variables). The empirical findings therefore supported the factor-endowments explanation for trade. Other results were mostly as expected: the effect of domestic military demand was negative for the short term, and positive – or not negative – for the long term; a rise in world demand for arms made it possible to expand Israel's defense exports; wars decreased defense exports, and the relation between civilian and defense exports was positive and statistically significant.

The study also examined the predictability power of the model by comparing the model forecast for 1992 with the actual defense exports for that year. The result was interpreted to suggest that following the major changes in the world arms market, Israel's defense export was launched on a new trend in which Israel was no longer a marginal player in the world arms trade. In particular, following the sharp decrease in the scope of world arms trade, the effect of outside demand on Israel's defense exports apparently became greater than in the past, while the effect of other factors was reduced (Tishler & Rotem, 1995: 489-490).

Arrangements relating to Israel's defense imports from the United States too contributed to the growth of defense exports (Kleiman, 1992: 261-271). Israel attempted through the years to make its special relationship with the United States, and in particular its defense imports from there, into a lever for enhancing defense exports. These efforts were on three levels: direct sales to the American Department of Defense, buybacks by American arms manufacturers selling military equipment to Israel, and joint ventures of Israeli and American defense companies. On the first level, mainly during the early 1980s, Israel's defense industries were allowed to participate in certain tenders of the American Department of Defense, but the results were modest, not exceeding tens of millions of dollars per year. The buybacks route apparently produced more substantial results. Commitments to this effect were explicitly included in import contracts, and although no valid data are available about actual realization, partial information suggests that sales within this framework reached sizable levels. Moreover, over the years, sales of subsystems and components extended beyond the weapons systems intended for Israel alone, and trade relations that began as a forced commitment for reciprocal acquisition were later carried out willingly, out of pure economic-technological considerations no longer connected to Israel's arms purchase transactions per se.

The close acquaintance that developed between the defense industries of the United States and Israel gave rise, especially since the late 1980s, to joint ventures of companies from both countries for the development, production and marketing of novel weapons systems. There were various models of cooperation: division of labor in development and production; co-opting American firms into the production of military systems developed in Israel, as a way of penetrating the American market; joint marketing to third parties, and others. The extent to which Israel's defense industries have penetrated the American market and attained cooperation with American counterparts, are another expression of the special status achieved by Israel, the "young" arms producer, on the world market.

1.2 Economic Implications of Defense Exports

The growing defense exports became an important source of foreign currency to the economy, and due to the nature of the goods involved usually carried a relatively high added value. In the world arms market of the 1970s and part of the 1980s, it also maintained averagely higher rates of profitability than civilian industrial exports. In addition, defense exports often paved the way for civilian goods: the former penetrated overseas markets previously unknown to Israeli industry, and made-in-Israel products gained a reputation that helped broaden customer circles, at first for plants producing civilian goods within the defense industry itself, and later – for other industries too. However, throughout the years, the relatively high share of defense exports in the balance of payments was subject to concerns and criticism. It was often asserted that an export-led economic growth policy based on defense exports is undesirable because of the risks implied in allocating large resources – as defense projects usually require – to one single industry. Moreover, defense exports are highly vulnerable; the world demand for military products is unstable, and the arms trade is widely influenced by political considerations (JIM, 1987: 47-48). There were other non-economic concerns too. From a political standpoint, it was argued that the search for economic benefits might limit the maneuverability of Israel's foreign policy. Military circles sometimes worried that exported military equipment might fall into enemy hands, or that early exposure for export purposes would give away secrets and do away with opportunities for surprise in the battlefield.

As far as micro-economic advantages are concerned, some were fully realized, others only partially. At the beginning of the 1970s, when the domestic defense industry grew rapidly, it was hoped that enlarging production series through export sales would reduce unit production costs, thereby reducing procurement expenditures of the IDF as well. Export prospects were factored into economic calculations, and in many instances provided a convincing argument for launching development programs for new weapons systems. Indeed, some advanced arms currently at the disposal of the IDF would not have been developed in the first place, were it not for the prospects of exporting them. However, the expectations for costs savings were not always realized. Firstly, for secrecy considerations, the export of certain types of military equipment was forbidden, or was permitted only after they were exposed in operational campaigns and in degraded versions. Therefore when export became possible, the IDF was already acquiring more advanced generations of equipment, and the extended production for export had little effect on the costs of the new items, if any. Secondly, Israel did not succeed in exporting some of the more expensive major weapons systems indigenously developed and produced, or they were exported in small, insignificant numbers (e.g., the "Kfir" combat aircraft). For one reason, American regulations forbid the sale of military equipment containing components acquired from the United States to a third party without prior consent, which is extremely hard to obtain. For another reason, potential customers are more politically sensitive as regards the source of major weapons systems, and were reluctant to acquire such systems from Israel, although they were sometimes traditional buyers of other Israeli military equipment. Thirdly, Israel's experience, like that of larger arms suppliers in industrial countries, demonstrated that competition on the world arms market often compelled producers to sell at marginal pricing, i.e. at prices covering variable production costs only, without any contribution to lowering prices for the domestic customer. In fact, in no few instances domestic sales subsidized exports rather than the other way around. Fourthly, in an increasing number of instances it became necessary to offer special models of equipment that complied with customers' requirements, and the investments involved, as well as the serial production that followed, made no direct contribution to domestic sales.

Notwithstanding the above reservations, the overall balance of defense exports in reducing the IDF's domestic procurement expenditures seems to be positive. In other words, if not for exports, the prices local producers charged the IDF would apparently have been higher. Two complementary effects should be considered in this context, both of which diminished over time. First, by allowing for economies of scale, in many cases export sales actually led to savings of variable production costs. However, this effect diminished as an increasing part of the domestic defense industry's activities became completely unrelated to local sales. Secondly, profitable export sales helped cover some of the fixed costs incurred by maintaining expensive development and production infrastructures, and this contribution increased substantially as the export share of overall sales grew. In the 1990s, however, following the end of the Cold War and the declining demand in the world arms market, export profitability decreased, thereby limiting the potential contribution to fixed costs too.

The defense exports have had other micro-economic advantages as well. For example, Israel provides a clear illustration of the role that exports may have in balancing out local demand fluctuations. As noted already, following the 1967 War and until the mid 1970s, the domestic defense industry grew due to the rapid increase in local demands, but it continued to expand later, until the mid 1980s, despite the leveling off in domestic sales, due to growing defense exports. Moreover, in the late 1980s, when domestic sales decreased sharply, the further growth of exports compensated for them, so that a moderate reduction only in the overall defense industry activity took place. The stabilizing effect of the defense exports made it possible, then, to avoid expensive adjustment processes of shutdowns, manpower turnover and the like for almost twenty years. However, the effectiveness of defense exports in balancing out local demand fluctuations depends not only on their overall volume, but also on their composition. The more export composition deviates from that of local sales, the more specific development and production infrastructure it requires, and over time labor and production equipment surpluses may emerge in particular lines and factories. Indeed, this kind of structural imbalances were discernible at the beginning of the 1990s in the Israeli defense industries. The defense exports did not necessarily replace the domestic sales of the same plants, and therefore while in some companies, or even in certain plants of the same company, production capacity surpluses developed, other parts of the industry maintained full employment. Because of the structural rigidities of defense industries, production capacity adjustments to changing demand patterns were slow and partial, thus failing to take full advantage of the potential contribution of defense exports to stable and regular industrial operation. These developments were crucial in leading up to the business crisis of the government defense companies in the early 1990s (see chapter 9, section 6). Nonetheless, even then, were it not for defense exports, the severity of the crisis and the adjustment costs needed would have been considerably greater.

A specific case of balancing out demand fluctuations, or of maintaining continuous operation of lines of production, relates to the inter-generational transition of weapons systems. The Middle East arms race, and the frequent wars that forced the exposure of secret weapons meant a relatively short life span for weapons systems, and made it necessary to acquire new, advanced and more sophisticated military equipment constantly. The development and production cycle of novel military systems often involves interim periods of idle capacity, which exports can fill by producing previous or in-between generations of systems, thereby maintaining "warm" production lines until the new model matures and is ready for current production.

From time to time, it has been suggested that defense exports may be considered a strategic substitute for emergency stockpiles of military equipment and ammunition. Following this line of thinking, the IDF could make do with lower levels of stocks, saving costs, and in time of need use the military products originally designated for export available in the defense industries' warehouses and production facilities. It has been suggested further that the large production capacity in itself, though initially built to meet export requirements, is a kind of replacement for emergency stocks; in wartime it can be used for supplying urgent needs of the IDF. In fact, in a few instances export-designated products were indeed diverted to internal use, and in emergency times the defense industries worked continuously to increase supplies to the IDF, but on the whole, the contribution to combat efforts was marginal. In the late 1980s, the importance of the indigenous industrial capacity as a substitute for emergency stocks diminished anyway, in view of the pre-positioning arrangements with the United States, which stored American military materials in Israel. Those stocks are primarily for American use, yet under certain circumstances Israel would be permitted to use them too.

For a long period, when defense exports grew rapidly and were highly profitable, they contributed significantly to the current financial management of the defense budget. The profits accumulated through exports, and in particular the substantial advance payments typical of export orders provided the defense companies with positive cash flows, and they could finance development projects for the IDF from their own sources, and were reimbursed upon delivery. Consequently, in procuring from the domestic defense industry, the Defense Ministry could do away with the traditional monthly installments, and turn instead to progress payments according to contractual milestones and to payments on delivery. The effect of the new financial arrangements was in practice similar to supplier credit, and as the defense budget is managed on a cash basis, it meant expanding, if temporarily, the effective purchasing power of the Defense Ministry. In any case, the fact that the defense industries' cash flow came to be depended primarily on export revenues, and not exclusively on the Defense Ministry funds, was a great relief for the current management of the defense budget.

1.3 Future Prospects of the Defense Exports

At the beginning of the 2000s, several key issues arose as regards the future of Israel's defense exports. From the perspective of the defense companies, for over a decade the share of export sales within their overall activities was greater than that of the domestic sales, and in quantitative terms at least, their survival came to depend on exports. Moreover, no meaningful change is expected in this respect in the near future, as there are no grounds to assume any substantial increase in the defense budget for domestic purchases. Over time, then, defense exports become less of a by-product of the sales to the IDF. The supply of military products developed with Defense Ministry financing for the IDF is diminishing, implying fewer products developed on IDF initiative, according to its specifications and through close direct contacts with the final user, are offered on the world arms market. The last point is enormously important; the tight interrelations between the IDF and the defense industries doubtlessly contributed to the outstanding achievements of Israel's defense development and production, and possibly underlie its economic advantages too, as they made possible shortcuts in development processes, and especially in field-testing, saving considerable costs. No less important, the reputation of the IDF as a user of such products clearly promoted their marketing overseas. Under the new circumstances, to successfully compete on the world arms market, Israel's defense companies would have to overcome several problems that are anything but trivial: raising the finances needed exclusively for export-specific R&D and production equipment; finding an appropriate substitute for the necessary feedback between the developer and producer of military systems and potential customers while maintaining reasonable costs, and establishing effective marketing strategies to offset the unavoidable disadvantage of selling equipment not in use by the IDF.

It has been argued that Israel occupies a middle position, between the large industrial countries, which concentrate on the development of sophisticated technologies and offer the most advanced weapons systems, and the smaller arms producers, mainly from Third World countries, which offer less advanced systems and standard items. There are, however, doubts as to Israel's ability to maintain this position in the future. Those doubts do not stem from possible shortage of scientific and engineering manpower; Israel abounds with high quality human resources, and her advantages in this area will presumably be prominent in the coming years as well. The doubts arise due to the vast investments in defense R&D currently required to keep up with the technological race, and because of the shrinking local market for defense systems. Furthermore, not only Israel's ability in this respect is doubtful, the desirability of such strategy is questionable since its political and economic cost-effectiveness are far from obvious.

Defense exports were instrumental in developing relationships with countries that did not maintain relations with Israel in any other sphere, relieved its diplomatic isolation, and promoted other political and strategic interests. With that, Israel's involvement in the international arms business carried a political price, and particularly damaged her image as a moral and peace-seeking nation. In the post-Cold War geopolitical environment, and especially in view of possible positive developments in the Middle East peace process, the arms business will not be as important as it was in the past for expanding Israel's foreign relations. The defense export policy may become more selective as regards customers and types of military equipment, and a question arises as to the effect of such an approach on the scope and composition of exports.

The relative weight of defense exports in the overall industrial exports of Israel decreased from a peak of 27 percent in the mid 1980s, and more than 20 percents during most of the 1980s, to approximately 10 percent at the beginning of the 2000s. Other relative measures of defense exports in the economy indicate a similar trend. Hence, though they are not unimportant, defense exports lost their place at the top among the major contributors to foreign currency revenues. Moreover, the keener competition on the world arms market apparently eroded export profitability. A serious question arises, then, regarding the long-term economic justification of defense exports, and it becomes even more acute given the competition for scarce human capital and other resources between the defense industries and the export-oriented, civilian high tech industries that grew substantially in the meantime.

Israel cannot do without an indigenous defense-industrial base, nor would it be possible for her to give it up in the foreseeable future even with better security conditions in the region. At the same time, the domestic defense-industrial base became increasingly dependent on exports, and as the discussion above indicates, the prospects of those exports are unclear. To contend successfully with this dilemma, Israel should adopt a defense-industrial policy that bases her defense exports on broader collaboration with defense industries in other countries. The trends of economic globalization, cross-borders mergers and joint ventures in arms development and production, with the de-politicization of arms transfers, give rise to new rules of the game, and defense companies that do not manage to join the circle of industrial collaboration might be left at the margins of the world arms market. At the same time, the technological achievements of Israel's defense industries, their proven ability in integrating advanced know-how with creative solutions for military operational problems, and no less, their marketing successes in the international arena, make them sought after industrial collaborators for defense companies in both industrial and Third World countries. Israel's defense industries are indeed familiar with the idea of cooperating with foreign defense industries. A prominent current example is the Arrow program, in which the United States and Israel cooperated in developing an anti-ballistic missile system, and it can serve as a model for additional joint development projects. In such projects not only are costs divided between the partners, but they also provide for division of labor and specialization that may economize on development and production costs. Besides, marketing prospects are enhanced because the involvement of another country brings in a prospective customer, and cost savings lead to competitive advantages in selling to third parties. This framework may be an appropriate answer to most of the problems confronting Israel's defense industry and exports. Furthermore, there is no reason to limit it to individual programs, and it could extend to establishing new companies under joint ownership, centered on selected technologies or broader families of defense systems. Finally, it may be expected that foreign companies will seek to join with Israeli industries in areas where the latter clearly have comparative advantages. Thus creating industrial alliances as suggested above will subject defense development and production activities in Israel to strict economic tests, thereby accelerating the restructure and rationalization of her defense-industrial base.

2. DEFENSE IMPORTS AND THEIR DETERMINANTS

The quantitative development of defense imports and their impact on Israel's balance of payments were discussed earlier (see chapter 6). Other characteristics – main supply sources, quality of arms imported, and the institutional and financial aspects of transactions – suggest three distinct periods of development. The first, until about 1955, was marked by the embargo on arms transfers to the Middle East. Initially established by a United Nations Security Council resolution in 1947, the embargo was later reaffirmed in the Tripartite Declaration of 1950, where Britain, France and the United States undertook to limit arms supplies to the belligerents in the region. Israel then turned to supply sources in the Soviet Bloc, particularly to the Czech Republic, as well as to some secondary sources in Europe (Schiff & Haber, 1976: 486). During those years, the defense imports were mainly secondhand equipment, most of it World War II surpluses, and acquisitions were made through arms traders and middlemen rather than directly from governments. Towards the end of 1953, the embargo was relaxed, and Britain agreed to sell Meteor combat aircraft to Israel, and a year later – two renovated destroyers.

The second period, from the mid 1950s to 1967, centered on the emerging supply relationship with France. During those years, Israel acquired large quantities of up-to-date weapons, chiefly for the air force, and most transactions were carried out between governments. France was the primary source, but important acquisitions were also made in Britain (renovated submarines, Centurion tanks and more) and in West Germany (including secondhand American weapons systems, mainly Patton tanks, with the blessing of the United States). The United States, although supporting occasional third party arms sales to Israel, still refrained from direct supply in the early 1960s. The change came in 1962, when the United States agreed, for the first time, to sell to Israel protective equipment – Hawk anti-aircraft missiles – and expanded in 1964 with the sale of M48 tanks. Another new development related to financing; some defense imports were financed with grants from West Germany. In the early 1960s those grants amounted to tens of millions of marks, and grew to 300 million marks in 1964 (Schiff & Haber, 1976: 488).

The third period started with the French embargo on military supplies to Israel, imposed on the eve of the 1967 War. Israel was left in effect without regular supply sources, and reinforced her acquisition efforts from the United States. The United States, on her part, at first attempted to come to terms with the Soviet Union so they might together restrict the Middle East arms race, but when those efforts failed the way was clear for the American chapter in Israel's defense imports. It began with several ad hoc transactions, but during the 1973 War the United States operated an intensive airlift, supplying Israel with unprecedented quantities of weapons, ammunition and other military equipment. After the War, the supply relations were enhanced, and for the first time the United States was willing to discuss multi-year acquisition programs, thus making her arms transfers to Israel regular and continuous. Furthermore, the United States agreed to equip Israel with state of the art weapons systems (e.g. F-15 and F-16 combat aircraft), and most important – extended large-scale annual military aid that made it possible for Israel to pay for them. Hence the United States replaced France as the major arms supplier, and since the 1970s American dominance in Israel's defense imports has been sustained.

It is not without grounds to assume that throughout most years the volume of defense imports would have been larger, if derived from the IDF programs and requirements by means of "ordinary" economic optimization. In the early years, political constraints limited available supply sources, thereby dictating non-optimal solutions, while later on it was the fixed amount of American military aid. As noted earlier, since the 1970s, in view of the continuous flow of military aid, defense budgeting was carried out almost independently under two separate constraints: the domestic resource constraint determined local defense expenditure, and the American military aid determined defense imports.

An econometric study pointed out the different behavioral patterns of defense imports in the years before and after the American military aid became significant. It was found that prior to 1967 the volume of defense imports was associated mainly with the defense expenditures of Arab countries, while in subsequent years the main explanatory variable was the level of American military aid (Mintz, Ward & Bichler, 1990: 185-186). To note the effectiveness of American military aid in constraining the volume of defense imports, it was demonstrated that the desired level of military aid, as estimated from the IDF plans, was higher than the actual aid extended (Mintz & Ward, 1989).

The impact of the Soviet and American arms transfers on the Middle East conflict was analyzed for the years 1954-1985 (Kinsella, 1994), and the findings led to three main conclusions. First, the arms provided by the Superpowers to the rival parties in the region did not create a stabilizing deterrence, but rather accelerated the conflict. Secondly, the accelerating effect of the Soviet arms transfers to the Arabs was more prominent than that of the American arms transfers to Israel. In fact, according to one measure at least, the American supplies did not reinforce the regional belligerence at all. Thirdly, the Superpowers' behavior followed an action-reaction pattern, although their responses were not symmetrical; the Soviets responded immediately, while American supplies to Israel lagged behind Soviet deliveries to the Arabs, frequently with relatively long delays. With respect to Israel's defense import behavior, the above findings might be interpreted as additional evidence of the dominant effect of supply constraints.

The relationship between the volume of defense imports and the level of American military aid became somewhat looser in the late 1980s. At that time, Israel sought to convert larger fractions of the annual aid appropriations to domestic uses, or alternatively, to expand their legitimate purposes so as to use them in importing components and other inputs for domestic defense production. It seems, then, that American willingness to supply Israel's defense needs continuously for almost two decades made it possible to overcome the traumatic experiences of the past, and subsequently reinforced the tendency to seek optimization in deciding about defense imports.

3. AMERICAN MILITARY AID AND ITS CONSEQUENCES

The size of the American military aid to Israel and its long duration made it a key issue on Israel's agenda, with important consequences both for national security and for the economy.

3.1 The Scope and Means of Military Aid

Military aid was extended principally through annual financial appropriations to pay for current acquisitions of military products from American suppliers. From time to time however, other modes of assistance augmented it. During the 1973 War, the United States operated an intensive airlift, supplying Israel with vast quantities of weapons, ammunition and spare parts withdrawn from her own emergency inventories for immediate use. At the end of the 1970s, following the peace treaty with Egypt and Israel's withdrawal from the Sinai peninsula, the United States, with her own Corps of Engineering, built new air ports for the Israeli Air Force to replace those left behind. Since the late 1980s, the United States has stored military equipment and ammunition intended primarily for her own use in Israel, yet it could also be available for the IDF under certain circumstances. During the 1991 Gulf War, Patriot missile systems manned by American soldiers protected Israel against Iraqi Scud missile attacks. After the Cold War ended, American military surpluses were transferred to Israel from Europe free of charge. The United States participates in financing the Israeli Arrow program for developing and producing an anti-ballistic missile system.

Due to the variety of its means, estimating the full value of American military aid to Israel is by no means a simple task. Moreover, with time, the ramifications of defense-economic relations have provided indirect benefits that also helped Israel cope with her defense burden. As noted already, acquisitions from American defense companies often involved offset agreements that eventually paved the way for increased exports of Israel's defense industries to the United States. In addition, the United States enabled Israel to access advanced military technologies, and to acquire from

American suppliers the critical components necessary for indigenous production of sophisticated defense systems.

As for the financial appropriations, they too underwent substantial changes. Firstly, their nominal value increased over the years. Until the end of the 1960s, the amounts were relatively small, but since then they advanced in leaps: 300 million dollars a year in the early 1970s, 1 billion dollars in the second half of that decade, 1.4 billion dollars in the first half of the 1980s, and 1.8 billion dollars a year ever since 1987. In the late 1990s, as part of a plan to eliminate economic aid gradually, it was agreed to increase the annual amount of military aid by 60 million dollars for ten consecutive years. Secondly, in the beginning military aid was extended in loans and later in a combination of grants and loans, but since 1985 in grants only. Thirdly, military aid was initially earmarked exclusively for financing Israel's acquisitions in the American defense market, and so it was actually used until 1984. Since then, however, Israel was allowed to use an agreed fraction of the annual appropriation as she sees fit, and the yearly convertible amounts increased from 250 million dollars in 1984 and 1985 to 475 million dollars since 1990. Fourthly, until the early 1990s, financial appropriations corresponded with actual payments to suppliers. Later on, the annual amount of military aid was made available to Israel at the beginning of each fiscal year, increasing its effective purchasing power.

3.2 The Strategic, Political and Economic Roles of the Military Aid

The American military aid to Israel fulfilled three main roles at one and the same time: it gave Israel access to supply sources of military products – the strategic or military role; it extended financial means to pay for the military acquisitions – the economic role; and it clearly manifested the support and commitment of the United States to Israel's survival as an independent state – the political role. Over the years, due to international and regional changes, there were also changes in the content and relative importance of these different roles.

Following the French embargo of 1967, Israel was left with virtually no military supply sources, and the emerging supply relations with the United States, as well as the financial means extended, were important primarily for their military role. The military role was especially emphasized by the emergency deliveries during the 1973 War, and continued to be of utmost significance for the rehabilitation and restructuring of the IDF immediately afterwards. However, it was not only a matter of availability and quantity, but also one of quality: the military role of the American aid was most significantly expressed in the advanced technological quality of the equipment offered to Israel. The military role certainly remained essential, yet its relative importance declined somewhat for three main reasons. First,

Israel developed her own defense-industrial base, including the capacity to produce major weapons systems and a large variety of missiles indigenously. Secondly, during the 1980s and the 1990s, the world arms market became more competitive, turning into a buyers' market, thereby expanding Israel's acquisition options to other than American sources. Thirdly, following fundamental international changes, the Arab-Israeli conflict ceased to be a battleground between Superpowers, the collapse of the Soviet Union deprived Israel's adversaries of their traditional supply sources, and the framework of the Arab-Israeli arms race changed substantially.

The economic role of the American military aid is discussed at length later on. At this stage, however, it is worth noting that over time its relative importance has changed too. From the mid 1980s to the end of the 1990s, the nominal value of aid appropriations remained unchanged, thus subject to substantial decline in real purchasing power. At the same time, Israel's relatively rapid economic growth in the 1990s reduced the relative share of foreign military aid in overall economic resources, and in foreign currency resources in particular.

In the political sphere, its persistence under both Republican and Democrat presidents and administrations for many years established military aid as the most conspicuous expression of United States support for Israel. The Arab states and their allies were greatly influenced by the American position, and American commitment and support for Israel became a major strategic factor in shaping the Israeli-Arab conflict during the last quarter of the twentieth century. As noted previously, the literature about arms transfers is not conclusive regarding the regional impact of Superpower military aid to Third World countries: it invoked instability and escalation in some cases, and had restraining effects in others. As for the impact of American military aid to Israel, it appears that even if in the earlier stages it had an escalating effect on the Israeli-Arab conflict, its restraining influence increased over time. In the post-Cold War era, the political role of American military aid may change: due to its restraining effect, it could guarantee the peace agreements between Israel and her neighbors, those already signed and others that might follow, to no less an extent than it guaranteed Israel's ability to survive the regional arms race and the actual armed conflicts of the past.

3.3 Macroeconomic Implications of Military Aid

American military aid to Israel had economic implications on three levels: on the national economy level, it affected macroeconomic developments – economic growth, public revenues and expenditures and the balance of payments – as well as overall defense expenditures and the defense burden; within the Defense Establishment, military aid influenced the composition of defense expenditures; at the industrial level, it had structural consequences reflected in the relative shares of the tradable and non-tradable sectors, and in particular, it influenced the development of the domestic defense industry.

As explained in the general discussion, foreign aid affects resource allocation in the economy directly and indirectly. The direct influence, in essence an income effect, results from the additional resources provided by foreign aid, while the indirect influence, a substitution effect, results from relative price change, as due to foreign aid, *ceteris paribus*, imports become cheaper relatively to locally produced goods and services. Both influences will be stronger for larger volumes of foreign aid, yet two additional features are important too: the stability of foreign aid over time, and the flexibility in diverting resources from one use to another. The last feature, on its part, depends, *inter alia*, on foreign aid fungibility.

Analyses of the income and substitution effects of American military aid to Israel pointed at both positive and negative macroeconomic consequences. On the positive side of the equation, as shown below, the military aid made it possible to allocate less from Israel's own resources for defense, and thus to maintain throughout the years a relatively higher level of civilian resource uses. In particular, compared with an alternative scenario, where Israel would not have had such military aid, it was continuously possible with the aid to maintain a rising standard of living, to appropriate a larger part of tax revenues for expanding civilian public services and increasing transfer payments to households, or to avoid further tax increases that could have reduced private savings and investments. Besides, military aid allowed for a lower level of free foreign currency defense expenditures, hence releasing resources to finance imports of raw materials for current production, and of capital goods to expand the overall industrial base.

All the influences above are income effects due to resource supplements. There were, however, some noteworthy specific contributions. First, regularity and stability over the years made the military aid a permanent element that could be relied upon while pursuing long-term allocation processes, thus obtaining optimality. Secondly, growing military aid was essential to mitigate the negative economic consequences of external shocks, such as the Yom Kippur War and the energy crisis of 1973 (Razin, 1984: 51-52; Syrquin, 1989: 34; Sussman, 1995). Thirdly, as a source of budgetary revenues, the military aid helped balance the state budget, thus avoiding the use of other, inflationary means of finance (Sussman, 1989). Fourth, military aid and the accompanying defense acquisitions provided leverage for counter trade, and buybacks of American producers from the Israeli industry enhanced industrial growth and exports. Fifth, large-scale, permanent foreign aid gave Israel a kind of financial guarantee that improved its creditworthiness in international capital markets (Razin, 1984: 52).

Negative conseqluences of foreign aid resemble those of the so-called "Dutch Disease" in many respects, i.e. the disadvantages of a temporary "gift of nature" that carries substantial, positive contributions in a limited time, yet causes structural damage, impairing the competitiveness of the economy in the longer term (Gafni, 1989: 5). The possibility of maintaining a relatively high level of civilian uses and of increasing defense expenditures simultaneously, created an illusion that both security and welfare banners could be raised at one and the same time (Barkai, 1980: 19-20). Moreover, the high level of military aid maintained for many years, gave rise to expectations that it would continue in the future, and such expectations, when internalized, influenced economic behavior. In particular, they inflated perceptions as to requirements in both defense and civilian spheres (Syrquin, 1989: 40). As a result, fiscal discipline deteriorated while implementation of essential economic reforms and other policy measures was put off. Indeed, public expenditures grew rapidly, the budgetary deficit expanded to unprecedented shares of GDP, the relative weight of the public sector in the economy increased, and soon high inflationary pressures emerged. At the same time, easy acceptance of inefficiencies and avoidance of fundamental reforms led inevitably to lower rates of factor productivity increase, and eventually to lower rates of economic growth. In addition, foreign aid slowed down economic growth by revaluating the rate of exchange. Compared with an alternative scenario of no foreign aid, the exchange rate was maintained at a lower level, thus reducing profits in the tradable sector and creating incentives to expand the non-tradable sector, where productivity growth is typically slower. This structural change in itself, then, had negative consequences for economic growth (Syrquin, 1989: 38).

Although military aid was not the direct trigger, and certainly not the only or principal cause for all the developments above, it mitigated their immediate negative consequences to a great extent, and thus indirectly exacerbated them. Moreover, since military aid created conditions for an exaggerated expansion of the non-tradable sector, retarding development of an economic basis for reducing import surplus in the future, dependency on foreign aid was, in fact, perpetuated. The military aid, then, induced structural processes that require it to continue in the long run.

3.4 The Effect of Military Aid on Resources Allocated for Defense

The influence on the scope of resources allocated for defense is contained principally in the question of how much American aid was actually devoted to increasing defense consumption, i.e. did defense consumption grow by the full amount of military aid, or ways were found to divert part of it to other uses? Diversion may take place directly, or by releasing and replacing resources otherwise allocated for defense at least partially. Alternatively, evaluation can be made of how the level of defense consumption would have changed had the additional resources from military aid not been available. Prevailing assumptions hold that defense consumption was actually increased by the full amount of aid extended, or near it, and that without military aid the defense budget would have been augmented by Israel's own resources in an attempt to cover part of the gap. For example, by one evaluation, in the absence of military aid the defense budget would have been increased to cover half to two thirds of the gap (Zusman, 1989: 14). It may be concluded, then, that even if military aid did not reduce the local defense burden, it presumably prevented its increase.

Two observations should be noted in this context. First, despite the nominal increase of military aid throughout the years, the share of defense consumption related to military aid in GDP was decreasing: some 13 percent in the mid 1970s, 8-9 percent in the 1980s, 2-4 percent in the first half of the 1990s, and some 2 percent at the end of the 1990s. Without military aid, then, an additional 2 percent of GDP would have to be allocated in recent years to defense to maintain defense consumption unimpaired. Secondly, some indication as to whether Israel would have increased defense consumption from its own resources in the no military aid scenario derives from the fact that though the value of American aid remained more or less the same in nominal terms since the late 1980s, implying substantial erosion of its real value, there were no significant compensatory additions of local resources. Certainly this is a partial indication only, since actual outlays represent the outcome of many different influences, not the separate effect of declining military aid in real terms.

To consider the effect of military aid on defense consumption separately, a more detailed analysis is required. Over the years aid was extended in two distinct circumstances: an exceptional and transitory surge of defense requirements on one hand, and continuous, current requirements within multi-year plans of force buildup and the long-term arms race on the other hand. When especially vast additional resources were needed within a short time – as in the intensive airlift during the 1973 War or following the peace treaty with Egypt and the withdrawal from Sinai – it is reasonable to assume that defense consumption increased by the full amount of military aid provided. Supposedly even without limitations on their use, the additional resources would have been used for their original purpose. It was simply not feasible to mobilize sufficient resources from within Israel's economy so promptly, and any attempt to do so would have caused severe economic turbulence. In these circumstances, military aid prevented temporary, massive dislocations of resources, thus minimizing deviations from the optimal, long-term economic development path.

A different conclusion might be reached regarding the more regular military aid. Since American military aid was extended at a stable level for a long period, it could be taken into account in the long range planning of force buildup, and thus release other resources marginally for civilian uses. However, an almost total separation developed between the local and free foreign currency budget on one hand, and the aid-financed budget on the other hand. Indeed, discussions on the size of the defense budget, as of its internal composition, treated the two budgetary constraints as completely independent. Since most military aid could be used for acquisitions in the United States only, aid funds were regarded as a given quantity suitable for certain specific defense requirements, and budgetary quarrels between the Defense Ministry and the Finance Ministry centered mainly on increasing local currency appropriations to meet other, different types of requirements. In terms of the issue under discussion here, it appears, then, that the dominant approach regarded aid as a net supplement to defense consumption, making it possible not to increase internal resource allocations for defense, yet by no means a substitute for internal resources already allocated.

Meanwhile, attempts were made constantly to arrive at more flexible methods of using aid funds. In this context, it was agreed to broaden the definition of legitimate uses to include imports of raw materials and intermediate inputs for the domestic defense production. In addition, it became possible for Israeli defense companies to develop and produce, on a subcontracting basis, components and subsystems for weapons systems the IDF acquired with aid funds. But especially there were constant pressures to allow partial conversion of aid funds into fungible resources, and to finance local defense expenses with them. Rigid proportions between aid and internal resources, a result of the restrictions imposed on spending aid funds, appeared to lead to inefficient production of national security in Israel, and in seeking a better input mix attempts were made to substitute military aid for internal resources. The American government allowed conversion of part of the aid, but most probably there were also some indirect ways, not easy to identify and even more difficult to quantify, that made the freely used aid funds greater than officially indicated. All in all, defense consumption did not ultimately grow by the full amount of military aid.

The hypothesis that military aid was in fact fungible, and thus was partially diverted to civilian uses, was studied for the years 1960-1979 (McGuire, 1982, 1987). The first study considered resource allocation among defense, civilian public consumption and private consumption, and concluded that military aid indeed substituted partially for Israel's own resources: it was found that 9 to 30 percent of the military aid was diverted to non-military uses. The second study added an equation representing demand for investments to the model, and the estimated share of freely used military aid was smaller.

Another study that examined the spending of American foreign aid by the eight largest recipient countries in 1972-1987 (Khilji & Zampelli, 1994) found, as noted before (chapter 10, section 6.1), that military aid was perfectly fungible: the estimate for the freely used share was, on the average, 0.91, and not statistically different from one. Since this is relevant to the Israeli case, it is worthwhile to repeat the explanation given for this finding: in countries that face acute threats, governments are likely to devote a substantial proportion of their own resources for defense purposes, even in the absence of military aid, and therefore military aid enables them to release some of their own resources that would have been otherwise allocated to defense for non-military uses. In that study it was estimated that of 100 dollars of American military aid Israel spent 21 dollars for defense and 14.5 dollars for civilian public consumption, while 64.5 dollars were transferred to the private sector, divided between private consumption, 40.5 dollars, and investment, 24 dollars.

Since defense consumption presumably did not increase by the full amount of military aid, it is plausible to assume that it would not decrease by the full amount if military aid were to be reduced either. In other words, marginal propensity to spend aid funds on defense is lower than one in both directions, when military aid increases as when it decreases. Moreover, as noted in the general discussion (see chapter 10, section 6.2), there might possibly be an asymmetry, or some sort of ratchet effect, and the marginal propensity to spend when military aid increases is higher than the marginal propensity to save expenses when military aid decreases. Hence, *ceteris paribus*, if American military aid decreased, Israel would inevitably have to increase allocations of its own resources for defense.

Finally, arrangements for more flexible spending of military aid notwithstanding, the unfavorable influences of the restrictions on use of aid funds as regards the allocation of resources in the economy, and particularly in the defense establishment, have not disappeared completely. In most years, direct defense imports were constrained by available military aid, thus preventing optimization in defense procurement from the standpoints of both military force buildup and foreign currency spending by the economy as a whole. Compared to a hypothetical situation of fully convertible foreign aid (not complete absence of military aid, of course), it appears that on the whole Israel had to make do with less defense and a lower level of civilian uses (Zusman, 1989: 17).

3.5 Military Aid Influence on Composition of Defense Expenditure

Planning defense spending under two separate, independent budgetary constraints is likely to result in a non-optimal expenditure structure. The following issues illustrate possible outcomes of a non-optimal expenditure allocation: -

i. When the perceived shadow price of items acquired in the United States with aid funds is low, or often – for a particular customer or even for the defense establishment as a whole – is equal to zero, a tendency to over-equip may develop, leading subsequently to a shortage of current financial means for properly absorbing, maintaining and operating that equipment. Imbalance between force buildup and current outlays may impair manpower capabilities and equipment usability, and eventually erode the overall capacity of the army to contend effectively with security missions. On the other hand, increasing current expenditures to accommodate satisfactorily larger forces

may require most of the affordable defense budgets, aid funds included, leaving insufficient resources for technological renewal. Imbalance between present and future force buildup in times of rapid technological progress may deprive the army of essential capabilities for contending effectively on the battlefield of the future.
ii. Available military aid spurring the IDF to build its force on of-the-shelf American defense systems might have ruled out indigenous development of equipment more operationally suitable to the theater and warfare doctrines. These concerns were a constant source of deliberations in the defense planning process, and from time to time gave rise to opinions advocating fundamental change in Israel's national security perception.

3.6 Microeconomic Implications of Military aid

At the industrial level, it is usually argued that foreign aid affects the economic structure, and in particular expands the non-tradable sector at the expense of the tradable sector. Besides, when most aid funds are channeled to the government, they may increase the share of the public sector in the economy at the expense of the private one. Studies of these possible effects in Israel are scant.

One study (Weinblatt, 1998) pointed at two findings. First, there was a positive, statistically significant association between the share of industrial output in GDP, and a negative, statistically significant association between the share of service industries' output in GDP, and the GDP share of foreign unilateral transfers. Secondly, attempts to explain changes in wage differentials among principal industries by changes in the relative size of foreign unilateral transfers did not yield any statistically significant results, although such statistically significant associations were found in a few cases, at sub-industry level within manufacturing industries. In particular, the GDP share of American aid was found to affect the relative wage in metal products and transport equipment industries (though not in electronic industries) positively and significantly. These findings may indeed be of limited significance, since they ignore many other factors that appear to influence the industrial structure. However, taken literally, they indicate that the income effect of foreign aid was greater than the substitution effect. The income effect on acquisitions in the domestic industry, and especially on acquisitions in some specific industries, was large enough to stimulate industrial activity at the expense of service industries, eventually leading to higher wage rises in certain segments of manufacturing industries than in others.

A more specific argument was often raised regarding possible retarding effects of the military aid on the growth of the domestic defense industry. Two levels of influence were noted. At the strategic level, the large-scale, continuous military aid inevitably led to reassessing the role of indigenous defense development and production within national security policy, and on the whole undermined its importance. At the defense expenditure planning level, since the opportunity costs of military acquisitions in the United States were perceived as low, and often even as equal to zero, imports were preferred to domestic production, thus imposing contraction, or at best hindering further growth of the domestic defense industry.

The factors that determined Israel's defense-industrial policy were discussed previously (see chapter 9). Here, however, it is worthwhile emphasizing again that though the military aid was an important stimulus for reformulating the defense-industrial policy in the 1980s, and particularly in relegating the goal of self-sufficiency to a secondary place, it was neither the sole nor the most important consideration. It may be in place, too, to note again that as aid grants are limited in amount and exploited in full, domestic production of certain items releases financial means for importing others, and hence the opportunity cost of aid grants is a positive quantity. The accepted economic rationale for import substitution is valid, then, for defense expenditure planning when targeted military aid is available as well.

With that, the demand for domestically produced equipment is definitely affected by the income and substitution effects of military aid. To analyze these effects, comparisons should be made with two alternative situations: absence (or reduced level) of military aid on one hand, and fully convertible, non-targeted aid grants on the other hand: -

i. Compared with a "no aid" situation, military aid at current levels generates a positive income effect. The resources at the economy's disposal are greater, and *ceteris paribus*, the tendency will be to increase defense consumption and overall acquisitions within it, including those from the domestic industry.

ii. From the national economy standpoint, changing levels of military aid affect the price of foreign currency as well. *Ceteris paribus*, increased (decreased) level of aid reduces (raises) the price of foreign currency, thus generating a substitution effect that might reduce (increase) demand for domestic products.

iii. If the same level of military aid were extended without any restrictions on spending, it is likely that a larger portion would be converted into local currency, and larger amounts of the overall budget spent on domestic acquisitions. Removing restrictions increase the opportunity costs of imports from the United States, with an effect like that of devaluating the exchange rate, thus stimulating domestic production at the expense of imports.

Hence the overall influence of military aid on the demand for domestic industrial product is not clear. Its volume affects domestic acquisitions in opposite directions, while restrictions imposed on spending aid funds reduce domestic acquisitions. In the balance, especially when taking into account the buybacks that went along with imports and the portion of aid funds converted to local currency and spent on domestic acquisitions, there are grounds to assume that American military aid contributed to the domestic industry rather than retarded its growth.

3.7 Some Concluding Remarks

The discussion so far reaches the following principal conclusions: -

i. While the military and economic roles of military aid depreciated somewhat in importance, the importance of its political role remained and may be even greater than in the past, though with a different content.

ii. Some elements of military aid did not help alleviate the economic burden of defense, but were crucial in preventing its increase. Other elements released internal resources for civilian uses, thereby reducing the defense burden. At the turn of the century, defense consumption financed by military aid is some 2 percent of GDP, apparently the upper limit of the additional defense burden Israel would have to bear in the absence of aid funds.

iii. The restrictions on using the military aid interfered with internal optimization of defense expenditure structure, yet the impact was limited since increasing amounts could be converted and spent locally.

iv. The military aid, combined with other considerations, led to reformulation of Israel's defense-industrial policy. Besides, the volume of military aid and the restrictions on spending aid funds, affected the demand for domestic defense industry products in opposite directions.

Four different aspects of military aid should be considered for the future: the real, physical aspect of military procurement in the United States versus financial aspects; one-time assistance in exceptional situations versus current aid; the volume of military aid versus degrees of freedom in spending it; and assistance in military procurement versus other expressions of support and commitment.

Since Israel military power will continue to be the main guarantee for the peace settlements between Israel and her neighbors, those already reached and others yet to be established, for many years to come, and since despite peace with her closer neighbors Israel's security might be threatened by more distant enemies, it is essential for Israel to have access to the American arms market. The technological gap between the American defense industry and those of other countries makes the latter an inadequate substitute as a source of arms for Israel, if a substitute at all. At the same time, the domestic defense industry too is a limited alternative: during recent decades it was restructured to provide the IDF with unique, technology-intensive means, designed to complement expensive platforms acquired in the United States, or operated in combination with such systems, and on the whole incapable of producing substitutes for the major platforms themselves. The domestic defense industry depends also on access to cutting-edge American military technology and to sources of critical components for producing defense systems indigenously. For all these reasons, severe restrictions on American military supplies to Israel would have by far more serious implications than any changes in the financial volume of military aid or its other terms.

The military and economic significance of one-time assistance in exceptional situations, i.e. of extending massive, vast resources within a relatively short period to assist a special military effort in emergency situations or in any other junctures, is very different from that of current aid over time. From an economic standpoint, the former makes it possible to avoid severe inflationary pressures or dangerous depletion of foreign currency reserves in the short term, and to limit deviations from the long-term optimal path of economic development. In the coming years, such needs may arise in connection with additional peace settlements that would require redeployment of forces and reinforced early warning capabilities. Therefore whatever the current changes regarding current aid, a clear distinction must be made between current aid and one-time requirements.

Military aid funds in real terms have decreased since the late 1980s, and if the trend continues, Israel will have to allocate more of its own resources for defense. The economic implications depend on the general economic situation, and since military aid amounts presently to merely few percentage points of the GDP, they seem to be bearable. From the standpoint of efficient production of national security, accommodating a lower level of military aid will require a different internal optimization in defense expenditures. In recent years, although approximately one fourth of aid grants are convertible, the dominant opinion in the Israeli Defense Establishment is that overall expenditure composition is non-optimal, and favors conversion of a larger portion of aid grants into local currency. Increasing the non-targeted amounts of aid grants might be a partial compensation for the declining volume in real terms. Even if the option to convert aid grants is not realized every year, and military procurement in the United States would require more spending on imports, removing the restrictions would make it possible to arrive at more optimal spending, and thus to produce defense more efficiently. In these circumstances, the additional own resources required to compensate for the lower level of military aid would also be kept to relatively smaller dimensions.

The decrease of military aid in real terms could be interpreted as withdrawal from American traditional political support for Israel. Therefore it is of the utmost importance to maintain alternative expressions of commitment. A higher level of defense cooperation, with practical content and clearly demonstrating long-term intentions, would be desirable. In this context, both sides may benefit from large-scale, collaborative programs for developing and producing novel defense systems, similar to the Arrow project. Joint programs of this nature are common in Western Europe. There, however, the principal motive is sharing the financial burden and technological risks, whereas in the American-Israeli case the United States would have to bear most of the financial burden. At the same time, as leading American high tech companies increasingly found an economic rationale in establishing R&D centers in Israel, so too the Federal government may benefit from Israel's comparative advantage in R&D. From the Israeli standpoint, collaborative programs will strengthen the defense-industrial base, and will also contribute to its technological progress and to the balance of payments. The long-term military contribution is also self-evident, as is the clear, unmistakable expression of long-term American political-military commitment to Israel. All these advantages will be more concrete and significant, if instead of one program only, a kind of episode, a comprehensive policy including several multi-year programs is adopted.

In conclusion, from Israel's standpoint it is essential that decreasing military aid should not interfere with the real side of procurement options; that even if current aid decreases, the United States continues to recognize the special needs that may arise in relation to redeployment of forces or other one-time, exceptional events; that expanding the non-targeted share of military aid be seriously considered; and finally, that alternative expressions of American long-term commitment be sought, such as an industrial alliance focused on defense development and production.

REFERENCES*

ACDA (U.S. Arms Control and Disarmament Agency). various years. *World Military Expenditures and Arms Transfers.* Washington D.C.: U.S. Government Printing Office.

ACDA. 1979. "Soviet Military Expenditures." In *World Military Expenditures and Arms Transfers 1968-1977.* Washington D.C.: U.S. Government Printing Office, pp. 13-15

ACDA. 1979a. "Arms Transfers 1968-1977: A Decade of Change". In *World Military Expenditures and Arms Transfers 1968-1977.* Washington D.C.: U.S. Government Printing Office, pp. 16-19.

ACDA. 1990. "Diversification of Arms Sources by Third World Nations." In *World Military Expenditures and Arms Transfers 1989.* Washington D.C.: U.S. Government Printing Office pp. 25-29.

Allon, J. 1988. *A Curtain of Sand.* Tel Aviv: Hakibbutz Hameuchad (H).

Amacher, R.C., Miller, J.C. III, Pauly, M.V., Tollison, R.D. and Willett, T.D. 1982. "The Economics of Military Draft." In *The Military Draft,* M. Anderson, ed. Washington D.C.: Hoover Institution Press, pp. 347-389.

Amidror, B. 1979. "Introduction." In B. Catton, *American Civil War.* Tel Aviv: Ministry of Defense Publishing House (H).

Anderton, C.H. 1995. "Economics of Arms Trade." In *Handbook of Defense Economics,* K. Hartley and T. Sandler, eds. Amsterdam: Elsevier, pp. 523-561.

Angrist, J.D. 1990. Lifetime Earnings and the Vietnam Era Draft Lottery: Evidence from Social Security Administrative Records. *American Economic Review* 80: 313-336.

Arms Sales Monitor. 15.12. 1995. *Arms Sales Monitor - Highlighting U.S. Government Policies on Conventional Arms Exports and Weapons Proliferation.* Federation of American Scientists Fund.

Ash, C., Udis, B. and McNown, R.F. 1983. Enlistment in the All-Volunteer Force: A Military Personnel Supply Model and its Forecasts. *American Economic Review* 73: 145-155.

Augustine, N.R. 1983. *Augustine's Laws.* New York: American Institute of Aeronautics and Astronautics.

Augustine, N.R. 1997. Reshaping an Industry: Lockheed Martin's Survival Story. *Harvard Business Review* May-June: 83-94.

Avnimelech, M. 1978. "The Influence of Defense Expenditures on the Redistribution of Income." In *Emdot: A Selection for 1975-1977,* M. Brinker, ed. Tel Aviv: Massada, pp. 109-106 (H).

Baldwin, D.A. 1985. *Economic Statecraft.* Princton: Princton University Press.

Ball, N. 1988. *Security and Economy in the Third World.* Princton: Princton University Press.

Bank of Israel. various years. *Annual Report.* Jerusalem. (H).

Barber, W.J. 1991. "British and American Economists and Attempts to Comprehend the Nature of War." In *Economics and National Security,* C.D. Goodwin, ed. Durham and London: Duke University Press, pp. 61-86.

Barber, W.J. 1991a. "From the Economics of Welfare to the Economics of Warfare (and Back) in the Thought of A.C.Pigou." In *Economics and National Security,* C.D. Goodwin, ed. Durham and London: Duke University Press, pp. 131-142.

Barkai, H. 1980. *The Cost of Security in Retrospect.* Jerusalem: Maurice Falk Institute for Economic Research in Israel (H).

Barkai, H. 1987. The Crossroads of the Defense Industry. *Monthly Survey* 9/34: 35-47 (H).

* H denotes Hebrew texts. H* indicates texts translated into Hebrew, and page numbers there refer to the Hebrew version.

Barkai, H. 1990. *The Beginnings of the Israeli Economy.* Jerusalem: Bialik Institute (H).

Barkai, H. 1996. Security Cost and the Level of Wages in the Economy. *Globs* 5-6 Nov. (H).

Beard, T.R. 1993. "Industrial Organization and the Defense Industry: A Research Agenda." In *Economics and National Security*, J. Leitzel, ed. Boulder: Westview Press, pp. 29-43.

Beenstock, M. 1998. Country Survey xi: Defence and the Israeli Economy. *Defence and Peace Economics* 9: 171-222.

Bell, M. 2000. Leaving Portsoken – Defence Procurement in the 1980s and 1990s. *Rusi Journal* 145/4: 30-36.

Ben-Gurion, D. 1971. *Uniqueness and Mission: Selected Speeches on the Security of Israel.* Tel Aviv: Ministry of Defense Publishing House (H).

Ben-Gurion, D. 1981. Army and State. *Ma'arachot* 279-280: 2-11 (H).

Benoit, E. 1968. The Monetary and Real Costs of National Defense. *American Economic Review* 58: 398-416.

Benoit, E. 1973. *Defense and Economic Growth in Developing Countries.* Lexington: Lexington Books.

Benoit, E. and Lubell, H. 1966. World Defense Expenditures. *Journal of Peace Research* 2: 97-113.

Ben-Zvi, S. 1993. *Defense Expenditures and the National Economy.* National Security College (H).

Berger, M.C. and Hirsch, B.T. 1983. The Civilian Earnings Experience of Vietnam-era Veterans. *Journal of Human Resources* 18: 455-479.

Berglas, E. 1970. Security, Standard of Living and External Debt. *Economic Quarterly* 67: 191-202 (H).

Berglas, E. 1983. *Defense and the economy: The Israeli experience.* Jerusalem: Maurice Falk Institute for Economic Research in Israel, Discussion Paper, 83.01.

Berman, E. and Halperin, A. 1990. "Skilled Labor Force, Security and Growth." In *Industrial-Technological Policy for Israel,* D. Brodet, M. Justman and M. Teubal, eds. Jerusalem: Jerusalem Institute for Israeli Studies, pp. 147-182 (H).

Berner, J.K. and Daula, T.V. 1993. Recruiting Goals, Regime Shifts, and the Supply of Labor to the Army. *Defence Economics* 4: 315-338.

Biddle, J.E. and Samuels, W.J. 1991. "Thorstein Veblen on War, Peace, and National Security." In *Economics and National Security,* C.D. Goodwin, ed. Durham and London: Duke University Press, pp. 87-130.

Blumental, N. 1985. "Investments in the Defense Industry and their Impact on the Economy." In *Israeli Security Planning in the 1980s,* Z. Lanir, ed. Tel Aviv: Ministry of Defense and Jaffee Center for Strategic Studies, Tel Aviv University, pp. 130-141 (H).

Bonen, Z. 1994. The Israeli Defence Industry: Past and Future. *Rusi Journal* June: 56-59.

Bonen, Z. 1995. "The Defense Industry: The Horse of Shalom Aleichem." In *The Defense Industry in Israel.* Ramat Gan: The Begin-Sadat Center for Strategic Studies, Bar-Ilan University, pp. 37-41 (H).

Borcherding, T.E. 1971. A Neglected Social Cost of a Voluntary Military. *American Economic Review* 61: 195-196.

Boston Study Group. 1981. *The Price of Defense.* Tel Aviv: Ministry of Defense Publishing House (H*).

Boulding, K.E. 1962. *Conflict and Defense: A General Theory.* New York: Harper and Row.

Boulding, K.E. 1986. The Economics and Noneconomics of the World War Industry. *Contemporary Policy Issues* 4: 12-21.

Bowlin, W.F. 1995. A Note on the Financial Condition of Defense Contractors. *Defence and Peace Economics* 6: 295-304.

Bradford, D.F. 1968. A Model of the Enlistment Decision under Draft Uncertainty. *Quarterly Journal of Economics,* 73: 621-638.

Brauer, J. 1991. Arms Production in Developing Nations: The Relation to Industrial Structure, Industrial Diversification, and Human Capital Formation. *Defence Economics* 2: 165-175.

Brauer, J. and Marlin, J.T. 1992. Converting Resources from Military to Non-Military Uses. *Journal of Economic Perspectives* 4: 145-164.

Brodie, B. 1980. *War and Politics.* Tel Aviv: Ministry of Defense Publishing House (H*).

Brown, C. 1985. Military Enlistments: What Can We Learn from Geographic Variations? *American Economic Review* 75: 228-234.

Bruno, M. 1975. "Guns, Butter and the Muses." In *Economists on Economic Policy in Israel,* N.T. Gross, ed. Tel Aviv: University's Publishing Enterprises, pp. 7-11 (H).

Bruno, M. 1989. The Economy's Recovery: A Historical Perspective. *Economic Quarterly* 141: 89-113 (H).

Bruno, M. 1993. *Crisis, Stabilization, and Economic Reform.* Oxford: Clarendon Press.

Brzoska, M. 1981. The Reporting of Military Expenditures. *Journal of Peace Research* 18: 261-275.

Brzoska, M. 1983. Research Communication: The Military Related External Debt of Third World Countries. *Journal of Peace Research* 20: 271-277.

Brzoska, M. 1994. The Financing Factor in Military Trade. *Defence and Peace Economics* 5: 67-80.

Brzoska, M. 1995. "World Military Expenditures." In *Handbook of Defense Economics,* K. Hartley and T. Sandler, eds. Amsterdam: Elsevier, pp. 45-67.

Brzoska, M. 1998. "Too Small to vanish, Too Large to Flourish: Dilemmas and Practices of Defence Industry Restructuring in West European Countries." In *The Politics and Economics of Defence Industries,* E. Inbar and B. Zilberfarb, eds. London: Frank Cass, pp. 71-94.

Buck, D., Hartley, K. and Hooper, N. 1993. Defence Research and Development, Crowding-out and the Peace Dividend. *Defence Economics* 4: 161-178.

Bullens, H. 1995. "Conversion and the Future of the German Defence Firm: A Systematic View." In *The Future of the Defence Firm: New Challenges, New Directions,* A. Latham and N. Hooper eds. Dordrecht: Kluwer Academic Publishers, pp. 161-174.

Burk, J. 1989. Debating the Draft in America. *Armed Forces and Society* 15: 431-438.

Business Week. 16.3.1992. "A Life Raft for Arms Makers", pp. 83-84.

Business Week. 5.7.1993. "From the Gulf War to the War on Crime", pp. 46-47.

Cahn, A.H. 1979. "The Economics of Arms Transfers." In *Arms Transfers in the Modern World,* S.G. Neuman and R.E. Harkavy, eds. New York: Prager, pp. 173-183.

Cashel-Cordo, P. and Craig, S.G. 1990. The Public Sector Impact of International Resource Transfers. *Journal of Development Economics* 32: 17-42.

Catrina, C. 1988. *Arms Transfers and Dependence.* United Nations Institute for Disarmament Research, New York: Taylor and Francis.

Catton, B. 1979. *American Civil War.* Tel Aviv: Ministry of Defense Publishing House (H*).

CBS (Central Bureau of Statistics). various years. *Statistical Abstract of Israel.* Jerusalem: Central Bureau of Statistics.

CBS. 1983. Defense Expenditures in Israel 1950-1982. *Statistical Monthly Supplement* 6: 59-68.

CBS. 1996. Defense Expenditures in Israel 1950-1995. *Current Briefings in Statistics* 23. Jerusalem: Central Bureau of Statistics.

Chakrabarti, A.K., Glisman, H.H. and Horn, E. 1992. Defence and Space Expenditures in the US: An Inter-firm Analysis. *Defence Economics* 3: 169-189.

Chan, S. 1985. The Impact of Defense Spending on Economic Performance: A Survey of Evidence and Problems. *Orbis* 29: 403-434.

Clark, J.J. 1966. "The new economics of national defense." reprinted In *The Economic Impact of the Cold War,* J.L. Clayton, ed. New York: Harcourt, Brace & World, Inc., 1970, pp. 7-26.

Clark, R. 1978. "Capital-Labor Ratios in a Military Service: A Putty Clay Application." In *Defense Manpower Policy: Presentations from the 1976 Rand Conference on Defense Manpower,* R.V.L. Cooper, ed. Santa Monica: Rand, pp. 11-23.

Clough, S. 1951. *The Rise and Fall of Civilization.* New York: McGraw Hill.

Cohen, S.A. 1995 *The IDF: From a "People's Army" to a "Professional Military" – Causes and Implications.* Ramat-Gan: The Begin-Sadat Center for Strategic Studies, Bar-Ilan University.

Cohen, S.A. 1997. *Towards a New Portrait of the (New) Israeli Soldier.* Ramat Gan: The Begin-Sadat Center for Strategic Studies, Bar-Ilan University.

Dale, C. and Gilroy, C. 1984. Determinants of Enlistments: A Macroeconomic Time-Series View. *Armed Forces and Society* 10: 192-210.

Dale, C. and Gilroy, C. 1985. Enlistments in the All-Volunteer Force: Note. *American Economic Review,* 75: 547-551.

Davis, R. and Palomba, N. 1968. On the Shifting of the Military Draft as a Progressive Tax in Kind. *Western Economic Journal* 6: 150-153.

Dayan, M. 1976. *Story of My Life.* Jerusalem: Edanim Publishers, and Tel Aviv: Dvir (H).

De Tray, D. 1982. Veteran Status as a Screening Device. *American Economic Review,* 72: 133-142.

DeBoer, L. and Blackley, P.R. 1990. The Structure of Defense Production in the United States, 1929-1987. *Defence Economics* 1: 85-95.

DeBoer, L. and Brorsen, B.W. 1989. The Demand for and Supply of Military Labor. *Southern Economic Journal* 55: 853-869.

Defense News. 14.6.1999. "Mergers Across Europe's Borders Spell Job Losses".

Defense News. 5.7.1999. "Warning to Washington".

Defense News. 18.10.1999. "Russian Defense Industry Shows Short-Term Rebound".

Defense News. 29.5.2000. "Protect the Industrial Base".

Defense News. 3.7.2000. "DoD Panel Wants to Lift Industry's Bottom Line".

Defense News. 29.1.2001. "Industry Consolidates, but Factories Stay Open".

Deger, S. and Sen, S. 1990. *Military Expenditure: The Political Economy of International Security.* Oxford: Sipri and Oxford University Press.

Deutsch, E. and Schopp, W. 1982. *Civil Versus Military R&D - Expenditures and Industrial Productivity.* Vienna: University of Technology of Vienna (unpublished).

Dowdall, P. and Braddon, D. 1995. "Puppets or Partners: The Defence Industry Supply Chain in Perspective." In *The Future of the Defence Firm: New Challenges, New Directions,* A. Latham and N. Hooper eds. Dordrecht: Kluwer Academic Publishers, pp. 103-119.

Dror, Y. 1987. Strategic-Political Considerations for Renewing the Warfare Doctrine and Force Structure. *Ma'arachot* 310: 8-12 (H).

Dror, Y. 1989. *A Grand Strategy for Israel.* Jerusalem: Academon (H).

Dun & Bradstreet. 2002. *Israel's Leading Enterprises 2001.* Israel: Dun & Bradstreet.

Dunne, J.P. 1993. The Changing Military Industrial Complex in the UK. *Defence Economics* 2: 91-111.

Dunne, J.P. 1995. "The Defense Industrial Base." In *Handbook of Defense Economics,* K. Hartley and T. Sandler, eds. Amsterdam: Elsevier, pp. 399-430.

Dussauge, P. and Garrette, B. 1993. Industrial Alliances in Aerospace and Defence: An Empirical Study of Strategic and Organizational Patterns. *Defence Economics* 4: 45-62.

Dussauge, P. and Garrette, B. 1995. "The Future of the Defence Firm: Collaboration, Co-operation and Strategic Alliance." In *The Future of the Defence Firm: New Challenges, New Directions,* A. Latham and N. Hooper eds. Dordrecht: Kluwer Academic Publishers, pp. 121-132.

Dvir, M. and Bar-Zackay, S. 1985. "Challenges for the Israeli Defense Industries." In *Israeli Security Planning in the 1980s,* Z. Lanir, ed. Tel Aviv: Ministry of Defense and Jaffee Center for Strategic Studies, Tel Aviv University, pp. 142-150 (H).

Economist. 2.10.1993. "Still Waiting for the Bang", pp. 69-70.

Economist. 13.1.1996. "Americans Monsters, European Minnows", pp. 65-66.

Economist. 18.1.1997. "Raytheon's Rise", p. 69.

Economist. 17.5.1997. "Missile to Queen's Rook Four", pp. 73-74.

Economist. 22.11.1997. "A Shrinking Arms Market", p. 18.

Electronic News. 3.7.1995. "Lockheed Martin Setting $ 1.7B Consolidation".

Elizur, Y. 1997. *Economic Warfare: The hundred-Year Economic Confrontation between Jews and Arabs.* Tel Aviv: Kinneret (H).

Enke, S. ed. 1967. *Defense management.* Englewood Cliffs: Prentice-Hall.

Enthoven, A.C. 1963. Economic Analysis in the Department of Defense. *American Economic Review* 53 (Supp.): 413-423.

Enthoven, A.C. and Smith, K. 1974. *How Much Is Enough.* Tel Aviv: Ministry of Defense Publishing House (H*).

Fallows, J. 1981. *National defense.* New York: Vintage Books.

Fergusson, J. 1995. "The Defence Firm in a Changing Politico-Strategic Environment." In *The Future of the Defence Firm: New Challenges, New Directions,* A. Latham and N. Hooper eds. Dordrecht: Kluwer Academic Publishers, pp. 23-35.

Financial Times. 9.1.1996. "From Swords into Cash".

Financial Times. 26.8.1998. "BAe May Close Royal Ordnance Munitions Plants".

Fisher, A.C. 1969. The Cost of the Draft and the Cost of Ending the Draft. *American Economic Review* 59: 239-254.

Flight International. 28.1.1998. "Dasa Achieves Record Figures", p. 16.

Flight International. 12.12.2000. "New Global Philosophy and Identity for Thomson-CSF", p. 25

Fontanel, J. 1994. The Economics of Disarmament: A Survey. *Defence and Peace Economics* 5: 87-120.

Fox, J.R. 1974. *Arming America: How the U.S. Buys Weapons.* Cambridge, MA.: Harvard University Press.

Fredland, J.E. and Little, R.D. 1980. Long-Term Returns to Vocational Training: Evidence from Military Sources. *Journal of Human Resources* 15: 49-66.

Frumkin, H. 1970. Effective Economic Planning for Security and Immigration. *Economic Quarterly* 67: 261-269 (H).

Gadish, Y. "A National Equilibrium." In *The Price of Power,* Z. Offer and A. Kober, eds. Tel Aviv: Ministry of Defense Publishing House, pp. 185-189 (H).

Gafni, A. 1989. "U.S. aid to Israel – Politics, Defense and Economics." In *U.S. Aid to Israel: Issues in Israel's Economic, Foreign and Defense Policies.* Tel-Aviv: Israeli International Institute for Applied Economic Policy Review, pp. 1-6.

Gansler, J.S. 1980. *The Defense Industry.* Cambridge, MA.: The MIT Press.

Gansler, J.S. 1989. *Affording Defense.* Cambridge, MA.: The MIT Press.

Gansler, J.S. 1993. Forging an Integrated Industrial Complex. *Technology Review* July: 24-27.

Gansler, J.S. 1995. "The Future of the Defence Firm: Integrating Civil and Military Technologies." In *The Future of the Defence Firm: New Challenges, New Directions,* A. Latham and N. Hooper eds. Dordrecht: Kluwer Academic Publishers, pp. 89-95.

Garfinkel, M.R. 1990. The Role of the Military Draft in Optimal Fiscal Policy. *Southern Economic Journal* 56: 718-731.

Gerner, D.J. 1983. Arms Transfers to the Third World: Research on Patterns, Causes and Effects. *International Interactions* 10: 5-37.

Gholz, E. and Sapolsky, H.M. 1999. Restructuring the U.S. Defense Industry. *International Security* 24/3: 5-51.

Gilshon, A. 1986. "Defense Expenditures and Economic Growth in Israel." In *Studies in Israel's Economy 1982, 1984,* M. Felber, M. Michaely and Z. Sussman, eds. Jerusalem: The Israeli Economic Association, pp. 13-33 (H).

Gilshon, A. and Beenstock, M. 1989. *Economic Significance of the Defense Burden in Israel.* (unpublished) (H).

Goldberg, M.S. and Warner, J.T. 1987. Military Experience, Civilian Experience, and the Earnings of Veterans. *Journal of Human Resources* 22: 62-81.

Goodwin, C.D. 1991. "National Security in Classical Political Economy." In *Economics and National Security,* C.D. Goodwin, ed. Durham and London: Duke University Press, pp. 23-35.

Gottlieb, D. 1994. *On the Economic Implications of the Peace Process for Israel.* Jerusalem: Bank of Israel (unpublished).

Greenberg, Y. 1991. *National Security and Military Power – Statesman versus Commander-in-Chief.* Tel Aviv: Ministry of Defense, History Department (H).
Greenberg, Y. 1993. The Ministry of Defense and the General Staff: The Controversy over the Management of the Defense Budget 1948-1967. *State, Government and International Relations* 38: 49-76 (H).
Greenberg, Y. 1997. *Defense Budgets and Military Power – The Case of Israel 1957-1967.* Tel Aviv: Ministry of Defense Publishing House (H).
Grimmett, R.F. 1995. *Conventional Arms Transfers to Developing Nations, 1987-1994.* Washington: Congressional Research Service (CRS) report.
Grobar, L.M., Stern, R.M. and Deardorff, A.V. 1990. The Economic Effects of International Trade in Armaments in the Major Western Industrialized and Developing Countries. *Defence Economics* 1: 97-120.
Gross, N.T. 1975. *Economists on Economic Policy in Israel.* Tel Aviv: University's Publishing Enterprises (H).
Gruneberg, D.S. 1995. "The Defence Firm and Trends in Civil and Military Technologies: Integration versus "Differentiation"." In *The Future of the Defence Firm: New Challenges, New Directions,* A. Latham and N. Hooper eds. Dordrecht: Kluwer Academic Publishers, pp. 97-101.
Gupta, A. 1993. Third World Militaries: New Suppliers, Deadlier Weapons. *Orbis,* 37: 57-68.
Hall, P. and Markowski, S. 1994. On the Normality and Abnormality of Offsets Obligations. *Defence and Peace Economics* 5: 173-188.
Halperin, A. 1986. *The Development of Military Capital Stocks in Israel and the Confrontation States.* Jerusalem: Maurice Falk Institute for Economic Research in Israel (H).
Halperin, A. 1987. Force Buildup and Economic Growth. *Economic Quarterly* 131: 990-1010 (H).
Halperin, A. 1989. *Military capital stock.* Massachusetts Institute of Technology, Working Paper No. 5-89.
Hansen, L.W. and Weisbrod, B.A. 1967. Economics of the Military Draft. *Quarterly Journal of Economics* 81: 395-421.
Harford, J.D. and Marcus, R.D. 1988. A General Equilibrium Model of the Volunteer Military. *Southern Economic Journal* 55: 472-484.
Harkabi, Y. 1990. *War and Strategy.* Tel Aviv: Ministry of Defense Publishing House (H).
Harkavy, R.E. 1975. *The Arms Trade and International Systems.* Cambridge MA.: Ballinger.
Hartley, K. 1995. "Industrial Policies in the Defense Sector." In *Handbook of Defense Economics,* K. Hartley and T. Sandler, eds. Amsterdam: Elsevier, pp. 460-489.
Hartley, K. 1996. Defence Industries Adjusting to Change - Review Article. *Defence and Peace Economics* 7: 169-184.
Hartley, K. and Hooper, N. 1990. *The Economics of Defense, Disarmament and Peace: An Annotated Bibliography.* Aldershot: Elgar.
Hartley, K. and Martin, S. 1993. Evaluating Collaborative Programmes. *Defence Economics* 4: 195-211.
Hartley, K. and Mclean, P. 1981. UK Defence Expenditure. *Public Finance* 36: 171-192.
Hasid, N. and Lesser, Y. 1981. Economic Resources for the Security of Israel. *Economic Quarterly* 110-111: 243-252 (H).
Hess, P.N. 1989. Force Ratios, Arms Imports and Foreign Aid Receipts in the Developing Nations. *Journal of Peace Research* 26: 399-412.
Hewitt, D.P. 1991. *Military Expenditure: International Comparison of Trends.* Washington D.C.: Fiscal Affairs Department, IMF Working Papers 91/54
Hildebrandt, G.G. 1990. Services and Wealth Measures of Military Capital. *Defence Economics* 1: 159-176.
Hilton, B. 1987. Defence Economics Review. *Scottish Journal of Political Economy* 34: 306-314.
Hitch, C.J. 1960. National Security Policy as a Field for Economics Research. *World Politics* 12: 434-449.

Hitch, C. J. and McKean, R. N. 1971. *The Economics of Defense in the Nuclear Age.* Tel Aviv: Ministry of Defense Publishing House (H*).

Holzman, F.D. 1980. Is There a Soviet-U.S. Military Spending Gap? *Challenge* Sep.-Oct.: 3-9.

Hooper, N. 1995. "The Future of the Defence Firm in the United Kingdom: The Impact of the Changing Politico-Commercial Environment." In *The Future of the Defence Firm: New Challenges, New Directions,* A. Latham and N. Hooper eds. Dordrecht: Kluwer Academic Publishers, pp. 57-71.

Horne, D.K. 1985. Modeling Army Enlistment Supply for the All-Volunteer Forces. *Monthly Labor Review* 108: 35-39.

Horowitz, D. 1984. "Israel's National Security Perception 1948-1972." In *Diplomacy and Confrontation: Selected Issues in Israel's Foreign Relations 1948-1978,* B. Neuberger, ed. Tel Aviv: Everyman's University, pp. 104-148 (H).

Horowitz, D. 1985. "Constant and Changing Factors in Israel's Security Perception." In *War by Choice,* J. Alpher, ed. Tel Aviv: Hakibbutz Hameuchad and Jaffee Center for Strategic Studies, Tel Aviv University, pp. 57-115 (H).

Howard, M. 1985. *War in European History.* Tel Aviv: Ministry of Defense Publishing House (H*).

Imbens, G. and Klaauw, W. 1993. *Evaluating the Cost of Conscription in the Netherlands.* Harvard Institute of Economic Research, Discussion Paper No. 1632.

Inbar, E. 1983. Israel Strategy since the Yom Kippur War. *Ma'arachot* 289-290: 14-28 (H).

Inbar, E. 1996. Contours of Israel's New Strategic Thinking. *Political Science Quarterly* 111: 41-64.

Interavia. April, 1996. "Can Europe Compete?", pp. 14-19.

Interavia. August/September, 1996. "UK Industry Fitter and Leaner", pp. 31-38.

Interavia. November, 1996. "Mega-Primes: More Deals Coming", pp. 20-23.

Interavia. January/February, 1997. "Can Europe Harmonise its Defence Requirements?", pp. 39-43.

Interavia. March, 1997. "Russia's French Connection", pp. 18-21.

Interavia. April, 1997. "Investors, Beware of Defence Spending Increases", pp. 16-17.

Intriligator, M.D. 1991. On the Nature and Scope of Defence Economics: A Reply to Judith Reppy's Comment. Defence Economics 2: 273-274.

Intriligator, M.D. and Brito, D.L. 1989. "Arms Races and Arms Control in the Middle East." In *Economic Cooperation in the Middle East,* G. Fishelson, ed. Boulder: Westview Press, pp. 105-122.

Intrillgator, M.D. 1998. "The Economics of Defence Conversion." In *The Politics and Economics of Defence Industries,* E. Inbar and B. Zilberfarb, eds. London: Frank Cass, pp.29-50.

Jane's Defence Weekly. 19.6.1996. "Lockheed Martin Set to Stay on its Toes", pp. 82-85.

Jane's Defence Weekly. 9.7.1997. "Lockheed Martin Poised to Grow Across Borders", p. 3.

Jane's Defence Weekly. 25.3.1998. "Lockheed Martin Heading for Showdown on Merger", p. 5.

Jane's Defence Weekly. 29.4.1998. "GEC Moves into Next Phase with $1.4b Tracor Bid", p. 18.

Jane's Defence Weekly. 4.11.1998. "Spain Moves towards Privatization", p. 20.

Jane's Defence Weekly. 3.5.2000. "Giat's Losses Continue to Built", p. 18.

Jane's Defence Weekly. 31.5.2000. "Air, Sea Projects Take Off as MBT Future Is in Doubt", pp. 34-35.

Jane's Defence Weekly. 28.6.2000. "New RO Defence Formed", p. 10.

Jane's Defence Weekly. 6.12.2000. "DoD Panel Suggest Steps to Boost US Industry", p. 18.

Jane's Defence Weekly. 17.1.2001. "France's SNPE Restructures", p.20.

Janowitz, M. 1973. *The US Forces and the Zero Draft.* London: International Institute for Strategic Studies, Adelphi Papers, No.94.

JIM. 1987. *Export-Led Growth Strategy for Israel (final report).* Tel-Aviv: JIM, Jerusalem Institute of Management.

Kanovsky, E. 1994. *Assessing the Mideast Peace Economic Dividend.* Ramat Gan: The Begin-Sadat Center for Strategic Studies, Bar-Ilan University (H).
Kapstein, E.B. 1992. *The Political Economy of National Security – A Global Perspective.* Columbia, South Carolina: University of South Carolina Press.
Katzir, Y. And Shadmi, Z. 1984. "Reducing Defense Expenditures – Is It the Key for Economic Recovery?" In *The Price of Power,* Z. Offer and A. Kober, eds. Tel Aviv: Ministry of Defense Publishing House, pp. 117-123 (H).
Kelly, R.C. Jr. 1977. "Military Manpower Costs and Manpower Policy: An Economic Assessment." In *Military Unions: US Trends and Issues,* W.J. Taylor, R.J. Arango and R.S. Lockwood, eds. Beverly Hills and London: Sage Publications, pp. 292-304.
Kennedy, G. 1975. *The Economics of Defence.* London: Faber and Faber.
Kennedy, P. 1992. *The Rise and Fall of the Great Powers.* Tel Aviv: Dvir (H*).
Kerstens, K. and Meyermans, E. 1993. The Draft versus an All-Volunteer Force: Issues of Efficiency and Equity in the Belgian Draft. *Defence Economics* 4: 271-284.
Keynes, J.M. 1919. *The Economic Consequences of the Peace.* London: Macmillan.
Khilji, N.M. and Zampelli, E.M. 1994. The Fungibility of U.S. Military and Non-Military Assistance and Impacts on Expenditures of Major Aid Recipients. *Journal of Development Economics* 43: 345-362.
Kindleberger, C.P. 1980. The Life of an Economist. *Banca Nazionale de Lavoro* 134: 231-245.
King, W.R. 1977. "The All-Volunteer Armed Forces and National Service: Alternatives for the Nation." In *Military Unions: US Trends and Issues,* W.J. Taylor, R.J. Arango and R.S. Lockwood, eds. Beverly Hills and London: Sage Publications, pp. 272-290.
Kinsella, D. 1994. The Impact of Superpower Arms Transfers on Conflict in the Middle East. *Defence and Peace Economics* 5: 19-36.
Kirkpatrick, D.L. 1995. The Rising Unit Cost of Defence Equipment – The Reasons and the Results. *Defence and Peace Economics* 6: 263-288.
Kleiman, E. 1970. The Consequences of the Six-Day War for the Structure of the Israeli Economy. *Banking Quarterly* 37: 49-57 (H).
Klieman, A. 1992. *Double-Edged Sword: Israel Defense Exports as an Instrument of Foreign Policy.* Tel Aviv: Am Oved (H).
Klieman, A. and Pedatzur, R. 1991. *Rearming Israel: Defense Procurement through the 1990s.* Tel-Aviv: Tel Aviv University, Jaffee Center for Strategic Studies.
Klinov, R. 1993. *Labor Force in Israel 1948-1990.* Jerusalem: Maurice Falk Institute for Economic Research in Israel (H).
Knapp, C.B. 1973. Human Capital Approach to the Burden of the Military Draft. *Journal of Human Resources* 8: 485-496.
Knorr, K. 1957. The Concept of Economic Potential for War. *World Politics* 10: 49-62.
Knorr, K. 1977. "Military Strength: Economic and Non-Economic Bases." In *Economic Issues and National Security,* K. Knorr and F.N. Trager, eds. Lawrence: Regent Press of Kansas, pp. 183-200.
Kohav, D. and Lifshitz, Y. 1973. Defense Expenditures and their Influence on the National Economy and Industry. *Economic Quarterly* 78-79: 3-17 (H).
Kolodziej, E.A. 1979. "Arms Transfers and International Politics: The Interdependence of Independence." In *Arms Transfers in the Modern World,* S.G. Neuman and R.E. Harkavy, eds. New York: Prager, pp. 3-26.
Landau, D. 1990. Public Choice and Economic Aid. *Economic Development and Cultural Change* 38: 559-576.
Latham, A. 1995. "The Structural Transformation of the US Defence Firm: Changes in Manufacturing Technology, Production Process and Principles of Corporate Organisation." In *The Future of the Defence Firm: New Challenges, New Directions,* A. Latham and N. Hooper eds. Dordrecht: Kluwer Academic Publishers, pp. 175-192.

Latham, A. and Hooper, N. 1995. "Introduction." In *The Future of the Defence Firm: New Challenges, New Directions,* A. Latham and N. Hooper eds. Dordrecht: Kluwer Academic Publishers, pp. 11-20.

Layne, C. and Metzger, R.S. 1995. Reforming Post-Cold War Arms Policy: The Crucial Link between Exports and the Defence Industrial Base. *Journal of Strategic Studies* 18: 1-32.

Leitzel, J. 1993. "Introduction." In *Economics and National Security,* J. Leitzel ed. Boulder: Westview Press, pp. ix-x.

Leonard, R.J. 1991. "War as a 'Simple Economic Problem': The Rise of an Economics of Defense" In *Economics and National Security,* C.D. Goodwin, ed. Durham and London: Duke University Press, pp. 261-283.

Lerner, J. 1992. The Mobility of Corporate Scientists and Engineers between Civil and Defense Activities: Implications for Economic Competitiveness in the post-Cold War Era. *Defence Economics* 3: 229-242.

Levine, P., Sen, S. and Smith, R. 1994. A Model of the International Arms Market. *Defence and Peace Economics* 5: 1-18.

Levite, A. 1988. *Offense and Defense in Israeli Military Doctrine.* Tel Aviv: Hakibbutz Hameuchad and Jaffee Center for Strategic Studies, Tel Aviv University (H).

Lichtenberg, F.R. 1995. "Economics of Defense R&D." In *Handbook of Defense Economics,* K. Hartley and T. Sandler, eds. Amsterdam: Elsevier, pp. 431-457.

Lifshitz, Y. 1974. "Defense Expenditure and Allocation of Resources." In *Issues in the Economy of Israel,* N. Halevi and Y. Kop, eds. Jerusalem: Maurice Falk Institute for Economic Research in Israel, pp. 88-111 (H).

Lifshitz, Y. 1991. "The Israeli Economy and the Gulf Crisis." In *War in the Gulf: Implications for Israel,* J. Alpher, ed. Tel Aviv: Jaffee Center for Strategic Studies, Tel Aviv University, pp. 314-335 (H).

Looney, R.E. 1989. The Influence of Arms Imports on Third World Debt. *Journal of Developing Areas* 23: 221-232.

Looney, R.E. and Frederiksen, P.C. 1993. "Arms Imports and Third World Growth in the 1980s." In *Defense Spending and Economic Growth,* J.P. Payne and A.P. Sahu, eds. Boulder: Westview Press, pp. 237-254.

Lovering, J. 1990. Military Expenditure and the Restructuring of Capitalism: The Military Industry in Britain. *Cambridge Journal of Economics* 14: 453-467.

Lovering, J. 1993. Restructuring the British Defence Industrial Base after the Cold War: Institutional and Geographical Perspectives. *Defence Economics* 2: 123-139.

Lundquist, J.T. 1992. Shrinking Fast and Smart in the Defense Industry. *Harvard Business Review* Nov.-Dec.: 74-85.

Luttwak, E. and Horowitz, D. 1975. *The Israeli Army.* London: Allen Lane.

Ma'ayan, J. 1985. "The Limits of Power." In *Quality and Quantity in Military Buildup,* Z. Offer and A. Kober, eds. Tel Aviv: Ministry of Defense Publishing House, pp. 295-303 (H).

Mahdavi, S. 1990. The Effects of Foreign Resource Inflows on Composition of Aggregate Expenditure in Developing Countries: A Seemingly Unrelated Model. *Kyklos* 43: 111-137.

Martin, S. and Hartley, K. 1995. UK Firms' Experience and Perceptions of Defence Offsets: Survey Results. *Defence and Peace Economics* 6: 123-139.

Martin, S., White, R. and Hartley, K. 1996. Defence and Firm Performance in the UK. *Defence and Peace Economics* 7: 325-337.

McGuire, M.C. 1982. U.S. Foreign Assistance, Israeli Resource Allocation, and the Arms Race in the Middle East. *Journal of Conflict Resolution* 26: 199-235.

McGuire, M.C. 1987. Foreign Assistance, Investment and Defense: A Methodological Study with Application to Israel, 1960-1979. *Economic Development and Cultural Change* 35: 847-873.

McGuire, M.C. 1987a. "Defence Economics." In *The New Palgrave Dictionary of Economics.* London: Macmillan, I: 760-762.

McGuire, M.C. 1995. "Defense Economics and International Security." In *Handbook of Defense Economics,* K. Hartley and T. Sandler, eds. Amsterdam: Elsevier, pp. 13-43.
Mehay, S.L. 1990. Determinants of Enlistments in the US Army Reserve. *Armed Forces and Society* 16: 351-367.
Mehay, S.L. 1991. Reserve Participation versus Moonlighting: Are They the Same? *Defence Economics* 2: 325-337.
Mellors, C. and Mckean, J. 1984. The Politics of Conscription in Western Europe. *West European Politics* 7: 25-42.
Melman, S. 1974. *The Permanent War Economy.* New York: Simon and Schuster.
Michaeli, M. 1975. "The Budget I would Propose." In *Economists on Economic Policy in Israel,* N.T. Gross, ed. Tel Aviv: University's Publishing Enterprises, pp. 32-38 (H).
Miller, J.C. III and Tollison, R.D. 1971. The Implicit Tax on Reluctant Military Recruits. *Social Science Quarterly* 51: 924-931.
Miller, J.C. III, Tollison, R.D. and Willett, T.D. 1968. Marginal Criteria and Draft Deferment Policy. *Quarterly Review of Economics and Business* 8: 69-73.
Ministry of Finance. 1984. "The State Budget Proposal for 1984 – Budget Principals." In *The Price of Power,* Z. Offer and A. Kober, eds. Tel Aviv: Ministry of Defense Publishing House, pp. 81-86 (H).
Ministry of Finance. various years. *The State Budget Proposal – Budget Principals.* Jerusalem (H).
Mintz, A. 1985. The Military-Industrial Complex: American Concepts and Israeli Realities. *Journal of Conflict Resolution* 29: 623-639.
Mintz, A. 1986. Arms Imports as an Action-Reaction Process: An Empirical Test of Six Pairs of Developing Nations. *International Interactions* 12: 229-243.
Mintz, A. 1992. "Introduction: Political Economy and National Security." In *The Political Economy of Military Spending in the United States,* A. Mintz, ed. London and New York: Routledge, pp. 1-11.
Mintz, A. and Ward, M.D. 1989. The Political Economy of Military Spending in Israel. *American Political Science Review* 83: 521-533.
Mintz, A., Ward, M.D. and Bichler, S. 1990. "Defence Spending in Israel." In *The Economics of Defence Spending,* K. Hartley and T. Sandler, eds. London and New York: Routledge, pp. 177-188.
Molas-Gallart Jordi. 1995. Missile Systems: "Flexible Modularity" and Incremental Technological Change in Military Production. *Defence and Peace Economics* 6: 141-158.
Moodie, M. 1979. "Defense Industries in the Third World: Problems and Promises." In *Arms Transfers in the Modern World,* S.G. Neuman and R.E. Harkavy, eds. New York: Praeger. pp. 294-311.
Neubach, A. 1984. "The Defense Burden on the Israeli Economy." In *The Price of Power,* Z. Offer and A. Kober, eds. Tel Aviv: Ministry of Defense Publishing House, pp. 161-184 (H).
Neuman, S.G. 1979. "Arms Transfers and Economic Development: Some Research and Policy Issues." In *Arms Transfers in the Modern World,* S.G. Neuman and R.E. Harkavy, eds. New York: Prager, pp. 219-245.
Neuman, S.G. and Harkavy, R.E. 1979. "The Road to Further Research and Theory in Arms Transfers." In *Arms Transfers in the Modern World,* S.G. Neuman and R.E. Harkavy, eds. New York: Prager, pp. 315-321.
Nevo, J. 1975. "On Guns and Butter in Times of War and Peace." In *Economists on Economic Policy in Israel,* N.T. Gross, ed. Tel Aviv: University's Publishing Enterprises, pp. 17-21 (H).
Oi, W.Y. 1967. The Economic Cost of the Draft. *American Economic Review, Papers and Proceeding* 57: 39-62.
Olson, M. 1989. *The Rise and Decline of Nations.* Tel Aviv: Am Oved (H*).
Olson, M. and Zeckhauser, R. 1966. The Economic Theory of Alliances. *Review of Economics and Statistics* 48: 266-279.

Oser, J. and Blanchfield, W. C. 1978. *The Evolution of Economic Thought.* Tel Aviv: Zmora-Bitan-Modan (H*).

Ostfeld, Z. 1994. *An Army Is Born.* Tel Aviv: Ministry of Defense Publishing House (H).

Owen, N. 1994. How Many Men Do Armed Forces Need? An International Comparison. *Defence and Peace Economics* 5: 269-288.

Patinkin, D. 1965. *The Israeli Economy in the First Decade.* Jerusalem: Maurice Falk Institute for Economic Research in Israel (H).

Pearson, F.S. 1989. The Correlates of Arms Importation. *Journal of Peace Research* 26: 153-164.

Peck, M.J. and Scherer F.M. 1962. *The Weapon Acquisition Process: An Economic Analysis.* Boston: Harvard University.

Pedatzur, R. 1990. Israel: Updating the Military Doctrine. *Ma'arachot* 319: 20-29 (H).

Peres, S. 1970. *David's Sling.* Jerusalem: Weidenfeld and Nicolson (H).

Peri, J. and Neubach, A. 1984. *The Military-Industrial Complex in Israel.* Tel Aviv: International Center for Peace in the Middle East (H).

Pierre, A.J. 1982. *The Global Politics of Arms Sales.* Princton: Princton University Press.

Pigou, A.C. 1921. *The Political Economy of War.* London: Macmillan.

Porter, M.E. 1990. *The Competitive Advantage of Nations.* New York: The Free Press

Pugh, P.G. 1993. The Procurement Nexus. *Defence Economics* 4: 179-194.

Rabin, I. 1979. Security Problems of Israel in the 1980s. *Ma'arachot* 270-271:18-20 (H).

Rabin, I. 1996. "Israel's Security Policy after the Gulf War." In *In Memoriam: Yitzhak Rabin and Israeli National Security* Ramat Gan: The Begin-Sadat Center for Strategic Studies, Bar-Ilan University (H).

Ratner, J. and Thomas, C. 1990. The Defence Industrial Base and Foreign Supply of Defence Goods. *Defence Economics* 2: 57-68.

Razin, A. 1984. "On Honey and on Sting – The Impact of American Aid." In *The Price of Power,* Z. Offer and A. Kober, eds. Tel Aviv: Ministry of Defense Publishing House, pp. 47-57 (H).

Reppy, J. 1991. On the Nature and Scope of Defence Economics: A Comment. *Defence Economics* 2: 269 - 271.

Richardson, L.F. 1960. *Arms and Insecurity: A Mathematical Study of the Causes and Origins of War.* Pittsburgh: Homewood.

Ridge, M. and Smith, R. 1991. UK Military Manpower and Substituability. *Defence Economics* 2: 283-293.

Rogerson, W.P. 1991. Excess Capacity in Weapon Production: An Empirical Analysis. *Defence Economics* 2: 235-249.

Rogerson, W.P. 1995. "Incentive Models of the Defense Procurement Process. In *Handbook of Defense Economics,* K. Hartley and T. Sandler, eds. Amsterdam: Elsevier, pp. 309-346.

Ronen, J. 1968. The Economy at the Service of Security. *Economic Quarterly* 58: 202-209 (H).

Rosh, R.M. 1990. Third World Arms Production and the Evolving Interstate System. *Journal of Conflict Resolution* 34: 57-73.

Rotem, Z. 1985. *The Israeli-Arab Conflict as Mirrored in Defense Expenditures.* Ramat Gan: Bar-Ilan University (M.A. Thesis) (H).

Rotem, Z. 1988. The Military Production Function and Changes in Israel-Arab Balance of Power 1950-1986. *Economic Quarterly* 138: 289-303 (H).

Rotem, Z. 1990. Changes in Israel-Arab Balance of Power 1950-1986. *Ma'arachot* 319: 30-39 (H).

Russet, B.M. 1970. *What Price Vigilance?* New Haven: Yale University Press.

Sadan, E. 1985. "National Security and National Economy." In *Israeli Security Planning in the 1980s,* Z. Lanir, ed. Tel Aviv: Ministry of Defense and Jaffee Center for Strategic Studies, Tel Aviv University, pp. 119-129 (H).

Safran, N. 1969. *From War to War.* Jerusalem: Keter Publishing House (H).

Sandler, T. and Cauley, J. 1975. On the Economic Theory of Alliances. *Journal of Conflict Resolution* 19: 320-348.
Sandler, T. and Hartley, K. 1995. *The Economics of Defense.* Cambridge: University Press of Cambridge.
Schelling, T.C. 1960. *The Strategy of Conflict.* Cambridge, MA.: Harvard University Press.
Scherer, F. 1964. *The weapons acquisition process: The economic incentives.* Boston: Harvard University Press.
Schiff, Z. and Haber, E. 1976. *Israel, Army and Defense – A Dictionary.* Tel Aviv: Zmora-Bitan-Modan (H).
Schwartz, S. 1986. The Relative Earnings of Vietnam and Korean-Era Veterans. *Industrial and Labor Relations Review* 39: 564-572.
Sharon, A. 1985. "The Security Perception of Ariel Sharon (A Speech not Delivered)." In *War by Choice,* J. Alpher, ed. Tel Aviv: Hakibbutz Hameuchad and Jaffee Center for Strategic Studies, Tel Aviv University, pp.157-163 (H).
Shefer, E. undated. *Israel-Arab: Economy and Security.* IDF (H).
Shefer, E. 1985. "The Economic Burden of the Arms Race between the Confrontation States and Israel." In *Israeli Security Planning in the 1980s,* Z. Lanir, ed. Tel Aviv: Ministry of Defense and Jaffee Center for Strategic Studies, Tel Aviv University, pp. 95-118 (H).
Shefer, E. 1987. The Economic Burden of the Arms Race in the Middle East. *Economic Quarterly* 131: 865-872 (H).
Sipri (Stockholm International Peace Research Institute). 1971. *The arms trade with the third world.* Stockholm: Almqvist & Wiksell.
Sipri. 1973. *The meaning and measurement of military expenditure.* Stockholm: Research report No. 10.
Sipri. various years. *SIPRI Yearbook.* New York: Oxford University Press.
Skons, E. 1983. "Military Prices." In *SIPRI Yearbook 1983.* London: Taylor & Francis, pp. 195-211.
Smith, Adam. 1776. *An Inquiry into the Nature and Causes of the Wealth of Nations* (printed for W. Strahan and T. Cadell).
Smith, R., Humm, A. and Fontanel, J. 1985. The Economics of Exporting Arms. *Journal of Peace Research* 22: 239-247.
Solnick, L.M., Henderson, D.R. and Capt. Kroeschel, J.W. 1991. Using Quit Rates to Estimate Compensating Wages Differentials in the Military. *Defence Economics* 2: 123-133.
Sorokin, K. 1993. Russia's 'New Look' Arms Sales Strategy. *Arms Control Today* October: 7-12.
Stigler, G.J. and Friedland, C. 1971. Profits of Defense Contractors. *American Economic Review* 61: 692-694.
Sussman, Z. 1989. "The Contribution of U.S. Aid to the Budget and the Success of the July 1985 Stabilization Program." In *U.S. Aid to Israel: Issues in Israel's Economic, Foreign and Defense Policies.* Tel-Aviv: Israeli International Institute for Applied Economic Policy Review, pp. 19-31.
Sussman, Z. 1995. From Crisis (1973) to Stabilization (1985): The Israeli Economy at the Mercy of External Shocks. *Economic Quarterly* 42: 683-698 (H).
Sutton, J.L. and Kemp, G. 1966. *Arms to Developing Countries, 1945-1965.* London: International Institute for Strategic Studies, Adelphi Papers, No. 28.
Syrquin, M. 1989. "U.S. Aid and the Structure of the Economy." In *U.S. Aid to Israel: Issues in Israel's Economic, Foreign and Defense Policies.* Tel-Aviv: Israeli International Institute for Applied Economic Policy Review, pp. 32-42.
Tal, I. 1996. *National Security: The Few against the Many.* Tel Aviv: Dvir (H).
Tarr, C.W. 1981. *By the Numbers: The Reform of the Selective Service System, 1970-1972.* National Defense University Press.
Taylor, T. 1993. West European Defence Industrial Issues for the 1990s. *Defence Economics* 4: 113-121

Tishler, A. and Rotem, Z. 1995. Factors Explaining the International Success of the Israeli Defense Industry. *Economic Quarterly* 3/95: 468-496 (H).

Toffler, A. and Toffler, H. 1994. *War and Anti-War.* Tel Aviv: Ma'ariv Book Guild (H*).

Tov, I. 1992. Defense Industries in Israel in Changing Conditions 1984-1990. *Economic Quarterly* 151: 636-659 (H).

Trevino, R. and Higgs, R. 1992. Profits of US Defense Contractors. *Defence Economic* 3: 211-218.

Tropp, Z. 1989. How Much Is 10.5 Percent of GNP? *Politics* 26: 52-53 (H).

Tuohy, W. 1991. "Conscript Armies May Be a Thing of the Past". *Los Angeles Times* (reprinted in *The Jerusalem Post,* 19.8.1991).

Udis, B. and Maskus, K.E. 1991. Offsets as Industrial Policy: Lessons from Aerospace. *Defence_Economics* 2: 151-164.

UNIDIR. (United Nations Institute for Disarmament Research). 1993. *Economic Aspects of Disarmament: Disarmament as an Investment Process.* New York: United Nations Publication.

van Creveld, M. 1985. "God Is on the Side of the Big Batallions?" In *Quality and Quantity in Military Buildup,* Z. Offer and A. Kober, eds. Tel Aviv: Ministry of Defense Publishing House, pp. 163-168 (H).

van Ypersele de Strihou, J. 1967. Sharing the Defense Burden Among Western Allies. *Review of Economics and Statistics* 49: 527-536.

Wald, E. 1987. *The Curse of the Broken Weapons.* Tel Aviv: Schocken Publishing House (H).

Walker, W. and Willett, S. 1993. Restructuring the European Defence Industrial Base. *Defence Economics* 4: 141-160.

Wallach, J.L. 1976. "Military." *Encyclopaedia Hebraica.* Ramat Gan: Massada, 28: 476-485 (H).

Wallach, J.L. 1980. *Military Doctrines: Their Development in the 19th and 20th Centuries.* Tel Aviv: Ministry of Defense Publishing House (H*).

Ward, M.D. and Mintz, A. 1987. Dynamics of Military Spending in Israel: A Computer Simulation. *Journal of Conflict Resolution* 31: 86-105.

Warner, J.T. and Asch, B. J. 1995. The Economics of Military Manpower. In *Handbook of Defense Economics,* K. Hartley and T. Sandler, eds. Amsterdam: Elsevier, pp. 347-398.

Warner, J.T. and Asch, B.J. 1996. The Economic Theory of Military Draft Reconsidered. *Defence and Peace Economics* 7: 297-312.

Weidenbaum, M. 1997. The U.S. Defense Industry after the Cold War. *Orbis* 41: 591-600.

Weinblatt, J. 1998. "The Influence of Stopping American Aid on the Israeli Economy." In *American Foreign Aid to Israel,* A. Gafni, ed. Jerusalem: The Center for Social Policy Research in Israel, pp. 13-33 (H).

Wheelock, T.R. 1978. Arms for Israel: The Limit of Leverage. *International Security* 3: 123-137.

Whitaker, J.K. 1991. "The Economics of Defence in British Political Economy, 1848-1914." In *Economics and National Security,* C.D. Goodwin, ed. Durham and London: Duke University Press, pp. 37-60.

Wiberg, H. 1983. Measuring Military Expenditures: Purposes, Methods, Sources. *Cooperation and Conflict* 18: 161- 177.

Willett, S. 1990. Conversion Policy in the UK. *Cambridge Journal of Economics* 14: 469-482.

Winter, J.M. 1975. "Introduction: The Economic and Social History of War." In *War and Economic Development,* J.M. Winter, ed. Cambridge: Cambridge University Press, pp. 1-10.

Withers, G.A. 1972. The Wage Costs of an All-Volunteer Army. *Economic Record* 48: 321-339.

Wulf, H. 1982. *Arms Production in Third World Countries: Effects on Industrialization.* Hamburg: University of Hamburg (unpublished).

Yaniv, A. 1994. *Politics and Strategy in Israel.* Tel Aviv: Sifriat Poalim (H).

Yariv, A. 1985. "A War by Choice: A War of No Choice." In *War by Choice,* J. Alpher, ed. Tel Aviv: Hakibbutz Hameuchad and Jaffee Center for Strategic Studies, Tel Aviv University, pp. 9-29 (H).
Zandberg, M. 1970. Security and Standard of Living. *Economic Quarterly* 66: 9-14 (H).
Zook, D.H. and Higham, R. 1981. *A Short History of Warfare.* Tel Aviv: Ministry of Defense Publishing House (H*).
Zusman, P. 1984. "Why is the Defense Burden in Israel so Oppressive?" In *The Price of Power,* Z. Offer and A. Kober, eds. Tel Aviv: Ministry of Defense Publishing House, pp. 17-25 (H).
Zusman, P. 1985. "The Dynamics of Economic Growth, Technological Progress and Power Augmentation: Some Strategic Impacts." In *Israeli Security Planning in the 1980s,* Z. Lanir, ed. Tel Aviv: Ministry of Defense and Jaffee Center for Strategic Studies, Tel Aviv University, pp. 182-203 (H).
Zusman, P. 1989. "U.S. Government Defense Aid to Israel." In *U.S. Aid to Israel: Issues in Israel's Economic, Foreign and Defense Policies.* Tel-Aviv: Israeli International Institute for Applied Economic Policy Review, pp. 7-17.

INDEX

MIX
Papier aus verantwortungsvollen Quellen
Paper from responsible sources
FSC® C105338

If you have any concerns about our products,
you can contact us on
ProductSafety@springernature.com

In case Publisher is established outside the EU,
the EU authorized representative is:
Springer Nature Customer Service Center GmbH
Europaplatz 3, 69115 Heidelberg, Germany

Printed by Libri Plureos GmbH
in Hamburg, Germany

Modern Analytical Electromagnetic Homogenization with Mathematica® (Second Edition)

Tom G Mackay
School of Mathematics, University of Edinburgh, Edinburgh, UK

Akhlesh Lakhtakia
Department of Engineering Science and Mechanics, Pennsylvania State University, University Park, PA, USA

IOP Publishing, Bristol, UK

ISBN 978-0-7503-3423-5 (ebook)
ISBN 978-0-7503-3421-1 (print)
ISBN 978-0-7503-3424-2 (myPrint)
ISBN 978-0-7503-3422-8 (mobi)

DOI 10.1088/978-0-7503-3423-5

Version: 20201201

IOP ebooks

British Library Cataloguing-in-Publication Data: A catalogue record for this book is available from the British Library.

Published by IOP Publishing, wholly owned by The Institute of Physics, London

IOP Publishing, Temple Circus, Temple Way, Bristol, BS1 6HG, UK

US Office: IOP Publishing, Inc., 190 North Independence Mall West, Suite 601, Philadelphia, PA 19106, USA

Dedicated to the worldwide eradication of poliomyelitis.

Contents

Preface to 1st edition

Electromagnetic homogenization is the process of estimating the effective electromagnetic properties of composite materials in the long-wavelength regime, wherein the length scales of nonhomogeneities are much smaller than the wavelengths involved. This is a venerable topic whose origins predate even the formulation of the Maxwell postulates of electromagnetism. Notable landmarks in its development include the establishment of the eponymous formalisms of Maxwell Garnett and Bruggeman, as well as of the strong-property-fluctuation theory. Our aim in this monograph is to provide a bird's-eye view of currently available homogenization formalisms for particulate composite materials. We present analytical methods only—the reader is referred elsewhere for descriptions of numerical approaches to homogenization.

Our focus is skewed towards the general settings of anisotropy and bianisotropy. The richness of electromagnetic phenomenons that accompany anisotropy and bianisotropy is of considerable technoscientific interest. Indeed, while nature provides us with a wide variety of anisotropic and bianisotropic crystals and minerals, it is through the process of homogenization that anisotropy and bianisotropy can be introduced in a controlled manner, fine tuned, and possibly exploited for technoscientific applications. Recent developments pertaining to nanostructured composite materials and thin films, for example, bear testament to this. In order to efficiently accommodate anisotropic and bianisotropic materials, the coordinate-free language of dyadics (i.e., second-rank Cartesian tensors) is adopted. A brief introduction to dyadic notation and algebra is provided in the appendix[1] for readers unfamiliar with dyadics.

We largely concentrate on 'effective' materials as opposed to 'equivalent' materials, and emphasize the fundamental (but sometimes overlooked) differences between these two categories of homogenized composite materials. The properties of an 'effective' material represents those of its composite material, regardless of the geometry and dimensions of the bulk materials and regardless of the orientations and polarization states of the illuminating electromagnetic fields. In contrast, the properties of 'equivalent' materials only represent those of their corresponding composite materials under certain restrictive circumstances.

The subjects covered are as follows: In chapter 1, an overview of homogenization in general is provided, including a brief account of the historical development of homogenization. The constitutive relations of isotropic, anisotropic, and bianisotropic materials—linear or nonlinear—are set forth in chapter 2, using the language of dyadics. The mathematical constructions of depolarization dyadics and polarizability density dyadics—upon which homogenization formalisms are built—are presented in chapter 3 for the most general categories of linear homogeneous materials. The homogenization formalisms themselves are delivered in chapters 4 and 5 for linear and weakly nonlinear composite materials, respectively. Some representative numerical

[1] Appendix A in the 2nd edition.

examples are included in chapters 4 and 5, to illustrate particularly noteworthy aspects. We finish off with a brief epilogue in chapter 6.[2] In this concise monograph, only key results are given, full details of derivations of these results being available in the references listed at the end of each chapter. The intended readership comprises graduate students and researchers, seeking a self-contained survey of the state-of-the-art in analytical homogenization. A familiarity with undergraduate-level electromagnetic theory is assumed. SI units are used throughout.

Tom G Mackay, Edinburgh, Scotland
Akhlesh Lakhtakia, University Park, PA
March 2015

[2] Chapter 7 in the 2nd edition.

Preface to 2nd edition

We have added a new chapter (chapter 6) on applications and examples, containing many numerical results. The Mathematica™ codes used to generate these results are provided in a new appendix (appendix B). Elsewhere, some sections have been updated and some typographical errors have been corrected.

Tom G Mackay, Edinburgh, Scotland

Akhlesh Lakhtakia, University Park, PA

November 2020

Acknowledgements

Our conceptions of electromagnetic homogenization have been shaped by discussions with Craig F Bohren, Jiajia Cui, Bernhard Michel, and (late) Werner S Weiglhofer. We thank our families for their loving support and Ashley Gasque as well as the staff of IOP Publishing for producing the ebook.

Author biographies

Tom G Mackay

Tom G Mackay graduated from the Universities of Edinburgh, Glasgow, and Strathclyde. His university education was supported, in part, by *The Carnegie Trust for The Universities of Scotland*. He is a reader in the School of Mathematics at the University of Edinburgh, and also an adjunct professor in the Department of Engineering Science and Mechanics at the Pennsylvania State University. In 2006/07 he held a *Royal Society of Edinburgh/Scottish Executive Support Research Fellowship* and in 2009/2010 a *Royal Academy of Engineering/Leverhulme Trust Senior Research Fellowship*. He is a Fellow of the Institute of Physics (UK) and SPIE–The International Society for Optics and Photonics. His research interests lie in the electromagnetic theory of complex mediums, including homogenized composite materials and analogs of curved spacetime.

Akhlesh Lakhtakia

Akhlesh Lakhtakia is Evan Pugh University Professor and Charles Godfrey Binder (Endowed) Professor of Engineering Science and Mechanics at The Pennsylvania State University. He received his BTech (1979) and DSc (2006) degrees in Electronics Engineering from the Institute of Technology, Banaras Hindu University, and his MS (1981) and PhD (1983) degrees in Electrical Engineering from the University of Utah. He was the Editor-in-Chief of the *Journal of Nanophotonics* from its inception in 2007 through 2013. He has been elected a Fellow of the American Association for the Advancement of Sciences, American Physical Society, Institute of Physics (UK), Optical Society of America, SPIE–The International Society for Optics and Photonics, Institute of Electrical and Electronics Engineers, Royal Society of Chemistry, and Royal Society of Arts. His current research interests relate to electromagnetic fields in complex mediums, sculptured thin films, multicontrollable metasurfaces, mimumes, surface multi-plasmonics and electromagnetic surface waves, thin-film solar cells, forensic science, biorepication, engineered biomimicry, and biologically inspired design for environment.

IOP Publishing

Modern Analytical Electromagnetic Homogenization with Mathematica® (Second Edition)

Tom G Mackay and Akhlesh Lakhtakia

Chapter 1

Introduction to homogenization

The concepts of *homogeneity* and *nonhomogeneity* underlying this ebook can be illustrated by way of two examples: (i) Take a bowl of vegetable soup, containing roughly chopped potatoes, carrots, onions, etc. The individual chunks of vegetables are easily seen. The soup is nonhomogeneous to the naked eye. Now pass the soup through a blender so that the vegetable chunks are broken down into tiny fragments. The soup now looks smooth throughout and is homogeneous to the naked eye. (ii) Take a large crystal that has been carefully grown to minimize the occurrence of defects. To the naked eye, the interior of the crystal seems uniform. We say that the crystal is homogeneous at visible wavelengths. However, under inspection with x-ray radiation, the crystal structure is revealed via diffraction patterns, showing that the crystal is non-homogeneous at x-ray wavelengths. We infer from the first example that the state of homogeneity at a fixed wavelength depends upon the size of the constituent particles, and we infer from the second example that the state of homogeneity of a fixed material depends upon the electromagnetic wavelength selected for its inspection.

In mathematical terms, a material is homogeneous if its constitutive parameters are independent of the spatial coordinates. Of course, this is an abstraction since every material is a composite material, made up from atoms and possibly larger building blocks. But it is an extremely useful abstraction, because in practical settings there are zillions of atoms contained within volumes of wavelength-sized dimensions in the optical and lower-frequency spectral regimes. In such instances, there is negligible loss in ignoring the particulate nature of the material under investigation, whereas the complexity that would be involved in taking into account each constituent particle of the material would be overwhelming [1].

doi:10.1088/978-0-7503-3423-5ch1

1.1 The notion of a homogenized composite material

The mathematical process by which a composite material may be regarded as an effectively homogeneous material is known as homogenization. The constitutive parameters of a homogenized composite material (HCM) may be estimated using a homogenization formalism. Such a formalism usually requires as inputs the constitutive parameters of the constituent materials of the composite material, as well as details concerning the shapes, orientations, sizes, and distributions of its constituent particles [2]. Homogenization is justifiable provided that the constituent particles are electrically small, i.e., much smaller than the wavelengths in the constituent materials and in the HCM [3, 4]. This spectral regime is called the long-wavelength regime.

Homogenization formalisms are useful for interpreting experimental data, since transducers typically provide measurements averaged over macroscopic length scales. Another important, and rapidly growing, area of application is the design of engineered composite materials. Relatively simple constituent materials may be mixed together to make a relatively complex HCM. The HCM may exhibit properties not exhibited by its constituent materials, or at least not exhibited to the same extent by its constituent materials [5, 6]. For example, HCMs exhibiting exotic properties such as negative-phase-velocity propagation may be conceptualized using simple constituent materials that do not themselves support negative-phase-velocity propagation [7]. As well as for material design, homogenization formalisms may be usefully employed in the design of devices such as optical sensors. For example, optical sensors are often based on the infiltration of a porous host material by an analyte fluid. The resulting changes in the optical response characteristics may be estimated using homogenization formalisms, and thereby the sensing process can be modeled, understood, and optimized [8, 9].

1.2 Salient features of homogenization formalisms

The process of homogenization engenders complexity. In particular, since HCMs inherit spacetime symmetries from each of their constituent materials, HCMs generally exhibit lower degrees of spacetime symmetry than do their constituent materials. Consequently, the mathematical characterization of the constitutive properties of HCMs generally involves a larger parameter space than that needed for their constituent materials, and HCMs generally exhibit a richer palette of electromagnetic responses than do their constituent materials. Thus, HCMs are frequently

- anisotropic, meaning that the electromagnetic phasors **D** and **E** are not mutually aligned and/or the electromagnetic phasors **H** and **B** are not mutually aligned, or
- bianisotropic, meaning that **D** is anisotropically coupled to both **E** and **B**, and **H** is anisotropically coupled to both **E** and **B**.

Furthermore, anisotropy or bianisotropy can arise even when the constituent materials themselves are isotropic. The anisotropic or bianisotropic nature of the

HCM may stem from one or more of the following sources: (i) the intrinsic electromagnetic properties of the constituent materials, (ii) the shapes and orientations of the constituent particles, and (iii) an external source such as an applied quasi-static electric or magnetic field. Dyadics (i.e., second-rank Cartesian tensors) rather than scalars are adopted to characterize the constitutive properties of anisotropic or bianisotropic materials.

In addition to spacetime symmetries, an HCM may inherit other attributes from its constituent materials. If one of the constituent materials has a nonlinear electromagnetic response, then the electromagnetic response of the HCM will also be nonlinear. If one of the constituent materials has a nonlocal electromagnetic response, with respect to space and/or time, then the electromagnetic response of the HCM will likewise be nonlocal. Appreciable manifestations of spatial nonlocality are highly uncommon. The same cannot be said of temporal nonlocality which manifests itself as dissipation. Thus, an HCM is dissipative if at least one of its constituent materials is dissipative, assuming that none of the other constituent materials is an active material. (If one of the constituent materials is a dissipative material and another is an active material, it is conceivable that the resulting HCM could be nondissipative.) However, an HCM can be effectively dissipative even though each of its constituent materials is nondissipative. The HCM dissipation in that case is attributed to scattering losses. Such scattering losses may be accommodated by homogenization formalisms when the constituent particles are assigned a nonzero size, or alternatively when a nonzero correlation length is assigned to the interactions between neighboring nonhomogeneities in the composite material.

Certain bounds can be placed on the constitutive parameter ranges for HCMs. For example, consider a mixture of two nondissipative constituent materials ‘a’ and ‘b’, each being an isotropic dielectric material with real-valued permittivities ε_a and ε_b, respectively. The constituent materials are distributed randomly as electrically small spheres, with the volume fraction of material ‘a’ being denoted by f_a while that of material ‘b’ being $f_b = 1 - f_a$. Using physical arguments, Wiener showed that the HCM’s permittivity ε_{HCM} should lie between the bounds [10, 11]

$$\left.\begin{aligned} W_\alpha &= f_a\varepsilon_a + f_b\varepsilon_b \\ W_\beta &= \left(\frac{f_a}{\varepsilon_a} + \frac{f_b}{\varepsilon_b}\right)^{-1} \end{aligned}\right\}; \tag{1.1}$$

that is, either $W_\alpha \leqslant \varepsilon_{HCM} \leqslant W_\beta$ or $W_\beta \leqslant \varepsilon_{HCM} \leqslant W_\alpha$. Generally tighter constraints on ε_{HCM} are represented by the Hashin–Shtrikman bounds

$$\left.\begin{aligned} HS_\alpha &= \varepsilon_b + \frac{3f_a\varepsilon_b(\varepsilon_a - \varepsilon_b)}{\varepsilon_a + 2\varepsilon_b - f_a(\varepsilon_a - \varepsilon_b)} \\ HS_\beta &= \varepsilon_a + \frac{3f_b\varepsilon_a(\varepsilon_b - \varepsilon_a)}{\varepsilon_b + 2\varepsilon_a - f_b(\varepsilon_b - \varepsilon_a)} \end{aligned}\right\}, \tag{1.2}$$

which were derived on the basis of energy considerations, using a variational method [12]. Extensions of both the Wiener bounds (1.1) and the Hashin–Shtrikman bounds (1.2) to dissipative materials, as represented by complex-valued ε_a and ε_b, have been established [13]. Both sets of bounds coincide as either $f_a \to 0$ or $f_a \to 1$. Thus, the HCM's constitutive parameters must coincide with those of one of its constituent materials in the limit as the volume fraction of that constituent material tends to unity —which is a physically reasonable expectation not only for the isotropic dielectric HCMs in consideration but also for anisotropic and bianisotropic HCMs.

Considerable simplifications in homogenization may arise depending upon the spatial arrangements of the constituent materials. If each constituent material takes the form of a planar sheet and the sheets are stacked one upon another to form a laminated composite material, then the nonhomogeneity is essentially one dimensional. A laminated composite material may be amenable to homogenization provided that the constituent layers are electrically thin. Two-dimensional nonhomogeneity can arise when the particles of one (or more) of the constituent materials take the form of electrically long cylinders, with all the cylinders being mutually aligned. The composite structure is essentially homogeneous in the direction parallel to the cylinder axis and nonhomogeneous in every direction orthogonal to that axis. Such a composite structure may be amenable to homogenization provided that the radiuses of the cylindrical particles are electrically small. A general mixture of materials is nonhomogeneous in all directions. This three-dimensional nonhomogeneity arises when the constituent particles are of finite dimensions in every direction. Such composite materials may be amenable to homogenization provided that the constituent particles are electrically small.

An HCM is a homogeneous material whose electromagnetic response in the long-wavelength spectral regime is effectively the same as that for the nonhomogeneous mixture of constituent materials from which the HCM arises. Therefore, sometimes HCMs are called 'effective' materials. Whether a slab, or a cylinder, or a spheroid, or some other object is made of a composite material, its long-wavelength response must be replicated by that of an identical object made of the HCM—regardless of the dimensions of the object as well as the spatial distributions of the Cartesian components of the illuminating electromagnetic field.

One should carefully distinguish effective materials from 'equivalent' materials. An equivalent material is a homogeneous material with only certain aspects of its electromagnetic response being the same as that of some composite material. This equivalence may hold only for electromagnetic illumination at a certain angle of incidence and/or a certain state of polarization. Also, the equivalent material might not have the same bulk dimensions as the composite material that it is related to. Furthermore, the nonhomogeneities in the corresponding composite material need not be small in relation to the wavelength—which contrasts starkly with the case for effective materials.

1.3 Brief history of homogenization formalisms

Electromagnetic homogenization has a long history [2], even predating the introduction of the Maxwell postulates of electromagnetism. In the early 19th century,

Biot and Arago proposed a very simple formula for estimating the refractive index of a mixture of mutually inert gases: the sum of the individual refractive indexes of the constituent gases, weighted according to their volume fractions in the mixture [14].

More rigorous approaches to electromagnetic homogenization emerged within a few decades. The much-used Maxwell Garnett formalism was put forward in 1904 to estimate the effective permittivity of an isotropic dielectric HCM [15] for a composite material made from a dispersal of spherical particles in a host material. A rigorous basis for this formalism was established by Faxén [16]. Although the name 'Maxwell Garnett' is the one that caught on, this formalism is in fact equivalent to other formalisms established earlier, namely the Mossotti–Clausius, the Lorenz–Lorentz, and the Rayleigh formalisms [17]. Much more recently, versions of the Maxwell Garnett formalism suitable for linear bianisotropic [18] and weakly nonlinear bianisotropic [19] HCMs, accommodating ellipsoidal particles, were developed. The popularity of the Maxwell Garnett formalism may be attributed to its relative simplicity and to its close connection to the Hashin–Shtrikman bounds (1.2) [12]. However, the Maxwell Garnett formalism is appropriate only for dilute distributions of the particulate materials in the host material. As such, it fails to predict behavior relating to the *percolation threshold.*

The phenomenon of percolation is intimately connected with the homogenization of particulate composite materials. Consider a composite material, with one of its constituent materials being labeled a. If the volume fraction f_{a} of this material is sufficiently small then the composite material cannot be traversed from one side to the other on a path that traverses only particles of the constituent material a. But as f_{a} is steadily increased, then a critical value—known as the percolation threshold—will be reached at which the composite material can be traversed from one side to the other on a path that traverses only particles of the constituent material a [20]. Depending on the nature of the materials involved, at the percolation threshold certain physical properties of the HCM, such as its quasi-static conductivity and permittivity, change rather abruptly.

When the particles in the Maxwell Garnett formalism are not spherical, physically unreasonable estimates of the HCM's constitutive parameters are generated as the volume fraction of the particles approaches unity. This shortcoming is overcome by the Bragg–Pippard formalism, first developed for anisotropic dielectric HCMs [21] and much later for bianisotropic ones [22]. This formalism emerges on replacing the depolarization dyadic of a spherical cavity in the Maxwell Garnett formalism with a depolarization dyadic of a cavity which is conformal with the particles, both cavities being infinitesimally small. Furthermore, the restriction on the Maxwell Garnett formalism to low values of f_{a} is overcome by two refinements. In the *incremental* Maxwell Garnett formalism, the particles are added to the host material in a series of discrete steps, with a homogenization process taking place at the end of each step [23]. The *differential* Maxwell Garnett formalism emerges from the incremental Maxwell Garnett formalism as the number of steps tends to infinity [24]. However, even with the Bragg–Pippard formalism, and the incremental and differential Maxwell Garnett formalisms, the constituent materials—i.e., the host and the particle materials—are treated asymmetrically.

The Bruggeman formalism [25] treats each of the constituent materials in a composite material on the same footing. No constituent material is designated as the one hosting the particles of other constituent materials. Instead, particles of all constituent materials are assumed to be dispersed in the HCM itself. Consequently, the Bruggeman formalism is suitable regardless of the values of the volume fractions, and the percolation threshold can be accommodated [26, 27], to a degree at least. The original version of this formalism, which applies to isotropic dielectric HCMs, predates equivalent formalisms put forward by Polder and van Santen [28] and by Böttcher [29, 30]. Much more recently, versions of the Bruggeman formalism were established for linear bianisotropic [18] and weakly nonlinear anisotropic [31] HCMs.

In the Maxwell Garnett and the Bruggeman formalisms, the spatial distributions of the constituent materials are characterized simply by their volume fractions. The strong-property-fluctuation theory (SPFT), although originally developed to investigate wave propagation in continuous random mediums [32, 33], provides a more sophisticated approach to homogenization. The SPFT has been adapted to estimate the constitutive parameters of isotropic dielectric[1] [34], anisotropic dielectric [35, 36], isotropic chiral [37], and bianisotropic [38] HCMs in the case of linear materials, as well as isotropic dielectric [39], anisotropic dielectric [40], and isotropic chiral [41] HCMs in the case of weakly nonlinear materials. The constituent materials in the composite material are treated as random fluctuations in the constitutive parameters of an ambient comparison medium. The spatial distributions of these fluctuations are characterized in terms of statistical moments. In principle, statistical moments of arbitrarily high order can be accommodated [42], but in practice the second-order SPFT formalism is usually implemented in which the spatial distributions are characterized by a two-point covariance function and its associated correlation length. In fact, the lowest-order SPFT formalism—which emerges from the second-order SPFT formalism in the limit as the correlation length tends to zero—is equivalent to the Bruggeman homogenization formalism and thus provides a rigorous basis for the Bruggeman formalism.

The Maxwell Garnett, Bruggeman, and SPFT formalisms share a common provenance [11]. In each of these formalisms, the electromagnetic responses of the constituent materials of the composite material are represented using depolarization dyadics and their associated polarizability density dyadics. Furthermore, in the standard implementations of these depolarization dyadics, each constituent particle is essentially represented by co-located electric and magnetic dipoles. Depolarization dyadics of cavities that are not infinitesimally small but are of the same sizes as the electrically small particles have been formulated for ellipsoidal particles [43]. This development led to extended versions of the Maxwell Garnett, Bruggeman, and SPFT formalisms, appropriate to linear bianisotropic HCMs [44] and weakly nonlinear anistropic HCMs [45], arising from ellipsoidal constituent particles of nonzero size. Similar extended versions of the Maxwell Garnett and Bruggeman

[1] In the case of dielectric composite materials, 'SPFT' stands for the strong-*permittivity*-fluctuation theory.

formalisms, not based on spectral representations of the depolarization dyadics, are available for isotropic HCMs [46–51]. A note of caution should be added here: extended homogenization formalisms generally do not scale in a physically intuitive manner [52], but the significance of this aspect is unclear.

An alternative approach to the homogenization of particulate composite materials leads to effective-wavenumber formalisms. These involve calculating the forward-scattering amplitudes of the constituent particles. Effective wavenumbers in the HCM are thereby estimated, wherefrom the effective constitutive parameters of the HCM may be inferred. Formalisms appropriate to isotropic dielectric [53], isotropic chiral [54], and certain anisotropic dielectric [55] HCMs are available. However, this approach cannot be extended to bianisotropic HCMs. The problem is essentially that there can exist at most only four independent wavenumbers for a bianisotropic material, whereas the bianisotropic material itself is characterized by as many as 36 complex-valued constitutive scalars [56].

Two-dimensional nonhomogeneous materials, composed of arrays of parallel and infinitely long cylinders, can be homogenized in a similar same way as three-dimensional particulate composite materials, provided that the cylinder radiuses are small relative to the wavelengths involved. Most notably, the properties of columnar thin films (CTFs) have been investigated by means of homogenization studies. A CTF is an array of parallel columnar nanowires, typically oriented at an acute angle to a planar substrate [57]. The array of nanowires may be grown on a substrate by means of physical vapor deposition. Within the homogenization context, a CTF may be regarded as a composite material with two constituent materials: (i) the material of which the nanowires are made of and (ii) air or some other fluid filling the spaces between the nanowires. The individual columns are envisaged as strings of conformal ellipsoidal particles, laid out end-to-end [58]. The Maxwell Garnett [59, 60], the Bragg–Pippard [61], the Bruggeman [8, 60, 62], and the SPFT [63] formalisms have all been implemented to estimate the effective constitutive dyadics of CTFs. Moreover, an inverse formulation of the Bruggeman formalism has been implemented to estimate the microscopic dielectric and morphological parameters of a CTF, using experimental measurements of the effective permittivity dyadic of the CTF [64].

In principle, the effective constitutive dyadics of one-dimensionally nonhomogeneous materials, i.e., laminated composite materials, can be estimated by solving the problem of reflection and transmission of plane waves [65]. The formulation involves exploiting the continuity of the tangential components of the electromagnetic phasors **E** and **H** across the boundaries of adjacent constituent layers. This process has been carried out for the case in which the constituent layers are made of different bianisotropic materials [66]. However, in the case of general bianisotropy, the resulting analytical expressions for the effective constitutive parameters of the HCMs are unwieldy. Related approaches have been adopted to estimate the equivalent constitutive dyadics of one-dimensionally nonhomogeneous materials. For example, a symmetrically laminated composite material, with its constituent layers made of isotropic dielectric materials, is electromagnetically equivalent to a certain homogeneous biaxial dielectric material [67, 68]. But the equivalence holds only for a specific angle of incidence—if the angle of incidence changes then the constitutive parameters of the equivalent homogeneous material must also change. There may also be a dependence on the polarization state of the illuminating plane wave [69].

Similar equivalent-homogeneous-material formalisms have been recently developed for engineered composite materials, designed to display exotic electromagnetic phenomenons such as negative refraction [70, 71]. Typically, such formalisms involve the calculation of scattering (S) parameters (in which instances they are called S-parameter retrieval schemes) [72]. As these calculations are based on experimental data, this approach is somewhat empirical. The composite materials considered generally have layers of nanostructures and the equivalent homogeneous materials may be bianisotropic, with magnetoelectric coupling playing an important role. For example, a stack of isotropic dielectric planar layers, under normal illumination, is electromagnetically equivalent to a certain type of homogeneous bianisotropic material [73]. While these formalisms only yield estimates relating to equivalent homogeneous materials and they lack the rigor of effective-homogeneous-material formalisms, they can be of considerable practical value as they facilitate a partial characterization of complex composite materials in terms of measurable quantities.

In addition to the aforementioned analytical approaches to homogenization, the homogenization problem can be approached numerically. Techniques based on the finite-element method [74, 75], the Monte Carlo method [76], and the finite-difference time-domain method [77, 78] have been developed for a variety of complex composite materials, including thin films [79] and some which exhibit bianisotropy [77, 78]. Generally good agreement has been reported between the predictions of these numerical techniques and the estimates yielded by analytical homogenization formalisms, albeit the constitutive parameters of only a limited range of HCMs have been considered. However, as these numerical techniques do not lie within the remit of this ebook, they will not be discussed henceforth; instead, the reader is referred to a comprehensive review article [78].

1.4 Organization of this ebook

In this ebook a modern approach is adopted in which dyadics (i.e., second-rank Cartesian tensors) rather than scalars are used to delineate the constitutive properties of HCMs. Dyadics provide a convenient and elegant means of treating the anisotropy or bianisotropy that is frequently exhibited by HCMs in practice. See appendix A for a succinct introduction to dyadics.

Chapter 2 provides a description of constitutive dyadics, including their common classifications and properties needed for the later chapters. The homogenization formalisms considered herein are largely based on depolarization dyadics and the associated polarizability density dyadics of individual particles relative to their neighborhoods. In chapter 3 the origins of these important theoretical constructions are described, with reference to the dyadic Green functions for the material taken to occupy the neighborhood. The homogenization formalisms themselves are delineated in chapter 4 for linear HCMs and chapter 5 for nonlinear HCMs. Some applications and illustrative numerical examples of the formalisms presented in chapters 4 and 5 are provided in chapter 6. While the focus lies largely on three-dimensional particulate composite materials, two-dimensional composite materials whose constituent materials present cylindrical symmetry are also considered. Mathematica™ codes for the numerical examples in chapter 6 are provided in appendix B. A brief epilogue is given in chapter 7.

As regards notation, 3-vectors are identified by boldface while 3 × 3 dyadics are double underlined. The ^ symbol in combination with boldface signifies a unit vector. Thus, Cartesian unit vectors aligned with the coordinate axes are written as $\hat{\mathbf{u}}_x$, $\hat{\mathbf{u}}_y$, and $\hat{\mathbf{u}}_z$, and the identity 3 × 3 dyadic may be expressed as $\underline{\underline{I}} = \hat{\mathbf{u}}_x\hat{\mathbf{u}}_x + \hat{\mathbf{u}}_y\hat{\mathbf{u}}_y + \hat{\mathbf{u}}_z\hat{\mathbf{u}}_z$. Boldface with single underlining denotes a 6-vector while boldface with double underlining denotes a 6 × 6 dyadic. Components of vectors and dyadics are identified by subscripted indexes. The 3 × 3 null dyadic is $\underline{\underline{0}}$ while the 6 × 6 null dyadic is $\underline{\underline{\mathbf{0}}}$ and the 6 × 6 identity dyadic is $\underline{\underline{\mathbf{I}}}$. The dyadic transpose and inverse are identified by the superscripts ^{T} and $^{-1}$, respectively. The adjoint and determinant of a dyadic are delivered by the dyadic operators 'adj' and 'det', respectively. The real and imaginary parts of complex quantities are delivered by the operators Re{·} and Im{·}, respectively. The complex conjugate of a complex-valued scalar z is denoted as z^*. The conjugate transpose of a 3 × 3 dyadic $\underline{\underline{X}}$ is written as $\underline{\underline{X}}^\dagger$; thus, $[\, \underline{\underline{X}}^\dagger \,]_{\ell j} = [\, \underline{\underline{X}} \,]^*_{j\ell}$ for $\ell \in \{1, 2, 3\}$ and $j \in \{1, 2, 3\}$. The symbols ε_0 and μ_0 represent the permittivity and permeability of free space, respectively. The free-space wavenumber is written as $k_0 = \omega\sqrt{\varepsilon_0\mu_0}$, where ω is the angular frequency.

References

[1] Van Kranendonk J and Sipe J E 1977 Foundations of the macroscopic electromagnetic theory of dielectric media *Prog. Optics* **15** 245–350

[2] Lakhtakia A 1996 *Selected Papers on Linear Optical Composite Materials* (Bellingham, WA: SPIE Optical Engineering Press)

[3] van de Hulst H C 1981 *Light Scattering by Small Particles* (New York: Dover)

[4] Mackay T G 2008 Lewin's homogenization formula revisited for nanocomposite materials *J. Nanophoton.* **2** 029503

[5] Walser R M 2003 Metamaterials: an introduction *Introduction to Complex Mediums for Optics and Electromagnetics* ed W S Weiglhofer and A Lakhtakia (Bellingham, WA: SPIE Press) pp 295–316

[6] Mackay T G 2005 Linear and nonlinear homogenized composite mediums as metamaterials *Electromagnetics* **25** 461–81

[7] Mackay T G and Lakhtakia A 2004 Plane wave with negative phase velocity in Faraday chiral mediums *Phys. Rev. E* **69** 026602

[8] Mackay T G and Lakhtakia A 2010 Modeling columnar thin films as platforms for surface-plasmonic-polaritonic optical sensing *Photon. Nanostruct.: Fundam. Appl.* **8** 140–9

[9] Mackay T G 2012 On the sensitivity of generic porous optical sensors *Appl. Opt.* **51** 2752–8

[10] Wiener O 1912 Die Theorie des Mischkörpers für das Feld der Stationären Strömung *Abh. Math. Phys. Kl. Sächs.* **32** 507–604 (Reproduced in [2], which also provides an English synopsis of [10] by B Michel)

[11] Aspnes D E 1982 Local-field effects and effective-medium theory: a microscopic perspective *Am. J. Phys.* **50** 704–9 (Reproduced in [2])
[12] Hashin Z and Shtrikman S 1962 A variational approach to the theory of the effective magnetic permeability of multiphase materials *J. Appl. Phys.* **33** 3125–31
[13] Milton G W 1980 Bounds on the complex dielectric constant of a composite material *Appl. Phys. Lett.* **37** 300–2
[14] Biot J-B and Arago F 1806 Mémoire sur les affinités des corps pour la lumière, et particulièrement sur les forces réfringentes des différens gaz *Mém. Inst. France* **7** 301–87
[15] Maxwell Garnett J C 1904 Colours in metal glasses and in metallic films *Phil. Trans. R. Soc. Lond. A* **203** 385–420 (Reproduced in [2])
[16] Faxén H 1920 Der Zusammenhang zwischen den Maxwellschen Gleichungen für Dielektrika und den atomistischen Ansätzen von H.A. Lorentz u.a. *Z. Phys.* **2** 218–29 (Reproduced in [2])
[17] Lakhtakia A 1996 Cakes and pastries, and linear optical composite materials too *Selected Papers on Linear Optical Composite Materials* ed A Lakhtakia (Bellingham, WA: SPIE Optical Engineering Press) pp xiii–xxiv
[18] Weiglhofer W S, Lakhtakia A and Michel B 1997 Maxwell Garnett and Bruggeman formalisms for a particulate composite with bianisotropic host medium *Microw. Opt. Technol. Lett.* **15** 263–6
Weiglhofer W S, Lakhtakia A and Michel B 1999 Correction to 'Maxwell Garnett and Bruggeman formalisms for a particulate composite with bianisotropic host medium' *Microw. Opt. Technol. Lett.* **22** 221 (correction)
[19] Lakhtakia A and Weiglhofer W S 2000 Maxwell Garnett formalism for weakly nonlinear, bianisotropic, dilute, particulate composite media *Int. J. Electron.* **87** 1401–8
[20] Stauffer D 1985 *Introduction to Percolation Theory* (London: Taylor & Francis)
[21] Bragg W L and Pippard A B 1953 The form birefringence of macromolecules *Acta Crystallogr.* **6** 865–7
[22] Sherwin J A and Lakhtakia A 2002 Bragg–Pippard formalism for bianisotropic particulate composites *Microw. Opt. Technol. Lett.* **33** 40–4
[23] Lakhtakia A 1998 Incremental Maxwell Garnett formalism for homogenizing particulate composite media *Microw. Opt. Technol. Lett.* **17** 276–9
[24] Michel B, Lakhtakia A, Weiglhofer W S and Mackay T G 2001 Incremental and differential Maxwell Garnett formalisms for bi-anisotropic composites *Compos. Sci. Technol.* **61** 13–8
[25] Bruggeman D A G 1935 Berechnung verschiedener physikalischer Konstanten von heterogenen Substanzen, I. Dielektrizitätskonstanten und Leitfähigkeiten der Mischkörper aus isotropen Substanzen *Ann. Phys. Lpz.* **24** 636–79 (Reproduced in [2])
[26] Goncharenko A V and Venger E F 2004 Percolation threshold for Bruggeman composites *Phys. Rev. E* **70** 057102
[27] Mackay T G and Lakhtakia A 2006 Percolation thresholds in the homogenization of spheroidal particles oriented in two directions *Opt. Commun.* **259** 727–37
[28] Polder D and van Santen J H 1946 The effective permeability of mixtures of solids *Physica* **12** 257–71
[29] Bohren C F and Battan L J 1981 Radar backscattering by inhomogeneous precipitation particles *J. Atmos. Sci.* **38** 459
[30] Böttcher C J F 1945 The dielectric constant of crystalline powders *Rec. Trav. Chim. Pays–Bas* **64** 47–51

[31] Lakhtakia M N and Lakhtakia A 2001 Anisotropic composite materials with intensity-dependent permittivity tensor: the Bruggeman approach *Electromagnetics* **21** 129–38

[32] Yu A, Ryzhov and Tamoikin V V 1970 Radiation and propagation of electromagnetic waves in randomly inhomogeneous media *Radiophys. Quantum Electron.* **14** 228–33

[33] Tatarskii V I and Zavorotnyi V U 1980 Strong fluctuations in light propagation in a randomly inhomogeneous medium *Prog. Optics* **18** 204–56

[34] Tsang L and Kong J A 1981 Scattering of electromagnetic waves from random media with strong permittivity fluctuations *Radio. Sci.* **16** 303–20 (Reproduced in [2])

[35] Genchev Z D 1992 Anisotropic and gyrotropic version of Polder and van Santen's mixing formula *Waves Random Media* **2** 99–110

[36] Zhuck N P 1994 Strong-fluctuation theory for a mean electromagnetic field in a statistically homogeneous random medium with arbitrary anisotropy of electrical and statistical properties *Phys. Rev. B* **50** 15636–45

[37] Michel B and Lakhtakia A 1995 Strong-property-fluctuation theory for homogenizing chiral particulate composites *Phys. Rev. E* **51** 5701–7

[38] Mackay T G, Lakhtakia A and Weiglhofer W S 2000 Strong-property-fluctuation theory for homogenization of bianisotropic composites: formulation *Phys. Rev. E* **62** 6052–64
Mackay T G, Lakhtakia A and Weiglhofer W S 2001 Strong-property-fluctuation theory for homogenization of bianisotropic composites: formulation *Phys. Rev. E* **63** 049901 (erratum)

[39] Lakhtakia A 2001 Application of strong permittivity fluctuation theory for isotropic, cubically nonlinear, composite mediums *Opt. Commun.* **192** 145–51

[40] Mackay T G 2003 Geometrically derived anisotropy in cubically nonlinear dielectric composites *J. Phys. D: Appl. Phys.* **36** 583–91

[41] Mackay T G, Lakhtakia A and Weiglhofer W S 2003 The strong-property-fluctuation theory for cubically nonlinear, isotropic chiral composite mediums *Electromagnetics* **23** 455–79

[42] Mackay T G, Lakhtakia A and Weiglhofer W S 2001 Third-order implementation and convergence of the strong-property-fluctuation theory in electromagnetic homogenisation *Phys. Rev. E* **64** 066616

[43] Mackay T G 2004 Depolarization volume and correlation length in the homogenization of anisotropic dielectric composites *Waves Random Media* **14** 485–98
Mackay T G 2006 Erratum: Depolarization volume and correlation length in the homog enization of anisotropic dielectric composites *Waves Random Complex Media* **16** 85 (erratum)

[44] Cui J and Mackay T G 2007 Depolarization regions of nonzero volume in bianisotropic homogenized composites *Waves Random Complex Media* **17** 269–81

[45] Cui J and Mackay T G 2007 Depolarization regions of nonzero volume for anisotropic, cubically nonlinear, homogenized nanocomposites *J. Nanophoton.* **1** 013506

[46] Doyle W T 1989 Optical properties of a suspension of metal spheres *Phys. Rev. B* **39** 9852–8

[47] Prinkey M T, Lakhtakia A and Shanker B 1994 On the extended Maxwell–Garnett and the extended Bruggeman approaches for dielectric-in-dielectric composites *Optik* **96** 25–30

[48] Shanker B and Lakhtakia A 1993 Extended Maxwell Garnett model for chiral-in-chiral composites *J. Phys. D: Appl. Phys.* **26** 1746–58

[49] Shanker B 1996 The extended Bruggeman approach for chiral-in-chiral mixtures *J. Phys. D: Appl. Phys.* **29** 281–8

[50] Lakhtakia A and Shanker B 1993 Beltrami fields within continuous source regions, volume integral equations, scattering algorithms, and the extended Maxwell–Garnett model *Int. J. Appl. Electromagn. Mater.* **4** 65–82 (Reproduced in [2])
[51] Mallet P, Guérin C A and Sentenac A 2005 Maxwell–Garnett mixing rule in the presence of multiple-scattering: Derivation and accuracy *Phys. Rev. B* **72** 014205
[52] Bohren C F 2009 Do extended effective-medium formulas scale properly? *J. Nanophoton.* **3** 039501
[53] Waterman P C and Truell R 1961 Multiple scattering of waves *J. Math Phys.* **2** 512–37
[54] Lakhtakia A 1993 Application of the Waterman–Truell approach for chiral composites *Int. J. Electron.* **75** 1243–9
[55] Twersky V 1978 Coherent electromagnetic waves in pair-correlated random distributions of aligned scatterers *J. Math. Phys.* **19** 215–30
[56] Mackay T G and Lakhtakia A 2019 *Electromagnetic Anisotropy and Bianisotropy: A Field Guide* 2nd edn (Singapore: World Scientific) https://doi.org/10.1142/7515
[57] Lakhtakia A and Messier R 2005 *Sculptured Thin Films: Nanoengineered Morphology and Optics* (Bellingham, WA: SPIE Press)
[58] Sherwin J A, Lakhtakia A and Hodgkinson I J 2002 On calibration of a nominal structure–property relationship model for chiral sculptured thin films by axial transmittance measurements *Opt. Commun.* **209** 369–75
[59] Suzuki M and Taga Y 1992 Anisotropy in the optical absorption of Ag–SiO_2 thin films with oblique columnar structures *J. Appl. Phys.* **71** 2848–54 (Reproduced in [2])
[60] Smith G B 1989 Effective medium theory and angular dispersion of optical constants in films with oblique columnar structure *Opt. Commun.* **71** 279–84 (Reproduced in [2])
[61] Hodgkinson I, Wu Q H and Collett S 2001 Dispersion equations for vacuum-deposited tilted-columnar biaxial media *Appl. Opt.* **40** 452–7
[62] Lakhtakia A 1998 On determining gas concentrations using dielectric thin-film helicoidal bianisotropic medium bilayers *Sens. Actuat. B: Chem.* **52** 243–50
[63] Jamaian S S and Mackay T G 2012 On columnar thin films as platforms for surface-plasmonic-polaritonic optical sensing: higher-order considerations *Opt. Commun.* **285** 5535–42
[64] Mackay T G and Lakhtakia A 2010 Determination of constitutive and morphological parameters of columnar thin films by inverse homogenization *J. Nanophoton.* **4** 041535
[65] Mackay T G and Lakhtakia A 2020 *The Transfer-Matrix Method in Electromagnetics and Optics* (San Rafael, CA: Morgan & Claypool) https://doi.org/10.2200/S00993ED1V01Y202002EMA001
[66] Lakhtakia A 2009 Constraints on effective constitutive parameters of certain bianisotropic laminated composite materials *Electromagnetics* **29** 508–14
[67] Abelès F 1950 Recherches sur la propagation des ondes électromagnétiques sinusoïdales dans les milieux stratifiés. Application aux couches minces (1re partie) *Ann. Phys. (Paris)* **5** 596–640 (Reproduced in [2])
[68] Lafait J, Yamaguchi T, Frigerio J M, Bichri A and Driss-Khodja K 1990 Effective medium equivalent to a symmetric multilayer at oblique incidence *Appl. Opt.* **29** 2460–5 (Reproduced in [2])
[69] Lakhtakia A and Krowne C M 2003 Restricted equivalence of paired epsilon-negative and mu-negative layers to a negative phase-velocity material (*alias* left-handed material) *Optik* **114** 305–7

[70] Chen X, Wu B-I, Kong J A and Grzegorczyk T M 2005 Retrieval of the effective constitutive parameters of bianisotropic metamaterials *Phys. Rev. E* **71** 046610
[71] Li Z, Aydin K and Ozbay E 2009 Determination of the effective constitutive parameters of bianisotropic metamaterials from reflection and transmission coefficients *Phys. Rev. E* **79** 026610
[72] Smith D R, Vier D C, Koschny Th and Soukoulis C M 2005 Electromagnetic parameter retrieval from inhomogeneous metamaterials *Phys. Rev. E* **71** 036617
[73] Kriegler C E, Rill M S, Linden S and Wegener M 2010 Bianisotropic photonic metamaterials *IEEE J. Sel. Top. Quantum Electron.* **16** 367–75
[74] Krakovsky I and Myroshnychenko V 2002 Modeling dielectric properties of composites by finite-element method *J. Appl. Phys.* **92** 6743–8
[75] Myroshnychenko V and Brosseau C 2005 Finite-element modeling method for the prediction of the complex effective permittivity of two-phase random statistically isotropic heterostructures *J. Appl. Phys.* **97** 044101
[76] Zhao X, Wu Y, Fan Z and Li F 2004 Three-dimensional simulations of the complex dielectric properties of random composites by finite element method *J. Appl. Phys.* **95** 8110–7
[77] Akyurtlu A and Werner D H 2004 Modeling of transverse propagation through a uniaxial bianisotropic medium using the finite-difference time-domain technique *IEEE Trans. Antennas Propagat.* **52** 3273–9
[78] Teixeira F L 2008 Time-domain finite-difference and finite-element methods for Maxwell equations in complex media *IEEE Trans. Antennas Propagat.* **56** 2150–66
[79] Hutchinson N J, Coquil T, Navid A and Pilon L 2010 Effective optical properties of highly ordered mesoporous thin films *Thin Solid Films* **518** 2141–6

IOP Publishing

Modern Analytical Electromagnetic Homogenization with Mathematica® (Second Edition)

Tom G Mackay and Akhlesh Lakhtakia

Chapter 2

Constitutive dyadics

Constitutive dyadics provide the vocabulary for characterizing and classifying the electromagnetic properties of materials at macroscopic length scales. Indeed, the use of the term 'material' implies matter whose macroscopic electromagnetic properties are describable in terms of constitutive dyadics. This chapter describes how constitutive dyadics arise in constitutive relations, and presents some of their basic properties and classifications. Concomitantly, the electromagnetic theory needed for later chapters is presented.

2.1 Microscopic and macroscopic electromagnetic perspectives

The focus of this chapter is on the electromagnetic properties of materials. The materials may be solids, liquids or gases. Furthermore, they may be either crystalline or amorphous, and either naturally occurring or engineered. All materials are spatially nonhomogeneous and temporally varying from the *microscopic* electromagnetic perspective, since they are collections of point charges. Since the standpoint adopted in this ebook is a classical one, uncertainties in the positions or velocities of these point charges are irrelevant. An ensemble of point charges q_ℓ, $\ell \in \{1, 2, 3, \ldots\}$, positioned at $\mathbf{r}_\ell(t)$ and moving with velocity $\mathbf{v}_\ell(t)$ at time t, gives rise to the microscopic charge density

$$\tilde{c}(\mathbf{r}, t) = \sum_\ell q_\ell \, \delta[\mathbf{r} - \mathbf{r}_\ell(t)] \tag{2.1}$$

and the microscopic current density

$$\tilde{\mathbf{j}}(\mathbf{r}, t) = \sum_\ell q_\ell \, \mathbf{v}_\ell \, \delta[\mathbf{r} - \mathbf{r}_\ell(t)], \tag{2.2}$$

doi:10.1088/978-0-7503-3423-5ch2

with $\delta(\cdot)$ being the Dirac delta. Both of these densities are the sources of two microscopic electromagnetic fields: the electric field $\tilde{\mathbf{e}}(\mathbf{r}, t)$ and the magnetic field $\tilde{\mathbf{b}}(\mathbf{r}, t)$.

The relationships between the microscopic source densities and fields are encapsulated by the microscopic Maxwell postulates [1]

$$\left.\begin{aligned} \nabla \times \tilde{\mathbf{e}}(\mathbf{r}, t) + \frac{\partial}{\partial t}\tilde{\mathbf{b}}(\mathbf{r}, t) &= \mathbf{0} \\ \nabla \times \tilde{\mathbf{b}}(\mathbf{r}, t) - \varepsilon_0\mu_0 \frac{\partial}{\partial t}\tilde{\mathbf{e}}(\mathbf{r}, t) &= \mu_0 \tilde{\mathbf{j}}(\mathbf{r}, t) \\ \nabla \bullet \tilde{\mathbf{e}}(\mathbf{r}, t) &= \frac{1}{\varepsilon_0}\tilde{c}(\mathbf{r}, t) \\ \nabla \bullet \tilde{\mathbf{b}}(\mathbf{r}, t) &= 0 \end{aligned}\right\}. \tag{2.3}$$

SI units are adopted in these equations as well as throughout this ebook. The permittivity and permeability of matter-free space are denoted by $\varepsilon_0 = 8.854 \times 10^{-12}$ F m^{-1} and $\mu_0 = 4\pi \times 10^{-7}$ H m^{-1}, respectively. The microscopic fields $\tilde{\mathbf{e}}(\mathbf{r}, t)$ and $\tilde{\mathbf{b}}(\mathbf{r}, t)$ possess spatial variations over distances $\lesssim 10^{-10}$ m and temporal variations over durations ranging from $\lesssim 10^{-13}$ s for nuclear vibrations to $\lesssim 10^{-17}$ s for electronic orbital motion [1].

The summation index ℓ in equations (2.1) and (2.2) ranges from 1 to an impractically large number in any volume occupied by a material that may be characterized from a macroscopic perspective. Consequently, it is desirable to consider the spatiotemporal averages of the microscopic quantities in equations (2.3), when entering the realm of *macroscopic* electromagnetism [2]. In fact, spatial averaging alone suffices since it implicitly involves temporal averaging, due to the universal maximum speed $(\varepsilon_0\mu_0)^{-1/2}$ being finite [1]. The spatial average should be taken over volumes which are sufficiently large as to contain many point charges but the linear extent of the averaging volume should be much smaller than the smallest electromagnetic wavelength in consideration. Thus, the macroscopic perspective is appropriate for solids and liquids in the extreme-ultraviolet regime and in longer-wavelength regimes. The appropriate wavelength range for the macroscopic perspective may have a much larger lower bound in rarefied gases.

Adoption of the macroscopic perspective leads to the replacement of equations (2.3) by

$$\left.\begin{aligned} \nabla \times \tilde{\mathbf{E}}(\mathbf{r}, t) + \frac{\partial}{\partial t}\tilde{\mathbf{B}}(\mathbf{r}, t) &= \mathbf{0} \\ \nabla \times \tilde{\mathbf{B}}(\mathbf{r}, t) - \varepsilon_0\mu_0 \frac{\partial}{\partial t}\tilde{\mathbf{E}}(\mathbf{r}, t) &= \mu_0 \tilde{\mathbf{J}}(\mathbf{r}, t) \\ \nabla \bullet \tilde{\mathbf{E}}(\mathbf{r}, t) &= \frac{1}{\varepsilon_0}\tilde{\rho}(\mathbf{r}, t) \\ \nabla \bullet \tilde{\mathbf{B}}(\mathbf{r}, t) &= 0 \end{aligned}\right\}. \tag{2.4}$$

The macroscopic fields $\tilde{\mathbf{E}}(\mathbf{r}, t)$ and $\tilde{\mathbf{B}}(\mathbf{r}, t)$ represent the spatial averages of $\tilde{\mathbf{e}}(\mathbf{r}, t)$ and $\tilde{\mathbf{b}}(\mathbf{r}, t)$, respectively; and the macroscopic charge and current densities $\tilde{\rho}(\mathbf{r}, t)$ and $\tilde{\mathbf{J}}(\mathbf{r}, t)$ are likewise related to $\tilde{c}(\mathbf{r}, t)$ and $\tilde{\mathbf{j}}(\mathbf{r}, t)$.

In a material, the macroscopic source densities may be regarded as either externally impressed or arising from internal mechanisms. Thus, these source densities can be partitioned as

$$\left.\begin{aligned} \tilde{\rho}(\mathbf{r}, t) &= \tilde{\rho}_{\text{ext}}(\mathbf{r}, t) + \tilde{\rho}_{\text{int}}(\mathbf{r}, t) \\ \tilde{\mathbf{J}}(\mathbf{r}, t) &= \tilde{\mathbf{J}}_{\text{ext}}(\mathbf{r}, t) + \tilde{\mathbf{J}}_{\text{int}}(\mathbf{r}, t) \end{aligned}\right\}, \tag{2.5}$$

wherein the subscripts 'ext' and 'int' identify the externally impressed and internally arising source densities, respectively. The macroscopic polarization $\tilde{\mathbf{P}}(\mathbf{r}, t)$ and magnetization $\tilde{\mathbf{M}}(\mathbf{r}, t)$ are introduced to characterize the internally arising source densities:

$$\left.\begin{aligned} \tilde{\rho}_{\text{int}}(\mathbf{r}, t) &= -\nabla \bullet \tilde{\mathbf{P}}(\mathbf{r}, t) \\ \tilde{\mathbf{J}}_{\text{int}}(\mathbf{r}, t) &= \frac{\partial}{\partial t}\tilde{\mathbf{P}}(\mathbf{r}, t) + \frac{1}{\mu_0}\nabla \times \tilde{\mathbf{M}}(\mathbf{r}, t) \end{aligned}\right\}. \tag{2.6}$$

However, a degree of ambiguity is associated with the internally arising source densities represented by equations (2.6). Thus, if $\tilde{\mathbf{P}}(\mathbf{r}, t)$ were substituted by $\tilde{\mathbf{P}}(\mathbf{r}, t) - \nabla \times \tilde{\mathbf{A}}(\mathbf{r}, t)$ and $\tilde{\mathbf{M}}(\mathbf{r}, t)$ substituted by $\tilde{\mathbf{M}}(\mathbf{r}, t) + \mu_0(\partial/\partial t)\tilde{\mathbf{A}}(\mathbf{r}, t)$, where $\tilde{\mathbf{A}}(\mathbf{r}, t)$ is some differentiable vector function, then $\tilde{\rho}_{\text{int}}(\mathbf{r}, t)$ and $\tilde{\mathbf{J}}_{\text{int}}(\mathbf{r}, t)$ given by equations (2.6) would remain unchanged.

The polarization and magnetization are subsumed into the following definitions of two macroscopic electromagnetic fields:[1]

$$\left.\begin{aligned} \tilde{\mathbf{D}}(\mathbf{r}, t) &= \varepsilon_0\, \tilde{\mathbf{E}}(\mathbf{r}, t) + \tilde{\mathbf{P}}(\mathbf{r}, t) \\ \tilde{\mathbf{H}}(\mathbf{r}, t) &= \frac{1}{\mu_0}\tilde{\mathbf{B}}(\mathbf{r}, t) - \frac{1}{\mu_0}\tilde{\mathbf{M}}(\mathbf{r}, t) \end{aligned}\right\}. \tag{2.7}$$

The fields $\tilde{\mathbf{D}}(\mathbf{r}, t)$ and $\tilde{\mathbf{H}}(\mathbf{r}, t)$ develop in a material in response to the fields $\tilde{\mathbf{E}}(\mathbf{r}, t)$ and $\tilde{\mathbf{B}}(\mathbf{r}, t)$. Consequently, $\tilde{\mathbf{D}}(\mathbf{r}, t)$ and $\tilde{\mathbf{H}}(\mathbf{r}, t)$ are regarded as *induction* fields, while $\tilde{\mathbf{E}}(\mathbf{r}, t)$ and $\tilde{\mathbf{B}}(\mathbf{r}, t)$ are regarded as *primitive* fields. Unlike the induction fields, the primitive fields may be measured directly via the Lorentz force [1]

$$\tilde{\mathbf{F}}_{\text{Lor}}(\mathbf{r}, t) = q(\mathbf{r}, t)\left[\tilde{\mathbf{E}}(\mathbf{r}, t) + \mathbf{v}(\mathbf{r}, t) \times \tilde{\mathbf{B}}(\mathbf{r}, t)\right] \tag{2.8}$$

acting on a point charge $q(\mathbf{r}, t)$ traveling at velocity $\mathbf{v}(\mathbf{r}, t)$.

[1] An alternative convention is in common use concerning the definitions of the vectors $\tilde{\mathbf{J}}_{\text{int}}(\mathbf{r}, t)$, $\tilde{\mathbf{M}}(\mathbf{r}, t)$, and $\tilde{\mathbf{H}}(\mathbf{r}, t)$. This convention involves equation $(2.6)_2$ being replaced by $\tilde{\mathbf{J}}_{\text{int}}(\mathbf{r}, t) = (\partial/\partial t)\tilde{\mathbf{P}}(\mathbf{r}, t) + \nabla \times \tilde{\mathbf{M}}(\mathbf{r}, t)$ and equation $(2.7)_2$ being replaced by $\tilde{\mathbf{H}}(\mathbf{r}, t) = \mu_0^{-1}\tilde{\mathbf{B}}(\mathbf{r}, t) - \tilde{\mathbf{M}}(\mathbf{r}, t)$. The alternative convention is not used in this ebook.

By substituting equations (2.5)–(2.7) into equations (2.4), the four macroscopic electromagnetic fields $\tilde{\mathbf{E}}(\mathbf{r}, t)$, $\tilde{\mathbf{D}}(\mathbf{r}, t)$, $\tilde{\mathbf{B}}(\mathbf{r}, t)$, and $\tilde{\mathbf{H}}(\mathbf{r}, t)$ are brought together in the macroscopic Maxwell postulates written in standard form as

$$\left.\begin{aligned} \nabla \times \tilde{\mathbf{H}}(\mathbf{r}, t) - \frac{\partial}{\partial t}\tilde{\mathbf{D}}(\mathbf{r}, t) &= \tilde{\mathbf{J}}_{\text{ext}}(\mathbf{r}, t) \\ \nabla \times \tilde{\mathbf{E}}(\mathbf{r}, t) + \frac{\partial}{\partial t}\tilde{\mathbf{B}}(\mathbf{r}, t) &= \mathbf{0} \\ \nabla \bullet \tilde{\mathbf{D}}(\mathbf{r}, t) &= \tilde{\rho}_{\text{ext}}(\mathbf{r}, t) \\ \nabla \bullet \tilde{\mathbf{B}}(\mathbf{r}, t) &= 0 \end{aligned}\right\}. \tag{2.9}$$

Thus, there are four linear differential equations: (i) two homogeneous differential equations $(2.9)_2$ and $(2.9)_4$ involving the primitive fields and (ii) two inhomogeneous differential equations $(2.9)_1$ and $(2.9)_3$ involving the induction fields.

The divergence postulate $(2.9)_4$ is completely consistent with the curl postulate $(2.9)_2$, as can be seen by taking the divergence of the left and right sides of the latter postulate. The consistency of the divergence postulate $(2.9)_3$ and the curl postulate $(2.9)_1$ mandates the reasonable constraint

$$\nabla \bullet \tilde{\mathbf{J}}_{\text{ext}}(\mathbf{r}, t) + \frac{\partial}{\partial t}\tilde{\rho}_{\text{ext}}(\mathbf{r}, t) = 0. \tag{2.10}$$

The two Maxwell curl postulates suffice for our purposes in the remainder of this ebook, the macroscopic continuity equation (2.10) being presumed to hold in practical situations.

2.2 Constitutive relations

While the Maxwell postulates (2.9) govern all electromagnetic phenomenons in materials, they cannot be solved (for the primitive or the induction fields) without further information being supplied. This further information is provided by constitutive relations, which relate the primitive fields to the induction fields in the material under consideration. Formally, these may be expressed in the general form

$$\left.\begin{aligned} \tilde{\mathbf{D}}(\mathbf{r}, t) &= \mathcal{F}\left\{\tilde{\mathbf{E}}(\mathbf{r}, t), \tilde{\mathbf{B}}(\mathbf{r}, t)\right\} \\ \tilde{\mathbf{H}}(\mathbf{r}, t) &= \mathcal{G}\left\{\tilde{\mathbf{E}}(\mathbf{r}, t), \tilde{\mathbf{B}}(\mathbf{r}, t)\right\} \end{aligned}\right\}, \tag{2.11}$$

with $\mathcal{F}$ and $\mathcal{G}$ being linear functions of $\tilde{\mathbf{E}}(\mathbf{r}, t)$ and $\tilde{\mathbf{B}}(\mathbf{r}, t)$ for linear materials, and nonlinear functions of $\tilde{\mathbf{E}}(\mathbf{r}, t)$ and $\tilde{\mathbf{B}}(\mathbf{r}, t)$ for nonlinear materials. The remainder of this chapter up to and including section 2.6 is devoted to linear materials, nonlinear materials being visited in section 2.7.

In general, a material's electromagnetic response is spatiotemporally nonlocal. Consequently, in full generality, the constitutive relations of a linear material are expressed as [3]

$$\left.\begin{aligned}\tilde{\mathbf{D}}(\mathbf{r},t) &= \int_{t'}\int_{\mathbf{r}'}\Big[\underline{\underline{\tilde{\varepsilon}}}_{\,\mathrm{EB}}(\mathbf{r}',t')\bullet\tilde{\mathbf{E}}(\mathbf{r}-\mathbf{r}',t-t')+\underline{\underline{\tilde{\xi}}}_{\,\mathrm{EB}}(\mathbf{r}',t')\bullet \\ &\qquad \tilde{\mathbf{B}}(\mathbf{r}-\mathbf{r}',t-t')\Big]\,d^3\mathbf{r}'\,dt' \\ \tilde{\mathbf{H}}(\mathbf{r},t) &= \int_{t'}\int_{\mathbf{r}'}\Big[\underline{\underline{\tilde{\zeta}}}_{\,\mathrm{EB}}(\mathbf{r}',t')\bullet\tilde{\mathbf{E}}(\mathbf{r}-\mathbf{r}',t-t')+\underline{\underline{\tilde{\nu}}}_{\,\mathrm{EB}}(\mathbf{r}',t')\bullet \\ &\qquad \tilde{\mathbf{B}}(\mathbf{r}-\mathbf{r}',t-t')\Big]\,d^3\mathbf{r}'\,dt'\end{aligned}\right\},\tag{2.12}$$

in terms of the four 3 × 3 constitutive dyadics $\underline{\underline{\tilde{\varepsilon}}}_{\,\mathrm{EB}}(\mathbf{r},t)$, $\underline{\underline{\tilde{\xi}}}_{\,\mathrm{EB}}(\mathbf{r},t)$, $\underline{\underline{\tilde{\zeta}}}_{\,\mathrm{EB}}(\mathbf{r},t)$, and $\underline{\underline{\tilde{\nu}}}_{\,\mathrm{EB}}(\mathbf{r},t)$. A guide to 3 × 3 dyadics is provided in appendix A.

The manifestation of nonlocality can vary greatly from one material to another. If a characteristic length scale in a material is similar to the electromagnetic wavelength, then spatial nonlocality may be significant [4]. However, the effects of spatial nonlocality are negligible in the vast majority of situations currently considered as practical. On the other hand, since electromagnetic signals travel rapidly, the effects of temporal nonlocality must not be ignored. Spatially local but temporally nonlocal linear materials are characterized by the constitutive relations

$$\left.\begin{aligned}\tilde{\mathbf{D}}(\mathbf{r},t) &= \int_{t'}\Big[\underline{\underline{\tilde{\varepsilon}}}_{\,\mathrm{EB}}(\mathbf{r},t')\bullet\tilde{\mathbf{E}}(\mathbf{r},t-t')+\underline{\underline{\tilde{\xi}}}_{\,\mathrm{EB}}(\mathbf{r},t')\bullet\tilde{\mathbf{B}}(\mathbf{r},t-t')\Big]\,dt' \\ \tilde{\mathbf{H}}(\mathbf{r},t) &= \int_{t'}\Big[\underline{\underline{\tilde{\zeta}}}_{\,\mathrm{EB}}(\mathbf{r},t')\bullet\tilde{\mathbf{E}}(\mathbf{r},t-t')+\underline{\underline{\tilde{\nu}}}_{\,\mathrm{EB}}(\mathbf{r},t')\bullet\tilde{\mathbf{B}}(\mathbf{r},t-t')\Big]\,dt'\end{aligned}\right\}.\tag{2.13}$$

2.3 Frequency domain

The time-domain constitutive relations (2.13) can be inconvenient as the convolution integrals therein usually lead to mathematical complications that are analytically intractable and computationally challenging. These complications are bypassed by switching to the frequency domain by taking the temporal Fourier transforms of all fields and sources appearing in equations (2.9) and (2.13) as follows:

$$\begin{aligned}\mathcal{Z}(\mathbf{r},\omega) &= \int_{-\infty}^{\infty}\tilde{\mathcal{Z}}(\mathbf{r},t)\exp(i\omega t)\,dt, \\ \mathcal{Z} &\in \left\{\underline{\underline{\varepsilon}}_{\,\mathrm{EB}},\,\underline{\underline{\xi}}_{\,\mathrm{EB}},\,\underline{\underline{\zeta}}_{\,\mathrm{EB}},\,\underline{\underline{\nu}}_{\,\mathrm{EB}},\,\mathbf{E},\,\mathbf{D},\,\mathbf{B},\,\mathbf{H},\,\mathbf{J}_{\mathrm{ext}},\,\rho_{\mathrm{ext}}\right\}.\end{aligned}\tag{2.14}$$

Here, ω is the angular frequency and $i=\sqrt{-1}$. When $\tilde{\mathcal{Z}}(\mathbf{r},t)$ is either a source or a field, its counterpart $\mathcal{Z}(\mathbf{r},\omega)$ is referred to as a *phasor*.

Accordingly, equations (2.9) transform to

$$\left.\begin{aligned}\nabla\times\mathbf{H}(\mathbf{r},\omega)+i\omega\mathbf{D}(\mathbf{r},\omega) &= \mathbf{J}_{\mathrm{ext}}(\mathbf{r},\omega) \\ \nabla\times\mathbf{E}(\mathbf{r},\omega)-i\omega\mathbf{B}(\mathbf{r},\omega) &= \mathbf{0} \\ \nabla\bullet\mathbf{D}(\mathbf{r},\omega) &= \rho_{\mathrm{ext}}(\mathbf{r},\omega) \\ \nabla\bullet\mathbf{B}(\mathbf{r},\omega) &= 0\end{aligned}\right\},\tag{2.15}$$

and application of the convolution theorem [5] to equations (2.13) delivers the frequency-domain constitutive relations

$$\left.\begin{aligned} \mathbf{D}(\mathbf{r},\omega) &= \underline{\underline{\varepsilon}}_{\,\mathrm{EB}}(\mathbf{r},\omega)\bullet\mathbf{E}(\mathbf{r},\omega)+\underline{\underline{\xi}}_{\,\mathrm{EB}}(\mathbf{r},\omega)\bullet\mathbf{B}(\mathbf{r},\omega)\\ \mathbf{H}(\mathbf{r},\omega) &= \underline{\underline{\zeta}}_{\,\mathrm{EB}}(\mathbf{r},\omega)\bullet\mathbf{E}(\mathbf{r},\omega)+\underline{\underline{\nu}}_{\,\mathrm{EB}}(\mathbf{r},\omega)\bullet\mathbf{B}(\mathbf{r},\omega)\end{aligned}\right\}. \tag{2.16}$$

Equations $(2.15)_3$ and $(2.15)_4$ are not needed for homogenization.

There is a price to be paid for the relative simplicity of the frequency-domain constitutive relations (2.16) as compared with the time-domain constitutive relations (2.13). The electromagnetic fields and constitutive dyadics in the time-domain description (2.13) are all real-valued quantities whereas their counterparts in the frequency-domain description (2.16) are all complex valued. Thus, inverse temporal Fourier transforms must be taken to convert the frequency-domain quantities into real-valued physical quantities.

The principle of causality leads to constraints on the real and imaginary parts of the frequency-domain constitutive parameters. To see this, first consider the implications of causality for the time-domain constitutive parameters. A passive material cannot respond to a stimulus until *after* the stimulus has been received. In view of the convolution integrals in the time-domain constitutive relations (2.13), therefore

$$\left.\begin{aligned} \underline{\underline{\tilde{\varepsilon}}}_{\,\mathrm{EB}}(\mathbf{r},t)-\varepsilon_0\delta(t)\underline{\underline{I}} &\equiv \underline{\underline{0}}\\ \underline{\underline{\tilde{\xi}}}_{\,\mathrm{EB}}(\mathbf{r},t) &\equiv \underline{\underline{0}}\\ \underline{\underline{\tilde{\zeta}}}_{\,\mathrm{EB}}(\mathbf{r},t) &\equiv \underline{\underline{0}}\\ \mu_0^{-1}\delta(t)\underline{\underline{I}}-\underline{\underline{\tilde{\nu}}}_{\,\mathrm{EB}}(\mathbf{r},t) &\equiv \underline{\underline{0}}\end{aligned}\right\}\quad \text{for}\quad t\leqslant 0, \tag{2.17}$$

where $\underline{\underline{I}}$ is the 3×3 identity dyadic. Now let the scalar function $\tilde{\mathcal{Z}}(\mathbf{r},t)$ represent an arbitrary component of any one of the dyadics $\underline{\underline{\tilde{\varepsilon}}}_{\,\mathrm{EB}}(\mathbf{r},t)-\varepsilon_0\delta(t)\underline{\underline{I}}$, $\underline{\underline{\tilde{\xi}}}_{\,\mathrm{EB}}(\mathbf{r},t)$, $\underline{\underline{\tilde{\zeta}}}_{\,\mathrm{EB}}(\mathbf{r},t)$, and $\mu_0^{-1}\delta(t)\underline{\underline{I}}-\underline{\underline{\tilde{\nu}}}_{\,\mathrm{EB}}(\mathbf{r},t)$. The analytic continuation of $\mathcal{Z}(\mathbf{r},\omega)$ in the upper half of the complex-ω plane then yields the Hilbert transforms

$$\left.\begin{aligned} \mathrm{Re}\,\{\mathcal{Z}(\mathbf{r},\omega)\} &= \frac{1}{\pi}\mathsf{P}\int_{-\infty}^{\infty}\frac{\mathrm{Im}\,\{\mathcal{Z}(\mathbf{r},s)\}}{s-\omega}\,ds\\ \mathrm{Im}\,\{\mathcal{Z}(\mathbf{r},\omega)\} &= -\frac{1}{\pi}\mathsf{P}\int_{-\infty}^{\infty}\frac{\mathrm{Re}\,\{\mathcal{Z}(\mathbf{r},s)\}}{s-\omega}\,ds\end{aligned}\right\}, \tag{2.18}$$

where P signifies the Cauchy principal value. Notice that the scalar quantity $\tilde{\mathcal{Z}}(\mathbf{r},t)$ is real valued. Accordingly, $\mathcal{Z}(\mathbf{r},\omega)$ is related to its complex conjugate $\mathcal{Z}^*(\mathbf{r},\omega)$ by

$$\mathcal{Z}(\mathbf{r},-\omega)=\mathcal{Z}^*(\mathbf{r},\omega). \tag{2.19}$$

By exploiting the symmetry condition (2.19), the Hilbert transforms (2.18) give rise to

$$\left.\begin{aligned} \text{Re}\,\{\mathcal{Z}(\mathbf{r},\,\omega)\} &= \frac{2}{\pi}\mathsf{P}\int_0^\infty \frac{s\,\text{Im}\,\{\mathcal{Z}(\mathbf{r},\,s)\}}{s^2-\omega^2}\,ds \\ \text{Im}\,\{\mathcal{Z}(\mathbf{r},\,\omega)\} &= -\frac{2}{\pi}\mathsf{P}\int_0^\infty \frac{\omega\,\text{Re}\,\{\mathcal{Z}(\mathbf{r},\,s)\}}{s^2-\omega^2}\,ds \end{aligned}\right\}. \tag{2.20}$$

Equations (2.20) are known as the *Kramers–Kronig relations* [5]. These relations are representative of general constraints on the frequency responses of causal linear systems [6].

The partnering of the primitive fields $\mathbf{E}(\mathbf{r},\,\omega)$ and $\mathbf{B}(\mathbf{r},\,\omega)$ leads to the *Boys–Post* formulation of the constitutive relations equations (2.16). The phasors $\mathbf{E}(\mathbf{r},\,\omega)$ and $\mathbf{B}(\mathbf{r},\,\omega)$ are paired because their inverse temporal Fourier transforms appear in the Lorentz force. Alternatively, it can be apposite for $\mathbf{E}(\mathbf{r},\,\omega)$ to be partnered with $\mathbf{H}(\mathbf{r},\,\omega)$, as in the specification of boundary conditions as well as in the definition of the time-averaged Poynting vector. That pairing leads to the *Tellegen* formulation of the frequency-domain constitutive relations

$$\left.\begin{aligned} \mathbf{D}(\mathbf{r},\,\omega) &= \underline{\underline{\varepsilon}}_{\,\text{EH}}(\mathbf{r},\,\omega)\bullet\mathbf{E}(\mathbf{r},\,\omega) + \underline{\underline{\xi}}_{\,\text{EH}}(\mathbf{r},\,\omega)\bullet\mathbf{H}(\mathbf{r},\,\omega) \\ \mathbf{B}(\mathbf{r},\,\omega) &= \underline{\underline{\zeta}}_{\,\text{EH}}(\mathbf{r},\,\omega)\bullet\mathbf{E}(\mathbf{r},\,\omega) + \underline{\underline{\mu}}_{\,\text{EH}}(\mathbf{r},\,\omega)\bullet\mathbf{H}(\mathbf{r},\,\omega) \end{aligned}\right\}. \tag{2.21}$$

The constitutive dyadics in the Boys–Post equations (2.16) are related to those in the Tellegen equations (2.21) as follows [3]:

$$\left.\begin{aligned} \underline{\underline{\varepsilon}}_{\,\text{EH}}(\mathbf{r},\,\omega) &= \underline{\underline{\varepsilon}}_{\,\text{EB}}(\mathbf{r},\,\omega) - \underline{\underline{\xi}}_{\,\text{EB}}(\mathbf{r},\,\omega)\bullet\underline{\underline{\nu}}_{\,\text{EB}}^{-1}(\mathbf{r},\,\omega)\bullet\underline{\underline{\zeta}}_{\,\text{EB}}(\mathbf{r},\,\omega) \\ \underline{\underline{\xi}}_{\,\text{EH}}(\mathbf{r},\,\omega) &= \underline{\underline{\xi}}_{\,\text{EB}}(\mathbf{r},\,\omega)\bullet\underline{\underline{\nu}}_{\,\text{EB}}^{-1}(\mathbf{r},\,\omega) \\ \underline{\underline{\zeta}}_{\,\text{EH}}(\mathbf{r},\,\omega) &= -\underline{\underline{\nu}}_{\,\text{EB}}^{-1}(\mathbf{r},\,\omega)\bullet\underline{\underline{\zeta}}_{\,\text{EB}}(\mathbf{r},\,\omega) \\ \underline{\underline{\mu}}_{\,\text{EH}}(\mathbf{r},\,\omega) &= \underline{\underline{\nu}}_{\,\text{EB}}^{-1}(\mathbf{r},\,\omega) \end{aligned}\right\}. \tag{2.22}$$

Implicitly, the constitutive dyadic $\underline{\underline{\nu}}_{\,\text{EB}}(\mathbf{r},\,\omega)$ is nonsingular. Caution should be exercised here, since singular constitutive dyadics are known to occur [7].

Henceforth in this ebook, the Tellegen constitutive relations (2.21) are used. For brevity, the subscript 'EH' on the constitutive dyadics is omitted. Homogenization being focused at a fixed angular frequency for linear materials, ω is omitted from the argument list for the electromagnetic fields and sources, and the Tellegen constitutive relations are written as

$$\left.\begin{aligned} \mathbf{D}(\mathbf{r}) &= \underline{\underline{\varepsilon}}(\mathbf{r})\bullet\mathbf{E}(\mathbf{r}) + \underline{\underline{\xi}}(\mathbf{r})\bullet\mathbf{H}(\mathbf{r}) \\ \mathbf{B}(\mathbf{r}) &= \underline{\underline{\zeta}}(\mathbf{r})\bullet\mathbf{E}(\mathbf{r}) + \underline{\underline{\mu}}(\mathbf{r})\bullet\mathbf{H}(\mathbf{r}) \end{aligned}\right\}. \tag{2.23}$$

If a material is homogeneous, the constitutive relations are written even more simply as

$$\left.\begin{aligned} \mathbf{D}(\mathbf{r}) &= \underline{\underline{\varepsilon}} \bullet \mathbf{E}(\mathbf{r}) + \underline{\underline{\xi}} \bullet \mathbf{H}(\mathbf{r}) \\ \mathbf{B}(\mathbf{r}) &= \underline{\underline{\zeta}} \bullet \mathbf{E}(\mathbf{r}) + \underline{\underline{\mu}} \bullet \mathbf{H}(\mathbf{r}) \end{aligned}\right\}. \tag{2.24}$$

2.4 A compact representation

As illustrated by equations (2.24), the most general linear homogeneous material is characterized by four 3 × 3 constitutive dyadics—namely, the permittivity dyadic $\underline{\underline{\varepsilon}}$, the permeability dyadic $\underline{\underline{\mu}}$, and the two magnetoelectric dyadics $\underline{\underline{\xi}}$ and $\underline{\underline{\zeta}}$. When dealing with such materials, it can be convenient to adopt a 6 × 6-dyadic/6-vector notation. This allows the Tellegen constitutive relations (2.24) to be written compactly as

$$\underline{\mathbf{C}}(\mathbf{r}) = \underline{\underline{\mathbf{K}}} \bullet \underline{\mathbf{F}}(\mathbf{r}), \tag{2.25}$$

wherein the four 3 × 3 constitutive dyadics are contained within the 6 × 6 constitutive dyadic

$$\underline{\underline{\mathbf{K}}} = \begin{bmatrix} \underline{\underline{\varepsilon}} & \underline{\underline{\xi}} \\ \underline{\underline{\zeta}} & \underline{\underline{\mu}} \end{bmatrix} \tag{2.26}$$

and the electromagnetic fields are represented by the 6-vectors

$$\left.\begin{aligned} \underline{\mathbf{C}}(\mathbf{r}) &= \begin{bmatrix} \mathbf{D}(\mathbf{r}) \\ \mathbf{B}(\mathbf{r}) \end{bmatrix} \\ \underline{\mathbf{F}}(\mathbf{r}) &= \begin{bmatrix} \mathbf{E}(\mathbf{r}) \\ \mathbf{H}(\mathbf{r}) \end{bmatrix} \end{aligned}\right\}. \tag{2.27}$$

In this notation, the result of combining the Maxwell curl postulates $(2.15)_{1,2}$ with the constitutive relations (2.24) may be expressed succinctly as

$$\left[\underline{\underline{\mathbf{L}}}(\nabla) + i\omega \underline{\underline{\mathbf{K}}} \right] \bullet \underline{\mathbf{F}}(\mathbf{r}) = \underline{\mathbf{Q}}(\mathbf{r}). \tag{2.28}$$

Here the linear differential operator

$$\underline{\underline{\mathbf{L}}}(\nabla) = \begin{bmatrix} \underline{\underline{0}} & \nabla \times \underline{\underline{I}} \\ -\nabla \times \underline{\underline{I}} & \underline{\underline{0}} \end{bmatrix}, \tag{2.29}$$

where $\underline{\underline{0}}$ is the 3 × 3 null dyadic, and

$$\underline{\mathbf{Q}}(\mathbf{r}) = \begin{bmatrix} \mathbf{J}_{\text{ext}}(\mathbf{r}) \\ \mathbf{0} \end{bmatrix} \tag{2.30}$$

is the current density 6-vector.

2.5 Dissipative and nondissipative materials

According to the principle of causality, a passive material cannot respond instantaneously to an applied electromagnetic field. Consequently, energy is dissipated when an electromagnetic field is applied to a passive material. We exclude free space from our consideration here, since free space is not a material at all from our classical viewpoint; indeed, the distinction between microscopic and macroscopic electromagnetism does not apply to free space.

In certain instances, the degree of dissipation may be negligible, especially over a narrow range of angular frequencies and limited length scales. For example, much of the field of crystal optics is predicated upon the absence of dissipation. The dissipative or nondissipative nature of a material is reflected in the constitutive dyadics for the material. Specifically, a material characterized by the Tellegen constitutive relations (2.24) is dissipative if both of the 3×3 dyadics [8, 9]

$$\left.\begin{aligned} \underline{\underline{d}}_1 &= i(\underline{\underline{\varepsilon}} - \underline{\underline{\varepsilon}}^\dagger) \\ \underline{\underline{d}}_2 &= i\left[\underline{\underline{\mu}} - \underline{\underline{\mu}}^\dagger - i\left(\underline{\underline{\zeta}} - \underline{\underline{\xi}}^\dagger\right) \bullet \underline{\underline{d}}_1^{-1} \bullet \left(\underline{\underline{\xi}} - \underline{\underline{\zeta}}^\dagger\right)\right] \end{aligned}\right\} \qquad (2.31)$$

are negative definite, or—equivalently—if both of the 3×3 dyadics

$$\left.\begin{aligned} \underline{\underline{d}}_3 &= i\left(\underline{\underline{\mu}} - \underline{\underline{\mu}}^\dagger\right) \\ \underline{\underline{d}}_4 &= i\left[\underline{\underline{\varepsilon}} - \underline{\underline{\varepsilon}}^\dagger - i\left(\underline{\underline{\xi}} - \underline{\underline{\zeta}}^\dagger\right) \bullet \underline{\underline{d}}_3^{-1} \bullet \left(\underline{\underline{\zeta}} - \underline{\underline{\xi}}^\dagger\right)\right] \end{aligned}\right\} \qquad (2.32)$$

are negative definite, where the superscript $\dagger$ denotes the conjugate transpose. Parenthetically, since the dyadics $\underline{\underline{d}}_{1,2,3,4}$ are all Hermitian (i.e., $\underline{\underline{d}}_\ell = \underline{\underline{d}}_\ell^\dagger$ for $\ell \in \{1, 2, 3, 4\}$), these generally complex-valued dyadics are guaranteed to have real-valued eigenvalues [10]. On the other hand, the material is nondissipative provided that [11]

$$\left.\begin{aligned} \underline{\underline{\varepsilon}} &= \underline{\underline{\varepsilon}}^\dagger \\ \underline{\underline{\xi}} &= \underline{\underline{\zeta}}^\dagger \\ \underline{\underline{\mu}} &= \underline{\underline{\mu}}^\dagger \end{aligned}\right\}, \qquad (2.33)$$

in which case $\underline{\underline{d}}_\ell = \underline{\underline{0}}$ for $\ell \in \{1, 2, 3, 4\}$.

An active material is one that is neither dissipative nor nondissipative. Accordingly, for an active material both dyadics $\underline{\underline{d}}_1$ and $\underline{\underline{d}}_2$ are positive definite or—equivalently— both of the dyadics $\underline{\underline{d}}_3$ and $\underline{\underline{d}}_4$ are positive definite.

The particular case of an isotropic dielectric–magnetic material, characterized by permittivity scalar ε and permeability scalar μ (i.e., $\underline{\underline{\varepsilon}} = \varepsilon \underline{\underline{I}}$, $\underline{\underline{\xi}} = \underline{\underline{\zeta}} = \underline{\underline{0}}$, and $\underline{\underline{\mu}} = \mu \underline{\underline{I}}$), is especially noteworthy here. According to equations (2.31) or, equivalently, (2.32), this material is active if one or both inequalities $\mathrm{Im}\{\varepsilon\} \geqslant 0$ and $\mathrm{Im}\{\mu\} \geqslant 0$ do not hold [12, 13].

The picture can be more complicated for anisotropic and bianisotropic materials. For certain anisotropic and bianisotropic materials, it may be that some but not all of the four dyadics $\underline{\underline{d}}_1$, $\underline{\underline{d}}_2$, $\underline{\underline{d}}_3$, and $\underline{\underline{d}}_4$ are negative definite while others are positive definite. Such materials cannot be classified as wholly dissipative or wholly active: fields propagating in one direction in such a material may be attenuated but fields propagating in another direction may be amplified [14]. This matter is discussed further and illustrated by a numerical example in section 6.9.2.

2.6 Linear materials

A general linear material is characterized by four 3 × 3 constitutive dyadics, which comprise 36 complex-valued constitutive scalars in total, per the Tellegen constitutive relations (2.24). Symmetries greatly reduce the dimensionality of the constitutive-parameter space for many materials of interest [15], as exemplified next.

2.6.1 Isotropic and biisotropic materials

In isotropic dielectric–magnetic materials, the induction fields are aligned wholly parallel to the primitive fields. Hence, their constitutive dyadics are of the form

$$\left.\begin{array}{lcl} \underline{\underline{\varepsilon}} & = & \varepsilon \underline{\underline{I}} \\ \underline{\underline{\xi}} & = & \underline{\underline{0}} \\ \underline{\underline{\zeta}} & = & \underline{\underline{0}} \\ \underline{\underline{\mu}} & = & \mu \underline{\underline{I}} \end{array}\right\}, \qquad (\varepsilon, \mu \in \mathbb{C}), \tag{2.34}$$

where $\mathbb{C}$ is the set of all complex numbers. Furthermore, $\mu = \mu_0$ for isotropic dielectric materials and $\varepsilon = \varepsilon_0$ for isotropic magnetic materials.

A biisotropic material displays both isotropy and magnetoelectric coupling. The constitutive dyadics of biisotropic materials are of the form

$$\left.\begin{array}{lcl} \underline{\underline{\varepsilon}} & = & \varepsilon \underline{\underline{I}} \\ \underline{\underline{\xi}} & = & \xi \underline{\underline{I}} \\ \underline{\underline{\zeta}} & = & \zeta \underline{\underline{I}} \\ \underline{\underline{\mu}} & = & \mu \underline{\underline{I}} \end{array}\right\}, \qquad (\varepsilon, \xi, \zeta, \mu \in \mathbb{C}). \tag{2.35}$$

An important special case occurs if $\xi = -\zeta$; then the material is an isotropic chiral material or a reciprocal biisotropic material [16]. The *optical activity* of such materials can be harnessed to discriminate between left-handed and right-handed electromagnetic fields [17]. The case $\xi \neq -\zeta$ is also a notable one, but the electromagnetic response characteristics of nonreciprocal biisotropic materials can be accounted for by setting $\xi = -\zeta$ [18] and prescribing an appropriate surface current density [19].

Although the 3 × 3 constitutive dyadics of biisotropic materials are simply scalar multiples of $\underline{\underline{I}}$, the induction fields are generally not aligned with the primitive fields in these materials, unlike in isotropic dielectric–magnetic materials. However, key electromagnetic characteristics, such as those pertaining to plane-wave propagation, are independent of direction in biisotropic materials.

2.6.2 Anisotropic and bianisotropic materials

Many naturally occurring and engineered materials exhibit anisotropy, i.e., direction-dependent characteristics [20]. Nontrivial dyadics—as opposed to dyadics that are simply scalar multiples of $\underline{\underline{I}}$—are needed to relate the induction fields to the primitive fields in such materials.

In an anisotropic dielectric–magnetic material, the induction field **D** is anisotropically coupled to the primitive field **E** and/or the induction field **H** is anisotropically coupled to the primitive field **B**. Hence, the constitutive dyadics obey the following constraints:

$$\left.\begin{array}{rcl} \underline{\underline{\varepsilon}} & \neq & \varepsilon\underline{\underline{I}} \\ \underline{\underline{\xi}} & = & \underline{\underline{0}} \\ \underline{\underline{\zeta}} & = & \underline{\underline{0}} \\ \underline{\underline{\mu}} & \neq & \mu\underline{\underline{I}} \end{array}\right\}, \qquad (\varepsilon, \mu \in \mathbb{C}). \tag{2.36}$$

Whereas $\underline{\underline{\mu}} = \mu_0\underline{\underline{I}}$ for an anisotropic dielectric material, $\underline{\underline{\varepsilon}} = \varepsilon_0\underline{\underline{I}}$ for an anisotropic magnetic material.

The natural generalization of anisotropy is bianisotropy. In a bianisotropic material, both induction fields **D** and **H** are anisotropically coupled to both primitive fields **E** and **B**. Thus, the constitutive dyadics for a bianisotropic material satisfy

$$\left.\begin{array}{l} \underline{\underline{\varepsilon}} \neq \varepsilon\underline{\underline{I}} \\ \underline{\underline{\xi}} \neq \xi\underline{\underline{I}} \\ \underline{\underline{\zeta}} \neq \zeta\underline{\underline{I}} \\ \underline{\underline{\mu}} \neq \mu\underline{\underline{I}} \end{array}\right\}, \qquad (\varepsilon, \xi, \zeta, \mu \in \mathbb{C}). \tag{2.37}$$

The structures of the constitutive dyadics in equations (2.36) and (2.37) reflect underlying spacetime symmetries of the materials being characterized. These symmetries are conventionally classified in terms of 122 discrete magnetic point groups and 21 continuous magnetic point groups (which encompass the symmetries of isotropic and biisotropic materials as degenerate cases) [15, 21–23]. The magnetic point group with the lowest symmetry, namely $\mathbf{C}_1$ in the Schoenflies notation, is especially noteworthy. For materials in this magnetic point group, the structures of the four constitutive dyadics $\underline{\underline{\varepsilon}}$, $\underline{\underline{\xi}}$, $\underline{\underline{\zeta}}$, and $\underline{\underline{\mu}}$ are completely arbitrary; i.e., a total of 36 complex-valued constitutive scalars are necessary. However, most commonly encountered anisotropic and bianisotropic materials exhibit substantial degrees of spacetime symmetry and accordingly far fewer constitutive scalars are needed to characterize them.

The simplest form of anisotropy is represented by uniaxial materials, for which the constitutive dyadics take the form

$$\underline{\underline{\sigma}} = \sigma_1 \underline{\underline{I}} + \sigma_2 \hat{\mathbf{u}}\hat{\mathbf{u}}, \qquad (\sigma_{1,2} \in \mathbb{C};\ \sigma \in \{\epsilon, \xi, \zeta, \mu\}). \tag{2.38}$$

There is just one distinguished direction, identified by the unit vector $\hat{\mathbf{u}}$. For example, within the realm of crystal optics, a uniaxial dielectric material is characterized by

$$\left.\begin{array}{lcl} \underline{\underline{\epsilon}} & = & \epsilon_1 \underline{\underline{I}} + \epsilon_2 \hat{\mathbf{u}}\hat{\mathbf{u}} \\ \underline{\underline{\xi}} & = & \underline{\underline{0}} \\ \underline{\underline{\zeta}} & = & \underline{\underline{0}} \\ \underline{\underline{\mu}} & = & \mu_0 \underline{\underline{I}} \end{array}\right\}, \qquad (\epsilon_{1,2} \in \mathbb{R}), \tag{2.39}$$

with $\hat{\mathbf{u}}$ pointing in the direction of the *optic axis* [11]. That is, $\pm\hat{\mathbf{u}}$ identify the special directions along which plane waves propagate with only one phase speed. Parenthetically, $\hat{\mathbf{u}}$ is also aligned with the *optic ray axis* that represents the special direction along which plane waves propagate with only one energy speed [24].

On generalizing the anisotropy represented in equation (2.38), constitutive dyadics of the form

$$\underline{\underline{\sigma}} = \sigma_1 \underline{\underline{I}} + \sigma_2 \hat{\mathbf{u}}\hat{\mathbf{u}} + \sigma_3 \hat{\mathbf{v}}\hat{\mathbf{v}}, \qquad (\sigma_{1,2,3} \in \mathbb{C};\ \sigma \in \{\epsilon, \xi, \zeta, \mu\}) \tag{2.40}$$

emerge. These constitutive dyadics characterize orthorhombic biaxial materials. Herein the unit vectors $\hat{\mathbf{u}}$ and $\hat{\mathbf{v}}$ are mutually orthogonal. There are two distinguished directions here but the formulation (2.40) is not particularly insightful, as a physical interpretation of $\hat{\mathbf{u}}$ and $\hat{\mathbf{v}}$ is not readily forthcoming. The distinguished directions may be better appreciated by considering the following alternative formulation: for example, an orthorhombic dielectric material within the realm of crystal optics may be characterized by [11, 25]

$$\left.\begin{array}{lcl} \underline{\underline{\epsilon}} & = & \epsilon_{\mathrm{p}} \underline{\underline{I}} + \epsilon_{\mathrm{q}} (\hat{\mathbf{m}}\hat{\mathbf{n}} + \hat{\mathbf{n}}\hat{\mathbf{m}}) \\ \underline{\underline{\xi}} & = & \underline{\underline{0}} \\ \underline{\underline{\zeta}} & = & \underline{\underline{0}} \\ \underline{\underline{\mu}} & = & \mu_0 \underline{\underline{I}} \end{array}\right\}, \qquad (\epsilon_{\mathrm{p,q}} \in \mathbb{R}). \tag{2.41}$$

The unit vectors $\hat{\mathbf{m}}$ and $\hat{\mathbf{n}}$ herein are aligned with the optic ray axes [26]; that is, they are aligned with the two directions along which plane waves propagate with only one energy speed. Notice that in the case of biaxial dielectric materials the two optic ray axes do not generally coincide with the two optic axes (which represent the directions along which plane waves propagate with only one phase speed) [15, 24].

The uniaxial and biaxial constitutive dyadics present in equations (2.38) and (2.40) are symmetric. Antisymmetric constitutive dyadics are also commonly encountered, notably in the context of gyrotropic materials which are characterized by constitutive dyadics of the form

$$\underline{\underline{\sigma}} = \sigma_1 \underline{\underline{I}} + \sigma_2 \hat{\mathbf{u}}\hat{\mathbf{u}} + \sigma_3 \hat{\mathbf{u}} \times \underline{\underline{I}}, \qquad (\sigma_{1,2,3} \in \mathbb{C};\ \sigma \in \{\epsilon, \xi, \zeta, \mu\}). \tag{2.42}$$

For example, a nondissipative magnetically biased ferrite may characterized by [27]

$$\left.\begin{aligned}\underline{\underline{\varepsilon}} &= \varepsilon_0 \underline{\underline{I}}\\ \underline{\underline{\xi}} &= \underline{\underline{0}}\\ \underline{\underline{\zeta}} &= \underline{\underline{0}}\\ \underline{\underline{\mu}} &= \mu_1 \underline{\underline{I}} + \mu_2 \hat{\mathbf{u}}\hat{\mathbf{u}} + i\mu_3 \hat{\mathbf{u}} \times \underline{\underline{I}}\end{aligned}\right\}, \qquad \left(\mu_{1,2,3} \in \mathbb{R}\right), \tag{2.43}$$

where the unit vector $\hat{\mathbf{u}}$ is aligned with the direction of the biasing magnetic field. An example of gyrotropy in a bianisotropic setting arises in the case of Faraday chiral materials. These are characterized by constitutive dyadics of the form [28]

$$\left.\begin{aligned}\underline{\underline{\varepsilon}} &= \varepsilon_1 \underline{\underline{I}} + \varepsilon_2 \hat{\mathbf{u}}\hat{\mathbf{u}} + i\varepsilon_3 \hat{\mathbf{u}} \times \underline{\underline{I}}\\ \underline{\underline{\xi}} &= i(\xi_1 \underline{\underline{I}} + \xi_2 \hat{\mathbf{u}}\hat{\mathbf{u}} + i\xi_3 \hat{\mathbf{u}} \times \underline{\underline{I}})\\ \underline{\underline{\zeta}} &= -\underline{\underline{\xi}}\\ \underline{\underline{\mu}} &= \mu_1 \underline{\underline{I}} + \mu_2 \hat{\mathbf{u}}\hat{\mathbf{u}} + i\mu_3 \hat{\mathbf{u}} \times \underline{\underline{I}}\end{aligned}\right\}, \qquad (\sigma_{1,2,3} \in \mathbb{R};\, \sigma \in \{\varepsilon, \xi, \mu\}), \tag{2.44}$$

in the absence of dissipation. For dissipative Faraday chiral materials, the constitutive parameters $\sigma_{1,2,3}$, ($\sigma \in \{\varepsilon, \xi, \mu\}$), are complex valued.

Naturally occurring materials which exhibit appreciable bianisotropic effects, under normal environmental conditions, are relatively scarce. However, artificial materials with substantial bianisotropic effects may be readily realized. Such materials can be conceptualized as homogenized composite materials (HCMs), arising from constituent materials which are not themselves bianisotropic (or even anisotropic in some cases), as is discussed in section 6.4.

Far from being an esoteric property, bianisotropy is actually a ubiquitous one [15]. From the perspective of special relativity, the classifications of isotropy and bianisotropy are not covariant under the Lorentz transformation.

For example, a material which is an isotropic dielectric material with respect to one inertial reference frame is a bianisotropic material with respect to another inertial reference frame [11]. Furthermore, from the perspective of general relativity, free space subjected to a gravitational field is electromagnetically equivalent to a nonhomogeneous bianisotropic material [29, 30].

2.7 Nonlinear materials

The electromagnetic response of a nonlinear material, at a given angular frequency generally arises due to the simultaneous excitation by an ensemble of primitive fields across a range of angular frequencies. For a general description of a homogeneous bianisotropic material, the starting point is provided by the frequency-domain counterparts of equations (2.7). Thus,

$$\left.\begin{aligned}\mathbf{D}(\mathbf{r}, \omega) &= \varepsilon_0\, \mathbf{E}(\mathbf{r}, \omega) + \mathbf{P}(\mathbf{r}, \omega)\\ \mathbf{H}(\mathbf{r}, \omega) &= \frac{1}{\mu_0}\, \mathbf{B}(\mathbf{r}, \omega) - \frac{1}{\mu_0}\, \mathbf{M}(\mathbf{r}, \omega)\end{aligned}\right\}, \tag{2.45}$$

with the frequency-domain polarization $\mathbf{P}(\mathbf{r}, \omega)$ and magnetization $\mathbf{M}(\mathbf{r}, \omega)$ being the temporal Fourier transforms of $\tilde{\mathbf{P}}(\mathbf{r}, t)$ and $\tilde{\mathbf{M}}(\mathbf{r}, t)$, respectively, per equation (2.14) with $\mathcal{Z} = \mathbf{P}$ and $\mathcal{Z} = \mathbf{M}$.

The 6-vector/6 × 6-dyadic formulation introduced in section 2.4 is now very convenient. Thereby, the nonlinear constitutive relations may be expressed as [31]

$$\underline{\mathbf{C}}(\mathbf{r}, \omega) = \underline{\underline{\mathbf{K}}}_0 \bullet \underline{\mathbf{F}}(\mathbf{r}, \omega) + \underline{\mathbf{N}}(\mathbf{r}, \omega), \tag{2.46}$$

with

$$\underline{\underline{\mathbf{K}}}_0 = \begin{bmatrix} \varepsilon_0 \underline{\underline{I}} & \underline{\underline{0}} \\ \underline{\underline{0}} & \mu_0 \underline{\underline{I}} \end{bmatrix} \tag{2.47}$$

being the 6 × 6 Tellegen constitutive dyadic of free space, and the 6-vectors $\underline{\mathbf{C}}(\mathbf{r}, \omega)$ and $\underline{\mathbf{F}}(\mathbf{r}, \omega)$ defined as in equations (2.27). The 6-vector

$$\underline{\mathbf{N}}(\mathbf{r}, \omega) = \begin{bmatrix} \mathbf{P}(\mathbf{r}, \omega) \\ \mathbf{M}(\mathbf{r}, \omega) \end{bmatrix} \tag{2.48}$$

is the sum of linear and nonlinear parts, i.e.,

$$\underline{\mathbf{N}}(\mathbf{r}, \omega) = \underline{\mathbf{N}}_{\mathrm{L}}(\mathbf{r}, \omega) + \underline{\mathbf{N}}_{\mathrm{NL}}(\mathbf{r}, \omega). \tag{2.49}$$

The linear part

$$\underline{\mathbf{N}}_{\mathrm{L}}(\mathbf{r}, \omega) = \left[\underline{\underline{\mathbf{K}}}^{\mathrm{L}}(\omega) - \underline{\underline{\mathbf{K}}}_0 \right] \bullet \underline{\mathbf{F}}(\mathbf{r}, \omega) \tag{2.50}$$

is specified in terms of $\underline{\underline{\mathbf{K}}}^{\mathrm{L}}(\omega)$, which is the 6 × 6 Tellegen constitutive dyadic that characterizes the linear bianisotropic response of the nonlinear material.

The nonlinear part $\underline{\mathbf{N}}_{\mathrm{NL}}(\mathbf{r}, \omega)$ provides the exclusively nonlinear response of the nonlinear material, subject to the simultaneous stimulation by an ensemble of fields $\underline{\mathbf{F}}(\mathbf{r}, \omega_m)$, where ω_m belongs to the set of angular frequencies

$$\mathcal{W} = \{ \omega_1, \omega_2, \dots, \omega_M \}, \qquad M > 1. \tag{2.51}$$

The members of the set $\mathcal{W}$ need not be distinct. The jth component of the 6-vector $\underline{\mathbf{N}}_{\mathrm{NL}}(\mathbf{r}, \omega_{\mathrm{NL}})$ evaluated at the angular frequency $\omega = \omega_{\mathrm{NL}}$ may be expressed as

$$\begin{aligned} &[\underline{\mathbf{N}}_{\mathrm{NL}}(\mathbf{r}, \omega_{\mathrm{NL}})]_j \\ &= \sum_{j_1=1}^{6} \sum_{j_2=1}^{6} \cdots \sum_{j_m=1}^{6} \cdots \sum_{j_M=1}^{6} \left\{ \chi^{\mathrm{NL}}_{jj_1j_2\cdots j_m\cdots j_M}(\omega_{\mathrm{NL}}; \mathcal{W}) \prod_{n=1}^{M} [\, \underline{\mathbf{F}}(\mathbf{r}, \omega_n) \,]_{j_n} \right\}, \end{aligned} \tag{2.52}$$

for $j \in [1, 6]$. The nonlinear susceptibility tensor $\chi^{\mathrm{NL}}_{jj_1j_2\cdots j_m\cdots j_M}(\omega_{\mathrm{NL}}; \mathcal{W})$ encapsulates the nonlinear constitutive properties of the material under consideration. The angular frequency ω_{NL} here is determined by members of the set of angular frequencies $\mathcal{W}$ per the sum

$$\omega_{\rm NL} = \sum_{m=1}^{M} a_m \omega_m, \qquad \omega_m \in \mathcal{W}, \; a_m \in \{-1, +1\}; \tag{2.53}$$

if $a_n = -1$ then $[\, \underline{\mathbf{F}}(\mathbf{r}, \omega_n) \,]_{j_n}$ on the right side of equation (2.52) should be replaced by its complex conjugate.

The constitutive relation (2.52) implies that, when an ensemble of primitive electromagnetic fields oscillating at angular frequencies $\omega_m \in \mathcal{W}$ are simultaneously launched into the nonlinear material, their combined effect is an electromagnetic field oscillating at angular frequency $\omega_{\rm NL}$ in the nonlinear material. A broad spectrum of nonlinear behaviors is encapsulated in equation (2.52), including frequency-mixing processes [32, 33] for dielectric materials and a variety of non-linear magnetic and magnetoelectric effects [34–36].

References

[1] Jackson J D 1999 *Classical Electrodynamics* 3rd edn (New York: Wiley)

[2] Buchwald J Z 1985 *From Maxwell to Microphysics: Aspects of Electromagnetic Theory in the Last Quarter of the Nineteenth Century* (Chicago, IL: University of Chicago Press)

[3] Weiglhofer W S 2003 Constitutive characterization of simple and complex mediums *Introduction to Complex Mediums for Optics and Electromagnetics* ed W S Weiglhofer and A Lakhtakia (Bellingham, WA: SPIE Press) pp 27–61

[4] Chern R-L 2013 Spatial dispersion and nonlocal effective permittivity for periodic layered metamaterials *Opt. Express* **21** 16514–27

[5] Arfken G B and Weber H J 1995 *Mathematical Methods for Physicists* 4th edn (London: Academic)

[6] Hilgevoord J 1962 *Dispersion Relations and Causal Descriptions* (Amsterdam: North-Holland)

[7] Lakhtakia A and Weiglhofer W S 2000 Electromagnetic waves in a material with simultaneous mirror-conjugated and racemic chirality characteristics *Electromagnetics* **20** 481–8 (The first negative sign on the right side of equation (26) of this paper should be replaced by a positive sign)

[8] Lindell I V and Dahl F M 2001 Conditions for the parameter dyadics of lossy bianisotropic media *Microw. Opt. Technol. Lett.* **9** 175–8

[9] Tan E L 2004 Reduced conditions for the constitutive parameters of lossy bi-anisotropic media *Microw. Opt. Technol. Lett.* **41** 133–5

[10] Whitelaw T A 1991 *Linear Algebra* 2nd edn (Glasgow: Blackie)

[11] Chen H C 1983 *Theory of Electromagnetic Waves: A Coordinate-Free Approach* (New York: McGraw-Hill)

[12] Depine R A and Lakhtakia A 2004 Comment I on "Resonant and antiresonant frequency dependence of the effective parameters of metamaterials" *Phys. Rev. E* **70** 048601

[13] Elfros A L 2004 Comment II on "Resonant and antiresonant frequency dependence of the effective parameters of metamaterials" *Phys. Rev. E* **70** 048602

[14] Mackay T G and Lakhtakia A 2015 Dynamically controllable anisotropic metamaterials with simultaneous attenuation and amplification *Phys. Rev. A* **92** 053847

[15] Mackay T G and Lakhtakia A 2019 *Electromagnetic Anisotropy and Bianisotropy: A Field Guide* 2nd edn (Singapore: World Scientific) https://doi.org/10.1142/7515

[16] Lakhtakia A 1994 *Beltrami Fields in Chiral Media* (Singapore: World Scientific)
[17] Lakhtakia A (ed) 1990 *Selected Papers on Natural Optical Activity* (Bellingham, WA: SPIE Optical Engineering Press)
[18] Lakhtakia A 2009 Remarks on the current status of the Post constraint *Optik* **120** 422–4
[19] Lakhtakia A and Mackay T G 2016 Classical electromagnetic model of surface states in topological insulators *J. Nanophoton.* **10** 033004
[20] Wakaki M 2012 *Optical Materials and Applications* (London: CRC Press)
[21] Dmitriev V 1999 Group theoretical approach to complex and bianisotropic media description *Eur. Phys. J. AP* **6** 49–55
[22] Litvin D B 2003 Point group symmetries *Introduction to Complex Mediums for Optics and Electromagnetics* ed W S Weiglhofer and A Lakhtakia (Bellingham, WA: SPIE Press) pp 79–102
[23] Lifshitz R 2005 Magnetic point groups and space groups *Encyclopedia of Condensed Matter Physics* vol 3 ed F Bassani, G L Liedl and P Wyder (Oxford: Elsevier) pp 219–26
[24] Born M and Wolf E 1999 *Principles of Optics* 7th (expanded) edn (Cambridge: Cambridge University Press)
[25] Weiglhofer W S and Lakhtakia A 1999 On electromagnetic waves in biaxial bianisotropic media *Electromagnetics* **19** 351–62
[26] Mackay T G and Weiglhofer W S 2000 Homogenization of biaxial composite materials: dissipative anisotropic properties *J. Opt. A: Pure Appl. Opt.* **2** 426–32
[27] Lax B and Button K J 1962 *Microwave Ferrites and Ferrimagnetics* (New York: McGraw-Hill)
[28] Weiglhofer W S and Lakhtakia A 1998 The correct constitutive relations of chiroplasmas and chiroferrites *Microw. Opt. Technol. Lett.* **17** 405–8
[29] Plébanski J 1960 Electromagnetic waves in gravitational fields *Phys. Rev.* **118** 1396–408
[30] Schleich W and Scully M O 1984 General relativity and modern optics *New Trends in Atomic Physics* ed G Grynberg and R Stora (Amsterdam: Elsevier) pp 995–1124
[31] Lakhtakia A and Weiglhofer W S 2000 Maxwell Garnett formalism for weakly nonlinear, bianisotropic, dilute, particulate composite media *Int. J. Electron.* **87** 1401–8
[32] Boyd R W 2008 *Nonlinear Optics* 3rd edn (London: Academic)
[33] Kobayashi T 1991 Introduction to nonlinear optical materials *Nonlin. Opt.* **1** 91–117
[34] Kielich S and Piekara A 1966 Frequency- and spatially variable electric and magnetic polarizations induced in nonlinear media by electromagnetic fields *Acta Physica Polo.* **29** 875–98
[35] Atkins P W and Barron L D 1968 Quantum field theory of optical birefringence phenomena I. Linear and nonlinear optical rotation *Proc. R. Soc. Lond. A* **304** 303–17
[36] Weiglhofer W S and Lakhtakia A 1996 Mediation of nonlinear polarization by the magnetic field in a composite medium with a chiral component *Microw. Opt. Technol. Lett.* **13** 285–7

IOP Publishing

Modern Analytical Electromagnetic Homogenization with Mathematica® (Second Edition)

Tom G Mackay and Akhlesh Lakhtakia

Chapter 3

Depolarization dyadics

The process of homogenization of a composite material often follows a two-step procedure. First, the response of a single electrically small particle, immersed in an unbounded homogeneous ambient material, to an applied electromagnetic field is calculated. Second, the responses of all particles in the composite material are combined to estimate the effective properties of the homogenized composite material (HCM). The first step is essentially the same for the Maxwell Garnett, Bruggeman, and SPFT formalisms, whereas these formalisms implement the second step in different ways. The first step is formulated in terms of depolarization dyadics and polarizability density dyadics, by which means the electrically small constituent particles are represented as co-located electric and magnetic dipoles.

This chapter begins with a description of how depolarization dyadics arise from dyadic Green functions (DGFs) for the unbounded homogeneous ambient materials in which the particles are immersed. Then a survey is presented of depolarization dyadics for isotropic, anisotropic, and bianisotropic materials. These depolarization dyadics, along with their corresponding polarizability density dyadics, are fundamental components in the homogenization formalisms presented in chapter 4. Only linear materials are considered in this chapter.

3.1 Dyadic Green functions

3.1.1 Defining properties

Let us recall equation (2.28), which compactly represents the combination of the frequency-domain Maxwell curl postulates and the frequency-domain constitutive relations of a homogeneous bianisotropic material characterized by the Tellegen 6×6 constitutive dyadic $\underline{\underline{\mathbf{K}}}$. This equation shows that the 6-vector field phasor $\underline{\mathbf{F}}(\mathbf{r})$ emerges from the 6-vector current density phasor $\underline{\mathbf{Q}}(\mathbf{r})$ through the agency of the differential operator $\underline{\underline{\mathbf{L}}}(\nabla) + i\omega\underline{\underline{\mathbf{K}}}$. Since the differential operator in equation (2.28) is linear, the solution to this equation may be formulated by means of a 6×6 DGF $\underline{\underline{G}}(\mathbf{r} - \mathbf{r}')$. That is, the field phasor at position $\mathbf{r}$ may be expressed as [1]

$$\underline{\mathbf{F}}(\mathbf{r}) = \underline{\mathbf{F}}^{\,\mathrm{cf}}(\mathbf{r}) + \int_{V'} \underline{\underline{\mathbf{G}}}(\mathbf{r} - \mathbf{r}') \bullet \underline{\mathbf{Q}}(\mathbf{r}')\, d^3\mathbf{r}'. \tag{3.1}$$

Here the integration region V' includes all locations where $\underline{\mathbf{Q}}(\mathbf{r})$ is not null-valued. The 6-vector $\underline{\mathbf{F}}^{\,\mathrm{cf}}(\mathbf{r})$ is the corresponding complementary function which satisfies everywhere the corresponding homogeneous differential equation

$$\left[\underline{\underline{\mathbf{L}}}(\nabla) + i\omega\underline{\underline{\mathbf{K}}}\right] \bullet \underline{\mathbf{F}}^{\,\mathrm{cf}}(\mathbf{r}) = \underline{\mathbf{0}}, \tag{3.2}$$

whereas the DGF satisfies everywhere the differential equation

$$\left[\underline{\underline{\mathbf{L}}}(\nabla) + i\omega\underline{\underline{\mathbf{K}}}\right] \bullet \underline{\underline{\mathbf{G}}}(\mathbf{r} - \mathbf{r}') = \delta(\mathbf{r} - \mathbf{r}')\, \underline{\underline{\mathbf{I}}}, \tag{3.3}$$

with

$$\delta(\mathbf{r}) = \left(\frac{1}{2\pi}\right)^3 \int_{\mathbf{q}} \exp\left(i\mathbf{q} \bullet \mathbf{r}\right) d^3\mathbf{q} \tag{3.4}$$

being the Dirac delta, expressed here in spectral form in anticipation of section 3.1.2. As well as the differential equation (3.3), the DGF must satisfy the Sommerfeld radiation condition as $|\,\mathbf{r} - \mathbf{r}'| \to \infty$ [1]. Observe that the DGF is not a function of $\mathbf{r}$ and $\mathbf{r}'$ independently, but is a function of the difference $\mathbf{r} - \mathbf{r}'$ instead; i.e., the DGF is translationally invariant [2].

DGFs are useful mathematical constructions for solving a multitude of electromagnetic problems involving sources [3, 4]. These functions play a key role in scattering theory, courtesy of the Huygens principle [5, 6]. In an informal sense, the DGF yields the 'response' of the material at $\mathbf{r}$ to a point 'source' at $\mathbf{r}'$.

3.1.2 Spectral representation

Closed-form representations of DGFs are available for isotropic materials and for some relatively simple anisotropic and bianisotropic materials, but not for general anisotropic and bianisotropic materials. As an alternative to closed-form representations, formal expansions in terms of eigenfunction solutions to equation (3.3) can be developed. For example, the *spectral representation* [3, 7]

$$\underline{\underline{\mathbf{G}}}(\mathbf{R}) = \left(\frac{1}{2\pi}\right)^3 \int_{\mathbf{q}} \underline{\underline{\breve{\mathbf{G}}}}(\mathbf{q}) \exp(i\,\mathbf{q} \bullet \mathbf{R})\, d^3\mathbf{q} \tag{3.5}$$

may be regarded as an expansion in terms of plane waves. Parenthetically, expansions in terms of cylindrical vector wavefunctions [8, 9] and spherical harmonics [10] have also been developed, but these formulations are cumbersome, restrictive in scope, and usually necessitate numerical implementation.

The spatial Fourier transform[1]

$$\check{\underline{\underline{G}}}(\mathbf{q}) = \int_{\mathbf{R}} \underline{\underline{G}}(\mathbf{R})\exp(-i\,\mathbf{q} \bullet \mathbf{R})\, d^3\mathbf{R} \tag{3.6}$$

of the DGF [7] follows simply from inverting equation (3.5). Furthermore, the spatial Fourier transform of equation (3.3) yields [3]

$$\check{\underline{\underline{G}}}(\mathbf{q}) = \frac{1}{i\omega}\frac{\operatorname{adj}\check{\underline{\underline{A}}}(\mathbf{q})}{\det \check{\underline{\underline{A}}}(\mathbf{q})}, \tag{3.7}$$

with the 6×6 dyadic

$$\check{\underline{\underline{A}}}(\mathbf{q}) = \begin{bmatrix} \underline{\underline{0}} & (\mathbf{q}/\omega) \times \underline{\underline{I}} \\ -(\mathbf{q}/\omega) \times \underline{\underline{I}} & \underline{\underline{0}} \end{bmatrix} + \underline{\underline{\mathbf{K}}}. \tag{3.8}$$

For the homogenization formalisms described in chapter 4, it transpires that only the singular part of the DGF (and the immediate vicinity of the singularity in the case of extended homogenization formalisms) is needed. Additionally, in order to extract the DGF's singularity (and its immediate vicinity), it is sufficient to work with the spatial Fourier transform of the DGF rather than the DGF itself. As may be appreciated from equation (3.7), the spatial Fourier transform of the DGF can be found for all homogeneous materials characterized by the constitutive dyadic $\underline{\underline{\mathbf{K}}}$, even those which are bianisotropic.

3.2 Depolarization dyadics

Consider a single particle made of a material characterized by the constitutive dyadic $\underline{\underline{\mathbf{K}}}_{\mathrm{p}}$. The particle occupies a region V'. It is immersed in an ambient material characterized by the constitutive dyadic $\underline{\underline{\mathbf{K}}}_{\mathrm{amb}}$. The particle and the ambient material are both homogeneous and, in general, both can be bianisotropic. The Maxwell curl postulates (2.28) everywhere can be written as [11]

$$\left[\, \underline{\underline{\mathbf{L}}}(\nabla) + i\omega\underline{\underline{\mathbf{K}}}_{\mathrm{amb}}\right] \bullet \underline{\mathbf{F}}(\mathbf{r}) = \underline{\mathbf{Q}}_{\mathrm{equiv}}(\mathbf{r}), \tag{3.9}$$

where the equivalent source current density

$$\underline{\mathbf{Q}}_{\mathrm{equiv}}(\mathbf{r}) = \begin{cases} i\omega\left(\underline{\underline{\mathbf{K}}}_{\mathrm{amb}} - \underline{\underline{\mathbf{K}}}_{\mathrm{p}}\right) \bullet \underline{\mathbf{F}}(\mathbf{r}), & \mathbf{r} \in V' \\ \underline{\mathbf{0}}, & \mathbf{r} \notin V' \end{cases} \tag{3.10}$$

has been introduced to represent the particle. The solution of equation (3.9) may be given in terms of the DGF for the ambient material, $\underline{\underline{\mathbf{G}}}_{\mathrm{amb}}(\mathbf{r} - \mathbf{r}')$, as

[1] Note the convention: for the spatial Fourier transform (3.6) a $-i$ factor is used in the exponential term but for the temporal Fourier transform (2.14) a $+i$ factor is used in the exponential term.

$$\underline{\mathbf{F}}(\mathbf{r}) = \underline{\mathbf{F}}^{\,\mathrm{cf}}(\mathbf{r}) + \int_{V'} \underline{\underline{\mathbf{G}}}_{\mathrm{amb}}(\mathbf{r} - \mathbf{r}') \bullet \underline{\mathbf{Q}}_{\mathrm{equiv}}(\mathbf{r}')\, d^3\mathbf{r}', \tag{3.11}$$

per equation (3.1).

Now suppose that the particle is sufficiently small that the equivalent source current density $\underline{\mathbf{Q}}_{\mathrm{equiv}}(\mathbf{r})$ is approximately uniform throughout the particle—this is the Rayleigh approximation. In the equivalent source region V', the solution (3.11) simplifies to

$$\underline{\mathbf{F}}(\mathbf{r}) \approx \underline{\mathbf{F}}^{\,\mathrm{cf}}(\mathbf{r}) + \left[\int_{V'} \underline{\underline{\mathbf{G}}}_{\mathrm{amb}}(\mathbf{r} - \mathbf{r}')\, d^3\mathbf{r}'\right] \bullet \underline{\mathbf{Q}}_{\mathrm{equiv}}(\mathbf{r}), \qquad (\mathbf{r} \in V'). \tag{3.12}$$

It is convenient to locate the centroid of the particle at the origin of the coordinate system. At the origin, the field phasor may be expressed as

$$\underline{\mathbf{F}}(\mathbf{0}) \approx \underline{\mathbf{F}}^{\,\mathrm{cf}}(\mathbf{0}) + \underline{\underline{\mathbf{D}}}_{\mathrm{amb}} \bullet \underline{\mathbf{Q}}_{\mathrm{equiv}}(\mathbf{0}), \tag{3.13}$$

where the *depolarization dyadic*

$$\underline{\underline{\mathbf{D}}}_{\mathrm{amb}} = \int_{V'} \underline{\underline{\mathbf{G}}}_{\mathrm{amb}}(-\mathbf{r}')\, d^3\mathbf{r}' \tag{3.14}$$

has been introduced.

Notice that the integration region V' for the depolarization dyadic has the same shape, orientation, and size as the particle immersed in the ambient material. In addition, $\underline{\underline{\mathbf{D}}}_{\mathrm{amb}}$ depends upon $\underline{\underline{\mathbf{K}}}_{\mathrm{amb}}$ but not upon $\underline{\underline{\mathbf{K}}}_{\mathrm{p}}$. Thus, the depolarization dyadic belongs to a region but not to a particle. Since the DGF for the ambient material is singular in the limit $|\,\mathbf{r} - \mathbf{r}'| \to 0$, caution must be exercised in evaluating depolarization dyadics [12].

3.2.1 Ellipsoidal region

Suppose now that the particle occupying the region V' has an ellipsoidal shape. Its surface is prescribed by the vector

$$\mathbf{r}_e(\theta, \phi) = \rho\, \underline{\underline{U}} \bullet \hat{\mathbf{r}}(\theta, \phi), \tag{3.15}$$

where $\hat{\mathbf{r}}(\theta, \phi)$ is the radial unit vector which prescribes the surface of the unit sphere, with θ and ϕ being the spherical polar coordinates. The shape and orientation of the ellipsoidal particle is encapsulated by the real symmetric dyadic $\underline{\underline{U}}$, this dyadic having positive eigenvalues and unit determinant [13]. The linear dimensions of the ellipsoidal particle are represented by the size parameter $\rho > 0$. Thus, the lengths of the ellipsoidal particle's semi-axes are provided by the eigenvalues of $\rho\underline{\underline{U}}$.

The depolarization dyadic's integral representation in equation (3.14) cannot be evaluated as it stands, because in general $\underline{\underline{\mathbf{G}}}_{\mathrm{amb}}(-\mathbf{r}')$ is unknown. To progress, the spectral representation (3.5) is introduced, which enables equation (3.14) to be written as [14, 15]

$$\underline{\underline{\mathbf{D}}}_{\,U/\mathrm{amb}}(\rho) = \frac{\rho}{2\pi^2} \int_{\mathbf{q}} \frac{1}{q^2}\left[\frac{\sin(q\rho)}{q\rho} - \cos(q\rho)\right] \underline{\underline{\check{\mathbf{G}}}}_{\mathrm{amb}}(\underline{\underline{U}}^{-1} \bullet \mathbf{q})\, d^3\mathbf{q}, \tag{3.16}$$

with $q = |\mathbf{q}|$ and $\check{\underline{\underline{\mathbf{G}}}}_{\text{amb}}$ representing the spatial Fourier transform of $\underline{\underline{\mathbf{G}}}_{\text{amb}}$ per equation (3.6). On the left side of equation (3.16), the notation for the depolarization dyadic has been augmented to emphasize the dependency upon $\underline{\underline{U}}$ and ρ.

It is instructive to represent the depolarization dyadic as the sum [16, 17]

$$\underline{\underline{\mathbf{D}}}_{U/\text{amb}}(\rho) = \underline{\underline{\mathbf{D}}}^{0}_{U/\text{amb}} + \underline{\underline{\mathbf{D}}}^{+}_{U/\text{amb}}(\rho), \tag{3.17}$$

where the 6×6 dyadic terms

$$\left.\begin{aligned} \underline{\underline{\mathbf{D}}}^{0}_{U/\text{amb}} &= \frac{\rho}{2\pi^2}\int_{\mathbf{q}} \frac{1}{q^2}\left[\frac{\sin(q\rho)}{q\rho} - \cos(q\rho)\right] \check{\underline{\underline{\mathbf{G}}}}^{\infty}_{\text{amb}}(\underline{\underline{U}}^{-1}\bullet\hat{\mathbf{q}})\, d^3\mathbf{q} \\ \underline{\underline{\mathbf{D}}}^{+}_{U/\text{amb}}(\rho) &= \frac{\rho}{2\pi^2}\int_{\mathbf{q}} \frac{1}{q^2}\left[\frac{\sin(q\rho)}{q\rho} - \cos(q\rho)\right] \check{\underline{\underline{\mathbf{G}}}}^{+}_{\text{amb}}(\underline{\underline{U}}^{-1}\bullet\mathbf{q})\, d^3\mathbf{q} \end{aligned}\right\}, \tag{3.18}$$

with

$$\left.\begin{aligned} \check{\underline{\underline{\mathbf{G}}}}^{\infty}_{\text{amb}}(\underline{\underline{U}}^{-1}\bullet\hat{\mathbf{q}}) &= \lim_{q\to\infty} \check{\underline{\underline{\mathbf{G}}}}_{\text{amb}}(\underline{\underline{U}}^{-1}\bullet\mathbf{q}) \\ &= \frac{1}{i\omega\, b_{U/\text{amb}}(\theta, \phi)} \\ &\quad \times \begin{bmatrix} \beta_{\mu}(\theta, \phi)\, \underline{\underline{U}}^{-1}\bullet\hat{\mathbf{q}}\hat{\mathbf{q}}\bullet\underline{\underline{U}}^{-1} & -\beta_{\xi}(\theta, \phi)\, \underline{\underline{U}}^{-1}\bullet\hat{\mathbf{q}}\hat{\mathbf{q}}\bullet\underline{\underline{U}}^{-1} \\ -\beta_{\zeta}(\theta, \phi)\, \underline{\underline{U}}^{-1}\bullet\hat{\mathbf{q}}\hat{\mathbf{q}}\bullet\underline{\underline{U}}^{-1} & \beta_{\varepsilon}(\theta, \phi)\, \underline{\underline{U}}^{-1}\bullet\hat{\mathbf{q}}\hat{\mathbf{q}}\bullet\underline{\underline{U}}^{-1} \end{bmatrix} \\ \check{\underline{\underline{\mathbf{G}}}}^{+}_{\text{amb}}(\underline{\underline{U}}^{-1}\bullet\mathbf{q}) &= \check{\underline{\underline{\mathbf{G}}}}_{\text{amb}}(\underline{\underline{U}}^{-1}\bullet\mathbf{q}) - \check{\underline{\underline{\mathbf{G}}}}^{\infty}_{\text{amb}}(\underline{\underline{U}}^{-1}\bullet\hat{\mathbf{q}}) \end{aligned}\right\} \tag{3.19}$$

.

Herein the scalar functions

$$\left.\begin{aligned} \beta_{\sigma}(\theta, \phi) &= \hat{\mathbf{q}}\bullet\underline{\underline{U}}^{-1}\bullet\underline{\underline{\sigma}}_{\text{amb}}\bullet\underline{\underline{U}}^{-1}\bullet\hat{\mathbf{q}}, \qquad (\sigma \in \{\varepsilon, \xi, \zeta, \mu\}) \\ b_{U/\text{amb}}(\theta, \phi) &= \beta_{\varepsilon}(\theta, \phi)\,\beta_{\mu}(\theta, \phi) - \beta_{\xi}(\theta, \phi)\,\beta_{\zeta}(\theta, \phi) \end{aligned}\right\}, \tag{3.20}$$

where the Tellegen constitutive dyadics $\underline{\underline{\varepsilon}}_{\text{amb}}$, $\underline{\underline{\xi}}_{\text{amb}}$, $\underline{\underline{\zeta}}_{\text{amb}}$, and $\underline{\underline{\mu}}_{\text{amb}}$ are the 3×3 components of $\underline{\underline{\mathbf{K}}}_{\text{amb}}$, per equation (2.26), and the unit vector

$$\hat{\mathbf{q}} = \frac{1}{q}\mathbf{q} = \sin\theta\cos\phi\,\hat{\mathbf{u}}_x + \sin\theta\sin\phi\,\hat{\mathbf{u}}_y + \cos\theta\,\hat{\mathbf{u}}_z. \tag{3.21}$$

For convenience later on, let us record that $\underline{\underline{\mathbf{D}}}^{0}_{U/\text{amb}}$ and $\underline{\underline{\mathbf{D}}}^{+}_{U/\text{amb}}(\rho)$ may written in terms of their 3×3 dyadic components as

$$\left.\begin{aligned} \underline{\underline{\mathbf{D}}}^{0}_{U/\text{amb}} &= \begin{bmatrix} {}^{\text{ee}}\underline{\underline{D}}^{0}_{U/\text{amb}} & {}^{\text{em}}\underline{\underline{D}}^{0}_{U/\text{amb}} \\ {}^{\text{me}}\underline{\underline{D}}^{0}_{U/\text{amb}} & {}^{\text{mm}}\underline{\underline{D}}^{0}_{U/\text{amb}} \end{bmatrix} \\ \underline{\underline{\mathbf{D}}}^{+}_{U/\text{amb}}(\rho) &= \begin{bmatrix} {}^{\text{ee}}\underline{\underline{D}}^{+}_{U/\text{amb}}(\rho) & {}^{\text{em}}\underline{\underline{D}}^{+}_{U/\text{amb}}(\rho) \\ {}^{\text{me}}\underline{\underline{D}}^{+}_{U/\text{amb}}(\rho) & {}^{\text{mm}}\underline{\underline{D}}^{+}_{U/\text{amb}}(\rho) \end{bmatrix} \end{aligned}\right\}. \tag{3.22}$$

The partition portrayed by equation (3.17) separates singular and nonsingular contributions to the depolarization dyadic. The dyadic $\underline{\underline{\mathbf{D}}}^{0}_{U/\text{amb}}$ is the depolarization contribution that arises from the infinitesimally small ellipsoidal region in the limit $\rho \to 0$. It is essentially given by the integrated singularity of the DGF for the ambient material. On the other hand, the dyadic $\underline{\underline{\mathbf{D}}}^{+}_{U/\text{amb}}(\rho)$ is the depolarization contribution which stems from the ellipsoidal region of nonzero volume, not including the singularity at the origin. The influence of the dyadic $\underline{\underline{\mathbf{D}}}^{0}_{U/\text{amb}}$ generally dominates over that of the dyadic $\underline{\underline{\mathbf{D}}}^{+}_{U/\text{amb}}(\rho)$ in homogenization formalisms. Indeed, often the contribution of $\underline{\underline{\mathbf{D}}}^{+}_{U/\text{amb}}(\rho)$ is neglected altogether and $\underline{\underline{\mathbf{D}}}^{0}_{U/\text{amb}}$ alone is taken as the depolarization dyadic [18]. However, $\underline{\underline{\mathbf{D}}}^{+}_{U/\text{amb}}(\rho)$ can add significantly to estimates of the constitutive parameters of the HCM, especially when the constituent materials are nondissipative or weakly dissipative. The nonzero spatial extent of depolarization regions has been investigated for isotropic [19–23], anisotropic [16], and bianisotropic [17] HCMs. Homogenization formalisms that take into account the nonzero spatial extent of depolarization regions are known as *extended* formalisms.

Singular contribution

The integral representation of $\underline{\underline{\mathbf{D}}}^{0}_{U/\text{amb}}$ in equation $(3.18)_1$ can be simplified as follows [18]. First, carrying out the integration with respect to the radial coordinate yields the ρ-independent double integral [14, 15]

$$\underline{\underline{\mathbf{D}}}^{0}_{U/\text{amb}} = \frac{1}{4\pi} \int_{\phi=0}^{2\pi} \int_{\theta=0}^{\pi} \underline{\underline{\breve{\mathbf{G}}}}^{\infty}(\underline{\underline{U}}^{-1} \bullet \hat{\mathbf{q}}) \sin\theta \, d\theta \, d\phi. \tag{3.23}$$

Second, the integrands in equation (3.23) are expanded using the spectral form $(3.19)_1$ to deliver

$${}^{\alpha\beta}\underline{\underline{D}}^{\,0}_{U/\text{amb}} = \frac{1}{4\pi i\omega} \underline{\underline{U}}^{-1} \bullet {}^{\alpha\beta}\underline{\underline{d}}^{\,0}_{U/\text{amb}} \bullet \underline{\underline{U}}^{-1}, \qquad (\alpha\beta \in \{\text{ee, em, me, mm}\}), \tag{3.24}$$

where the 3×3 dyadics

$$\left.\begin{aligned}
{}^{\text{ee}}\underline{\underline{d}}^{\,0}_{U/\text{amb}} &= \int_{\phi=0}^{2\pi} \int_{\theta=0}^{\pi} \frac{\beta_\mu(\theta, \phi)}{b_{U/\text{amb}}(\theta, \phi)} \hat{\mathbf{q}}\hat{\mathbf{q}} \sin\theta \, d\theta \, d\phi \\
{}^{\text{em}}\underline{\underline{d}}^{\,0}_{U/\text{amb}} &= -\int_{\phi=0}^{2\pi} \int_{\theta=0}^{\pi} \frac{\beta_\xi(\theta, \phi)}{b_{U/\text{amb}}(\theta, \phi)} \hat{\mathbf{q}}\hat{\mathbf{q}} \sin\theta \, d\theta \, d\phi \\
{}^{\text{me}}\underline{\underline{d}}^{\,0}_{U/\text{amb}} &= -\int_{\phi=0}^{2\pi} \int_{\theta=0}^{\pi} \frac{\beta_\zeta(\theta, \phi)}{b_{U/\text{amb}}(\theta, \phi)} \hat{\mathbf{q}}\hat{\mathbf{q}} \sin\theta \, d\theta \, d\phi \\
{}^{\text{mm}}\underline{\underline{d}}^{\,0}_{U/\text{amb}} &= \int_{\phi=0}^{2\pi} \int_{\theta=0}^{\pi} \frac{\beta_\varepsilon(\theta, \phi)}{b_{U/\text{amb}}(\theta, \phi)} \hat{\mathbf{q}}\hat{\mathbf{q}} \sin\theta \, d\theta \, d\phi
\end{aligned}\right\}. \tag{3.25}$$

Everywhere that a 3×3 constitutive dyadic term appears in equations (3.25), it is pre- and post-multiplied by $\underline{\underline{U}}^{-1}$. Therefore, the evaluation of $\underline{\underline{\mathbf{D}}}^{0}_{U/\text{amb}}$ relevant to an ellipsoidal particle immersed in an isotropic ambient material is isomorphic to the evaluation for a spherical particle immersed in a certain anisotropic or bianisotropic ambient material. Accordingly, explicit evaluations of $\underline{\underline{\mathbf{D}}}^{0}_{U/\mathrm{amb}}$ are postponed until section 3.2.2 wherein the case of spherical particles is taken up. Furthermore, particles of certain cylindrical shapes, namely needle-shaped particles and disk-shaped particles, may be accommodated as limiting cases of the ellipsoidal particle, as discussed in section 3.2.3.

Nonsingular contribution
Let us turn to the depolarization contribution stemming from the nonzero spatial extent of the $\underline{\underline{U}}$-shaped particle, i.e., the depolarization contribution represented by the 6 × 6 dyadic $\underline{\underline{\mathbf{D}}}^{+}_{U/\mathrm{amb}}(\rho)$. In order to make analytical progress with the integration of the right side of equation $(3.18)_2$, attention is restricted to Lorentz-reciprocal ambient materials [24]—i.e., the symmetries [25]

$$\left.\begin{aligned} \underline{\underline{\varepsilon}}_{\mathrm{amb}} &= \underline{\underline{\varepsilon}}^{\mathrm{T}}_{\mathrm{amb}} \\ \underline{\underline{\xi}}_{\mathrm{amb}} &= -\underline{\underline{\zeta}}^{\mathrm{T}}_{\mathrm{amb}} \\ \underline{\underline{\mu}}_{\mathrm{amb}} &= \underline{\underline{\mu}}^{\mathrm{T}}_{\mathrm{amb}} \end{aligned}\right\} \tag{3.26}$$

are presumed to hold. A wide range of commonly encountered anisotropic and bianisotropic materials, including uniaxial and biaxial materials, are Lorentz-reciprocal, but there are certain important classes of materials that are not, such as gyrotropic materials [5]. In view of equations (3.26), the triple integration in equation $(3.18)_2$ reduces to a double integration [26]; hence,

$$\begin{aligned} \underline{\underline{\mathbf{D}}}^{+}_{U/\mathrm{amb}}(\rho) = \frac{\omega^4}{4\pi} \int_{\phi=0}^{2\pi} \int_{\theta=0}^{\pi} \frac{1}{b_{U/\mathrm{amb}}(\theta,\phi)} \Bigg[& \frac{1}{\kappa_+ - \kappa_-} \Bigg(\frac{\exp(i\rho q)}{2q^2} (1 - i\rho q) \\ & \times \Big\{ \det\Big[\underline{\underline{\breve{\mathbf{A}}}}_{\mathrm{amb}}(\underline{\underline{U}}^{-1} \bullet \mathbf{q})\Big] \underline{\underline{\breve{\mathbf{G}}}}^{+}_{\mathrm{amb}}(\underline{\underline{U}}^{-1} \bullet \mathbf{q}) \\ & + \det\Big[\underline{\underline{\breve{\mathbf{A}}}}_{\mathrm{amb}}(-\underline{\underline{U}}^{-1} \bullet \mathbf{q})\Big] \underline{\underline{\breve{\mathbf{G}}}}^{+}_{\mathrm{amb}}(-\underline{\underline{U}}^{-1} \bullet \mathbf{q}) \Big\} \Bigg)_{q=\sqrt{\kappa_-}}^{q=\sqrt{\kappa_+}} \\ & + \frac{\det\Big[\underline{\underline{\breve{\mathbf{A}}}}_{\mathrm{amb}}(\mathbf{0})\Big]}{\kappa_+ \kappa_-} \underline{\underline{\breve{\mathbf{G}}}}^{+}_{\mathrm{amb}}(\mathbf{0}) \Bigg] \sin\theta \, d\theta \, d\phi, \end{aligned} \tag{3.27}$$

with

$$\underline{\underline{\breve{\mathbf{A}}}}_{\mathrm{amb}}(\mathbf{q}) = \frac{1}{i\omega} \Big[\underline{\underline{\breve{\mathbf{G}}}}_{\mathrm{amb}}(\mathbf{q})\Big]^{-1} \tag{3.28}$$

and $\kappa_{\pm}$ representing the roots of the quadratic (in q^2) equation $\det\Big[\underline{\underline{\breve{\mathbf{A}}}}_{\mathrm{amb}}(\underline{\underline{U}}^{-1} \bullet \mathbf{q})\Big] = 0$, i.e.,

$$\det\left[\underline{\underline{\breve{\mathbf{A}}}}_{\text{amb}}(\underline{\underline{U}}^{-1} \bullet \mathbf{q})\right] \propto (q^2 - \kappa_+)(q^2 - \kappa_-) = 0. \tag{3.29}$$

The notation $(f(\nu))_a^b$ stands for the difference $f(b) - f(a)$.

Explicit evaluations of $\underline{\underline{\mathbf{D}}}^+_{U/\text{amb}}(\rho)$ may be derived for certain isotropic and uniaxial ambient materials, as presented in section 3.2.2. Such derivations follow from approximating $\exp(i\rho q)(1 - i\rho q)$ in the integrand on the right side of equation (3.27) by $1 + (\rho q)^2/2 + i(\rho q)^3/3$. This step is justified in the long-wavelength regime because therein $\rho\sqrt{\kappa_\pm} \ll 1$ [26].

3.2.2 Spherical region

In general, numerical techniques are needed to evaluate the double integrals on the right sides of equations (3.25) and (3.27). But for certain highly symmetric ambient materials, analytic progress can be made. Since spherical particles are most commonly encountered in homogenization formalisms, and the formalisms themselves are less cumbersome for spherical particles, in this section we focus on the case $\underline{\underline{U}} = \underline{\underline{I}}$. As noted in section 3.2.1, depolarization dyadics relevant to ellipsoidal particles immersed in isotropic ambient materials are isomorphic to those relevant to spherical particles immersed in certain anisotropic or bianisotropic ambient materials.

Isotropic ambient materials

Suppose that the ambient material is isotropic dielectric–magnetic, i.e., described by the constitutive relations (2.34). Then the particle-size-independent contributions to the depolarization dyadic reduce to the familiar form [12]

$$\left.\begin{aligned} {}^{\text{ee}}\underline{\underline{D}}^{\,0}_{I/\text{amb}} &= \frac{1}{3i\omega\varepsilon}\,\underline{\underline{I}} \\ {}^{\text{em}}\underline{\underline{D}}^{\,0}_{I/\text{amb}} &= \underline{\underline{0}} \\ {}^{\text{me}}\underline{\underline{D}}^{\,0}_{I/\text{amb}} &= \underline{\underline{0}} \\ {}^{\text{mm}}\underline{\underline{D}}^{\,0}_{I/\text{amb}} &= \frac{1}{3i\omega\mu}\,\underline{\underline{I}} \end{aligned}\right\}, \tag{3.30}$$

while the particle-size-dependent contributions are given as [26]

$$\left.\begin{aligned} {}^{\text{ee}}\underline{\underline{D}}^{\,+}_{I/\text{amb}}(\rho) &= \frac{i\omega\rho^2\mu}{9}\left(3 + i2\rho\omega\sqrt{\varepsilon\mu}\right)\underline{\underline{I}} \\ {}^{\text{em}}\underline{\underline{D}}^{\,+}_{I/\text{amb}}(\rho) &= \underline{\underline{0}} \\ {}^{\text{me}}\underline{\underline{D}}^{\,+}_{I/\text{amb}}(\rho) &= \underline{\underline{0}} \\ {}^{\text{mm}}\underline{\underline{D}}^{\,+}_{I/\text{amb}}(\rho) &= \frac{i\omega\rho^2\varepsilon}{9}\left(3 + i2\rho\omega\sqrt{\varepsilon\mu}\right)\underline{\underline{I}} \end{aligned}\right\}. \tag{3.31}$$

Clearly from equations (3.30), the ρ-independent depolarization dyadic $\underline{\underline{\mathbf{D}}}^0_{I/\text{amb}}$ is undefined for the theoretical concept termed *nihility* [27], wherein both ε and μ are null valued.

Now let us generalize to the case where the ambient material is isotropic chiral, i.e., described by the constitutive relations (2.35) with $\xi = -\zeta$. The 3×3 dyadic contributions to the 6×6 depolarization dyadic are all scalar multiples of the 3×3 identity dyadic. The particle size-independent contributions are [26]

$$\left.\begin{aligned} {}^{\rm ee}\underline{\underline{D}}^{\,0}_{\,I/{\rm amb}} &= \frac{\mu}{3i\omega\,\Upsilon_+}\,\underline{\underline{I}} \\ {}^{\rm em}\underline{\underline{D}}^{\,0}_{\,I/{\rm amb}} &= -\frac{\xi}{3i\omega\,\Upsilon_+}\,\underline{\underline{I}} \\ {}^{\rm me}\underline{\underline{D}}^{\,0}_{\,I/{\rm amb}} &= \frac{\xi}{3i\omega\,\Upsilon_+}\,\underline{\underline{I}} \\ {}^{\rm mm}\underline{\underline{D}}^{\,0}_{\,I/{\rm amb}} &= \frac{\varepsilon}{3i\omega\,\Upsilon_+}\,\underline{\underline{I}} \end{aligned}\right\}, \tag{3.32}$$

while the particle-size-dependent contributions are

$$\left.\begin{aligned} {}^{\rm ee}\underline{\underline{D}}^{\,+}_{\,I/{\rm amb}}(\rho) &= \frac{i\omega\rho^2}{3}\left(\mu + \frac{2i\rho\,\omega\,\Upsilon_-}{3}\sqrt{\frac{\mu}{\varepsilon}}\right)\underline{\underline{I}} \\ {}^{\rm em}\underline{\underline{D}}^{\,+}_{\,I/{\rm amb}}(\rho) &= \frac{i\omega\xi\rho^2}{3}\left(1 + \frac{4i\rho\,\omega}{3}\sqrt{\varepsilon\,\mu}\right)\underline{\underline{I}} \\ {}^{\rm me}\underline{\underline{D}}^{\,+}_{\,I/{\rm amb}}(\rho) &= -\frac{i\omega\xi\rho^2}{3}\left(1 + \frac{4i\rho\,\omega}{3}\sqrt{\varepsilon\,\mu}\right)\underline{\underline{I}} \\ {}^{\rm mm}\underline{\underline{D}}^{\,+}_{\,I/{\rm amb}}(\rho) &= \frac{i\omega\rho^2}{3}\left(\varepsilon + \frac{2i\rho\,\omega\,\Upsilon_-}{3}\sqrt{\frac{\varepsilon}{\mu}}\right)\underline{\underline{I}} \end{aligned}\right\}, \tag{3.33}$$

with

$$\Upsilon_\pm = \varepsilon\,\mu \pm \xi^2. \tag{3.34}$$

Anisotropic ambient materials

We begin our consideration of anisotropic ambient materials with the uniaxial dielectric–magnetic case. That is, the permittivity and permeability dyadics $\underline{\underline{\varepsilon}}_{\rm amb}$ and $\underline{\underline{\mu}}_{\rm amb}$ both have the form specified in equation (2.38) whereas the magnetoelectric dyadics $\underline{\underline{\xi}}_{\rm amb}$ and $\underline{\underline{\zeta}}_{\rm amb}$ are both identical to the null dyadic. The symmetries of the depolarization dyadics mirror those of the constitutive dyadics. The particle-size-independent contributions to the depolarization dyadic may be expressed as [14]

$$\left.\begin{aligned} {}^{\rm ee}\underline{\underline{D}}^{\,0}_{\,I/{\rm amb}} &= D^0_\varepsilon(\,\underline{\underline{I}} - \hat{\mathbf{u}}\,\hat{\mathbf{u}}) + D^{0,u}_\varepsilon\,\hat{\mathbf{u}}\,\hat{\mathbf{u}} \\ {}^{\rm em}\underline{\underline{D}}^{\,0}_{\,I/{\rm amb}} &= \underline{\underline{0}} \\ {}^{\rm me}\underline{\underline{D}}^{\,0}_{\,I/{\rm amb}} &= \underline{\underline{0}} \\ {}^{\rm mm}\underline{\underline{D}}^{\,0}_{\,I/{\rm amb}} &= D^0_\mu(\,\underline{\underline{I}} - \hat{\mathbf{u}}\,\hat{\mathbf{u}}) + D^{0,u}_\mu\,\hat{\mathbf{u}}\,\hat{\mathbf{u}} \end{aligned}\right\}. \tag{3.35}$$

The dimensionless scalar

$$\gamma_\sigma = \frac{\sigma_1 + \sigma_2}{\sigma_1}, \qquad (\sigma \in \{\varepsilon, \mu\}) \tag{3.36}$$

determines the explicit forms that D_σ^0 and $D_\sigma^{0,u}$, $(\sigma \in \{\varepsilon, \mu\})$, in equations (3.35) take. For a nondissipative ambient material [14, 28],

$$\left.\begin{aligned} D_\sigma^0 &= \frac{1}{2i\omega\sigma_2}\left(\frac{\gamma_\sigma \sinh^{-1}\sqrt{\frac{1-\gamma_\sigma}{\gamma_\sigma}}}{\sqrt{1-\gamma_\sigma}} - 1\right) \\ D_\sigma^{0,u} &= \frac{1}{i\omega\sigma_2}\left(1 - \frac{\sinh^{-1}\sqrt{\frac{1-\gamma_\sigma}{\gamma_\sigma}}}{\sqrt{1-\gamma_\sigma}}\right) \end{aligned}\right\}, \qquad (\sigma \in \{\varepsilon, \mu\}), \tag{3.37}$$

if $0 < \gamma_\sigma < 1$, but

$$\left.\begin{aligned} D_\sigma^0 &= \frac{1}{2i\omega\sigma_2}\left(\frac{\gamma_\sigma \sec^{-1}\sqrt{\gamma_\sigma}}{\sqrt{\gamma_\sigma - 1}} - 1\right) \\ D_\sigma^{0,u} &= \frac{1}{i\omega\sigma_2}\left(1 - \frac{\sec^{-1}\sqrt{\gamma_\sigma}}{\sqrt{\gamma_\sigma}}\right) \end{aligned}\right\}, \qquad (\sigma \in \{\varepsilon, \mu\}), \tag{3.38}$$

if $\gamma_\sigma > 1$. The condition $\gamma_\sigma < 0$ holds for a nondissipative uniaxial ambient material with indefinite constitutive dyadics [29–31], and the components of $\underline{\underline{\mathbf{D}}}^0$ are undefined for this anomalous case. In the general case involving a dissipative uniaxial ambient material, the parameter γ_σ is complex-valued and either equations (3.37) or equations (3.38) may be employed.

Depolarization dyadics of the form (3.35) with components equivalent to those in equations (3.37) and (3.38) are also appropriate for a spheroidal particle, immersed in an isotropic dielectric–magnetic ambient material, with its rotational axis aligned with $\hat{\mathbf{u}}$ [28]. In addition, the depolarization dyadic components for uniaxial dielectric ambient materials and uniaxial magnetic ambient materials arise from equations (3.35)–(3.38), directly upon making the substitutions $\underline{\underline{\mu}}_{\text{amb}} = \mu_0 \underline{\underline{I}}$ and $\underline{\underline{\varepsilon}}_{\text{amb}} = \varepsilon_0 \underline{\underline{I}}$, respectively.

In order to consider the particle-size-dependent contributions to the depolarization dyadic we turn to uniaxial dielectric ambient materials, i.e., those characterized by 3×3 constitutive dyadics of the form given in equations (2.39). The explicit form [26]

$${}^{\text{ee}}\underline{\underline{D}}^{+}_{I/\text{amb}}(\rho) = D_\varepsilon^{+}(\rho)(\underline{\underline{I}} - \hat{\mathbf{u}}\,\hat{\mathbf{u}}) + D_\varepsilon^{+,u}(\rho)\,\hat{\mathbf{u}}\,\hat{\mathbf{u}} \tag{3.39}$$

is determined by the dimensionless scalar

$$\tau_\varepsilon = -\frac{\varepsilon_2}{\varepsilon_1}. \tag{3.40}$$

For a nondissipative ambient material,

$$\left.\begin{aligned} D_\varepsilon^+(\rho) &= \frac{i\omega\mu_0\rho^2}{8}\left[\frac{1-\tau_\varepsilon}{\tau_\varepsilon} + i\rho\frac{4\omega(4\varepsilon_1+\varepsilon_2)}{9}\sqrt{\frac{\mu_0}{\varepsilon_1}} - \left(\frac{1-\tau_\varepsilon}{\tau_\varepsilon}\right)^2\sqrt{\tau_\varepsilon}\tanh^{-1}\sqrt{\tau_\varepsilon}\right] \\ D_\varepsilon^{+,u}(\rho) &= \frac{i\omega\mu_0\rho^2}{4}\left[i\rho\frac{4\omega\sqrt{\varepsilon_1\mu_0}}{9} + \frac{1}{\tau_\varepsilon}\left(-1+\frac{1+\tau_\varepsilon}{\tau_\varepsilon}\sqrt{\tau_\varepsilon}\tanh^{-1}\sqrt{\tau_\varepsilon}\right)\right] \end{aligned}\right\} \tag{3.41}$$

if $0 < \tau_\varepsilon < 1$, and

$$\left.\begin{aligned} D_\varepsilon^+(\rho) &= \frac{i\omega\mu_0\rho^2}{8}\left[\frac{1-\tau_\varepsilon}{\tau_\varepsilon} + i\rho\frac{4\omega(4\varepsilon_1+\varepsilon_2)}{9}\sqrt{\frac{\mu_0}{\varepsilon_1}} + \left(\frac{1-\tau_\varepsilon}{\tau_\varepsilon}\right)^2\sqrt{-\tau_\varepsilon}\tan^{-1}\sqrt{-\tau_\varepsilon}\right] \\ D_\varepsilon^{+,u}(\rho) &= \frac{i\omega\mu_0\rho^2}{4}\left[i\rho\frac{4\omega\sqrt{\varepsilon_1\mu_0}}{9} + \frac{1}{\tau_\varepsilon}\left(1-\frac{1+\tau_\varepsilon}{\tau_\varepsilon}\sqrt{-\tau_\varepsilon}\tan^{-1}\sqrt{-\tau_\varepsilon}\right)\right] \end{aligned}\right\} \tag{3.42}$$

if $\tau_\varepsilon < 0$. The condition $\tau_\varepsilon > 1$ holds for nondissipative uniaxial dielectric ambient materials with indefinite constitutive dyadics [29–31]. The components of $\underline{\underline{\mathbf{D}}}^+(\rho)$ are not defined for this anomalous case (nor are the components of $\underline{\underline{\mathbf{D}}}^0$). In the general case involving a dissipative uniaxial dielectric ambient material, the parameter τ_ε is complex-valued and either equations (3.41) or equations (3.42) may be implemented.

If the ambient material were to be a uniaxial magnetic material then the particle-size-dependent depolarization dyadic expressions would be exactly analogous to those presented in equations (3.39)–(3.42), with the roles of ε and μ interchanged.

A consequence of the symmetries of the integrals on the right sides of equations (3.25) is that skew-symmetric components of any 3×3 constitutive dyadics make no contribution to the corresponding 3×3 component dyadics of $\underline{\underline{\mathbf{D}}}^0$. Therefore, the particle-size-independent 3×3 depolarization dyadics for gyrotropic dielectric ambient materials and gyrotropic magnetic ambient materials (as characterized by the constitutive dyadics in equation (2.42) with $\sigma = \varepsilon$ and μ, respectively) are the same as those for uniaxial dielectric ambient materials and uniaxial magnetic ambient materials given in equations (3.35)–(3.38).

Lastly in this section, suppose that the ambient material is orthorhombic dielectric–magnetic, i.e., $\underline{\underline{\varepsilon}}_{\text{amb}}$ and $\underline{\underline{\mu}}_{\text{amb}}$ are characterized as in equation (2.40) while $\underline{\underline{\xi}}_{\text{amb}}$ and $\underline{\underline{\zeta}}_{\text{amb}}$ are null dyadics. It is convenient then to express $\underline{\underline{\varepsilon}}_{\text{amb}}$ and $\underline{\underline{\mu}}_{\text{amb}}$ in diagonal form per

$$\underline{\underline{\sigma}}_{\text{amb}} = \sigma_x\,\hat{\mathbf{u}}_x\,\hat{\mathbf{u}}_x + \sigma_y\,\hat{\mathbf{u}}_y\,\hat{\mathbf{u}}_y + \sigma_z\,\hat{\mathbf{u}}_z\,\hat{\mathbf{u}}_z, \qquad (\sigma \in \{\varepsilon, \mu\}). \tag{3.43}$$

The particle-size-independent contributions to the depolarization dyadic have the diagonal representations

$$\left.\begin{aligned}
{}^{\mathrm{ee}}\underline{\underline{D}}\,{}^{0}_{I/\mathrm{amb}} &= D_x^{\varepsilon}\,\hat{\mathbf{u}}_x\,\hat{\mathbf{u}}_x + D_y^{\varepsilon}\,\hat{\mathbf{u}}_y\,\hat{\mathbf{u}}_y + D_z^{\varepsilon}\,\hat{\mathbf{u}}_z\,\hat{\mathbf{u}}_z \\
{}^{\mathrm{em}}\underline{\underline{D}}\,{}^{0}_{I/\mathrm{amb}} &= \underline{\underline{0}} \\
{}^{\mathrm{me}}\underline{\underline{D}}\,{}^{0}_{I/\mathrm{amb}} &= \underline{\underline{0}} \\
{}^{\mathrm{mm}}\underline{\underline{D}}\,{}^{0}_{I/\mathrm{amb}} &= D_x^{\mu}\,\hat{\mathbf{u}}_x\,\hat{\mathbf{u}}_x + D_y^{\mu}\,\hat{\mathbf{u}}_y\,\hat{\mathbf{u}}_y + D_z^{\mu}\,\hat{\mathbf{u}}_z\,\hat{\mathbf{u}}_z
\end{aligned}\right\}. \tag{3.44}$$

For nondissipative ambient materials, explicit evaluations of the dyadics in equations $(3.44)_{1,4}$ depend upon the relative magnitudes of the constitutive parameters $\sigma_{x,y,z}$, $(\sigma \in \{\varepsilon, \mu\})$. For $\sigma_{x,y,z} > 0$, these may be written as [32, 33]

$$D_x^{\sigma} = \begin{cases}
\dfrac{1 + i\sqrt{\dfrac{\sigma_y}{\sigma_z - \sigma_x}}\,E\!\left(i \sinh^{-1}\sqrt{\dfrac{\sigma_z}{\sigma_x} - 1},\ \sigma_1\right)}{i\omega(\sigma_x - \sigma_y)}, & \sigma_z > \sigma_y > \sigma_x,\ \ \sigma_y > \sigma_z > \sigma_x,\ \ \sigma_z > \sigma_x > \sigma_y; \\[2ex]
\dfrac{1 - \sqrt{\dfrac{\sigma_y}{\sigma_x - \sigma_z}}\,E\!\left(\sec^{-1}\sqrt{\dfrac{\sigma_x}{\sigma_z}},\ \sigma_1\right)}{i\omega(\sigma_x - \sigma_y)}, & \sigma_y > \sigma_x > \sigma_z,\ \ \sigma_x > \sigma_z > \sigma_y,\ \ \sigma_x > \sigma_y > \sigma_z;
\end{cases} \tag{3.45}$$

$$D_y^{\sigma} = \begin{cases}
\dfrac{-1 - i\sqrt{\dfrac{\sigma_x}{\sigma_z - \sigma_y}}\,E\!\left(i \sinh^{-1}\sqrt{\dfrac{\sigma_z}{\sigma_y} - 1},\ \sigma_2\right)}{i\omega(\sigma_x - \sigma_y)}, & \sigma_z > \sigma_y > \sigma_x,\ \ \sigma_x > \sigma_z > \sigma_y,\ \ \sigma_z > \sigma_x > \sigma_y; \\[2ex]
\dfrac{-1 + \sqrt{\dfrac{\sigma_x}{\sigma_y - \sigma_z}}\,E\!\left(\sec^{-1}\sqrt{\dfrac{\sigma_y}{\sigma_z}},\ \sigma_2\right)}{i\omega(\sigma_x - \sigma_y)}, & \sigma_y > \sigma_x > \sigma_z,\ \ \sigma_y > \sigma_z > \sigma_x,\ \ \sigma_x > \sigma_y > \sigma_z;
\end{cases} \tag{3.46}$$

$$D_z^{\sigma} = \begin{cases}
\dfrac{F\!\left(\sec^{-1}\sqrt{\dfrac{\sigma_x}{\sigma_z}},\ \sigma_1\right) - E\!\left(\sec^{-1}\sqrt{\dfrac{\sigma_x}{\sigma_z}},\ \sigma_1\right)}{i\omega(\sigma_y - \sigma_z)\dfrac{\sqrt{\sigma_x - \sigma_z}}{\sqrt{\sigma_y}}}, & \sigma_y > \sigma_x > \sigma_z,\ \ \sigma_x > \sigma_z > \sigma_y,\ \ \sigma_x > \sigma_y > \sigma_z; \\[2ex]
\dfrac{F\!\left(i \sinh^{-1}\sqrt{\dfrac{\sigma_z - \sigma_y}{\sigma_y}},\ \sigma_2\right) - E\!\left(i \sinh^{-1}\sqrt{\dfrac{\sigma_z - \sigma_y}{\sigma_y}},\ \sigma_2\right)}{i\omega(\sigma_z - \sigma_x)\dfrac{\sqrt{\sigma_z - \sigma_y}}{i\sqrt{\sigma_x}}}, & \sigma_z > \sigma_y > \sigma_x,\ \ \sigma_z > \sigma_x > \sigma_y; \\[2ex]
\dfrac{E\!\left(\sin^{-1}\sqrt{\dfrac{\sigma_z - \sigma_x}{\sigma_x}},\ \sigma_3\right) - F\!\left(\sin^{-1}\sqrt{\dfrac{\sigma_z - \sigma_x}{\sigma_x}},\ \sigma_3\right)}{i\omega(\sigma_z - \sigma_y)\dfrac{\sqrt{\sigma_z - \sigma_x}}{\sqrt{\sigma_y}}}, & \sigma_y > \sigma_z > \sigma_x,
\end{cases} \tag{3.47}$$

where

$$\left.\begin{aligned} \sigma_1 &= \sqrt{\frac{\sigma_x(\sigma_y - \sigma_z)}{\sigma_y(\sigma_x - \sigma_z)}} \\ \sigma_2 &= \sqrt{\frac{\sigma_y(\sigma_x - \sigma_z)}{\sigma_x(\sigma_y - \sigma_z)}} \\ \sigma_3 &= \sqrt{\frac{\sigma_x(\sigma_y - \sigma_z)}{\sigma_y(\sigma_z - \sigma_x)}} \end{aligned}\right\}, \tag{3.48}$$

and $\sigma \in \{\varepsilon, \mu\}$. Herein the scalar functions $F(\varphi, k)$ and $E(\varphi, k)$ are the incomplete elliptic integrals of the first and second kind, respectively, defined as [34]

$$\left.\begin{aligned} F(\varphi, k) &= \int_0^{\varphi} \frac{dt}{\sqrt{1 - k^2 \sin^2 t}} \\ E(\varphi, k) &= \int_0^{\varphi} \sqrt{1 - k^2 \sin^2 t}\, dt \end{aligned}\right\}, \tag{3.49}$$

with φ being the *amplitude*. Although some imaginary-valued amplitudes are used in equations (3.45)–(3.47) for compact representation, all quantities $i\omega D^{\sigma}_{x,y,z}$ $(\sigma \in \{\varepsilon, \mu\})$ are real valued for nondissipative ambient materials. Standard identities for elliptic integrals [35] can be implemented to re-express equations (3.45)–(3.47) in terms of real-valued quantities; however, the resulting expressions are rather more cumbersome than those in equations (3.45)–(3.47). For dissipative ambient materials, we have $\sigma_{x,y,z} \in \mathbb{C}$, $(\sigma \in \{\varepsilon, \mu\})$, and any of the expressions in equations (3.45)–(3.47) may be used.

By means of the substitutions $\underline{\underline{\mu}}_{\text{amb}} = \mu_0 \underline{\underline{I}}$ and $\underline{\underline{\varepsilon}}_{\text{amb}} = \varepsilon_0 \underline{\underline{I}}$, respectively, the corresponding 3×3 dyadics in the 6×6 depolarization dyadic for orthorhombic dielectric ambient materials and orthorhombic magnetic ambient materials follow directly from equations (3.44). For an ellipsoidal particle, with rotational axes aligned with the Cartesian coordinate axes, immersed in an isotropic dielectric–magnetic ambient material, the form of the particle-size-independent depolarization dyadic $\underline{\underline{\mathbf{D}}}^0$ is analogous to that represented by equations (3.44) [32, 36].

Bianisotropic ambient materials

Generally, explicit expressions are not available for depolarization dyadics for bianisotropic ambient materials. One notable exception is that of uniaxial bianisotropy, i.e., an ambient material whose constitutive dyadics $\underline{\underline{\varepsilon}}_{\text{amb}}$, $\underline{\underline{\xi}}_{\text{amb}}$, $\underline{\underline{\zeta}}_{\text{amb}}$, and $\underline{\underline{\mu}}_{\text{amb}}$ all have the form represented in equation (2.38). The particle-size-independent 3×3 depolarization dyadics have the same symmetry as the 3×3 constitutive dyadics. That is, they may be expressed as [15]

$$\left.\begin{aligned} {}^{\text{ee}}\underline{\underline{D}}^{\,0}_{I/\text{amb}} &= D^0_{\varepsilon}(\underline{\underline{I}} - \hat{\mathbf{u}}\,\hat{\mathbf{u}}) + D^{0,u}_{\varepsilon}\,\hat{\mathbf{u}}\,\hat{\mathbf{u}} \\ {}^{\text{em}}\underline{\underline{D}}^{\,0}_{I/\text{amb}} &= D^0_{\xi}(\underline{\underline{I}} - \hat{\mathbf{u}}\,\hat{\mathbf{u}}) + D^{0,u}_{\xi}\,\hat{\mathbf{u}}\,\hat{\mathbf{u}} \\ {}^{\text{me}}\underline{\underline{D}}^{\,0}_{I/\text{amb}} &= D^0_{\zeta}(\underline{\underline{I}} - \hat{\mathbf{u}}\,\hat{\mathbf{u}}) + D^{0,u}_{\zeta}\,\hat{\mathbf{u}}\,\hat{\mathbf{u}} \\ {}^{\text{mm}}\underline{\underline{D}}^{\,0}_{I/\text{amb}} &= D^0_{\mu}(\underline{\underline{I}} - \hat{\mathbf{u}}\,\hat{\mathbf{u}}) + D^{0,u}_{\mu}\,\hat{\mathbf{u}}\,\hat{\mathbf{u}} \end{aligned}\right\}, \tag{3.50}$$

wherein the scalar components are

$$\left.\begin{aligned} D_\varepsilon^0 &= \frac{1}{2i\omega}\left[\mu_1 I_0 + (\mu_2 - \mu_1)I_1 - \mu_2 I_2\right] \\ D_\varepsilon^{0,u} &= \frac{1}{i\omega}(\mu_1 I_1 + \mu_2 I_2) \\ D_\xi^0 &= -\frac{1}{2i\omega}\left[\xi_1 I_0 + (\xi_2 - \xi_1)I_1 - \xi_2 I_2\right] \\ D_\xi^{0,u} &= -\frac{1}{i\omega}(\xi_1 I_1 + \xi_2 I_2) \\ D_\zeta^0 &= -\frac{1}{2i\omega}\left[\zeta_1 I_0 + (\zeta_2 - \zeta_1)I_1 - \zeta_2 I_2\right] \\ D_\zeta^{0,u} &= -\frac{1}{i\omega}(\zeta_1 I_1 + \zeta_2 I_2) \\ D_\mu^0 &= \frac{1}{2i\omega}\left[\varepsilon_1 I_0 + (\varepsilon_2 - \varepsilon_1)I_1 - \varepsilon_2 I_2\right] \\ D_\mu^{0,u} &= \frac{1}{i\omega}(\varepsilon_1 I_1 + \varepsilon_2 I_2) \end{aligned}\right\}. \tag{3.51}$$

The integrals

$$I_n = \int_0^1 \frac{t^{2n}}{A + Bt^2 + Ct^4}\, dt, \qquad (n \in \{0, 1, 2\}), \tag{3.52}$$

involve the coefficients

$$\left.\begin{aligned} A &= \varepsilon_1\mu_1 - \xi_1\zeta_1 \\ B &= \varepsilon_1\mu_2 + \varepsilon_2\mu_1 - \zeta_1\xi_2 - \zeta_2\xi_1 \\ C &= \varepsilon_2\mu_2 - \xi_2\zeta_2 \end{aligned}\right\}. \tag{3.53}$$

For dissipative ambient materials, the constitutive parameters are all generally complex valued and the foregoing integrals are evaluated as

$$\left.\begin{aligned} I_0 &= \frac{C}{h}(I_{0+} + I_{0-}) \\ I_1 &= -\frac{1}{h}(D_+ I_{0+} + D_- I_{0-}) \\ I_2 &= \frac{1 - AI_0 - BI_1}{C} \end{aligned}\right\}, \tag{3.54}$$

wherein

$$\left.\begin{aligned} h &= \sqrt{B^2 - 4AC} \\ D_\pm &= \frac{1}{2}(B \mp h) \\ I_{0\pm} &= \frac{1}{\sqrt{CD_\pm}} \tan^{-1}\sqrt{\frac{C}{D_\pm}} \end{aligned}\right\}. \tag{3.55}$$

While the 3 × 3 depolarization dyadics in equations (3.50) relate to a spherical particle immersed in a uniaxial bianisotropic ambient material, they are analogous to the depolarization dyadics relevant to a spheroidal particle immersed in a biisotropic ambient material, or indeed a spheroidal particle immersed in a uniaxial bianisotropic ambient material provided that the rotational axis of the spheroid is aligned with the distinguished axis of the uniaxial material.

Finally, let us add that in the case of a bianisotropic ambient material whose magnetoelectric dyadics $\underline{\underline{\xi}}_{\text{amb}}$ and $\underline{\underline{\zeta}}_{\text{amb}}$ are wholly skew-symmetric, the particle-size-independent depolarization dyadics are simply those of the corresponding anisotropic ambient material, because the skew-symmetric terms make no contribution to the corresponding depolarization dyadics. For example, such skew-symmetric magnetoelectric constitutive dyadics can arise in materials which are electromagnetic analogs of certain curved spacetimes [37, 38]. Furthermore, homogenization may provide a viable method of realizing these analog materials [39, 40].

3.2.3 Cylindrical region

The ellipsoidal shape and orientation encapsulated by the real symmetric dyadic $\underline{\underline{U}}$ introduced in equation (3.15) can describe a wide range of particles of practical interest, including spheroidal and spherical particles. But the $\underline{\underline{U}}$-dyadic representation can be extended further: by considering the limiting dimensional values for a spheroidal particle, disk-shaped and needle-shaped particles can be accommodated too.

Attention is restricted now to ambient materials that are orthorhombic biaxial dielectric–magnetic, i.e., characterized by the permittivity and permeability dyadics given in equation (3.43). Consider a spheroidal particle whose rotational axis is aligned with $\hat{\mathbf{u}}_z$, as specified by

$$\underline{\underline{U}} = \frac{1}{\sqrt{\Gamma}}\left(\hat{\mathbf{u}}_x\,\hat{\mathbf{u}}_x + \hat{\mathbf{u}}_y\,\hat{\mathbf{u}}_y\right) + \Gamma\,\hat{\mathbf{u}}_z\,\hat{\mathbf{u}}_z, \qquad (\Gamma > 0). \tag{3.56}$$

The limit $\Gamma \to 0$ corresponds to a disk-shaped particle (oriented parallel to the xy plane) while the limit $\Gamma \to \infty$ corresponds to a needle-shaped particle (oriented parallel to the z axis).

On applying these limits to the integrals on the right sides of equations (3.25), the particle-size-independent depolarization dyadics emerge as [41]

$$\left.\begin{aligned} {}^{\text{ee}}\underline{\underline{D}}^{\,0}_{I/\text{amb}} &= \frac{1}{i\omega\varepsilon_z}\,\hat{\mathbf{u}}_z\,\hat{\mathbf{u}}_z \\ {}^{\text{em}}\underline{\underline{D}}^{\,0}_{I/\text{amb}} &= \underline{\underline{0}} \\ {}^{\text{me}}\underline{\underline{D}}^{\,0}_{I/\text{amb}} &= \underline{\underline{0}} \\ {}^{\text{mm}}\underline{\underline{D}}^{\,0}_{I/\text{amb}} &= \frac{1}{i\omega\mu_z}\,\hat{\mathbf{u}}_z\,\hat{\mathbf{u}}_z \end{aligned}\right\} \tag{3.57}$$

for a disk-shaped region, and

$$\left.\begin{aligned}
{}^{\mathrm{ee}}\underline{\underline{D}}^{\,0}_{\,I/\mathrm{amb}} &= \frac{1}{i\omega(\varepsilon_x-\varepsilon_y)}\left[\left(1-\sqrt{\frac{\varepsilon_y}{\varepsilon_x}}\right)\hat{\mathbf{u}}_x\,\hat{\mathbf{u}}_x-\left(1-\sqrt{\frac{\varepsilon_x}{\varepsilon_y}}\right)\hat{\mathbf{u}}_y\,\hat{\mathbf{u}}_y\right]\\
{}^{\mathrm{em}}\underline{\underline{D}}^{\,0}_{\,I/\mathrm{amb}} &= \underline{\underline{0}}\\
{}^{\mathrm{me}}\underline{\underline{D}}^{\,0}_{\,I/\mathrm{amb}} &= \underline{\underline{0}}\\
{}^{\mathrm{mm}}\underline{\underline{D}}^{\,0}_{\,I/\mathrm{amb}} &= \frac{1}{i\omega(\mu_x-\mu_y)}\left[\left(1-\sqrt{\frac{\mu_y}{\mu_x}}\right)\hat{\mathbf{u}}_x\,\hat{\mathbf{u}}_x-\left(1-\sqrt{\frac{\mu_x}{\mu_y}}\right)\hat{\mathbf{u}}_y\,\hat{\mathbf{u}}_y\right]
\end{aligned}\right\} \tag{3.58}$$

for a needle-shaped region. The results (3.57) and (3.58) can also be derived by a quite different route which requires expansion of the corresponding DGF in terms of eigenfunctions in cylindrical coordinates [42].

3.3 Polarizability density

The *polarizability density* dyadic of an electrically small particle immersed in the ambient material is closely related to the depolarization dyadic of a region of the same orientation, shape, and size as the particle. The polarizability density dyadic plays a central part in the homogenization formalisms presented in chapter 4. Consideration of this quantity sheds light on the physical aspects of the depolarization dyadic and its role in homogenization.

Let us focus on a $\underline{\underline{U}}^{\ell}$-shaped ellipsoidal particle made of a bianisotropic material characterized by the 6 × 6 Tellegen constitutive dyadic $\underline{\underline{\mathbf{K}}}_{\,\ell}$ and located at the origin of the coordinate system. The particle is immersed in a bianisotropic ambient material characterized by the 6 × 6 Tellegen constitutive dyadic $\underline{\underline{\mathbf{K}}}_{\mathrm{amb}}$. As in section 3.2, in the presence of the field phasor $\underline{\mathbf{F}}(\mathbf{r})$, the particle may be represented by an equivalent source current density

$$\underline{\mathbf{Q}}_{\mathrm{equiv}}(\mathbf{0}) = i\omega\left(\underline{\underline{\mathbf{K}}}_{\mathrm{amb}}-\underline{\underline{\mathbf{K}}}_{\,\ell}\right)\bullet\underline{\mathbf{F}}(\mathbf{0}), \tag{3.59}$$

per equation (3.10). This gives rise to the following implicit definition of the polarizability density dyadic $\underline{\underline{\boldsymbol{\alpha}}}_{\ell/\mathrm{amb}}(\rho)$ [18]:

$$\underline{\underline{\boldsymbol{\alpha}}}_{\ell/\mathrm{amb}}(\rho)\bullet\underline{\mathbf{F}}^{\mathrm{cf}}(\mathbf{0}) = -\frac{1}{i\omega}\underline{\mathbf{Q}}_{\mathrm{equiv}}(\mathbf{0}), \tag{3.60}$$

with $\underline{\mathbf{F}}^{\mathrm{cf}}(\mathbf{0})$ being the field phasor that would exist at the origin *if* the particle were to be made of the ambient material. By combining equation (3.13) with equations (3.59) and (3.60), an explicit expression for the polarizability density dyadic is found as [18]

$$\underline{\underline{\boldsymbol{\alpha}}}_{\ell/\mathrm{amb}}(\rho) = \left(\underline{\underline{\mathbf{K}}}_{\,\ell}-\underline{\underline{\mathbf{K}}}_{\mathrm{amb}}\right)\bullet\left[\,\underline{\underline{\mathbf{I}}}+i\omega\underline{\underline{\mathbf{D}}}_{\,U^{\ell}/\mathrm{amb}}(\rho)\bullet\left(\underline{\underline{\mathbf{K}}}_{\,\ell}-\underline{\underline{\mathbf{K}}}_{\mathrm{amb}}\right)\right]^{-1}. \tag{3.61}$$

The particle-size-independent polarizability density dyadic follows upon replacing $\underline{\underline{\mathbf{D}}}_{U^\ell/\text{amb}}(\rho)$ in equation (3.61) with $\underline{\underline{\mathbf{D}}}^0_{U^\ell/\text{amb}}$; i.e.,

$$\underline{\underline{\boldsymbol{\alpha}}}^0_{\ell/\text{amb}} = \left(\underline{\underline{\mathbf{K}}}_\ell - \underline{\underline{\mathbf{K}}}_\text{amb}\right) \bullet \left[\underline{\underline{\mathbf{I}}} + i\omega \underline{\underline{\mathbf{D}}}^0_{U^\ell/\text{amb}} \bullet \left(\underline{\underline{\mathbf{K}}}_\ell - \underline{\underline{\mathbf{K}}}_\text{amb}\right)\right]^{-1}. \tag{3.62}$$

In essence, the polarizability density dyadic may be regarded as a measure of the electromagnetic response of an electrically small ellipsoidal particle to an applied electromagnetic field. Notice that the response represented by $\underline{\underline{\boldsymbol{\alpha}}}_{\ell/\text{amb}}(\rho)$ can become extremely large in the limit as the determinant of the dyadic term $\underline{\underline{\mathbf{I}}} + i\omega \underline{\underline{\mathbf{D}}}_{U^\ell/\text{amb}}(\rho) \bullet \left(\underline{\underline{\mathbf{K}}}_\ell - \underline{\underline{\mathbf{K}}}_\text{amb}\right)$ on the right side of equation (3.61) becomes extremely small. This may be better appreciated by considering the simplest case wherein the particle is composed of an isotropic dielectric–magnetic material characterized by the permittivity ε_ℓ and permeability μ_ℓ, and the ambient material is also isotropic dielectric–magnetic, characterized by the permittivity ε_amb and permeability μ_amb. If, in addition, the particle is spherical (i.e., $\underline{\underline{U}}^\ell = \underline{\underline{I}}$) and particle-size effects are neglected, then the corresponding polarizability density dyadic has the form

$$\underline{\underline{\boldsymbol{\alpha}}}^0_{\ell/\text{amb}} = \begin{bmatrix} \alpha^{0,\varepsilon}_{\ell/\text{amb}}\, \underline{\underline{I}} & \underline{\underline{0}} \\ \underline{\underline{0}} & \alpha^{0,\mu}_{\ell/\text{amb}}\, \underline{\underline{I}} \end{bmatrix}, \tag{3.63}$$

with

$$\alpha^{0,\sigma}_{\ell/\text{amb}} = \frac{3\, \sigma_\text{amb}(\sigma_\ell - \sigma_\text{amb})}{2\sigma_\text{amb} + \sigma_\ell}, \qquad (\sigma \in \{\varepsilon, \mu\}). \tag{3.64}$$

In the limit $\sigma_\ell \to -2\sigma_\text{amb}$, the scalar polarizability densities $\alpha^\sigma_{U^\ell/\text{amb}}$ become unbounded. This limit cannot be practically attained because real materials exhibit dissipation. Nevertheless, it is quite possible for $\alpha^\sigma_{U^\ell/\text{amb}}$ to become extremely large if the particle material ℓ and the ambient material are only weakly dissipative and $\text{Re}\{\sigma_\ell\} \approx -2\,\text{Re}\{\sigma_\text{amb}\}$. In fact, this phenomenon—which is often referred to as a Fröhlich mode [43]—gives rise to *localized surface plasmon resonance*, as exhibited by an electrically small metallic particle immersed in a dielectric ambient material [44, 45]. The excitation of this resonance is a delicate process that is acutely sensitive to changes in the constitutive properties of the particle and the ambient material, as well as their interfacial morphologies [46]. Consequently, localized surface plasmon resonance has been extensively utilized for optical sensing applications, especially at nanometer length scales [47–49].

References

[1] Felsen L B and Marcuvitz N 1984 *Radiation and Scattering of Waves* (Piscataway, NJ: IEEE)

[2] Weiglhofer W S 1995 Frequency-dependent dyadic Green functions for bianisotropic media *Advanced Electromagnetism: Foundations, Theory and Applications* ed T W Barrett and D M Grimes (Singapore: World Scientific) pp 376–89

[3] Faryad M and Lakhtakia A 2018 *Infinite-Space Dyadic Green Functions in Electromagnetism* (San Rafael, CA: Morgan & Claypool) https://doi.org/10.1088/978-1-6817-4557-2

[4] Brau C A 2004 *Modern Problems in Classical Electrodynamics* (New York: Oxford University Press)
[5] Mackay T G and Lakhtakia A 2019 *Electromagnetic Anisotropy and Bianisotropy: A Field Guide* 2nd edn (Singapore: World Scientific) https://doi.org/10.1142/7515
[6] Lakhtakia A 2018 The Ewald–Oseen extinction theorem and the extended boundary condition method *The World of Applied Electromagnetics: In Appreciation of Magdy Fahmy Iskander* ed A Lakhtakia and C M Furse (Cham: Springer) pp 481–513
[7] Kong J A 1972 Theorems of bianisotropic media *Proc. IEEE* **60** 1036–46
[8] Chew W C 1999 *Waves and Fields in Inhomogenous Media* (Piscataway, NJ: IEEE)
[9] Li L-W, Kang X K and Leong M-S 2001 *Spheroidal Wave Functions in Electromagnetic Theory* (Hoboken, NJ: Wiley)
[10] Monzon J C 1989 Three-dimensional field expansion in the most general rotationally symmetric anisotropic material: application to scattering by a sphere *IEEE Trans. Antennas Propagat.* **37** 728–35
[11] Weiglhofer W S, Lakhtakia A and Michel B 1997 Maxwell Garnett and Bruggeman formalisms for a particulate composite with bianisotropic host medium *Microw. Opt. Technol. Lett.* **15** 263–6
Weiglhofer W S, Lakhtakia A and Michel B 1999 Corrections to 'Maxwell Garnett and Bruggeman formalisms for a particulate composite with bianisotropic host medium' *Microw. Opt. Technol. Lett.* **22** 221 (correction)
[12] Van Bladel J 1991 *Singular Electromagnetic Fields and Sources* (Oxford: Oxford University Press) (reissued in association with IEEE, Piscataway, NJ, 1995)
[13] Lakhtakia A 2000 Orthogonal symmetries of polarizability dyadics of bianisotropic ellipsoids *Microw. Opt. Technol. Lett.* **27** 175–7
[14] Michel B 1997 A Fourier space approach to the pointwise singularity of an anisotropic dielectric medium *Int. J. Appl. Electromagn Mech.* **8** 219–27
[15] Michel B and Weiglhofer W S 1997 Pointwise singularity of dyadic Green function in a general bianisotropic medium *Arch. Elektron. Übertrag.* **51** 219–23
Michel B and Weiglhofer W S 1998 Pointwise singularity of dyadic Green function in a general bianisotropic medium *Arch. Elektron. Übertrag* **52** 310 (erratum)
[16] Mackay T G 2004 Depolarization volume and correlation length in the homogenization of anisotropic dielectric composites *Waves Random Media* **14** 485–98
Mackay T G 2006 Erratum: Depolarization volume and correlation length in the homoge nization of anisotropic dielectric composites *Waves Random Complex Media* **16** 85 (erratum)
[17] Cui J and Mackay T G 2007 Depolarization regions of nonzero volume in bianisotropic homogenized composites *Waves Random Complex Media* **17** 269–81
[18] Michel B 2000 Recent developments in the homogenization of linear bianisotropic composite materials *Electromagnetic Fields in Unconventional Materials and Structures* ed O N Singh and A Lakhtakia (New York: Wiley) pp 39–82
[19] Doyle W T 1989 Optical properties of a suspension of metal spheres *Phys. Rev.* B **39** 9852–8
[20] Dungey C E and Bohren C F 1991 Light scattering by nonspherical particles: a refinement to the coupled-dipole method *J. Opt. Soc. Am.* A **8** 81–7
[21] Shanker B and Lakhtakia A 1993 Extended Maxwell Garnett model for chiral-in-chiral composites *J. Phys. D: Appl. Phys.* **26** 1746–58
[22] Prinkey M T, Lakhtakia A and Shanker B 1994 On the extended Maxwell–Garnett and the extended Bruggeman approaches for dielectric-in-dielectric composites *Optik* **96** 25–30

[23] Shanker B 1996 The extended Bruggeman approach for chiral-in-chiral mixtures *J. Phys. D: Appl. Phys.* **29** 281–8

[24] Altman C and Suchy K 1991 *Reciprocity, Spatial Mapping and Time Reversal in Electromagnetics* (Dordrecht: Kluwer Academic Publishers)

[25] Krowne C M 1984 Electromagnetic theorems for complex anisotropic media *IEEE Trans. Antennas Propagat.* **32** 1224–30

[26] Mackay T G 2008 On extended homogenization formalisms for nanocomposites *J. Nanophoton.* **2** 021850

[27] Lakhtakia A 2002 On perfect lenses and nihility *Int. J. Infrared Millim. Waves* **23** 339–43

[28] Mackay T G and Lakhtakia A 2005 Anisotropic enhancement of group velocity in a homogenized dielectric composite medium *J. Opt. A: Pure Appl. Opt.* **7** 669–74

[29] Mackay T G, Lakhtakia A and Depine R A 2006 Uniaxial dielectric media with hyperbolic dispersion relations *Microw. Opt. Technol. Lett.* **48** 363–7

[30] Li G X, Tam H L, Wang F Y and Cheah K W 2007 Superlens from complementary anisotropic metamaterials *J. Appl. Phys.* **102** 116101

[31] Smith D R and Schurig D 2003 Electromagnetic wave propagation in media with indefinite permittivity and permeability tensors *Phys. Rev. Lett.* **90** 077405

[32] Weiglhofer W S 1998 Electromagnetic depolarization dyadics and elliptic integrals *J. Phys. A: Math. Gen.* **31** 7191–6

[33] Mackay T G 2014 Towards metamaterials with giant dielectric anisotropy via homogenization: An analytical study *Photon. Nanostruct.: Fundam. Appl.* **13** 8–19

[34] Gradshteyn I S and Ryzhik I M 2007 *Table of Integrals, Series, and Products* 7th edn ed A Jeffrey and D Zwillinger (Burlington, MA: Academic)

[35] Abramowitz M and Stegun I A 1972 *Handbook of Mathematical Functions with Formulas, Graphs, and Mathematical Tables* (New York: Dover)

[36] Fricke H 1953 The Maxwell–Wagner dispersion in a suspension of ellipsoids *J. Phys. Chem.* **57** 934–7

[37] Plébanski J 1960 Electromagnetic waves in gravitational fields *Phys. Rev.* **118** 1396–408

[38] Schleich W and Scully M O 1984 General relativity and modern optics *New Trends in Atomic Physics* ed G Grynberg and R Stora (Amsterdam: Elsevier) pp 995–1124

[39] Mackay T G and Lakhtakia A 2011 Towards a realization of Schwarzschild-(anti-)de Sitter spacetime as a particulate metamaterial *Phys. Rev.* B **83** 195424

[40] Mackay T G and Lakhtakia A 2014 Towards a piecewise-homogeneous metamaterial model of the collision of two linearly polarized gravitational plane waves *IEEE Trans. Antennas Propagat.* **62** 6149–54

[41] Weiglhofer W S and Mackay T G 2002 Needles and pillboxes in anisotropic mediums *IEEE Trans. Antennas Propagat.* **50** 85–6

[42] Cottis P G, Vazouras C N and Spyrou C 1999 Green's function for an unbounded biaxial medium in cylindrical coordinates *IEEE Trans. Antennas Propagat.* **47** 195–9

[43] Bohren C F and Huffman D R 1983 *Absorption and Scattering of Light by Small Particles* (New York: Wiley)

[44] Kalele S A, Tiwari N R, Gosavi S W and Kulkarni S K 2007 Plasmon-assisted photonics at the nanoscale *J. Nanophoton.* **1** 012501

[45] Maier S A 2007 *Plasmonics: Fundamentals and Applications* (New York: Springer)

[46] Polo J A Jr, Mackay T A and Lakhtakia A 2013 *Electromagnetic Surface Waves: A Modern Perspective* (Waltham, MA: Elsevier) https://doi.org/10.1016/C2011-0-07510-5

[47] Hutter E and Fendler J H 2004 Exploitation of localized surface plasmon resonance *Adv. Mater.* **16** 1685–706
[48] Homola J (ed) 2006 *Surface Plasmon Resonance Based Sensors* (Heidelberg: Springer)
[49] Zourob M and Lakhtakia A (ed) 2010 *Optical Guided-Wave Chemical and Biosensors I* (Heidelberg: Springer)

Tom G Mackay and Akhlesh Lakhtakia

Chapter 4

Homogenization formalisms: linear materials

As discussed in section 1.3, numerous homogenization formalisms have been developed over the past two centuries to estimate the constitutive parameters of homogenized composite materials (HCMs). However, only a handful of formalisms have been rigorously established for bianisotropic HCMs—most notably, the Maxwell Garnett formalism (and variants thereof), the Bruggeman formalism, and the strong-property-fluctuation theory (SPFT) formalism. These formalisms were first developed for isotropic dielectric (and dielectric–magnetic) composite materials, and subsequently generalized to anisotropic and bianisotropic composite materials after the emergence of appropriate expressions for the relevant depolarization dyadics. Each homogenization formalism has its own strengths and weaknesses. While some formalisms may provide more realistic estimates of the effective constitutive parameters than others for particular types of composite materials, the acid test of any formalism is perhaps provided only by comparison with carefully conducted experimental measurements. However, in many instances, suitable experimental data are simply not available.

After a preliminary discussion of constituent materials and homogenization, the Maxwell Garnett, Bruggeman, and SPFT formalisms are presented for linear HCMs in this chapter. (Analogous developments for weakly nonlinear HCMs are presented in chapter 5.) Both the standard versions (wherein the constituent particles are treated as infinitesimally small entities) and the extended versions (wherein the constituent particles are treated as electrically small but not infinitesimally small entities) of the formalisms are presented.

4.1 Preliminaries

4.1.1 Constituent materials

Let our attention be focused on HCMs arising from two constituent materials labeled 'a' and 'b'. Extensions to three or more constituent materials are typically

doi:10.1088/978-0-7503-3423-5ch4

straightforward. Each constituent material is itself homogeneous. The most general linear case is considered in this chapter wherein both constituent materials are bianisotropic materials, characterized by the 6 × 6 Tellegen constitutive dyadics $\underline{\underline{\mathbf{K}}}_{\,\mathrm{a}}$ and $\underline{\underline{\mathbf{K}}}_{\,\mathrm{b}}$.

For compatibility with the notion of homogenization, the nonhomogeneity of the composite material must be at length scales much smaller than the electromagnetic wavelengths in the constituent materials and the HCM [2, 3]. In the spectral regime of visible wavelengths (380–780 nm), for example, the linear dimensions of particles making up the constituent materials cannot exceed 38–78 nm. There are also restrictions upon the minimum sizes of the constituent particles because, at sufficiently small length scales, quantum phenomenons invalidate the classical macroscopic viewpoint upon which the characterization of the constituent materials in terms of their constitutive dyadics $\underline{\underline{\mathbf{K}}}_{\ell}$ is based. The lower bound on the linear dimensions of constituent particles is material dependent. For example, spectroscopic ellipsometry has confirmed that an extended version of the Maxwell Garnett formalism satisfactorily characterizes a composite material containing silver particles as small as 2.3 nm in diameter [4].

Even though the sizes of the constituent particles may fall into the appropriate range to facilitate the process of homogenization, care must be exercised in selecting values for the constitutive parameters of the constituent materials. This is because the constitutive parameters of the material in a tiny particle can be significantly different from those of the corresponding bulk material. Furthermore, these differences may be sensitive to the shape and size of the constituent particle [5]. Particular care is needed for metallic particles with diameters smaller than the mean free path of conduction electrons in the bulk metal, since in this case the mean free path in the particulate material may be determined by collisions at the particle's boundary [6].

4.1.2 Homogenized composite materials

As delineated in section 3.3, a single ellipsoidal particle composed of a material characterized by the constitutive dyadic $\underline{\underline{\mathbf{K}}}_{\ell}$, immersed in an ambient material characterized by the constitutive dyadic $\underline{\underline{\mathbf{K}}}_{\mathrm{amb}}$, may be represented electromagnetically by an equivalent source current density given by equation (3.59). At a coarser length scale, the electromagnetic response of a great number of such particles may be characterized by a phasor $\underline{\mathbf{N}}_{\mathrm{exc}}(\mathbf{r})$. Representing the excess polarization and/or magnetization relative to the ambient material, $\underline{\mathbf{N}}_{\mathrm{exc}}(\mathbf{r})$ arises from spatially averaging over many constituent particles. Thus, the coarser-length-scale many-particle equivalent of equation (3.9) is given by the differential equation

$$\left[\, \underline{\underline{\mathbf{L}}}(\nabla) + i\omega \underline{\underline{\mathbf{K}}}_{\mathrm{amb}} \right] \bullet \underline{\mathbf{F}}(\mathbf{r}) = -i\omega \underline{\mathbf{N}}_{\mathrm{exc}}(\mathbf{r}). \tag{4.1}$$

In mathematical terms, the aim of homogenization formalisms is to provide an estimate of the constitutive dyadic $\underline{\underline{\mathbf{K}}}_{\mathrm{HCM}}$ that satisfies the differential equation

$$\left[\, \underline{\underline{\mathbf{L}}}(\nabla) + i\omega \underline{\underline{\mathbf{K}}}_{\mathrm{HCM}} \right] \bullet \underline{\mathbf{F}}(\mathbf{r}) = \mathbf{0} \tag{4.2}$$

and thereby characterizes the HCM. Key properties of the HCM are implicit in equation (4.2), namely: the HCM provides a long-wavelength representation of the composite material that applies for all angles and polarization states of illumination, and the HCM has the same bulk dimensions as its corresponding composite material. Thus, the HCM should be distinguished from 'equivalent' materials whose constitutive parameters may be inferred using the S-parameter retrieval scheme [7], for example. Such equivalent materials may only represent the corresponding composite material for certain angles of incidence or for certain polarization states; furthermore, neither may they share the same bulk dimensions as the corresponding composite materials nor may they be necessarily composed of electrically small constituent particles.

Subtraction of equation (4.1) from equation (4.2) yields

$$\left(\underline{\underline{\mathbf{K}}}_{\mathrm{HCM}} - \underline{\underline{\mathbf{K}}}_{\mathrm{amb}}\right) \bullet \underline{\mathbf{F}}(\mathbf{r}) = \underline{\mathbf{N}}_{\mathrm{exc}}(\mathbf{r}). \tag{4.3}$$

Therefore, the problem of deriving an expression for the HCM's constitutive dyadic $\underline{\underline{\mathbf{K}}}_{\mathrm{HCM}}$ is, in effect the problem of finding the excess polarization/magnetization $\underline{\mathbf{N}}_{\mathrm{exc}}(\mathbf{r})$ in terms of $\underline{\mathbf{F}}(\mathbf{r})$.

Consider an HCM whose constituent materials are characterized by the constitutive dyadics $\underline{\underline{\mathbf{K}}}_{\ell}$ $(\ell = a, b)$ and which are distributed with volume fraction $f_\ell \in (0, 1)$. A general constraint on $\underline{\underline{\mathbf{K}}}_{\mathrm{HCM}}$ is that

$$\underline{\underline{\mathbf{K}}}_{\mathrm{HCM}} \to \underline{\underline{\mathbf{K}}}_{\ell} \tag{4.4}$$

as $f_\ell \to 1$.

In sections 4.2, 4.3, and 4.4, the estimates of $\underline{\underline{\mathbf{K}}}_{\mathrm{HCM}}$ delivered by the Maxwell Garnett, Bruggeman, and SPFT formalisms, respectively, are presented for bianisotropic HCMs. In these sections the standard versions are provided, the effects of the sizes of the constituent particle being neglected. Extended versions of the Maxwell Garnett, Bruggeman, and SPFT formalisms, in which the sizes of the constituent particles are accommodated, are described in section 4.5.

4.2 Maxwell Garnett formalism

In the Maxwell Garnett approach to the homogenization of two constituent materials, labeled as 'a' and 'b', one of the constituent materials, say 'b', is regarded as a host material and particles of constituent material 'a' are regarded as particles randomly dispersed throughout the host material. In the simplest case, all particles have the same shape and orientation and are described by $\underline{\underline{U}}^{\mathrm{a}}$, per equation (3.15). No particular topology is assigned to the host material, which may or may not be particulate. The particles are assumed to be well separated in the host material. Accordingly, the volume fraction of constituent material 'a' is taken to be $f_{\mathrm{a}} \lesssim 0.3$, with $f_{\mathrm{b}} = 1 - f_{\mathrm{a}}$ being the volume fraction of the other constituent material. A schematic representation of the modeling approach adopted is shown in figure 4.1.

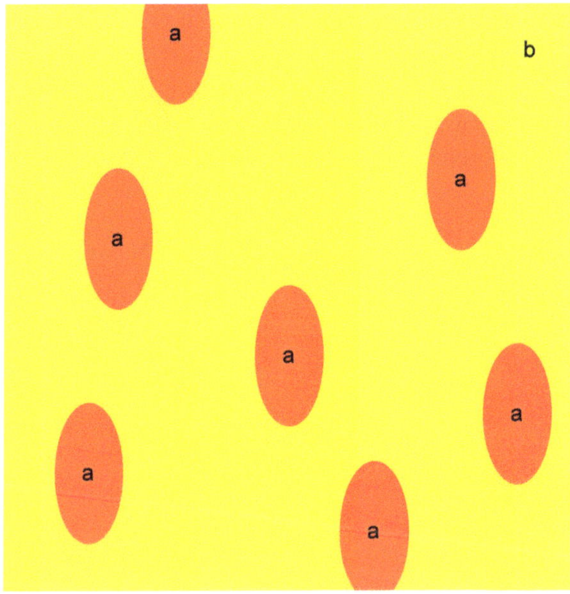

Figure 4.1. Schematic representation of the constituent materials for the Maxwell Garnett formalism. Ellipsoidal particles of constituent material 'a' are randomly distributed in constituent material 'b'. The ellipsoidal particles all have the same shape and orientation, as specified by $\underline{\underline{U}}^{\mathrm{a}}$. The volume fraction of constituent material 'a' must be relatively low.

4.2.1 Formulas

In view of the defining equation (3.60) for the polarizability density dyadic, the excess polarization and/or magnetization due to the particles in the host material may be expressed as

$$\underline{\mathbf{N}}_{\mathrm{exc}}(\mathbf{r}) = f_{\mathrm{a}}\, \underline{\underline{\boldsymbol{\alpha}}}^{0}_{\mathrm{a/b}} \bullet \underline{\mathbf{F}}_{\mathrm{loc}}(\mathbf{r}), \tag{4.5}$$

wherein the local phasor $\underline{\mathbf{F}}_{\mathrm{loc}}(\mathbf{r})$ represents the electromagnetic field which illuminates the particle located at $\mathbf{r}$. In order to estimate $\underline{\mathbf{F}}_{\mathrm{loc}}(\mathbf{r})$, attention is focused on a single particle which is considered as being contained inside a cavity occupying the region V_{Lor} [8]. This cavity is known as the *Lorentzian cavity*. The particles are assumed to be distributed sufficiently sparsely in the host material that a Lorentzian cavity contains only one particle. Furthermore, since the particles are assumed to be distributed isotropically in the host material, V_{Lor} is chosen to be spherical.

Suppose that the entire space occupied by the composite material is partitioned as $V_{\mathrm{Lor}} \cup V_{\mathrm{out}}$. Then, in light of equation (3.13), the solution of equation (4.1) may be expressed formally as

$$\underline{\mathbf{F}}(\mathbf{r}) = \underline{\mathbf{F}}^{\mathrm{cf}}(\mathbf{r}) - i\omega \int_{V_{\mathrm{out}}} \underline{\underline{\mathbf{G}}}_{\mathrm{b}}(\mathbf{r} - \mathbf{r}') \bullet \underline{\mathbf{N}}_{\mathrm{exc}}(\mathbf{r}')\, d^3\mathbf{r}' - i\omega \underline{\underline{\mathbf{D}}}^{0}_{I/\mathrm{b}} \bullet \underline{\mathbf{N}}_{\mathrm{exc}}(\mathbf{r}), \tag{4.6}$$

where $\underline{\mathbf{F}}^{\,\mathrm{cf}}(\mathbf{r})$ is the electromagnetic field phasor that would be present in the absence of the particle, $\underline{\underline{\mathbf{G}}}_{\,\mathrm{b}}(\mathbf{r}-\mathbf{r}')$ is the DGF of the host material, and $\underline{\underline{\mathbf{D}}}^{0}_{\,I/\mathrm{b}}$ is the depolarization dyadic for a spherical region in the host material.

The corresponding integral equation for the local field phasor follows similarly, but with one key difference: by construction, the excess polarization and/or magnetization are null valued inside the Lorentzian cavity. Thus

$$\underline{\mathbf{F}}_{\,\mathrm{loc}}(\mathbf{r}) = \underline{\mathbf{F}}^{\,\mathrm{cf}}(\mathbf{r}) - i\omega \int_{V_{\mathrm{out}}} \underline{\underline{\mathbf{G}}}_{\,\mathrm{b}}(\mathbf{r}-\mathbf{r}') \bullet \underline{\mathbf{N}}_{\,\mathrm{exc}}(\mathbf{r}')\; d^3\mathbf{r}'. \tag{4.7}$$

Subtraction of equation (4.6) from equation (4.7) delivers

$$\underline{\mathbf{F}}_{\,\mathrm{loc}}(\mathbf{r}) - \underline{\mathbf{F}}\,(\mathbf{r}) = i\omega \underline{\underline{\mathbf{D}}}^{0}_{\,I/\mathrm{b}} \bullet \underline{\mathbf{N}}_{\,\mathrm{exc}}(\mathbf{r}). \tag{4.8}$$

Consequently, after combining equations (4.3), (4.5), and (4.8), the Maxwell Garnett estimate $\underline{\underline{\mathbf{K}}}_{\,\mathrm{MG}}$ of $\underline{\underline{\mathbf{K}}}_{\mathrm{HCM}}$ emerges as [9]

$$\underline{\underline{\mathbf{K}}}_{\,\mathrm{MG}} = \underline{\underline{\mathbf{K}}}_{\,\mathrm{b}} + f_{\mathrm{a}}\, \underline{\underline{\boldsymbol{\alpha}}}^{\,0}_{\,\mathrm{a/b}} \bullet \left(\underline{\underline{\mathbf{I}}} - i\omega f_{\mathrm{a}}\, \underline{\underline{\mathbf{D}}}^{0}_{\,I/\mathrm{b}} \bullet \underline{\underline{\boldsymbol{\alpha}}}^{\,0}_{\,\mathrm{a/b}} \right)^{-1}. \tag{4.9}$$

The estimate (4.9) is readily extended to accommodate three or more constituent materials. Suppose that the constituent materials $\mathrm{a}_1, \mathrm{a}_2, \cdots, \mathrm{a}_n$, each one characterized by the Tellegen constitutive dyadic $\underline{\underline{\mathbf{K}}}_{\,\mathrm{a}_j}$, $(j = 1, 2, \ldots, n)$, and each comprising ellipsoid particles whose shape and orientation were characterized by $\underline{\underline{U}}^{\mathrm{a}_j}$, $(j = 1, 2, \ldots, n)$, are randomly dispersed in the host material ‘b’. Then the corresponding Maxwell Garnett estimate of $\underline{\underline{\mathbf{K}}}_{\mathrm{HCM}}$ is given by

$$\underline{\underline{\mathbf{K}}}_{\,\mathrm{MG}} = \underline{\underline{\mathbf{K}}}_{\,\mathrm{b}} + \sum_{j=1}^{n} \left[f_{\mathrm{a}_j}\, \underline{\underline{\boldsymbol{\alpha}}}^{\,0}_{\,\mathrm{a}_j/\mathrm{b}} \bullet \left(\underline{\underline{\mathbf{I}}} - i\omega f_{\mathrm{a}_j}\, \underline{\underline{\mathbf{D}}}^{0}_{\,I/\mathrm{b}} \bullet \underline{\underline{\boldsymbol{\alpha}}}^{\,0}_{\,\mathrm{a}_j/\mathrm{b}} \right)^{-1} \right], \tag{4.10}$$

where f_{a_j} denotes the volume fraction of constituent material a_j and the restriction $\sum_{j=1}^{n} f_{\mathrm{a}_j} \lesssim 0.3$ is imposed.

The spherical shape of the Lorentzian cavity contrasts with the generally ellipsoidal shape of the particles. This can lead to physically unreasonable estimates of the constitutive parameters of the HCM in cases where the ellipsoidal particles are relatively elongated and present in relatively high concentration. In an attempt to overcome this difficulty, the Bragg–Pippard formalism was introduced [10]. The Bragg–Pippard estimate of $\underline{\underline{\mathbf{K}}}_{\mathrm{HCM}}$ arises as the right side of equation (4.9) but with the depolarization term $\underline{\underline{\mathbf{D}}}^{0}_{\,I/\mathrm{b}}$ therein replaced by $\underline{\underline{\mathbf{D}}}^{0}_{\,U^{\mathrm{a}}/\mathrm{b}}$ [11].

The restriction $f_{\mathrm{a}} \lesssim 0.3$ which applies to equation (4.9) can be countered to an extent by interchanging the roles of the constituent materials ‘a’ and ‘b’. That is, if constituent ‘a’ were regarded as the host material and constituent ‘b’ as ellipsoidal particles randomly dispersed in the host material, then the corresponding Maxwell Garnett estimate

$$\underline{\underline{\mathbf{K}}}_{\,\mathrm{MG}} = \underline{\underline{\mathbf{K}}}_{\,\mathrm{a}} + f_{\mathrm{b}}\, \underline{\underline{\boldsymbol{\alpha}}}^{\,0}_{\,\mathrm{b/a}} \bullet \left(\underline{\underline{\mathbf{I}}} - i\omega f_{\mathrm{b}}\, \underline{\underline{\mathbf{D}}}^{0}_{\,I/\mathrm{a}} \bullet \underline{\underline{\boldsymbol{\alpha}}}^{\,0}_{\,\mathrm{b/a}} \right)^{-1}, \tag{4.11}$$

applies for $f_a \gtrsim 0.7$. However, the standard Maxwell Garnett formalism is not appropriate for the regime of intermediate values of f_a.

4.2.2 Inverse formalism

Homogenization formalisms are conventionally used to estimate the constitutive parameters of HCMs from a knowledge of their constituent materials. However, for certain applications one may wish to estimate the constitutive parameters of one of the constituent materials from a knowledge of the other constituent material and the HCM [12]. By virtue of equation (4.9), the *inverse* Maxwell Garnett formalism yields [13]

$$\underline{\underline{\mathbf{K}}}_{a} = \underline{\underline{\mathbf{K}}}_{b} + \frac{1}{f_a}\left(\underline{\underline{\mathbf{I}}} - i\omega\frac{1}{f_a}\,\underline{\underline{\boldsymbol{\alpha}}}^{0,\ddagger}_{MG/b} \bullet \underline{\underline{\mathbf{D}}}^{0}_{U^a/b}\right)^{-1} \bullet \underline{\underline{\boldsymbol{\alpha}}}^{0,\ddagger}_{MG/b}, \tag{4.12}$$

where the 6 × 6 dyadic

$$\underline{\underline{\boldsymbol{\alpha}}}^{0,\ddagger}_{MG/b} = \left[\underline{\underline{\mathbf{I}}} + i\omega\left(\underline{\underline{\mathbf{K}}}_{MG} - \underline{\underline{\mathbf{K}}}_{b}\right) \bullet \underline{\underline{\mathbf{D}}}^{0}_{I/b}\right]^{-1} \bullet \left(\underline{\underline{\mathbf{K}}}_{MG} - \underline{\underline{\mathbf{K}}}_{b}\right). \tag{4.13}$$

In a similar fashion, $\underline{\underline{\mathbf{K}}}_{b}$ may be expressed in terms of $\underline{\underline{\mathbf{K}}}_{a}$ and $\underline{\underline{\mathbf{K}}}_{MG}$.

4.2.3 Incremental and differential formalisms

Two refinements of the Maxwell Garnett formalism allow the restriction on the dilute concentration of particles to be overcome. These refinements are known as the *incremental* and *differential* Maxwell Garnett formalisms.

The incremental formalism involves adding the constituent material 'a' to the constituent material 'b' in a fixed number N of stages rather than all at once [14]. After each incremental addition, the composite material is homogenized using the Maxwell Garnett formalism and the resulting HCM is used as the host material for next step in the incremental process. The iteration scheme is given by

$$\left.\begin{aligned} &\underline{\underline{\mathbf{K}}}(0) = \underline{\underline{\mathbf{K}}}_{b} \\ &\underline{\underline{\mathbf{K}}}(n+1) = \underline{\underline{\mathbf{K}}}(n) + \left(1 - f_b^{1/N}\right)\underline{\underline{\boldsymbol{\alpha}}}^{0}_{a/n} \\ &\bullet\left[\underline{\underline{\mathbf{I}}} - i\omega\left(1 - f_b^{1/N}\right)\underline{\underline{\mathbf{D}}}^{0}_{I/n} \bullet \underline{\underline{\boldsymbol{\alpha}}}^{0}_{a/n}\right]^{-1} \quad \text{for} \quad n \in \{0, 1, \ldots, N-1\} \end{aligned}\right\}. \tag{4.14}$$

The polarizability density dyadic $\underline{\underline{\boldsymbol{\alpha}}}^{0}_{a/n}$ herein is that of a particle of material 'a' with shape and orientation specified by $\underline{\underline{U}}^{a}$ and immersed in a homogeneous material characterized by the Tellegen constitutive dyadic $\underline{\underline{\mathbf{K}}}(n)$. Thus, the incremental Maxwell Garnett estimate of $\underline{\underline{\mathbf{K}}}_{HCM}$ emerges as

$$\underline{\underline{\mathbf{K}}}_{IMG} = \underline{\underline{\mathbf{K}}}(N). \tag{4.15}$$

The incremental formalism yields the differential formalism in the limit $N \to \infty$ [15]. That is, the differential Maxwell Garnett estimate $\underline{\underline{\mathbf{K}}}_{DMG}$ of $\underline{\underline{\mathbf{K}}}_{HCM}$ arises as the solution of the initial-value problem

$$\frac{d}{ds}\underline{\underline{\mathbf{K}}}(s) = \frac{1}{1-s}\underline{\underline{\boldsymbol{\alpha}}}^{0}_{a/s}, \qquad \underline{\underline{\mathbf{K}}}(0) = \underline{\underline{\mathbf{K}}}_{b}. \tag{4.16}$$

The volume fraction of constituent material 'a' added to material 'b' is denoted by the continuous variable s in equation (4.16), while the polarizability density dyadic $\underline{\underline{\boldsymbol{\alpha}}}^{0}_{a/s}$ is that of a particle of material 'a' of shape and orientation specified by $\underline{\underline{U}}^{a}$ and immersed in a material characterized by the constitutive dyadic $\underline{\underline{\mathbf{K}}}(s)$. Notice that the ordinary differential equation (4.16) is nonlinear, courtesy of the dependency of $\underline{\underline{\boldsymbol{\alpha}}}^{0}_{a/s}$ upon $\underline{\underline{\mathbf{K}}}(s)$. The solution of equation (4.16) evaluated at $s = f_a$, namely

$$\underline{\underline{\mathbf{K}}}_{DMG} = \underline{\underline{\mathbf{K}}}(f_a), \tag{4.17}$$

is the differential Maxwell Garnett estimate of $\underline{\underline{\mathbf{K}}}_{HCM}$.

Variations on the recursive approach that forms the basis of the incremental and differential Maxwell Garnett formalisms, as delineated here for bianisotropic HCMs, have been generally utilized in the study of isotropic dielectric HCMs [16–19].

4.3 Bruggeman formalism

Unlike the Maxwell Garnett formalism, as well as its incremental and differential variants, the Bruggeman formalism is founded on a democratic principle: all constituent materials are treated equally. That is, for the homogenization of constituent materials 'a' and 'b', neither of the two is considered as a 'host' material in which particles of the other material are randomly dispersed. Instead, both constituent materials are regarded as being randomly dispersed in the HCM itself. The shape and orientation of the ellipsoidal particles of the constituent material 'a' are specified by $\underline{\underline{U}}^{a}$ per equation (3.15), and similarly the shape and orientation of the ellipsoidal particles of the constituent material 'b' are specified by $\underline{\underline{U}}^{b}$. In general, $\underline{\underline{U}}^{a} \neq \underline{\underline{U}}^{b}$ but the possibility that $\underline{\underline{U}}^{a} = \underline{\underline{U}}^{b}$ is permitted. The volume fractions of the constituent materials 'a' and 'b' are f_a and f_b, respectively. These can take any value in the range 0–1, subject to the constraint $f_a + f_b = 1$. This constraint presents a conceptual problem: how can all space be completely filled by ellipsoidal particles alone? In order to deal with this issue, a fractal-like distribution of the constituent materials may be envisaged, as is schematically represented in figure 4.2.

4.3.1 Formulas

Let $\underline{\underline{\mathbf{K}}}_{Br}$ be the estimate of the HCM's constitutive dyadic $\underline{\underline{\mathbf{K}}}_{HCM}$ delivered by the Bruggeman formalism. In view of the defining equation (3.60) for the polarizability density dyadic, the excess polarization and/or magnetization due to all constituent particles relative to the HCM may be expressed as

$$\underline{\mathbf{N}}_{exc}(\mathbf{r}) = f_a\, \underline{\underline{\boldsymbol{\alpha}}}^{0}_{a/Br} \bullet \underline{\mathbf{F}}_{loc}(\mathbf{r}) + f_b\, \underline{\underline{\boldsymbol{\alpha}}}^{0}_{b/Br} \bullet \underline{\mathbf{F}}_{loc}(\mathbf{r}). \tag{4.18}$$

The Bruggeman formalism rests on the assertion that

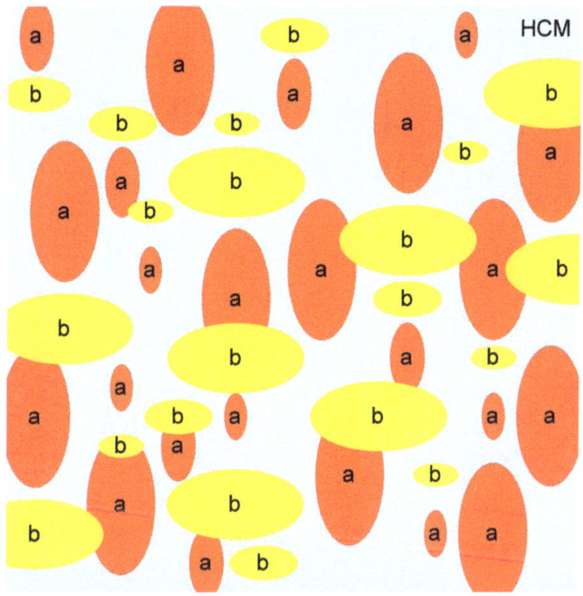

Figure 4.2. Schematic representation of the random dispersal of particles of the constituent materials 'a' and 'b' in the HCM per the Bruggeman formalism. All 'a' particles have the same shape and orientation, as specified by $\underline{\underline{U}}^{\mathrm{a}}$; likewise all 'b' particles have the same shape and orientation, as specified by $\underline{\underline{U}}^{\mathrm{b}}$. In general, $\underline{\underline{U}}^{\mathrm{a}} \neq \underline{\underline{U}}^{\mathrm{b}}$. A fractal-like distribution of the constituent particles may be envisaged, in order for the two constituent materials to completely occupy all space.

$$\underline{\mathbf{N}}_{\mathrm{exc}}(\mathbf{r}) = \underline{\mathbf{0}} \tag{4.19}$$

for all $\mathbf{r}$. Therefore, the dyadic Bruggeman equation emerges as [9]

$$f_{\mathrm{a}}\, \underline{\underline{\boldsymbol{\alpha}}}^{\,0}_{\mathrm{a/Br}} + f_{\mathrm{b}}\, \underline{\underline{\boldsymbol{\alpha}}}^{\,0}_{\mathrm{b/Br}} = \underline{\underline{\mathbf{0}}}, \tag{4.20}$$

which is wholly consistent with equation (4.3) with the HCM playing the role of the ambient material.

Equation (4.20) represents a nonlinear implicit expression for $\underline{\underline{\mathbf{K}}}_{\mathrm{Br}}$. Numerical techniques are generally needed to extract $\underline{\underline{\mathbf{K}}}_{\mathrm{Br}}$ from this equation. For example, the following Jacobi scheme has been extensively used [20]: The sought-after solution may be found via the recursive formula

$$\underline{\underline{\mathbf{K}}}_{\mathrm{Br}}(n) = \mathcal{T}\left\{ \underline{\underline{\mathbf{K}}}_{\mathrm{Br}}(n-1) \right\}, \qquad (n \in \{1, 2, \ldots\}). \tag{4.21}$$

In practice, the initial value $\underline{\underline{\mathbf{K}}}_{\mathrm{Br}}(0) = \underline{\underline{\mathbf{K}}}_{\mathrm{MG}}$ has been found to work well. The dyadic operator $\mathcal{T}$ is found by rearranging equation (4.20) and utilizing the expression (3.61) for the polarizability density dyadic; i.e.,

$$\begin{aligned}\mathcal{T}\{\underline{\underline{\mathbf{K}}}_{\,\mathrm{Br}}\} = &\Big\{ f_{\mathrm{a}}\,\underline{\underline{\mathbf{K}}}_{\,\mathrm{a}} \bullet \Big[\,\underline{\underline{\mathbf{I}}} + i\omega \underline{\underline{\mathbf{D}}}^{0}_{U^{\mathrm{a}}/\mathrm{Br}} \bullet \big(\underline{\underline{\mathbf{K}}}_{\,\mathrm{a}} - \underline{\underline{\mathbf{K}}}_{\,\mathrm{Br}}\big)\Big]^{-1} \\ &+ f_{\mathrm{b}}\,\underline{\underline{\mathbf{K}}}_{\,\mathrm{b}} \bullet \Big[\,\underline{\underline{\mathbf{I}}} + i\omega \underline{\underline{\mathbf{D}}}^{0}_{U^{\mathrm{b}}/\mathrm{Br}} \bullet \big(\underline{\underline{\mathbf{K}}}_{\,\mathrm{b}} - \underline{\underline{\mathbf{K}}}_{\,\mathrm{Br}}\big)\Big]^{-1} \Big\} \\ &\bullet \Big\{ f_{\mathrm{a}} \Big[\,\underline{\underline{\mathbf{I}}} + i\omega \underline{\underline{\mathbf{D}}}^{0}_{U^{\mathrm{a}}/\mathrm{Br}} \bullet \big(\underline{\underline{\mathbf{K}}}_{\,\mathrm{a}} - \underline{\underline{\mathbf{K}}}_{\,\mathrm{Br}}\big)\Big]^{-1} \\ &+ f_{\mathrm{b}} \Big[\,\underline{\underline{\mathbf{I}}} + i\omega \underline{\underline{\mathbf{D}}}^{0}_{U^{\mathrm{b}}/\mathrm{Br}} \bullet \big(\underline{\underline{\mathbf{K}}}_{\,\mathrm{b}} - \underline{\underline{\mathbf{K}}}_{\,\mathrm{Br}}\big)\Big]^{-1} \Big\}^{-1}. \end{aligned} \tag{4.22}$$

Equation (4.20) is readily extended to accommodate the constituent materials a_1, a_2, ⋯, a_n, each one characterized by the Tellegen constitutive dyadic $\underline{\underline{\mathbf{K}}}_{\,\mathrm{a}_j}$, $(j = 1, 2, \dots, n)$, and each comprising ellipsoidal particles whose shape and orientation are characterized by $\underline{\underline{U}}^{\mathrm{a}_j}$, $(j = 1, 2, \dots, n)$. Each constituent material in turn is regarded as being randomly dispersed in the HCM. Then the corresponding Bruggeman estimate of the HCM's constitutive dyadic is given implicitly by

$$\sum_{j=1}^{n}\Big(f_{\mathrm{a}_j}\, \underline{\underline{\boldsymbol{\alpha}}}^{0}_{\,\mathrm{a}_j/\mathrm{Br}} \Big) = \underline{\underline{\mathbf{0}}}, \tag{4.23}$$

with f_{a_j} being the volume fraction of constituent material a_j and $\sum_{j=1}^{n} f_{\mathrm{a}_j} = 1$. The Jacobi scheme represented by equations (4.21) and (4.22) for use with equation (4.20) can be similarly applied to equation (4.23) to extract $\underline{\underline{\mathbf{K}}}_{\,\mathrm{Br}}$.

4.3.2 Inverse formalism

The Bruggeman formalism may be implemented in an inverse sense, in a similar manner to that described for the Maxwell Garnett formalism in section 4.2.2. For example, equation (4.20) can be rearranged to provide an estimate of the constitutive dyadic $\underline{\underline{\mathbf{K}}}_{\,\mathrm{a}}$ in terms of the constitutive dyadics $\underline{\underline{\mathbf{K}}}_{\,\mathrm{b}}$ and $\underline{\underline{\mathbf{K}}}_{\,\mathrm{Br}}$, and the particle shape/orientation dyadics $\underline{\underline{U}}^{\mathrm{a}}$ and $\underline{\underline{U}}^{\mathrm{b}}$. One finds [13]

$$\underline{\underline{\mathbf{K}}}_{\,\mathrm{a}} = \underline{\underline{\mathbf{K}}}_{\,\mathrm{Br}} - \left(\frac{f_{\mathrm{a}}}{f_{\mathrm{b}}}\underline{\underline{\mathbf{I}}} + i\omega\, \underline{\underline{\boldsymbol{\alpha}}}^{0}_{\,\mathrm{b}/\mathrm{Br}} \bullet \underline{\underline{\mathbf{D}}}^{0}_{U^{\mathrm{a}}/\mathrm{Br}} \right)^{-1} \bullet \underline{\underline{\boldsymbol{\alpha}}}^{0}_{\,\mathrm{b}/\mathrm{Br}}. \tag{4.24}$$

Significantly, the inverse Bruggeman formalism represented by equation (4.24) delivers an explicit expression $\underline{\underline{\mathbf{K}}}_{\,\mathrm{a}}$. This is unlike the forward Bruggeman formalism represented by equation (4.20) which delivers only an implicit expression for $\underline{\underline{\mathbf{K}}}_{\,\mathrm{Br}}$.

4.4 Strong-property-fluctuation theory

4.4.1 Introduction

As described in section 4.3 for the Bruggeman formalism, in the SPFT approach to homogenization the total space occupied by the two constituent materials is partitioned as $V_{\mathrm{a}} \cup V_{\mathrm{b}}$, where the region V_{a} is occupied by randomly distributed ellipsoidal particles made of the constituent material 'a' and the region V_{b} is likewise

Figure 4.3. Schematic representation of the constituent materials for the SPFT formalism. Ellipsoidal particles of constituent materials 'a' and 'b' are randomly distributed in the HCM. All ellipsoidal particles have the same shape and orientation, as specified by $\underline{\underline{U}}$. In order for the two constituent materials to completely occupy all space, a fractal-like distribution of the constituent particles is envisaged.

occupied by randomly distributed ellipsoidal particles of the constituent material 'b'. Unlike the case for the Bruggeman formalism, the ellipsoidal particles of both constituent materials are required to have the same shape and orientation, specified by $\underline{\underline{U}}$ per equation (3.15). As for the Bruggeman formalism, a fractal-like distribution of constituent materials 'a' and 'b' ensures that the total space is completely filled with ellipsoidal particles. A schematic representation of the constituent materials for the SPFT is provided in figure 4.3.

The characteristic functions

$$\Phi_\ell(\mathbf{r}) = \begin{cases} 1, & \mathbf{r} \in V_\ell, \\ \\ 0, & \mathbf{r} \not\in V_\ell, \end{cases} \qquad (\ell \in \{\mathrm{a,b}\}) \tag{4.25}$$

are introduced to describe the distributions of the constituent particles throughout $V_\mathrm{a} \cup V_\mathrm{b}$. The ensemble average $\langle\, \Phi_\ell(\mathbf{r}_1)\cdots\Phi_\ell(\mathbf{r}_n)\, \rangle_\mathrm{e}$, which is known as the nth moment of $\Phi_\ell(\mathbf{r})$, represents the probability of $\mathbf{r}_1, \ldots, \mathbf{r}_n \in V_\ell$, $(\ell \in \{\mathrm{a,b}\})$. Accordingly, the volume fraction of constituent material ℓ is given as

$$f_\ell = \langle\, \Phi_\ell(\mathbf{r})\, \rangle_\mathrm{e}, \qquad (\ell \in \{\mathrm{a,b}\}), \tag{4.26}$$

and the partition of total space $V_\mathrm{a} \cup V_\mathrm{b}$ implies that $\langle\, \Phi_\mathrm{a}(\mathbf{r})\, \rangle_\mathrm{e} + \langle\, \Phi_\mathrm{b}(\mathbf{r})\, \rangle_\mathrm{e} = 1$. In effect, only the first moments of $\Phi_\mathrm{a,b}(\mathbf{r})$ are employed in the Maxwell Garnett and

Bruggeman formalisms to characterize the distribution of the constituent materials, whereas arbitrarily high-order moments can be accommodated in the SPFT, at least in principle [23, 24]. Accordingly, losses due to coherent scattering can be taken into account in the SPFT, whereas they are not accounted for in the Maxwell Garnett and Bruggeman formalisms.

A homogeneous ambient material, characterized by the Tellegen constitutive dyadic $\underline{\underline{\mathbf{K}}}_{\text{amb}}$ and its corresponding DGF $\underline{\underline{\mathbf{G}}}_{\text{amb}}(\mathbf{r}-\mathbf{r}')$, is central to homogenization via the SPFT. A recursive process of refinement is undertaken, in which a Feynman-diagrammatic method is employed to ensemble average a Born-series representation of the electromagnetic field phasor. A direct approach accommodates only weak spatial fluctuations in the infinity norm of [21]

$$\underline{\underline{\mathbf{K}}}_{\text{a}}\,\Phi_{\text{a}}(\mathbf{r}) + \underline{\underline{\mathbf{K}}}_{\text{b}}\,\Phi_{\text{b}}(\mathbf{r}) - \underline{\underline{\mathbf{K}}}_{\text{amb}}. \tag{4.27}$$

However, a reformulation that accommodates strong spatial fluctuations follows by carefully taking the singularity of $\underline{\underline{\mathbf{G}}}_{\text{amb}}(\mathbf{r}-\mathbf{r}')$ into account [22, 23].

When electromagnetic wavelengths are much larger than the statistical correlation lengths between constituent particles in the composite material, the nth-order SPFT yields a macroscopic description of the composite material as a spatially local and effectively homogeneous material specified by the Tellegen constitutive dyadic $\underline{\underline{\mathbf{K}}}^{[n]}_{\text{SPFT}}$. That is,

$$\underline{\underline{\mathbf{K}}}^{[n]}_{\text{SPFT}} = \underline{\underline{\mathbf{K}}}_{\text{amb}} - \frac{1}{i\omega}\left(\underline{\underline{\mathbf{I}}} + \underline{\underline{\Sigma}}^{[n]} \bullet \underline{\underline{\mathbf{D}}}^{0}_{U/\text{amb}}\right)^{-1} \bullet \underline{\underline{\Sigma}}^{[n]} \tag{4.28}$$

is the nth-order SPFT estimate of $\underline{\underline{\mathbf{K}}}_{\text{HCM}}$ [24, 25]. Herein the *mass operator* term $\underline{\underline{\Sigma}}^{[n]}$ is expressible as an infinite series in terms of the integrated DGF for the ambient material and the corresponding polarizability density dyadics $\underline{\underline{\boldsymbol{\alpha}}}^{0}_{\text{a,b/amb}}$. The truncation of this series is determined by the positive integer n.

In order to eliminate secular terms from the Born-series representation, it is necessary to require that the ensemble average [26]

$$\langle\, \Phi_{\text{a}}(\mathbf{r})\,\underline{\underline{\boldsymbol{\alpha}}}^{0}_{\text{a/amb}}(\omega) + \Phi_{\text{b}}(\mathbf{r})\,\underline{\underline{\boldsymbol{\alpha}}}^{0}_{\text{b/amb}}(\omega)\,\rangle_{\text{e}} = \underline{\underline{\mathbf{0}}}. \tag{4.29}$$

4.4.2 Lowest-order approximation

The zeroth-order and first-order truncations of the mass operator are both null-valued; i.e.,

$$\underline{\underline{\Sigma}}^{[0]} = \underline{\underline{\Sigma}}^{[1]} = \underline{\underline{\mathbf{0}}}. \tag{4.30}$$

Therefore, the zeroth-order and first-order estimates of $\underline{\underline{\mathbf{K}}}_{\text{HCM}}$ emerge trivially from equation (4.28) as

$$\underline{\underline{\mathbf{K}}}^{[0]}_{\text{SPFT}} = \underline{\underline{\mathbf{K}}}^{[1]}_{\text{SPFT}} = \underline{\underline{\mathbf{K}}}_{\text{amb}}. \tag{4.31}$$

Since the non-secularity condition (4.29) is effectively equivalent to equation (4.20), the constitutive dyadic $\underline{\underline{\mathbf{K}}}_{\text{amb}}$ may be extracted therefrom using the Jacobi scheme described in section 4.3.1. Indeed, since the lowest-order SPFT estimates of $\underline{\underline{\mathbf{K}}}_{\text{HCM}}$ are identical to the Bruggeman estimate $\underline{\underline{\mathbf{K}}}_{\text{Br}}$, the lowest-order SPFT provides a rigorous basis for the Bruggeman formalism.

4.4.3 Second-order approximation

Usually the second-order level of approximation, called the *bilocal* approximation, is implemented when calculating the SPFT estimate of $\underline{\underline{\mathbf{K}}}_{\text{HCM}}$ [23, 24, 27–30]. Here the two-point covariance function

$$\begin{aligned}\Lambda(\mathbf{r}-\mathbf{r}') &= \langle\Phi_{\text{a}}(\mathbf{r})\Phi_{\text{a}}(\mathbf{r}')\rangle_{\text{e}} - \langle\Phi_{\text{a}}(\mathbf{r})\rangle_{\text{e}}\langle\Phi_{\text{a}}(\mathbf{r}')\rangle_{\text{e}} \\ &= \langle\Phi_{\text{b}}(\mathbf{r})\Phi_{\text{b}}(\mathbf{r}')\rangle_{\text{e}} - \langle\Phi_{\text{b}}(\mathbf{r})\rangle_{\text{e}}\langle\Phi_{\text{b}}(\mathbf{r}')\rangle_{\text{e}}\end{aligned} \tag{4.32}$$

is employed to characterize the distributional statistics of the constituent materials 'a' and 'b'. The associated correlation length L_{cor} is taken to be small relative to the electromagnetic wavelengths involved. The correlated scattering responses of constituent particles contained within an electrically small region of linear cross-sectional dimensions given by L_{cor}, and of shape dictated by the covariance function, result in an attenuation of the macroscopic coherent field. On the other hand, those scattering interactions between constituent particles that are separated by distances much greater than L_{cor} are taken to be statistically independent.

The second-order truncation of the mass operator is [24]

$$\underline{\underline{\Sigma}}^{[2]}(L_{\text{cor}}) = -\omega^2\left(\underline{\underline{\boldsymbol{\alpha}}}^{0}_{\text{a/amb}} - \underline{\underline{\boldsymbol{\alpha}}}^{0}_{\text{b/amb}}\right) \bullet \underline{\underline{\mathbf{P}}}_{\Lambda}(L_{\text{cor}}) \bullet \left(\underline{\underline{\boldsymbol{\alpha}}}^{0}_{\text{a/amb}} - \underline{\underline{\boldsymbol{\alpha}}}^{0}_{\text{b/amb}}\right). \tag{4.33}$$

Herein, the two-point covariance function and the associated correlation length feature via the principal-value integral defined as

$$\underline{\underline{\mathbf{P}}}_{\Lambda}(L_{\text{cor}}) = \mathsf{P}\int_{\mathbf{R}} \Lambda(\mathbf{R})\, \underline{\underline{\mathbf{G}}}_{\text{amb}}(\mathbf{R})\, d^3\mathbf{R}. \tag{4.34}$$

The second-order SPFT estimate $\underline{\underline{\mathbf{K}}}^{[2]}_{\text{SPFT}}$ clearly depends upon choice of covariance function [31, 32]. The step covariance function

$$\Lambda_{\text{step}}(\mathbf{r}-\mathbf{r}') = \begin{cases} f_{\text{a}} f_{\text{b}}, & |\, \underline{\underline{U}}^{-1} \bullet (\mathbf{r}-\mathbf{r}')\,| \leqslant L_{\text{cor}}, \\ 0, & |\, \underline{\underline{U}}^{-1} \bullet (\mathbf{r}-\mathbf{r}')\,| > L_{\text{cor}}, \end{cases} \tag{4.35}$$

yielding

$$\langle\Phi_\ell(\mathbf{r})\Phi_\ell(\mathbf{r}')\rangle_{\text{e}} = \begin{cases} f_\ell, & |\, \underline{\underline{U}}^{-1} \bullet (\mathbf{r}-\mathbf{r}')\,| \leqslant L_{\text{cor}}, \\ f_\ell^2, & |\, \underline{\underline{U}}^{-1} \bullet (\mathbf{r}-\mathbf{r}')\,| > L_{\text{cor}}, \end{cases} \qquad (\ell \in \{\text{a,b}\}), \tag{4.36}$$

is extensively used. In fact, it transpires that the form of the covariance function has only a weak influence upon the SPFT estimate $\underline{\underline{\mathbf{K}}}^{[2]}_{\mathrm{SPFT}}$, as has been confirmed by theoretical and numerical studies for a variety of physically motivated forms of the covariance function [32]. Notice that in the limit $L_{\mathrm{cor}} \to 0$, the step function $\Lambda_{\mathrm{step}}(\mathbf{r} - \mathbf{r}')$ becomes null valued and thereby $\underline{\underline{\Sigma}}^{[2]}(L_{\mathrm{cor}})$ also becomes null valued. Therefore, the second-order SPFT reduces to the zeroth-order SPFT as the correlation length vanishes, as would be expected on physical grounds.

With the choice $\Lambda_{\mathrm{step}}(\mathbf{r} - \mathbf{r}')$ for the two-point covariance function, the principal-value integral in equation (4.34) may be expressed as

$$\underline{\underline{\mathbf{P}}}_{\Lambda_{\mathrm{step}}}(L_{\mathrm{cor}}) = f_{\mathrm{a}} f_{\mathrm{b}}\, \underline{\underline{\mathbf{D}}}^{+}_{U/\mathrm{amb}}(L_{\mathrm{cor}}). \tag{4.37}$$

The non-principal-value integral represented by $\underline{\underline{\mathbf{D}}}^{+}_{U/\mathrm{amb}}(L_{\mathrm{cor}})$ on the right side of equation (4.37), which is defined in equation $(3.18)_2$, can be simplified in certain cases. That is, for certain anisotropic [28, 29] and bianisotropic [24] ambient materials, $\underline{\underline{\mathbf{D}}}^{+}_{U/\mathrm{amb}}(L_{\mathrm{cor}})$ reduces to a double integral which is equivalent to that given in equation (3.27); however, numerical techniques are usually needed to evaluate this double integral. For isotropic chiral and uniaxial dielectric (or magnetic) ambient materials, explicit expressions for $\underline{\underline{\mathbf{P}}}_{\Lambda_{\mathrm{step}}}(L_{\mathrm{cor}})$ can be found by means of equations (3.33) and (3.39).

4.4.4 Third-order approximation

For the third-order level of approximation, also known as the *trilocal* approximation, a three-point covariance function [33, 34] is needed. By generalizing the two-point covariance function represented in equation (4.36), one finds

$$\langle\, \Phi_{\mathrm{a}}(\mathbf{r})\, \Phi_{\mathrm{a}}(\mathbf{r}')\, \Phi_{\mathrm{a}}(\mathbf{r}'')\, \rangle_{\mathrm{e}} = \begin{cases} f_{\mathrm{a}}^{3}, & \min\{L_{12}, L_{13}, L_{23}\} > L_{\mathrm{cor}} \\ f_{\mathrm{a}}, & \max\{L_{12}, L_{13}, L_{23}\} \leqslant L_{\mathrm{cor}} \\ \dfrac{1}{3}\,(f_{\mathrm{a}} + 2 f_{\mathrm{a}}^{3}), & \text{one of } L_{12}, L_{13}, L_{23} \leqslant L_{\mathrm{cor}} \\ \dfrac{1}{3}\,(2 f_{\mathrm{a}} + f_{\mathrm{a}}^{3}), & \text{two of } L_{12}, L_{13}, L_{23} \leqslant L_{\mathrm{cor}} \end{cases}, \tag{4.38}$$

with the lengths

$$\left.\begin{aligned} L_{12} &= |\, \underline{\underline{U}}^{-1} \bullet (\mathbf{r} - \mathbf{r}')| \\ L_{13} &= |\, \underline{\underline{U}}^{-1} \bullet (\mathbf{r} - \mathbf{r}'')| \\ L_{23} &= |\, \underline{\underline{U}}^{-1} \bullet (\mathbf{r}' - \mathbf{r}'')| \end{aligned}\right\}. \tag{4.39}$$

The three-point covariance function (4.38) gives rise to the third-order truncation of the mass operator [35]

$$\underline{\underline{\Sigma}}^{[3]}(L_{\text{cor}}) = \underline{\underline{\Sigma}}^{[2]}(L_{\text{cor}}) + \frac{i\omega^3 f_{\text{a}}(1 - 2f_{\text{a}})}{3(1 - f_{\text{a}})^2} \underline{\underline{\boldsymbol{\alpha}}}^{0}_{\text{a/amb}} \bullet \Big[\underline{\underline{\mathbf{V}}} \bullet \underline{\underline{\boldsymbol{\alpha}}}^{0}_{\text{a/amb}} \bullet \underline{\underline{\mathbf{P}}}_{\Lambda_{\text{step}}}(L_{\text{cor}}) + \underline{\underline{\mathbf{P}}}_{\Lambda_{\text{step}}}(L_{\text{cor}}) \bullet \underline{\underline{\boldsymbol{\alpha}}}^{0}_{\text{a/amb}} \bullet \underline{\underline{\mathbf{V}}} + \underline{\underline{\mathbf{P}}}_{\Lambda_{\text{step}}}(L_{\text{cor}}) \bullet \underline{\underline{\boldsymbol{\alpha}}}^{0}_{\text{a/amb}} \bullet \underline{\underline{\mathbf{P}}}_{\Lambda_{\text{step}}}(L_{\text{cor}}) \Big] \bullet \underline{\underline{\boldsymbol{\alpha}}}^{0}_{\text{a/amb}}, \tag{4.40}$$

where

$$\underline{\underline{\mathbf{V}}} = \frac{1}{i\omega} \underline{\underline{\mathbf{K}}}^{-1}_{\text{amb}} - \underline{\underline{\mathbf{D}}}^{0}_{U/\text{amb}}. \tag{4.41}$$

Superficially, it may appear from equation (4.40) that $\underline{\underline{\Sigma}}^{[3]}(L_{\text{cor}}) - \underline{\underline{\Sigma}}^{[2]}(L_{\text{cor}})$ does not depend upon $\underline{\underline{\mathbf{K}}}_{\text{b}}$. However, this is not the case: the dependency of the third-order mass operator upon the constitutive properties of constituent material 'b' enters via the non-secularity condition (4.29) that $\underline{\underline{\boldsymbol{\alpha}}}^{0}_{\text{a/amb}}$ must satisfy.

The derivation of equation (4.40) holds rigorously for isotropic chiral ambient materials associated with spherical constituent particles [35]. However, it is likely that this result extends to weakly bianisotropic materials characterized by diagonally dominant 3×3 constitutive dyadics. Furthermore, on the basis of equation (4.40), numerical experiments have revealed that the SPFT estimates of $\underline{\underline{\mathbf{K}}}_{\text{HCM}}$ converge at the second-order level of approximation in the cases of certain isotropic chiral HCMs and certain bianisotropic HCMs of the Faraday chiral type that are simultaneously weakly uniaxial and weakly gyrotropic [35]. An illustration of SPFT convergence is presented in section 6.2.

4.5 Extended formalisms

As described in sections 4.2, 4.3, and 4.4, the Maxwell Garnett, Bruggeman, and SPFT approaches to homogenization are all based on depolarization dyadics. These dyadics are required to represent the scattering responses of the electrically small particles of which the composite materials are composed. The standard convention was followed in sections 4.2–4.4 whereby the spatial extent of the constituent particles is neglected, thereby allowing $\underline{\underline{\mathbf{D}}}^{0}_{U/\text{amb}}$ to be taken as the depolarization dyadic, per equation (3.23).

By taking $\underline{\underline{\mathbf{D}}}_{U/\text{amb}}(\rho)$ as the depolarization dyadic, per equation (3.23), instead of $\underline{\underline{\mathbf{D}}}^{0}_{U/\text{amb}}$, and $\underline{\underline{\Sigma}}_{\ell/\text{amb}}(\rho)$ instead of $\underline{\underline{\Sigma}}^{0}_{\ell/\text{amb}}$, per equation (3.61), the extended versions of the Maxwell Garnett (including its incremental and differential variants), Bruggeman, and SPFT formalisms emerge [36]. These extended formalisms take the spatial extent of the constituent particles into account, via the particle size-parameter ρ. Such extended versions have been established for bianisotropic HCMs, arising from constituent materials composed of ellipsoidal particles [37]. In addition, numerical studies have revealed that the extended SPFT formalism converges at the bilocal level of approximation for certain bianisotropic HCMs [38]. A variety of alternative extended homogenization formalisms, not explicitly involving the spectral representation of the depolarization dyadic but still based on the Maxwell Garnett and Bruggeman formalisms, have been established for isotropic HCMs [39–44].

While the constituent particles catered for in extended homogenization formalisms are not regarded as infinitesimally small, they must nevertheless be much smaller than the wavelengths in the constituent materials and in the HCM [3]; otherwise, the process of homogenization would be nonsensical [45]. In the depolarization-dyadic approach adopted in the Maxwell Garnett, Bruggeman, and SPFT formalisms, the constituent particles are effectively represented as co-located electric and magnetic dipoles. Higher-order multipolar effects do not feature at all. Indeed, higher-order multipolar approaches to homogenization are fraught with difficulties [46]. Furthermore, extended approaches to homogenization generally do not scale in a physically intuitive manner. However, the physical significance of this feature is uncertain [47]. For consistency with the fractal-like distribution of constituent particles that underlies the Bruggeman and SPFT formalisms (see figure 4.2 and 4.3), the size parameter ρ should be regarded as an upper bound on particles sizes for these formalisms.

Since more detailed information concerning the constituent materials is taken into account in extended homogenization formalisms, it is a reasonable expectation that these formalisms will deliver better estimates of $\underline{\underline{\mathbf{K}}}_{\text{HCM}}$ than their standard counterparts. However, whether or not this expectation is realized for a particular composite material can only be confirmed by checking against carefully conducted experimental measurements.

Parenthetically, notice that details about the sizes of constituent particles may be inferred by implementing extended formalisms in the inverse sense, as outlined in section 4.2.2 and 4.2.3. For example, this may be exploited for particle-sizing applications for nanocomposite materials [4].

References

[1] Lakhtakia A (ed) 1996 *Selected Papers on Linear Optical Composite Materials* (Bellingham, WA: SPIE Optical Engineering Press)

[2] van de Hulst H C 1981 *Light Scattering by Small Particles* (New York: Dover)

[3] Mackay T G 2008 Lewin's homogenization formula revisited for nanocomposite materials *J. Nanophoton.* **2** 029503

[4] Oates T W H 2006 Real time spectroscopic ellipsometry of nanoparticle growth *Appl. Phys. Lett.* **88** 213115

[5] Bohren C F and Huffman D R 1983 *Absoption and Scattering of Light by Small Particles* (New York: Wiley)

[6] Kreibig U and von Fragstein C 1969 The limitation of electron mean free path in small silver particles *Z. Physik* **224** 307–23

[7] Smith D R, Vier D C, Koschny Th and Soukoulis C M 2005 Electromagnetic parameter retrieval from inhomogeneous metamaterials *Phys. Rev.* E **71** 036617

[8] Aspnes D E 1982 Local-field effects and effective-medium theory: a microscopic perspective *Am. J. Phys.* **50** 704–9 (Reproduced in [1])

[9] Weiglhofer W S, Lakhtakia A and Michel B 1997 Maxwell Garnett and Bruggeman formalisms for a particulate composite with bianisotropic host medium *Microw. Opt. Technol. Lett.* **15** 263–6
Weiglhofer W S, Lakhtakia A and Michel B 1999 Correction to 'Maxwell Garnett and Bruggeman formalisms for a particulate composite with bianisotropic host medium' *Microw. Opt. Technol. Lett.* **22** 221 (correction)
[10] Bragg W L and Pippard A B 1953 The form birefringence of macromolecules *Acta Crystallogr.* **6** 865–7
[11] Sherwin J A and Lakhtakia A 2002 Bragg-Pippard formalism for bianisotropic particulate composites *Microw. Opt. Technol. Lett.* **33** 40–4
[12] Cherkaev E 2001 Inverse homogenization for evaluation of effective properties of a mixture *Inverse Problems* **17** 1203–18
[13] Weiglhofer W S 2001 On the inverse homogenization problem of linear composite materials *Microw. Opt. Technol. Lett.* **28** 421–3
[14] Lakhtakia A 1998 Incremental Maxwell Garnett formalism for homogenizing particulate composite media *Microw. Opt. Technol. Lett.* **17** 276–9
[15] Michel B, Lakhtakia A, Weiglhofer W S and Mackay T G 2001 Incremental and differential Maxwell Garnett formalisms for bi-anisotropic composites *Compos. Sci. Technol.* **61** 13–8
[16] Ghosh K and Fuchs R 1991 Critical behavior in the dielectric properties of random self-similar composites *Phys. Rev.* B **44** 7330–43
[17] Barrera R G and Fuchs R 1995 Theory of electron energy loss in a random system of spheres *Phys. Rev.* B **52** 3256–73
[18] Fuchs R, Barrera R G and Carrillo J L 1996 Spectral representations of the electron energy loss in composite media *Phys. Rev.* B **54** 12824–34
[19] Sosa I O, Mendoza C I and Barrera R G 2001 Calculation of electron-energy-loss spectra of composites and self-similar structures *Phys. Rev.* E **63** 144201
[20] Michel B, Lakhtakia A and Weiglhofer W S 1998 Homogenization of linear bianisotropic particulate composite media–Numerical studies *Int. J. Appl. Electromag. Mech.* **9** 167–78
Michel B, Lakhtakia A and Weiglhofer W S 1999 Homogenization of linear bianisotropic particulate composite media—Numerical studies *Int. J. Appl. Electromag. Mech.* **10** 537–8 (erratum)
[21] Ryzhov Yu A and Tamoikin V V 1970 Radiation and propagation of electromagnetic waves in randomly inhomogeneous media (review) *Radiophys. Quantum Electron.* **14** 228–33
[22] Tatarskii V I and Zavorotnyi V U 1980 Strong fluctuations in light propagation in a randomly inhomogeneous medium *Prog. Optics* **18** 204–56
[23] Tsang L and Kong J A 1981 Scattering of electromagnetic waves from random media with strong permittivity fluctuations *Radio Sci.* **16** 303–20 (Reproduced in [1])
[24] Mackay T G, Lakhtakia A and Weiglhofer W S 2000 Strong-property-fluctuation theory for homogenization of bianisotropic composites: formulation *Phys. Rev.* E **62** 6052–64
Mackay T G, Lakhtakia A and Weiglhofer W S 2001 Strong-property-fluctuation theory for homogenization of bianisotropic composites: formulation Phys. Rev. *E* **63** 049901 (erratum)
[25] Mackay T G, Lakhtakia A and Weiglhofer W S 2001 Ellipsoidal topology, orientation diversity and correlation length in bianisotropic composite mediums *Arch. Elektron. Übertrag.* **55** 243–51
[26] Frisch U 1970 Wave propagation in random media *Probabilistic Methods in Applied Mathematics* vol 1 ed A T Bharucha-Reid (London: Academic) pp 75–198

[27] Stogryn A 1983 The bilocal approximation for the electric field in strong fluctuation theory *IEEE Trans. Antennas Propagat.* **31** 985–6
[28] Genchev Z D 1992 Anisotropic and gyrotropic version of Polder and van Santenas mixing formula *Waves Random Media* **2** 99–110
[29] Zhuck N P 1994 Strong-fluctuation theory for a mean electromagnetic field in a statistically homogeneous random medium with arbitrary anisotropy of electrical and statistical properties *Phys. Rev.* B **50** 15636–45
[30] Michel B and Lakhtakia A 1995 Strong-property-fluctuation theory for homogenizing chiral particulate composites *Phys. Rev.* E **51** 5701–7
[31] Tsang L, Kong J A and Newton R W 1982 Application of strong fluctuation random medium theory to scattering of electromagnetic waves from a half-space of dielectric mixture *IEEE Trans. Antennas Propagat.* **30** 292–302
[32] Mackay T G, Lakhtakia A and Weiglhofer W S 2001 Homogenisation of similarly oriented, metallic, ellipsoidal inclusions using the bilocal-approximated strong-property-fluctuation theory *Opt. Commun.* **197** 89–95
[33] Miller M N 1969 Bounds for effective electrical, thermal, and magnetic properties of heterogeneous materials *J. Math. Phys.* **10** 1988–2004
[34] Milton G W and Phan-Thien N 1982 New bounds on effective elastic moduli of two-component materials *Proc. R. Soc. Lond.* A **380** 305–31
[35] Mackay T G, Lakhtakia A and Weiglhofer W S 2001 Third-order implementation and convergence of the strong-property-fluctuation theory in electromagnetic homogenisation *Phys. Rev.* E **64** 066616
[36] Mackay T G 2004 Depolarization volume and correlation length in the homogenization of anisotropic dielectric composites *Waves Random Media* **14** 485–98
Mackay T G 2006 Erratum: Depolarization volume and correlation length in the homoge nization of anisotropic dielectric composites *Waves Random Complex Media* **16** 85 (erratum)
[37] Cui J and Mackay T G 2007 Depolarization regions of nonzero volume in bianisotropic homogenized composites *Waves Random Complex Media* **17** 269–81
[38] Cui J and Mackay T G 2007 On convergence of the extended strong-property-fluctuation theory for bianisotropic homogenized composites *Electromagnetics* **27** 495–506
[39] Doyle W T 1989 Optical properties of a suspension of metal spheres *Phys. Rev.* B **39** 9852–8
[40] Prinkey M T, Lakhtakia A and Shanker B 1994 On the extended Maxwell-Garnett and the extended Bruggeman approaches for dielectric-in-dielectric composites *Optik* **96** 25–30
[41] Shanker B and Lakhtakia A 1993 Extended Maxwell Garnett model for chiral-in-chiral composites *J. Phys. D: Appl. Phys.* **26** 1746–58
[42] Shanker B 1996 The extended Bruggeman approach for chiral-in-chiral mixtures *J. Phys. D: Appl. Phys.* **29** 281–8
[43] Lakhtakia A and Shanker B 1993 Beltrami fields within continuous source regions, volume integral equations, scattering algorithms, and the extended Maxwell-Garnett model *Int. J. Appl. Electromagn. Mater.* **4** 65–82
[44] Mallet P, Guérin C A and Sentenac A 2005 Maxwell-Garnett mixing rule in the presence of multiple scattering: Derivation and accuracy *Phys. Rev.* B **72** 014205
[45] Lakhtakia A 1996 Cakes and pastries, and linear optical composite materials too *Selected Papers on Linear Optical Composite Materials* ed A Lakhtakia (Bellingham, WA: SPIE Optical Engineering Press) pp xiii–xxiv

[46] Bohren C F 1986 Applicability of effective-medium theories to problems of scattering and absorption by nonhomogeneous atmospheric particles *J. Atmos. Sci.* **43** 468–75
[47] Bohren C F 2009 Do extended effective-medium formulas scale properly? *J. Nanophoton.* **3** 039501

Tom G Mackay and Akhlesh Lakhtakia

Chapter 5

Homogenization formalisms: nonlinear materials

The presentation of homogenization formalisms for bianisotropic materials in chapter 4 is extended in this chapter into the realm of nonlinear materials, as far as it is practicable. The introduction of nonlinearity greatly enriches the palette of electromagnetic phenomenons that are displayed, especially in the context of anisotropic and bianisotropic materials. However, the introduction of nonlinearity is also accompanied by a substantial increase in the degree of mathematical complexity. Consequently, homogenization formalisms are available only for weakly nonlinear materials. Even with this restriction, there is considerable scope for complex nonlinear behavior—such as frequency-mixing phenomenons, the optical Kerr effect, and stimulated Brillouin scattering [1, 2].

After a brief discussion of some preliminary matters, the Maxwell Garnett formalism is set forth in this chapter for weakly nonlinear bianisotropic materials. This is followed by a presentation of the weakly nonlinear strong-property-fluctuation theory (SPFT) for isotropic dielectric, isotropic chiral, and anisotropic dielectric materials. Recall that the SPFT formalism contains the Bruggeman formalism as its lowest-order variant.

5.1 Preliminaries

The foundation for the homogenization of nonlinear composite materials is provided by the constitutive relation (2.46) which characterizes a generally non-linear, homogeneous bianisotropic material. The linear and nonlinear contributions to the 6-vector $\underline{\mathbf{N}}(\mathbf{r}, \omega)$ therein are provided by $\underline{\mathbf{N}}_{\mathrm{L}}(\mathbf{r}, \omega)$ and $\underline{\mathbf{N}}_{\mathrm{NL}}(\mathbf{r}, \omega)$, as defined in equations (2.50) and (2.52), respectively, per the partition represented by equation (2.49). Let us reiterate: from the constitutive relation (2.46) it may be inferred that when an ensemble of primitive electromagnetic fields oscillating at angular

doi:10.1088/978-0-7503-3423-5ch5

frequencies $\omega_m \in \mathcal{W}$, $(m \in \{1,2,...,M\})$, are simultaneously launched into the nonlinear material, their combined effect in the nonlinear material is an electromagnetic field oscillating at angular frequency ω_{NL}, as given in equation (2.53). For consistency with the concept of homogenization, the particles of the constituent materials are much smaller in size than the electromagnetic wavelengths at ω_{NL} as well as at every angular frequency $\omega_m \in \mathcal{W}$.

The mathematical intractability that can be associated with nonlinearity, particularly when accompanied by anisotropy or bianisotropy, has largely restricted theoretical progress on homogenization to weakly nonlinear materials. Henceforth in this chapter, attention is limited to the regime represented by satisfaction of the inequality

$$| \underline{\mathbf{N}}_{\mathrm{L}}(\mathbf{r}, \omega) | \gg | \underline{\mathbf{N}}_{\mathrm{NL}}(\mathbf{r}, \omega) |. \tag{5.1}$$

Essentially, within this regime the linear and nonlinear contributions to the constitutive relations of the HCM may be handled independently [3–5]. In the following sections, the Maxwell Garnett and SPFT formalisms for weakly nonlinear HCMs are delineated in their most general formulations presently available, with the lowest-order SPFT formalism being equivalent to the Bruggeman formalism, as described in section 4.4.2.

5.2 Maxwell Garnett formalism

As in section 4.1.1, the homogenization of a random mixture of two constituent materials, labeled ‘a’ and ‘b’, is considered. As in section 4.2, the constituent material ‘b’ is viewed as the host material and particles of constituent material ‘a’ are viewed as particles which are randomly distributed throughout the host material. The particles are generally regarded as ellipsoidal particles, whose shape and orientation are specified by $\underline{\underline{U}}^{\mathrm{a}}$, per equation (3.15). The particles all have the same shape and orientation. No particular topology is assigned to the host material, which may or may not be particulate. Since the particles are assumed to be dilutely dispersed in the host material, the volume fraction f_{a} of the constituent material ‘a’ is restricted to $f_{\mathrm{a}} \lesssim 0.3$. The schematic representation of the modeling approach adopted, which is provided in figure 4.3 for linear materials, also applies here for nonlinear materials.

The constituent material ‘a’ is taken to be a weakly nonlinear bianisotropic material characterized by constitutive relations expressed compactly as equation (2.46). The linear contribution to the 6-vector $\underline{\mathbf{N}}(\mathbf{r}, \omega)$ is provided by the right side of equation (2.50) wherein the Tellegen constitutive dyadic $\underline{\underline{\mathbf{K}}}^{\mathrm{L}}(\omega) = \underline{\underline{\mathbf{K}}}^{\mathrm{L}}_{\mathrm{a}}(\omega)$[1]. The components of the weakly nonlinear contribution to $\underline{\mathbf{N}}(\mathbf{r}, \omega)$ are provided by the right side of equation (2.52) wherein the nonlinear susceptibility tensor $\chi^{\mathrm{NL}}_{jj_1j_2\cdots j_m\cdots j_M}(\omega_{\mathrm{NL}}; \mathcal{W}) = {}^{\mathrm{a}}\chi^{\mathrm{NL}}_{jj_1j_2\cdots j_m\cdots j_M}(\omega_{\mathrm{NL}}; \mathcal{W})$. In contrast, the constituent material

[1] Since frequency-mixing phenomenons are accommodated in this section, it is necessary to explicitly state the angular-frequency dependencies in the arguments of all quantities concerned.

'b' is taken to be a linear bianisotropic material characterized by the Tellegen constitutive dyadic $\underline{\underline{\mathbf{K}}}_{\,\mathrm{b}}(\omega)$.

In the weakly nonlinear regime represented by the inequality (5.1), the Maxwell Garnett formalism yields constitutive relations of the HCM as [5]

$$\underline{\mathbf{C}}(\mathbf{r},\,\omega) = \underline{\underline{\mathbf{K}}}^{\mathrm{L}}_{\,\mathrm{MG}}(\omega) \bullet \underline{\mathbf{F}}(\mathbf{r},\,\omega) + s(\omega,\,\omega_{\mathrm{NL}})\, \underline{\mathbf{N}}^{\mathrm{NL}}_{\,\mathrm{MG}}(\mathbf{r},\,\omega),\ \omega \in \mathcal{W} \cup \{\omega_{\mathrm{NL}}\}. \quad (5.2)$$

Here the 6-vector

$$\underline{\mathbf{N}}^{\mathrm{NL}}_{\,\mathrm{MG}}(\mathbf{r},\,\omega_{\mathrm{NL}}) = \left\{\underline{\underline{\mathbf{I}}} - i\omega_{\mathrm{NL}}\left[\,\underline{\underline{\mathbf{I}}} - i\omega_{\mathrm{NL}}\, f_{\mathrm{a}}\, \underline{\underline{\mathbf{D}}}^{0}_{\,I/\mathrm{b}}(\omega_{\mathrm{NL}}) \bullet \underline{\underline{\boldsymbol{\alpha}}}^{\,0}_{\,\mathrm{a/b}}(\omega_{\mathrm{NL}})\,\right]^{-1} \bullet \underline{\underline{\mathbf{D}}}^{0}_{\,I/\mathrm{b}}(\omega_{\mathrm{NL}})\right\} \bullet \underline{\check{\mathbf{N}}}^{\mathrm{NL}}_{\,\mathrm{MG}}(\mathbf{r},\,\omega_{\mathrm{NL}}), \quad (5.3)$$

while the components of the 6-vector $\underline{\check{\mathbf{N}}}^{\mathrm{NL}}_{\,\mathrm{MG}}(\mathbf{r},\,\omega_{\mathrm{NL}})$ therein may be expressed as

$$[\,\underline{\check{\mathbf{N}}}_{\,\mathrm{NL}}(\mathbf{r},\,\omega_{\mathrm{NL}})]_j = \sum_{j_1=1}^{6}\sum_{j_2=1}^{6}\cdots\sum_{j_m=1}^{6}\cdots\sum_{j_M=1}^{6}\left({}^{\mathrm{a}}\chi^{\mathrm{NL}}_{jj_1j_2\cdots j_m\cdots j_M}(\omega_{\mathrm{NL}};\,\mathcal{W}) \times \prod_{n=1}^{M}\left\{\left[\,\underline{\underline{\mathbf{K}}}^{\mathrm{L}}_{\,\mathrm{a}}(\omega_n) - \underline{\underline{\mathbf{K}}}^{\mathrm{L}}_{\,\mathrm{MG}}(\omega_n)\,\right]^{-1} \bullet \underline{\underline{\boldsymbol{\alpha}}}^{\,0}_{\,\mathrm{a/MG}}(\omega_n) \bullet \underline{\mathbf{F}}(\mathbf{r},\,\omega_n)\right\}_{j_n}\right). \quad (5.4)$$

The Tellegen constitutive dyadic $\underline{\underline{\mathbf{K}}}^{\mathrm{L}}_{\,\mathrm{MG}}(\omega_n)$ representing the linear contribution to the HCM constitutive properties is delivered as in equation (4.9) but with $\underline{\underline{\mathbf{K}}}^{\mathrm{L}}_{\,\mathrm{a}}(\omega_n)$ in lieu of $\underline{\underline{\mathbf{K}}}_{\,\mathrm{a}}$. The polarizability density dyadic $\underline{\underline{\boldsymbol{\alpha}}}^{\,0}_{\,\mathrm{a/MG}}(\omega_n)$ is defined as in equation (3.61), but with $\underline{\underline{\mathbf{K}}}^{\mathrm{L}}_{\,\mathrm{a}}(\omega_n)$ in lieu of $\underline{\underline{\mathbf{K}}}_{\,\mathrm{a}}$ and $\underline{\underline{\mathbf{K}}}^{\mathrm{L}}_{\,\mathrm{MG}}(\omega_n)$ in lieu of $\underline{\underline{\mathbf{K}}}_{\,\mathrm{MG}}$. Notice that the presence of $\underline{\check{\mathbf{N}}}_{\,\mathrm{NL}}(\mathbf{r},\,\omega_{\mathrm{NL}})$ renders equation (5.2) an implicit nonlinear relation in terms of $\underline{\underline{\mathbf{K}}}^{\mathrm{L}}_{\,\mathrm{MG}}(\omega_n)$. Accordingly, numerical techniques are needed to extract the effective constitutive parameters from equation (5.2). The switching function

$$s(\omega,\,\omega_{\mathrm{NL}}) = \begin{cases} 1, & \omega = \omega_{\mathrm{NL}} \\ 0, & \omega \neq \omega_{\mathrm{NL}} \end{cases} \quad (5.5)$$

ensures that linear processes are accommodated at all angular frequencies in the set $\mathcal{W} \cup \{\omega_{\mathrm{NL}}\}$, whereas nonlinear processes are accommodated only when $\omega = \omega_{\mathrm{NL}}$.

The nonlinear bianisotropic Maxwell Garnett formalism represented by equation (5.2) is built upon the foundations laid by the corresponding nonlinear isotropic dielectric formalism, incorporating spherical constituent particles [4], and extensions of this appropriate to biisotropic [6] and gyrotropic [7] HCMs.

The sizes of the particles are not taken into account in the Maxwell Garnett formalism underlying equation (5.2), since this result incorporates the depolarization dyadics $\underline{\underline{\mathbf{D}}}^{0}_{\,U^{a/b},\,MG}(\omega_{\mathrm{NL}})$ and the polarizability density dyadics $\underline{\underline{\boldsymbol{\alpha}}}^{\,0}_{\,\mathrm{a/b,MG}}(\omega_{\mathrm{NL}})$ which are independent of the size of the particles. In a fashion analogous to that described in section 4.5 for linear HCMs, the extended version of the Maxwell Garnett formalism does take the size of the particles into account through the parameter ρ. This extended formalism arises on replacing $\underline{\underline{\mathbf{D}}}^{0}_{U^{a/b},\,MG}(\omega_{\mathrm{NL}})$ by $\underline{\underline{\mathbf{D}}}_{U^{a/b},MG}(\rho,\,\omega_{\mathrm{NL}})$ and $\underline{\underline{\boldsymbol{\alpha}}}^{0}_{\mathrm{a/b,MG}}(\omega_{\mathrm{NL}})$ by $\underline{\underline{\boldsymbol{\alpha}}}_{\mathrm{a/b,MG}}(\rho,\,\omega_{\mathrm{NL}})$, per equations (3.17) and (3.61), respectively.

5.3 Strong-property-fluctuation theory

Next, consider the second-order SPFT for the homogenization of two constituent materials, labeled as 'a' and 'b', both of which are taken to be homogeneous. Recall from section 4.4 that the shapes and orientations of all constituent particles are required to be the same for the SPFT; i.e., $\underline{\underline{U}}^{\mathrm{a}} = \underline{\underline{U}}^{\mathrm{b}} = \underline{\underline{U}}$. The lowest-order versions (i.e., the zeroth- and first-order versions) represent the Bruggeman formalism, in same way as described for linear composite materials in section 4.4.2.

Suppose that the constituent material 'a' is an isotropic dielectric material, characterized by an intensity-dependent permittivity scalar of the form

$$\varepsilon_{\mathrm{a}} = \varepsilon_{\mathrm{a}}^{\mathrm{L}} + \varepsilon_{\mathrm{a}}^{\mathrm{NL}}\ |\,\mathbf{E}_{\mathrm{a}}\,|^{2}, \tag{5.6}$$

with $\varepsilon_{\mathrm{a}}^{\mathrm{L}}$ and $\varepsilon_{\mathrm{a}}^{\mathrm{NL}}$ being the linear and weakly nonlinear contributions to the permittivity, respectively[2]. Particles of the constituent material 'a' are assumed to be sufficiently small that the electric field phasor $\mathbf{E}_{\mathrm{a}}$ is uniform inside each particle. The nonlinearity represented by the intensity-dependent permittivity in equation (5.6) gives rise to the optical Kerr effect [1], which in turn gives rise to technoscientifically important phenomenons such as self-focusing and self-trapping of light [8, 9]. Whereas nature provides us with many materials whose permittivities are characterized by equation (5.6), engineered HCMs with intensity-dependent permittivities are especially noteworthy because the process of homogenization can enhance the nonlinearity [10], as elaborated upon in section 6.10. While the materials considered here with intensity-dependent permittivity are weakly nonlinear, per inequality (5.1), such materials have the propensity to exhibit the often strong process of stimulated Brillouin scattering by means of electrostriction [1].

The nature of the constituent material 'b' depends upon the homogenization scenario under consideration. The second-order SPFT has been formulated for the following scenarios:

- Nonlinear isotropic dielectric HCMs [11, 12] are considered in section 5.3.1. The constituent material 'b' is simply a linear isotropic dielectric material, and both constituent materials are randomly distributed as spherical particles.
- Nonlinear isotropic chiral HCMs [13] are considered in section 5.3.2. The constituent material 'b' is a linear isotropic chiral material, and both constituent materials are randomly distributed spherical particles.
- Nonlinear anisotropic dielectric HCMs [14–16] are considered in section 5.3.1. The constituent material 'b' is an isotropic dielectric material characterized by an intensity-dependent permittivity of the form represented in

[2] Since frequency-mixing phenomenons are not accommodated in this section, the explicit dependency on angular frequency is omitted from the arguments of all ω-dependent quantities.

equation (5.6), and anisotropy arises from the ellipsoidal shape of the oriented constituent particles.

5.3.1 Isotropic dielectric composite materials

The most straightforward weakly nonlinear homogenization scenario, at least from a mathematical perspective, is that wherein both constituent materials are isotropic dielectric materials, and both are randomly distributed as spherical particles. That is, the constituent material 'a' is characterized by the intensity-dependent permittivity given in equation (5.6), while the constituent material 'b' is simply a linear material characterized by the scalar permittivity ε_b, and the shapes of the constituent particles are such that $\underline{\underline{U}} = \underline{\underline{I}}$.

The ambient material—which represents the starting point for the iterative refinement process in the SPFT, as described in section 4.4.1—inherits the spacetime symmetries and nonlinear properties of the constituent materials. Thus, it is a nonlinear, isotropic dielectric material characterized by the intensity-dependent permittivity

$$\varepsilon_{\text{amb}} = \varepsilon_{\text{amb}}^{\text{L}} + \varepsilon_{\text{amb}}^{\text{NL}} \mid \mathbf{E}_{\text{HCM}} \mid^2, \tag{5.7}$$

with $\mathbf{E}_{\text{HCM}}$ being the spatially averaged electric field in the HCM. The linear contribution to ε_{amb} is $\varepsilon_{\text{amb}}^{\text{L}}$. It arises as the solution to the isotropic dielectric version of the scalar Bruggeman equation (6.10), with ε_{Br} and ε_{a} therein being substituted by $\varepsilon_{\text{amb}}^{\text{L}}$ and $\varepsilon_{\text{a}}^{\text{L}}$, respectively. The nonlinear contribution to ε_{amb} is delivered by [11]

$$\varepsilon_{\text{amb}}^{\text{NL}} = g_{\varepsilon}^{\text{FF}} \, \varepsilon_{\text{a}}^{\text{NL}} \frac{\partial \varepsilon_{\text{amb}}^{\text{L}}}{\partial \varepsilon_{\text{a}}^{\text{L}}}. \tag{5.8}$$

From equation (6.10), the partial derivative term on the right side of equation (5.8) is given by

$$\frac{\partial \varepsilon_{\text{amb}}^{\text{L}}}{\partial \varepsilon_{\text{a}}^{\text{L}}} = \frac{f_{\text{a}} \varepsilon_{\text{amb}}^{\text{L}} (\varepsilon_{\text{b}} + 2\varepsilon_{\text{amb}}^{\text{L}})^2}{f_{\text{a}} \varepsilon_{\text{a}}^{\text{L}} (\varepsilon_{\text{b}} + 2\varepsilon_{\text{amb}}^{\text{L}})^2 + f_{\text{b}} \varepsilon_{\text{b}} (\varepsilon_{\text{a}}^{\text{L}} + 2\varepsilon_{\text{amb}}^{\text{L}})^2}. \tag{5.9}$$

Various plausible estimates of the local field factor

$$g_{\varepsilon}^{\text{FF}} = \frac{d \mid \mathbf{E}_{\text{a}} \mid^2}{d \mid \mathbf{E}_{\text{HCM}} \mid^2} \tag{5.10}$$

have been proposed [11, 17], among which the estimate

$$g_{\varepsilon}^{\text{FF}} = \left| \frac{3\varepsilon_{\text{amb}}^{\text{L}}}{\varepsilon_{\text{a}}^{\text{L}} + 2\varepsilon_{\text{amb}}^{\text{L}}} \right|^2 \tag{5.11}$$

has been implemented [18].

The second-order SPFT estimate $\varepsilon_{\text{SPFT}}^{[2]}$ of the permittivity of the HCM shares the nonlinearity and spacetime symmetries of the ambient material; i.e.,

$$\varepsilon_{\text{SPFT}}^{[2]} = \varepsilon_{\text{SPFT}}^{[2],\text{L}} + \varepsilon_{\text{SPFT}}^{[2],\text{NL}} \mid \mathbf{E}_{\text{HCM}} \mid^2. \tag{5.12}$$

For the step covariance function $\Lambda_{\text{step}}(\mathbf{r} - \mathbf{r}')$ prescribed in equation (4.36) with $\underline{\underline{U}} = \underline{\underline{I}}$, the linear contribution is given by [12]

$$\begin{aligned}\varepsilon_{\text{SPFT}}^{[2],\text{L}} = \varepsilon_{\text{amb}}^{\text{L}} + \mu_0(\omega L_{\text{cor}} \varepsilon_{\text{amb}}^{\text{L}})^2 \\ \times\left[3 + i2\omega L_{\text{cor}}\left(\mu_0 \varepsilon_{\text{amb}}^{\text{L}}\right)^{1/2}\right]\left[f_{\text{a}}\left(\frac{\varepsilon_{\text{a}}^{\text{L}} - \varepsilon_{\text{amb}}^{\text{L}}}{\varepsilon_{\text{a}}^{\text{L}} + 2\varepsilon_{\text{amb}}^{\text{L}}}\right)^2\right. \\ \left.+ f_{\text{b}}\left(\frac{\varepsilon_{\text{b}} - \varepsilon_{\text{amb}}^{\text{L}}}{\varepsilon_{\text{b}} + 2\varepsilon_{\text{amb}}^{\text{L}}}\right)^2\right]\end{aligned} \tag{5.13}$$

and the nonlinear contribution by

$$\begin{aligned}\varepsilon_{\text{SPFT}}^{[2],\text{NL}} = \varepsilon_{\text{amb}}^{\text{NL}} + \mu_0(\omega L_{\text{cor}})^2\varepsilon_{\text{amb}}^{\text{L}}\left\{2\varepsilon_{\text{amb}}^{\text{L}}\left[3 + i2\omega L_{\text{cor}}\left(\mu_0 \varepsilon_{\text{amb}}^{\text{L}}\right)^{1/2}\right]\right. \\ \times\left[f_{\text{a}}\left(\frac{\varepsilon_{\text{a}}^{\text{L}} - \varepsilon_{\text{amb}}^{\text{L}}}{\varepsilon_{\text{a}}^{\text{L}} + 2\varepsilon_{\text{amb}}^{\text{L}}}\right)^2\left(\frac{g_\varepsilon^{\text{FF}}\varepsilon_{\text{a}}^{\text{NL}} - \varepsilon_{\text{amb}}^{\text{NL}}}{\varepsilon_{\text{a}}^{\text{L}} - \varepsilon_{\text{amb}}^{\text{L}}} - \frac{g_\varepsilon^{\text{FF}}\varepsilon_{\text{a}}^{\text{NL}} + 2\varepsilon_{\text{amb}}^{\text{NL}}}{\varepsilon_{\text{a}}^{\text{L}} + 2\varepsilon_{\text{amb}}^{\text{L}}}\right)\right. \\ \left.- f_{\text{b}}\left(\frac{\varepsilon_{\text{b}} - \varepsilon_{\text{amb}}^{\text{L}}}{\varepsilon_{\text{b}} + 2\varepsilon_{\text{amb}}^{\text{L}}}\right)^2\left(\frac{\varepsilon_{\text{amb}}^{\text{NL}}}{\varepsilon_{\text{b}} - \varepsilon_{\text{amb}}^{\text{L}}} + \frac{2\varepsilon_{\text{amb}}^{\text{NL}}}{\varepsilon_{\text{b}} + 2\varepsilon_{\text{amb}}^{\text{L}}}\right)\right] \\ + \varepsilon_{\text{amb}}^{\text{NL}}\left[6 + i5\omega L_{\text{cor}}\left(\mu_0 \varepsilon_{\text{amb}}^{\text{L}}\right)^{1/2}\right]\left[f_{\text{a}}\left(\frac{\varepsilon_{\text{a}}^{\text{L}} - \varepsilon_{\text{amb}}^{\text{L}}}{\varepsilon_{\text{a}}^{\text{L}} + 2\varepsilon_{\text{amb}}^{\text{L}}}\right)^2\right. \\ \left.\left.+ f_{\text{b}}\left(\frac{\varepsilon_{\text{b}} - \varepsilon_{\text{amb}}^{\text{L}}}{\varepsilon_{\text{b}} + 2\varepsilon_{\text{amb}}^{\text{L}}}\right)^2\right]\right\}.\end{aligned} \tag{5.14}$$

As is the case for the linear SPFT estimates [19], the weakly nonlinear SPFT estimates are not especially sensitive to the choice of covariance function, so long as the covariance function chosen is physically reasonable. For example, the choice of a Gaussian covariance function, as opposed to the step covariance function chosen here, results in only very modest differences to the SPFT estimate $\varepsilon_{\text{SPFT}}^{[2]}$ [11].

Corresponding expressions for the third-order SPFT estimate of the HCM permittivity, derived using the three-point covariance function (4.38), can be found elsewhere [12]. In practice, the third-order SPFT estimates may not be required since convergence at the second-order level has been established by means of numerical investigations in the regime of weak nonlinearity [12].

5.3.2 Isotropic chiral composite materials

Magnetoelectric coupling can play a substantive role in weakly nonlinear scenarios. As in section 5.3.1, suppose that the constituent material 'a' is characterized by the intensity-dependent permittivity given in equation (5.6). But, now the constituent material 'b' is a linear isotropic chiral material characterized by 3 × 3 Tellegen constitutive dyadics of the form given in equations (2.34) with $\xi = -\zeta$. It is convenient here to adopt the 6 × 6 constitutive dyadic notation of section 2.2. The 6 × 6 constitutive dyadic of the constituent material 'a' is then written as

$$\underline{\underline{\mathbf{K}}}_{\,\mathrm{a}} = \begin{bmatrix} \varepsilon_{\mathrm{a}}^{\mathrm{L}} \underline{\underline{I}} & \underline{\underline{0}} \\ \underline{\underline{0}} & \mu_0 \underline{\underline{I}} \end{bmatrix} + \begin{bmatrix} \varepsilon_{\mathrm{a}}^{\mathrm{NL}} \underline{\underline{I}} & \underline{\underline{0}} \\ \underline{\underline{0}} & \underline{\underline{0}} \end{bmatrix} |\, \mathbf{E}_{\mathrm{a}} \,|^2 = \underline{\underline{\mathbf{K}}}_{\,\mathrm{a}}^{\mathrm{L}} + \underline{\underline{\mathbf{K}}}_{\,\mathrm{a}}^{\mathrm{NL}} \,|\, \mathbf{E}_{\mathrm{a}} \,|^2 \qquad (5.15)$$

and the 6 × 6 constitutive dyadic of the constituent material 'b' is written as

$$\underline{\underline{\mathbf{K}}}_{\,\mathrm{b}} = \begin{bmatrix} \varepsilon_{\mathrm{b}} \underline{\underline{I}} & \xi_{\mathrm{b}} \underline{\underline{I}} \\ -\xi_{\mathrm{b}} \underline{\underline{I}} & \mu_{\mathrm{b}} \underline{\underline{I}} \end{bmatrix}. \qquad (5.16)$$

Both constituent materials are randomly distributed as electrically small spherical particles (i.e., $\underline{\underline{U}} = \underline{\underline{I}}$). Application of the second-order SPFT formalism then yields the following results [13].

Since the HCM inherits both the weak nonlinearity of the constituent material 'a' and the isotropic chirality of the constituent material 'b', the 6 × 6 Tellegen constitutive dyadic of the ambient medium may be expressed as

$$\begin{aligned} \underline{\underline{\mathbf{K}}}_{\,\mathrm{amb}} &= \begin{bmatrix} \varepsilon_{\mathrm{amb}}^{\mathrm{L}} \underline{\underline{I}} & \xi_{\mathrm{amb}}^{\mathrm{L}} \underline{\underline{I}} \\ -\xi_{\mathrm{amb}}^{\mathrm{L}} \underline{\underline{I}} & \mu_{\mathrm{amb}}^{\mathrm{L}} \underline{\underline{I}} \end{bmatrix} + \begin{bmatrix} \varepsilon_{\mathrm{amb}}^{\mathrm{NL}} \underline{\underline{I}} & \xi_{\mathrm{amb}}^{\mathrm{NL}} \underline{\underline{I}} \\ -\xi_{\mathrm{amb}}^{\mathrm{NL}} \underline{\underline{I}} & \mu_{\mathrm{amb}}^{\mathrm{NL}} \underline{\underline{I}} \end{bmatrix} |\mathbf{E}_{\mathrm{HCM}}|^2 \\ &= \underline{\underline{\mathbf{K}}}_{\,\mathrm{amb}}^{\mathrm{L}} + \underline{\underline{\mathbf{K}}}_{\,\mathrm{amb}}^{\mathrm{NL}} \,|\, \mathbf{E}_{\mathrm{HCM}} \,|^2, \end{aligned} \qquad (5.17)$$

with $\mathbf{E}_{\mathrm{HCM}}$ being the spatially averaged electric field in the HCM. In a similar vein, the form of the depolarization dyadic for an infinitesimally small spherical region in the ambient material mirrors that of $\underline{\underline{\mathbf{K}}}_{\,\mathrm{amb}}$; i.e.,

$$\underline{\underline{\mathbf{D}}}_{\,\mathrm{I/amb}}^{0} = \underline{\underline{\mathbf{D}}}_{\,\mathrm{I/amb}}^{0,\mathrm{L}} + \underline{\underline{\mathbf{D}}}_{\,\mathrm{I/amb}}^{0,\mathrm{NL}} \,|\, \mathbf{E}_{\mathrm{HCM}} \,|^2. \qquad (5.18)$$

Since the nonlinearity is assumed to be weak per equation (5.1), the linear contribution to the depolarization dyadic is identical to that which would arise in the case where $\varepsilon_{\mathrm{a}}^{\mathrm{NL}} = 0$. Accordingly, $\underline{\underline{\mathbf{D}}}_{\,\mathrm{I/amb}}^{0,\mathrm{L}}$ is composed of four 3 × 3 dyadics of the form specified in equations (3.32), with $\varepsilon_{\mathrm{amb}}^{\mathrm{L}}$, $\xi_{\mathrm{amb}}^{\mathrm{L}}$, and $\mu_{\mathrm{amb}}^{\mathrm{L}}$ used in lieu of ε, ξ, and μ, respectively, therein. The nonlinear contribution to the depolarization dyadic takes the form

$$\underline{\underline{\mathbf{D}}}_{\,\mathrm{I/amb}}^{0,\mathrm{NL}} = \frac{1}{3i\omega\,\Upsilon_{+}} \begin{bmatrix} {}^{ee}d_{\,\mathrm{I/amb}}^{\,0,\mathrm{NL}} \underline{\underline{I}} & {}^{em}d_{\,\mathrm{I/amb}}^{\,0,\mathrm{NL}} \underline{\underline{I}} \\ {}^{me}d_{\,\mathrm{I/amb}}^{\,0,\mathrm{NL}} \underline{\underline{I}} & {}^{mm}d_{\,\mathrm{I/amb}}^{\,0,\mathrm{NL}} \underline{\underline{I}} \end{bmatrix}, \qquad (5.19)$$

where the scalar functions

$$\left.\begin{aligned}
{}^{ee}d^{\,0,\mathrm{NL}}_{\mathrm{l/amb}} &= \mu^{\mathrm{NL}}_{\mathrm{amb}} - \mu^{\mathrm{L}}_{\mathrm{amb}}\breve{\Upsilon}\\
{}^{em}d^{\,0,\mathrm{NL}}_{\mathrm{l/amb}} &= \xi^{\mathrm{NL}}_{\mathrm{amb}}\breve{\Upsilon} - \xi^{\mathrm{L}}_{\mathrm{amb}}\\
{}^{me}d^{\,0,\mathrm{NL}}_{\mathrm{l/amb}} &= \xi^{\mathrm{L}}_{\mathrm{amb}} - \xi^{\mathrm{NL}}_{\mathrm{amb}}\breve{\Upsilon}\\
{}^{mm}d^{\,0,\mathrm{NL}}_{\mathrm{l/amb}} &= \varepsilon^{\mathrm{NL}}_{\mathrm{amb}} - \varepsilon^{\mathrm{L}}_{\mathrm{amb}}\breve{\Upsilon}
\end{aligned}\right\} \tag{5.20}$$

are defined in terms of the scalar parameter

$$\breve{\Upsilon} = \frac{\varepsilon^{\mathrm{L}}_{\mathrm{amb}}\,\mu^{\mathrm{NL}}_{\mathrm{amb}} + 2\,\xi^{\mathrm{L}}_{\mathrm{amb}}\,\xi^{\mathrm{NL}}_{\mathrm{amb}}}{\Upsilon_+}. \tag{5.21}$$

The scalar Υ_+ is specified as in equation (3.34) but with ε, ξ, and μ therein replaced by $\varepsilon^{\mathrm{L}}_{\mathrm{amb}}$, $\xi^{\mathrm{L}}_{\mathrm{amb}}$, and $\mu^{\mathrm{L}}_{\mathrm{amb}}$, respectively.

The form of the depolarization dyadics given in equation (5.18) is shared by the corresponding 6 × 6 polarizability density dyadics; i.e.,

$$\underline{\underline{\boldsymbol{\alpha}}}^{\,0}_{\ell/\mathrm{amb}} = \underline{\underline{\boldsymbol{\alpha}}}^{\,0,\mathrm{L}}_{\ell/\mathrm{amb}} + \underline{\underline{\boldsymbol{\alpha}}}^{\,0,\mathrm{NL}}_{\ell/\mathrm{amb}}\,|\,\mathbf{E}_{\mathrm{HCM}}\,|^2, \qquad (\ell \in \{\mathrm{a,b}\}). \tag{5.22}$$

The linear contributions $\underline{\underline{\boldsymbol{\alpha}}}^{\,0,\mathrm{L}}_{\ell/\mathrm{amb}}$ are specified as in equation (3.34), with $\underline{\underline{\mathbf{K}}}^{\mathrm{L}}_{\mathrm{a}}$ standing in for $\underline{\underline{\mathbf{K}}}_{\mathrm{a}}$, $\underline{\underline{\mathbf{K}}}^{\mathrm{L}}_{\mathrm{amb}}$ standing in for $\underline{\underline{\mathbf{K}}}_{\mathrm{amb}}$, and $\underline{\underline{\mathbf{D}}}^{0,\mathrm{L}}_{\mathrm{l/amb}}$ standing in for $\underline{\underline{\mathbf{D}}}^{0}_{U^\ell/\mathrm{amb}}$ therein. The nonlinear contributions are specified by

$$\left.\begin{aligned}
\underline{\underline{\boldsymbol{\alpha}}}^{\,0,\mathrm{NL}}_{\mathrm{a/amb}} =& \left(g^{\mathrm{FF}}_{\mathrm{ICM}}\,\underline{\underline{\mathbf{K}}}^{\mathrm{NL}}_{\mathrm{a}} - \underline{\underline{\mathbf{K}}}^{\mathrm{NL}}_{\mathrm{amb}}\right)\bullet\left(\underline{\underline{\boldsymbol{\aleph}}}^{\mathrm{L}}_{\mathrm{a/amb}}\right)^{-1} + \left(\underline{\underline{\mathbf{K}}}^{\mathrm{L}}_{\mathrm{a}} - \underline{\underline{\mathbf{K}}}^{\mathrm{L}}_{\mathrm{amb}}\right)\bullet\\
&\mathcal{N}_{\mathrm{ICM}}\left\{\underline{\underline{\boldsymbol{\aleph}}}_{\mathrm{a/amb}}\right\}\\
\underline{\underline{\boldsymbol{\alpha}}}^{\,0,\mathrm{NL}}_{\mathrm{b/amb}} =& -\underline{\underline{\mathbf{K}}}^{\mathrm{NL}}_{\mathrm{amb}}\bullet\left(\underline{\underline{\boldsymbol{\aleph}}}^{\mathrm{L}}_{\mathrm{b/amb}}\right)^{-1} + \left(\underline{\underline{\mathbf{K}}}_{\mathrm{b}} - \underline{\underline{\mathbf{K}}}^{\mathrm{L}}_{\mathrm{amb}}\right)\bullet\mathcal{N}_{\mathrm{ICM}}\left\{\underline{\underline{\boldsymbol{\aleph}}}_{\mathrm{b/amb}}\right\}
\end{aligned}\right\}. \tag{5.23}$$

Herein the 6 × 6 dyadic $\underline{\underline{\boldsymbol{\aleph}}}_{\ell/\mathrm{amb}}$ may be written as

$$\underline{\underline{\boldsymbol{\aleph}}}_{\ell/\mathrm{amb}} = \underline{\underline{\boldsymbol{\aleph}}}^{\mathrm{L}}_{\ell/\mathrm{amb}} + \underline{\underline{\boldsymbol{\aleph}}}^{\mathrm{NL}}_{\ell/\mathrm{amb}}\,|\,\mathbf{E}_{\mathrm{HCM}}\,|^2, \qquad (\ell \in \{\mathrm{a,b}\}), \tag{5.24}$$

with linear contributions

$$\underline{\underline{\boldsymbol{\aleph}}}^{\mathrm{L}}_{\ell/\mathrm{amb}} = \underline{\underline{\mathbf{I}}} + i\omega\underline{\underline{\mathbf{D}}}^{0,\mathrm{L}}_{\mathrm{l/amb}}\bullet\left(\underline{\underline{\mathbf{K}}}^{\mathrm{L}}_{\ell} - \underline{\underline{\mathbf{K}}}^{\mathrm{L}}_{\mathrm{amb}}\right), \qquad (\ell \in \{a, b\}), \tag{5.25}$$

and nonlinear contributions

$$\left.\begin{aligned}
\underline{\underline{\boldsymbol{\aleph}}}^{\mathrm{NL}}_{\mathrm{a/amb}} =&\; i\omega\left[\underline{\underline{\mathbf{D}}}^{0,\mathrm{L}}_{\mathrm{l/amb}}\bullet\left(g^{\mathrm{FF}}_{\mathrm{ICM}}\,\underline{\underline{\mathbf{K}}}^{\mathrm{NL}}_{\mathrm{a}} - \underline{\underline{\mathbf{K}}}^{\mathrm{NL}}_{\mathrm{amb}}\right) + \underline{\underline{\mathbf{D}}}^{0,\mathrm{NL}}_{\mathrm{l/amb}}\bullet\left(\underline{\underline{\mathbf{K}}}^{\mathrm{L}}_{\mathrm{a}} - \underline{\underline{\mathbf{K}}}^{\mathrm{L}}_{\mathrm{amb}}\right)\right]\\
\underline{\underline{\boldsymbol{\aleph}}}^{\mathrm{NL}}_{\mathrm{b/amb}} =&\; i\omega\left[-\underline{\underline{\mathbf{D}}}^{0,\mathrm{L}}_{\mathrm{l/amb}}\bullet\underline{\underline{\mathbf{K}}}^{\mathrm{NL}}_{\mathrm{amb}} + \underline{\underline{\mathbf{D}}}^{0,\mathrm{NL}}_{\mathrm{l/amb}}\bullet\left(\underline{\underline{\mathbf{K}}}_{\mathrm{b}} - \underline{\underline{\mathbf{K}}}^{\mathrm{L}}_{\mathrm{amb}}\right)\right]
\end{aligned}\right\}. \tag{5.26}$$

Equations (5.23) contain the 6 × 6 dyadic operator $\mathcal{N}_{\mathrm{ICM}}\{\,\bullet\,\}$. Its action

$$\mathcal{N}_{\text{ICM}}\left\{\underline{\underline{\aleph}}_{\ell/\text{amb}}\right\} = \frac{1}{\det \underline{\underline{\aleph}}^{\text{L}}_{\ell/\text{amb}}}\left[\text{adj}\left(\underline{\underline{\aleph}}^{\text{NL}}_{\ell/\text{amb}}\right) - \mathcal{M}_{\text{ICM}}\left\{\underline{\underline{\aleph}}_{\ell/\text{amb}}\right\}\left(\underline{\underline{\aleph}}^{\text{L}}_{\ell/\text{amb}}\right)^{-1}\right] \quad (5.27)$$

is specified in terms of the scalar operator

$$\begin{aligned}\mathcal{M}_{\text{ICM}}\left\{\underline{\underline{\aleph}}_{\ell/\text{amb}}\right\} = {}& {}^{ee}\aleph^{\text{L}}_{\ell/\text{amb}}\ {}^{mm}\aleph^{\text{NL}}_{\ell/\text{amb}} - {}^{em}\aleph^{\text{L}}_{\ell/\text{amb}}\ {}^{me}\aleph^{\text{NL}}_{\ell/\text{amb}} \\ & + {}^{ee}\aleph^{\text{NL}}_{\ell/\text{amb}}\ {}^{mm}\aleph^{\text{L}}_{\ell/\text{amb}} - {}^{em}\aleph^{\text{NL}}_{\ell/\text{amb}}\ {}^{me}\aleph^{\text{L}}_{\ell/\text{amb}},\end{aligned} \quad (5.28)$$

with the components of the 6×6 dyadics $\underline{\underline{\aleph}}^{\text{L,NL}}_{\ell/\text{amb}}$ being prescribed via

$$\underline{\underline{\aleph}}^{\text{j}}_{\ell/\text{amb}} = \begin{bmatrix} {}^{ee}\aleph^{\text{j}}_{\ell/\text{amb}}\,\underline{\underline{I}} & {}^{em}\aleph^{\text{j}}_{\ell/\text{amb}}\,\underline{\underline{I}} \\ {}^{me}\aleph^{\text{j}}_{\ell/\text{amb}}\,\underline{\underline{I}} & {}^{mm}\aleph^{\text{j}}_{\ell/\text{amb}}\,\underline{\underline{I}} \end{bmatrix}, \qquad (\text{j} \in \{\text{L, NL}\}). \quad (5.29)$$

The local field factor $g^{\text{FF}}_{\text{ICM}}$ is introduced in equation $(5.23)_1$. This quantity represents the derivative on the right side of equation (5.10). A plausible estimate is given by [20, 21]

$$g^{\text{FF}}_{\text{ICM}} = \left| \left\{\left[\underline{\underline{\aleph}}^{\text{L}}_{\text{a/amb}}\right]^{-1}\right\}_{11} \right|^2. \quad (5.30)$$

The ambient material—which is equivalent to the HCM that arises from the Bruggeman formalism, as described in section 4.4.2— is specified via the nonlinear equations

$$f_{\text{a}}\,\underline{\underline{\boldsymbol{\alpha}}}^{0,\text{j}}_{a/\text{amb}} + f_{\text{b}}\,\underline{\underline{\boldsymbol{\alpha}}}^{0,\text{j}}_{b/\text{amb}} = \underline{\underline{\mathbf{0}}}, \qquad (\text{j} \in \{\text{L, NL}\}). \quad (5.31)$$

The constitutive parameters of the ambient material can be extracted iteratively from equations (5.31) by the Jacobi scheme [22] represented by

$$\underline{\underline{\mathbf{K}}}^{\text{j}}_{\text{amb}}[n] = \mathcal{S}^{\text{j}}_{\text{ICM}}\left\{\underline{\underline{\mathbf{K}}}^{\text{j}}_{\text{amb}}[n-1]\right\}, \qquad (\text{j} \in \{\text{L, NL}\};\ n \in \{1, 2, \dots\}). \quad (5.32)$$

The dyadic operator $\mathcal{S}^{\text{L}}_{\text{ICM}}\{\,\bullet\,\}$ which yields the linear contribution to $\underline{\underline{\mathbf{K}}}_{\text{amb}}$ is defined as in equation (4.22), but with $\underline{\underline{\mathbf{K}}}^{\text{L}}_{\text{a}}$ in lieu of $\underline{\underline{\mathbf{K}}}_{\text{a}}$, $\underline{\underline{\mathbf{K}}}^{\text{L}}_{\text{amb}}$ in lieu of $\underline{\underline{\mathbf{K}}}_{\text{Br}}$, and $\underline{\underline{\mathbf{D}}}^{0,\text{L}}_{\text{l/amb}}$ in lieu of $\underline{\underline{\mathbf{D}}}^{0}_{U^{\text{a}}/\text{Br}}$ and $\underline{\underline{\mathbf{D}}}^{0}_{U^{\text{b}}/\text{Br}}$ therein. The dyadic operator $\mathcal{S}^{\text{NL}}_{\text{ICM}}\{\,\bullet\,\}$ which yields the nonlinear contribution to $\underline{\underline{\mathbf{K}}}_{\text{amb}}$ is given by

$$\begin{aligned}\mathcal{S}^{\text{NL}}_{\text{ICM}}\left\{\underline{\underline{\mathbf{K}}}^{\text{NL}}_{\text{amb}}\right\} = {}& \left\{f_{\text{a}}\left[g^{\text{FF}}_{\text{ICM}}\,\underline{\underline{\mathbf{K}}}^{\text{NL}}_{\text{a}} \bullet \left(\underline{\underline{\aleph}}^{\text{L}}_{\text{a/amb}}\right)^{-1} + \left(\underline{\underline{\mathbf{K}}}^{\text{L}}_{\text{a}} - \underline{\underline{\mathbf{K}}}^{\text{L}}_{\text{amb}}\right) \bullet \mathcal{N}_{\text{ICM}}\left\{\underline{\underline{\aleph}}_{\text{a/amb}}\right\}\right]\right. \\ & \left. + f_{\text{b}}\left(\underline{\underline{\mathbf{K}}}_{\text{b}} - \underline{\underline{\mathbf{K}}}^{\text{L}}_{\text{amb}}\right) \bullet \mathcal{N}_{\text{ICM}}\left\{\underline{\underline{\aleph}}_{\text{b/amb}}\right\}\right\} \\ & \bullet \left[f_{\text{a}}\left(\underline{\underline{\aleph}}^{\text{L}}_{\text{a/amb}}\right)^{-1} + f_{\text{b}}\left(\underline{\underline{\aleph}}^{\text{L}}_{\text{b/amb}}\right)^{-1}\right]^{-1}.\end{aligned} \quad (5.33)$$

As outlined in section 4.4.1, the mass operator is a keystone of the SPFT, and the principal-value integral represented by $\underline{\underline{\mathbf{P}}}_{\Lambda}(L_{\text{cor}})$ in equation (4.34) is central to the calculation of the mass operator. Consider the linear and nonlinear contributions to $\underline{\underline{\mathbf{P}}}_{\Lambda}(L_{\text{cor}})$ per

$$\underline{\underline{\mathbf{P}}}_{\Lambda}(L_{\text{cor}}) = \underline{\underline{\mathbf{P}}}^{\text{L}}_{\Lambda}(L_{\text{cor}}) + \underline{\underline{\mathbf{P}}}^{\text{NL}}_{\Lambda}(L_{\text{cor}}) \mid \mathbf{E}_{\text{HCM}} \mid^2. \tag{5.34}$$

For the step covariance function $\Lambda_{\text{step}}(\mathbf{r} - \mathbf{r}')$, as given in equation (4.36) with $\underline{\underline{U}} = \underline{\underline{I}}$, the linear contribution to $\underline{\underline{\mathbf{P}}}_{\Lambda}(L_{\text{cor}})$ is as shown in equation (4.37), with the components of $\underline{\underline{\mathbf{D}}}^{+}_{I/\text{amb}}(L_{\text{cor}})$ being specified as in equations (3.33) with ε, ξ, and μ being substituted by $\varepsilon^{\text{L}}_{\text{amb}}$, $\xi^{\text{L}}_{\text{amb}}$, and $\mu^{\text{L}}_{\text{amb}}$, respectively. The nonlinear contribution to $\underline{\underline{\mathbf{P}}}_{\Lambda}(L_{\text{cor}})$ can be written in the form

$$\underline{\underline{\mathbf{P}}}^{\text{NL}}_{\Lambda_{\text{step}}}(L_{\text{cor}}) = \frac{i\omega L_{\text{cor}}^2 f_a f_b}{3} \begin{bmatrix} {}^{\text{ee}}P^{\text{NL}}_{\Lambda_{\text{step}}} \underline{\underline{I}} & {}^{\text{em}}P^{\text{NL}}_{\Lambda_{\text{step}}} \underline{\underline{I}} \\ {}^{\text{me}}P^{\text{NL}}_{\Lambda_{\text{step}}} \underline{\underline{I}} & {}^{\text{mm}}P^{\text{NL}}_{\Lambda_{\text{step}}} \underline{\underline{I}} \end{bmatrix}. \tag{5.35}$$

Herein, the component scalar functions

$$\left.\begin{aligned}
{}^{\text{ee}}P^{\text{NL}}_{\Lambda_{\text{step}}} &= \mu^{\text{NL}}_{\text{amb}} + i\frac{2}{3}\omega L_{\text{cor}}\left(\mu^{\text{NL}}_{\text{amb}}\, p_1 + \mu^{\text{L}}_{\text{amb}}\, p_2 \right) \\
{}^{\text{em}}P^{\text{NL}}_{\Lambda_{\text{step}}} &= \xi^{\text{NL}}_{\text{amb}} + i\frac{4}{3}\omega L_{\text{cor}}\left(\xi^{\text{NL}}_{\text{amb}}\, p_3 + \xi^{\text{L}}_{\text{amb}}\, p_4 \right) \\
{}^{\text{me}}P^{\text{NL}}_{\Lambda_{\text{step}}} &= -\xi^{\text{NL}}_{\text{amb}} - i\frac{4}{3}\omega L_{\text{cor}}\left(\xi^{\text{NL}}_{\text{amb}}\, p_3 + \xi^{\text{L}}_{\text{amb}}\, p_4 \right) \\
{}^{\text{mm}}P^{\text{NL}}_{\Lambda_{\text{step}}} &= \varepsilon^{\text{NL}}_{\text{amb}} + i\frac{2}{3}\omega L_{\text{cor}}\left(\varepsilon^{\text{NL}}_{\text{amb}}\, p_1 + \varepsilon^{\text{L}}_{\text{amb}}\, p_2 \right)
\end{aligned}\right\} \tag{5.36}$$

are expressed in terms of the scalar parameters

$$\left.\begin{aligned}
p_1 &= \frac{\varepsilon^{\text{L}}_{\text{amb}}\, \mu^{\text{L}}_{\text{amb}} - (\xi^{\text{L}}_{\text{amb}})^2}{p_3} \\
p_2 &= \left[1 + \left(\frac{\xi^{\text{L}}_{\text{amb}}}{p_3} \right)^2 \right] p_4 - \frac{2\, \xi^{\text{L}}_{\text{amb}}\, \xi^{\text{NL}}_{\text{amb}}}{p_3} \\
p_3 &= \left(\varepsilon^{\text{L}}_{\text{amb}}\, \mu^{\text{L}}_{\text{amb}} \right)^{1/2} \\
p_4 &= \frac{\varepsilon^{\text{L}}_{\text{amb}}\, \mu^{\text{NL}}_{\text{amb}} + \varepsilon^{\text{NL}}_{\text{amb}}\, \mu^{\text{L}}_{\text{amb}}}{2\, p_3}
\end{aligned}\right\}. \tag{5.37}$$

The second-order mass operator is written as the sum of its linear and nonlinear contributions as

$$\underline{\underline{\mathbf{\Sigma}}}^{[2]}(L_{\text{cor}}) = \underline{\underline{\mathbf{\Sigma}}}^{[2],\,\text{L}}(L_{\text{cor}}) + \underline{\underline{\mathbf{\Sigma}}}^{[2],\,\text{NL}}(L_{\text{cor}}) \mid \mathbf{E}_{\text{HCM}} \mid^2. \tag{5.38}$$

The linear contribution to $\underline{\underline{\Sigma}}^{[2]}(L_{\text{cor}})$ is given by the right side of equation (4.33), but with $\underline{\underline{\alpha}}^{0,\text{L}}_{\text{a,b/amb}}$ in lieu of $\underline{\underline{\alpha}}^{0}_{\text{a,b/amb}}$, and $\underline{\underline{\mathbf{P}}}^{\text{L}}_{\Lambda_{\text{step}}}(L_{\text{cor}})$ in lieu of $\underline{\underline{\mathbf{P}}}_{\Lambda}(L_{\text{cor}})$. The nonlinear contribution to $\underline{\underline{\Sigma}}^{[2]}(L_{\text{cor}})$ is

$$
\begin{aligned}
\underline{\underline{\Sigma}}^{[2],\,\text{NL}}(L_{\text{cor}}) = -\,\omega^2 \Bigg\{ & \left[\left(\underline{\underline{\alpha}}^{\text{L}}_{\text{a/amb}} - \underline{\underline{\alpha}}^{\text{L}}_{\text{b/amb}} \right) \bullet \underline{\underline{\mathbf{P}}}^{\text{L}}_{\Lambda_{\text{step}}}(L_{\text{cor}}) \bullet \left(\underline{\underline{\alpha}}^{\text{NL}}_{\text{a/amb}} - \underline{\underline{\alpha}}^{\text{NL}}_{\text{b/amb}} \right) \right] \\
& + \left[\left(\underline{\underline{\alpha}}^{\text{L}}_{\text{a/amb}} - \underline{\underline{\alpha}}^{\text{L}}_{\text{b/amb}} \right) \bullet \underline{\underline{\mathbf{P}}}^{\text{NL}}_{\Lambda_{\text{step}}}(L_{\text{cor}}) \bullet \left(\underline{\underline{\alpha}}^{\text{L}}_{\text{a/amb}} - \underline{\underline{\alpha}}^{\text{L}}_{\text{b/amb}} \right) \right] \\
& + \left[\left(\underline{\underline{\alpha}}^{\text{NL}}_{\text{a/amb}} - \underline{\underline{\alpha}}^{\text{NL}}_{\text{b/amb}} \right) \bullet \underline{\underline{\mathbf{P}}}^{\text{L}}_{\Lambda_{\text{step}}}(L_{\text{cor}}) \bullet \left(\underline{\underline{\alpha}}^{\text{L}}_{\text{a/amb}} - \underline{\underline{\alpha}}^{\text{L}}_{\text{b/amb}} \right) \right] \Bigg\}.
\end{aligned} \tag{5.39}
$$

By assembling the foregoing expressions, the second-order SPFT estimate of the HCM's constitutive dyadic emerges as

$$
\underline{\underline{\mathbf{K}}}^{[2]}_{\text{SPFT}} = \underline{\underline{\mathbf{K}}}^{[2],\text{L}}_{\text{SPFT}} + \underline{\underline{\mathbf{K}}}^{[2],\text{NL}}_{\text{SPFT}} \,|\, \mathbf{E}_{\text{HCM}} \,|^2. \tag{5.40}
$$

The linear contribution $\underline{\underline{\mathbf{K}}}^{[2],\text{L}}_{\text{SPFT}}$ is delivered by equation (4.28) evaluated for $n = 2$, with $\underline{\underline{\mathbf{K}}}_{\text{amb}}$, $\underline{\underline{\mathbf{D}}}^{0}_{U/\text{amb}}$, and $\underline{\underline{\Sigma}}^{[2]}(L_{\text{cor}})$ therein substituted by $\underline{\underline{\mathbf{K}}}^{\text{L}}_{\text{amb}}$, $\underline{\underline{\mathbf{D}}}^{0,\text{L}}_{\text{I/amb}}$, and $\underline{\underline{\Sigma}}^{[2],\,\text{L}}(L_{\text{cor}})$, respectively. The nonlinear contribution $\underline{\underline{\mathbf{K}}}^{[2],\text{NL}}_{\text{SPFT}}$ is given by

$$
\underline{\underline{\mathbf{K}}}^{[2],\text{NL}}_{\text{SPFT}} = \underline{\underline{\mathbf{K}}}^{\text{NL}}_{\text{amb}} - \frac{1}{i\omega}[(\underline{\underline{\mathbf{e}}}^{\text{L}})^{-1} \bullet \underline{\underline{\Sigma}}^{[2],\,\text{NL}}(L_{\text{cor}}) + \mathcal{N}_{\text{ICM}}\{\underline{\underline{\mathbf{e}}}\} \bullet \underline{\underline{\Sigma}}^{[2],\,\text{L}}(L_{\text{cor}})], \tag{5.41}
$$

with the linear and nonlinear contributions to the 6×6 dyadic

$$
\underline{\underline{\mathbf{e}}} = \underline{\underline{\mathbf{e}}}^{\text{L}} + \underline{\underline{\mathbf{e}}}^{\text{NL}} \,|\, \mathbf{E}_{\text{HCM}} \,|^2 \tag{5.42}
$$

being specified as

$$
\left.
\begin{aligned}
\underline{\underline{\mathbf{e}}}^{\text{L}} &= \underline{\underline{I}} + \underline{\underline{\Sigma}}^{[2],\,\text{L}}(L_{\text{cor}}) \bullet \underline{\underline{D}}^{0,\text{L}}_{\text{I/amb}} \\
\underline{\underline{\mathbf{e}}}^{\text{NL}} &= \underline{\underline{\Sigma}}^{[2],\,\text{L}}(L_{\text{cor}}) \bullet \underline{\underline{D}}^{0,\text{NL}}_{\text{I/amb}} + \underline{\underline{\Sigma}}^{[2],\,\text{NL}}(L_{\text{cor}}) \bullet \underline{\underline{D}}^{0,\text{L}}_{\text{I/amb}}
\end{aligned}
\right\}. \tag{5.43}
$$

The corresponding third-order SPFT estimate of the HCM permittivity has been established for the three-point covariance function (4.38) [13]. As is the case for weakly nonlinear isotropic dielectric HCMs detailed in section 5.3.1, the third-order SPFT estimates are unlikely to be needed in practice because numerical studies have confirmed convergence at the second-order level for weakly nonlinear HCMs [13].

5.3.3 Anisotropic dielectric composite materials

Weakly nonlinear HCMs can be anisotropic, and the second-order SPFT formalism has been established for this scenario [14]. As always with the SPFT, the derivation includes the corresponding Bruggeman formalism as the lowest-order version [15]. Moreover, an extended version has been formulated to accommodate the size of the constituent particles [16].

As in section 5.3.1, the constituent material 'a' is characterized by the intensity-dependent permittivity given by equation (5.6). The constituent material 'b' is also characterized by an intensity-dependent permittivity, as given by equation (5.6) but with the subscript 'a' therein replaced with the subscript 'b'. All constituent particles are randomly distributed as identically oriented, conformal, ellipsoidal particles whose shapes and orientations are captured by the dyadic $\underline{\underline{U}}$, per equation (3.15). The schematic representation of the modeling approach displayed in figure 4.3 for linear composite materials is also relevant here for nonlinear composite materials. For mathematical convenience, and without loss of generality, the principal axes of the ellipsoidal particles are taken to be aligned with the axes of the Cartesian coordinate system. Thus, $\underline{\underline{U}}$ is isomorphic to a diagonal matrix.

The HCM is anisotropic—in fact, in general it is an orthorhombic biaxial material—because of the shape of the constituent particles. Similarly to equations (5.7) and (5.17), the permittivity dyadic of the ambient material may be expressed as

$$\begin{aligned}\underline{\underline{\varepsilon}}_{\text{amb}} &= \begin{pmatrix} \varepsilon_{\text{amb}}^{x,\text{L}} & 0 & 0 \\ 0 & \varepsilon_{\text{amb}}^{y,\text{L}} & 0 \\ 0 & 0 & \varepsilon_{\text{amb}}^{z,\text{L}} \end{pmatrix} + \begin{pmatrix} \varepsilon_{\text{amb}}^{x,\text{NL}} & 0 & 0 \\ 0 & \varepsilon_{\text{amb}}^{y,\text{NL}} & 0 \\ 0 & 0 & \varepsilon_{\text{amb}}^{z,\text{NL}} \end{pmatrix} |\,\mathbf{E}_{\text{HCM}}\,|^2 \\ &= \underline{\underline{\varepsilon}}_{\text{amb}}^{\text{L}} + \underline{\underline{\varepsilon}}_{\text{amb}}^{\text{NL}} |\,\mathbf{E}_{\text{HCM}}\,|^2,\end{aligned} \tag{5.44}$$

where $\mathbf{E}_{\text{HCM}}$ is the spatially averaged electric field in the HCM.

The 3×3 depolarization dyadic relevant to an electrically small ellipsoidal particle immersed in the ambient material prescribed by equations (5.44) is written as ${}^{\text{ee}}\underline{\underline{D}}_{U/\text{amb}}^{0}$ per equation $(3.22)_1$. This dyadic can be stated as

$${}^{\text{ee}}\underline{\underline{D}}_{U/\text{amb}}^{0} = {}^{\text{ee}}\underline{\underline{D}}_{U/\text{amb}}^{0,\text{L}} + {}^{\text{ee}}\underline{\underline{D}}_{U/\text{amb}}^{0,\text{NL}} |\,\mathbf{E}_{\text{HCM}}\,|^2, \tag{5.45}$$

wherein the linear and nonlinear contributions are

$${}^{\text{ee}}\underline{\underline{D}}_{U/\text{amb}}^{0,\text{L}} = \frac{1}{4\pi i\omega} \int_{\phi=0}^{2\pi} \int_{\theta=0}^{\pi} \frac{\sin\theta}{\text{tr}\left[\underline{\underline{\varepsilon}}_{\text{amb}}^{\text{L}} \bullet \underline{\underline{\Omega}}(\theta, \phi)\right]} \underline{\underline{\Omega}}(\theta, \phi)\, d\theta\, d\phi \tag{5.46}$$

and

$${}^{\text{ee}}\underline{\underline{D}}_{U/\text{amb}}^{0,\text{NL}} = -\frac{1}{4\pi i\omega} \int_{\phi=0}^{2\pi} \int_{\theta=0}^{\pi} \frac{\text{tr}\left[\underline{\underline{\varepsilon}}_{\text{amb}}^{\text{NL}} \bullet \underline{\underline{\Omega}}(\theta, \phi)\right] \sin\theta}{\left\{\text{tr}\left[\underline{\underline{\varepsilon}}_{\text{amb}}^{\text{L}} \bullet \underline{\underline{\Omega}}(\theta, \phi)\right]\right\}^2} \underline{\underline{\Omega}}(\theta, \phi)\, d\theta\, d\phi, \tag{5.47}$$

respectively. In these equations,

$$\underline{\underline{\Omega}}(\theta, \phi) = \underline{\underline{U}}^{-1} \bullet \begin{pmatrix} \sin^2\theta\cos^2\phi & 0 & 0 \\ 0 & \sin^2\theta\sin^2\phi & 0 \\ 0 & 0 & \cos^2\theta \end{pmatrix} \bullet \underline{\underline{U}}^{-1} \tag{5.48}$$

is a 3 × 3 diagonal dyadic. Parenthetically, the surface integral on the right side of equation (5.46) can be written as an incomplete elliptic integral, as provided in equations (3.45)–(3.47).

In a similar vein, the corresponding 3 × 3 polarizability density dyadic may be written as the sum of linear and nonlinear contributions as

$$ {}^{\text{ee}}\underline{\underline{\alpha}}^{\,0}_{\ell/\text{amb}} = {}^{\text{ee}}\underline{\underline{\alpha}}^{\,0,\text{L}}_{\ell/\text{amb}} + {}^{\text{ee}}\underline{\underline{\alpha}}^{\,0,\text{NL}}_{\ell/\text{amb}} \,|\, \mathbf{E}_{\text{HCM}} \,|^2, \qquad (\ell \in \{\text{a,b}\}), \tag{5.49} $$

with the linear contribution being

$$ {}^{\text{ee}}\underline{\underline{\alpha}}^{\,0,\text{L}}_{\ell/\text{amb}} = \left(\varepsilon_\ell^{\text{L}}\, \underline{\underline{I}} - \underline{\underline{\varepsilon}}^{\,\text{L}}_{\text{amb}} \right) \bullet \left({}^{\text{ee}}\underline{\underline{\aleph}}_{\ell/\text{amb}} \right)^{-1} \tag{5.50} $$

and the nonlinear contribution being

$$ \begin{aligned} {}^{\text{ee}}\underline{\underline{\alpha}}^{\,0,\text{NL}}_{\ell/\text{amb}} = & \left(g^{\text{FF}}_{\ell,\,\underline{\underline{\varepsilon}}}\, \varepsilon_\ell^{\text{NL}}\, \underline{\underline{I}} - \underline{\underline{\varepsilon}}^{\,\text{NL}}_{\text{amb}} \right) \bullet \left({}^{\text{ee}}\underline{\underline{\aleph}}_{\ell/\text{amb}} \right)^{-1} \\ & + \left(\varepsilon_\ell^{\text{L}}\, \underline{\underline{I}} - \underline{\underline{\varepsilon}}^{\,\text{L}}_{\text{amb}} \right) \bullet \mathcal{N}_{\underline{\underline{\varepsilon}}} \left\{ {}^{\text{ee}}\underline{\underline{\aleph}}_{\ell/\text{amb}} \right\}. \end{aligned} \tag{5.51} $$

In equations (5.50) and (5.51), the 3 × 3 dyadic

$$ {}^{\text{ee}}\underline{\underline{\aleph}}_{\ell/\text{amb}} = {}^{\text{ee}}\underline{\underline{\aleph}}^{\,\text{L}}_{\ell/\text{amb}} + {}^{\text{ee}}\underline{\underline{\aleph}}^{\,\text{NL}}_{\ell/\text{amb}} \,|\, \mathbf{E}_{\text{HCM}} \,|^2, \qquad (\ell \in \{\text{a, b}\}), \tag{5.52} $$

has the diagonal form

$$ {}^{\text{ee}}\underline{\underline{\aleph}}^{\,\text{j}}_{\ell/\text{amb}} = \begin{pmatrix} {}^{\text{ee}}\aleph^{x,\text{j}}_{\ell/\text{amb}} & 0 & 0 \\ 0 & {}^{\text{ee}}\aleph^{y,\text{j}}_{\ell/\text{amb}} & 0 \\ 0 & 0 & {}^{\text{ee}}\aleph^{z,\text{j}}_{\ell/\text{amb}} \end{pmatrix}, \qquad (\text{j} \in \{\text{L, NL}\}). \tag{5.53} $$

Its linear part is given as

$$ {}^{\text{ee}}\underline{\underline{\aleph}}^{\,\text{NL}}_{\ell/\text{amb}} = \underline{\underline{I}} + i\omega\, {}^{\text{ee}}\underline{\underline{D}}^{\,0,\text{L}}_{U/\text{amb}} \bullet \left(\varepsilon_\ell^{\text{L}}\, \underline{\underline{I}} - \underline{\underline{\varepsilon}}^{\,\text{L}}_{\text{amb}} \right), \tag{5.54} $$

while its nonlinear part involves

$$ \begin{aligned} {}^{\text{ee}}\underline{\underline{\aleph}}^{\,\text{L}}_{\ell/\text{amb}} = i\omega \Big[& {}^{\text{ee}}\underline{\underline{D}}^{\,0,\text{L}}_{U/\text{amb}} \bullet \left(g^{\text{FF}}_{\ell,\,\underline{\underline{\varepsilon}}}\, \varepsilon_\ell^{\text{NL}}\, \underline{\underline{I}} - \underline{\underline{\varepsilon}}^{\,\text{NL}}_{\text{amb}} \right) + {}^{\text{ee}}\underline{\underline{D}}^{\,0,\text{NL}}_{U/\text{amb}} \\ & \bullet \left(\varepsilon_\ell^{\text{L}}\, \underline{\underline{I}} - \underline{\underline{\varepsilon}}^{\,\text{L}}_{\text{amb}} \right) \Big]. \end{aligned} \tag{5.55} $$

The action of the 3 × 3 dyadic operator $\mathcal{N}_{\underline{\underline{\varepsilon}}}\{\,\bullet\,\}$ in equation (5.51) is specified by

$$ \begin{aligned} \mathcal{N}_{\underline{\underline{\varepsilon}}} \left\{ {}^{\text{ee}}\underline{\underline{\aleph}}_{\ell/\text{amb}} \right\} = \frac{1}{\det {}^{\text{ee}}\underline{\underline{\aleph}}^{\,\text{L}}_{\ell/\text{amb}}} & \left[\begin{pmatrix} \Pi^{x}_{\ell/\text{amb}} & 0 & 0 \\ 0 & \Pi^{y}_{\ell/\text{amb}} & 0 \\ 0 & 0 & \Pi^{z}_{\ell/\text{amb}} \end{pmatrix} \right. \\ & \left. - \mathcal{M}_{\underline{\underline{\varepsilon}}} \left\{ {}^{\text{ee}}\underline{\underline{\aleph}}_{\ell/\text{amb}} \right\} \left({}^{\text{ee}}\underline{\underline{\aleph}}^{\,\text{L}}_{\ell/\text{amb}} \right)^{-1} \right], \end{aligned} \tag{5.56} $$

with the scalar operator $\mathcal{M}_{\underline{\underline{\varepsilon}}}\{ \bullet \}$ therein being specified by

$$\begin{aligned}\mathcal{M}_{\underline{\underline{\varepsilon}}}\left\{ {}^{ee}\underline{\underline{\aleph}}_{\,\ell/\mathrm{amb}} \right\} = {} & {}^{ee}\aleph^{x,\mathrm{L}}_{\ell/\mathrm{amb}}\; {}^{ee}\aleph^{y,\mathrm{L}}_{\ell/\mathrm{amb}}\; {}^{ee}\aleph^{z,\mathrm{NL}}_{\ell/\mathrm{amb}} + {}^{ee}\aleph^{x,\mathrm{L}}_{\ell/\mathrm{amb}}\; {}^{ee}\aleph^{y,\mathrm{NL}}_{\ell/\mathrm{amb}}\; {}^{ee}\aleph^{z,\mathrm{L}}_{\ell/\mathrm{amb}} \\ & + {}^{ee}\aleph^{x,\mathrm{NL}}_{\ell/\mathrm{amb}}\; {}^{ee}\aleph^{y,\mathrm{L}}_{\ell/\mathrm{amb}}\; {}^{ee}\aleph^{z,\mathrm{L}}_{\ell/\mathrm{amb}},\end{aligned} \tag{5.57}$$

and the scalar parameters

$$\left.\begin{aligned} \Pi^{x}_{\ell/\mathrm{amb}} &= {}^{ee}\aleph^{y,\mathrm{NL}}_{\ell/\mathrm{amb}}\; {}^{ee}\aleph^{z,\mathrm{L}}_{\ell/\mathrm{amb}} + {}^{ee}\aleph^{y,\mathrm{L}}_{\ell/\mathrm{amb}}\; {}^{ee}\aleph^{z,\mathrm{NL}}_{\ell/\mathrm{amb}} \\ \Pi^{y}_{\ell/\mathrm{amb}} &= {}^{ee}\aleph^{z,\mathrm{NL}}_{\ell/\mathrm{amb}}\; {}^{ee}\aleph^{x,\mathrm{L}}_{\ell/\mathrm{amb}} + {}^{ee}\aleph^{z,\mathrm{L}}_{\ell/\mathrm{amb}}\; {}^{ee}\aleph^{x,\mathrm{NL}}_{\ell/\mathrm{amb}} \\ \Pi^{z}_{\ell/\mathrm{amb}} &= {}^{ee}\aleph^{y,\mathrm{NL}}_{\ell/\mathrm{amb}}\; {}^{ee}\aleph^{x,\mathrm{L}}_{\ell/\mathrm{amb}} + {}^{ee}\aleph^{y,\mathrm{L}}_{\ell/\mathrm{amb}}\; {}^{ee}\aleph^{x,\mathrm{NL}}_{\ell/\mathrm{amb}} \end{aligned}\right\}. \tag{5.58}$$

The local field factors $g^{\mathrm{FF}}_{\ell,\,\underline{\underline{\varepsilon}}}$ ($\ell \in \{\mathrm{a,b}\}$) in equation (5.51) represent the derivative on the right side of equation (5.10) with the subscript ℓ in lieu of the subscript 'a' therein. Plausible estimates of these factors are provided by

$$g^{\mathrm{FF}}_{\ell,\,\underline{\underline{\varepsilon}}} = \frac{1}{3}\left|\ \mathrm{tr}\left({}^{ee}\underline{\underline{\aleph}}^{\mathrm{L}}_{\,\ell/\mathrm{amb}}\right)^{-1}\ \right|, \qquad (\ell \in \{\mathrm{a,b}\}). \tag{5.59}$$

The permittivity dyadic of the ambient material, $\underline{\underline{\varepsilon}}_{\,\mathrm{amb}}$ of the form specified in equation (5.44), is given implicitly by nonlinear equations isomorphic to those in equations (5.31), but with the 6×6 polarizability density dyadics $\underline{\underline{\boldsymbol{\alpha}}}^{\,\mathrm{L,NL}}_{\,\ell/\mathrm{amb}}$ substituted by the 3×3 polarizability density dyadics ${}^{ee}\underline{\underline{\alpha}}^{\,\mathrm{L,NL}}_{\,\ell/\mathrm{amb}}$. By means of the following Jacobi scheme [22], the linear and nonlinear contributions to $\underline{\underline{\varepsilon}}_{\,\mathrm{amb}}$ emerge as

$$\underline{\underline{\varepsilon}}^{\,\mathrm{j}}_{\,\mathrm{amb}}[n] = \mathcal{S}^{\mathrm{j}}_{\underline{\underline{\varepsilon}}}\left\{\underline{\underline{\varepsilon}}^{\,\mathrm{j}}_{\,\mathrm{amb}}\,[n-1]\right\}, \qquad (\mathrm{j} \in \{\mathrm{L,\ NL}\};\ n \in \{1,\ 2,\ \ldots\}). \tag{5.60}$$

Herein, the action of the 3×3 dyadic operator $\mathcal{S}^{\mathrm{L}}_{\underline{\underline{\varepsilon}}}\{ \bullet \}$ is yielded by the 3×3 dielectric specialization of equation (4.22), with the replacements: $\underline{\underline{\varepsilon}}^{\,\mathrm{L}}_{\,\mathrm{a}}$ for $\underline{\underline{\mathbf{K}}}_{\,\mathrm{a}}$, $\underline{\underline{\varepsilon}}^{\,\mathrm{L}}_{\,\mathrm{amb}}$ for $\underline{\underline{\mathbf{K}}}_{\,\mathrm{Br}}$, ${}^{ee}\underline{\underline{D}}^{\,0,\mathrm{L}}_{\,\mathrm{U/amb}}$ for $\underline{\underline{\mathbf{D}}}^{0}_{\,U^{\mathrm{a}}/\mathrm{Br}}$ and $\underline{\underline{\mathbf{D}}}^{0}_{\,U^{\mathrm{b}}/\mathrm{Br}}$, and $\underline{\underline{I}}$ for $\underline{\underline{\mathbf{I}}}$. The action of the 3×3 dyadic operator $\mathcal{S}^{\mathrm{NL}}_{\underline{\underline{\varepsilon}}}\{ \bullet \}$ is given by

$$\begin{aligned}\mathcal{S}^{\mathrm{NL}}_{\underline{\underline{\varepsilon}}}\left\{\underline{\underline{\varepsilon}}^{\,\mathrm{NL}}_{\,\mathrm{amb}}\right\} = {} & \Bigg\{ f_{\mathrm{a}}\Big[g^{\mathrm{FF}}_{\mathrm{a},\,\underline{\underline{\varepsilon}}}\, \varepsilon^{\mathrm{NL}}_{\mathrm{a}} \left({}^{ee}\underline{\underline{\aleph}}^{\mathrm{L}}_{\,\mathrm{a/amb}}\right)^{-1} + \left(\varepsilon^{\mathrm{L}}_{\mathrm{a}}\,\underline{\underline{I}} - \underline{\underline{\varepsilon}}^{\,\mathrm{L}}_{\,\mathrm{amb}}\right) \bullet \mathcal{N}_{\underline{\underline{\varepsilon}}}\left\{{}^{ee}\underline{\underline{\aleph}}_{\,\mathrm{a/amb}}\right\}\Big] \\ & + f_{\mathrm{b}}\Big[g^{\mathrm{FF}}_{\mathrm{b},\,\underline{\underline{\varepsilon}}}\, \varepsilon^{\mathrm{NL}}_{\mathrm{b}} \left({}^{ee}\underline{\underline{\aleph}}^{\mathrm{L}}_{\,b/\mathrm{amb}}\right)^{-1} + \left(\varepsilon^{\mathrm{L}}_{\mathrm{b}}\,\underline{\underline{I}} - \underline{\underline{\varepsilon}}^{\,\mathrm{L}}_{\,\mathrm{amb}}\right) \bullet \mathcal{N}_{\underline{\underline{\varepsilon}}}\left\{{}^{ee}\underline{\underline{\aleph}}_{\,\mathrm{b/amb}}(\omega)\right\}\Big]\Bigg\} \\ & \bullet \left[f_{\mathrm{a}}\left({}^{ee}\underline{\underline{\aleph}}^{\mathrm{L}}_{\,\mathrm{a/amb}}\right)^{-1} + f_{\mathrm{b}}\left({}^{ee}\underline{\underline{\aleph}}^{\mathrm{L}}_{\,\mathrm{b/amb}}\right)^{-1}\right]^{-1}.\end{aligned} \tag{5.61}$$

Since the ambient material is equivalent to the HCM that arises in the Bruggeman formalism, equations (5.60) effectively yield the Bruggeman estimate of the HCM's intensity-dependent permittivity dyadic.

To proceed towards the second-order SPFT estimate of the HCM's intensity-dependent permittivity dyadic, the mass operator is required, as described in section

4.4.1. The mass operator builds upon the principal-value integral represented by $\underline{\underline{\mathbf{P}}}_{\Lambda}(L_{\text{cor}})$ in equation (4.34). Since there is no magnetoelectric coupling here, only the components $\left[\,\underline{\underline{\mathbf{P}}}_{\Lambda}(L_{\text{cor}})\,\right]_{mn}$ with $m, n \in \{1, 2, 3\}$ are required. Together these components constitute the 3×3 dyadic ${}^{\text{ee}}\underline{\underline{P}}_{\Lambda}(L_{\text{cor}})$, which is defined such that $\left[\,\underline{\underline{\mathbf{P}}}_{\Lambda}(L_{\text{cor}})\,\right]_{mn} = \left[\,{}^{\text{ee}}\underline{\underline{P}}_{\Lambda}(L_{\text{cor}})\,\right]_{mn}$ for $m, n \in \{1, 2, 3\}$. The dyadic ${}^{\text{ee}}\underline{\underline{P}}_{\Lambda}(L_{\text{cor}})$ may be partitioned as a sum of linear and nonlinear contributions, as in equation (5.34) but with ${}^{\text{ee}}\underline{\underline{P}}_{\Lambda}(L_{\text{cor}})$ in lieu of $\underline{\underline{\mathbf{P}}}_{\Lambda}(L_{\text{cor}})$. For the step covariance function $\Lambda_{\text{step}}(\mathbf{r} - \mathbf{r}')$ given in equation (4.36), the linear contribution to ${}^{\text{ee}}\underline{\underline{P}}_{\Lambda_{\text{step}}}(L_{\text{cor}})$ may be expressed as

$$
\begin{aligned}
{}^{\text{ee}}\underline{\underline{P}}^{\,\text{L}}_{\Lambda_{\text{step}}}(L_{\text{cor}}) = \frac{f_{\text{a}} f_{\text{b}} L_{\text{cor}}^{3}}{12\pi i\omega} \int_{\phi=0}^{2\pi} \int_{\theta=0}^{\pi} \frac{\sin\theta}{\Delta_0(\theta,\phi)} \\
\left[\,\tau_{\text{u}}(L_{\text{cor}},\theta,\phi)\,\underline{\underline{u}}_{\,0}(\theta,\phi) + \tau_{\text{v}}(\theta,\phi)\,\underline{\underline{v}}_{\,0}(\theta,\phi)\right] d\theta\, d\phi.
\end{aligned}
\tag{5.62}
$$

Herein, the 3×3 dyadic functions

$$
\left.
\begin{aligned}
\underline{\underline{u}}_{\,0}(\theta,\phi) &= \left[2\,\underline{\underline{\varepsilon}}^{\,\text{L}}_{\,\text{amb}} - \left(\text{tr}\,\underline{\underline{\varepsilon}}^{\,\text{L}}_{\,\text{amb}}\right)\underline{\underline{I}}\,\right] \bullet \underline{\underline{\Omega}}(\theta,\phi) - \left\{\text{tr}\left[\underline{\underline{\varepsilon}}^{\,\text{L}}_{\,\text{amb}} \bullet \underline{\underline{\Omega}}(\theta,\phi)\right]\right\}\underline{\underline{I}} \\
&\quad - \frac{\text{tr}\left[\left(\text{adj}\,\underline{\underline{\varepsilon}}^{\,\text{L}}_{\,\text{amb}}\right) \bullet \underline{\underline{\Omega}}(\theta,\phi)\right] - \left(\text{tr adj}\,\underline{\underline{\varepsilon}}^{\,\text{L}}_{\,\text{amb}}\right)\left[\text{tr}\,\underline{\underline{\Omega}}(\theta,\phi)\right]}{\text{tr}\left[\,\underline{\underline{\varepsilon}}^{\,\text{L}}_{\,\text{amb}} \bullet \underline{\underline{\Omega}}(\theta,\phi)\,\right]}\,\underline{\underline{\Omega}}(\theta,\phi) \\
\underline{\underline{v}}_{\,0}(\theta,\phi) &= \text{adj}\,\underline{\underline{\varepsilon}}^{\,\text{L}}_{\,\text{amb}} - \frac{\det \underline{\underline{\varepsilon}}^{\,\text{L}}_{\,\text{amb}}}{\text{tr}\left[\,\underline{\underline{\varepsilon}}^{\,\text{L}}_{\,\text{amb}} \bullet \underline{\underline{\Omega}}(\theta,\phi)\,\right]}\,\underline{\underline{\Omega}}(\theta,\phi)
\end{aligned}
\right\}
\tag{5.63}
$$

and the scalar functions

$$
\left.
\begin{aligned}
&\tau_{\text{u}}(L_{\text{cor}},\theta,\phi) \\
&\qquad = \frac{3[\kappa_{0+}(\theta,\phi) - \kappa_{0-}(\theta,\phi)]}{2L_{\text{cor}}} + i\left[\kappa_{0+}^{\frac{3}{2}}(\theta,\phi) - \kappa_{0-}^{\frac{3}{2}}(\theta,\phi)\right] \\
&\tau_{\text{v}}(\theta,\phi) = i\omega^2\mu_0\left[\kappa_{0+}^{\frac{1}{2}}(\theta,\phi) - \kappa_{0-}^{\frac{1}{2}}(\theta,\phi)\right] \\
&\Delta_0(\theta,\phi) = \left[t_{\text{B0}}^2(\theta,\phi) - 4t_{\text{A0}}(\theta,\phi)\,t_{\text{C0}}(\theta,\phi)\right]^{1/2} \\
&\kappa_{0\pm}(\theta,\phi) = -\mu_0\omega^2\,\frac{t_{\text{B0}}(\theta,\phi) \mp \Delta_0(\theta,\phi)}{2t_{\text{C0}}(\theta,\phi)}
\end{aligned}
\right\},
\tag{5.64}
$$

with

$$\left.\begin{aligned} t_{A0} &= \det \underline{\underline{\varepsilon}}^{\,L}_{\,amb} \\ t_{B0}(\theta, \phi) &= \left[\operatorname{tr} \left(\operatorname{adj} \underline{\underline{\varepsilon}}^{\,L}_{\,amb}\right) \bullet \underline{\underline{\Omega}}(\theta, \phi)\right] - \left(\operatorname{tr} \operatorname{adj} \underline{\underline{\varepsilon}}^{\,L}_{\,amb}\right) \operatorname{tr} \underline{\underline{\Omega}}(\theta, \phi) \\ t_{C0}(\theta, \phi) &= \left\{\operatorname{tr}\left[\underline{\underline{\varepsilon}}^{\,L}_{\,amb} \bullet \underline{\underline{\Omega}}(\theta, \phi) \right]\right\} \operatorname{tr} \underline{\underline{\Omega}}(\theta, \phi) \end{aligned}\right\}. \tag{5.65}$$

The corresponding nonlinear contribution to ${}^{ee}\underline{\underline{P}}_{\,\Lambda_{step}}(L_{cor})$ may be expressed as

$$\begin{aligned} {}^{ee}\underline{\underline{P}}^{\,NL}_{\,\Lambda_{step}}(L_{cor}) = \; & \frac{f_a f_b L_{cor}^3}{12\pi i \omega} \int_{\phi=0}^{2\pi} \int_{\theta=0}^{\pi} \frac{\sin\theta}{\Delta_0(\theta, \phi)} \Bigg(\tau_u(L_{cor}, \theta, \phi) \\ & \times \left[\underline{\underline{u}}_{\,1}(\theta, \phi) - \frac{\Delta_1(\theta, \phi)}{\Delta_0(\theta, \phi)} \underline{\underline{u}}_{\,0}(\theta, \phi) \right] \\ & + \tau_v(L_{cor}, \theta, \phi) \left[\underline{\underline{v}}_{\,1}(\theta, \phi) - \frac{\Delta_1(\theta, \phi)}{\Delta_0(\theta, \phi)} \underline{\underline{v}}_{\,0}(\theta, \phi) \right] \\ & + \frac{i}{2} \left[\frac{\kappa_{1+}(\theta, \phi)}{\kappa_{0+}^{\frac{1}{2}}(\theta, \phi)} - \frac{\kappa_{1-}(\theta, \phi)}{\kappa_{0-}^{\frac{1}{2}}(\theta, \phi)} \right] \underline{\underline{v}}_{\,0}(\theta, \phi) \\ & + \frac{3}{2} \Bigg\{ \left[\frac{1}{L_{cor}} + i\kappa_{0+}^{\frac{1}{2}}(\theta, \phi) \right] \kappa_{1+}(\theta, \phi) \\ & - \left[\frac{1}{L_{cor}} + i\kappa_{0-}^{\frac{1}{2}}(\theta, \phi) \right] \kappa_{1-}(\theta, \phi) \Bigg\} \underline{\underline{u}}_{\,0}(\theta, \phi) \Bigg) \, d\theta \, d\phi. \end{aligned} \tag{5.66}$$

Herein, the 3×3 dyadic functions

$$\left.\begin{aligned} \underline{\underline{u}}_{\,1}(\theta, \phi) = & \left\{ 2\, \underline{\underline{\varepsilon}}^{\,NL}_{\,amb} - \left[\frac{t_{B1}(\theta, \phi) t_{C0}(\theta, \phi) - t_{B0}(\theta, \phi) t_{C1}(\theta, \phi)}{t_{C0}(\theta, \phi) \left\{ \operatorname{tr}\left[\underline{\underline{\varepsilon}}^{\,L}_{\,amb} \bullet \underline{\underline{\Omega}}(\theta, \phi) \right]\right\}} + \operatorname{tr} \underline{\underline{\varepsilon}}^{\,NL}_{\,amb} \right] \underline{\underline{I}} \right\} \\ & \bullet \underline{\underline{\Omega}}(\theta, \phi) - \left\{ \operatorname{tr}\left[\underline{\underline{\varepsilon}}^{\,NL}_{\,amb} \bullet \underline{\underline{\Omega}}(\theta, \phi) \right]\right\} \underline{\underline{I}} \\ \underline{\underline{v}}_{\,1}(\theta, \phi) = & \; \underline{\underline{\Psi}} - \frac{t_{B1}(\theta, \phi) t_{C0}(\theta, \phi) - t_{B0}(\theta, \phi) t_{C1}(\theta, \phi)}{t_{C0}(\theta, \phi) \left\{ \operatorname{tr}\left[\underline{\underline{\varepsilon}}^{\,L}_{\,amb} \bullet \underline{\underline{\Omega}}(\theta, \phi) \right]\right\}} \underline{\underline{\Omega}}(\theta, \phi) \end{aligned}\right\}, \tag{5.67}$$

where the diagonal dyadic

$$\underline{\underline{\Psi}} = \begin{pmatrix} \Psi^x & 0 & 0 \\ 0 & \Psi^y & 0 \\ 0 & 0 & \Psi^z \end{pmatrix}, \tag{5.68}$$

has components

$$\left.\begin{aligned}\Psi^x &= \varepsilon_{\text{amb}}^{y,\text{NL}}\,\varepsilon_{\text{amb}}^{z,\text{L}} + \varepsilon_{\text{amb}}^{y,\text{L}}\,\varepsilon_{\text{amb}}^{z,\text{NL}}\\ \Psi^y &= \varepsilon_{\text{amb}}^{z,\text{NL}}\,\varepsilon_{\text{amb}}^{x,\text{L}} + \varepsilon_{\text{amb}}^{z,\text{L}}\,\varepsilon_{\text{amb}}^{x,\text{NL}}\\ \Psi^z &= \varepsilon_{\text{amb}}^{x,\text{NL}}\,\varepsilon_{\text{amb}}^{y,\text{L}} + \varepsilon_{\text{amb}}^{x,\text{L}}\,\varepsilon_{\text{amb}}^{y,\text{NL}}\end{aligned}\right\};\qquad(5.69)$$

and the scalar functions

$$\left.\begin{aligned}\Delta_1(\theta,\phi) &= \frac{t_{\text{B}0}(\theta,\phi)t_{\text{B}1}(\theta,\phi) - 2\left[\,t_{\text{A}1}t_{\text{C}0}(\theta,\phi) + t_{\text{A}0}t_{\text{C}1}(\theta,\phi)\,\right]}{\Delta_0(\theta,\phi)}\\ \kappa_{1\pm}(\theta,\phi) &= \frac{\omega^2[\,-t_{\text{B}1}(\theta,\phi) \pm \Delta_1(\theta,\phi)\,] - 2t_{\text{C}1}(\theta,\phi)\,\kappa_{0\pm}(\theta,\phi)}{2\,t_{\text{C}0}(\theta,\phi)}\end{aligned}\right\},\qquad(5.70)$$

with

$$\left.\begin{aligned}t_{\text{A}1} &= \varepsilon_{\text{amb}}^{x,\text{NL}}\,\varepsilon_{\text{amb}}^{y,\text{L}}\,\varepsilon_{\text{amb}}^{z,\text{L}} + \varepsilon_{\text{amb}}^{x,\text{L}}\,\varepsilon_{\text{amb}}^{y,\text{NL}}\,\varepsilon_{\text{amb}}^{z,\text{L}} + \varepsilon_{\text{amb}}^{x,\text{L}}\,\varepsilon_{\text{amb}}^{y,\text{L}}\,\varepsilon_{\text{amb}}^{z,\text{NL}}\\ t_{\text{B}1}(\theta,\phi) &= \text{tr}\,[\,\underline{\underline{\Psi}}\bullet\underline{\underline{\Omega}}(\theta,\phi)] - (\text{tr}\,\underline{\underline{\Psi}}\,)\ \text{tr}\,\underline{\underline{\Omega}}(\theta,\phi)\\ t_{\text{C}1}(\theta,\phi) &= \text{tr}\left[\,\underline{\underline{\varepsilon}}_{\,\text{amb}}^{\,\text{NL}}\bullet\underline{\underline{\Omega}}(\theta,\phi)\right]\text{tr}\,\underline{\underline{\Omega}}(\theta,\phi)\end{aligned}\right\}.\qquad(5.71)$$

Since the HCM is an anisotropic dielectric material, only the components $[\,\underline{\underline{\Sigma}}^{[2]}(L_{\text{cor}})\,]_{mn}$ with $m, n \in \{1, 2, 3\}$ of the second-order mass operator are relevant. These components constitute the 3×3 dyadic ${}^{\text{ee}}\underline{\underline{\Sigma}}^{[2]}(L_{\text{cor}})$ which is defined such that $[\,\underline{\underline{\Sigma}}^{[2]}(L_{\text{cor}})\,]_{mn} = [\,{}^{\text{ee}}\underline{\underline{\Sigma}}^{[2]}(L_{\text{cor}})\,]_{mn}$ for $m, n \in \{1, 2, 3\}$. In the weakly nonlinear regime, the dyadic ${}^{\text{ee}}\underline{\underline{\Sigma}}^{[2]}(L_{\text{cor}})$ may be partitioned as

$${}^{\text{ee}}\underline{\underline{\Sigma}}^{[2]}(L_{\text{cor}}) = {}^{\text{ee}}\underline{\underline{\Sigma}}^{[2],\,\text{L}}(L_{\text{cor}}) + {}^{\text{ee}}\underline{\underline{\Sigma}}^{[2],\,\text{NL}}(L_{\text{cor}})\,|\,\mathbf{E}_{\,\text{HCM}}\,|^2,\qquad(5.72)$$

wherein the linear contribution is

$${}^{\text{ee}}\underline{\underline{\Sigma}}^{[2],\,\text{L}}(L_{\text{cor}}) = -\omega^2\left(\underline{\underline{\alpha}}_{\,\text{a/amb}}^{\,0,\text{L}} - \underline{\underline{\alpha}}_{\,\text{b/amb}}^{\,0,\text{L}}\right)\bullet{}^{\text{ee}}\underline{\underline{P}}_{\,\Lambda_{\text{step}}}^{\,\text{L}}(L_{\text{cor}})\bullet\left(\underline{\underline{\alpha}}_{\,\text{a/amb}}^{\,0,\text{L}} - \underline{\underline{\alpha}}_{\,\text{b/amb}}^{\,0,\text{L}}\right)\qquad(5.73)$$

and the nonlinear contribution is

$$\begin{aligned}{}^{\text{ee}}\underline{\underline{\Sigma}}^{[2],\,\text{NL}}(L_{\text{cor}}) = -\,\omega^2\Big[2\left(\underline{\underline{\alpha}}_{\,\text{a/amb}}^{\,0,\text{L}} - \underline{\underline{\alpha}}_{\,\text{b/amb}}^{\,0,\text{L}}\right)\bullet{}^{\text{ee}}\underline{\underline{P}}_{\,\Lambda_{\text{step}}}^{\,\text{L}}(L_{\text{cor}})\bullet\left(\underline{\underline{\alpha}}_{\,\text{a/amb}}^{\,0,\text{NL}} - \underline{\underline{\alpha}}_{\,\text{b/amb}}^{\,0,\text{NL}}\right)\\ + \left(\underline{\underline{\alpha}}_{\,\text{a/amb}}^{\,0,\text{L}} - \underline{\underline{\alpha}}_{\,\text{b/amb}}^{\,0,\text{L}}\right)\bullet{}^{\text{ee}}\underline{\underline{P}}_{\,\Lambda_{\text{step}}}^{\,\text{NL}}(L_{\text{cor}})\bullet\left(\underline{\underline{\alpha}}_{\,\text{a/amb}}^{\,0,\text{L}} - \underline{\underline{\alpha}}_{\,\text{b/amb}}^{\,0,\text{L}}\right)\Big].\end{aligned}\qquad(5.74)$$

After assembling the foregoing expressions, the second-order SPFT estimate of the constitutive dyadic of the HCM emerges as

$$\underline{\underline{\varepsilon}}_{\,\text{SPFT}}^{\,[2]} = \underline{\underline{\varepsilon}}_{\,\text{SPFT}}^{\,[2],\text{L}} + \underline{\underline{\varepsilon}}_{\,\text{SPFT}}^{\,[2],\text{NL}}\,|\,\mathbf{E}_{\,\text{HCM}}\,|^2.\qquad(5.75)$$

Thus, the anisotropic dielectric specialization of equation (4.28) for $n = 2$ yields the linear contribution

$$\underline{\underline{\varepsilon}}^{[2],\mathrm{L}}_{\mathrm{SPFT}} = \underline{\underline{\varepsilon}}^{\mathrm{L}}_{\mathrm{amb}} - \frac{1}{i\omega}\left(\underline{\underline{e}}^{\mathrm{L}}\right)^{-1} \bullet {}^{ee}\underline{\underline{\Sigma}}^{[2],\,\mathrm{L}}(L_{\mathrm{cor}}) \tag{5.76}$$

and the nonlinear contribution

$$\underline{\underline{\varepsilon}}^{[2],\mathrm{NL}}_{\mathrm{SPFT}} = \underline{\underline{\varepsilon}}^{\mathrm{NL}}_{\mathrm{amb}} - \frac{1}{i\omega}\Big[\left(\underline{\underline{e}}^{\mathrm{L}}\right)^{-1} \bullet {}^{ee}\underline{\underline{\Sigma}}^{[2],\,\mathrm{NL}}(L_{\mathrm{cor}}) + \mathcal{N}_{\underline{\underline{\varepsilon}}}\{\underline{\underline{e}}\} \bullet {}^{ee}\underline{\underline{\Sigma}}^{[2],\,\mathrm{L}}(L_{\mathrm{cor}})\Big]. \tag{5.77}$$

In these expressions, $\underline{\underline{e}}$ represents the 3 × 3 dyadic analogue of the 6 × 6 dyadic $\underline{\underline{\mathbf{e}}}$ defined by equation (5.42). The linear and nonlinear contributions to $\underline{\underline{e}}$—written as $\underline{\underline{e}}^{\mathrm{L}}$ and $\underline{\underline{e}}^{\mathrm{NL}}$, respectively—are provided by equations (5.43) but with the 3 × 3 dyadics ${}^{ee}\underline{\underline{\Sigma}}^{[2],\,\mathrm{L,NL}}(L_{\mathrm{cor}})$ and ${}^{ee}\underline{\underline{D}}^{0,\mathrm{L,NL}}_{\mathrm{U/amb}}$ in lieu of the 6 × 6 dyadics $\underline{\underline{\mathbf{\Sigma}}}^{[2],\,\mathrm{L,NL}}(L_{\mathrm{cor}})$ and $\underline{\underline{\mathbf{D}}}^{0,\mathrm{L,NL}}_{\mathrm{I/comp}}$, respectively.

The second-order SPFT estimate of the HCM's intensity-dependent permittivity dyadic provided by equation (5.75) does not take into account the size of the constituent particles. However, the corresponding estimate of the extended second-order SPFT, which does take the constituent particle size into account [16], arises from equations (5.76) and (5.77) by replacing ${}^{ee}\underline{\underline{D}}^{0}$ with ${}^{ee}\underline{\underline{D}}(\rho)$ and $\underline{\underline{\alpha}}^{0,\mathrm{L,NL}}_{\mathrm{a,b/amb}}$ with $\underline{\underline{\alpha}}^{\mathrm{L,NL}}_{\mathrm{a,b/amb}}(\rho)$ in the definitions of $\underline{\underline{e}}$ and ${}^{ee}\underline{\underline{\Sigma}}^{[2],}(L_{\mathrm{cor}})$, in a manner analogous to that described for linear HCMs in section 4.5.

References

[1] Boyd R W 2008 *Nonlinear Optics* 3rd edn (London: Academic)

[2] Kobayashi T 1991 Introduction to nonlinear optical materials *Nonlin. Opt.* **1** 91–117

[3] Stroud D and Hui P M 1988 Nonlinear susceptibilities of granular matter *Phys. Rev.* B **37** 8719–24

[4] Sipe J E and Boyd R W 1992 Nonlinear susceptibility of composite optical materials in the Maxwell Garnett model *Phys. Rev.* A **46** 1614–29

[5] Lakhtakia A and Weiglhofer W S 2000 Maxwell Garnett formalism for weakly nonlinear, bianisotropic, dilute, particulate composite media *Int. J. Electron.* **87** 1401–8

[6] Lakhtakia A and Weiglhofer W S 1996 Maxwell Garnett approach for nonlinear dilute particulate composites with bi-isotropic host media *Int. J. Electron.* **80** 665–76

[7] Lakhtakia A and Weiglhofer W S 2000 Maxwell Garnett formalism for cubically nonlinear, gyrotropic, composite media *Int. J. Electron.* **84** 285–94

[8] Brandt H E (ed) 1991 *Selected Papers on Nonlinear Optics* (Bellingham, WA: Optical Engineering Press)

[9] Cumberbatch E 1970 Self-focusing in non-linear optics *IMA J. Appl. Maths.* **6** 250–62

[10] Levy O, Yagil Y and Bergman D J 1994 Field-induced tuning of the optical properties of nonlinear composites near resonance *J. Appl. Phys.* **76** 1431–5

[11] Lakhtakia A 2001 Application of strong permittivity fluctuation theory for isotropic, cubically nonlinear, composite mediums *Opt. Commun.* **192** 145–51

[12] Mackay T G, Lakhtakia A and Weiglhofer W S 2002 Homogenisation of isotropic, cubically nonlinear, composite mediums by the strong-permittivity-fluctuation theory: third-order considerations *Opt. Commun.* **204** 219–28

[13] Mackay T G, Lakhtakia A and Weiglhofer W S 2003 The strong-property-fluctuation theory for cubically nonlinear, isotropic chiral composite mediums *Electromagnetics* **23** 455–79

[14] Mackay T G 2003 Geometrically derived anisotropy in cubically nonlinear dielectric composites *J. Phys. D: Appl. Phys.* **36** 583–91

[15] Lakhtakia M N and Lakhtakia A 2001 Anisotropic composite materials with intensity-dependent permittivity tensor: the Bruggeman approach *Electromagnetics* **21** 129–38

[16] Cui J and Mackay T G 2007 Depolarization regions of nonzero volume for anisotropic, cubically nonlinear, homogenized nanocomposites *J. Nanophoton.* **1** 013506

[17] Zeng X C, Bergman D J, Hui P M and Stroud D 1988 Effective-medium theory for weakly nonlinear composites *Phys. Rev.* B **38** 10970–3

[18] Lakhtakia A 1996 Extended Maxwell Garnett formulae for weakly nonlinear dielectric-in-dielectric composites *Optik* **103** 85–7

[19] Mackay T G, Lakhtakia A and Weiglhofer W S 2001 Homogenisation of similarly oriented, metallic, ellipsoidal inclusions using the bilocal-approximated strong-property-fluctuation theory *Opt. Commun.* **197** 89–95

[20] Slepyan G Ya, Maksimenko S A, Bass F G and Lakhtakia A 1995 Nonlinear electromagnetics in chiral media: self-action of waves *Phys. Rev.* E **52** 1049–58

[21] Slepyan G Ya, Lakhtakia A and Maksimenko S A 1996 Bruggeman and Maxwell Garnett models of a chiral composite with weak cubic nonlinearities *Microw. Opt. Technol. Lett.* **12** 342–46

[22] Rao S S 2001 *Applied Numerical Methods for Engineers and Scientists* (Cambridge: Pearson)

Chapter 6

Applications and numerical examples

The conceptualization of a homogenizable composite material (HCM) revolves around its ability to replace a homogenous material [1]. Therefore, the homogenization formalisms for linear and weakly nonlinear HCMs presented in chapters 4 and 5, respectively, have long been applied in a wide range of electromagnetic scenarios. In recent years, many applications have involved HCMs that exhibit properties that are either not exhibited at all by their constituent materials or not exhibited to the same extent by their constituent materials [2].

In this chapter a range of applications of the Maxwell Garnett, Bruggeman, and strong-property-fluctuation theory (SPFT) formalisms are presented. Where appropriate, these applications are illustrated by means of numerical examples. Mathematica™ codes for the numerical examples are provided in appendix B. Also discussed and illustrated are important limitations of these formalisms that restrict their application to certain segments of the space spanned by the constitutive parameters.

6.1 Refinements to the Maxwell Garnett formalism

Let us begin with the Maxwell Garnett formalism. As discussed in section 4.2, this formalism is suitable only for a dilute distribution of a particulate material 'a' in a host material 'b'. A broad but not universal consensus restricts the volume fraction of constituent material 'a' to $f_a \lessapprox 0.3$. The incremental and differential refinements [3, 4] of this formalism are designed to overcome this restriction.

The effects of the incremental and differential refinements may be illustrated numerically as follows. Suppose the constituent materials are an isotropic dielectric material 'a' and a uniaxial dielectric material 'b'. The permittivity dyadic of constituent material 'a' is $\underline{\underline{\varepsilon}}_{\,a} = \varepsilon_a \, \underline{\underline{I}}$ with $\varepsilon_a = (12 + 4i)\varepsilon_0$, whereas the permittivity dyadic of constituent material 'b' is $\underline{\underline{\varepsilon}}_{\,b} = \varepsilon_\perp^b(\underline{\underline{I}} - \hat{\mathbf{u}}_z\hat{\mathbf{u}}_z) + \varepsilon_\parallel^b \hat{\mathbf{u}}_z\hat{\mathbf{u}}_z$ with $\varepsilon_\perp^b = (6 + i)\varepsilon_0$

doi:10.1088/978-0-7503-3423-5ch6

and $\varepsilon_{\parallel}^{\mathrm{b}} = (2.5 + 0.5i)\varepsilon_0$. The shape and orientation of the particles of constituent material 'a' are described by $\underline{\underline{U}}^{\mathrm{a}} = [(\underline{\underline{I}} - \hat{\mathbf{u}}_z\,\hat{\mathbf{u}}_z) + 5\hat{\mathbf{u}}_z\,\hat{\mathbf{u}}_z]/\sqrt[3]{5}$, per equation (3.15). The permittivity dyadic of the corresponding HCM has the uniaxial form

$$\underline{\underline{\varepsilon}}_{\mathrm{HCM}} = \varepsilon_{\perp}^{\mathrm{HCM}}(\underline{\underline{I}} - \hat{\mathbf{u}}_z\hat{\mathbf{u}}_z) + \varepsilon_{\parallel}^{\mathrm{HCM}}\hat{\mathbf{u}}_z\hat{\mathbf{u}}_z. \tag{6.1}$$

Estimates of the real and imaginary parts of the permittivity parameters $\varepsilon_{\perp}^{\mathrm{HCM}}$ and $\varepsilon_{\parallel}^{\mathrm{HCM}}$, as delivered by the Maxwell Garnett formalism, the incremental Maxwell Garnett formalism with $N = 3$, and the differential Maxwell Garnett formalism, are plotted as functions of the volume fraction f_{a} in figure 6.1. For comparison, the estimates of $\varepsilon_{\perp}^{\mathrm{HCM}}$ and $\varepsilon_{\parallel}^{\mathrm{HCM}}$ delivered by the Bruggeman formalism are also presented. In order to calculate the Bruggeman estimates, the particles of constituent material 'b' were taken to have the same shape and orientation as those of constituent material 'a' (i.e., $\underline{\underline{U}}^{\mathrm{b}} = \underline{\underline{U}}^{\mathrm{a}}$), although other choices for $\underline{\underline{U}}^{\mathrm{b}}$ are admissible.

Over the range $0 < f_{\mathrm{a}} < 0.3$, the Maxwell Garnett, the incremental Maxwell Garnett, the differential Maxwell Garnett, and the Bruggeman estimates are all in close agreement, but the estimates diverge markedly as f_{a} increases beyond 0.3. As f_{a} approaches unity, the HCM's permittivity parameters should converge to those of constituent material 'a', i.e., $\varepsilon_{\perp}^{\mathrm{HCM}} \to \varepsilon_{\perp}^{\mathrm{a}}$ and $\varepsilon_{\parallel}^{\mathrm{HCM}} \to \varepsilon_{\parallel}^{\mathrm{a}}$ as $f_{\mathrm{a}} \to 1$. These conditions are satisfied by the differential Maxwell Garnett estimates and the Bruggeman estimates. However, they are not quite satisfied by the incremental Maxwell Garnett estimates and they are not even close to being satisfied by the Maxwell Garnett estimates [5, 6]. For all values of f_{a}, the estimates of the differential Maxwell Garnett and the Bruggeman formalisms are almost identical.

6.2 Convergence of the strong-property-fluctuation theory

Next let us illustrate numerically the convergence of the SPFT by presenting the zeroth-order, second-order, and third-order estimates of the constitutive parameters that specify an isotropic chiral HCM. Per equations (2.35), the constitutive dyadics of the HCM are expressed as

$$\left.\begin{array}{lcl} \underline{\underline{\varepsilon}}_{\mathrm{SPFT}}^{[\ell]} & = & \varepsilon_{\mathrm{SPFT}}^{[\ell]}\,\underline{\underline{I}} \\ \underline{\underline{\xi}}_{\mathrm{SPFT}}^{[\ell]} & = & \xi_{\mathrm{SPFT}}^{[\ell]}\,\underline{\underline{I}} \\ \underline{\underline{\zeta}}_{\mathrm{SPFT}}^{[\ell]} & = & -\,\underline{\underline{\xi}}_{\mathrm{SPFT}}^{[\ell]} \\ \underline{\underline{\mu}}_{\mathrm{SPFT}}^{[\ell]} & = & \mu_{\mathrm{SPFT}}^{[\ell]}\,\underline{\underline{I}} \end{array}\right\}, \tag{6.2}$$

with $\ell = 0$ for the zeroth-order estimates, $\ell = 2$ for the second-order estimates, and $\ell = 3$ for the third-order estimates. As described in section 4.4, the zeroth-order estimate is identical to the first-order estimate and to the estimate delivered by Bruggeman homogenization formalism [7]. Suppose that constituent material 'a' is an isotropic chiral material characterized by the constitutive dyadics $\underline{\underline{\varepsilon}}_{\mathrm{a}} = (5 + 0.04i)\varepsilon_0\underline{\underline{I}}$, $\underline{\underline{\xi}}_{\mathrm{a}} = -\underline{\underline{\zeta}}_{\mathrm{a}} = (0.2 + 0.003i)i\sqrt{\varepsilon_0\mu_0}\,\underline{\underline{I}}$, and $\underline{\underline{\mu}}_{\mathrm{a}} = (2 + 0.02i)\mu_0\,\underline{\underline{I}}$,

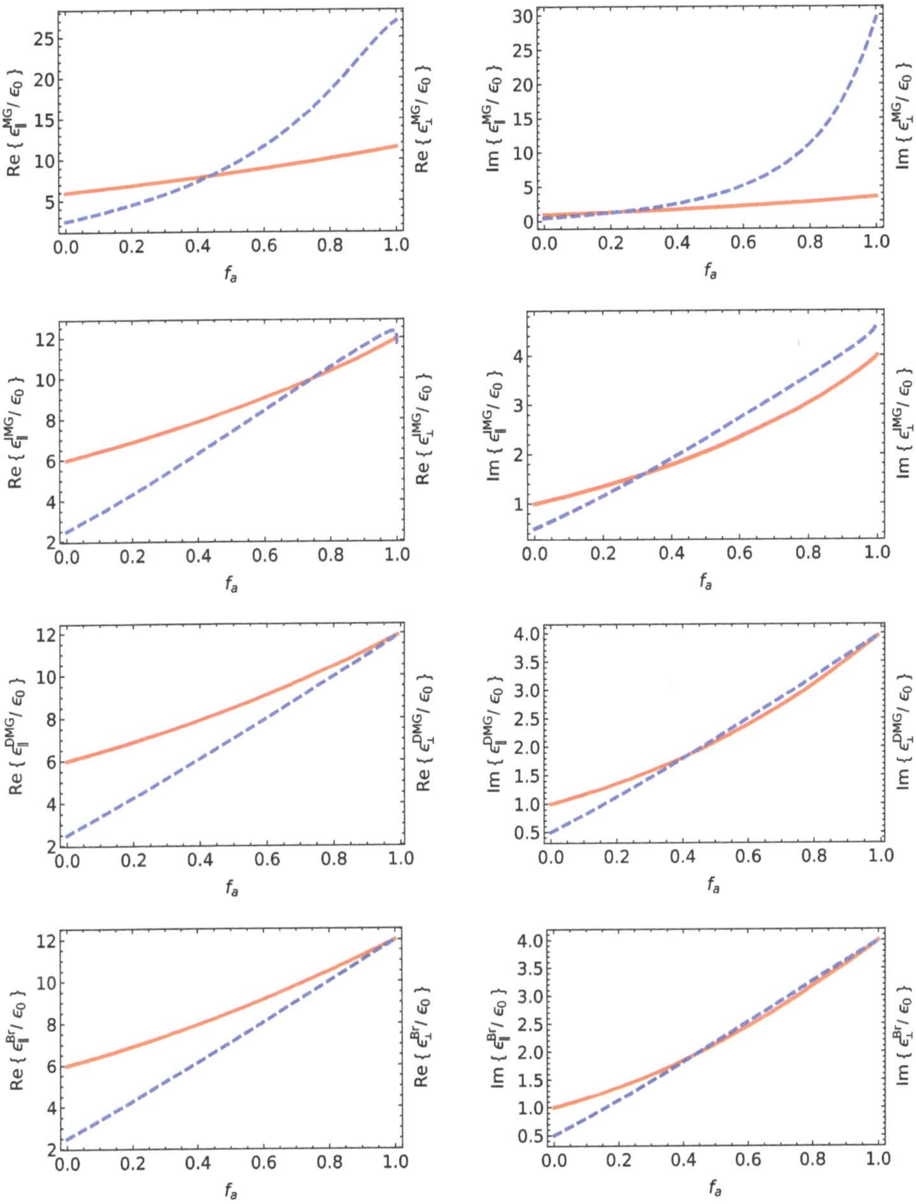

Figure 6.1. Real and imaginary parts of $\varepsilon_{\perp}^{\mathrm{HCM}}/\varepsilon_0$ (red solid curves) and $\varepsilon_{\parallel}^{\mathrm{HCM}}/\varepsilon_0$ (blue dashed curves) plotted against f_{a}, as estimated by the Maxwell Garnett formalism (HCM = MG), the incremental Maxwell Garnett formalism (HCM = IMG) with $N = 3$, the differential Maxwell Garnett formalism (HCM = DMG), and the Bruggeman formalism (HCM = Br). Input parameters: $\underline{\underline{\varepsilon}}_{\mathrm{a}} = (12 + 4i)\varepsilon_0\, \underline{\underline{I}}$, $\underline{\underline{\varepsilon}}_{\mathrm{b}} = \left[(6 + i)(\underline{\underline{I}} - \hat{\mathbf{u}}_z\, \hat{\mathbf{u}}_z) + (2.5 + 0.5i)\hat{\mathbf{u}}_z\, \hat{\mathbf{u}}_z\right]\varepsilon_0$, $\underline{\underline{U}}^{\mathrm{a}} = [(\underline{\underline{I}} - \hat{\mathbf{u}}_z\, \hat{\mathbf{u}}_z) + 5\hat{\mathbf{u}}_z\, \hat{\mathbf{u}}_z]/\sqrt[3]{5}$, and in the case of the Bruggeman formalism $\underline{\underline{U}}^{\mathrm{b}} = \underline{\underline{U}}^{\mathrm{a}}$.

while constituent material 'b' is simply free space, i.e., $\underline{\underline{\varepsilon}}_{\rm b} = \varepsilon_0 \underline{\underline{I}}$, $\underline{\underline{\xi}}_{\rm b} = \underline{\underline{\zeta}}_{\rm b} = \underline{\underline{0}}$, and $\underline{\underline{\mu}}_{\rm b} = \mu_0 \underline{\underline{I}}$. The particles of constituent materials 'a' and 'b' are spherical, so that $\underline{\underline{U}}^{\rm a} = \underline{\underline{U}}^{\rm b} = \underline{\underline{I}}$. The volume fraction is fixed at $f_{\rm a} = 0.4$ and the step covariance functions (4.35) and (4.38) are adopted for the second-order and third-order SPFT estimates, respectively.

Real and imaginary parts of the estimates of the HCM's constitutive parameters are plotted against the normalized correlation length $k_0 L_{\rm cor}$ in figure 6.2 for $\ell \in \{0, 2, 3\}$. The $\ell = 0$ estimates are independent of correlation length, but the $\ell = 2$ and 3 estimates are not. The $\ell = 2$ and 3 estimates for each constitutive

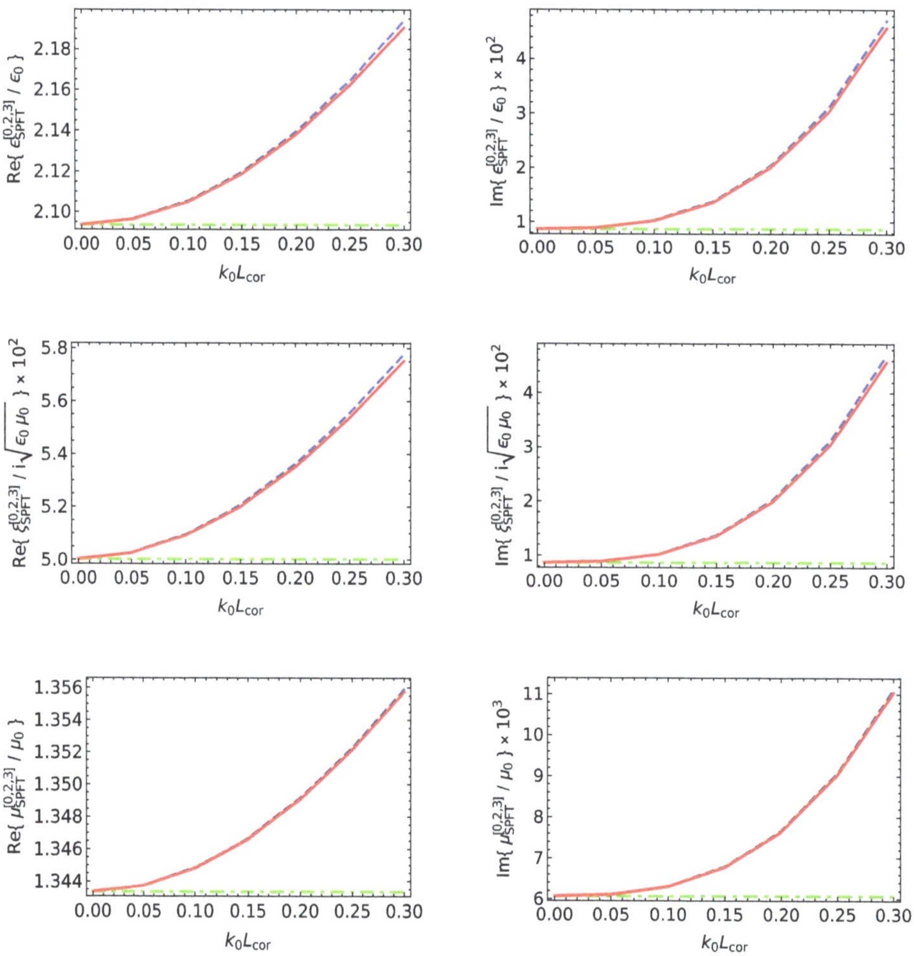

Figure 6.2. Real and imaginary parts of $\varepsilon_{\rm SPFT}^{[\ell]}/\varepsilon_0$, $\xi_{\rm SPFT}^{[\ell]}/(i\sqrt{\varepsilon_0\mu_0})$, and $\mu_{\rm SPFT}^{[\ell]}/\mu_0$ plotted against $k_0 L_{\rm cor}$, as computed using the zeroth-order ($\ell = 0$: green broken dashed curves), second-order ($\ell = 2$: blue dashed curves), and third-order ($\ell = 3$: red solid curves) SPFT formalisms. Step covariance functions (4.38) and (4.38) were used for $\ell = 2$ and 3, respectively. Input parameters: $f_a = 0.4$, $\underline{\underline{\varepsilon}}_{\rm a} = (5 + 0.04i)\varepsilon_0\,\underline{\underline{I}}$, $\underline{\underline{\xi}}_{\rm a} = -\underline{\underline{\zeta}}_{\rm a} = (0.2 + 0.003i)i\sqrt{\varepsilon_0\mu_0}\,\underline{\underline{I}}$, $\underline{\underline{\mu}}_{\rm a} = (2 + 0.02i)\mu_0\,\underline{\underline{I}}$, $\underline{\underline{\varepsilon}}_{\rm b} = \varepsilon_0 \underline{\underline{I}}$, $\underline{\underline{\xi}}_{\rm b} = \underline{\underline{\zeta}}_{\rm b} = \underline{\underline{0}}$, $\underline{\underline{\mu}}_{\rm b} = \mu_0 \underline{\underline{I}}$, and $\underline{\underline{U}}^{\rm a} = \underline{\underline{U}}^{\rm b} = \underline{\underline{I}}$.

parameter of the HCM are identical to the $\ell = 0$ estimates when $L_{cor} = 0$. As the correlation length increases, the real and imaginary parts of the $\ell = 2$ and 3 estimates diverge markedly from the $\ell = 0$ estimate but the $\ell = 2$ and 3 estimates remain very close to each other over the range $0 < k_0 L_{cor} < 0.3$. As the third-order SPFT estimate delivers only a tiny modification to the second-order SPFT estimate, convergence at the second-order level of approximation is demonstrated.

6.3 Extended formalisms: the isotropic dielectric HCM

The extended homogenization formalisms described in section 4.4 are substantially more cumbersome than their standard counterparts. Numerical methods are usually needed for their implementations. Indeed, explicit expressions for the extended depolarization dyadic $\underline{\underline{\mathbf{D}}}_{U/\text{amb}}(\rho)$ exist only for isotropic ambient materials and certain rather simple uniaxial ambient materials, as presented in section 3.2.2. In order to highlight the essential features of extended formalisms, it is helpful to consider the simplest imaginable homogenization scenario.

Suppose that the two constituent materials are both isotropic dielectric materials[1], characterized by permittivities ε_a and ε_b, and both are distributed as spherical particles. Accordingly, the HCM is also an isotropic dielectric material with permittivity ε_{HCM}. The estimates of ε_{HCM} delivered by the extended Maxwell Garnett, Bruggeman, and SPFT formalisms, namely ε_{MG}, ε_{Br}, and $\varepsilon_{\text{SPFT}}$, respectively, are as follows.

6.3.1 Extended Maxwell Garnett estimate

Only the 3×3 components ${}^{ee}\underline{\underline{D}}^{\,0}_{I/b}$ and ${}^{ee}\underline{\underline{D}}^{\,+}_{I/b}(\rho)$ of the 6×6 depolarization dyadic $\underline{\underline{\mathbf{D}}}_{I/b}(\rho)$, per equations $(3.30)_1$ and $(3.31)_1$, are pertinent here. Since ${}^{ee}\underline{\underline{D}}^{\,0}_{I/b}$ and ${}^{ee}\underline{\underline{D}}^{\,+}_{I/b}(\rho)$ are simply scalar multiples of $\underline{\underline{I}}$, the depolarization scalar

$$d_b(\rho) = d_b^0 + d_b^+(\rho) \tag{6.3}$$

emerges. The ρ-independent contribution to $d_b(\rho)$ is

$$d_b^0 = \frac{1}{i3\omega\varepsilon_b} \tag{6.4}$$

by virtue of equation $(3.30)_1$, whereas the ρ-dependent contribution

$$d_b^+(\rho) = \frac{i\omega\rho^2\mu_0}{9}\left(3 + i2\rho\omega\sqrt{\varepsilon_b\mu_0}\right) \tag{6.5}$$

to $d_b(\rho)$ is delivered by equation $(3.31)_1$. In a similar fashion, the pertinent component of the corresponding polarizability density dyadic $\underline{\underline{\alpha}}_{a/b}(\rho)$ is simply a scalar multiple of $\underline{\underline{I}}$; thus, the polarizability density scalar

[1] The case for isotropic magnetic materials is analogous to that for isotropic dielectric materials.

$$\alpha_{\mathrm{a/b}}(\rho) = \frac{\varepsilon_{\mathrm{a}} - \varepsilon_{\mathrm{b}}}{1 + i\omega d_{\mathrm{b}}(\rho)(\varepsilon_{\mathrm{a}} - \varepsilon_{\mathrm{b}})} \tag{6.6}$$

turns out to be the isotropic-dielectric specialization of equation (3.61). Notice that equation (6.6) becomes isomorphic to equation (3.64) in the limit $\rho \to 0$. Then, the extended Maxwell Garnett estimate of $\varepsilon_{\mathrm{HCM}}$ emerges from the isotropic-dielectric specialization of equation (3.61) as

$$\varepsilon_{\mathrm{MG}} = \varepsilon_{\mathrm{b}} + \frac{f_{\mathrm{a}}\alpha_{\mathrm{a/b}}(\rho)}{1 - i\omega f_{\mathrm{a}} d_{\mathrm{b}}(\rho)\, \alpha_{\mathrm{a/b}}(\rho)}. \tag{6.7}$$

When evaluated at $\rho = 0$, the formula (6.7) is precisely the same as that for the Hashin–Shtrikman bound given in equation $(1.2)_1$. However, the Maxwell Garnett estimate is only appropriate for $f_{\mathrm{a}} \lessapprox 0.3$ whereas there is no such restriction on the Hashin–Shtrikman bounds.

6.3.2 Extended Bruggeman estimate

The polarizability density scalars for the Bruggeman formalism are given as

$$\alpha_{\ell/\mathrm{Br}}(\rho) = \frac{\varepsilon_{\ell} - \varepsilon_{\mathrm{Br}}}{1 + i\omega d_{\mathrm{Br}}(\rho)(\varepsilon_{\ell} - \varepsilon_{\mathrm{Br}})}, \qquad (\ell \in \{\mathrm{a}, \mathrm{b}\}), \tag{6.8}$$

where the depolarization scalar $d_{\mathrm{Br}}(\rho)$ has the same form as that given in equations (6.3)–(6.5) but with ε_{b} therein replaced by $\varepsilon_{\mathrm{Br}}$. Thus, the extended version of the isotropic-dielectric specialization of equation (4.20) arises as

$$\frac{f_{\mathrm{a}}(\varepsilon_{\mathrm{a}} - \varepsilon_{\mathrm{Br}})}{1 + i\omega d_{\mathrm{Br}}(\rho)(\varepsilon_{\mathrm{a}} - \varepsilon_{\mathrm{Br}})} + \frac{f_{\mathrm{b}}(\varepsilon_{\mathrm{b}} - \varepsilon_{\mathrm{Br}})}{1 + i\omega d_{\mathrm{Br}}(\rho)(\varepsilon_{\mathrm{b}} - \varepsilon_{\mathrm{Br}})} = 0. \tag{6.9}$$

When evaluated at $\rho = 0$, equation (6.9) reduces to a quadratic equation in terms of $\varepsilon_{\mathrm{Br}}$ and the solution

$$\begin{aligned}\varepsilon_{\mathrm{Br}} = \frac{1}{4}\Big\{&-\left[\varepsilon_{\mathrm{a}}(f_{\mathrm{b}} - 2f_{\mathrm{a}}) + \varepsilon_{\mathrm{b}}(f_{\mathrm{a}} - 2f_{\mathrm{b}})\right] \\ &+ \sqrt{\left[\varepsilon_{\mathrm{a}}(f_{\mathrm{b}} - 2f_{\mathrm{a}}) + \varepsilon_{\mathrm{b}}(f_{\mathrm{a}} - 2f_{\mathrm{b}})\right]^2 + 8\varepsilon_{\mathrm{a}}\varepsilon_{\mathrm{b}}}\Big\}\end{aligned} \tag{6.10}$$

follows. The sign of the square root term in equation (6.10) may be determined by imposing the constraint (4.4); i.e., $\varepsilon_{\mathrm{Br}} \to \varepsilon_{\ell}$ as $f_{\ell} \to 1$ for $\ell \in \{\mathrm{a}, \mathrm{b}\}$.

6.3.3 Extended SPFT estimate

For the SPFT formalism, the ambient comparison material is an isotropic dielectric material with permittivity $\varepsilon_{\mathrm{amb}}$. In fact, $\varepsilon_{\mathrm{amb}} = \varepsilon_{\mathrm{Br}}$, as given in equation (6.10), courtesy of the non-secularity condition (4.29). Under the bilocal approximation, the SPFT estimate of the permittivity $\varepsilon_{\mathrm{HCM}}$ is delivered as [8]

$$\varepsilon_{\text{SPFT}}^{[2]} = \varepsilon_{\text{amb}} - \frac{1}{i\omega}\left[\frac{\Sigma^{[2]}(\rho, L_{\text{cor}})}{1 + \Sigma^{[2]}(\rho, L_{\text{cor}})\, d_{\text{amb}}(\rho)}\right], \tag{6.11}$$

which represents the isotropic-dielectric specialization of equation (4.28) for $n = 2$. Here the depolarization scalar $d_{\text{amb}}(\rho)$ has the same form as that given in equations (6.3)–(6.5) but with ε_{b} therein replaced by ε_{amb}. The scalar mass operator term in equation (6.11) is provided as

$$\Sigma^{[2]}(\rho, L_{\text{cor}}) = -f_{\text{a}} f_{\text{b}}\, \omega^2 [\alpha_{\text{a/amb}}(\rho) - \alpha_{\text{b/amb}}(\rho)]^2 d^{+}_{\text{amb}}(L_{\text{cor}}), \tag{6.12}$$

which represents the isotropic-dielectric specialization of equation (4.33) with the step function (4.36) taken as the choice of two-point covariance function. The scalar polarizability density scalars $\alpha_{\text{a/amb}}(\rho)$ and $\alpha_{\text{b/amb}}(\rho)$ in the foregoing equation are defined as in equation (6.8), but with ε_{Br} therein replaced by ε_{amb}.

6.3.4 Scattering loss

A particular feature of the extended Maxwell Garnett and Bruggeman formalisms—not shared by their standard counterparts—is that their estimates of ε_{HCM} can be complex valued even when both ε_{a} and ε_{b} are real valued. This may be appreciated from an inspection of equations (6.4) and (6.5): the terms $i\omega\, d^{0}_{\text{b, Br}}$ are real valued when $\varepsilon_{\text{b,Br}}$ are real valued, while the terms $i\omega\, d^{+}_{\text{b, Br}}(\rho)$ have nonzero imaginary parts whenever the size parameter $\rho > 0$. Hence, the HCM is predicted to be dissipative even when its constituent materials are nondissipative. The dissipation may be regarded as a manifestation of a scattering loss from the coherent field, arising due to the non-zero spatial extent of the constituent particles. Furthermore, for non-dissipative constituent materials the bilocally approximated and the trilocally approximated SPFT formalisms predict a dissipative HCM even when the size parameter $\rho = 0$ (i.e., even for the standard SPFT formulation). This is a consequence of $d^{+}_{\text{amb}}(L_{\text{cor}})$ being complex valued even when ε_{amb} is real valued. Again, dissipation in the HCM may be attributed to a scattering loss from the coherent field, in this case brought about by scattering interactions between constituent particles separated by distances less than the correlation length L_{cor}.

The manifestation of scattering loss that is associated with extended homogenization formalisms, as well as the standard higher-order SPFT formalisms, is a general feature that also occurs for anisotropic and bianisotropic HCMs, including nonlinear HCMs.

6.3.5 Numerical illustration

As an illustration, suppose that $\varepsilon_{\text{a}} = 0.2\,\varepsilon_0$ and $\varepsilon_{\text{b}} = 5\,\varepsilon_0$. Plots of the real and imaginary parts of $\varepsilon_{\text{MG}}/\varepsilon_0$ (per equation (6.7)), $\varepsilon_{\text{Br}}/\varepsilon_0$ (per equation (6.7)), and $\varepsilon^{[2]}_{\text{SPFT}}$ (per equation (6.11)) versus f_{a} and $k_0\rho$ are presented in figure 6.3. For $\varepsilon^{[2]}_{\text{SPFT}}$, the correlation length $L_{\text{cor}} = 0.3/k_0$. Since $k_0 = \omega\sqrt{\varepsilon_0\mu_0}$ is the free-space wavenumber, the quantity $k_0\rho$ is a dimensionless measure of particle size relative to wavelength. The constraint (4.4) ensures that for all three formalisms, the estimates of ε_{HCM} are

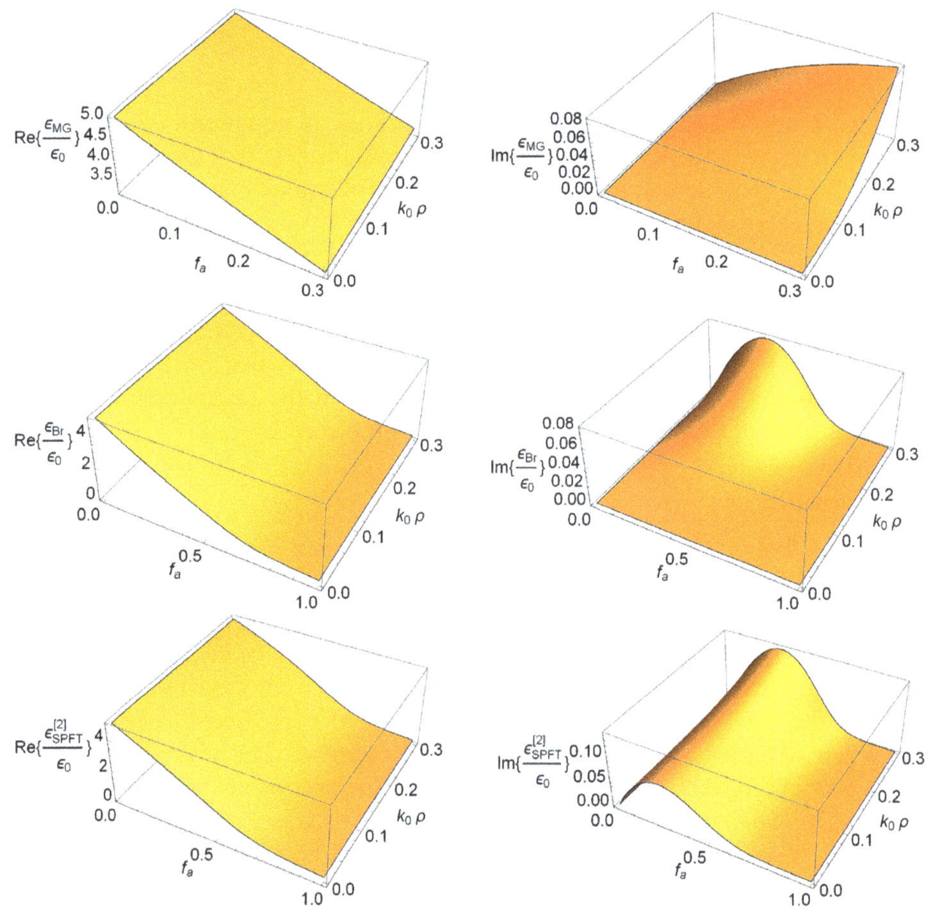

Figure 6.3. Real and imaginary parts of the estimates of $\varepsilon_{\text{HCM}}/\varepsilon_0$ plotted versus f_{a} and $k_0\rho$ for the extended Maxwell Garnett, Bruggeman, and second-order SPFT formalisms. The isotropic dielectric constituent materials are specified by $\varepsilon_{\text{a}} = 0.2\varepsilon_0$ and $\varepsilon_{\text{b}} = 5\varepsilon_0$, and are randomly distributed as spherical particles. The correlation length $L_{\text{cor}} = 0.3/k_0$ for $\varepsilon_{\text{SPFT}}^{[2]}$.

such that $\text{Re}\{\varepsilon_{\text{HCM}}\} \to \varepsilon_\ell$ and $\text{Im}\{\varepsilon_{\text{HCM}}\} \to 0$ as $f_\ell \to 1$, for $\ell \in \{a, b\}$. At a fixed value of $k_0\rho$, the real parts of the estimates of ε_{HCM} decrease approximately linearly as f_{a} increases over the range (0, 0.3) for all three formalisms. At larger values of f_{a} (where the Maxwell Garnett formalism is not appropriate), the estimates of ε_{HCM} from the Bruggeman and SPFT formalisms decrease in a more nonlinear fashion as f_{a} increases. There are only modest differences in the estimates of $\text{Re}\{\varepsilon_{\text{HCM}}\}$ provided by the three formalisms. The estimates of $\text{Im}\{\varepsilon_{\text{HCM}}\}$ delivered by the Maxwell Garnett and Bruggeman formalisms are similar, qualitatively and even quantitatively, for $f_{\text{a}} \in (0, 0.3)$: at any fixed value of $k_0\rho$ the imaginary parts of ε_{HCM} increase rapidly as f_{a} increases from 0, and at any fixed value of f_{a} the imaginary parts of ε_{HCM} increase rapidly as $k_0\rho$ increases from 0. The latter observation is consistent with greater scattering losses being associated with larger constituent particles. As f_a varies while the value of $k_0\rho$ is fixed, the Bruggeman and SPFT estimates of $\mathrm{Im}\{\varepsilon_{\mathrm{HCM}}\}$ have a local maximum in the vicinity of $f_a \approx 0.3$. The imaginary part of the SPFT estimate of $\varepsilon_{\mathrm{HCM}}$ is qualitatively different from the corresponding estimates provided by the Maxwell Garnett and Bruggeman formalisms. Specifically, $\mathrm{Im}\{\varepsilon^{[2]}_{\mathrm{SPFT}}\} \not\to 0$ in the limit $\rho \to 0$, whereas both $\mathrm{Im}\{\varepsilon_{\mathrm{MG}}\} \to 0$ and $\mathrm{Im}\{\varepsilon_{\mathrm{Br}}\} \to 0$ in the limit $\rho \to 0$. Furthermore, the values of $\mathrm{Im}\{\varepsilon^{[2]}_{\mathrm{SPFT}}\}$ are greater than those of $\mathrm{Im}\{\varepsilon_{\mathrm{MG}}\}$ and $\mathrm{Im}\{\varepsilon_{\mathrm{Br}}\}$ for all values of f_a and $k_0\rho$ considered. This is because the extended second-order SPFT formalism, as compared to the extended Maxwell Garnett and Bruggeman formalisms, incorporates an additional mechanism for taking into account scattering loss between neighboring constituent particles, namely that mediated via the correlation length L_{cor}.

The relationship between the size parameter ρ and the correlation length L_{cor} is further delineated in figure 6.4, wherein the real and imaginary parts of $\varepsilon^{[2]}_{\mathrm{SPFT}}$ are plotted versus ρ/L_{cor} and k_0L_{cor}. Here the volume fraction $f_a = 0.3$. At a fixed value of ρ/L_{cor}, both the real and imaginary parts of $\varepsilon^{[2]}_{\mathrm{SPFT}}$ increase nonlinearly as k_0L_{cor} increases from 0. The increase in $\mathrm{Im}\{\varepsilon^{[2]}_{\mathrm{SPFT}}\}$ may be explained as follows: as the correlation length increases, the interactions of an increasing number of scattering centers become correlated; accordingly, the scattering loss increases and a concomitant increase in the imaginary part of $\varepsilon^{[2]}_{\mathrm{SPFT}}$ occurs. Notice that the imaginary part of $\varepsilon^{[2]}_{\mathrm{SPFT}}$ is null valued and the SPFT estimate of $\varepsilon_{\mathrm{HCM}}$ coincides with that from the standard Bruggeman formalism when $\rho = L_{\mathrm{cor}} = 0$.

6.4 Realization of anisotropy and bianisotropy

Through the process of homogenization, the spacetime symmetries of constituent materials (and other properties such as nonlinearity and nonlocality) are combined and passed on to HCMs. Thus, relatively complex HCMs can arise from relatively simple constituent materials. In particular, anisotropic and bianisotropic HCMs may be straightforwardly conceptualized, when the constituent materials themselves are not necessarily anisotropic or bianisotropic.

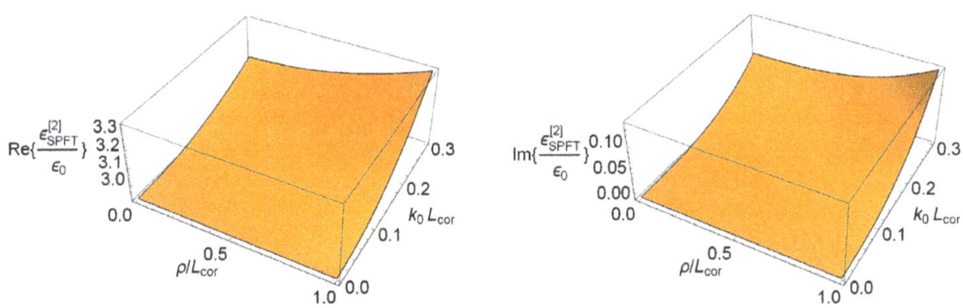

Figure 6.4. Real and imaginary parts of $\varepsilon^{[2]}_{\mathrm{SPFT}}/\varepsilon_0$ plotted versus ρ/L_{cor} and k_0L_{cor}. The isotropic dielectric constituent materials are specified by $\varepsilon_a = 0.2\varepsilon_0$ and $\varepsilon_b = 5\varepsilon_0$, and are randomly distributed as spherical particles. The volume fraction $f_a = 0.3$.

6.4.1 Biaxial materials

The manifestation of biaxial anisotropy [9, 10] and biaxial bianisotropy [11] by means of homogenization has been widely investigated. In general, the constituent materials must singly or jointly possess two non-collinear distinguished axes for a biaxial HCM to arise. If the constituent materials together possess only one distinguished axis then the corresponding HCM is uniaxial, in general. The distinguished axes of the constituent materials may derive from spacetime symmetries (e.g., one or more of the constituent materials may be uniaxial) or they may derive from the shape of the constituent particles (e.g., one or more of the constituent materials may be composed of oriented ellipsoidal particles).

For example, consider the homogenization of two dielectric constituent materials characterized by the permittivity dyadics $\underline{\underline{\varepsilon}}_{\,\mathrm{a}}$ and $\underline{\underline{\varepsilon}}_{\,\mathrm{b}}$, both dispersed randomly as ellipsoidal particles whose shapes and orientations are specified by $\underline{\underline{U}}^{\mathrm{a}}$ and $\underline{\underline{U}}^{\mathrm{b}}$ per equation (3.15). The corresponding HCM may be a biaxial dielectric material in any one of the following instances:

- both $\underline{\underline{\varepsilon}}_{\,\mathrm{a}}$ and $\underline{\underline{\varepsilon}}_{\,\mathrm{b}}$ have uniaxial symmetry, and both $\underline{\underline{U}}^{\mathrm{a}}$ and $\underline{\underline{U}}^{\mathrm{b}}$ describe spherical particles;
- both $\underline{\underline{\varepsilon}}_{\,\mathrm{a}}$ and $\underline{\underline{\varepsilon}}_{\,\mathrm{b}}$ have isotropic symmetry, and both $\underline{\underline{U}}^{\mathrm{a}}$ and $\underline{\underline{U}}^{\mathrm{b}}$ describe spheroidal particles;
- both $\underline{\underline{\varepsilon}}_{\,\mathrm{a}}$ and $\underline{\underline{\varepsilon}}_{\,\mathrm{b}}$ have isotropic symmetry, with $\underline{\underline{U}}^{\mathrm{a}}$ describing ellipsoidal particles and $\underline{\underline{U}}^{\,\mathrm{b}}$ describing spherical particles;
- $\underline{\underline{\varepsilon}}_{\,\mathrm{a}}$ has uniaxial symmetry and $\underline{\underline{\varepsilon}}_{\,\mathrm{b}}$ has isotropic symmetry, with $\underline{\underline{U}}^{\mathrm{a}}$ describing spherical particles and $\underline{\underline{U}}^{\mathrm{b}}$ describing spheroidal particles;
- $\underline{\underline{\varepsilon}}_{\,\mathrm{a}}$ has uniaxial symmetry and $\underline{\underline{\varepsilon}}_{\,\mathrm{b}}$ has isotropic symmetry, with $\underline{\underline{U}}^{\mathrm{a}}$ describing spheroidal particles and $\underline{\underline{U}}^{\mathrm{b}}$ describing spherical particles; and
- $\underline{\underline{\varepsilon}}_{\,\mathrm{a}}$ has biaxial symmetry and $\underline{\underline{\varepsilon}}_{\,\mathrm{b}}$ has isotropic symmetry, with both $\underline{\underline{U}}^{\mathrm{a}}$ and $\underline{\underline{U}}^{\mathrm{b}}$ describing spherical particles.

Provided that the distinguished axes presented by the constituent materials are mutually perpendicular, the biaxiality exhibited by the HCM is of the orthorhombic type, as characterized by constitutive dyadics of the form (2.40). More general forms of biaxiality, namely monoclinic and triclinic biaxiality, can arise when the distinguished axes presented by the constituent materials are not mutually perpendicular [10, 11].

Let us consider a specific example: an orthorhombic dielectric HCM. The Bruggeman estimate of the HCM's permittivity dyadic is written as

$$\underline{\underline{\varepsilon}}_{\mathrm{Br}} = \varepsilon_{\mathrm{x}}^{\mathrm{Br}}\, \hat{\mathbf{u}}_x\, \hat{\mathbf{u}}_x + \varepsilon_{\mathrm{y}}^{\mathrm{Br}}\, \hat{\mathbf{u}}_y\, \hat{\mathbf{u}}_y + \varepsilon_{\mathrm{z}}^{\mathrm{Br}}\, \hat{\mathbf{u}}_z\, \hat{\mathbf{u}}_z. \tag{6.13}$$

Constituent material 'a' is taken to be isotropic dielectric with permittivity dyadic $\underline{\underline{\varepsilon}}_{\mathrm{a}} = (12 + 4i)\varepsilon_0\, \underline{\underline{I}}$ while constituent material 'b' is taken as a uniaxial dielectric material with permittivity dyadic $\underline{\underline{\varepsilon}}_{\mathrm{b}} = [(6 + i)(\underline{\underline{I}} - \hat{\mathbf{u}}_z\, \hat{\mathbf{u}}_z) + (2.5 + 0.5i)\hat{\mathbf{u}}_z\, \hat{\mathbf{u}}_z]\varepsilon_0$. Both constituent materials are distributed randomly as spheroidal particles. The shape and orientation of the constituent particles are specified by $\underline{\underline{U}}^{\mathrm{a}} = [(\underline{\underline{I}} - \hat{\mathbf{u}}_z\, \hat{\mathbf{u}}_z)$

$+U\hat{\mathbf{u}}_z\,\hat{\mathbf{u}}_z]/\sqrt[3]{U}$ and $\underline{\underline{U}}^{\rm b} = [(\underline{\underline{I}} - \hat{\mathbf{u}}_x\,\hat{\mathbf{u}}_x) + U\hat{\mathbf{u}}_x\,\hat{\mathbf{u}}_x]/\sqrt[3]{U}$ for materials 'a' and 'b', respectively.

Plots of the real and imaginary parts of $\varepsilon_{\rm x}^{\rm Br}$, $\varepsilon_{\rm y}^{\rm Br}$, and $\varepsilon_{\rm z}^{\rm Br}$ as functions of volume fraction $f_{\rm a}$ are displayed in figure 6.5 for the shape parameter $U \in \{0.2, 1, 0.5\}$. The real and imaginary parts of the HCM's permittivity parameters increase uniformly as $f_{\rm a}$ increases from zero to unity. The HCM permittivity dyadic $\underline{\underline{\varepsilon}}_{\rm Br}$ is identical to $\underline{\underline{\varepsilon}}_{\rm b}$ at $f_{\rm a} = 0$ and to $\underline{\underline{\varepsilon}}_{\rm a}$ at $f_{\rm a} = 1$. For the degenerate case $U = 1$, the constituent particles are spherical and the corresponding HCM is uniaxial with $\varepsilon_{\rm x}^{\rm Br} = \varepsilon_{\rm y}^{\rm Br}$ for all values of $f_{\rm a}$.

In order for a biaxial bianisotropic HCM to arise, magnetoelectric coupling must be exhibited by one (or more) of the constituent materials. For example, consider the homogenization of a composite material comprising:

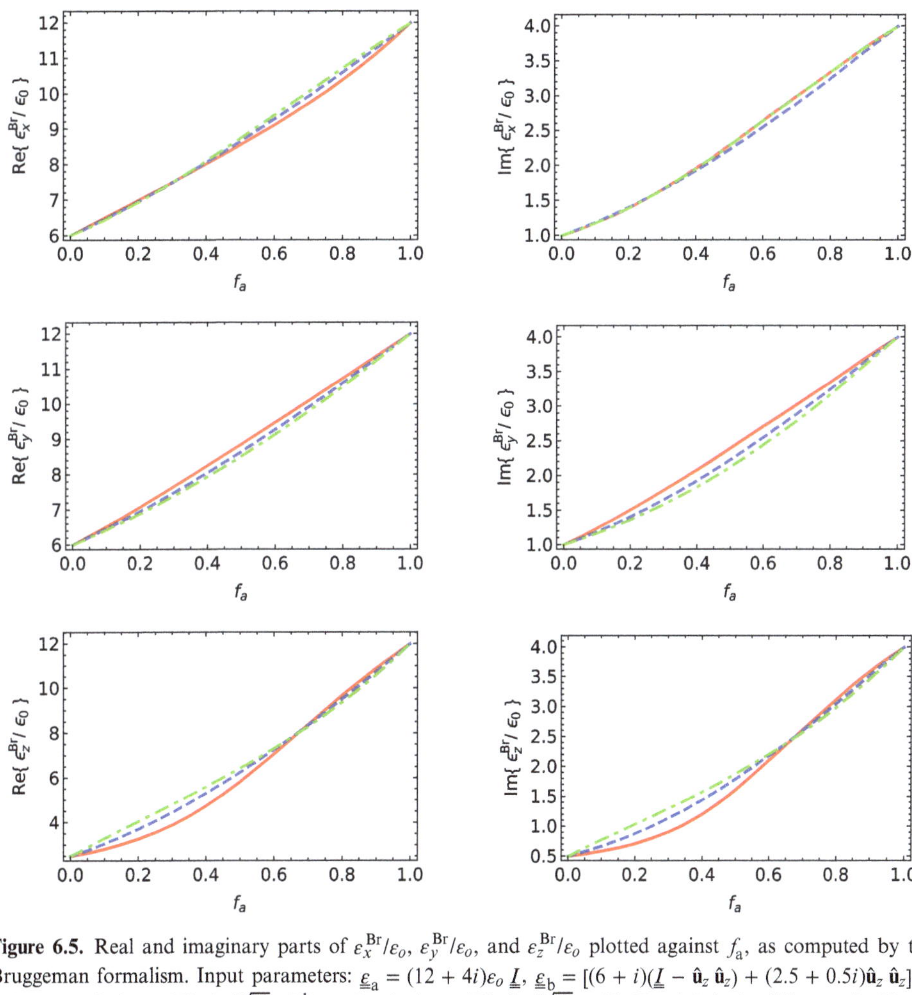

Figure 6.5. Real and imaginary parts of $\varepsilon_x^{\rm Br}/\varepsilon_o$, $\varepsilon_y^{\rm Br}/\varepsilon_o$, and $\varepsilon_z^{\rm Br}/\varepsilon_o$ plotted against $f_{\rm a}$, as computed by the Bruggeman formalism. Input parameters: $\underline{\underline{\varepsilon}}_{\rm a} = (12 + 4i)\varepsilon_o\,\underline{\underline{I}}$, $\underline{\underline{\varepsilon}}_{\rm b} = [(6 + i)(\underline{\underline{I}} - \hat{\mathbf{u}}_z\,\hat{\mathbf{u}}_z) + (2.5 + 0.5i)\hat{\mathbf{u}}_z\,\hat{\mathbf{u}}_z]\varepsilon_o$, $\underline{\underline{U}}^{\rm a} = [(\underline{\underline{I}} - \hat{\mathbf{u}}_z\,\hat{\mathbf{u}}_z) + U\hat{\mathbf{u}}_z\,\hat{\mathbf{u}}_z]/\sqrt[3]{U}$, $\underline{\underline{U}}^{\rm b} = [(\underline{\underline{I}} - \hat{\mathbf{u}}_x\,\hat{\mathbf{u}}_x) + U\hat{\mathbf{u}}_x\,\hat{\mathbf{u}}_x]/\sqrt[3]{U}$, with $U = 0.2$ (red solid curves), $U = 1$ (blue dashed curves), and $U = 5$ (green broken dashed curves).

- an isotropic chiral material 'a' described by the constitutive dyadics $\underline{\underline{\varepsilon}}_a = (5 + 0.04i)\varepsilon_0\, \underline{\underline{I}}$, $\underline{\underline{\xi}}_a = -\underline{\underline{\zeta}}_a = (0.2 + 0.003i)i\sqrt{\varepsilon_0\mu_0}\, \underline{\underline{I}}$, and $\underline{\underline{\mu}}_a = (2 + 0.02i)\mu_0\, \underline{\underline{I}}$, and
- a uniaxial dielectric–magnetic material 'b' described by the constitutive dyadics $\underline{\underline{\varepsilon}}_b = [1.5(\underline{\underline{I}} - \hat{\mathbf{u}}_x\, \hat{\mathbf{u}}_x) + 2.5\hat{\mathbf{u}}_x\, \hat{\mathbf{u}}_x]\varepsilon_0$, $\underline{\underline{\xi}}_b = \underline{\underline{\zeta}}_b = \underline{\underline{0}}$, and $\underline{\underline{\mu}}_b = [1.1(\underline{\underline{I}} - \hat{\mathbf{u}}_x\, \hat{\mathbf{u}}_x) + 1.4\hat{\mathbf{u}}_x\, \hat{\mathbf{u}}_x]\mu_0$.

Both constituent materials are randomly distributed as spheroidal particles whose shape and orientation are specified by $\underline{\underline{U}}^a = \underline{\underline{U}}^b = [(\underline{\underline{I}} - \hat{\mathbf{u}}_z\, \hat{\mathbf{u}}_z) + U\hat{\mathbf{u}}_z\, \hat{\mathbf{u}}_z]/\sqrt[3]{U}$, with $U = 5$. The corresponding HCM is orthorhombic bianisotropic. For the purposes of illustration, the HCM's constitutive parameters are estimated by the second-order SPFT formalism with the step covariance function (4.35). The constitutive dyadics of the HCM are written as

$$\underline{\underline{\sigma}}_{\text{SPFT}} = \sigma_{\text{x}}^{\text{SPFT}}\, \hat{\mathbf{u}}_x\, \hat{\mathbf{u}}_x + \sigma_{\text{y}}^{\text{SPFT}}\, \hat{\mathbf{u}}_y\, \hat{\mathbf{u}}_y + \sigma_{\text{z}}^{\text{SPFT}}\, \hat{\mathbf{u}}_z\, \hat{\mathbf{u}}_z, \qquad (\sigma \in \{\varepsilon, \xi, \mu\}), \tag{6.14}$$

with $\underline{\underline{\zeta}}_{\text{SPFT}} = -\underline{\underline{\xi}}_{\text{SPFT}}$. The volume fraction is fixed at $f_a = 0.4$.

Plots of the real and imaginary parts of the HCM's constitutive parameters against normalized correlation length $k_0 L_{\text{cor}}$ are presented in figure 6.6. These plots show that both the real and imaginary parts of the HCM's constitutive parameters increase uniformly as the correlation length increases from zero. The relative increases in the imaginary parts as the correlation length increases are substantially larger than the relative increases in the real parts.

6.4.2 Faraday chiral materials

Faraday chiral materials, i.e., materials characterized by constitutive dyadics of the form (2.44), may be conceptualized as homogenized composite materials [12]. As such, these materials exemplify those bianisotropic HCMs that arise as mixtures of non-bianisotropic constituent materials. Bianisotropy in this instance is jointly due to one constituent material exhibiting natural optical activity [13] and the other constituent material exhibiting Faraday rotation [14–16].

For example, a Faraday chiral material may be conceptualized as a *chiroferrite* material which arises from the homogenization of an isotropic chiral material with a magnetically biased ferrite [17]. Alternatively, the magnetically biased ferrite may be substituted by a magnetically biased plasma to create a *chiroplasma* [18]. More general forms of Faraday chiral material than that characterized via equation (2.44) can arise if one (or more) of the constituent materials is dispersed as particles of nonspherical shapes [18, 19].

Let us illustrate by means of a numerical example. Consider the homogenization of a composite material comprising:

- an isotropic chiral material 'a' characterized by the constitutive dyadics $\underline{\underline{\varepsilon}}_a = (5 + 0.04i)\varepsilon_0\, \underline{\underline{I}}$, $\underline{\underline{\xi}}_a = -\underline{\underline{\zeta}}_a = (0.2 + 0.003i)i\sqrt{\varepsilon_0\mu_0}\, \underline{\underline{I}}$, and $\underline{\underline{\mu}}_a = (2 + 0.02i)\mu_0\, \underline{\underline{I}}$, and

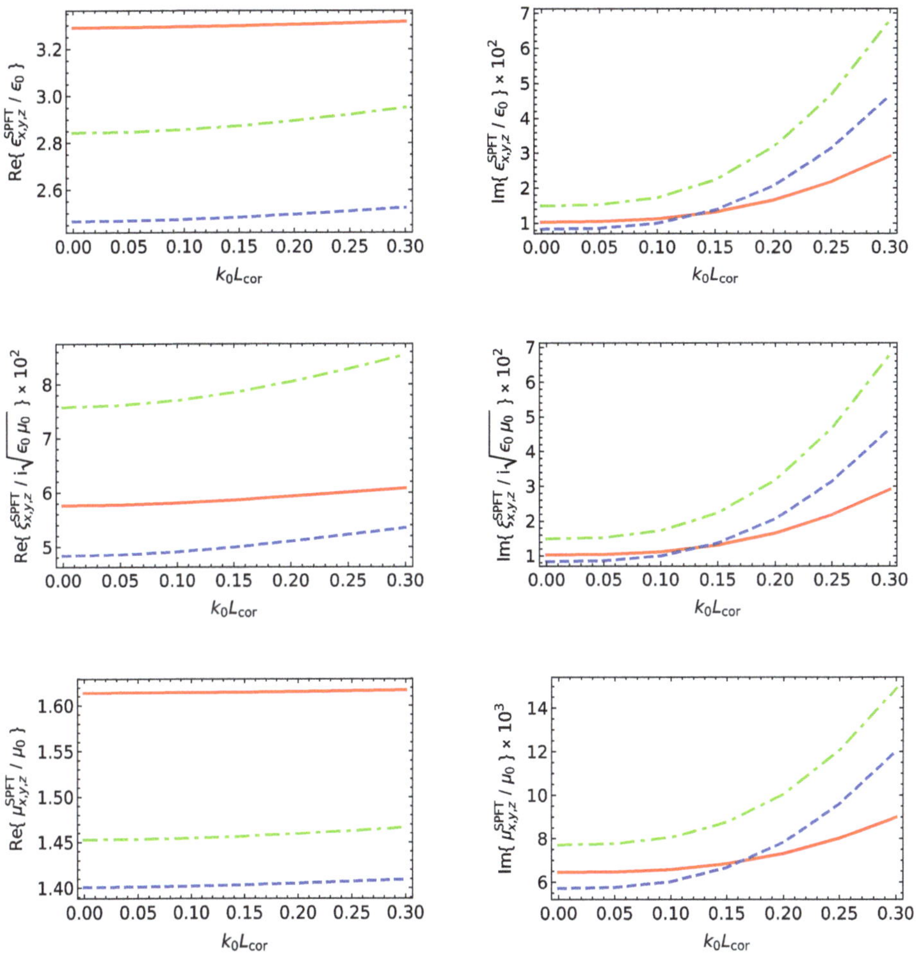

Figure 6.6. Real and imaginary parts of $\varepsilon_\ell^{\rm SPFT}/\varepsilon_0$, $\xi_\ell^{\rm SPFT}/(i\sqrt{\varepsilon_0\mu_0})$, and $\mu_\ell^{\rm SPFT}/\mu_0$ plotted against $k_0 L_{\rm cor}$, as computed by the second-order SPFT formalism with the step covariance function (4.35). Key: $\ell = x$ (red solid curves), $\ell = y$ (blue dashed curves), and $\ell = z$ (green broken dashed curves). Input parameters: $f_a = 0.4$, $\underline{\underline{\varepsilon}}_{\rm a} = (5 + 0.04i)\varepsilon_0\, \underline{\underline{I}}$, $\underline{\underline{\xi}}_{\rm a} = -\underline{\underline{\zeta}}_{\rm a} = (0.2 + 0.003i)i\sqrt{\varepsilon_0\mu_0}\, \underline{\underline{I}}$, $\underline{\underline{\mu}}_{\rm a} = (2 + 0.02i)\mu_0\, \underline{\underline{I}}$, $\underline{\underline{\varepsilon}}_{\rm b} = [1.5(\underline{\underline{I}} - \hat{\mathbf{u}}_x\, \hat{\mathbf{u}}_x) + 2.5\hat{\mathbf{u}}_x\, \hat{\mathbf{u}}_x]\varepsilon_0$, $\underline{\underline{\xi}}_{\rm b} = \underline{\underline{\zeta}}_{\rm b} = \underline{\underline{0}}$, $\underline{\underline{\mu}}_{\rm b} = [1.1(\underline{\underline{I}} - \hat{\mathbf{u}}_x\, \hat{\mathbf{u}}_x) + 1.4\hat{\mathbf{u}}_x\, \hat{\mathbf{u}}_x]\mu_0$, $\underline{\underline{U}}^{\rm a} = \underline{\underline{U}}^{\rm b} = [(\underline{\underline{I}} - \hat{\mathbf{u}}_z\, \hat{\mathbf{u}}_z) + U\hat{\mathbf{u}}_z\, \hat{\mathbf{u}}_z]/\sqrt[3]{U}$, with $U = 5$.

- a magnetically biased ferrite material 'b' characterized by the constitutive dyadics $\underline{\underline{\varepsilon}}_{\rm b} = (1.5 + 0.01i)\varepsilon_0\, \underline{\underline{I}}$, $\underline{\underline{\xi}}_{\rm b} = \underline{\underline{\zeta}}_{\rm b} = \underline{\underline{0}}$, and $\underline{\underline{\mu}}_{\rm b} = [(3.5 + 0.02i)(\underline{\underline{I}} - \hat{\mathbf{u}}_z\, \hat{\mathbf{u}}_z)] - i(2 + 0.01i)\hat{\mathbf{u}}_z \times \underline{\underline{I}} + (1 + 0.005i)\hat{\mathbf{u}}_z\, \hat{\mathbf{u}}_z]\mu_0$.

Both constituent materials are randomly distributed as spheroidal particles whose shape and orientation are specified by $\underline{\underline{U}}^{\rm a} = \underline{\underline{U}}^{\rm b} = [(\underline{\underline{I}} - \hat{\mathbf{u}}_z\, \hat{\mathbf{u}}_z) + U\hat{\mathbf{u}}_z\, \hat{\mathbf{u}}_z]/\sqrt[3]{U}$, with $U = 5$. The corresponding HCM is a chiroferrite. Use of the Bruggeman formalism delivers the constitutive dyadics of the HCM as

$$\left.\begin{aligned}
\underline{\underline{\varepsilon}}_{\rm Br} &= \varepsilon_1^{\rm Br}\underline{\underline{I}} + \varepsilon_2^{\rm Br}\hat{\mathbf{u}}_z\hat{\mathbf{u}}_z + i\varepsilon_3^{\rm Br}\hat{\mathbf{u}}_z \times \underline{\underline{I}} \\
\underline{\underline{\xi}}_{\rm Br} &= i(\xi_1^{\rm Br}\underline{\underline{I}} + \xi_2^{\rm Br}\hat{\mathbf{u}}_z\hat{\mathbf{u}}_z + i\xi_3^{\rm Br}\hat{\mathbf{u}}_z \times \underline{\underline{I}}\,) \\
\underline{\underline{\zeta}}_{\rm Br} &= -\,\underline{\underline{\xi}}_{\rm Br} \\
\underline{\underline{\mu}}_{\rm Br} &= \mu_1^{\rm Br}\underline{\underline{I}} + \mu_2^{\rm Br}\hat{\mathbf{u}}_z\hat{\mathbf{u}}_z + i\mu_3^{\rm Br}\hat{\mathbf{u}}_z \times \underline{\underline{I}}
\end{aligned}\right\}. \tag{6.15}$$

The real and imaginary parts of the HCM's constitutive parameters are plotted against volume fraction $f_{\rm a}$ in figure 6.7. The HCM's constitutive parameters vary smoothly as $f_{\rm a}$ increases from 0 to 1. The values of the HCM's constitutive

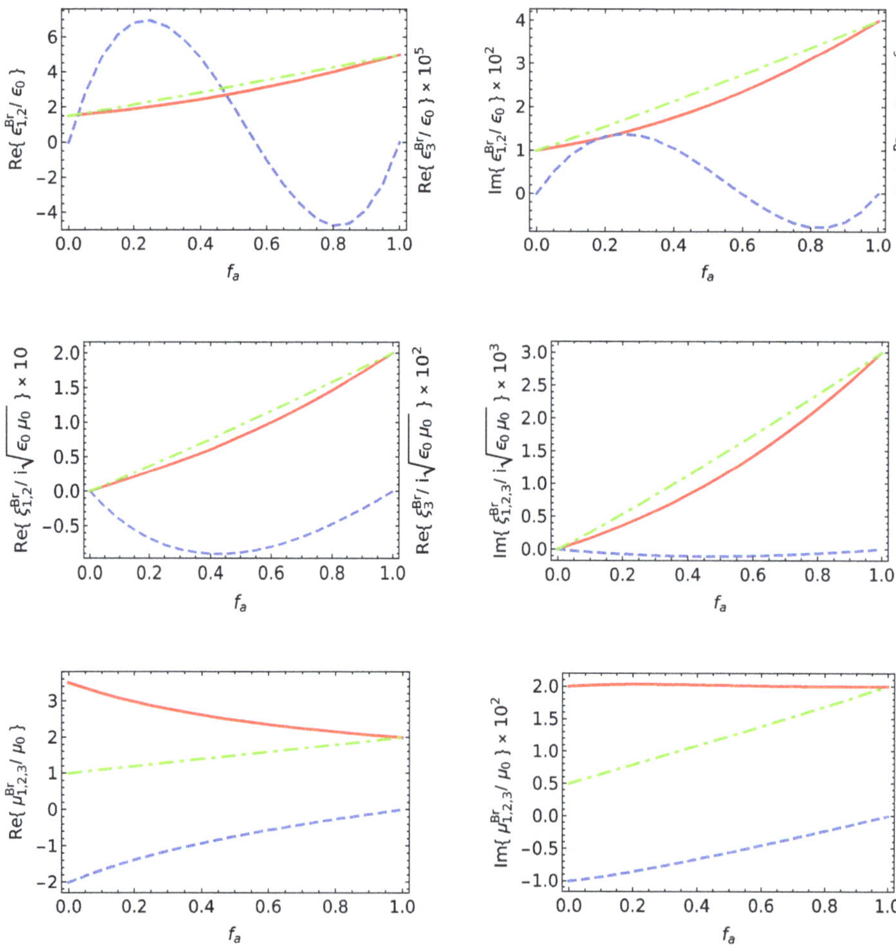

Figure 6.7. Real and imaginary parts of $\varepsilon_\ell^{\rm Br}/\varepsilon_0$, $\xi_\ell^{\rm Br}/(i\sqrt{\varepsilon_0\mu_0})$, and $\mu_\ell^{\rm Br}/\mu_0$ plotted against $f_{\rm a}$, as computed by the Bruggeman formalism, with $\ell = 1$ (red solid curves), $\ell = 2$ (green broken dashed curves), and $\ell = 3$ (blue dashed curves). Input parameters: $\underline{\underline{\varepsilon}}_{\rm a} = (5 + 0.04i)\varepsilon_0\,\underline{\underline{I}}$, $\underline{\underline{\xi}}_{\rm a} = -\underline{\underline{\zeta}}_{\rm a} = (0.2 + 0.003i)i\sqrt{\varepsilon_0\mu_0}\,\underline{\underline{I}}$, $\underline{\underline{\mu}}_{\rm a} = (2 + 0.02i)\mu_0\,\underline{\underline{I}}$, $\underline{\underline{\varepsilon}}_{\rm b} = (1.5 + 0.01i)\varepsilon_0\,\underline{\underline{I}}$, $\underline{\underline{\xi}}_{\rm b} = \underline{\underline{\zeta}}_{\rm b} = \underline{\underline{0}}$, $\underline{\underline{\mu}}_{\rm b} = [(3.5 + 0.02i)(\underline{\underline{I}} - \hat{\mathbf{u}}_z\,\hat{\mathbf{u}}_z) - i(2 + 0.01i)\hat{\mathbf{u}}_z \times \underline{\underline{I}} + (1 + 0.005i)\hat{\mathbf{u}}_z\,\hat{\mathbf{u}}_z]\mu_0$, $\underline{\underline{U}}^{\rm a} = \underline{\underline{U}}^{\rm b} = [(\underline{\underline{I}} - \hat{\mathbf{u}}_z\,\hat{\mathbf{u}}_z) + U\hat{\mathbf{u}}_z\,\hat{\mathbf{u}}_z]/\sqrt[3]{U}$, with $U = 5$.

parameters coincide with those of constituent materials ‘b’ and ‘a’, respectively, at $f_a = 0$ and 1.

6.5 Disk-shaped and needle-shaped particles

As described in section 3.2.3, the characterization of ellipsoidal constituent particles in terms of the dyadic $\underline{\underline{U}}$ defined via equation (3.15) can accommodate disk-shaped and needle-shaped constituent particles. The depolarization dyadics for disk-shaped and needle-shaped particles embedded in an orthorhombic dielectric–magnetic ambient material are given in equations (3.57) and (3.58), respectively. The homogenization of constituent materials dispersed as disk-shaped and/or needle-shaped particles in a host material can result in HCMs which exhibit very high degrees of anisotropy [20, 21]. Materials with higher degrees of anisotropy than are commonly encountered in natural materials [22, 23] are central to a variety of recent applications, involving negative refraction [24], optical cloaking [25, 26], null reflection [27], omnidirectional radiation [28], and so on. Indeed, such dielectric anisotropy can facilitate negative refraction in non-magnetic composite materials [20, 29].

Furthermore, high degrees of anisotropy are needed for the development of material analogs for the electromagnetic properties of certain curved-spacetime scenarios, such as rotating black holes [30, 31], Schwarzschild-(anti-)de Sitter spacetime [32], cosmic spinning strings [33], and colliding gravitational plane waves [34], as well as material analogs of quantum electrodynamic vacuum [35]. Most notably, regions of large spacetime curvature, especially in the vicinities of singularities or event horizons, require extremely high degrees of anisotropy. Additional details of these material analogs are provided in section 6.7.

The realization of high degrees of anisotropy that can be achieved via homogenization may be conveniently illustrated by means of the Bruggeman formalism. Consider the homogenization of a composite material with two constituent isotropic dielectric materials characterized by the permittivities ε_a and ε_b. For simplicity, both constituent materials are assumed to be nondissipative, i.e., $\varepsilon_a \in \mathbb{R}$ and $\varepsilon_b \in \mathbb{R}$. Suppose also that both materials are randomly distributed as disk-shaped particles oriented parallel to the xy plane. Then the corresponding HCM is a uniaxial dielectric material whose permittivity dyadic $\underline{\underline{\varepsilon}}_{\,\mathrm{HCM}}$ has the form given in equation (2.39) with $\hat{\mathbf{u}} = \hat{\mathbf{u}}_z$ and the Bruggeman estimate of $\underline{\underline{\varepsilon}}_{\,\mathrm{HCM}}$ may be expressed as

$$\underline{\underline{\varepsilon}}_{\,\mathrm{Br}} = \varepsilon_\perp^{\mathrm{Br}}(\underline{\underline{I}} - \hat{\mathbf{u}}_z\hat{\mathbf{u}}_z) + \varepsilon_\parallel^{\mathrm{Br}}\hat{\mathbf{u}}_z\hat{\mathbf{u}}_z. \tag{6.16}$$

After using the depolarization dyadic (3.57), with $\varepsilon_\parallel^{\mathrm{Br}}$ replacing ε_z and μ_0 replacing μ_z therein, equation (4.20) yields [21]

$$\left.\begin{aligned} \varepsilon_\perp^{\mathrm{Br}} &= f_a\varepsilon_a + f_b\varepsilon_b \\ \varepsilon_\parallel^{\mathrm{Br}} &= \frac{\varepsilon_a\varepsilon_b}{f_b\varepsilon_a + f_a\varepsilon_b}. \end{aligned}\right\}. \tag{6.17}$$

The ratio $\varepsilon_{\perp}^{\mathrm{Br}}/\varepsilon_{\parallel}^{\mathrm{Br}}$ is a measure of the degree of anisotropy of the HCM. When evaluated at $f_{\mathrm{a}} = f_{\mathrm{b}} = 0.5$, this ratio grows unboundedly as ε_{b} increasingly deviates from ε_{a}, which becomes evident as follows:

$$\frac{\varepsilon_{\perp}^{\mathrm{Br}}}{\varepsilon_{\parallel}^{\mathrm{Br}}} = \frac{1}{4}\frac{(\varepsilon_{\mathrm{a}} + \varepsilon_{\mathrm{b}})^2}{\varepsilon_{\mathrm{a}}\varepsilon_{\mathrm{b}}} \to \infty \quad \text{as} \quad \varepsilon_{\mathrm{b}} \to \begin{cases} 0 \\ \infty \end{cases} \quad \text{for fixed } \varepsilon_{\mathrm{a}}. \tag{6.18}$$

Furthermore, the estimated degree of anisotropy is proportional to ε_{b} for $\varepsilon_{\mathrm{b}} > \varepsilon_{\mathrm{a}}$ and proportional to $1/\varepsilon_{\mathrm{b}}$ for $\varepsilon_{\mathrm{b}} < \varepsilon_{\mathrm{a}}$.

The ratio $\varepsilon_{\perp}^{\mathrm{Br}}/\varepsilon_{\parallel}^{\mathrm{Br}}$ is plotted versus $\varepsilon_{\mathrm{b}}/\varepsilon_0$ and f_{a} in figure 6.8 for $\varepsilon_{\mathrm{a}} = 2\varepsilon_0$. This figure illustrates that the estimated degree of anisotropy of the HCM is greatest for some mid-range values of the volume fractions. Also, the estimated degree of anisotropy increases steadily as ε_{b} increases or decreases from the value of ε_{a}. The range of values of $\varepsilon_{\mathrm{b}}/\varepsilon_0$ covered in figure 6.8, namely (0.1, 200), may seem extreme when considered in the context of commonly encountered natural materials. However, such a huge range is not beyond the grasp of contemporary materials science and engineering. Indeed, in recent years much progress been made towards realizing engineered composite materials with very high permittivities [36] and with permittivities very close to zero [37].

Now suppose that the homogenization illustrated numerically in figure 6.8 is repeated using needle-shaped constituent particles instead of disk-shaped ones. All of the needle-shaped particles are taken to be aligned parallel to the z-axis. The estimate of $\underline{\underline{\varepsilon}}_{\mathrm{HCM}}$ delivered by the Bruggeman formalism again has the form provided by equation (6.16). By utilizing the depolarization dyadic (3.58), with $\varepsilon_{\perp}^{\mathrm{Br}}$ replacing $\varepsilon_{x,y}$, $\varepsilon_{\parallel}^{\mathrm{Br}}$ replacing ε_z, and μ_0 replacing $\mu_{x,y,z}$ therein, the eigenvalues of $\underline{\underline{\varepsilon}}_{\mathrm{Br}}$ are derived from equation (4.20) as [21]

$$\left.\begin{aligned} \varepsilon_{\perp}^{\mathrm{Br}} &= \frac{(f_{\mathrm{b}} - f_{\mathrm{a}})(\varepsilon_{\mathrm{b}} - \varepsilon_{\mathrm{a}}) + \sqrt{\left[(f_{\mathrm{b}} - f_{\mathrm{a}})(\varepsilon_{\mathrm{b}} - \varepsilon_{\mathrm{a}})\right]^2 + 4\varepsilon_{\mathrm{a}}\varepsilon_{\mathrm{b}}}}{2} \\ \varepsilon_{\parallel}^{\mathrm{Br}} &= f_{\mathrm{a}}\varepsilon_{\mathrm{a}} + f_{\mathrm{b}}\varepsilon_{\mathrm{b}} \end{aligned}\right\}. \tag{6.19}$$

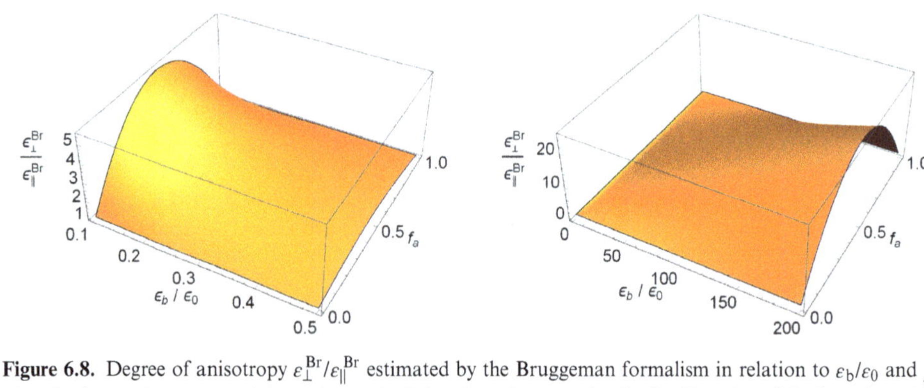

Figure 6.8. Degree of anisotropy $\varepsilon_{\perp}^{\mathrm{Br}}/\varepsilon_{\parallel}^{\mathrm{Br}}$ estimated by the Bruggeman formalism in relation to $\varepsilon_{\mathrm{b}}/\varepsilon_0$ and f_{a}, when both constituent materials are isotropic dielectric and are randomly distributed as disk-shaped particles oriented parallel to the xy plane. The isotropic dielectric constituent material 'a' has permittivity $\varepsilon_{\mathrm{a}} = 2\varepsilon_0$.

The constraint (4.4) can be used to determine the sign of the square root term in equation $(6.19)_1$; i.e., $\varepsilon_\perp^{\mathrm{Br}} \to \varepsilon_\ell$ as $f_\ell \to 1$ for $\ell \in \{\mathrm{a}, \mathrm{b}\}$. When evaluated at $f_\mathrm{a} = f_\mathrm{b} = 0.5$, the ratio $\varepsilon_\parallel^{\mathrm{Br}} / \varepsilon_\perp^{\mathrm{Br}}$ derived from equations (6.19) becomes

$$\frac{\varepsilon_\parallel^{\mathrm{Br}}}{\varepsilon_\perp^{\mathrm{Br}}} = \frac{1}{2}\left(\sqrt{\frac{\varepsilon_\mathrm{a}}{\varepsilon_\mathrm{b}}} + \sqrt{\frac{\varepsilon_\mathrm{b}}{\varepsilon_\mathrm{a}}}\right) \tag{6.20}$$

and grows unboundedly as either $(\varepsilon_\mathrm{b}/\varepsilon_\mathrm{a}) \to 0$ or $(\varepsilon_\mathrm{b}/\varepsilon_\mathrm{a}) \to \infty$. Therefore, as is the case for the HCM arising from disk-shaped constituent particles, the estimated degree of anisotropy increases without limit as ε_b increasingly deviates from ε_a. However, in the case of needle-shaped constituent particles the estimated degree of anisotropy is proportional to $\sqrt{\varepsilon_b}$ for $\varepsilon_\mathrm{b} > \varepsilon_\mathrm{a}$ and proportional to $1/\sqrt{\varepsilon_b}$ for $\varepsilon_\mathrm{b} < \varepsilon_\mathrm{a}$.

The ratio $\varepsilon_\parallel^{\mathrm{Br}} / \varepsilon_\perp^{\mathrm{Br}}$ is plotted versus $\varepsilon_\mathrm{b}/\varepsilon_0$ and f_a in figure 6.9 for $\varepsilon_\mathrm{a} = 2\varepsilon_0$. This figure is qualitatively rather similar to figure 6.8. In both figures, the degree of anisotropy estimated by the Bruggeman formalism achieves its greatest value at some mid-range value of f_a, albeit for needle-shaped constituent particles the maximum values of $\varepsilon_\parallel^{\mathrm{Br}} / \varepsilon_\perp^{\mathrm{Br}}$ are skewed towards lower values of f_a for $\varepsilon_\mathrm{b} < \varepsilon_\mathrm{a}$, and skewed towards higher values of f_a for $\varepsilon_\mathrm{b} > \varepsilon_\mathrm{a}$. It is also noteworthy that the magnitudes of $\varepsilon_\perp^{\mathrm{Br}} / \varepsilon_\parallel^{\mathrm{Br}}$ in figure 6.9 are generally lower than the corresponding magnitudes of $\varepsilon_\perp^{\mathrm{Br}} / \varepsilon_\parallel^{\mathrm{Br}}$ in figure 6.8.

6.6 Plane-wave phenomenons

By means of homogenization, constitutive-parameter regimes may become accessible which may not be accessible with the constituent materials. Thereby, HCMs may exhibit plane-wave phenomenons that their constituent materials do not exhibit at all, or at least certain plane-wave characteristics of HCMs may be enhanced relative to the corresponding plane-wave characteristics of the constituent materials. Generally, anisotropy and bianisotropy allow greater scope for the emergence or

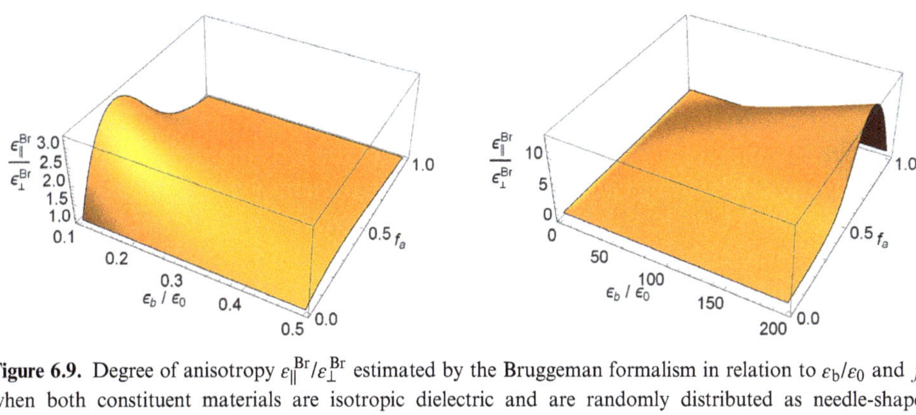

Figure 6.9. Degree of anisotropy $\varepsilon_\parallel^{\mathrm{Br}} / \varepsilon_\perp^{\mathrm{Br}}$ estimated by the Bruggeman formalism in relation to $\varepsilon_\mathrm{b}/\varepsilon_0$ and f_a, when both constituent materials are isotropic dielectric and are randomly distributed as needle-shaped particles aligned parallel to the z axis. The isotropic dielectric constituent material 'a' has permittivity $\varepsilon_\mathrm{a} = 2\varepsilon_0$.

enhancement of such plane-wave phenomenons, as becomes evident from the following examples.

6.6.1 Birefringence

As described in section 6.4, anisotropic HCMs can arise from constituent materials that are isotropic. Some anisotropic HCMs can display plane-wave characteristics that are completely different from those of their isotropic constituent materials. For example, anisotropic dielectric–magnetic HCMs can arise from the homogenization of isotropic dielectric–magnetic materials which are dispersed as nonspherical particles. Anisotropic dielectric–magnetic materials are *birefringent* in general [14, 38], whereas isotropic dielectric–magnetic materials are *unirefringent*.

In a similar vein, bianisotropic HCMs can arise from constituent materials that are anisotropic or even isotropic. It is possible for a bianisotropic HCM to exhibit plane-wave characteristics that are completely different from those of its isotropic or anisotropic constituent materials. For example, certain bianisotropic materials that are not Lorentz-reciprocal (i.e., characterized by constitutive dyadics which do not conform to the constraints given in equations (3.26)) can arise from the homogenization of anisotropic dielectric–magnetic materials and isotropic chiral materials. The wavenumbers of plane waves propagating in a certain direction do not change upon the reversal of direction in a Lorentz-reciprocal material, but that is not true for a material that does not satisfy equations (3.26) [39, 40]. A Faraday chiral material is a specific example [17, 18].

6.6.2 Negative phase velocity and negative reflection

In conventional materials at rest, uniform plane waves generally propagate with positive phase velocity (PPV). That is, the phase velocity casts a positive projection onto the time-averaged Poynting vector. However, appropriately engineered composite materials may exhibit negative-phase-velocity (NPV) propagation, wherein the phase velocity casts a negative projection onto the time-averaged Poynting vector [41]. In recent years there has been considerable research interest in materials that exhibit NPV propagation [42], because they may facilitate sub-wavelength imaging or may be used as planar lenses, for example. To date, attempts to construct commercially viable devices based on NPV propagation have been thwarted by unacceptably high dissipation [43, 44]. This has led to the consideration of active constituent materials for such composite materials [45, 46].

Homogenization offers a relatively simple route towards the realization of NPV-exhibiting materials, using constituent materials that themselves do not exhibit NPV propagation [47]. Consider the homogenization of two isotropic dielectric–magnetic materials, characterized by the permittivities ε_{a} and ε_{b}, and permeabilities μ_{a} and μ_{b}, respectively, and each dispersed randomly as spherical particles. If $\varepsilon_{\mathrm{a,b}}$ and $\mu_{\mathrm{a,b}}$ lie within appropriate parameter ranges, with $\mathrm{Re}\{\varepsilon_{\mathrm{a,b}}\} < 0$ and $\mathrm{Re}\{\mu_{\mathrm{a,b}}\} > 0$ (or vice versa), then the HCM can exhibit NPV propagation while the constituent materials do not. The factors that determine whether or not the HCM exhibits NPV propagation depend upon the sizes [48] and spatial distributions [49] of the constituent particles, as well as the volume fractions and constitutive properties of the constituent materials [47].

Anisotropic and bianisotropic materials are generally (but not always) associated with constitutive-parameter spaces of higher dimensions than are isotropic dielectric–magnetic materials. Consequently, anisotropic and bianisotropic materials present greater opportunities for NPV propagation [50]. For example, a Faraday chiral material allowing up to four different modes of NPV propagation may be conceptualized from constituent materials that do not allow NPV propagation [51].

Caution should be exercised in distinguishing NPV propagation from the often associated phenomenon of negative refraction. Negative refraction occurs at the planar interface of two dissimilar materials when the real part of the wavevector of a refracted plane wave is directed towards the interface [40]. While NPV propagation and negative refraction go hand in hand for uniform plane-wave propagation in passive isotropic dielectric materials [52], this is not true more generally. In particular, the two phenomenons are quite independent if nonuniform plane waves are considered [53], and likewise in the case of anisotropic [54] and/or active materials [55–57].

In a similar fashion, NPV propagation (and negative refraction) should also be distinguished from negative reflection [58–60]. Negative reflection occurs at the planar interface of two dissimilar materials when the real part of the wavevector of a reflected plane wave is directed towards the interface. As is the case for NPV propagation (and negative refraction), negative reflection may be harnessed for planar lenses, at least in principle [58]. Furthermore, HCMs which support the exhibition of negative reflection may be conceptualized as having constituent materials that do not themselves support negative reflection. For example, a Faraday chiral material, arising from isotropic chiral and anisotropic dielectric (or anisotropic magnetic) constituent materials that do not support negative reflection, can display negative reflection provided that the magnetoelectric constitutive parameter of the isotropic chiral constituent material is sufficiently large [61]. Furthermore, when a plane wave with PPV in such a Faraday chiral material is incident on a planar interface with another material, one of the two reflected plane waves may be a negatively reflected plane wave with NPV and the other a positively reflected plane wave with PPV; on the other hand, if the incident plane wave has NPV, then the positively reflected plane wave may have NPV and the negatively reflected plane wave may have PPV.

Parenthetically, negative refraction/NPV propagation/negative reflection are not solely the preserve of artificially produced materials. Negative refraction is exhibited by certain metallic ferromagnetic materials [62]. Furthermore, neither negative refraction nor NPV propagation are invariant under a Lorentz transformation [63]. Thus, materials which do not exhibit negative refraction or NPV propagation from the perspective of a co-moving observer may exhibit both negative refraction [64] and NPV [65] propagation from the perspective of a uniformly moving observer. Indeed, even free space itself, subject to a sufficiently strong gravitational field, can exhibit NPV propagation [66, 67]. This conclusion follows from the noncovariant electromagnetic characterization of free space subjected to an gravitational field, in terms of a nonhomogeneous bianisotropic material [68, 69], as outlined in section 2.6.2.

6.6.3 Hyperbolic dispersion relations

Consider plane-wave propagation in a uniaxial dielectric material characterized by a permittivity dyadic $\underline{\underline{\varepsilon}}$ of the form given in equation $(2.39)_1$. For definiteness, suppose that the optic axis is directed parallel to the x-axis, i.e., the unit vector $\hat{\mathbf{u}} = \hat{\mathbf{u}}_x$, and that the direction of propagation lies wholly in the xz plane. For simplicity, suppose that the eigenvalues ε_1 and $\varepsilon_1 + \varepsilon_2$ of $\underline{\underline{\varepsilon}}$ are real valued; i.e., the material is nondissipative.

Consider an ordinary plane wave with wave vector $\mathbf{k}_{\text{or}} = k_x\hat{\mathbf{u}}_x + k_z^{\text{or}}\hat{\mathbf{u}}_z$ and an extraordinary plane wave with wave vector $\mathbf{k}_{\text{ex}} = k_x\hat{\mathbf{u}}_x + k_z^{\text{ex}}\hat{\mathbf{u}}_z$. The dispersion relation for the ordinary plane wave is [14, 40]

$$k_x^2 + (k_z^{\text{or}})^2 = \frac{\varepsilon_1}{\varepsilon_0}\, k_0^2. \tag{6.21}$$

From a geometric perspective, the components of $\mathbf{k}_{\text{or}}$ describe a circle in the (k_x, k_z^{or}) space. The dispersion relation for the extraordinary plane wave is [14, 40]

$$\frac{k_x^2}{\varepsilon_1} + \frac{(k_z^{\text{ex}})^2}{\varepsilon_1 + \varepsilon_2} = \frac{k_0^2}{\varepsilon_0}. \tag{6.22}$$

The components of $\mathbf{k}_{\text{ex}}$ describe an ellipse in the (k_x, k_z^{ex}) space when the ratio $(\varepsilon_1 + \varepsilon_2)/\varepsilon_1 > 0$, i.e., when the permittivity dyadic $\underline{\underline{\varepsilon}}$ is either positive definite or negative definite. But the same components describe a hyperbola in the (k_x, k_z^{ex}) space when the ratio $(\varepsilon_1 + \varepsilon_2)/\varepsilon_1 < 0$, i.e., when $\underline{\underline{\varepsilon}}$ is indefinite. The concept of hyperbolic dispersion relations extends to dissipative materials, through the real part of the permittivity dyadic $\underline{\underline{\varepsilon}}$ being indefinite. Such materials are called hyperbolic materials.

Hyperbolic materials can exhibit exotic electromagnetic characteristics that are of considerable technoscientific interest. For example, diffraction gratings made of hyperbolic materials can direct light into an unlimited number of refraction channels [70]. Furthermore, negative refraction can be achieved without concomitant NPV propagation, with a cascade of two slabs of different hyperbolic materials [71]. Also, hyperbolic materials are associated with a strong enhancement in the density of optical states [72]. The relatively large wave numbers associated with hyperbolic materials can lead to broadband thermal emission exceeding that predicted by Planck's law [73], as well as heat transfer with efficiencies exceeding those predicted by the Stefan–Boltzmann law [74]. Greater opportunities for exotic behavior, especially negative refraction, are provided by hyperbolic materials which are anisotropic dielectric–magnetic, i.e., materials characterized by indefinite permittivity and permeability dyadics (or materials characterized by permittivity and permeability dyadics whose real parts are indefinite, in the case of dissipative materials) [75, 76].

A relatively simple route to hyperbolic materials is offered by homogenization. For example, the homogenization of a composite material comprising two isotropic dielectric materials, characterized by the permittivities ε_a and ε_b such that $\text{Re}\{\varepsilon_a\}\text{Re}\{\varepsilon_b\} < 0$ and randomly dispersed as oriented spheroidal particles, can give rise to a uniaxial dielectric HCM whose dispersion relations are of the hyperbolic type [77]. A note of caution is required here: in order to implement the Bruggeman formalism when $\text{Re}\{\varepsilon_a\}\text{Re}\{\varepsilon_b\} < 0$, it is necessary that the imaginary part of at least one of ε_a and ε_b be relatively large [78, 79]. This issue is elaborated upon in section 6.8.

6.6.4 Voigt waves

In an anisotropic material, generally two independent plane waves—with mutually orthogonal polarization states and different phase speeds—propagate in any given direction [38]. However, in certain circumstances, these two plane waves can coalesce to form a single plane wave called a Voigt wave [80–83]. A key characteristic of a Voigt wave is that its amplitude has a linear dependence upon the propagation distance—this feature distinguishes a Voigt wave from a standard plane wave. In mathematical terms, the requirement for Voigt wave propagation (VWP) is satisfied when an eigenvalue of the corresponding plane-wave propagation matrix has an algebraic multiplicity greater than its geometric multiplicity [84, 85]. VWP is possible in both dissipative and active materials, but not in nondissipative inactive materials [86].

Certain naturally occurring pleochroic minerals, exemplified by iolite, exhibit VWP [80]. Opportunities for realizing this singular form of wave propagation are also presented by engineered materials [87, 88]. Notably, VWP is possible in certain HCMs arising from constituent materials that do not support VWP [89]. Whether or not a particular HCM displays VWP depends on the orientations, shapes, sizes, and distributions of its constituent particles, as well as on the volume fractions and constitutive properties of the constituent materials [90, 91]. For example, an HCM that exhibits VWP may be conceptualized as arising from two constituent materials, both of which are made from uniaxial dielectric materials and randomly distributed as spherical particles [89], or both of which are made from isotropic dielectric materials and distributed as oriented spheroidal particles [91]. Therefore, through judicious design of the particulate structure and careful selection of the constitutive properties of the constituent materials, the VWP phenomenon may be engineered.

The exhibition of VWP in HCMs is highly sensitive to the volume fraction of the constituent materials. This presents possible opportunities of harnessing VWP in HCMs for optical-sensing applications. For example, as a basis for optical sensing one may envisage a porous material that does not support VWP but when this material is infiltrated by an analyte fluid the resulting composite material does support VWP [92]. Furthermore, the process of monitoring the directions that support VWP in such an HCM-based optical sensor could be facilitated by using electro-optic materials [93], for example.

6.6.5 Group-speed enhancement

The group velocity of a wave packet—which may be regarded as an assembly of plane waves—is conventionally defined as the gradient of the angular frequency with respect to the components of the wave vector. In the absence of dissipation, the group velocity in an anisotropic dielectric–magnetic material is the same as the velocity of energy transport [14]. The magnitude of group velocity, i.e., group speed, may be enhanced via homogenization. The group speed in a HCM can be higher than in the constituent materials [94, 95]. Anisotropic HCMs offer greater scope for group-speed enhancement, as compared to isotropic dielectric–magnetic HCMs [96]. In a similar fashion, the group speed can also be diminished by the process of homogenization [97].

6.7 Inverse homogenization

Usually homogenization formalisms are implemented in the forward sense, whereby the constitutive parameters of an HCM are estimated from a knowledge of the constitutive parameters and the distributional statistics of its constituent materials. As outlined in section 4.2.2 and 4.3.2, this process can be inverted in order to estimate the constitutive parameters (or other types of parameters) of one of the constituent materials, from a knowledge of the other constituent material(s) and the HCM [12]. Two recent examples of the applications of the inverse Bruggeman formalism are as follows.

6.7.1 Columnar thin films

A columnar thin film (CTF) consists of an array of identical parallel nanorods, which may be grown at an acute angle χ on a substrate by physical vapor deposition [98]. Each nanorod may be assumed to be homogeneous. While the nanorod array is often made by depositing a material that is an isotropic dielectric material in bulk form, from a macroscopic viewpoint the CTF is generally a homogeneous biaxial dielectric material whose anisotropy stems from the cross-sectional shape and orientation of the nanorods. To justify this viewpoint, the spaces between the nanorods as well as the cross-sectional dimensions of the individual nanorods must be much smaller than the wavelengths involved. Thus, a CTF may be considered macroscopically as an HCM arising from two constituent materials, one being the nanorods themselves and the other being the material that fills the interstitial spaces between the nanorods [99–103]. The homogenization process can proceed, for example, by regarding each nanorod as being composed of elongated electrically small ellipsoids, strung together end-to-end along the length of the nanorod [98]. The interstitial spaces may be regarded as an Apollonian distribution of electrically-small spheres [104]. A schematic representation of a CTF and the modeling approach taken in homogenizing this structure are represented in figure 6.10.

Usually the interstitial spaces are taken to be filled simply with air, but for optical-sensing applications these spaces may be infiltrated with an analyte-containing fluid. The changes induced in the optical response of the CTF by the analyte-containing

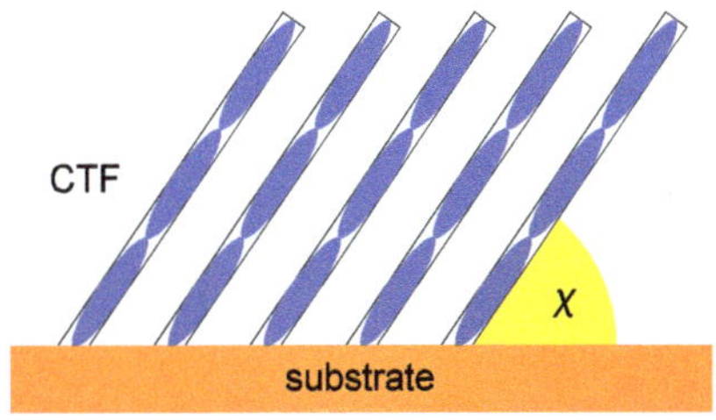

Figure 6.10. Schematic representation of a columnar thin film and the modeling approach adopted when homogenizing the constituent nanorods and interstitial spaces into a biaxial HCM.

fluid provide the basis for sensing. In principle, this optical-sensing scenario can be modeled using a homogenization formalism, but to do so requires accurate data on the shape of the ellipsoidal particles representing the nanorods, as well as the porosity of the CTF and the dielectric properties of the material that the deposited nanorods are made of [105]. However, such data are usually unavailable. In order to generate estimates of these data relating to the nanoscopic structure of the CTF, an inverse homogenization process can be implemented. As inputs to the inverse homogenization process, one can use experimentally measured values of the constitutive parameters of the macroscopic CTF with the interstitial spaces filled with air. These data may be inferred from transmission and/or reflection measurements [106, 107]. Such an inverse homogenization has been carried out with the Bruggeman formalism [108], and the results used to subsequently model CTF-based optical sensing [109, 110]. Furthermore, a similar inverse approach has been used to model optical sensing in more elaborate sculptured thin films [111–113].

6.7.2 Analogs of curved spacetime

The electromagnetic response of free space subjected to a gravitational field is formally equivalent to that of a certain nonhomogeneous bianisotropic material [68, 69], as mentioned in section 2.6.2. Consequently, material analogs for the electromagnetic properties of a wide variety of curved spacetime scenarios have been proposed, as outlined in section 6.5.

The question arises: how can one formulate practical designs for such material analogs? In principle, this question can be addressed by homogenization as follows. The curved spacetime of interest may be divided up into subregions which are sufficiently small that the material analog for each subregion can be taken to be approximately homogeneous. Then, by regarding the homogeneous material analog for each subregion of spacetime as an HCM, the inverse homogenization method can be used to estimate the required properties of constituent materials that give rise to an HCM that represents the homogeneous material analog for each subregion of spacetime.

Based on the inverse Bruggeman formalism, this process has been used to formulate designs for HCM-based analogs of Schwarzschild-(anti-)de Sitter spacetime [32], cosmic spinning strings [33], and colliding gravitational plane waves [34]. As discussed in section 6.5, these HCM analogs can require rather extreme degrees of anisotropy, especially in the vicinity of singularities or event horizons. However, HCMs belonging to these parameter regimes can, at least in principle, be attained through homogenizing highly-elongated particles, such as disk- and needle-shaped particles.

6.8 Limitations for linear materials

In this section we strike a note of caution. For certain ranges of the constitutive parameters of the constituent materials, a homogenization formalism may provide estimates of the effective constitutive dyadics that are not physically plausible. This is best illustrated by returning to the simple isotropic dielectric HCM considered in section 6.3.

Consider therefore the homogenization of a composite material made of two isotropic dielectric materials distributed randomly as spherical particles. The permittivities are denoted by ε_a and ε_b, and the volume fractions by f_a and f_b. The standard Bruggeman estimate ε_{Br} of the HCM permittivity is provided in equation (6.10). The discriminant term

$$\Delta_{Br} = \left[\varepsilon_a\left(f_b - 2f_a\right) + \varepsilon_b\left(f_a - 2f_b\right)\right]^2 + 8\varepsilon_a\varepsilon_b \tag{6.23}$$

therein is central when the applicability of the Bruggeman formalism is considered [78]. This becomes clear if attention is restricted to nondissipative constituent materials (i.e., $\varepsilon_{a,b} \in \mathbb{R}$). If $\varepsilon_a\varepsilon_b < 0$, then there is scope for Δ_{Br} to be negative valued. For example, $\Delta_{Br} < 0$ when $f_a = f_b$, provided that $(\varepsilon_a + \varepsilon_b)^2 < -32\varepsilon_a\varepsilon_b$. Hence, the Bruggeman estimate of the HCM permittivity may be complex valued with $\mathrm{Im}\{\varepsilon_{Br}\} \neq 0$, even though neither constituent material is dissipative. As the standard Bruggeman formalism has no explicit mechanism for taking scattering loss into account (unlike the extended Bruggeman formalism or the second-order SPFT formalism), this result is physically implausible. On the other hand, no such problem arises if $\varepsilon_a\varepsilon_b > 0$ because $\Delta_{Br} > 0$ then.

It is illuminating to examine the issues that arise here by way of representative numerical examples. To put the condition $\varepsilon_a\varepsilon_b < 0$ into context, consider first the case of $\varepsilon_a\varepsilon_b > 0$ which corresponds to the homogenization of a composite material comprising two nondissipative dielectric materials. For $\varepsilon_a = 6\,\varepsilon_0$ and $\varepsilon_b = 1.5\,\varepsilon_0$, the Bruggeman estimate ε_{Br} is plotted against the volume fraction f_a in figure 6.11. Also plotted are the Hashin–Shtrikman bounds $HS_{\alpha,\beta}$ defined in equations (1.2), and the Wiener bounds $W_{\alpha,\beta}$ defined in equations (1.1). The estimate ε_{Br} decreases uniformly, but nonlinearly, as f_a increases from 0 to 1. For all values of $f_a \in (0, 1)$, the estimate ε_{Br} is bounded above by HS_β and below by HS_α. Furthermore, the envelope formed by the Hashin–Shtrikman bounds is wholly contained within the envelope formed by the Wiener bounds. Also, $\varepsilon_{Br} \to \varepsilon_\ell$, $HS_\alpha \to \varepsilon_\ell$, $HS_\beta \to \varepsilon_\ell$, $W_\alpha \to \varepsilon_\ell$, and $W_\beta \to \varepsilon_\ell$ in the limit $f_\ell \to 1$ for $\ell \in \{a, b\}$, in conformity with the constraint (4.4).

In the realm of dissipative materials, a counterpart to the example illustrated in figure 6.11 is represented by the case $\varepsilon_a = (6 + 2i)\,\varepsilon_0$ and $\varepsilon_b = (1.5 + 0.5i)\,\varepsilon_0$, i.e., $\mathrm{Re}\{\varepsilon_a\}\,\mathrm{Re}\{\varepsilon_b\} > 0$. The corresponding real and imaginary parts of the Bruggeman estimate ε_{Br} are plotted against f_a in figure 6.12. The real and imaginary parts of the

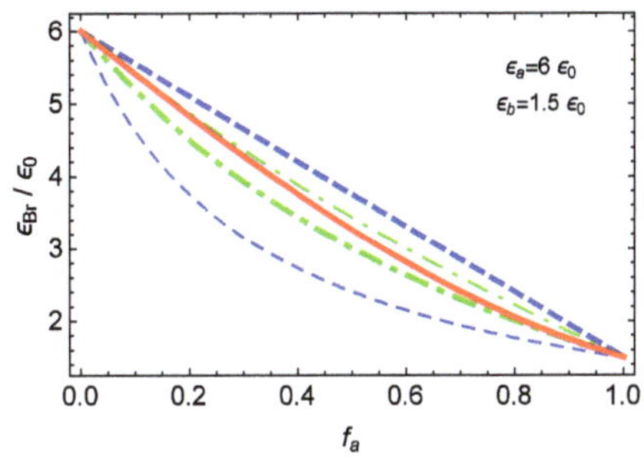

Figure 6.11. Normalized Bruggeman estimate $\varepsilon_{Br}/\varepsilon_0$ (red solid curve) plotted against volume fraction f_a for an isotropic dielectric HCM arising from isotropic dielectric constituent materials with $\varepsilon_a = 6\,\varepsilon_0$ and $\varepsilon_b = 1.5\,\varepsilon_0$. The constituent materials are distributed randomly as electrically small spheres. Also plotted are the normalized Hashin–Shtrikman bounds HS_α/ε_0 (green, thick broken dashed curve) and HS_β/ε_0 (green, thin broken dashed curve), and the normalized Wiener bounds W_α/ε_0 (blue, thick dashed curve) and W_β/ε_0 (blue, thin dashed curve).

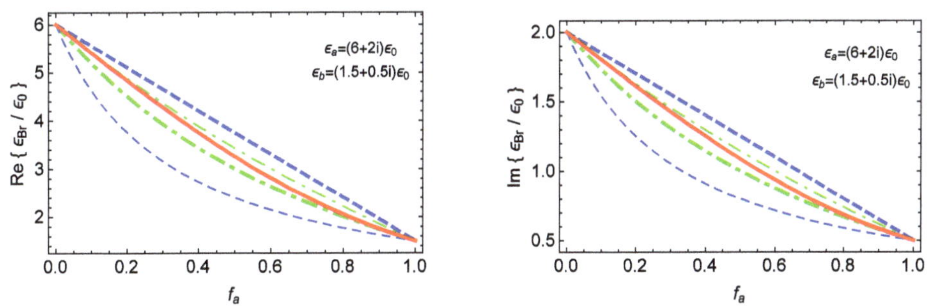

Figure 6.12. Real and imaginary parts of the normalized Bruggeman estimate $\varepsilon_{Br}/\varepsilon_0$ (red solid curve) plotted against f_a for an isotropic dielectric HCM arising from isotropic dielectric constituent materials with $\varepsilon_a = (6 + 2i)\,\varepsilon_0$ and $\varepsilon_b = (1.5 + 0.5i)\,\varepsilon_0$. The constituent materials are distributed randomly as electrically small spheres. Also plotted are the real and imaginary parts of the normalized Hashin–Shtrikman bounds HS_α/ε_0 (green, thick broken dashed curve) and HS_β/ε_0 (green, thin broken dashed curve), and the normalized Wiener bounds W_α/ε_0 (blue, thick dashed curve) and W_β/ε_0 (blue, thin dashed curve).

Hashin–Shtrikman and Wiener bounds are also plotted. The graph of $\text{Re}\{\varepsilon_{Br}\}$ in figure 6.12 is very similar to the graph of ε_{Br} in figure 6.11. Indeed, the graphs of $\text{Re}\{HS_{\alpha,\beta}\}$ and $\text{Re}\{W_{\alpha,\beta}\}$ in figure 6.12 are also very similar to corresponding graphs in figure 6.11. Qualitatively, the graph of $\text{Im}\{\varepsilon_{Br}\}$ in figure 6.12 is similar to the graph of $\text{Re}\{\varepsilon_{Br}\}$; in particular, $\text{Im}\{\varepsilon_{Br}\}$ is wholly contained within an envelope formed by $\text{Im}\{HS_{\alpha,\beta}\}$ and this envelope, in turn, is wholly contained within an envelope formed by $\text{Im}\{W_{\alpha,\beta}\}$.

Now let us turn to the case of $\varepsilon_a\varepsilon_b < 0$. For $\varepsilon_a = -6\,\varepsilon_0$ and $\varepsilon_b = 1.5\,\varepsilon_0$, the real and imaginary parts of the Bruggeman estimate ε_{Br} are plotted against volume fraction f_a in figure 6.13. The corresponding plots of the Hashin–Shtrikman bounds and Wiener bounds are shown alongside the graph of $\text{Re}\{\varepsilon_{Br}\}$. These bounds are entirely real valued here. While the graph of $\text{Re}\{\varepsilon_{Br}\}$ is not obviously remarkable, the same cannot be said of the graph of $\text{Im}\{\varepsilon_{Br}\}$ which exhibits a broad positive-

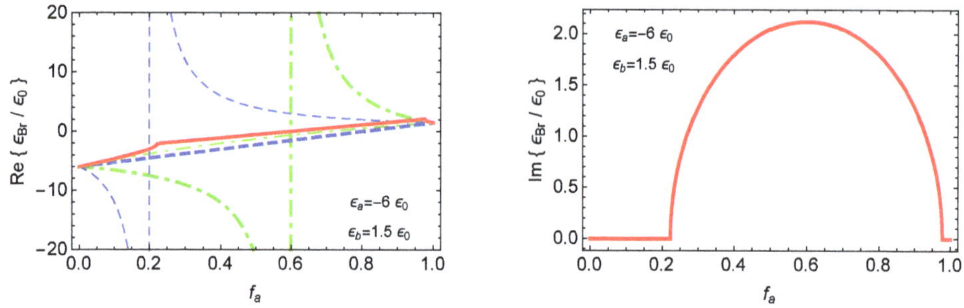

Figure 6.13. Real and imaginary parts of the normalized Bruggeman estimate $\varepsilon_{\text{Br}}/\varepsilon_0$ (red solid curve) plotted against the volume fraction f_a for an isotropic dielectric HCM arising from isotropic dielectric constituent materials with $\varepsilon_a = -6\varepsilon_0$ and $\varepsilon_b = 1.5\varepsilon_0$. The constituent materials are distributed randomly as electrically small spheres. Also plotted are the real parts of the normalized Hashin–Shtrikman bounds HS_α/ε_0 (green, thick broken dashed curve) and HS_β/ε_0 (green, thin broken dashed curve), and the normalized Wiener bounds W_α/ε_0 (blue, thick dashed curve) and W_β/ε_0 (blue, thin dashed curve). Since the imaginary parts of HS_α, HS_β, W_α, and W_β are null valued for all values of f_a, these quantities are not plotted.

valued peak over the range $0.22 < f_a < 0.97$. The nonzero values of $\text{Im}\{\varepsilon_{\text{Br}}\}$ are not physically plausible, given that the constituent materials are both nondissipative. Furthermore, the graphs of one of the Hashin–Shtrikman bounds and of one of the Wiener bounds exhibit resonances: $|HS_\alpha| \to \infty$ as $f_a \to 0.6$, as may be anticipated from equation $(1.2)_1$, and $|W_\beta| \to \infty$ as $f_a \to 0.2$, as may be anticipated from equation $(1.1)_2$.

Finally, consider the case $\varepsilon_a = (-6 + 2\delta i)\,\varepsilon_0$ and $\varepsilon_b = (1.5 + 0.5\delta i)\,\varepsilon_0$, i.e., $\text{Re}\{\varepsilon_a\}\,\text{Re}\{\varepsilon_b\} > 0$. The dissipation parameter $\delta > 0$. The corresponding real and imaginary parts of the Bruggeman estimate ε_{Br} are plotted against the volume fraction f_a in figure 6.14 for $\delta \in \{0.1, 1, 10\}$. As in figure 6.12, the real and imaginary parts of the Hashin–Shtrikman and Wiener bounds are also plotted. For $\delta = 0.1$, the graphs of $\text{Re}\{\varepsilon_{\text{Br}}\}$, $\text{Re}\{HS_{\alpha,\beta}\}$, and $\text{Re}\{W_{\alpha,\beta}\}$ are very similar to those for the corresponding nondissipative case presented in figure 6.13, with $\text{Re}\{HS_\alpha\}$ and $\text{Re}\{W_\beta\}$ exhibiting strong resonances in the vicinities of $f_a = 0.6$ and $f_a = 0.2$, respectively. The graph of $\text{Im}\{\varepsilon_{\text{Br}}\}$ is likewise very similar to that in figure 6.13, while the graphs of $\text{Im}\{HS_\alpha\}$ and $\text{Re}\{W_\beta\}$ exhibit strong resonances in the vicinities of $f_a = 0.6$ and $f_a = 0.2$, respectively. At $\delta = 1$, the peak of $\text{Im}\{\varepsilon_{\text{Br}}\}$ is diminished and the resonances exhibited by the real and imaginary parts of both HS_α and W_β are less pronounced, as compared to those for $\delta = 0.1$. This trend continues as the dissipation parameter increases further. At $\delta = 10$, there is no obvious sign of resonances exhibited by the real and imaginary parts of both HS_α and W_β, and both the real and imaginary parts of the Bruggeman estimate ε_{Br} are contained within moderately small envelopes formed by the real and imaginary parts of the Hashin–Shtrikman bounds. Also, the Hashin–Shtrikman bounds are contained within slightly larger envelopes formed by the real and imaginary parts of the Wiener bounds.

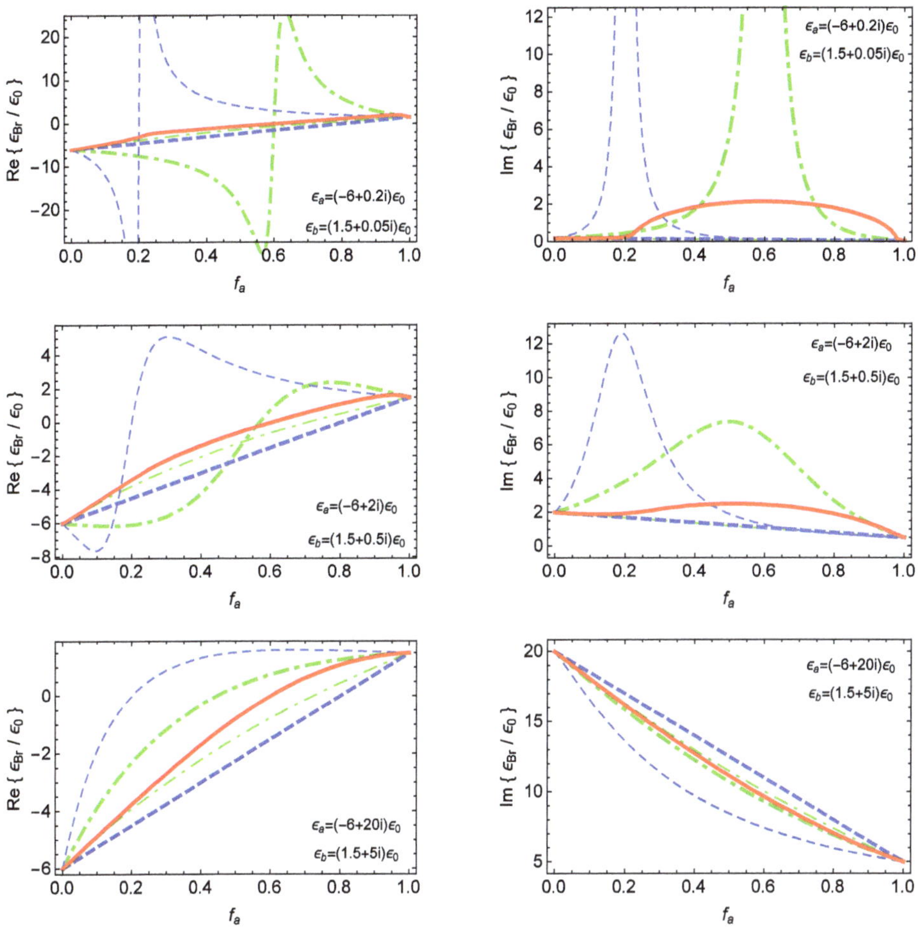

Figure 6.14. Real and imaginary parts of the normalized Bruggeman estimate $\varepsilon_{\rm Br}/\varepsilon_0$ (red solid curve) plotted against $f_{\rm a}$ for an isotropic dielectric HCM arising from isotropic dielectric constituent materials with $\varepsilon_{\rm a} = (-6 + 2i\delta)\,\varepsilon_0$ and $\varepsilon_{\rm b} = (1.5 + 0.5i\delta)\,\varepsilon_0$, for $\delta \in \{0.1, 1, 10\}$. The constituent materials are distributed randomly as electrically small spheres. Also plotted are the real and imaginary parts of the normalized Hashin–Shtrikman bounds HS_α/ε_0 (green, thick broken dashed curve) and HS_β/ε_0 (green, thin broken dashed curve), and the normalized Wiener bounds W_α/ε_0 (blue, thick dashed curve) and W_β/ε_0 (blue, thin dashed curve).

The graphs in figure 6.11–6.14 are representative examples which demonstrate that both $HS_{\alpha,\beta}$ and $W_{\alpha,\beta}$ can provide quite tight bounds on the Bruggeman estimate $\varepsilon_{\rm Br}$ for the cases $\varepsilon_{\rm a}\varepsilon_{\rm b} > 0$ and $\mathrm{Re}\{\varepsilon_{\rm a}\}\,\mathrm{Re}\{\varepsilon_{\rm b}\} > 0$, as well as the case of $\mathrm{Re}\{\varepsilon_{\rm a}\}\,\mathrm{Re}\{\varepsilon_{\rm b}\} < 0$ provided that the materials involved are at least moderately dissipative. However, for the cases $\varepsilon_{\rm a}\varepsilon_{\rm b} < 0$ and $\mathrm{Re}\{\varepsilon_{\rm a}\}\,\mathrm{Re}\{\varepsilon_{\rm b}\} < 0$ where the materials involved are only weakly dissipative, the Bruggeman estimates are physically implausible and both the Hashin–Shtrikman and Wiener bounds exhibit strong resonances—which thereby renders these bounds of little or no practical use.

Weak dissipation accompanying the case of $\mathrm{Re}\{\varepsilon_{\rm a}\}\,\mathrm{Re}\{\varepsilon_{\rm b}\} < 0$ is quite topical these days. This regime characterizes certain metal-in-insulator composites which are regarded as possible platforms for negative-refraction applications. Silver is often considered for these applications, as it exhibits particularly low dissipation at optical and near-infrared frequencies. While low dissipation is desirable for many applications, this property renders silver-in-insulator composites inappropriate for the application of the Bruggeman formalism at optical and near-infrared frequencies [79]. This limitation has not always been taken into account [114–116].

The restriction concerning the application of the standard Bruggeman formalism represented by equation (6.10) for (i) $\varepsilon_a \varepsilon_b < 0$ and (ii) the weakly dissipative $\mathrm{Re}\{\varepsilon_a\}\,\mathrm{Re}\{\varepsilon_b\} < 0$ case has a wider scope. As the extended Bruggeman formalism is built upon the standard formalism, the former too is restricted. The restriction applies to both passive and active HCMs [117]. The inverse Bruggeman formalism is similarly restricted [118]. Since the Bruggeman formalism is equivalent to the lowest-order SPFT formalism, the second-order (and higher-order) SPFT is likewise restricted.

The Maxwell Garnett formula given by equation (6.7), when evaluated at $\rho = 0$, is identical to the Hashin–Shtrikman formula $(1.2)_1$. Therefore, the applicability of the Maxwell Garnett formalism in the $\varepsilon_a \varepsilon_b < 0$ regime and the weakly dissipative $\mathrm{Re}\{\varepsilon_a\}\,\mathrm{Re}\{\varepsilon_b\} < 0$ regime is questionable. Furthermore, in a similar manner to the Hashin–Shtrikman bounds and Wiener bounds, the Bergman–Milton bounds [119–121] on the constitutive parameters of the HCM become exceedingly large for these problematic cases and thereby lose their practical utility [122].

6.9 Gain and loss

Homogenization provides opportunities for enhancing gain or loss. That is, an HCM may be conceptualized to exhibit a greater degree of gain or loss than is exhibited by its constituent materials [123]. Furthermore, if anisotropy or bianisotropy is involved, HCMs may be conceptualized that exhibit gain for certain propagation directions and loss for other propagation directions [124]. These possibilities are illustrated in this section by means of representative numerical examples.

6.9.1 Enhancement of gain and loss

The enhancement in gain or loss that can be achieved through homogenization is illustrated by means of a simple example involving an isotropic dielectric HCM [123].

Let us begin with the case in which both constituent materials of a composite material are active isotropic dielectric materials. Constituent material ‘a’ has permittivity $\varepsilon_a = (2 - 0.12i)\,\varepsilon_0$ whereas constituent material ‘b’ has permittivity $\varepsilon_b = (5 - 0.1i)\,\varepsilon_0$. These permittivity values lie within the range typically employed for active components of metamaterials in the visible wavelength regime [125]. Both constituent materials are randomly distributed as spherical particles. Plots of the real and imaginary parts of the Bruggeman estimate of the HCM's permittivity scalar ε_{Br} against volume fraction f_a are shown in figure 6.15. For comparison, the Wiener and Hashin–Shtrikman bounds, introduced in equations (1.1) and (1.2), respectively, are also presented. The imaginary part of ε_{Br} is less than the imaginary parts of both ε_a

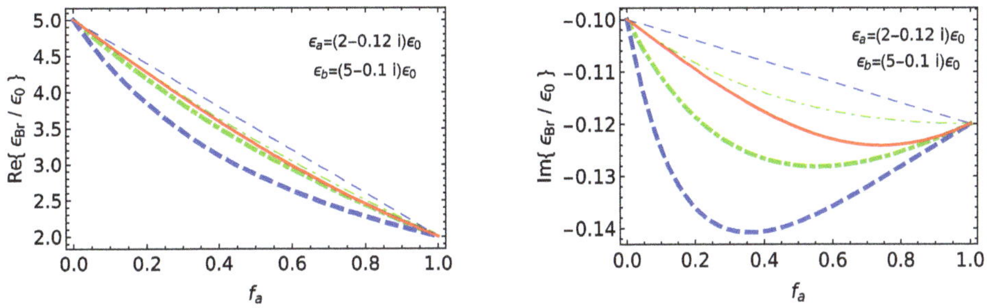

Figure 6.15. Real and imaginary parts of the normalized Bruggeman estimate $\varepsilon_{Br}/\varepsilon_0$ (red solid curve) plotted against volume fraction f_a for an isotropic dielectric HCM arising from active isotropic dielectric constituent materials with $\varepsilon_a = (2 - 0.12i)\,\varepsilon_0$ and $\varepsilon_b = (5 - 0.1i)\,\varepsilon_0$. The constituent materials are distributed randomly as spherical particles. Also plotted are the normalized Hashin–Shtrikman bounds HS_α/ε_0 (green, thick broken dashed curve) and HS_β/ε_0 (green, thin broken dashed curve), and the normalized Wiener bounds W_α/ε_0 (blue, thick dashed curve) and W_β/ε_0 (blue, thin dashed curve).

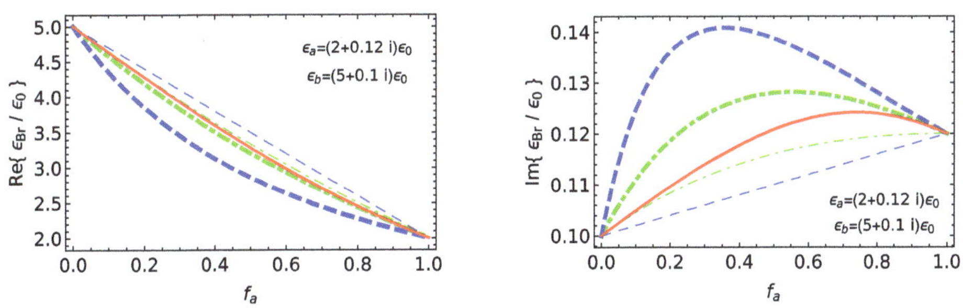

Figure 6.16. As figure 6.15 except that both constituent materials are dissipative with $\varepsilon_a = (2 + 0.12i)\,\varepsilon_0$ and $\varepsilon_b = (5 + 0.1i)\,\varepsilon_0$.

and ε_b across the range $0.47 < f_a < 1$. Thus, gain enhancement is exhibited for this range of volume fraction f_a.

In a similar vein, loss enhancement is demonstrated: Suppose now that both constituent materials are dissipative materials with permittivities $\varepsilon_a = (2 + 0.12i)\,\varepsilon_0$ and $\varepsilon_b = (5 + 0.1i)\,\varepsilon_0$. In figure 6.16 plots are provided of the real and imaginary parts of the Bruggeman estimate ε_{Br}, and the Wiener and Hashin–Shtrikman bounds, against volume fraction f_a. Now the imaginary part of ε_{Br} is greater than the imaginary parts of both ε_a and ε_b for $0.47 < f_a < 1$. Thus, loss enhancement is exhibited for this range of volume fraction.

In essence, gain enhancement and loss enhancement represent the two different sides of the same coin. Therefore, concentrating on gain enhancement, we introduce the gain-enhancement index

$$g_\rho = \frac{\mathrm{Im}\{\varepsilon_{Br}\}}{\min\{\mathrm{Im}\{\varepsilon_a\}, \mathrm{Im}\{\varepsilon_b\}\}}, \tag{6.24}$$

with $g_\rho > 1$ signifying gain enhancement.

For $\text{Re}\{\varepsilon_a\} = 2\,\varepsilon_0$ and $\varepsilon_b = (5 - 0.1i)\,\varepsilon_0$, g_ρ is plotted against f_a and $\text{Im}\{\varepsilon_a\}/\text{Im}\{\varepsilon_b\}$ in figure 6.17. The gain enhancement is greatest for mid-range values of f_a when $\text{Im}\{\varepsilon_a\}/\text{Im}\{\varepsilon_b\}$ is close to unity. In particular, no gain enhancement is possible if $\text{Im}\{\varepsilon_a\}/\text{Im}\{\varepsilon_b\}$ is either much greater or much less than unity, regardless of the value of the volume fraction f_a.

In figure 6.18, g_ρ is plotted against f_a and $\text{Re}\{\varepsilon_a\}/\text{Re}\{\varepsilon_b\}$ for $\text{Im}\{\varepsilon_a\} = -0.12\,\varepsilon_0$ and $\varepsilon_b = (5 - 0.1i)\,\varepsilon_0$. The gain enhancement is greatest for mid-range values of f_a when $\text{Re}\{\varepsilon_a\}/\text{Re}\{\varepsilon_b\}$ is either much less or much greater than unity. In particular, no

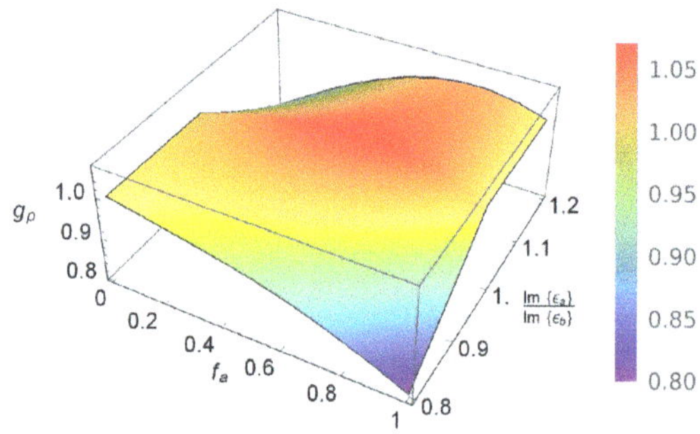

Figure 6.17. Gain-enhancement index g_ρ plotted against f_a and $\text{Im}\{\varepsilon_a\}/\text{Im}\{\varepsilon_b\}$ when $\text{Re}\{\varepsilon_a\} = 2\,\varepsilon_0$ and $\varepsilon_b = (5 - 0.1i)\,\varepsilon_0$.

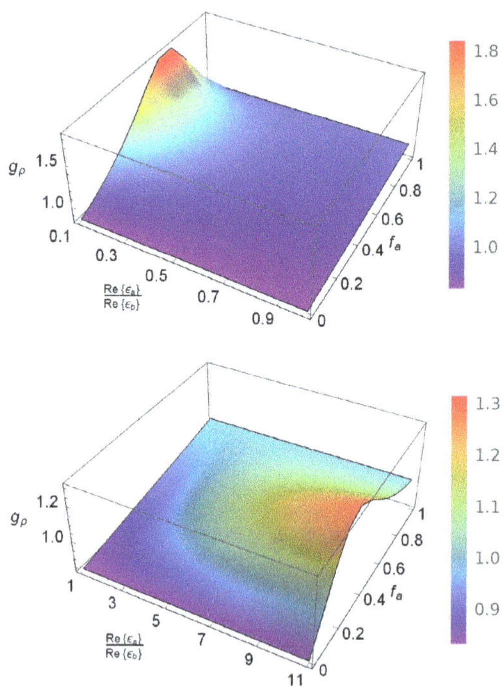

Figure 6.18. Gain-enhancement index g_ρ plotted against f_a and $\text{Re}\{\varepsilon_a\}/\text{Re}\{\varepsilon_b\}$ when $\text{Im}\{\varepsilon_a\} = -0.12\,\varepsilon_0$ and $\varepsilon_b = (5 - 0.1i)\,\varepsilon_0$.

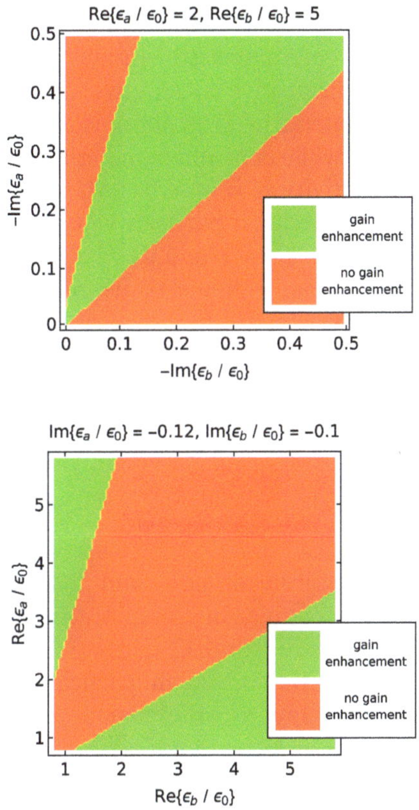

Figure 6.19. Top: gain-enhancement subspace in the $(-\mathrm{Im}\{\varepsilon_a/\varepsilon_0\}, -\mathrm{Im}\{\varepsilon_b/\varepsilon_0\})$ parameter space when $\mathrm{Re}\{\varepsilon_a\} = 2\varepsilon_0$ and $\mathrm{Re}\{\varepsilon_b\} = 5\varepsilon_0$. Bottom: gain-enhancement subspace in the $(\mathrm{Re}\{\varepsilon_a/\varepsilon_0\}, \mathrm{Re}\{\varepsilon_b\}/\varepsilon_0)$ parameter space when $\mathrm{Im}\{\varepsilon_a\} = -0.12\varepsilon_0$ and $\mathrm{Im}\{\varepsilon_b\} = -0.1\varepsilon_0$.

gain enhancement is possible if $\mathrm{Re}\{\varepsilon_a\}/\mathrm{Re}\{\varepsilon_b\}$ is close to unity, regardless of the volume faction.

The parameter space $(-\mathrm{Im}\{\varepsilon_a/\varepsilon_0\}, -\mathrm{Im}\{\varepsilon_b/\varepsilon_0\}) \in (0, 0.5) \times (0, 0.5)$ is mapped in figure 6.19(top) to identify the gain-enhancement subspace when $\mathrm{Re}\{\varepsilon_a\} = 2\varepsilon_0$ and $\mathrm{Re}\{\varepsilon_b\} = 5\varepsilon_0$. The gain-enhancement subspace is a window containing $\mathrm{Im}\{\varepsilon_a\} = \mathrm{Im}\{\varepsilon_b\}$ that becomes broader as the magnitudes of $\mathrm{Im}\{\varepsilon_a\}$ and $\mathrm{Im}\{\varepsilon_b\}$ are increased. Likewise, the parameter space $(\mathrm{Re}\{\varepsilon_a/\varepsilon_0\}, \mathrm{Re}\{\varepsilon_b\}/\varepsilon_0) \in (0.8, 6) \times (0.8, 6)$ is mapped in figure 6.19(bottom) to identify the two disjoint gain-enhancement subspaces that exist when $\mathrm{Im}\{\varepsilon_a\} = -0.12\varepsilon_0$ and $\mathrm{Im}\{\varepsilon_b\} = -0.1\varepsilon_0$. Within these two gain-enhancement subspaces, $\mathrm{Re}\{\varepsilon_a\}$ and $\mathrm{Re}\{\varepsilon_b\}$ are dissimilar in magnitude. The patterns revealed in figure 6.19 are wholly consistent with those displayed in figure 6.17 and 6.18.

6.9.2 Simultaneous gain and loss

Anisotropy and/or bianisotropy provide opportunities for simultaneous gain and loss, i.e., plane waves may be simultaneously amplified for one propagation direction but attenuated for another propagation direction. A composite material that exhibits this characteristic may be readily conceptualized via the homogenization of an active constituent material 'a' and a dissipative constitutive material 'b' [124].

Suppose that constituent material 'a' is an active isotropic dielectric material with permittivity $\varepsilon_a = (2 - 0.1i)\,\varepsilon_0$ while constituent material 'b' is a dissipative isotropic dielectric material with permittivity $\varepsilon_b = (4 + 0.005i)\,\varepsilon_0$. Both constituent materials are randomly distributed as spheriodal particles whose shape and orientation are specified by $\underline{\underline{U}}^a = \underline{\underline{U}}^b = [(\underline{\underline{I}} - \hat{\mathbf{u}}_z\,\hat{\mathbf{u}}_z) + U\hat{\mathbf{u}}_z\,\hat{\mathbf{u}}_z]/\sqrt[3]{U}$. The corresponding HCM is uniaxial dielectric, the Bruggeman estimate of its permittivity dyadic having the form given in equation (6.16).

The real and imaginary parts of the Bruggeman estimates of the HCM's permittivity parameters are plotted in figure 6.20 against volume fraction f_a for $U = 5$. As f_a increases, the real parts of the permittivity parameters decrease smoothly, as do the imaginary parts also. Significantly, $\text{Im}\{\varepsilon_\perp^{Br}\}$ decreases in a more nonlinear fashion than $\text{Im}\{\varepsilon_\parallel^{Br}\}$ does, as f_a increases. Consequently, $\text{Im}\{\varepsilon_\perp^{Br}\} < 0$ but $\text{Im}\{\varepsilon_\parallel^{Br}\} > 0$ for $f_a \in (0.22, 0.32)$. In this f_a range, the HCM is neither wholly dissipative nor wholly active; instead, gain and loss in different propagation directions are simultaneously possible. In contrast, the HCM is wholly active for $0 < f_a < 0.22$ and wholly dissipative for $0.32 < f_a < 1$.

For this example, the shape of the constituent particles has a strong influence over the volume-fraction range that supports simultaneous gain and loss. In figure 6.21, the real and imaginary parts of $\varepsilon_\perp^{Br}/\varepsilon_0$ and $\varepsilon_\parallel^{Br}/\varepsilon_0$ are plotted against U with $f_a = 0.225$ fixed. The constituent particles are oblate spheroids for $U < 1$ but prolate spheroids for $U > 1$. In the degenerate case $U = 1$, the constituent particles are spheres and the corresponding HCM is isotropic. As U increases, both $\text{Re}\{\varepsilon_\perp^{Br}\}$ and $\text{Im}\{\varepsilon_\perp^{Br}\}$ decrease nonlinearly, whereas both $\text{Re}\{\varepsilon_\parallel^{Br}\}$ and $\text{Im}\{\varepsilon_\parallel^{Br}\}$ increase nonlinearly. For $U < 0.68$, $\text{Im}\{\varepsilon_\parallel^{Br}\} < 0$ but $\text{Im}\{\varepsilon_\perp^{Br}\} > 0$. Also, for $U > 2.8$, $\text{Im}\{\varepsilon_\parallel^{Br}\} > 0$ but $\text{Im}\{\varepsilon_\perp^{Br}\} < 0$. Hence, gain and loss in different propagation

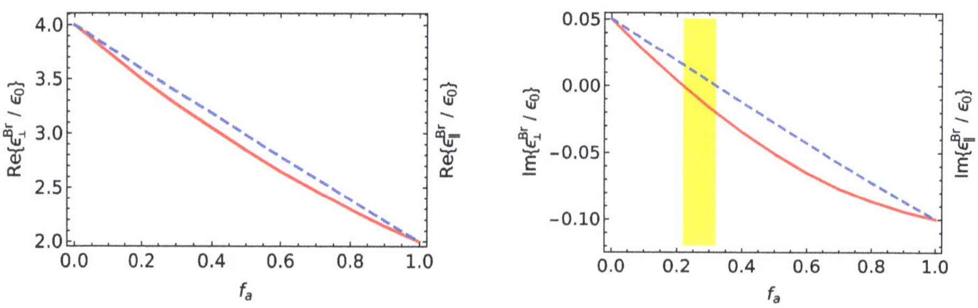

Figure 6.20. Real and imaginary parts of $\varepsilon_\perp^{Br}/\varepsilon_0$ (red solid curves) and $\varepsilon_\parallel^{Br}/\varepsilon_0$ (blue dashed curves) plotted against f_a, as delivered by the Bruggeman formalism. Input parameters: $\underline{\underline{\varepsilon}}_a = (2 - 0.1i)\varepsilon_0\,\underline{\underline{I}}$, $\underline{\underline{\varepsilon}}_b = (4 + 0.05i)\varepsilon_0\,\underline{\underline{I}}$, and $\underline{\underline{U}}^a = \underline{\underline{U}}^b = [(\underline{\underline{I}} - \hat{\mathbf{u}}_z\,\hat{\mathbf{u}}_z) + 5\hat{\mathbf{u}}_z\,\hat{\mathbf{u}}_z]/\sqrt[3]{5}$. The shaded (yellow) rectangle indicates the range of f_a for which $\text{Im}\{\varepsilon_\perp^{Br}\} < 0$ but $\text{Im}\{\varepsilon_\parallel^{Br}\} > 0$.

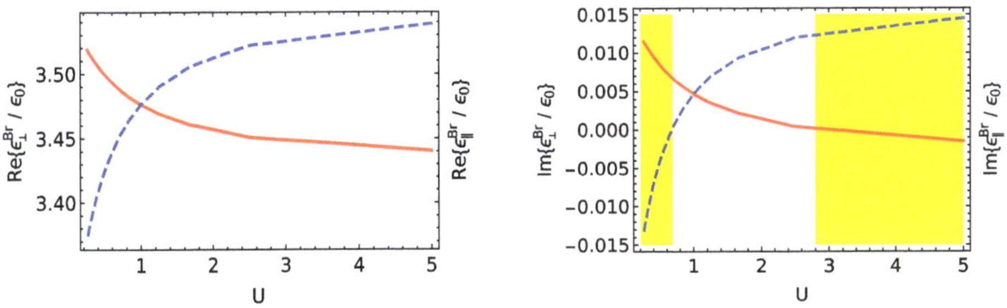

Figure 6.21. Real and imaginary parts of $\varepsilon_{\perp}^{\mathrm{Br}}/\varepsilon_0$ (red solid curves) and $\varepsilon_{\parallel}^{\mathrm{Br}}/\varepsilon_0$ (blue dashed curves) plotted against U, as delivered by the Bruggeman formalism. Input parameters: $f_{\mathrm{a}} = 0.225$, $\underline{\underline{\varepsilon}}_{\mathrm{a}} = (2 - 0.1i)\varepsilon_0\, \underline{\underline{I}}$, $\underline{\underline{\varepsilon}}_{\mathrm{b}} = (4 + 0.05i)\varepsilon_0\, \underline{\underline{I}}$, and $\underline{\underline{U}}^{\mathrm{a}} = \underline{\underline{U}}^{\mathrm{b}} = [(\underline{\underline{I}} - \hat{\mathbf{u}}_z\, \hat{\mathbf{u}}_z) + U\hat{\mathbf{u}}_z\, \hat{\mathbf{u}}_z]/\sqrt[3]{U}$. The shaded (yellow) rectangles indicate the range of U for which the product $\mathrm{Im}\{\varepsilon_{\perp}^{\mathrm{Br}}\}\mathrm{Im}\{\varepsilon_{\parallel}^{\mathrm{Br}}\} < 0$.

directions are simultaneously possible, provided that the constituent spheroidal particles are either sufficiently flattened or sufficiently elongated.

The influence of the permittivity of the active constituent material on the manifestation of simultaneous gain and loss in different propagation directions is illustrated figure 6.22, wherein the real and imaginary parts of $\varepsilon_{\perp}^{\mathrm{Br}}/\varepsilon_0$ and $\varepsilon_{\parallel}^{\mathrm{Br}}/\varepsilon_0$ are plotted against the real and (negative) imaginary parts of $\varepsilon_a/\varepsilon_0$ for $\varepsilon_{\mathrm{b}} = (4 + 0.05i)\varepsilon_0$, $f_{\mathrm{a}} = 0.225$, and $U = 5$. Also plotted in figure 6.22 is the product $\mathrm{Im}\{\varepsilon_{\perp}^{\mathrm{Br}}/\varepsilon_0\}\mathrm{Im}\{\varepsilon_{\parallel}^{\mathrm{Br}}/\varepsilon_0\}$, this product being negative only when $\mathrm{Im}\{\varepsilon_{\perp}^{\mathrm{Br}}\}$ and $\mathrm{Im}\{\varepsilon_{\parallel}^{\mathrm{Br}}\}$ have opposite signs. Gain and loss in different propagation directions are simultaneously possible for relatively small values of $\mathrm{Re}\{\varepsilon_{\mathrm{a}}\}$ and relatively large values of $-\mathrm{Im}\{\varepsilon_{\mathrm{a}}\}$. In particular, as $-\mathrm{Im}\{\varepsilon_{\mathrm{a}}\}$ approaches 0, the product $\mathrm{Im}\{\varepsilon_{\perp}^{\mathrm{Br}}\}\mathrm{Im}\{\varepsilon_{\parallel}^{\mathrm{Br}}\} > 0$ for all values of $\mathrm{Re}\{\varepsilon_{\mathrm{a}}\}$ considered.

Figure 6.23 shows the result of repeating the computations of figure 6.22 using the Maxwell Garnett formalism. The Maxwell Garnett estimate of the HCM's permittivity dyadic has the form given in equation (6.16) but with the subscript/superscript 'Br' therein replaced by the subscript/superscript 'MG'. For the Maxwell Garnett formalism, the value of the shape parameter U only applies to constituent material 'a', because no particle shape is assigned to the host material in this formalism. The ε_{a}-regime wherein the product $\mathrm{Im}\{\varepsilon_{\perp}^{\mathrm{MG}}\}\mathrm{Im}\{\varepsilon_{\parallel}^{\mathrm{MG}}\} < 0$ in figure 6.23 is slightly smaller than the ε_{a}-regime wherein the product $\mathrm{Im}\{\varepsilon_{\perp}^{\mathrm{Br}}\}\mathrm{Im}\{\varepsilon_{\parallel}^{\mathrm{Br}}\} < 0$ in figure 6.22. Aside from this relatively modest difference, the plots presented in figure 6.22 and 6.23 are very similar.

6.10 Nonlinearity enhancement

The nonlinear properties of materials can be enhanced through the process of homogenization. Thus, HCMs can arise which exhibit stronger nonlinear properties than their constituent mediums. This aspect of homogenization extends beyond the issue of nonlinearity: a range of physical properties may be similarly enhanced via homogenization [126, 127], as alluded to in section 1.1 and described in sections 6.6.5 and 6.9.1. Nonlinearity enhancement may be exploited in applications

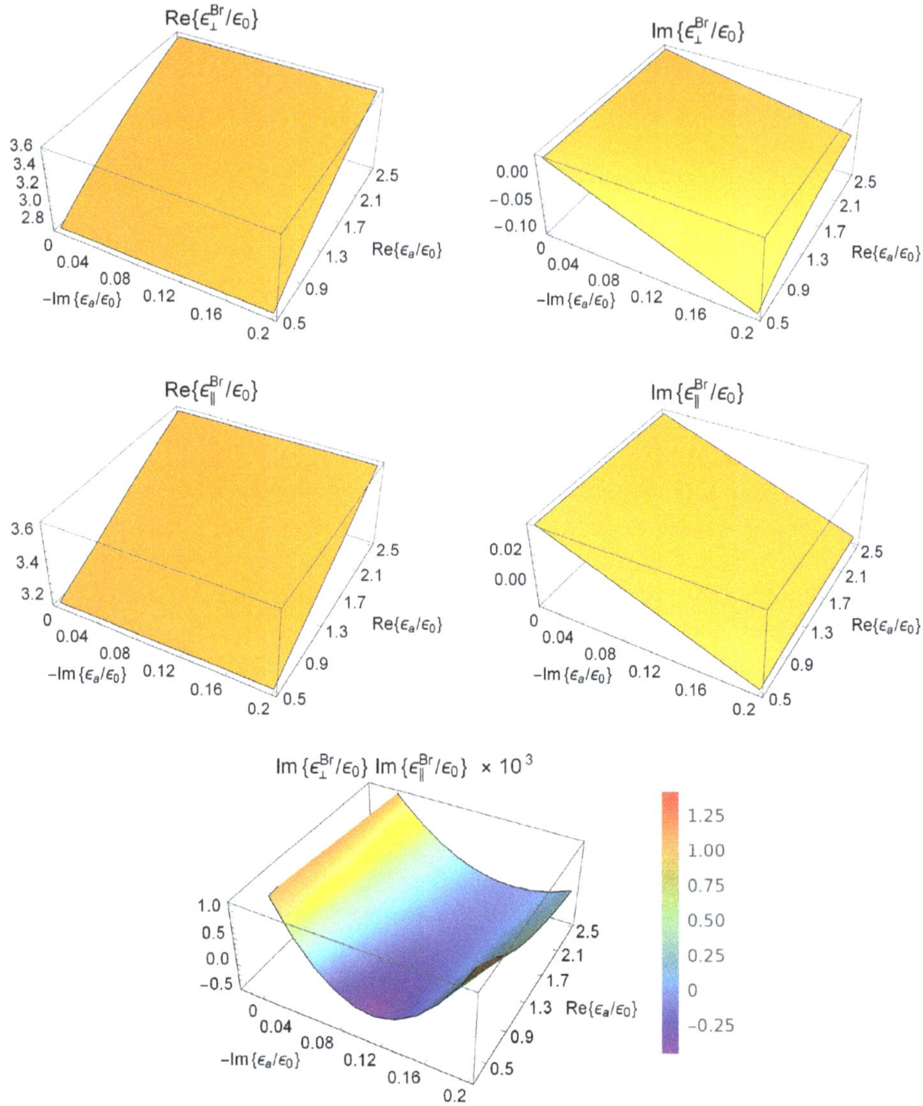

Figure 6.22. Real and imaginary parts of $\varepsilon_{\perp}^{\rm Br}/\varepsilon_0$ and $\varepsilon_{\parallel}^{\rm Br}/\varepsilon_0$, and the product $\mathrm{Im}\{\varepsilon_{\perp}^{\rm Br}/\varepsilon_0\}\mathrm{Im}\{\varepsilon_{\parallel}^{\rm Br}/\varepsilon_0\}$, as delivered by the Bruggeman formalism, plotted against the real and (negative) imaginary parts of $\varepsilon_{\rm a}/\varepsilon_0$. Input parameters: $f_{\rm a} = 0.225$, $\underline{\underline{\varepsilon}}_{\rm a} = \varepsilon_{\rm a}\, \underline{\underline{I}}$, $\underline{\underline{\varepsilon}}_{\rm b} = (4 + 0.05i)\varepsilon_0\, \underline{\underline{I}}$, and $\underline{\underline{U}}^{\rm a} = \underline{\underline{U}}^{\rm b} = [(\underline{\underline{I}} - \hat{\mathbf{u}}_z\,\hat{\mathbf{u}}_z) + 5\hat{\mathbf{u}}_z\,\hat{\mathbf{u}}_z]/\sqrt[3]{5}$.

pertaining to optical computing, real-time holography, optical correlators, and phase conjugators, among others [128].

In particular, for composite materials comprising materials with intensity-dependent permittivity, nonlinearity enhancement has been comprehensively explored by means of numerical studies, in the settings of isotropic dielectric [129–131], isotropic chiral [132], and anisotropic dielectric [133–135] HCMs. Let us illustrate this phenomenon here by means of a representative numerical example.

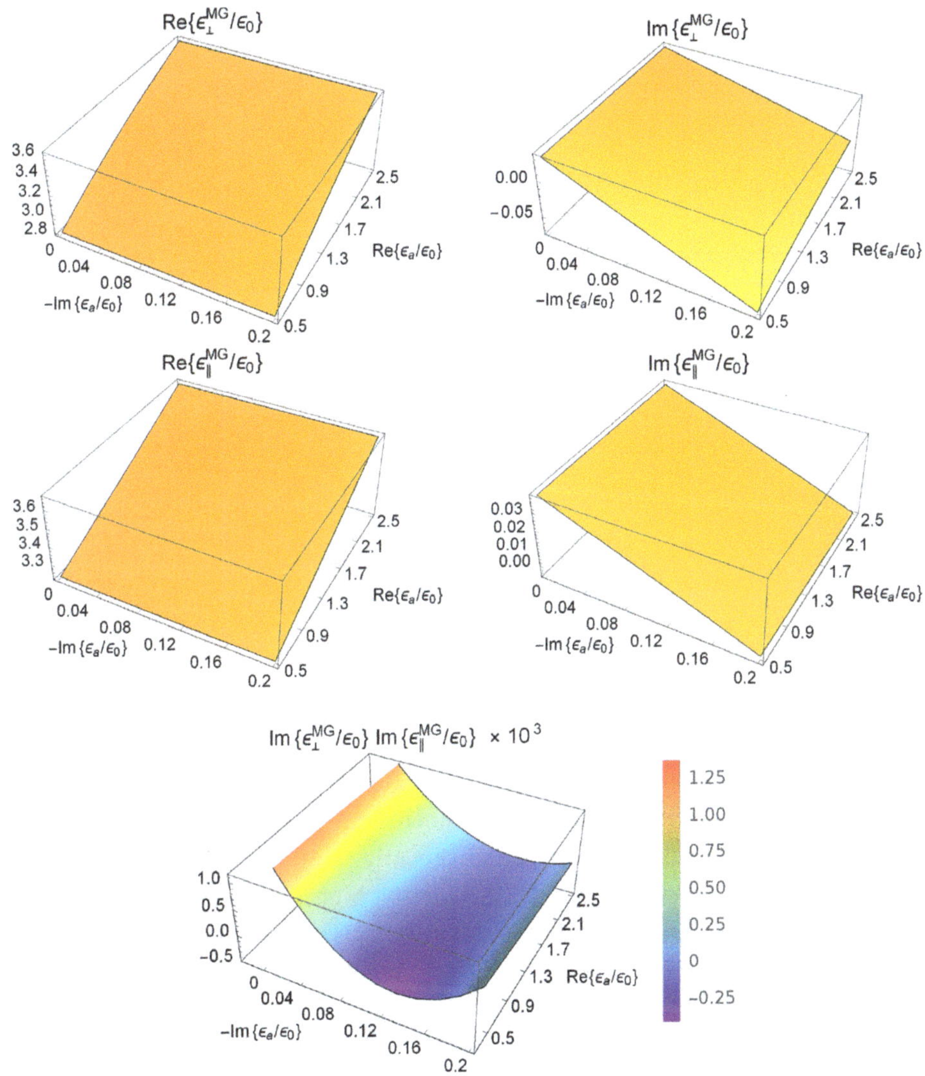

Figure 6.23. Real and imaginary parts of $\varepsilon_\perp^{\mathrm{MG}}/\varepsilon_0$ and $\varepsilon_\parallel^{\mathrm{MG}}/\varepsilon_0$, and the product $\mathrm{Im}\{\varepsilon_\perp^{\mathrm{MG}}/\varepsilon_0\}\mathrm{Im}\left\{\varepsilon_\parallel^{\mathrm{MG}}/\varepsilon_0\right\}$, as delivered by the Maxwell Garnett formalism, plotted against the real and (negative) imaginary parts of $\varepsilon_\mathrm{a}/\varepsilon_0$. Input parameters: $f_\mathrm{a} = 0.225$, $\underline{\underline{\varepsilon}}_\mathrm{a} = \varepsilon_\mathrm{a}\,\underline{\underline{I}}$, $\underline{\underline{\varepsilon}}_\mathrm{b} = (4 + 0.05i)\varepsilon_0\,\underline{\underline{I}}$, and $\underline{\underline{U}}^\mathrm{a} = [(\underline{\underline{I}} - \hat{\mathbf{u}}_z\,\hat{\mathbf{u}}_z) + 5\hat{\mathbf{u}}_z\,\hat{\mathbf{u}}_z]/\sqrt[3]{5}$.

For simplicity of presentation, let us focus on the most straightforward case, namely that of a weakly nonlinear isotropic dielectric HCM. The corresponding second-order SPFT formalism is provided in section 5.3.1. Recall that the ambient material, whose intensity-dependent permittivity has the form given in equation (5.7), is equivalent to the HCM that arises from the Bruggeman formalism (which is also equivalent to the HCM that arises from the zeroth- and first-order SPFT formalisms). Hence, the intensity-dependent permittivity of the Bruggeman HCM, namely

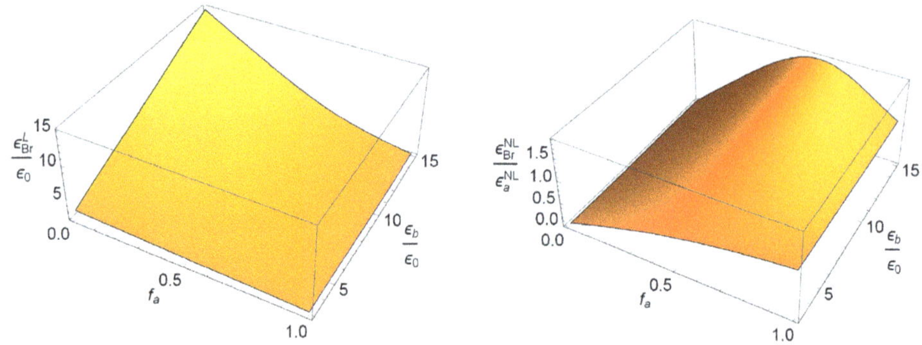

Figure 6.24. The linear contribution $\varepsilon_{\mathrm{Br}}^{\mathrm{L}}$ (normalized relative to ε_0) and the nonlinear contribution $\varepsilon_{\mathrm{Br}}^{\mathrm{NL}}$ (normalized relative to $\varepsilon_{\mathrm{a}}^{\mathrm{NL}}$) to the Bruggeman estimate of the intensity-dependent HCM permittivity plotted against f_{a} and $\varepsilon_{\mathrm{b}}/\varepsilon_0$, for $\varepsilon_{\mathrm{a}}^{\mathrm{L}} = 2\varepsilon_0$.

$\varepsilon_{\mathrm{Br}}$, has the form given in equation (5.7) but with the subscript 'amb' replaced by the subscript 'Br'.

Suppose that the linear contribution to the permittivity of the constituent material 'a' is $\varepsilon_{\mathrm{a}}^{\mathrm{L}} = 2\varepsilon_0$ and the permittivity of the constituent material 'b' is $\varepsilon_{\mathrm{b}} \in (3\varepsilon_0, 15\varepsilon_0)$. The linear contribution $\varepsilon_{\mathrm{Br}}^{\mathrm{L}}$ (normalized relative to ε_0) and the nonlinear contribution $\varepsilon_{\mathrm{Br}}^{\mathrm{NL}}$ (normalized relative to $\varepsilon_{\mathrm{a}}^{\mathrm{NL}}$) to $\varepsilon_{\mathrm{Br}}$ are plotted against f_{a} and $\varepsilon_{\mathrm{b}}/\varepsilon_0$ in figure 6.24. The graph of $\varepsilon_{\mathrm{Br}}^{\mathrm{L}}$ is precisely the same as the graph that would arise in the corresponding case of a purely linear composite material (i.e., when $\varepsilon_{\mathrm{a}}^{\mathrm{NL}} = 0$). Whereas $\varepsilon_{\mathrm{Br}}^{\mathrm{L}}$ decreases approximately linearly as f_{a} increases when ε_{b} is small, $\varepsilon_{\mathrm{Br}}^{\mathrm{L}}$ decreases in a more nonlinear fashion as f_{a} increases when ε_{b} is large. Regardless of the value of f_{a}, $\varepsilon_{\mathrm{Br}}^{\mathrm{L}}$ increases approximately linearly as ε_{b} increases. More interesting behavior is observed in the graph of $\varepsilon_{\mathrm{Br}}^{\mathrm{NL}}/\varepsilon_{\mathrm{a}}^{\mathrm{NL}}$. At higher values of ε_{b}, the graph of $\varepsilon_{\mathrm{Br}}^{\mathrm{NL}}/\varepsilon_{\mathrm{a}}^{\mathrm{NL}}$ versus f_{a} displays a local maximum in the vicinity of $f_{\mathrm{a}} \approx 0.6$. This local maximum is greatest at $\varepsilon_{\mathrm{b}} = 15\varepsilon_0$, where $\varepsilon_{\mathrm{Br}}^{\mathrm{NL}} \approx 1.8\, \varepsilon_{\mathrm{a}}^{\mathrm{NL}}$. That is, the Bruggeman estimate of the magnitude of the HCM's nonlinear permittivity exceeds that of constituent material 'a' by as much as approximately 80%. Parenthetically, even greater degrees of nonlinearity enhancement may be achieved by focusing on anisotropic HCMs [133, 134]. Notice that: (i) $\varepsilon_{\mathrm{Br}}^{\mathrm{L}}$ is independent of $\varepsilon_{\mathrm{a}}^{\mathrm{NL}}$ since the nonlinear permittivity does not feature in the linear scalar Bruggeman equation (6.10) and (ii) $\varepsilon_{\mathrm{Br}}^{\mathrm{NL}}/\varepsilon_{\mathrm{a}}^{\mathrm{NL}}$ is independent of $\varepsilon_{\mathrm{a}}^{\mathrm{NL}}$ according to equation (5.8).

The differences between the second-order SPFT estimates of the HCM permittivity and the corresponding Bruggeman estimates are plotted in figure 6.25. This figure contains the real and imaginary parts of $\varepsilon_{\mathrm{SPFT}}^{[2],\mathrm{L}} - \varepsilon_{\mathrm{Br}}^{\mathrm{L}}$ (normalized relative to ε_0) and $\varepsilon_{\mathrm{SPFT}}^{[2],\mathrm{NL}} - \varepsilon_{\mathrm{Br}}^{\mathrm{NL}}$ (normalized relative to $\varepsilon_{\mathrm{a}}^{\mathrm{NL}}$) plotted against f_{a} and $k_0 L_{\mathrm{cor}}$, for $\varepsilon_{\mathrm{a}}^{\mathrm{L}} = 2\varepsilon_0$ and $\varepsilon_{\mathrm{b}} = 9\varepsilon_0$. Both the real and imaginary parts of $\varepsilon_{\mathrm{SPFT}}^{[2],\mathrm{L}}$ are larger than the real and imaginary parts of $\varepsilon_{\mathrm{Br}}^{\mathrm{L}}$, respectively, for all values of L_{cor} considered and all possible values of f_{a}. Both the real and imaginary parts of the difference $\varepsilon_{\mathrm{SPFT}}^{[2],\mathrm{L}} - \varepsilon_{\mathrm{Br}}^{\mathrm{L}}$ increase as L_{cor} increases, and have maximum values in

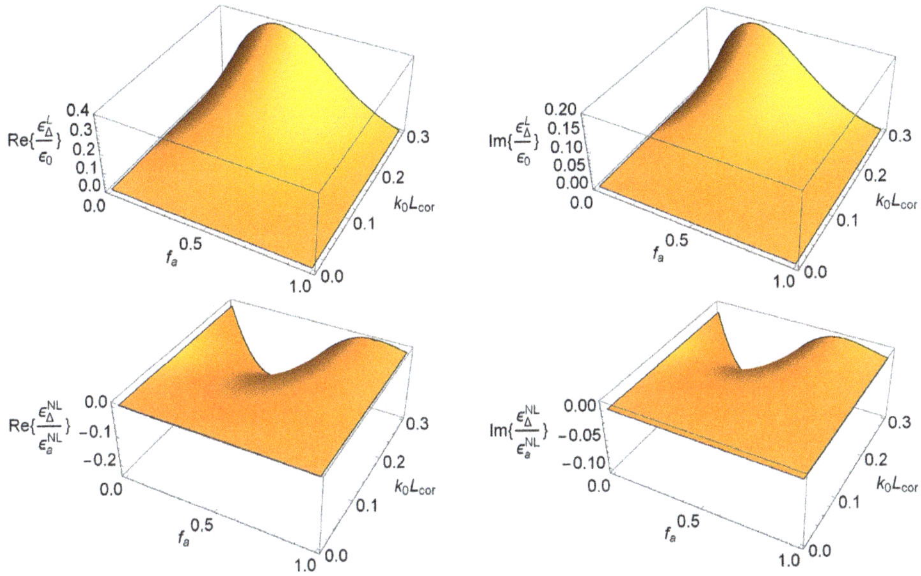

Figure 6.25. Real and imaginary parts of $\varepsilon_\Delta^{\rm L} = \varepsilon_{\rm SPFT}^{[2],\rm L} - \varepsilon_{\rm Br}^{\rm L}$ (normalized relative to ε_0) and $\varepsilon_\Delta^{\rm NL} = \varepsilon_{\rm SPFT}^{[2],\rm NL} - \varepsilon_{\rm Br}^{\rm NL}$ (normalized relative to $\varepsilon_{\rm a}^{\rm NL}$) plotted against $f_{\rm a}$ and $k_0 L_{\rm cor}$, for $\varepsilon_{\rm a}^{\rm L} = 2\varepsilon_0$ and $\varepsilon_{\rm b} = 9\varepsilon_0$.

the vicinity of $f_{\rm a} \approx 0.3$. The real and imaginary parts of $\varepsilon_{\rm SPFT}^{[2],\rm NL}$ are generally smaller than the real and imaginary parts of $\varepsilon_{\rm Br}^{\rm NL}$, respectively, across the ranges of $L_{\rm cor}$ and $f_{\rm a}$ considered. Both the real and imaginary parts of the difference $\varepsilon_{\rm SPFT}^{[2],\rm L} - \varepsilon_{\rm Br}^{\rm L}$ generally decrease as $L_{\rm cor}$ increases, and have minimum values in the vicinity of $f_{\rm a} \approx 0.3$. Hence, the degree of nonlinearity enhancement estimated using the second-order SPFT is slightly less than that estimated using the Bruggeman formalism, and the degree of nonlinearity enhancement diminishes slightly as the correlation length increases. Indeed, a similar diminishment of nonlinearity enhancement with respect to increasing correlation length has been observed for isotropic chiral [132] and anisotropic dielectric [134] HCMs. Furthermore, an analogous diminishment of nonlinearity enhancement with respect to increasing constituent particle size has been observed using an extended version of the SPFT [135]. Additionally, the difference $(\varepsilon_{\rm SPFT}^{[2],\rm L} - \varepsilon_{\rm Br}^{\rm L})$ is independent of $\varepsilon_{\rm a}^{\rm NL}$ since the nonlinear permittivity does not feature in the linear second-order SPFT equation (5.13), and the difference $(\varepsilon_{\rm SPFT}^{[2],\rm L} - \varepsilon_{\rm Br}^{\rm L})$ is independent of $\varepsilon_{\rm a}^{\rm NL}$ according to equation (5.14).

References

[1] Lakhtakia A 1996 Cakes and pastries, and linear optical composite materials too *Selected Papers on Linear Optical Composite Materials* ed A Lakhtakia (Bellingham, WA: SPIE Optical Engineering Press) pp xiii–xxiv

[2] Lakhtakia A and Mackay T G 2007 Meet the metamaterials *Opt. Photon. News* **18** 32–9

[3] Lakhtakia A 1998 Incremental Maxwell Garnett formalism for homogenizing particulate composite media *Microw. Opt. Technol. Lett.* **17** 276–9

[4] Michel B, Lakhtakia A, Weiglhofer W S and Mackay T G 2001 Incremental and differential Maxwell Garnett formalisms for bi-anisotropic composites *Compos. Sci. Technol.* **61** 13–8
[5] Bragg W L and Pippard A B 1953 The form birefringence of macromolecules *Acta Crystallogr.* **6** 865–7
[6] Sherwin J A and Lakhtakia A 2002 Bragg–Pippard formalism for bianisotropic particulate composites *Microw. Opt. Technol. Lett.* **33** 40–4
[7] Michel B and Lakhtakia A 1995 Strong-property-fluctuation theory for homogenizing chiral particulate composites *Phys. Rev. E* **51** 5701–7
[8] Mackay T G 2008 On extended homogenization formalisms for nanocomposites *J. Nanophoton.* **2** 021850
[9] Mackay T G and Weiglhofer W S 2000 Homogenization of biaxial composite materials: dissipative anisotropic properties *J. Opt. A: Pure Appl. Opt.* **2** 426–32
[10] Mackay T G and Weiglhofer W S 2001 Homogenization of biaxial composite materials: nondissipative dielectric properties *Electromagnetics* **21** 15–26
[11] Mackay T G and Weiglhofer W S 2001 Homogenization of biaxial composite materials: bianisotropic properties *J. Opt. A: Pure Appl. Opt.* **3** 45–52
[12] Engheta N, Jaggard D L and Kowarz M W 1992 Electromagnetic waves in Faraday chiral media *IEEE Trans. Antennas Propagat.* **40** 367–74
[13] Lakhtakia A 1994 *Beltrami Fields in Chiral Media* (Singapore: World Scientific)
[14] Chen H C 1983 *Theory of Electromagnetic Waves: A Coordinate-Free Approach* (New York: McGraw-Hill)
[15] Lax B and Button K J 1962 *Microwave Ferrites and Ferrimagnetics* (New York: McGraw-Hill)
[16] Collin R E 1996 *Foundations for Microwave Engineering* (New York: McGraw-Hill)
[17] Weiglhofer W S, Lakhtakia A and Michel B 1998 On the constitutive parameters of a chiroferrite composite medium *Microw. Opt. Technol. Lett.* **18** 342–5
[18] Weiglhofer W S and Mackay T G 2000 Numerical studies of the constitutive parameters of a chiroplasma composite medium *Arch. Elektron. Übertrag.* **54** 259–65
[19] Mackay T G, Lakhtakia A and Weiglhofer W S 2001 Ellipsoidal topology, orientation diversity and correlation length in bianisotropic composite mediums *Arch. Elektron. Übertrag.* **55** 243–51
[20] Peng L, Ran L and Mortensen N A 2010 Achieving anisotropy in metamaterials made of dielectric cylindrical rods *Appl. Phys. Lett.* **96** 241108
[21] Mackay T G 2015 Towards metamaterials with giant dielectric anisotropy via homogenization: An analytical study *Photon. Nanostruct.: Fundam. Appl.* **13** 8–19
[22] Givargizov E I 1987 *Highly Anisotropic Crystals* (New York: Springer)
[23] Wakaki M 2012 *Optical Materials and Applications* (London: CRC Press)
[24] Jiang W X, Chin J Y and Cui T J 2009 Anisotropic metamaterial devices *Mater. Today* **12** 26–33
[25] Liu X, Li C, Yao K, Meng X and Li F 2009 Invisibility cloaks modeled by anisotropic metamaterials based on inductor-capacitor networks *IEEE Antennas Wireless Propagat. Lett.* **8** 1154–7
[26] Anderson T H, Mackay T G and Lakhtakia A 2012 Toward a cylindrical cloak via inverse homogenization *J. Opt. Soc. Am. A* **29** 239–43

[27] Mirala A and Abdolali A 2013 A zero reflection from layered anisotropic metamaterial structures *Waves Random Complex Media* **23** 152–68

[28] Cheng Q, Jiang W X and Cui T J 2012 Spatial power combination for omnidirectional radiation via anisotropic metamaterials *Phys. Rev. Lett.* **108** 213903

[29] Podolskiy V A and Narimanov E E 2005 Strongly anisotropic waveguide as a nonmagnetic left-handed system *Phys. Rev. B* **71** 201101R

[30] Smolyaninov I I 2003 Surface plasmon toy model of a rotating black hole *New. J. Phys.* **5** 147

[31] Lu W, Jin J, Lin Z and Chen H 2010 A simple design of an artificial electromagnetic black hole *J. Appl. Phys.* **108** 064517

[32] Mackay T G and Lakhtakia A 2011 Towards a realization of Schwarzschild-(anti-)de Sitter spacetime as a particulate metamaterial *Phys. Rev. B* **83** 195424

[33] Mackay T G and Lakhtakia A 2010 Towards a metamaterial simulation of a spinning cosmic string *Phys. Lett. A* **374** 2305–8

[34] Mackay T G and Lakhtakia A 2014 Towards a piecewise-homogeneous metamaterial model of the collision of two linearly polarized gravitational plane waves *IEEE Trans. Antennas Propagat.* **62** 6149–54

[35] Mackay T G and Lakhtakia A 2012 Towards an experimental realization of affinely transformed linearized quantum electrodynamics vacuum via inverse homogenization *J. Opt. Soc. America B* **29** 1680–4

[36] Shin J, Shen J-T and Fan S 2009 Three-dimensional metamaterials with an ultrahigh effective refractive index over a broad bandwidth *Phys. Rev. Lett.* **102** 093903

[37] Navarro-Cia M, Beruete M, Campillo I and Sorolla M 2011 Enhanced lens by ϵ and μ near-zero metamaterial boosted by extraordinary optical transmission *Phys. Rev. B* **83** 115112

[38] Born M and Wolf E 1999 *Principles of Optics* 7th (expanded) edn (Cambridge: Cambridge University Press)

[39] Altman C and Suchy K 1991 *Reciprocity, Spatial Mapping and Time Reversal in Electromagnetics* (Dordrecht: Kluwer Academic Publishers)

[40] Mackay T G and A Lakhtakia A 2019 *Electromagnetic Anisotropy and Bianisotropy: A Field Guide* 2nd edn (Singapore: World Scientific) https://doi.org/10.1142/7515

[41] Lakhtakia A, McCall M W and Weiglhofer W S 2003 Negative phase-velocity mediums *Introduction to Complex Mediums for Optics and Electromagnetics* ed W S Weiglhofer and A Lakhtakia (Bellingham, WA: SPIE Press) pp 347–63

[42] Ramakrishna S A 2005 Physics of negative refractive index materials *Rep. Prog. Phys.* **68** 449–521

[43] Dolling G, Wegener M, Soukoulis C M and Linden S 2007 Negative-index metamaterial at 780 nm wavelength *Opt. Lett.* **32** 53–5

[44] Shalaev V M 2007 Optical negative-index metamaterials *Nature Photon.* **1** 41–8

[45] Noginov M A 2008 Compensation of surface plasmon loss by gain in dielectric medium *J. Nanophoton.* **2** 021855

[46] Bratkovsky A, Ponizovskaya E, Wang S-Y, Holmström P, Thylén L, Fu Y and Ågren H 2008 A metal-wire/quantum-dot composite metamaterial with negative ε and compensated optical loss *Appl. Phys. Lett.* **93** 193106

[47] Mackay T G and Lakhtakia A 2005 Negative phase velocity in isotropic dielectric-magnetic media via homogenization *Microw. Opt. Technol. Lett.* **47** 313–5

[48] Lakhtakia A and Mackay T G 2006 Negative phase velocity in isotropic dielectric–magnetic media via homogenization: Part II *Microw. Opt. Technol. Lett.* **48** 709–12
[49] Mackay T G and Lakhtakia A 2006 Correlation length and negative phase velocity in isotropic dielectric–magnetic materials *J. Appl. Phys.* **100** 063533
[50] Hu L and Lin Z 2003 Imaging properties of uniaxially anisotropic negative refractive index materials *Phys. Lett. A* **313** 316–24
[51] Mackay T G and Lakhtakia A 2004 Plane wave with negative phase velocity in Faraday chiral mediums *Phys. Rev. E* **69** 026602
[52] McCall M W, Lakhtakia A and Weiglhofer W S 2002 The negative index of refraction demystified *Eur. J. Phys.* **23** 353–9
[53] Mackay T G and Lakhtakia A 2009 Negative refraction, negative phase velocity and counterposition in bianisotropic materials and metamaterials *Phys. Rev. B* **79** 235121
[54] Belov P A 2003 Backward waves and negative refraction in uniaxial dielectrics with negative dielectric permittivity along the anisotropy axis *Microw. Opt. Technol. Lett.* **37** 259–63
[55] Lakhtakia A, Mackay T G and Geddes J B III 2009 On the inapplicability of a negative-phase-velocity condition as a negative-refraction condition for active materials *Microw. Opt. Technol. Lett.* **51** 1230
[56] Kinsler P 2010 Comment on: On the inapplicability of a negative-phase-velocity condition as a negative refraction condition for active materials *Microw. Opt. Technol. Lett.* **52** 247
[57] Lakhtakia A, Mackay T G and Geddes J B III 2010 Response to 'Comment on: On the inapplicability of a negative-phase-velocity condition as a negative-refraction condition for active materials' *Microw. Opt. Technol. Lett.* **52** 1681
[58] Zhang C and Cui T J 2007 Negative reflections of electromagnetic waves in a strong chiral medium *Appl. Phys. Lett.* **91** 194101
[59] Faryad M and Naqvi Q A 2008 Cylindrical reflector in chiral medium supporting simultaneously positive phase velocity and negative phase velocity *J. Electromag. Waves Applic.* **22** 563–72
[60] Wang Y, Zha X and Yan J 2005 Reflection and refraction of light at the interface of a uniaxial bicrystal *Europhys. Lett.* **72** 830–6
[61] Mackay T G and Lakhtakia A 2008 Negative reflection in a Faraday chiral medium *Microw. Opt. Technol. Lett.* **50** 1368–71
[62] Pimenov A, Loidl A, Gehrke K, Moshnyaga V and Samwer K 2007 Negative refraction observed in a metallic ferromagnet in the gigahertz frequency range *Phys. Rev. Lett.* **98** 197401
[63] Mackay T G and Lakhtakia A 2009 Negative refraction and positive refraction are not Lorentz covariant *Phys. Lett. A* **374** 101–5
[64] Mackay T G and Lakhtakia A 2007 Concealment by uniform motion *J. Eur. Opt. Soc. Rapid. Pub.* **2** 07003
[65] Mackay T G and Lakhtakia A 2004 Negative phase velocity in a uniformly moving, homogeneous, isotropic, dielectric–magnetic medium *J. Phys. A: Math. Gen.* **37** 5697–711
[66] Lakhtakia A and Mackay T G 2004 Towards gravitationally assisted negative refraction of light by vacuum *J. Phys. A: Math. Gen.* **37** L505–10
Lakhtakia A and Mackay T G 2004 Towards gravitationally assisted negative refraction of light by vacuum *J. Phys. A: Math. Gen.* **37** 12093 (corrigendum)

[67] Mackay T G, Lakhtakia A and Setiawan S 2005 Global and local perspectives of gravitationally assisted negative-phase-velocity propagation of electromagnetic waves in vacuum *Phys. Lett. A* **336** 89–96
[68] Plébanski J 1960 Electromagnetic waves in gravitational fields *Phys. Rev.* **118** 1396–408
[69] Schleich W and Scully M O 1984 General relativity and modern optics *New Trends in Atomic Physics* ed G Grynberg and R Stora (Amsterdam: Elsevier) pp 995–1124
[70] Depine R A and Lakhtakia A 2005 Diffraction by a grating made of a uniaxial dielectric–magnetic medium exhibiting negative refraction *New J. Phys.* **7** 158
[71] Li G X, Tam H L, Wang F Y and Cheah K W 2007 Superlens from complementary anisotropic metamaterials *J. Appl. Phys.* **102** 116101
[72] Poddubny A, Iorsh I, Belov P and Kivshar Y 2013 Hyperbolic metamaterials *Nature Photon.* **7** 958–67
[73] Guo Y, Cortes C L, Molesky S and Jacob Z 2012 Broadband super-Planckian thermal emission from hyperbolic metamaterials *Appl. Phys. Lett.* **101** 131106
[74] Guo Y and Jacob Z 2013 Thermal hyperbolic metamaterials *Opt. Express* **21** 15014–9
[75] Smith D R and Schurig D 2003 Electromagnetic wave propagation in media with indefinite permittivity and permeability tensors *Phys. Rev. Lett.* **90** 077405
[76] Smith D R, Kolinko P and Schurig D 2004 Negative refraction in indefinite media *J. Opt. Soc. Am. B* **21** 1032–43
[77] Mackay T G, Lakhtakia A and Depine R A 2006 Uniaxial dielectric media with hyperbolic dispersion relations *Microw. Opt. Technol. Lett.* **48** 363–7
[78] Mackay T G and Lakhtakia A 2004 A limitation of the Bruggeman formalism for homogenization *Opt. Commun.* **234** 35–42
Mackay T G and Lakhtakia A 2009 Erratum to: A limitation of the Bruggeman formalism for homogenization *Opt. Commun.* **282** 4028
[79] Mackay T G 2007 On the effective permittivity of silver–insulator nanocomposites *J. Nanophoton.* **1** 019501
[80] Voigt W 1902 On the behaviour of pleochroitic crystals along directions in the neighbourhood of an optic axis *Phil. Mag.* **4** 90–7
[81] Pancharatnam S 1958 The optical interference figures of amethystine quartz—Part II *Proc. Ind. Acad. Sci. A* **47** 210–29
[82] Fedorov F I and Goncharenko A M 1963 Propagation of light along the circular optical axes of absorbing crystals *Opt. Spectrosc. (USSR)* **14** 51–3
[83] Ranganath G S 1994 Optics of absorbing anisotropic media *Curr. Sci. (India)* **67** 231–7
[84] Gerardin J and Lakhtakia A 2001 Conditions for Voigt wave propagation in linear, homogeneous, dielectric mediums *Optik.* **112** 493–5
[85] Mackay T G and Lakhtakia A 2008 Electromagnetic fields in linear bianisotropic mediums *Prog. Optics* **51** 121–209
[86] Mackay T G and Lakhtakia A 2016 On the propagation of Voigt waves in energetically active materials *Eur. J. Phys.* **37** 064002
[87] Lakhtakia A 1998 Anomalous axial propagation in helicoidal bianisotropic media *Opt. Commun.* **157** 193–201
[88] Berry M V 2005 The optical singularities of bianisotropic crystals *Proc. R. Soc. Lond. A* **461** 2071–98
[89] Mackay T G and Lakhtakia A 2003 Voigt wave propagation in biaxial composite materials *J. Opt. A: Pure Appl. Opt.* **5** 91–5

[90] Mackay T G and Lakhtakia A 2004 Correlation length facilitates Voigt wave propagation *Waves Random Media* **14** L1–11
[91] Mackay T G 2011 Voigt waves in homogenized particulate composites based on isotropic dielectric components *J. Opt. (United Kingdom)* **13** 105702
[92] Mackay T G 2014 On the sensitivity of directions that support Voigt wave propagation in infiltrated biaxial dielectric materials *J. Nanophoton.* **8** 083993
[93] Mackay T G 2015 Controlling Voigt waves by the Pockels effect *J. Nanophoton.* **9** 093599
[94] Sølna K and Milton G W 2002 Can mixing materials make electromagnetic signals travel faster? *SIAM J. Appl. Math.* **62** 2064–91
[95] Mackay T G and Lakhtakia A 2004 Enhanced group velocity in metamaterials *J. Phys. A: Math. Gen.* **37** L19–24
[96] Mackay T G and Lakhtakia A 2005 Anisotropic enhancement of group velocity in a homogenized dielectric composite medium *J. Opt. A: Pure Appl. Opt.* **7** 669–74
[97] Gao L 2006 Decreased group velocity in compositionally graded films *Phys. Rev. E* **73** 036602
[98] Lakhtakia A and Messier R 2005 *Sculptured Thin Films: Nanoengineered Morphology and Optics* (Bellingham, WA: SPIE Press)
[99] Hodgkinson I J and Wu Q-h 1997 *Birefringent Thin Films and Polarizing Elements* (Singapore: World Scientific)
[100] Smith G B 1989 Effective medium theory and angular dispersion of optical constants in films with oblique columnar structure *Opt. Commun.* **71** 279–84
[101] Wang H 1995 Assessment of optical constants of multilayer thin films with columnar-structure-induced anisotropy *J. Phys. D: Appl. Phys.* **28** 571–5
[102] Lakhtakia A and Weiglhofer W S 1999 Significance of cross-sectional morphology for Motohiro–Taga interfaces *Optik* **110** 33–6
[103] Ward L 2000 *The Optical Constants of Bulk Materials and Films* 2nd edn (Bristol: Institute of Physics)
[104] Aste T 1996 Circle, sphere, and drop packings *Phys. Rev. E* **53** 2571–9
[105] Sherwin J A, Lakhtakia A and Hodgkinson I J 2002 On calibration of a nominal structure-property relationship model for chiral sculptured thin films by axial transmittance measurements *Opt. Commun.* **209** 369–75
[106] Hodgkinson I, Wu Q h and Hazel J 1998 Empirical equations for the principal refractive indices and column angle of obliquely deposited films of tantalum oxide, titanium oxide, and zirconium oxide *Appl. Opt.* **37** 2653–9
[107] Schmidt D, Schubert E and Schubert M 2013 Generalized ellipsometry characterization of sculptured thin films made by glancing angle deposition *Ellipsometry at the Nanoscale* ed M Losurdo and K Hingerl (Berlin: Springer) pp 341–410
[108] Mackay T G and Lakhtakia A 2010 Determination of constitutive and morphological parameters of columnar thin films by inverse homogenization *J. Nanophoton.* **4** 041535
[109] Mackay T G and Lakhtakia A 2010 Modeling columnar thin films as platforms for surface-plasmonic-polaritonic optical sensing *Photon. Nanostruct.: Fundam. Appl.* **8** 140–9
[110] Jamaian S S and Mackay T G 2012 On columnar thin films as platforms for surface-plasmonic-polaritonic optical sensing: higher-order considerations *Opt. Commun.* **285** 5535–42
[111] Mackay T G and Lakhtakia A 2012 Modeling chiral sculptured thin films as platforms for surface-plasmonic-polaritonic optical sensing *IEEE Sensors J.* **12** 273–80

[112] Lakhtakia A and Faryad M 2014 Theory of optical sensing with Dyakonov–Tamm waves *J. Nanophoton.* **8** 083072

[113] Abbas F, Lakhtakia A, Naqvi Q A and Faryad M 2015 An optical-sensing modality that exploits Dyakonov–Tamm waves *Photon. Res.* **3** 5–8

[114] Sancho-Parramon J, Janicki V and Zorc H 2010 On the dielectric function tuning of random metal-dielectric nanocomposites for metamaterial applications *Opt. Express* **18** 26915–28

[115] Chen W, Thoreson M D, Kildishev A V and Shalaev V M 2010 Fabrication and optical characterizations of smooth silver–silica nanocomposite films *Laser Phys. Lett.* **7** 677–84

[116] Sundari S T, Chandra S and Tyagi A K 2013 Temperature dependent optical properties of silver from spectroscopic ellipsometry and density functional theory calculations *J. Appl. Phys.* **114** 033515

[117] Mackay T G and Lakhtakia A 2009 On the application of homogenization formalisms to active dielectric composite materials *Opt. Commun.* **282** 2470–5

[118] Jamaian S S and Mackay T G 2010 On limitations of the Bruggeman formalism for inverse homogenization *J. Nanophoton.* **4** 043510

[119] Bergman D J 1981 Bounds for the complex dielectric constant of a two-component material *Phys. Rev. B* **23** 3058–65

[120] Milton G W 1981 Bounds on the complex permittivity of a two-component composite material *J. Appl. Phys.* **52** 5286–93

[121] Bergman D J 1980 Exactly solvable microscopic geometries and rigorous bounds for the complex dielectric constant of a two-component composite material *Phys. Rev. Lett.* **44** 1285–7

Bergman D J 1980 Exactly solvable microscopic geometries and rigorous bounds for the complex dielectric constant of a two-component composite material *Phys. Rev. Lett.* **45** 148 (correction)

[122] Duncan A J, Mackay T G and Lakhtakia A 2007 On the Bergman–Milton bounds for the homogenization of dielectric composite materials *Opt. Commun.* **271** 470–4

[123] Mackay T G and Lakhtakia A 2016 Gain and loss enhancement in active and passive particulate composite materials *Waves Random Complex Media* **26** 553–63

[124] Mackay T G and Lakhtakia A 2015 Dynamically controllable anisotropic metamaterials with simultaneous attenuation and amplification *Phys. Rev. A* **92** 053847

[125] Sun L, Yang X and Gao J 2013 Loss-compensated broadband epsilon-near-zero metamaterials with gain media *Appl. Phys. Lett.* **103** 201109

[126] Walser R M 2003 Metamaterials: an introduction *Introduction to Complex Mediums for Optics and Electromagnetics* ed W S Weiglhofer and A Lakhtakia (Bellingham, WA: SPIE Press) pp 295–316

[127] Mackay T G 2005 Linear and nonlinear homogenized composite mediums as metamaterials *Electromagnetics* **25** 461–81

[128] Liao H B, Xiao R F, Wang H, Wong K S and Wong G K L 1998 Large third-order optical nonlinearity in Au:TiO_2 composite films measured on a femtosecond time scale *Appl. Phys. Lett.* **72** 1817–9

[129] Boyd R W, Gehr R J, Fischer G L and Sipe J E 1996 Nonlinear optical properties of nanocomposite materials *Pure Appl. Opt.* **5** 505–12

[130] Lakhtakia A 2001 Application of strong permittivity fluctuation theory for isotropic, cubically nonlinear, composite mediums *Opt. Commun.* **192** 145–51

[131] Mackay T G, Lakhtakia A and Weiglhofer W S 2002 Homogenisation of isotropic, cubically nonlinear, composite mediums by the strong-permittivity-fluctuation theory: third-order considerations *Opt. Commun.* **204** 219–28
[132] Mackay T G, Lakhtakia A and Weiglhofer W S 2003 The strong-property-fluctuation theory for cubically nonlinear, isotropic chiral composite mediums *Electromagnetics* **23** 455–79
[133] Lakhtakia M N and Lakhtakia A 2001 Anisotropic composite materials with intensity-dependent permittivity tensor: the Bruggeman approach *Electromagnetics* **21** 129–38
[134] Mackay T G 2003 Geometrically derived anisotropy in cubically nonlinear dielectric composites *J. Phys. D: Appl. Phys.* **36** 583–91
[135] Cui J and Mackay T G 2007 Depolarization regions of nonzero volume for anisotropic, cubically nonlinear, homogenized nanocomposites *J. Nanophoton.* **1** 013506

Tom G Mackay and Akhlesh Lakhtakia

Chapter 7

Epilogue

A bird's-eye view on the state-of-the-art in analytical electromagnetic homogenization of particulate composite materials has been provided in the previous chapters of this ebook. Homogenization formalisms applicable to both linear and weakly nonlinear materials have been presented in their most general forms The formalisms named after Maxwell Garnett and Bruggeman can generally accommodate bianisotropic constituent materials that are randomly distributed as electrically small ellipsoidal particles. Furthermore, higher-order distributional statistics can be accommodated by the strong-property-fluctuation-theory (SPFT) formalism, and the actual sizes of the electrically small constituent particles can be accommodated by extended versions of the Maxwell Garnett, Bruggeman, and SPFT formalisms. While the standard versions of the Maxwell Garnett and Bruggeman formalisms, relevant to isotropic dielectric composite materials containing spherical constituent particles, have been available since the early 20th century [1], the emergence of homogenization formalisms for constitutively more complex materials followed on the heels of the establishment of depolarization dyadics appropriate to general anisotropic and bianisotropic materials in the late 1990s [2, 3].

The emergence of homogenization formalisms suitable for bianisotropic materials has coincided with a period of rapid advancement in the field of engineered and nanoengineered composite materials, both theoretically and experimentally. For example, the development in the early 21st century of engineered composite materials that can exhibit negative refraction spawned an explosion of research activity, motivated in part by the prospects of achieving sub-wavelength imaging and planar lenses [4], as outlined in section 6.6.2.

The scope for engineered composite materials with judiciously tailored properties continues to expand. An excellent example is provided by material analogs for the electromagnetic properties of curved spacetime [5], as described in section 6.7.2. Typically, materials with highly complex anisotropy and even bianisotropy are

doi:10.1088/978-0-7503-3423-5ch7

needed to simulate spacetimes of interest. In particular, high degrees of anisotropy and/or bianisotropy are needed (beyond those degrees of anisotropy and/or biansotropy exhibited by naturally occurring materials) to represent regions in the vicinities of spacetime singularities and event horizons. Homogenization formalisms can play a vital role in the design of appropriate material analogs, with highly elongated constituent particles being used to achieve the required high degrees of anisotropy and/or bianisotropy [6]. A major challenge in the case of composite-material analogs of curved spacetime—which also applies in the case of engineered composite materials that exhibit negative refraction—is the elimination of dissipation. These composite materials generally require metallic (or similar) constituent materials, and metals are generally dissipative at the wavelengths of interest. The use of metals such as silver at visible wavelengths helps to minimize dissipation, but it may be that active constituent materials will provide a more satisfactory solution [7].

While opportunities for homogenization formalisms abound, especially in exotic scenarios such as those involving negative refraction and analogs of curved spacetime, caution must always be exercised to ensure physical reasonableness. As discussed in section 6.8, there are limitations to the constitutive parameter space that is appropriate for certain homogenization formalisms [8, 9]. Ultimately, one must be mindful that homogenization formalisms yield *estimates* of effective constitutive parameters and checks with careful experimental measurements should be sought wherever practicable.

References

[1] Lakhtakia A (ed) 1996 *Selected Papers on Linear Optical Composite Materials* (SPIE Optical Engineering Press: Bellingham, WA)

[2] Michel B 1997 A Fourier space approach to the pointwise singularity of an anisotropic dielectric medium *Int. J. Appl. Electromagn. Mech.* **8** 219–27

[3] Michel B and Weiglhofer W S 1997 Pointwise singularity of dyadic Green function in a general bianisotropic medium *Arch. Elektron. Übertrag.* **51** 219–23
Michel B and Weiglhofer W S 1998 Pointwise singularity of dyadic Green function in a general bianisotropic medium *Arch. Elektron. Übertrag.* **52** 310 (erratum)

[4] Ramakrishna S A 2005 Physics of negative refractive index materials *Rep. Prog. Phys.* **68** 449–521

[5] Mackay T G and Lakhtakia A 2018 Toward the construction of parts of the universe on tabletops *The World of Applied Electromagnetics: In Appreciation of Magdy Fahmy Iskander* ed A Lakhtakia and C M Furse (Cham: Springer) pp 631–54

[6] Mackay T G 2015 Towards metamaterials with giant dielectric anisotropy via homogenization: An analytical study *Photon. Nanostruct.: Fundam. Appl.* **13** 8–19

[7] Noginov M A 2008 Compensation of surface plasmon loss by gain in dielectric medium *J. Nanophoton.* **2** 021855

[8] Mackay T G and Lakhtakia A 2004 A limitation of the Bruggeman formalism for homogenization *Opt. Commun.* **234** 35–42
Mackay T G and Lakhtakia A 2009 Erratum to: A limitation of the Bruggeman formalism for homogenization *Opt. Commun.* **282** 4028

[9] Mackay T G 2007 On the effective permittivity of silver-insulator nanocomposites *J. Nanophoton.* **1** 019501

IOP Publishing

Modern Analytical Electromagnetic Homogenization with Mathematica® (Second Edition)

Tom G Mackay and Akhlesh Lakhtakia

Appendix A

3 × 3 Dyadics

A 3 × 3 dyad is composed of two 3-vectors:

$$\underline{\underline{d}} = \mathbf{a}\,\mathbf{b}. \tag{A.1}$$

While $\underline{\underline{d}} \bullet \mathbf{p} = \mathbf{a}\,(\mathbf{b} \bullet \mathbf{p})$ and $\mathbf{p} \bullet \underline{\underline{d}} = (\mathbf{p} \bullet \mathbf{a})\,\mathbf{b}$ are vectors, $\underline{\underline{d}} \times \mathbf{p} = \mathbf{a}(\mathbf{b} \times \mathbf{p})$ and $\mathbf{p} \times \underline{\underline{d}} = (\mathbf{p} \times \mathbf{a})\mathbf{b}$ are dyads. The transpose of a dyad $\mathbf{a}\,\mathbf{b}$ is the dyad $\mathbf{b}\,\mathbf{a}$. Sometimes, dyads are called bivectors. Some authors prefer to write the dyad $\mathbf{a}\,\mathbf{b}$ as $\mathbf{b} \otimes \mathbf{a}$.

A 3 × 3 dyadic $\underline{\underline{M}}$ is a linear mapping from $\mathbf{a}$ to $\mathbf{b}$:

$$\mathbf{b} = \underline{\underline{M}} \bullet \mathbf{a}. \tag{A.2}$$

The 3 × 3 identity dyadic $\underline{\underline{I}}$ is such that $\mathbf{a} \bullet \underline{\underline{I}} = \underline{\underline{I}} \bullet \mathbf{a} = \mathbf{a}$; likewise, the 3 × 3 null dyadic $\underline{\underline{0}}$ is defined so that $\mathbf{a} \bullet \underline{\underline{0}} = \underline{\underline{0}} \bullet \mathbf{a}$ equals the null vector $\mathbf{0}$. The general representation of a dyadic is as a sum of dyads; i.e.,

$$\underline{\underline{M}} = \sum_{\ell=1,2,\ldots} \underline{\underline{d}}_{\,\ell} = \sum_{\ell=1,2,\ldots} M_\ell\, \hat{\mathbf{u}}_\ell \hat{\mathbf{v}}_\ell, \tag{A.3}$$

where M_ℓ are some scalar coefficients, while $\hat{\mathbf{u}}_\ell$ and $\hat{\mathbf{v}}_\ell$ are vectors of unit magnitude.

All 3-vectors can be written using matrix notation. The vector $\mathbf{a} = \hat{\mathbf{u}}_x a_x + \hat{\mathbf{u}}_y\, a_y + \hat{\mathbf{u}}_z\, a_z$ in a Cartesian coordinate system is equivalent to the column 3-vector

$$\mathbf{a} \equiv \begin{bmatrix} a_x \\ a_y \\ a_z \end{bmatrix}, \tag{A.4}$$

and likewise for the vector $\mathbf{b}$. Thus, the dyad $\underline{\underline{d}} = \mathbf{a}\,\mathbf{b}$ is equivalent to the 3 × 3 matrix

$$\underline{\underline{d}} \equiv \begin{bmatrix} a_x b_x & a_x b_y & a_x b_z \\ a_y b_x & a_y b_y & a_y b_z \\ a_z b_x & a_z b_y & a_z b_z \end{bmatrix}. \tag{A.5}$$

Hence, dyadics in electromagnetics can be written as 3 × 3 matrixes. The identity dyadic $\underline{\underline{I}}$ is equivalent to the 3 × 3 identity matrix, and the null dyadic $\underline{\underline{0}}$ to the 3 × 3 null matrix. The usual algebra of matrixes can thus be used for dyadics as well.

The trace of a dyadic is the sum of the diagonal elements in its matrix representation. Likewise, the determinant of a dyadic is the same as the determinant of its equivalent matrix. A dyadic can be transposed in the same way as a matrix. A dyadic can be inverted if it is nonsingular (i.e., if its determinant is not equal to zero).

The antisymmetric dyadic

$$\mathbf{a} \times \underline{\underline{I}} = \underline{\underline{I}} \times \mathbf{a}, \tag{A.6}$$

$$= a_x\left(\hat{\mathbf{u}}_z\hat{\mathbf{u}}_y - \hat{\mathbf{u}}_y\hat{\mathbf{u}}_z\right) + a_y(\hat{\mathbf{u}}_x\hat{\mathbf{u}}_z - \hat{\mathbf{u}}_z\hat{\mathbf{u}}_x) + a_z\left(\hat{\mathbf{u}}_y\hat{\mathbf{u}}_x - \hat{\mathbf{u}}_x\hat{\mathbf{u}}_y\right), \tag{A.7}$$

$$\equiv \begin{bmatrix} 0 & -a_z & a_y \\ a_z & 0 & -a_x \\ -a_y & a_x & 0 \end{bmatrix} \tag{A.8}$$

is often useful to denote gyrotropic constitutive properties that are characteristic of ferrites and plasmas. The simplest antisymmetric dyadic is $\hat{\mathbf{u}} \times \underline{\underline{I}}$, where $\hat{\mathbf{u}}$ is any vector of unit magnitude. The trace of any antisymmetric dyadic is zero.

Even vector differential operators can be thought of as 3 × 3 dyadics. Thus, the *curl* operator is written as $\nabla \times \underline{\underline{I}}$ and the *divergence* operator as $\nabla \bullet \underline{\underline{I}}$, with

$$\nabla = \hat{\mathbf{u}}_x\frac{\partial}{\partial x} + \hat{\mathbf{u}}_y\frac{\partial}{\partial y} + \hat{\mathbf{u}}_z\frac{\partial}{\partial z} \tag{A.9}$$

written as a vector in a Cartesian coordinate system.

Tom G Mackay and Akhlesh Lakhtakia

Appendix B

Mathematica™ codes

The Mathematica™ codes used to generate the numerical results presented in chapter 6 are provided in this appendix.

B.1 Figure 6.1

Codes for the Maxwell Garnett, incremental Maxwell Garnett, differential Maxwell Garnett, and Bruggeman formalisms in the case of a uniaxial dielectric HCM, as used for figure 6.1, follow. Straightforward modifications of the Bruggeman code were used for figures 6.20, 6.21, and 6.22, and straightforward modifications of the Maxwell Garnett code were used for figure 6.23.

```
(* Inputs *)

I33 = IdentityMatrix[3];

ϵa = (12 + 4I)I33;

ϵb = DiagonalMatrix[{6 + I, 6 + I, 2.5 + 0.5I}];

Ua = DiagonalMatrix[{1, 1, 5}] / ∛5 ;
```



```
(* Depolarization dyadic *)
depol[ϵhost_, Uinc_]:=Module[{tϵhost, γ, d11, d33},
tϵhost = Inverse[Uinc].ϵhost.Inverse[Uinc];
γ = tϵhost[[3,3]]/tϵhost[[1,1]];
d11 = 1/(2(tϵhost[[1,1]]−tϵhost[[3,3]])) (1 − γArcSec[√γ]/√(γ−1));
d33 = 1/(tϵhost[[1,1]]−tϵhost[[3,3]]) (ArcSec[√γ]/√(γ−1) − 1);
Inverse[Uinc].{{d11, 0, 0}, {0, d11, 0}, {0, 0, d33}}.Inverse[Uinc]]
```

```
(* Maxwell Garnett : Uniaxial dielectric HCM *)
αab = (ϵa − ϵb).Inverse[I33 + (depol[ϵb, Ua].(ϵa − ϵb))];
ϵMG = ϵb + (fa αab.Inverse[I33 − (fa depol[ϵb, I33].αab)]);
pr = Plot[{Re[ϵMG[[1, 1]]], Re[ϵMG[[3, 3]]]}, {fa, 0, 1},
PlotStyle → {{Thickness[0.01], Red}, {Dashing[{0.02, 0.02}], Thickness[0.01], Blue}}, Frame → True,
Axes → False, FrameLabel → {"fa", "Re { ϵ‖MG/ ϵ0 }", "", "Re { ϵ⊥MG/ ϵ0 }"},
BaseStyle → {FontSize → 14, FontFamily → "Myriad"}]
```

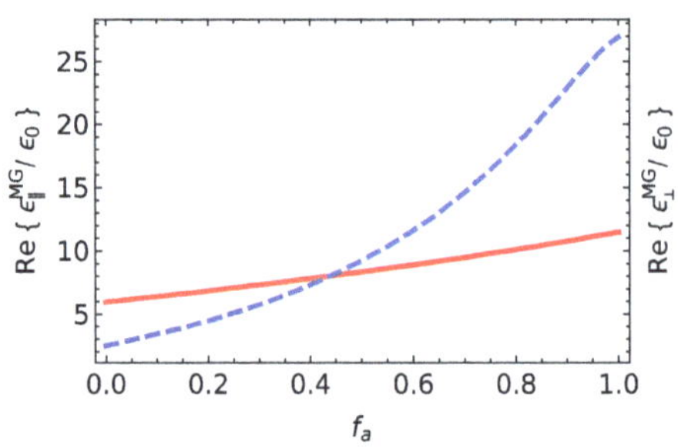


```
pi = Plot[{Im[ϵMG[[1, 1]]], Im[ϵMG[[3, 3]]]}, {fa, 0, 1}, PlotRange → All,
PlotStyle → {{Thickness[0.01], Red}, {Dashing[{0.02, 0.02}], Thickness[0.01], Blue}}, Frame → True,
Axes → False, FrameLabel → {"fa", "Im { ϵ‖MG/ ϵ0 }", "", "Im { ϵ⊥MG/ ϵ0 }"},
BaseStyle → {FontSize → 14, FontFamily → "Myriad"}]
```

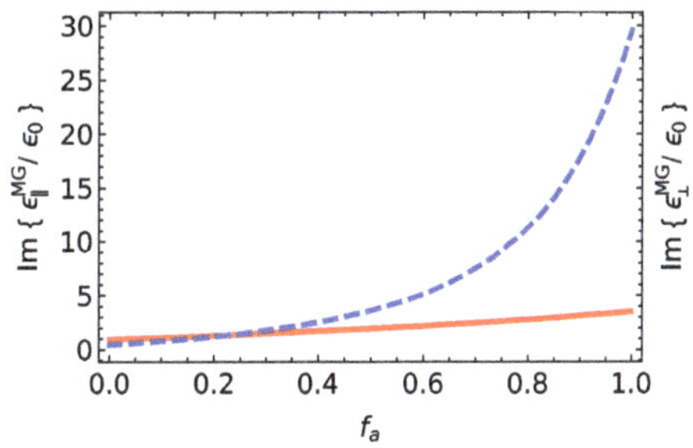


```
(* Incremental Maxwell Garnett : Uniaxial dielectric HCM *)
Niter = 3;
ϵIMG[fa_]:=Module[{fb, ϵn}, fb = 1 − fa;
ϵn = ϵb;
For[n = 0, n < Niter, n++, αan = (ϵa − ϵn).Inverse[I33 + (depol[ϵn, Ua].(ϵa − ϵn))];
ϵn = ϵn + (1 − fb^(1/Niter)) αan.Inverse[I33 − ((1 − fb^(1/Niter)) depol[ϵn, I33].αan)]];
ϵn]
```

```
pr = Plot[{Re[ϵIMG[fa][[1, 1]]], Re[ϵIMG[fa][[3, 3]]]}, {fa, 0, 1},
PlotStyle → {{Thickness[0.01], Red}, {Dashing[{0.02, 0.02}], Thickness[0.01], Blue}},
Frame → True, Axes → False, FrameLabel → {"fa", "Re { ϵ∥IMG/ ϵ0 }", "", "Re { ϵ⊥IMG/ ϵ0 }"},
BaseStyle → {FontSize → 14, FontFamily → "Myriad"}]
```

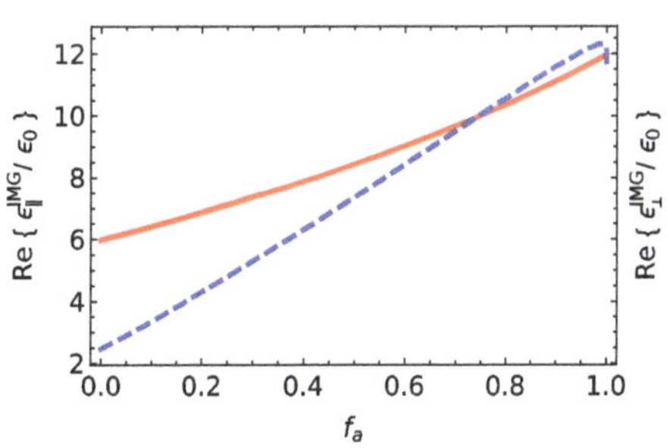


```
pi = Plot[{Im[ϵIMG[fa][[1, 1]]], Im[ϵIMG[fa][[3, 3]]]}, {fa, 0, 1},
PlotStyle → {{Thickness[0.01], Red}, {Dashing[{0.02, 0.02}], Thickness[0.01], Blue}},
Frame → True, Axes → False, FrameLabel → {"fa", "Im { ϵ∥IMG/ ϵ0 }", "", "Im { ϵ⊥IMG/ ϵ0 }"},
BaseStyle → {FontSize → 14, FontFamily → "Myriad"}]
```

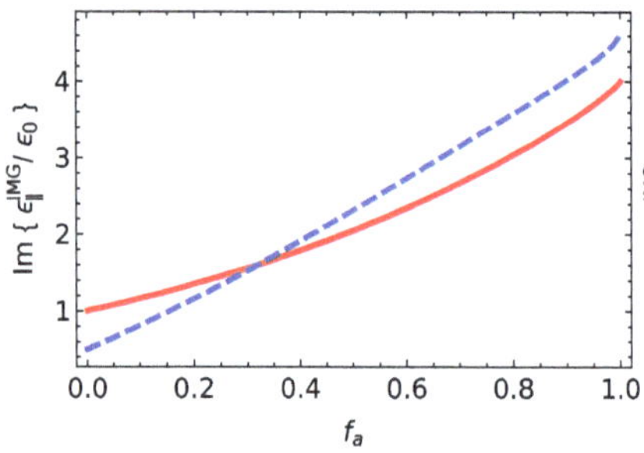


```
(* Differential Maxwell Garnett : Uniaxial dielectric HCM *)
αas[ϵs_]:=(ϵa − ϵs).Inverse[I33 + (depol[ϵs, Ua].(ϵa − ϵs))];
ϵs = NDSolve [{ϵsx'[s]== 1/(1−s) αas[DiagonalMatrix[{ϵsx[s], ϵsz[s], ϵsz[s]}]][[1, 1]],
ϵsz'[s]== 1/(1−s) αas[DiagonalMatrix[{ϵsx[s], ϵsz[s], ϵsz[s]}]][[3, 3]], ϵsx[0]==ϵb[[1, 1]],
ϵsz[0]==ϵb[[3, 3]]}, {ϵsx, ϵsz}, {s, 0.99}];
```

```
pr = Plot[{Re[Evaluate[ϵsx[s]/.ϵs[[1, 1]]]], Re[Evaluate[ϵsz[s]/.ϵs[[1, 2]]]]}, {s, 0, 0.99},
PlotStyle → {{Thickness[0.01], Red}, {Dashing[{0.02, 0.02}], Thickness[0.01], Blue}},
Frame → True, Axes → False, FrameLabel → {"fa", "Re { ϵ∥DMG/ ϵ0 }", "", "Re { ϵ⊥DMG/ ϵ0 }"},
BaseStyle → {FontSize → 14, FontFamily → "Myriad"}]
```

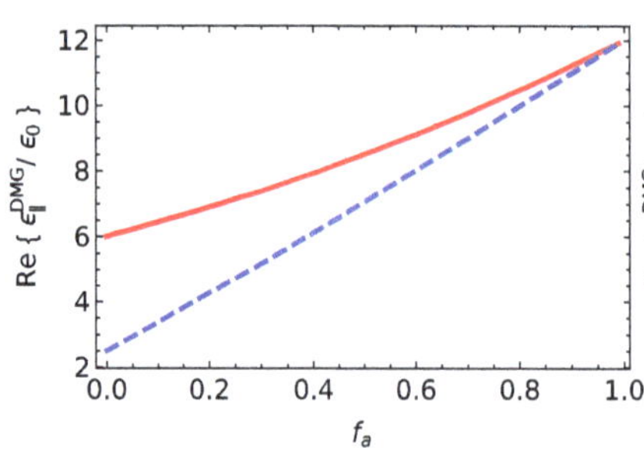


```
pi = Plot[{Im[Evaluate[ϵsx[s]/.ϵs[[1, 1]]]], Im[Evaluate[ϵsz[s]/.ϵs[[1, 2]]]]}, {s, 0, 0.99},
PlotStyle → {{Thickness[0.01], Red}, {Dashing[{0.02, 0.02}], Thickness[0.01], Blue}},
Frame → True, Axes → False, FrameLabel → {"fa", "Im { ϵ∥DMG/ ϵ0 }", "", "Im { ϵ⊥DMG/ ϵ0 }"},
BaseStyle → {FontSize → 14, FontFamily → "Myriad"}]
```

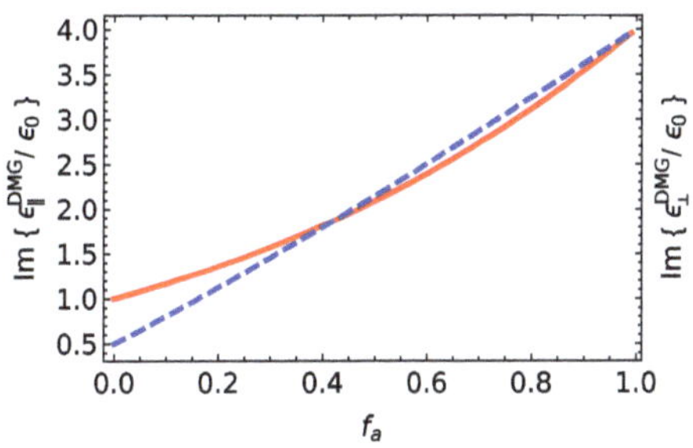


```
(* Bruggeman : Uniaxial dielectric HCM *)
ϵBr = DiagonalMatrix[{ϵBrx, ϵBrx, ϵBrz}];
αaBr[ϵBr_]:=(ϵa − ϵBr).Inverse[I33 + (depol[ϵBr, Ua].(ϵa − ϵBr))];
αbBr[ϵBr_]:=(ϵb − ϵBr).Inverse[I33 + (depol[ϵBr, Ua].(ϵb − ϵBr))];
tϵBr[fa_]:=
FindRoot[{faαaBr[ϵBr][[1, 1]] + (1 − fa)αbBr[ϵBr][[1, 1]] == 0,
faαaBr[ϵBr][[3, 3]] + (1 − fa)αbBr[ϵBr][[3, 3]] == 0}, {ϵBrx, ϵa[[1, 1]]}, {ϵBrz, ϵa[[1, 1]]}]

pr = Plot[{Re[tϵBr[fa][[1, 2]]], Re[tϵBr[fa][[2, 2]]]}, {fa, 0, 1},
PlotStyle → {{Thickness[0.01], Red}, {Dashing[{0.02, 0.02}], Thickness[0.01], Blue}},
Frame → True, Axes → False, FrameLabel → {"fa", "Re { ϵ∥Br/ ϵ0 }", "", "Re { ϵ⊥Br/ ϵ0 }"},
BaseStyle → {FontSize → 14, FontFamily → "Myriad"}]
```

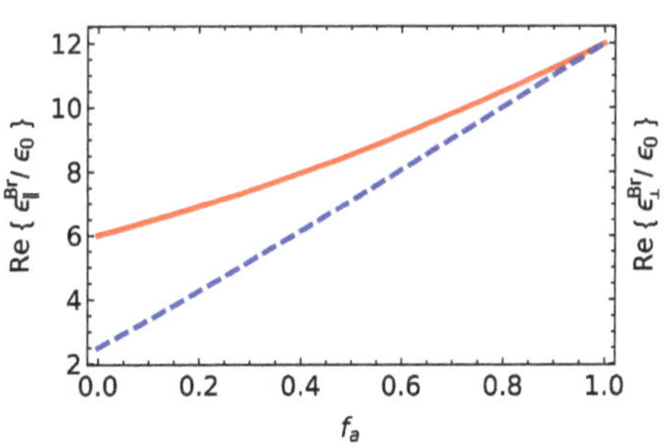

```
pi = Plot[{Im[tϵBr[fa][[1, 2]]], Im[tϵBr[fa][[2, 2]]]}, {fa, 0, 1},
PlotStyle → {{Thickness[0.01], Red}, {Dashing[{0.02, 0.02}], Thickness[0.01], Blue}},
Frame → True, Axes → False, FrameLabel → {"f_a", "Im { ϵ_∥^Br/ ϵ0 }", "", "Im { ϵ_⊥^Br/ ϵ0 }"},
BaseStyle → {FontSize → 14, FontFamily → "Myriad"}]
```

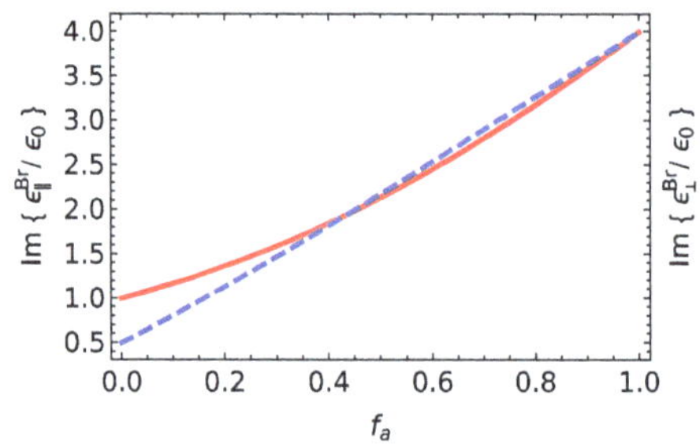

B.2 Figure 6.2

Here follow the codes for the zeroth-, second-, and third-order SPFT in the case of an isotropic chiral HCM, as used for figure 6.2.

```
(* Inputs : 0th-, 2nd-, 3rd-order SPFT for isotropic chiral HCM *)
I33 = IdentityMatrix[3]; I66 = IdentityMatrix[6];
ϵa = (5 + 0.04I) I33; ξa = (0.2 + 0.003I)I I33; ζa = −ξa; μa = (2 + 0.02I) I33;
ϵb = I33; ξb = 0 I33; ζb = 0 I33; μb = I33;
Ua = I33; Ub = I33;
Ka = {{ϵa[[1, 1]], ϵa[[1, 2]], ϵa[[1, 3]], ξa[[1, 1]], ξa[[1, 2]], ξa[[1, 3]]},
{ϵa[[2, 1]], ϵa[[2, 2]], ϵa[[2, 3]], ξa[[2, 1]], ξa[[2, 2]], ξa[[2, 3]]},
{ϵa[[3, 1]], ϵa[[3, 2]], ϵa[[3, 3]], ξa[[3, 1]], ξa[[3, 2]], ξa[[3, 3]]},
{ζa[[1, 1]], ζa[[1, 2]], ζa[[1, 3]], μa[[1, 1]], μa[[1, 2]], μa[[1, 3]]},
{ζa[[2, 1]], ζa[[2, 2]], ζa[[2, 3]], μa[[2, 1]], μa[[2, 2]], μa[[2, 3]]},
{ζa[[3, 1]], ζa[[3, 2]], ζa[[3, 3]], μa[[3, 1]], μa[[3, 2]], μa[[3, 3]]}};
Kb = I66;
fa = 0.4; fb = 1 − fa;
```

```
(* Depolarization dyadic : D0 *)
depol[K_, U_]:=Module[{ϵ, ξ, ζ, μ, iU, qh, nϵ, nξ, nζ, nμ, b, θ, ϕ, dϵx, dϵy, dϵz, dξx, dξy, dξz, dμx, dμy, dμz,
eed, emd, med, mmd},
ϵ = DiagonalMatrix[{K[[1, 1]], K[[2, 2]], K[[3, 3]]}];
ξ = DiagonalMatrix[{K[[1, 4]], K[[2, 5]], K[[3, 6]]}];
ζ = −ξ;
μ = DiagonalMatrix[{K[[4, 4]], K[[5, 5]], K[[6, 6]]}];
iU = Inverse[U];
qh = {Sin[θ] Cos[ϕ], Sin[θ] Sin[ϕ], Cos[θ]};
nϵ = qh.(iU.ϵ.iU).qh;
nξ = qh.(iU.ξ.iU).qh;
nζ = qh.(iU.ζ.iU).qh;
nμ = qh.(iU.μ.iU).qh;
b = (nϵ nμ) − (nξ nζ);
dϵx = NIntegrate[(nμ qh[[1]] qh[[1]] Sin[θ])/b, {θ, 0, Pi}, {ϕ, 0, 2Pi}];
dϵy = NIntegrate[(nμ qh[[2]] qh[[2]] Sin[θ])/b, {θ, 0, Pi}, {ϕ, 0, 2Pi}];
dϵz = NIntegrate[(nμ qh[[3]] qh[[3]] Sin[θ])/b, {θ, 0, Pi}, {ϕ, 0, 2Pi}];
eed = iU.{{dϵx, 0, 0}, {0, dϵy, 0}, {0, 0, dϵz}}.iU/(4Pi);
dξx = −NIntegrate[(nξ qh[[1]] qh[[1]] Sin[θ])/b, {θ, 0, Pi}, {ϕ, 0, 2Pi}];
dξy = −NIntegrate[(nξ qh[[2]] qh[[2]] Sin[θ])/b, {θ, 0, Pi}, {ϕ, 0, 2Pi}];
dξz = −NIntegrate[(nξ qh[[3]] qh[[3]] Sin[θ])/b, {θ, 0, Pi}, {ϕ, 0, 2Pi}];
emd = iU.{{dξx, 0, 0}, {0, dξy, 0}, {0, 0, dξz}}.iU/(4Pi);
med = −emd;
dμx = NIntegrate[(nϵ qh[[1]] qh[[1]] Sin[θ])/b, {θ, 0, Pi}, {ϕ, 0, 2Pi}];
dμy = NIntegrate[(nϵ qh[[2]] qh[[2]] Sin[θ])/b, {θ, 0, Pi}, {ϕ, 0, 2Pi}];
dμz = NIntegrate[(nϵ qh[[3]] qh[[3]] Sin[θ])/b, {θ, 0, Pi}, {ϕ, 0, 2Pi}];
mmd = iU.{{dμx, 0, 0}, {0, dμy, 0}, {0, 0, dμz}}.iU/(4Pi);
{{eed[[1, 1]], 0, 0, emd[[1, 1]], 0, 0}, {0, eed[[2, 2]], 0, 0, emd[[2, 2]], 0}, {0, 0, eed[[3, 3]], 0, 0, emd[[3, 3]]},
{med[[1, 1]], 0, 0, mmd[[1, 1]], 0, 0}, {0, med[[2, 2]], 0, 0, mmd[[2, 2]], 0}, {0, 0, med[[3, 3]], 0, 0, mmd[[3, 3]]}}]
```

```
(* Zeroth-order SPFT *)
KBr = fa Ka + fb Kb;
tKBr[K_]:=Module[{ta, tb},
ta = fa Inverse[I66 + depol[K, Ua].(Ka − K)]; tb = fb Inverse[I66 + depol[K, Ub].(Kb − K)];
(Ka.ta + Kb.tb).Inverse[ta + tb]];
tol = 0.000001; delta = 1; n = 1;
For[ncount = 1, ncount < 25, ncount++,
If[delta > tol,
newKBr = tKBr[KBr];
dK = newKBr − KBr;
delta =
Max[{Abs[dK][[1, 1]]/Abs[KBr][[1, 1]], Abs[dK][[2, 2]]/Abs[KBr][[2, 2]], Abs[dK][[3, 3]]/Abs[KBr][[3, 3]],
Abs[dK][[1, 4]]/Abs[KBr][[1, 4]], Abs[dK][[2, 5]]/Abs[KBr][[2, 5]], Abs[dK][[3, 6]]/Abs[KBr][[3, 6]],
Abs[dK][[4, 4]]/Abs[KBr][[4, 4]], Abs[dK][[5, 5]]/Abs[KBr][[5, 5]], Abs[dK][[6, 6]]/Abs[KBr][[6, 6]]}];
KBr = newKBr;
n++; ]];
Print ["fa = ", N[fa], ", no. of interations = ", n, ", delta = ", delta]

(* Initialization of arrays for plotting *)
Lmaxcount = 7;
plotRϵSPFT0 = Array[0&, {Lmaxcount, 2}];
plotRϵSPFT2 = Array[0&, {Lmaxcount, 2}];
plotRϵSPFT3 = Array[0&, {Lmaxcount, 2}];
plotIϵSPFT0 = Array[0&, {Lmaxcount, 2}];
plotIϵSPFT2 = Array[0&, {Lmaxcount, 2}];
plotIϵSPFT3 = Array[0&, {Lmaxcount, 2}];
```

$$\text{plotR}\epsilon\text{SPFT0[[1]]} = \{0, \text{Re[KBr[[1, 1]]]}\};$$

$$\text{plotR}\epsilon\text{SPFT2[[1]]} = \{0, \text{Re[KBr[[2, 2]]]}\};$$

$$\text{plotR}\epsilon\text{SPFT3[[1]]} = \{0, \text{Re[KBr[[3, 3]]]}\};$$

$$\text{plotI}\epsilon\text{SPFT0[[1]]} = \{0, 10\text{^}2\,\text{Im[KBr[[1, 1]]]}\};$$

$$\text{plotI}\epsilon\text{SPFT2[[1]]} = \{0, 10\text{^}2\,\text{Im[KBr[[2, 2]]]}\};$$

$$\text{plotI}\epsilon\text{SPFT3[[1]]} = \{0, 10\text{^}2\,\text{Im[KBr[[3, 3]]]}\};$$

$$\text{plotR}\xi\text{SPFT0} = \text{Array[0\&}, \{\text{Lmaxcount}, 2\}];$$

$$\text{plotR}\xi\text{SPFT2} = \text{Array[0\&}, \{\text{Lmaxcount}, 2\}];$$

$$\text{plotR}\xi\text{SPFT3} = \text{Array[0\&}, \{\text{Lmaxcount}, 2\}];$$

$$\text{plotI}\xi\text{SPFT0} = \text{Array[0\&}, \{\text{Lmaxcount}, 2\}];$$

$$\text{plotI}\xi\text{SPFT2} = \text{Array[0\&}, \{\text{Lmaxcount}, 2\}];$$

$$\text{plotI}\xi\text{SPFT3} = \text{Array[0\&}, \{\text{Lmaxcount}, 2\}];$$

$$\text{plotR}\xi\text{SPFT0[[1]]} = \{0, 10\text{^}2\,\text{Re[KBr[[1, 4]]}/I]\};$$

$$\text{plotR}\xi\text{SPFT2[[1]]} = \{0, 10\text{^}2\,\text{Re[KBr[[2, 5]]}/I]\};$$

$$\text{plotR}\xi\text{SPFT3[[1]]} = \{0, 10\text{^}2\,\text{Re[KBr[[3, 6]]}/I]\};$$

$$\text{plotI}\xi\text{SPFT0[[1]]} = \{0, 10\text{^}2\,\text{Im[KBr[[1, 4]]}/I]\};$$

$$\text{plotI}\xi\text{SPFT2[[1]]} = \{0, 10\text{^}2\,\text{Im[KBr[[2, 5]]}/I]\};$$

$$\text{plotI}\xi\text{SPFT3[[1]]} = \{0, 10\text{^}2\,\text{Im[KBr[[3, 6]]}/I]\};$$

$$\text{plotR}\mu\text{SPFT0} = \text{Array[0\&}, \{\text{Lmaxcount}, 2\}];$$

$$\text{plotR}\mu\text{SPFT2} = \text{Array[0\&}, \{\text{Lmaxcount}, 2\}];$$

```
plotRμSPFT3 = Array[0&, {Lmaxcount, 2}];
plotIμSPFT0 = Array[0&, {Lmaxcount, 2}];
plotIμSPFT2 = Array[0&, {Lmaxcount, 2}];
plotIμSPFT3 = Array[0&, {Lmaxcount, 2}];
plotRμSPFT0[[1]] = {0, Re[KBr[[4, 4]]]};
plotRμSPFT2[[1]] = {0, Re[KBr[[5, 5]]]};
plotRμSPFT3[[1]] = {0, Re[KBr[[6, 6]]]};
plotIμSPFT0[[1]] = {0, 10^3 Im[KBr[[4, 4]]]};
plotIμSPFT2[[1]] = {0, 10^3 Im[KBr[[5, 5]]]};
plotIμSPFT3[[1]] = {0, 10^3 Im[KBr[[6, 6]]]};

(* Depolarization dyadic : D+ *)
dplus[K_, U_, L_]:=
Module[{iU, iUq, q, θ, ϕ, iUqxI,
Amat, adjA, detA, Gmat, ϵ, ξ, ζ, μ, qh, nϵ, nξ, nζ, nμ, b, iUqqiU, Ginfmat, Gplusmat, κplus,
κminus, t4, t2, t0, T0, detA0, adjAplus, adjAminus, cplus, cminus, c, ex, ey, ez, zx, zy, zz, mx, my, mz},
iU = Inverse[U];
iUq = qiU.{Sin[θ] Cos[ϕ], Sin[θ] Sin[ϕ], Cos[θ]};
iUqxI = {{0, −iUq[[3]], iUq[[2]]}, {iUq[[3]], 0, −iUq[[1]]}, {−iUq[[2]], iUq[[1]], 0}};
Amat = {{0, 0, 0, iUqxI[[1, 1]], iUqxI[[1, 2]], iUqxI[[1, 3]]}, {0, 0, 0, iUqxI[[2, 1]], iUqxI[[2, 2]], iUqxI[[2, 3]]},
{0, 0, 0, iUqxI[[3, 1]], iUqxI[[3, 2]], iUqxI[[3, 3]]}, {−iUqxI[[1, 1]], −iUqxI[[1, 2]], −iUqxI[[1, 3]], 0, 0, 0},
{−iUqxI[[2, 1]], −iUqxI[[2, 2]], −iUqxI[[2, 3]], 0, 0, 0}, {−iUqxI[[3, 1]], −iUqxI[[3, 2]], −iUqxI[[3, 3]], 0, 0, 0}}
+K;
adjA = Map[Reverse, Minors[Transpose[Amat], Length[Amat] − 1], {0, 1}]
Table[(−1)^(i + j), {i, Length[Amat]}, {j, Length[Amat]}];
detA = Det[Amat];
Gmat = Inverse[Amat];
ϵ = {{K[[1, 1]], K[[1, 2]], K[[1, 3]]}, {K[[2, 1]], K[[2, 2]], K[[2, 3]]}, {K[[3, 1]], K[[3, 2]], K[[3, 3]]}};
ξ = {{K[[1, 4]], K[[1, 5]], K[[1, 6]]}, {K[[2, 4]], K[[2, 5]], K[[2, 6]]}, {K[[3, 4]], K[[3, 5]], K[[3, 6]]}};
```

```
ζ = {{K[[4, 1]], K[[4, 2]], K[[4, 3]]}, {K[[5, 1]], K[[5, 2]], K[[5, 3]]}, {K[[6, 1]], K[[6, 2]], K[[6, 3]]}};
μ = {{K[[4, 4]], K[[4, 5]], K[[4, 6]]}, {K[[5, 4]], K[[5, 5]], K[[5, 6]]}, {K[[6, 4]], K[[6, 5]], K[[6, 6]]}};
qh = {Sin[θ] Cos[ϕ], Sin[θ] Sin[ϕ], Cos[θ]};
nϵ = qh.(iU.ϵ.iU).qh;
nξ = qh.(iU.ξ.iU).qh;
nζ = qh.(iU.ζ.iU).qh;
nμ = qh.(iU.μ.iU).qh;
b = (nϵ nμ) − (nξ nζ);
iUqqiU = iU.{{qh[[1]] qh[[1]], qh[[1]] qh[[2]], qh[[1]] qh[[3]]}, {qh[[2]] qh[[1]], qh[[2]] qh[[2]], qh[[2]] qh[[3]]},
{qh[[3]] qh[[1]], qh[[3]] qh[[2]], qh[[3]] qh[[3]]}}.iU;
Ginfmat = {{nμ iUqqiU[[1, 1]], nμ iUqqiU[[1, 2]], nμ iUqqiU[[1, 3]], −nξ iUqqiU[[1, 1]], −nξ iUqqiU[[1, 2]],
−nξ iUqqiU[[1, 3]]}, {nμ iUqqiU[[2, 1]], nμ iUqqiU[[2, 2]], nμ iUqqiU[[2, 3]], −nξ iUqqiU[[2, 1]],
−nξ iUqqiU[[2, 2]], −nξ iUqqiU[[2, 3]]}, {nμ iUqqiU[[3, 1]], nμ iUqqiU[[3, 2]], nμ iUqqiU[[3, 3]],
−nξ iUqqiU[[3, 1]], −nξ iUqqiU[[3, 2]], −nξ iUqqiU[[3, 3]]}, {−nζ iUqqiU[[1, 1]], −nζ iUqqiU[[1, 2]],
−nζ iUqqiU[[1, 3]], nϵ iUqqiU[[1, 1]], nϵ iUqqiU[[1, 2]], nϵ iUqqiU[[1, 3]]}, {−nζ iUqqiU[[2, 1]],
−nζ iUqqiU[[2, 2]], −nζ iUqqiU[[2, 3]], nϵ iUqqiU[[2, 1]], nϵ iUqqiU[[2, 2]], nϵ iUqqiU[[2, 3]]},
{−nζ iUqqiU[[3, 1]], −nζ iUqqiU[[3, 2]], −nζ iUqqiU[[3, 3]], nϵ iUqqiU[[3, 1]], nϵ iUqqiU[[3, 2]],
nϵ iUqqiU[[3, 3]]}}/b;
Gplusmat = Gmat − Ginfmat;
t4 = Coefficient[detA, q, 4];
t2 = Coefficient[detA, q, 2];
t0 = Coefficient[detA, q, 0];
T0 = Gmat/.q → 0;
detA0 = detA/.q → 0;
κplus = (−t2 + √(t2^2 − 4 t4 t0))/(2 t4); κminus = (−t2 − √(t2^2 − 4 t4 t0))/(2 t4);
adjAplus = (adjA/.q → √κplus) + (adjA/.q → −√κplus);
adjAminus = (adjA/.q → √κminus) + (adjA/.q → −√κminus);
cplus = Exp[I q L]/(2q^2) (1 − (I q L))/.q → √κplus; cminus = Exp[I q L]/(2q^2) (1 − (I q L))/.q → √κminus;
c = Sin[θ]/(4 Pi b) (1/(κplus − κminus) ((cplus adjAplus) − (cminus adjAminus)) + detA0/(κplus κminus) (T0 − Ginfmat));
ex = NIntegrate[c[[1, 1]], {θ, 0, Pi}, {ϕ, 0, 2 Pi}];
ey = NIntegrate[c[[2, 2]], {θ, 0, Pi}, {ϕ, 0, 2 Pi}];
ez = NIntegrate[c[[3, 3]], {θ, 0, Pi}, {ϕ, 0, 2 Pi}];
```

```
zx = NIntegrate[c[[1, 4]], {θ, 0, Pi}, {ϕ, 0, 2 Pi}];
zy = NIntegrate[c[[2, 5]], {θ, 0, Pi}, {ϕ, 0, 2 Pi}];
zz = NIntegrate[c[[3, 6]], {θ, 0, Pi}, {ϕ, 0, 2 Pi}];
mx = NIntegrate[c[[4, 4]], {θ, 0, Pi}, {ϕ, 0, 2 Pi}];
my = NIntegrate[c[[5, 5]], {θ, 0, Pi}, {ϕ, 0, 2 Pi}];
mz = NIntegrate[c[[6, 6]], {θ, 0, Pi}, {ϕ, 0, 2 Pi}];
{{ex, 0, 0, zx, 0, 0}, {0, ey, 0, 0, zy, 0}, {0, 0, ez, 0, 0, zz},
{−zx, 0, 0, mx, 0, 0}, {0, −zy, 0, 0, my, 0}, {0, 0, −zz, 0, 0, mz}}]

(* Mass operator terms *)
alpha[K_, Kamb_, U_]:=(K − Kamb).Inverse[I66 + depol[Kamb, U].(K − Kamb)];
Pmat[L_]:=fa fb dplus[KBr, Ua, L];
Vmat[K_, U_]:=Inverse[K] − depol[K, U];
massop2[L_]:=(alpha[Ka, KBr, Ua] − alpha[Kb, KBr, Ub]).Pmat[L].
(alpha[Ka, KBr, Ua] − alpha[Kb, KBr, Ub]);
massop3[L_]:= massop2[L]−
fa(fb−fa)/(3 fb^2) alpha[Ka, KBr, Ua].((Vmat[KBr, Ua].alpha[Ka, KBr, Ua].Pmat[L])
+(Pmat[L].alpha[Ka, KBr, Ua].Vmat[KBr, Ua])+
(Pmat[L].alpha[Ka, KBr, Ua].Pmat[L])).alpha[Ka, KBr, Ua];

For[Lcount = 1, Lcount < Lmaxcount, Lcount++,
L = Lcount/20;
KSPFT0 = KBr;
KSPFT2 = KBr − Inverse[I66 + massop2[L].depol[KBr, Ua]].massop2[L];
KSPFT3 = KBr − Inverse[I66 + massop3[L].depol[KBr, Ua]].massop3[L];
Print["L = ", N[L]];
plotRϵSPFT0[[Lcount + 1]] = {L, Re[KSPFT0[[1, 1]]]};
plotRϵSPFT2[[Lcount + 1]] = {L, Re[KSPFT2[[2, 2]]]};
plotRϵSPFT3[[Lcount + 1]] = {L, Re[KSPFT3[[3, 3]]]};
plotIϵSPFT0[[Lcount + 1]] = {L, 10^2 Im[KSPFT0[[1, 1]]]};
plotIϵSPFT2[[Lcount + 1]] = {L, 10^2 Im[KSPFT2[[2, 2]]]};
plotIϵSPFT3[[Lcount + 1]] = {L, 10^2 Im[KSPFT3[[3, 3]]]};
```

```
plotRξSPFT0[[Lcount + 1]] = {L, 10^2 Re[KSPFT0[[1, 4]]/I]};
plotRξSPFT2[[Lcount + 1]] = {L, 10^2 Re[KSPFT2[[2, 5]]/I]};
plotRξSPFT3[[Lcount + 1]] = {L, 10^2 Re[KSPFT3[[3, 6]]/I]};
plotIξSPFT0[[Lcount + 1]] = {L, 10^2 Im[KSPFT0[[1, 4]]/I]};
plotIξSPFT2[[Lcount + 1]] = {L, 10^2 Im[KSPFT2[[2, 5]]/I]};
plotIξSPFT3[[Lcount + 1]] = {L, 10^2 Im[KSPFT3[[3, 6]]/I]};
plotRμSPFT0[[Lcount + 1]] = {L, Re[KSPFT0[[4, 4]]]};
plotRμSPFT2[[Lcount + 1]] = {L, Re[KSPFT2[[5, 5]]]};
plotRμSPFT3[[Lcount + 1]] = {L, Re[KSPFT3[[6, 6]]]};
plotIμSPFT0[[Lcount + 1]] = {L, 10^3 Im[KSPFT0[[4, 4]]]};
plotIμSPFT2[[Lcount + 1]] = {L, 10^3 Im[KSPFT2[[5, 5]]]};
plotIμSPFT3[[Lcount + 1]] = {L, 10^3 Im[KSPFT3[[6, 6]]]};]

pRe0 = ListPlot[plotRϵSPFT0, Joined → True, PlotRange → All,
PlotStyle → {Green, Dashing[{0.035, 0.025, 0.005, 0.025}], Thickness[0.008]}];
pRe2 = ListPlot[plotRϵSPFT2, Joined → True, PlotRange → All,
PlotStyle → {Blue, Dashing[{0.02, 0.02}], Thickness[0.008]}];
pRe3 = ListPlot[plotRϵSPFT3, Joined → True, PlotRange → All,
PlotStyle → {Red, Thickness[0.008]}];
pR = Show[Graphics[pRe0], Graphics[pRe2], Graphics[pRe3],
Frame → True, Axes → False, PlotRange → All,
FrameLabel → {"k0Lcor", "Re{ ϵ[0]SPFT / ϵ0 }", "Re{ ϵ[2]SPFT / ϵ0 }", "Re{ ϵ[3]SPFT / ϵ0 }"},
BaseStyle → {FontSize → 14, FontFamily → "Myriad"}]
```

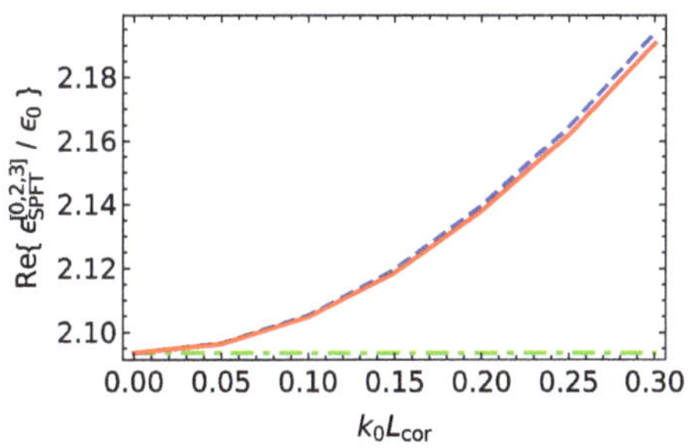

```
pIe0 = ListPlot[plotIϵSPFT0, Joined → True, PlotRange → All,
PlotStyle → {Green, Dashing[{0.035, 0.025, 0.005, 0.025}], Thickness[0.008]}];
pIe2 = ListPlot[plotIϵSPFT2, Joined → True, PlotRange → All,
PlotStyle → {Blue, Dashing[{0.02, 0.02}], Thickness[0.008]}];
pIe3 = ListPlot[plotIϵSPFT3, Joined → True, PlotRange → All,
PlotStyle → {Red, Thickness[0.008]}];
pI = Show[Graphics[pIe0], Graphics[pIe2], Graphics[pIe3],
Frame → True, Axes → False, PlotRange → All, FrameLabel →
{"k0Lcor", "Im{ ϵ[0]SPFT / ϵ0 } × 10^2", "Im{ ϵ[2]SPFT / ϵ0 } × 10^2", "Im{ ϵ[3]SPFT / ϵ0 } × 10^2"},
BaseStyle → {FontSize → 14, FontFamily → "Myriad"}]
```

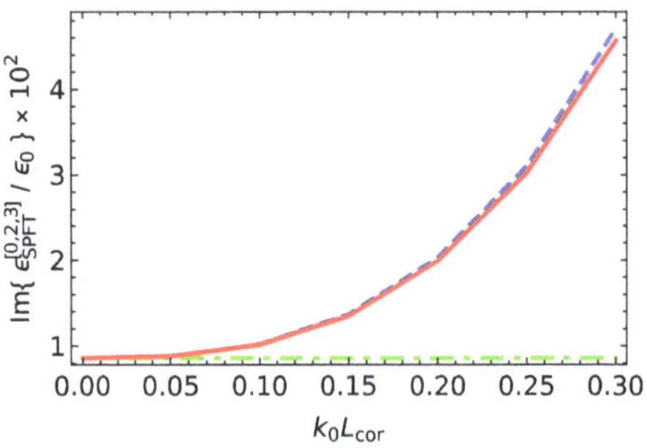


```
pRz0 = ListPlot[plotRξSPFT0, Joined → True, PlotRange → All,
PlotStyle → {Green, Dashing[{0.035, 0.025, 0.005, 0.025}], Thickness[0.008]}];
pRz2 = ListPlot[plotRξSPFT2, Joined → True, PlotRange → All,
PlotStyle → {Blue, Dashing[{0.02, 0.02}], Thickness[0.008]}];
pRz3 = ListPlot[plotRξSPFT3, Joined → True, PlotRange → All,
PlotStyle → {Red, Thickness[0.008]}];
pR = Show[Graphics[pRz0], Graphics[pRz2], Graphics[pRz3],
Frame → True, Axes → False, PlotRange → All, FrameLabel →
{"k0Lcor", "Re{ ξ[0]SPFT / i√(ϵ0μ0) } × 10^2", "Re{ ξ[2]SPFT / i√(ϵ0μ0) } × 10^2",
"Re{ ξ[3]SPFT / i√(ϵ0μ0) } × 10^2"}, BaseStyle → {FontSize → 14, FontFamily → "Myriad"}]
```

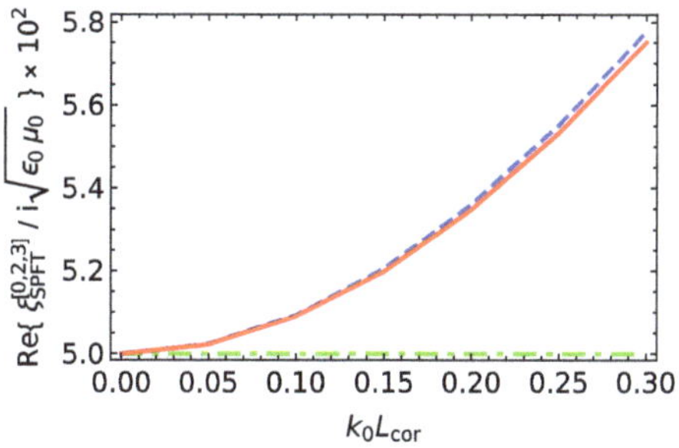

pIz0 = ListPlot[plotIξSPFT0, Joined → True, PlotRange → All,

PlotStyle → {Green, Dashing[{0.035, 0.025, 0.005, 0.025}], , Thickness[0.008]}];

pIz2 = ListPlot[plotIξSPFT2, Joined → True, PlotRange → All,

PlotStyle → {Blue, Dashing[{0.02, 0.02}], Thickness[0.008]}];

pIz3 = ListPlot[plotIξSPFT3, Joined → True, PlotRange → All,

PlotStyle → {Red, Thickness[0.008]}];

pI = Show[Graphics[pIe0], Graphics[pIe2], Graphics[pIe3],

Frame → True, Axes → False, PlotRange → All, FrameLabel →

{"$k_0 L_{\rm cor}$", "Im{ $\xi^{[0]}_{\rm SPFT}$ / i$\sqrt{\epsilon_0 \mu_0}$ } × 10^2", "Im{ $\xi^{[2]}_{\rm SPFT}$ / i$\sqrt{\epsilon_0 \mu_0}$ } × 10^2",

"Im{ $\xi^{[3]}_{\rm SPFT}$ / i$\sqrt{\epsilon_0 \mu_0}$ } × 10^2"}, BaseStyle → {FontSize → 14, FontFamily → "Myriad"}]

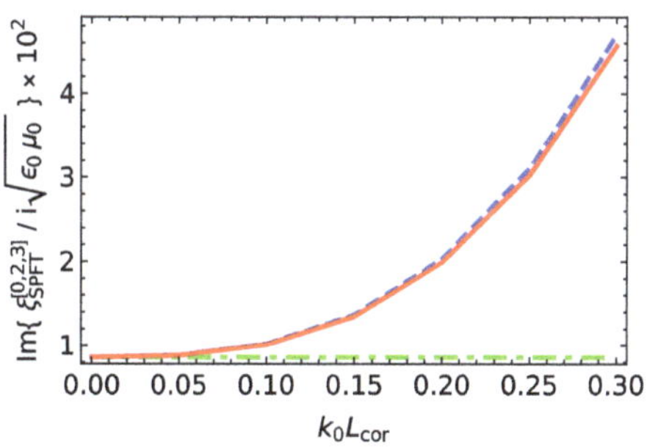

```
pRm0 = ListPlot[plotRμSPFT0, Joined → True, PlotRange → All,
PlotStyle → {Green, Dashing[{0.035, 0.025, 0.005, 0.025}], Thickness[0.008]}];
pRm2 = ListPlot[plotRμSPFT2, Joined → True, PlotRange → All,
PlotStyle → {Blue, Dashing[{0.02, 0.02}], Thickness[0.008]}];
pRm3 = ListPlot[plotRμSPFT3, Joined → True, PlotRange → All,
PlotStyle → {Red, Thickness[0.008]}];
pR = Show[Graphics[pRm0], Graphics[pRm2], Graphics[pRm3],
Frame → True, Axes → False, PlotRange → All, FrameLabel →
{"k0Lcor", "Re{ μ[0]SPFT / μ0 }", "Re{ μ[2]SPFT / μ0 }", "Re{ μ[3]SPFT / μ0 }"},
BaseStyle → {FontSize → 14, FontFamily → "Myriad"}]
```

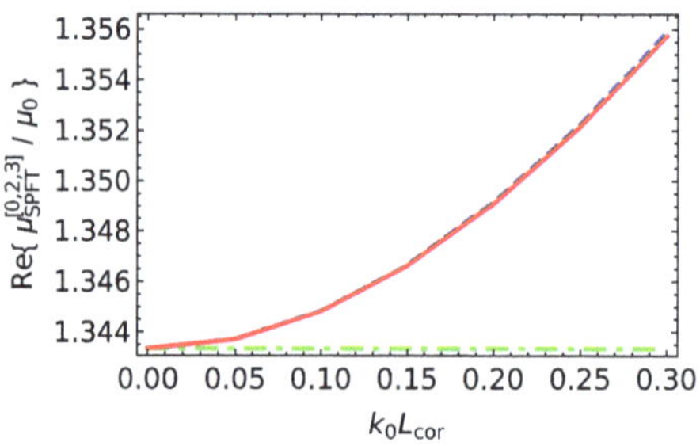


```
pIm0 = ListPlot[plotIμSPFT0, Joined → True, PlotRange → All,
PlotStyle → {Green, Dashing[{0.035, 0.025, 0.005, 0.025}], Thickness[0.008]}];
pIm2 = ListPlot[plotIμSPFT2, Joined → True, PlotRange → All,
PlotStyle → {Blue, Dashing[{0.02, 0.02}], Thickness[0.008]}];
pIm3 = ListPlot[plotIμSPFT3, Joined → True, PlotRange → All,
PlotStyle → {Red, Thickness[0.008]}];
pI = Show[Graphics[pIm0], Graphics[pIm2], Graphics[pIm3],
Frame → True, Axes → False, PlotRange → All, FrameLabel →
{"k0Lcor", "Im{ μ[0]SPFT / μ0 } × 10^3", "Im{ μ[2]SPFT / μ0 } × 10^3", "Im{ μ[3]SPFT / μ0 } × 10^3"},
BaseStyle → {FontSize → 14, FontFamily → "Myriad"}]
```

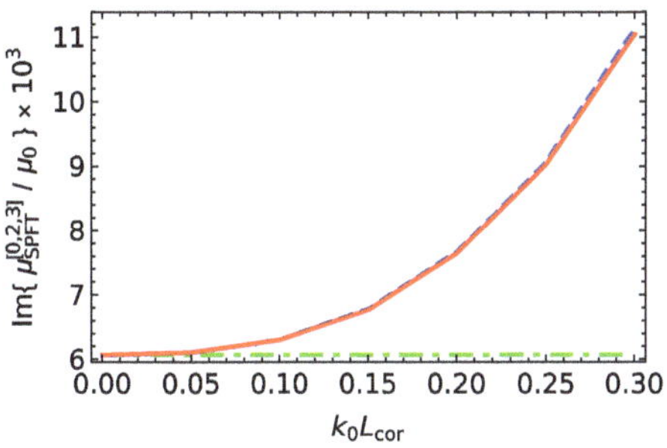

B.3 Figures 6.3 and 6.4

Here follow the codes for extended versions of the Maxwell Garnett, Bruggeman, and second-order SPFT formalisms in the case of an isotropic dielectric HCM, as used for figures 6.3 and 6.4.

```
(* Extended Maxwell Garnett : isotropic dielectric HCM *)

MaxwellGarnett[epsa_, epsb_, vola_, rho_]:=Module[{alpha, depol, depol0, depolp},

depol0 = 1/(3 epsb);

depolp = -(rho^2/9) (3 + (2 I rho Sqrt[epsb]));

depol = depol0 + depolp;

alpha = (epsa-epsb)/(1+(depol (epsa-epsb)));

epsb + (vola*alpha)/(1-(vola depol alpha))]
```

```
rp = Plot3D[Re[MaxwellGarnett[0.2, 5, fa, ρ]], {fa, 0, 0.3}, {ρ, 0, 0.3}, MaxRecursion → 0,

PlotPoints → 30, Mesh → None, AxesLabel → {"fa ", ρk0, "Re{εMG/ε0} "},

BaseStyle → {FontSize → 14, FontFamily → "Myriad"}]
```

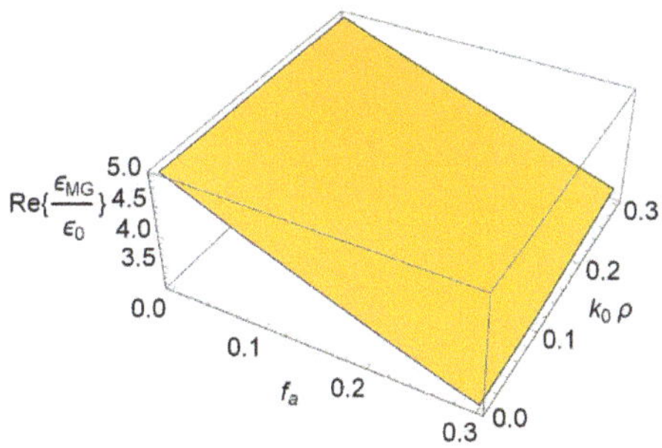

```
ip = Plot3D[Im[MaxwellGarnett[0.2, 5, fa, ρ]], {fa, 0, 0.3}, {ρ, 0, 0.3}, MaxRecursion → 0,
PlotPoints → 30, Mesh → None, PlotRange → Full, AxesLabel → {"fa ", ρk0, "Im{εMG/ε0} "},
BaseStyle → {FontSize → 14, FontFamily → "Myriad"}]
```

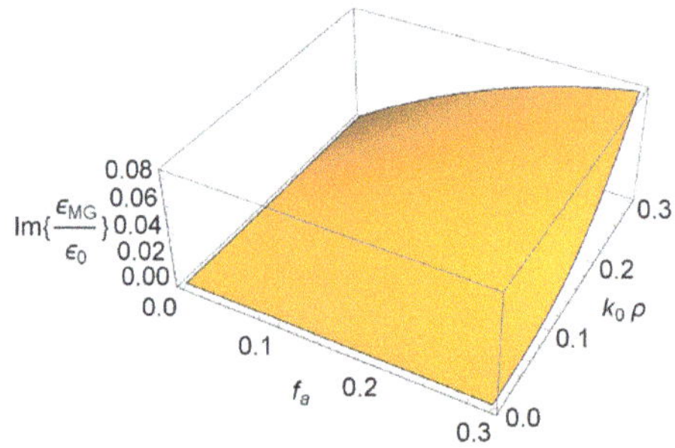


```
(* Extended Bruggeman : isotropic dielectric HCM *)
mBruggeman[epsa_, epsb_, vola_, rho_, Num_]:=
Module[{volb, εBr0, fεBr, epsBr}, volb = 1 − vola; εBr0 = (vola epsa) + (volb epsb);
fεBr[epsBr_]:=mBr[epsa, epsb, vola, epsBr, rho];
εBrN = NestList[fεBr, εBr0, Num][[Num + 1]]];
mBr[epsa_, epsb_, vola_, epsBr_, rho_]:=Module[{volb, depol, depol0, depolp, a, b, c}, volb = 1 − vola;
depol0 = 1/(3 epsBr);
depolp = −(rho^2/9) (3 + (2 I rho √epsBr));
depol = depol0 + depolp;
a = depol;
b = −1 − (depol (epsa + epsb));
c = (depol epsa epsb) + (vola epsa) + (volb epsb);
(−b − √(b^2 − (4 a c)))/(2a)];

rp = Plot3D[Re[mBruggeman[0.2, 5, fa, ρ, 25]], {fa, 0, 1}, {ρ, 0, 0.3}, MaxRecursion → 0,
PlotPoints → 30, Mesh → None, AxesLabel → {"fa ", ρk0, "Re{εBr/ε0}"},
BaseStyle → {FontSize → 14, FontFamily → "Myriad"}]
```

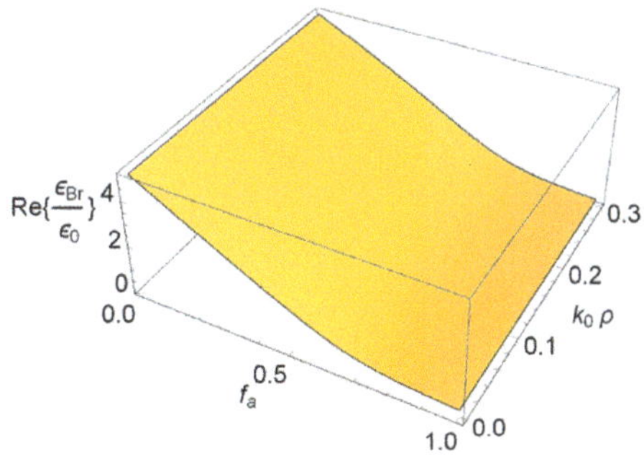

ip = Plot3D[Im[mBruggeman[0.2, 5, fa, ρ, 25]], {fa, 0, 1}, {ρ, 0, 0.3}, MaxRecursion → 0,

PlotPoints → 30, Mesh → None, PlotRange → Full, AxesLabel → $\left\{"f_a\ ", \rho k_0, "\mathrm{Im}\{\frac{\epsilon_{Br}}{\epsilon_0}\}\ "\right\}$,

BaseStyle → {FontSize → 14, FontFamily → "Myriad"}]

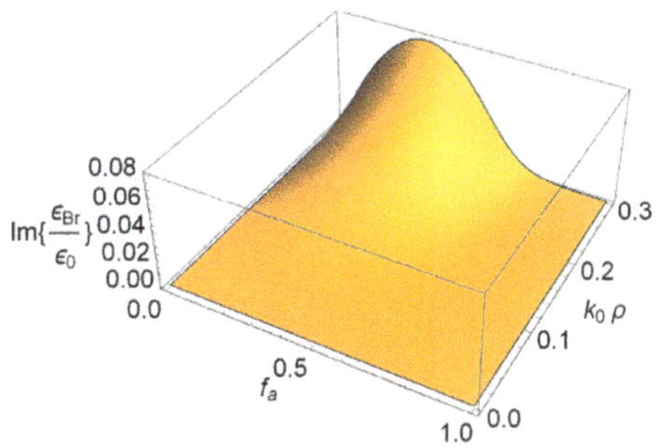

(* Extended second − order SPFT : isotropic dielectric HCM *)

SPFT[epsa_, epsb_, vola_, rho_, Lcor_, Num_]:=

Module[{sigma, alphaa, alphab, volb, depolpD, depolpL, depol0, depol, epsBr},

$\text{volb} = 1 - \text{vola};$

$\text{epsBr} = \text{mBruggeman}[\text{epsa}, \text{epsb}, \text{vola}, \text{rho}, \text{Num}];$

$\text{depol0} = \frac{1}{3\,\text{epsBr}};$

$\text{depolpD} = -\frac{\text{rho}^\wedge 2}{9}\left(3 + \left(2\,I\,\text{rho}\sqrt{\text{epsBr}}\right)\right);$

$\text{depol} = \text{depol0} + \text{depolpD};$

$\text{alphaa} = \frac{\text{epsa}-\text{epsBr}}{1+(\text{depol}\,(\text{epsa}-\text{epsBr}))};$

$\text{alphab} = \frac{\text{epsb}-\text{epsBr}}{1+(\text{depol}\,(\text{epsb}-\text{epsBr}))};$

$\text{depolpL} = \frac{\text{Lcor}^\wedge 2}{9}\left(3 + \left(2\,I\,\text{Lcor}\sqrt{\text{epsBr}}\right)\right);$

$\text{sigma} = -\text{vola}\,\text{volb}\,(\text{alphaa} - \text{alphab})\,(\text{alphaa} - \text{alphab})\,\text{depolpL};$

$\text{epsBr} - \frac{\text{sigma}}{1+(\text{sigma}\,\text{depol})}\Big]$

$\text{rp} = \text{Plot3D}[\text{Re}[\text{SPFT}[0.2, 5, \text{fa}, \rho, 0.3, 25]], \{\text{fa}, 0, 1\}, \{\rho, 0, 0.3\}, \text{MaxRecursion} \to 0,$
$\text{PlotPoints} \to 30, \text{Mesh} \to \text{None}, \text{AxesLabel} \to \left\{\text{" } f_a \text{ "}, \rho k_0, \text{"Re}\{\frac{\epsilon^{[2]}_{\text{SPFT}}}{\epsilon_0}\}\text{"}\right\},$
$\text{BaseStyle} \to \{\text{FontSize} \to 14, \text{FontFamily} \to \text{“Myriad”}\}]$

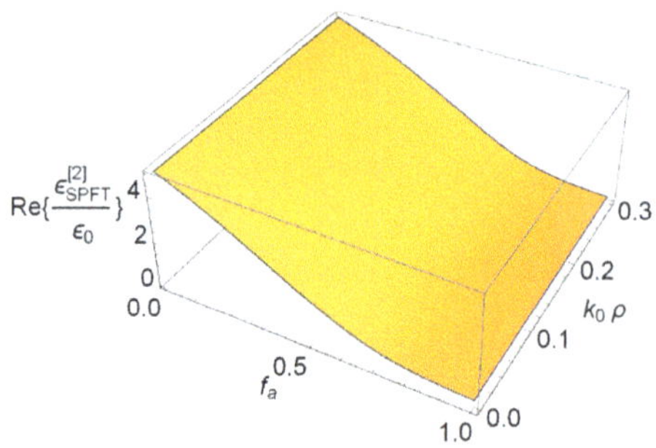

$\text{ip} = \text{Plot3D}[\text{Im}[\text{SPFT}[0.2, 5, \text{fa}, \rho, 0.3, 25]], \{\text{fa}, 0, 1\}, \{\rho, 0, 0.3\}, \text{MaxRecursion} \to 0,$
$\text{PlotPoints} \to 30, \text{Mesh} \to \text{None}, \text{PlotRange} \to \text{Full}, \text{AxesLabel} \to \left\{\text{" } f_a \text{ "}, \rho k_0, \text{"Im}\{\frac{\epsilon^{[2]}_{\text{SPFT}}}{\epsilon_0}\}\text{ "}\right\},$
$\text{BaseStyle} \to \{\text{FontSize} \to 14, \text{FontFamily} \to \text{“Myriad”}\}]$

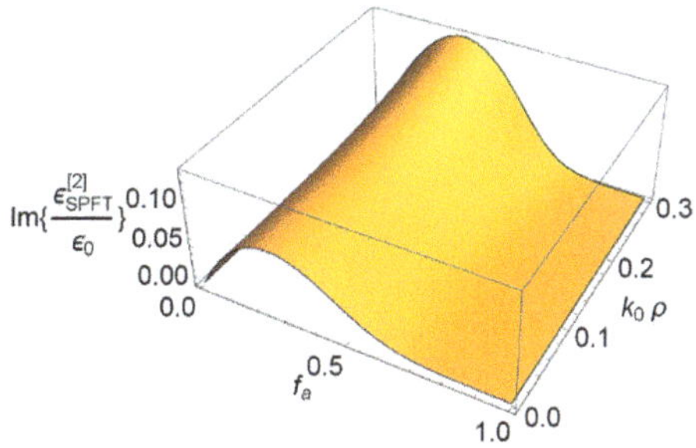


```
SPFTx[epsa_, epsb_, vola_, rhoL_, Lcor_, Num_]:=
Module[{sigma, alphaa, alphab, volb, depolpD, depolpL, depol0, depol, epsBr, rho},
rho = rhoL Lcor;
volb = 1 - vola;
epsBr = mBruggeman[epsa, epsb, vola, rho, Num];
depol0 = 1/(3 epsBr);
depolpD = -(rho^2)/9 (3 + (2 I rho Sqrt[epsBr]));
depol = depol0 + depolpD;
alphaa = (epsa-epsBr)/(1+(depol (epsa-epsBr)));
alphab = (epsb-epsBr)/(1+(depol (epsb-epsBr)));
depolpL = (Lcor^2)/9 (3 + (2 I Lcor Sqrt[epsBr]));
sigma = -vola volb (alphaa - alphab) (alphaa - alphab) depolpL;
epsBr - sigma/(1+(sigma depol))]
```

```
rp = Plot3D[Re[SPFTx[0.2, 5, 0.3, ρL, Lc, 15]], {ρL, 0, 1}, {Lc, 0, 0.3}, MaxRecursion → 0,
PlotPoints → 30, Mesh → None, AxesLabel → {"ρ/L_cor", L_cor k_0, "Re{ε_SPFT^[2]/ε_0} "},
BaseStyle → {FontSize → 14, FontFamily → "Myriad"}]
```

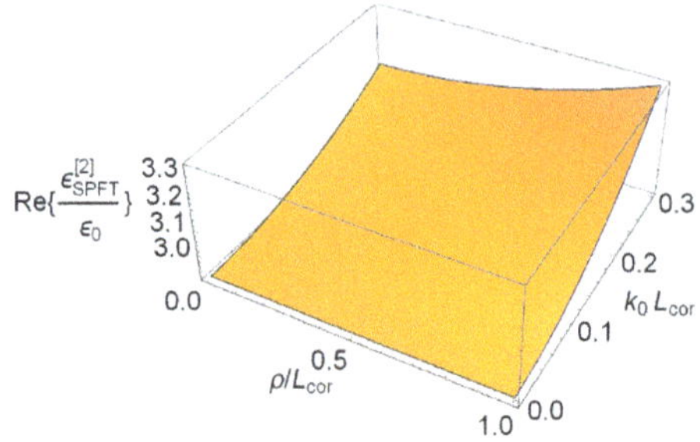

ip = Plot3D[Im[SPFTx[0.2, 5, 0.3, ρL, Lc, 15]], {ρL, 0, 1}, {Lc, 0, 0.3}, MaxRecursion → 0, PlotPoints → 30, Mesh → None, AxesLabel → $\left\{"\rho/L_{\rm cor}", L_{\rm cor}k_0, "\mathrm{Im}\{\frac{\epsilon^{[2]}_{\rm SPFT}}{\epsilon_0}\}"\right\}$, BaseStyle → {FontSize → 14, FontFamily → "Myriad"}]

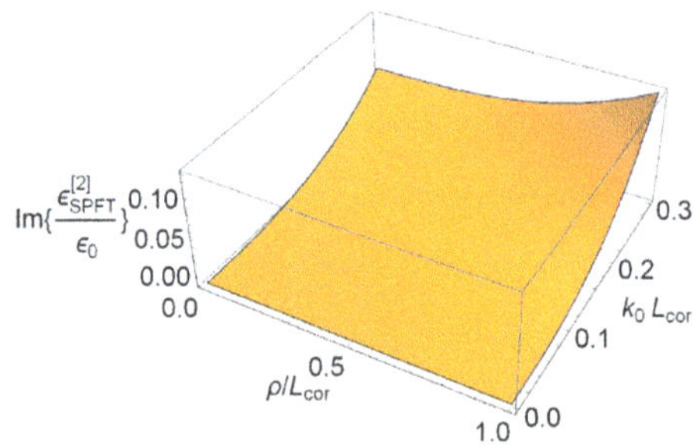

B.4 Figure 6.5

Here follows the code for the Bruggeman formalism in the case of a biaxial dielectric HCM, as used for figure 6.5.

```
(* Inputs : Bruggeman for biaxial dielectric HCM *)
I33 = IdentityMatrix[3];
εa = (12 + 4I)I33; εb = DiagonalMatrix[{6 + I, 6 + I, 2.5 + 0.5I}];
Ua = DiagonalMatrix[{1, 1, U}] / ∛U ;
Ub = DiagonalMatrix[{U, 1, 1}] / ∛U ;
```

```
(* Initialization of arrays for plotting *)
fmaxcount = 21;
plotRϵBrx = Array[0&, {fmaxcount, 2}];
plotRϵBry = Array[0&, {fmaxcount, 2}];
plotRϵBrz = Array[0&, {fmaxcount, 2}];
plotIϵBrx = Array[0&, {fmaxcount, 2}];
plotIϵBry = Array[0&, {fmaxcount, 2}];
plotIϵBrz = Array[0&, {fmaxcount, 2}];
plotRϵBrx[[1]] = {0, Re[ϵb[[1, 1]]]};
plotRϵBry[[1]] = {0, Re[ϵb[[2, 2]]]};
plotRϵBrz[[1]] = {0, Re[ϵb[[3, 3]]]};
plotIϵBrx[[1]] = {0, Im[ϵb[[1, 1]]]};
plotIϵBry[[1]] = {0, Im[ϵb[[2, 2]]]};
plotIϵBrz[[1]] = {0, Im[ϵb[[3, 3]]]};
plotRϵBrx[[fmaxcount]] = {1, Re[ϵa[[1, 1]]]};
plotRϵBry[[fmaxcount]] = {1, Re[ϵa[[2, 2]]]};
plotRϵBrz[[fmaxcount]] = {1, Re[ϵa[[3, 3]]]};
plotIϵBrx[[fmaxcount]] = {1, Im[ϵa[[1, 1]]]};
plotIϵBry[[fmaxcount]] = {1, Im[ϵa[[2, 2]]]};
plotIϵBrz[[fmaxcount]] = {1, Im[ϵa[[3, 3]]]};
```

```
(* Depolarization dyadic *)
depol[ϵ_, U_]:=Module[{b, θ, ϕ, qhat, iU, dx, dy, dz},
qhat = {Sin[θ] Cos[ϕ], Sin[θ] Sin[ϕ], Cos[θ]};
iU = Inverse[U];
b = qhat.(iU.ϵ.iU).qhat;
dx = NIntegrate[(qhat[[1]] qhat[[1]] Sin[θ]/b), {θ, 0, Pi}, {ϕ, 0, 2 Pi}];
dy = NIntegrate[(qhat[[2]] qhat[[2]] Sin[θ]/b), {θ, 0, Pi}, {ϕ, 0, 2 Pi}];
dz = NIntegrate[(qhat[[3]] qhat[[3]] Sin[θ]/b), {θ, 0, Pi}, {ϕ, 0, 2 Pi}];
iU.{{dx, 0, 0}, {0, dy, 0}, {0, 0, dz}}.iU/(4 Pi)]

tϵBr[ϵ_]:=Module[{ta, tb},
ta = fa Inverse[I33 + depol[ϵ, Ua].(ϵa − ϵ)]; tb = fb Inverse[I33 + depol[ϵ, Ub].(ϵb − ϵ)];
(ϵa.ta + ϵb.tb).Inverse[ta + tb]];
tol = 0.000001;

U = 1/5;
For[fcount = 1, fcount < fmaxcount − 1, fcount++,
fa = fcount/(fmaxcount − 1);
fb = 1 − fa;
ϵBr = (fa ϵa) + (fb ϵb);
delta = 1;
n = 1;
For[ncount = 1, ncount < 25, ncount++,
If[delta > tol,
newϵBr = tϵBr[ϵBr];
delta = Max[{Abs[newϵBr − ϵBr][[1, 1]]/Abs[ϵBr][[1, 1]], Abs[newϵBr − ϵBr][[2, 2]]/Abs[ϵBr][[2, 2]],
Abs[newϵBr − ϵBr][[3, 3]]/Abs[ϵBr][[3, 3]]}];
ϵBr = newϵBr;
n++; ]];
```

```
Print ["fa = ", N[fa], ", no. of interations = ", n, ", delta = ", delta] ;
plotRϵBrx[[fcount + 1]] = {fa, Re[ϵBr[[1, 1]]]};
plotRϵBry[[fcount + 1]] = {fa, Re[ϵBr[[2, 2]]]};
plotRϵBrz[[fcount + 1]] = {fa, Re[ϵBr[[3, 3]]]};
plotIϵBrx[[fcount + 1]] = {fa, Im[ϵBr[[1, 1]]]};
plotIϵBry[[fcount + 1]] = {fa, Im[ϵBr[[2, 2]]]};
plotIϵBrz[[fcount + 1]] = {fa, Im[ϵBr[[3, 3]]]}; ]

U1Rx = ListPlot[plotRϵBrx, Joined → True, PlotStyle → {Red, Thickness[0.008]}];
U1Ry = ListPlot[plotRϵBry, Joined → True, PlotStyle → {Red, Thickness[0.008]}];
U1Rz = ListPlot[plotRϵBrz, Joined → True, PlotStyle → {Red, Thickness[0.008]}];
U1Ix = ListPlot[plotIϵBrx, Joined → True, PlotStyle → {Red, Thickness[0.008]}];
U1Iy = ListPlot[plotIϵBry, Joined → True, PlotStyle → {Red, Thickness[0.008]}];
U1Iz = ListPlot[plotIϵBrz, Joined → True, PlotStyle → {Red, Thickness[0.008]}];

U = 1;
For[fcount = 1, fcount < fmaxcount − 1, fcount++,
fa = fcount/(fmaxcount − 1);
fb = 1 − fa;
ϵBr = (fa ϵa) + (fb ϵb);
delta = 1;
n = 1;
For[ncount = 1, ncount < 25, ncount++,
If[delta > tol,
newϵBr = tϵBr[ϵBr];
delta = Max[{Abs[newϵBr − ϵBr][[1, 1]]/Abs[ϵBr][[1, 1]], Abs[newϵBr − ϵBr][[2, 2]]/Abs[ϵBr][[2, 2]],
Abs[newϵBr − ϵBr][[3, 3]]/Abs[ϵBr][[3, 3]]}];
ϵBr = newϵBr;
n++; ]];
Print ["fa = ", N[fa], ", no. of interations = ", n, ", delta = ", delta] ;
plotRϵBrx[[fcount + 1]] = {fa, Re[ϵBr[[1, 1]]]};
plotRϵBry[[fcount + 1]] = {fa, Re[ϵBr[[2, 2]]]};
```

```
plotRϵBrz[[fcount + 1]] = {fa, Re[ϵBr[[3, 3]]]};
plotIϵBrx[[fcount + 1]] = {fa, Im[ϵBr[[1, 1]]]};
plotIϵBry[[fcount + 1]] = {fa, Im[ϵBr[[2, 2]]]};
plotIϵBrz[[fcount + 1]] = {fa, Im[ϵBr[[3, 3]]]};]

U2Rx = ListPlot[plotRϵBrx, Joined → True,
PlotStyle → {Blue, Dashing[{0.02, 0.02}], Thickness[0.008]}];
U2Ry = ListPlot[plotRϵBry, Joined → True,
PlotStyle → {Blue, Dashing[{0.02, 0.02}], Thickness[0.008]}];
U2Rz = ListPlot[plotRϵBrz, Joined → True,
PlotStyle → {Blue, Dashing[{0.02, 0.02}], Thickness[0.008]}];
U2Ix = ListPlot[plotIϵBrx, Joined → True,
PlotStyle → {Blue, Dashing[{0.02, 0.02}], Thickness[0.008]}];
U2Iy = ListPlot[plotIϵBry, Joined → True,
PlotStyle → {Blue, Dashing[{0.02, 0.02}], Thickness[0.008]}];
U2Iz = ListPlot[plotIϵBrz, Joined → True,
PlotStyle → {Blue, Dashing[{0.02, 0.02}], Thickness[0.008]}];

U = 5;
For[fcount = 1, fcount < fmaxcount − 1, fcount++,
fa = fcount/(fmaxcount − 1);
fb = 1 − fa;
ϵBr = (fa ϵa) + (fb ϵb);
delta = 1;
n = 1;
For[ncount = 1, ncount < 25, ncount++,
If[delta > tol,
newϵBr = tϵBr[ϵBr];
delta = Max[{Abs[newϵBr − ϵBr][[1, 1]]/Abs[ϵBr][[1, 1]], Abs[newϵBr − ϵBr][[2, 2]]/Abs[ϵBr][[2, 2]],
Abs[newϵBr − ϵBr][[3, 3]]/Abs[ϵBr][[3, 3]]}];
ϵBr = newϵBr;
n++; ]];
```

```
Print ["fa = ", N[fa], ", no. of interations = ", n, ", delta = ", delta];
plotRϵBrx[[fcount + 1]] = {fa, Re[ϵBr[[1, 1]]]};
plotRϵBry[[fcount + 1]] = {fa, Re[ϵBr[[2, 2]]]};
plotRϵBrz[[fcount + 1]] = {fa, Re[ϵBr[[3, 3]]]};
plotIϵBrx[[fcount + 1]] = {fa, Im[ϵBr[[1, 1]]]};
plotIϵBry[[fcount + 1]] = {fa, Im[ϵBr[[2, 2]]]};
plotIϵBrz[[fcount + 1]] = {fa, Im[ϵBr[[3, 3]]]};]

U3Rx = ListPlot[plotRϵBrx, Joined → True,
PlotStyle → {Green, Dashing[{0.035, 0.025, 0.005, 0.025}], Thickness[0.008]}];
U3Ry = ListPlot[plotRϵBry, Joined → True,
PlotStyle → {Green, Dashing[{0.035, 0.025, 0.005, 0.025}], Thickness[0.008]}];
U3Rz = ListPlot[plotRϵBrz, Joined → True,
PlotStyle → {Green, Dashing[{0.035, 0.025, 0.005, 0.025}], Thickness[0.008]}];
U3Ix = ListPlot[plotIϵBrx, Joined → True,
PlotStyle → {Green, Dashing[{0.035, 0.025, 0.005, 0.025}], Thickness[0.008]}];
U3Iy = ListPlot[plotIϵBry, Joined → True,
PlotStyle → {Green, Dashing[{0.035, 0.025, 0.005, 0.025}], Thickness[0.008]}];
U3Iz = ListPlot[plotIϵBrz, Joined → True,
PlotStyle → {Green, Dashing[{0.035, 0.025, 0.005, 0.025}], Thickness[0.008]}];

Rx = Show[Graphics[U1Rx], Graphics[U2Rx], Graphics[U3Rx],
Frame → True, Axes → False, PlotRange → All, FrameLabel → {"fa", "Re{ ϵx^Br/ ϵ0 }", "", ""},
BaseStyle → {FontSize → 14, FontFamily → "Myriad"}]
```

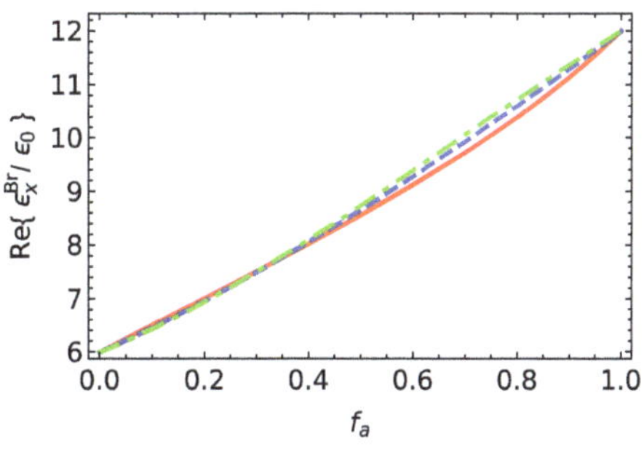

```
Ry = Show[Graphics[U1Ry], Graphics[U2Ry], Graphics[U3Ry],
Frame → True, Axes → False, PlotRange → All, FrameLabel → {"fa", "Re{ ϵy^Br/ ϵ0 }", "", ""},
BaseStyle → {FontSize → 14, FontFamily → "Myriad"}]
```

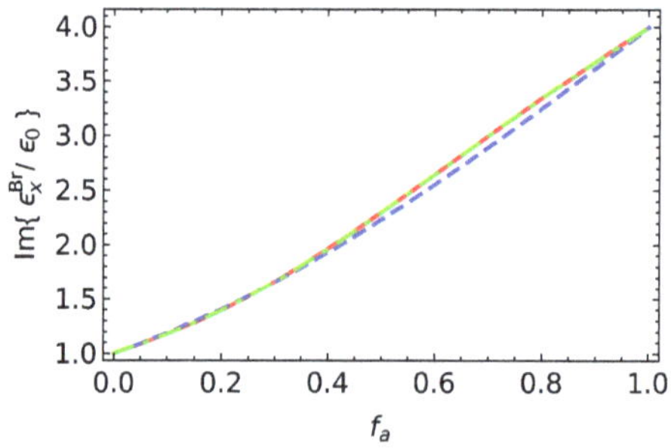


```
Rz = Show[Graphics[U1Rz], Graphics[U2Rz], Graphics[U3Rz],
Frame → True, Axes → False, PlotRange → All, FrameLabel → {"fa", "Re{ ϵz^Br/ ϵ0 }", "", ""},
BaseStyle → {FontSize → 14, FontFamily → "Myriad"}]
```

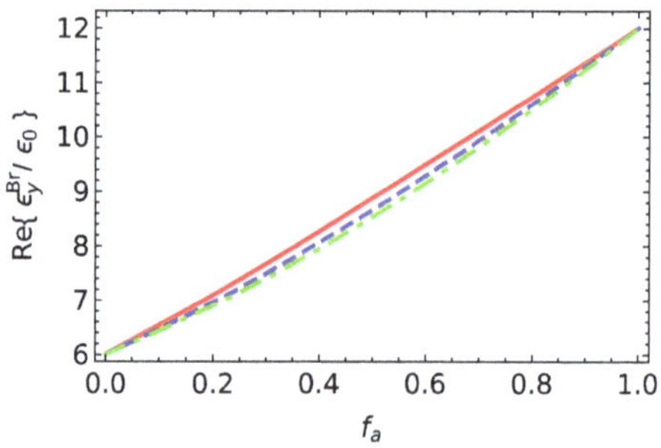


```
Ix = Show[Graphics[U1Ix], Graphics[U2Ix], Graphics[U3Ix],
Frame → True, Axes → False, PlotRange → All, FrameLabel → {"fa", "Im{ ϵx^Br/ ϵ0 }", "", ""},
BaseStyle → {FontSize → 14, FontFamily → "Myriad"}]
```

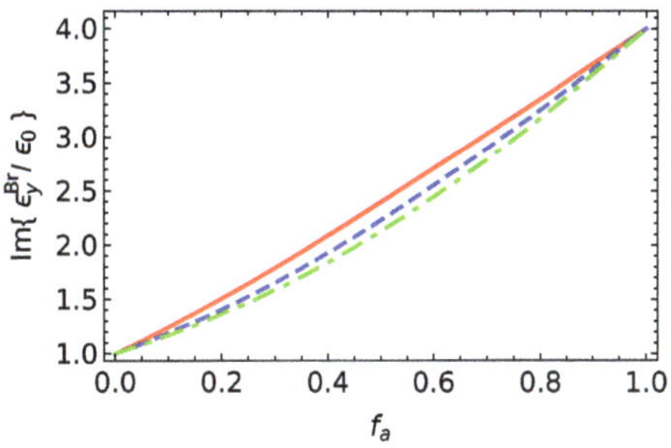

Iy = Show[Graphics[U1Iy], Graphics[U2Iy], Graphics[U3Iy],

Frame → True, Axes → False, PlotRange → All, FrameLabel → {"f_a", "Im{ ϵ_y^{Br}/ ϵ_0 }", "", ""},

BaseStyle → {FontSize → 14, FontFamily → "Myriad"}]

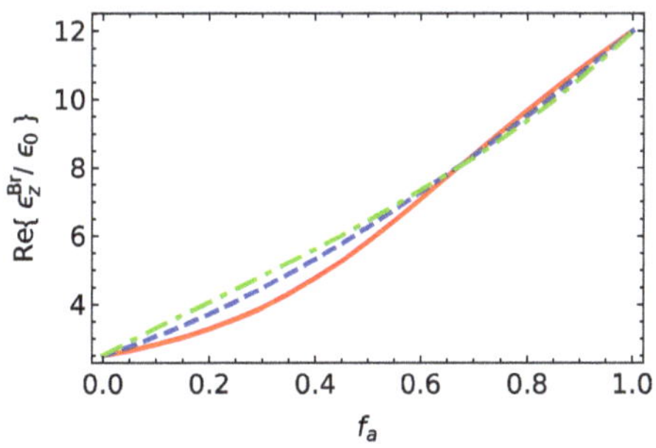

Iz = Show[Graphics[U1Iz], Graphics[U2Iz], Graphics[U3Iz],

Frame → True, Axes → False, PlotRange → All, FrameLabel → {"f_a", "Im{ ϵ_z^{Br}/ ϵ_0 }", "", ""},

BaseStyle → {FontSize → 14, FontFamily → "Myriad"}]

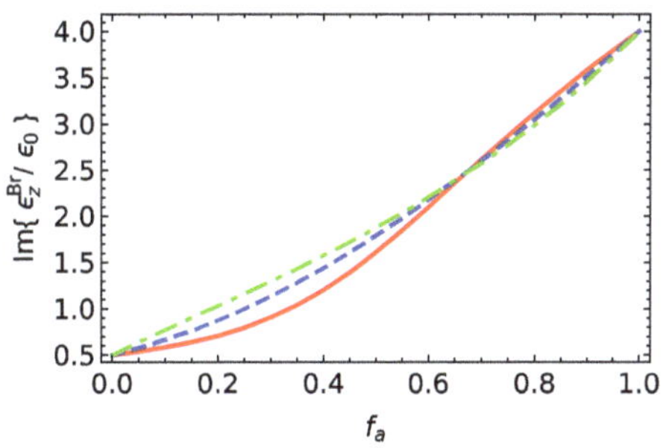

B.5 Figure 6.6

Here follows the code for the second-order SPFT in the case of a biaxial bianisotropic HCM, as used for figure 6.6.

```
(* Inputs : Second-order SPFT for biaxial bianisotropic HCM *)

I33 = IdentityMatrix[3]; I66 = IdentityMatrix[6];

ϵa = (5 + 0.04I)I33; ξa = (0.2 + 0.003I) I I33; ζa = −ξa; μa = (2 + 0.02I)I33;

ϵb = DiagonalMatrix[{2.5, 1.5, 1.5}];

ξb = 0 I33; ζb = 0 I33; μb = DiagonalMatrix[{1.4, 1.1, 1.1}];

U = 5;

Ua = DiagonalMatrix[{1, 1, U}] / ∛U ;

Ub = DiagonalMatrix[{1, 1, U}] / ∛U ;

Ka = {{ϵa[[1, 1]], ϵa[[1, 2]], ϵa[[1, 3]], ξa[[1, 1]], ξa[[1, 2]], ξa[[1, 3]]},
{ϵa[[2, 1]], ϵa[[2, 2]], ϵa[[2, 3]], ξa[[2, 1]], ξa[[2, 2]], ξa[[2, 3]]},
{ϵa[[3, 1]], ϵa[[3, 2]], ϵa[[3, 3]], ξa[[3, 1]], ξa[[3, 2]], ξa[[3, 3]]},
{ζa[[1, 1]], ζa[[1, 2]], ζa[[1, 3]], μa[[1, 1]], μa[[1, 2]], μa[[1, 3]]},
{ζa[[2, 1]], ζa[[2, 2]], ζa[[2, 3]], μa[[2, 1]], μa[[2, 2]], μa[[2, 3]]},
{ζa[[3, 1]], ζa[[3, 2]], ζa[[3, 3]], μa[[3, 1]], μa[[3, 2]], μa[[3, 3]]}};

Kb = {{ϵb[[1, 1]], ϵb[[1, 2]], ϵb[[1, 3]], ξb[[1, 1]], ξb[[1, 2]], ξb[[1, 3]]},
{ϵb[[2, 1]], ϵb[[2, 2]], ϵb[[2, 3]], ξb[[2, 1]], ξb[[2, 2]], ξb[[2, 3]]},
{ϵb[[3, 1]], ϵb[[3, 2]], ϵb[[3, 3]], ξb[[3, 1]], ξb[[3, 2]], ξb[[3, 3]]},
{ζb[[1, 1]], ζb[[1, 2]], ζb[[1, 3]], μb[[1, 1]], μb[[1, 2]], μb[[1, 3]]},
{ζb[[2, 1]], ζb[[2, 2]], ζb[[2, 3]], μb[[2, 1]], μb[[2, 2]], μb[[2, 3]]},
{ζb[[3, 1]], ζb[[3, 2]], ζb[[3, 3]], μb[[3, 1]], μb[[3, 2]], μb[[3, 3]]}};

fa = 0.4;

fb = 1 − fa;
```

```
(* Depolarization dyadic D0 *)
depol[K_, U_]:=
Module[{ϵ, ξ, ζ, μ, iU, qh, nϵ, nξ, nζ, nμ, b, θ, ϕ, dϵx, dϵy, dϵz, dξx, dξy, dξz, dμx, dμy, dμz,
eed, emd, med, mmd},
ϵ = DiagonalMatrix[{K[[1, 1]], K[[2, 2]], K[[3, 3]]}];
ξ = DiagonalMatrix[{K[[1, 4]], K[[2, 5]], K[[3, 6]]}];
ζ = −ξ;
μ = DiagonalMatrix[{K[[4, 4]], K[[5, 5]], K[[6, 6]]}];
iU = Inverse[U];
qh = {Sin[θ] Cos[ϕ], Sin[θ] Sin[ϕ], Cos[θ]};
nϵ = qh.(iU.ϵ.iU).qh;
nξ = qh.(iU.ξ.iU).qh;
nζ = qh.(iU.ζ.iU).qh;
nμ = qh.(iU.μ.iU).qh;
b = (nϵ nμ) − (nξ nζ);
dϵx = NIntegrate[(nμ qh[[1]] qh[[1]] Sin[θ])/b, {θ, 0, Pi}, {ϕ, 0, 2 Pi}];
dϵy = NIntegrate[(nμ qh[[2]] qh[[2]] Sin[θ])/b, {θ, 0, Pi}, {ϕ, 0, 2 Pi}];
dϵz = NIntegrate[(nμ qh[[3]] qh[[3]] Sin[θ])/b, {θ, 0, Pi}, {ϕ, 0, 2 Pi}];
eed = iU.{{dϵx, 0, 0}, {0, dϵy, 0}, {0, 0, dϵz}}.iU/(4 Pi);
dξx = −NIntegrate[(nξ qh[[1]] qh[[1]] Sin[θ])/b, {θ, 0, Pi}, {ϕ, 0, 2 Pi}];
dξy = −NIntegrate[(nξ qh[[2]] qh[[2]] Sin[θ])/b, {θ, 0, Pi}, {ϕ, 0, 2 Pi}];
dξz = −NIntegrate[(nξ qh[[3]] qh[[3]] Sin[θ])/b, {θ, 0, Pi}, {ϕ, 0, 2 Pi}];
emd = iU.{{dξx, 0, 0}, {0, dξy, 0}, {0, 0, dξz}}.iU/(4 Pi);
med = −emd;
dμx = NIntegrate[(nϵ qh[[1]] qh[[1]] Sin[θ])/b, {θ, 0, Pi}, {ϕ, 0, 2 Pi}];
dμy = NIntegrate[(nϵ qh[[2]] qh[[2]] Sin[θ])/b, {θ, 0, Pi}, {ϕ, 0, 2 Pi}];
dμz = NIntegrate[(nϵ qh[[3]] qh[[3]] Sin[θ])/b, {θ, 0, Pi}, {ϕ, 0, 2 Pi}];
mmd = iU.{{dμx, 0, 0}, {0, dμy, 0}, {0, 0, dμz}}.iU/(4 Pi);
{{eed[[1, 1]], 0, 0, emd[[1, 1]], 0, 0}, {0, eed[[2, 2]], 0, 0, emd[[2, 2]], 0}, {0, 0, eed[[3, 3]], 0, 0, emd[[3
```

```
{med[[1, 1]], 0, 0, mmd[[1, 1]], 0, 0}, {0, med[[2, 2]], 0, 0, mmd[[2, 2]], 0}, {0, 0, med[[3, 3]], 0, 0, mmd[[3, 3]]}}]

(* Zeroth-order SPFT *)
KBr = fa Ka + fb Kb;
tKBr[K_]:=Module[{ta, tb},
ta = fa Inverse[I66 + depol[K, Ua].(Ka − K)]; tb = fb Inverse[I66 + depol[K, Ub].(Kb − K)];
(Ka.ta + Kb.tb).Inverse[ta + tb]];
tol = 0.000001;
delta = 1;
n = 1;
For[ncount = 1, ncount < 25, ncount++,
If[delta > tol,
newKBr = tKBr[KBr];
dK = newKBr − KBr;
delta =
Max[{Abs[dK][[1, 1]]/Abs[KBr][[1, 1]], Abs[dK][[2, 2]]/Abs[KBr][[2, 2]], Abs[dK][[3, 3]]/Abs[KBr][[3, 3]],
Abs[dK][[1, 4]]/Abs[KBr][[1, 4]], Abs[dK][[2, 5]]/Abs[KBr][[2, 5]], Abs[dK][[3, 6]]/Abs[KBr][[3, 6]],
Abs[dK][[4, 4]]/Abs[KBr][[4, 4]], Abs[dK][[5, 5]]/Abs[KBr][[5, 5]], Abs[dK][[6, 6]]/Abs[KBr][[6, 6]]}];
KBr = newKBr;
n++; ]];
Print ["fa = ", N[fa], ", no. of interations = ", n, ", delta = ", delta]

(* Initialization of arrays for plotting *)
Lmaxcount = 7;
plotRεSPFTx = Array[0&, {Lmaxcount, 2}];
plotRεSPFTy = Array[0&, {Lmaxcount, 2}];
plotRεSPFTz = Array[0&, {Lmaxcount, 2}];
plotIεSPFTx = Array[0&, {Lmaxcount, 2}];
plotIεSPFTy = Array[0&, {Lmaxcount, 2}];
plotIεSPFTz = Array[0&, {Lmaxcount, 2}];
plotRεSPFTx[[1]] = {0, Re[KBr[[1, 1]]]};
plotRεSPFTy[[1]] = {0, Re[KBr[[2, 2]]]};
```

```
plotRϵSPFTz[[1]] = {0, Re[KBr[[3, 3]]]};
plotIϵSPFTx[[1]] = {0, 10^2 Im[KBr[[1, 1]]]};
plotIϵSPFTy[[1]] = {0, 10^2 Im[KBr[[2, 2]]]};
plotIϵSPFTz[[1]] = {0, 10^2 Im[KBr[[3, 3]]]};
plotRξSPFTx = Array[0&, {Lmaxcount, 2}];
plotRξSPFTy = Array[0&, {Lmaxcount, 2}];
plotRξSPFTz = Array[0&, {Lmaxcount, 2}];
plotIξSPFTx = Array[0&, {Lmaxcount, 2}];
plotIξSPFTy = Array[0&, {Lmaxcount, 2}];
plotIξSPFTz = Array[0&, {Lmaxcount, 2}];
plotRξSPFTx[[1]] = {0, 10^2 Re[KBr[[1, 4]]/I]};
plotRξSPFTy[[1]] = {0, 10^2 Re[KBr[[2, 5]]/I]};
plotRξSPFTz[[1]] = {0, 10^2 Re[KBr[[3, 6]]/I]};
plotIξSPFTx[[1]] = {0, 10^2 Im[KBr[[1, 4]]/I]};
plotIξSPFTy[[1]] = {0, 10^2 Im[KBr[[2, 5]]/I]};
plotIξSPFTz[[1]] = {0, 10^2 Im[KBr[[3, 6]]/I]};
plotRμSPFTx = Array[0&, {Lmaxcount, 2}];
plotRμSPFTy = Array[0&, {Lmaxcount, 2}];
plotRμSPFTz = Array[0&, {Lmaxcount, 2}];
plotIμSPFTx = Array[0&, {Lmaxcount, 2}];
plotIμSPFTy = Array[0&, {Lmaxcount, 2}];
plotIμSPFTz = Array[0&, {Lmaxcount, 2}];
plotRμSPFTx[[1]] = {0, Re[KBr[[4, 4]]]};
plotRμSPFTy[[1]] = {0, Re[KBr[[5, 5]]]};
plotRμSPFTz[[1]] = {0, Re[KBr[[6, 6]]]};
plotIμSPFTx[[1]] = {0, 10^3 Im[KBr[[4, 4]]]};
plotIμSPFTy[[1]] = {0, 10^3 Im[KBr[[5, 5]]]};
plotIμSPFTz[[1]] = {0, 10^3 Im[KBr[[6, 6]]]};

(* Depolarization dyadic D+ *)
dplus[K_, U_, L_]:=
Module[{iU, iUq, q, θ, ϕ, iUqxI, Amat, adjA, detA, Gmat, ϵ, ξ, ζ, μ, qh, nϵ, nξ, nζ, nμ, b, iUqqiU, Ginfmat,
```

```
Gplusmat, κplus, κminus, t4, t2, t0, T0, detA0, adjAplus, adjAminus, cplus, cminus, c,
ex, ey, ez, zx, zy, zz, mx, my, mz},
iU = Inverse[U];
iUq = qiU.{Sin[θ] Cos[φ], Sin[θ] Sin[φ], Cos[θ]};
iUqxI = {{0, −iUq[[3]], iUq[[2]]}, {iUq[[3]], 0, −iUq[[1]]}, {−iUq[[2]], iUq[[1]], 0}};
Amat = {{0, 0, 0, iUqxI[[1, 1]], iUqxI[[1, 2]], iUqxI[[1, 3]]}, {0, 0, 0, iUqxI[[2, 1]], iUqxI[[2, 2]], iUqxI[[2, 3]]},
{0, 0, 0, iUqxI[[3, 1]], iUqxI[[3, 2]], iUqxI[[3, 3]]}, {−iUqxI[[1, 1]], −iUqxI[[1, 2]], −iUqxI[[1, 3]], 0, 0, 0},
{−iUqxI[[2, 1]], −iUqxI[[2, 2]], −iUqxI[[2, 3]], 0, 0, 0}, {−iUqxI[[3, 1]], −iUqxI[[3, 2]], −iUqxI[[3, 3]], 0, 0, 0}}
+K;
adjA = Map[Reverse, Minors[Transpose[Amat], Length[Amat] − 1], {0, 1}]
Table[(−1)^(i + j), {i, Length[Amat]}, {j, Length[Amat]}];
detA = Det[Amat];
Gmat = Inverse[Amat];
ϵ = {{K[[1, 1]], K[[1, 2]], K[[1, 3]]}, {K[[2, 1]], K[[2, 2]], K[[2, 3]]}, {K[[3, 1]], K[[3, 2]], K[[3, 3]]}};
ξ = {{K[[1, 4]], K[[1, 5]], K[[1, 6]]}, {K[[2, 4]], K[[2, 5]], K[[2, 6]]}, {K[[3, 4]], K[[3, 5]], K[[3, 6]]}};
ζ = {{K[[4, 1]], K[[4, 2]], K[[4, 3]]}, {K[[5, 1]], K[[5, 2]], K[[5, 3]]}, {K[[6, 1]], K[[6, 2]], K[[6, 3]]}};
μ = {{K[[4, 4]], K[[4, 5]], K[[4, 6]]}, {K[[5, 4]], K[[5, 5]], K[[5, 6]]}, {K[[6, 4]], K[[6, 5]], K[[6, 6]]}};
qh = {Sin[θ] Cos[φ], Sin[θ] Sin[φ], Cos[θ]};
nϵ = qh.(iU.ϵ.iU).qh;
nξ = qh.(iU.ξ.iU).qh;
nζ = qh.(iU.ζ.iU).qh;
nμ = qh.(iU.μ.iU).qh;
b = (nϵ nμ) − (nξ nζ);
iUqqiU = iU.{{qh[[1]] qh[[1]], qh[[1]] qh[[2]], qh[[1]] qh[[3]]}, {qh[[2]] qh[[1]], qh[[2]] qh[[2]], qh[[2]] qh[[3]]},
{qh[[3]] qh[[1]], qh[[3]] qh[[2]], qh[[3]] qh[[3]]}}.iU;
Ginfmat = {{nμ iUqqiU[[1, 1]], nμ iUqqiU[[1, 2]], nμ iUqqiU[[1, 3]], −nξ iUqqiU[[1, 1]], −nξ iUqqiU[[1, 2]],
−nξ iUqqiU[[1, 3]]}, {nμ iUqqiU[[2, 1]], nμ iUqqiU[[2, 2]], nμ iUqqiU[[2, 3]], −nξ iUqqiU[[2, 1]],
−nξ iUqqiU[[2, 2]], −nξ iUqqiU[[2, 3]]}, {nμ iUqqiU[[3, 1]], nμ iUqqiU[[3, 2]], nμ iUqqiU[[3, 3]],
−nξ iUqqiU[[3, 1]], −nξ iUqqiU[[3, 2]], −nξ iUqqiU[[3, 3]]}, {−nζ iUqqiU[[1, 1]], −nζ iUqqiU[[1, 2]],
−nζ iUqqiU[[1, 3]], nϵ iUqqiU[[1, 1]], nϵ iUqqiU[[1, 2]], nϵ iUqqiU[[1, 3]]}, {−nζ iUqqiU[[2, 1]],
−nζ iUqqiU[[2, 2]], −nζ iUqqiU[[2, 3]], nϵ iUqqiU[[2, 1]], nϵ iUqqiU[[2, 2]], nϵ iUqqiU[[2, 3]]},
{−nζ iUqqiU[[3, 1]], −nζ iUqqiU[[3, 2]], −nζ iUqqiU[[3, 3]], nϵ iUqqiU[[3, 1]], nϵ iUqqiU[[3, 2]],
```

```
nϵ iUqqiU[[3, 3]]}}/b;
Gplusmat = Gmat − Ginfmat;
t4 = Coefficient[detA, q, 4];
t2 = Coefficient[detA, q, 2];
t0 = Coefficient[detA, q, 0];
T0 = Gmat/.q → 0;
detA0 = detA/.q → 0;
κplus = (−t2 + √(t2^2 − 4 t4 t0))/(2 t4); κminus = (−t2 − √(t2^2 − 4 t4 t0))/(2 t4);
adjAplus = (adjA/.q → √κplus) + (adjA/.q → −√κplus);
adjAminus = (adjA/.q → √κminus) + (adjA/.q → −√κminus);
cplus = Exp[I q L]/(2q^2) (1 − (I q L))/.q → √κplus; cminus = Exp[I q L]/(2q^2) (1 − (I q L))/.q → √κminus;
c = Sin[θ]/(4 Pi b) (1/(κplus − κminus) ((cplus adjAplus) − (cminus adjAminus)) + detA0/(κplus κminus) (T0 − Ginfmat));
ex = NIntegrate[c[[1, 1]], {θ, 0, Pi}, {ϕ, 0, 2 Pi}];
ey = NIntegrate[c[[2, 2]], {θ, 0, Pi}, {ϕ, 0, 2 Pi}];
ez = NIntegrate[c[[3, 3]], {θ, 0, Pi}, {ϕ, 0, 2 Pi}];
zx = NIntegrate[c[[1, 4]], {θ, 0, Pi}, {ϕ, 0, 2 Pi}];
zy = NIntegrate[c[[2, 5]], {θ, 0, Pi}, {ϕ, 0, 2 Pi}];
zz = NIntegrate[c[[3, 6]], {θ, 0, Pi}, {ϕ, 0, 2 Pi}];
mx = NIntegrate[c[[4, 4]], {θ, 0, Pi}, {ϕ, 0, 2 Pi}];
my = NIntegrate[c[[5, 5]], {θ, 0, Pi}, {ϕ, 0, 2 Pi}];
mz = NIntegrate[c[[6, 6]], {θ, 0, Pi}, {ϕ, 0, 2 Pi}];
{{ex, 0, 0, zx, 0, 0}, {0, ey, 0, 0, zy, 0}, {0, 0, ez, 0, 0, zz}, {−zx, 0, 0, mx, 0, 0}, {0, −zy, 0, 0, my, 0},
{0, 0, −zz, 0, 0, mz}}]

(* Mass operator *)
alpha[K_, Kamb_, U_]:=(K − Kamb).Inverse[I66 + depol[Kamb, U].(K − Kamb)];
massop[L_]:=fa fb (alpha[Ka, KBr, Ua] − alpha[Kb, KBr, Ub]).dplus[KBr, Ua, L].(alpha[Ka, KBr, Ua]
−alpha[Kb, KBr, Ub]);

For[Lcount = 1, Lcount < Lmaxcount, Lcount++,
L = Lcount/20;
KSPFT = KBr − Inverse[I66 + massop[L].depol[KBr, Ua]].massop[L];
```

```
Print["L = ", N[L]];
plotRϵSPFTx[[Lcount + 1]] = {L, Re[KSPFT[[1, 1]]]};
plotRϵSPFTy[[Lcount + 1]] = {L, Re[KSPFT[[2, 2]]]};
plotRϵSPFTz[[Lcount + 1]] = {L, Re[KSPFT[[3, 3]]]};
plotIϵSPFTx[[Lcount + 1]] = {L, 10^2 Im[KSPFT[[1, 1]]]};
plotIϵSPFTy[[Lcount + 1]] = {L, 10^2 Im[KSPFT[[2, 2]]]};
plotIϵSPFTz[[Lcount + 1]] = {L, 10^2 Im[KSPFT[[3, 3]]]};
plotRξSPFTx[[Lcount + 1]] = {L, 10^2 Re[KSPFT[[1, 4]]/I]};
plotRξSPFTy[[Lcount + 1]] = {L, 10^2 Re[KSPFT[[2, 5]]/I]};
plotRξSPFTz[[Lcount + 1]] = {L, 10^2 Re[KSPFT[[3, 6]]/I]};
plotIξSPFTx[[Lcount + 1]] = {L, 10^2 Im[KSPFT[[1, 4]]/I]};
plotIξSPFTy[[Lcount + 1]] = {L, 10^2 Im[KSPFT[[2, 5]]/I]};
plotIξSPFTz[[Lcount + 1]] = {L, 10^2 Im[KSPFT[[3, 6]]/I]};
plotRμSPFTx[[Lcount + 1]] = {L, Re[KSPFT[[4, 4]]]};
plotRμSPFTy[[Lcount + 1]] = {L, Re[KSPFT[[5, 5]]]};
plotRμSPFTz[[Lcount + 1]] = {L, Re[KSPFT[[6, 6]]]};
plotIμSPFTx[[Lcount + 1]] = {L, 10^3 Im[KSPFT[[4, 4]]]};
plotIμSPFTy[[Lcount + 1]] = {L, 10^3 Im[KSPFT[[5, 5]]]};
plotIμSPFTz[[Lcount + 1]] = {L, 10^3 Im[KSPFT[[6, 6]]]};]

pRex = ListPlot[plotRϵSPFTx, Joined → True,
PlotRange → All, PlotStyle → {Red, Thickness[0.008]}];
pRey = ListPlot[plotRϵSPFTy, Joined → True, PlotRange → All,
PlotStyle → {Blue, Dashing[{0.02, 0.02}], Thickness[0.008]}];
pRez = ListPlot[plotRϵSPFTz, Joined → True, PlotRange → All,
PlotStyle → {Green, Dashing[{0.035, 0.025, 0.005, 0.025}], Thickness[0.008]}];
pR = Show[Graphics[pRex], Graphics[pRey], Graphics[pRez],
Frame → True, Axes → False, PlotRange → All,
FrameLabel → {"k0Lcor", "Re{ ϵx^SPFT / ϵ0 }", "Re{ ϵy^SPFT / ϵ0 }", "Re{ ϵz^SPFT / ϵ0 }"},
BaseStyle → {FontSize → 14, FontFamily → "Myriad"}]
```

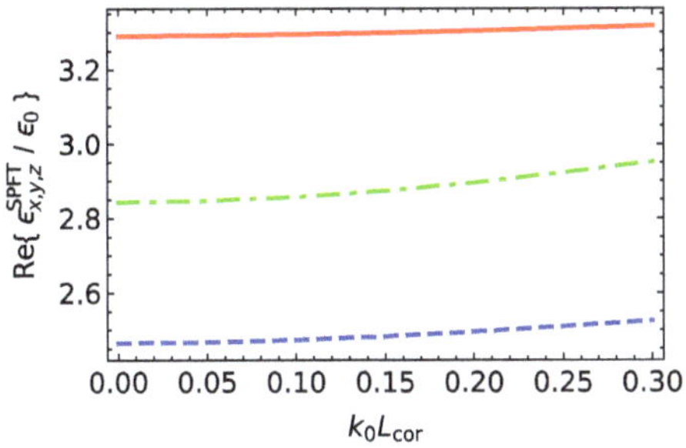


```
pIex = ListPlot[plotIεSPFTx, Joined → True, PlotRange → All, PlotStyle → {Red, Thickness[0.008]}];
pIey = ListPlot[plotIεSPFTy, Joined → True, PlotRange → All,
PlotStyle → {Blue, Dashing[{0.02, 0.02}], Thickness[0.008]}];
pIez = ListPlot[plotIεSPFTz, Joined → True, PlotRange → All,
PlotStyle → {Green, Dashing[{0.035, 0.025, 0.005, 0.025}], Thickness[0.008]}];
pI = Show[Graphics[pIex], Graphics[pIey], Graphics[pIez],
Frame → True, Axes → False, PlotRange → All,
FrameLabel → {"k0Lcor", "Im{ εx^SPFT / ε0 } × 10^2", "Im{ εy^SPFT / ε0 } × 10^2",
"Im{ εz^SPFT / ε0 } × 10^2"}, BaseStyle → {FontSize → 14, FontFamily → "Myriad"}]
```

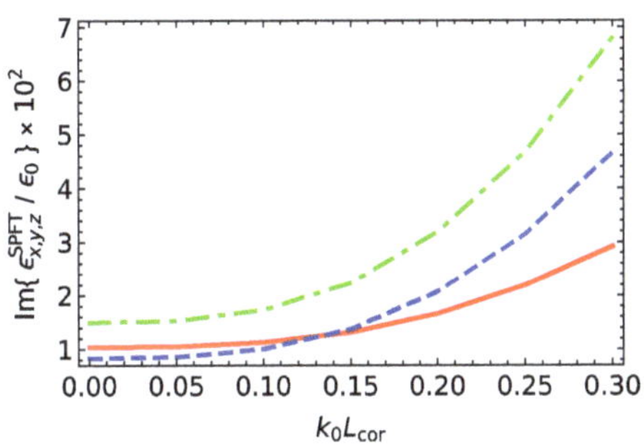


```
pRzx = ListPlot[plotRξSPFTx, Joined → True, PlotRange → All, PlotStyle → {Red, Thickness[0.008]}];
pRzy = ListPlot[plotRξSPFTy, Joined → True, PlotRange → All,
PlotStyle → {Blue, Dashing[{0.02, 0.02}], Thickness[0.008]}];
pRzz = ListPlot[plotRξSPFTz, Joined → True, PlotRange → All,
PlotStyle → {Green, Dashing[{0.035, 0.025, 0.005, 0.025}], Thickness[0.008]}];
pR = Show[Graphics[pRzx], Graphics[pRzy], Graphics[pRzz],
Frame → True, Axes → False, PlotRange → All,
FrameLabel → {"k0Lcor", "Re{ ξx^SPFT / i√(ε0μ0) } × 10^2", "Re{ ξy^SPFT / i√(ε0μ0) } × 10^2",
"Re{ ξz^SPFT / i√(ε0μ0) } × 10^2"}, BaseStyle → {FontSize → 14, FontFamily → "Myriad"}]
```

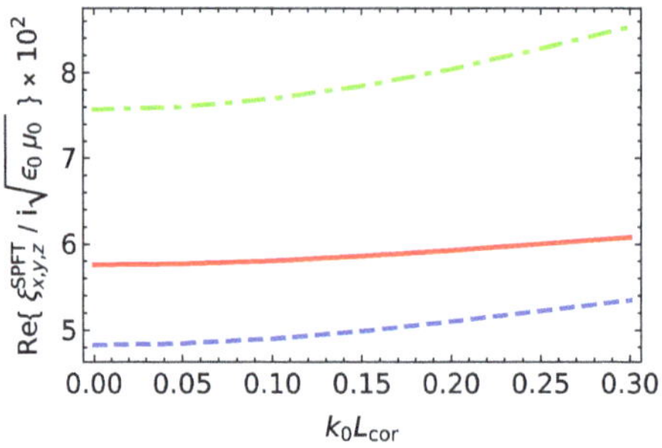

pIzx = ListPlot[plotIξSPFTx, Joined → True, PlotRange → All, PlotStyle → {Red, Thickness[0.008]}];

pIzy = ListPlot[plotIξSPFTy, Joined → True, PlotRange → All,

PlotStyle → {Blue, Dashing[{0.02, 0.02}], Thickness[0.008]}];

pIzz = ListPlot[plotIξSPFTz, Joined → True, PlotRange → All,

PlotStyle → {Green, Dashing[{0.035, 0.025, 0.005, 0.025}], Thickness[0.008]}];

pI = Show[Graphics[pIex], Graphics[pIey], Graphics[pIez],

Frame → True, Axes → False, PlotRange → All,

FrameLabel → {"$k_0 L_{\text{cor}}$", "Im{ ξ_x^{SPFT} / i$\sqrt{\epsilon_0\mu_0}$ } × 10^2", "Im{ ξ_y^{SPFT} / i$\sqrt{\epsilon_0\mu_0}$ } × 10^2",

"Im{ ξ_z^{SPFT} / i$\sqrt{\epsilon_0\mu_0}$ } × 10^2"}, BaseStyle → {FontSize → 14, FontFamily → "Myriad"}]

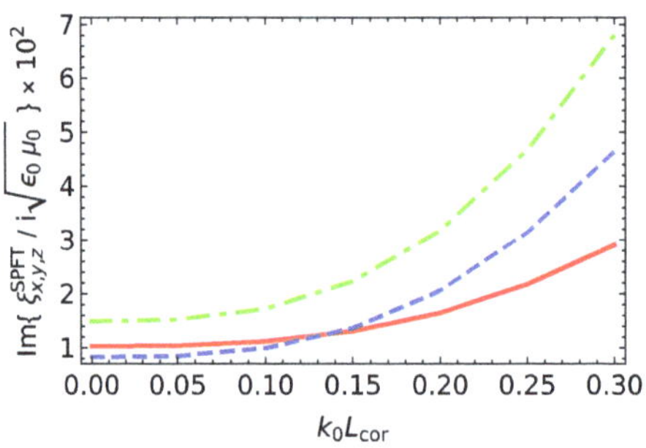

pRmx = ListPlot[plotRμSPFTx, Joined → True, PlotRange → All,

PlotStyle → {Red, Thickness[0.008]}];

pRmy = ListPlot[plotRμSPFTy, Joined → True, PlotRange → All,

PlotStyle → {Blue, Dashing[{0.02, 0.02}], Thickness[0.008]}];

pRmz = ListPlot[plotRμSPFTz, Joined → True, PlotRange → All,

PlotStyle → {Green, Dashing[{0.035, 0.025, 0.005, 0.025}], Thickness[0.008]}];

pR = Show[Graphics[pRmx], Graphics[pRmy], Graphics[pRmz],

Frame → True, Axes → False, PlotRange → All,

FrameLabel → {"$k_0 L_{\text{cor}}$", "Re{ μ_x^{SPFT} / μ_0 }", "Re{ μ_y^{SPFT} / μ_0 }", "Re{ μ_z^{SPFT} / μ_0 }"},

BaseStyle → {FontSize → 14, FontFamily → "Myriad"}]

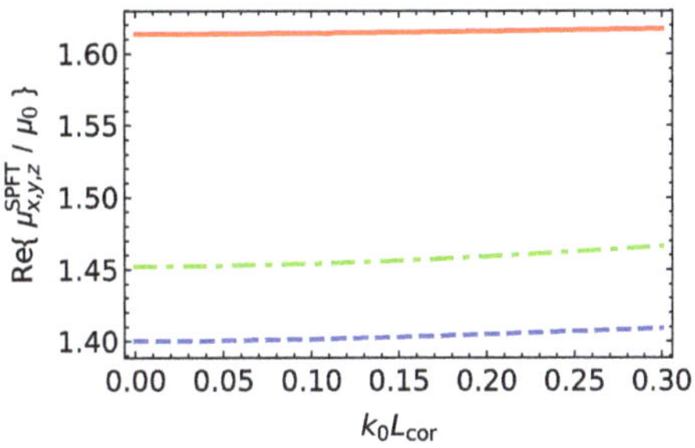


```
pImx = ListPlot[plotIμSPFTx, Joined → True, PlotRange → All, PlotStyle → {Red, Thickness[0.008]}];
pImy = ListPlot[plotIμSPFTy, Joined → True, PlotRange → All,
PlotStyle → {Blue, Dashing[{0.02, 0.02}], Thickness[0.008]}];
pImz = ListPlot[plotIμSPFTz, Joined → True, PlotRange → All,
PlotStyle → {Green, Dashing[{0.035, 0.025, 0.005, 0.025}], Thickness[0.008]}];
pI = Show[Graphics[pImx], Graphics[pImy], Graphics[pImz],
Frame → True, Axes → False, PlotRange → All,
FrameLabel → {"k0Lcor", "Im{ μx^SPFT / μ0 } × 10^3", "Im{ μy^SPFT / μ0 } × 10^3",
"Im{ μz^SPFT / μ0 } × 10^3"}, BaseStyle → {FontSize → 14, FontFamily → "Myriad"}]
```

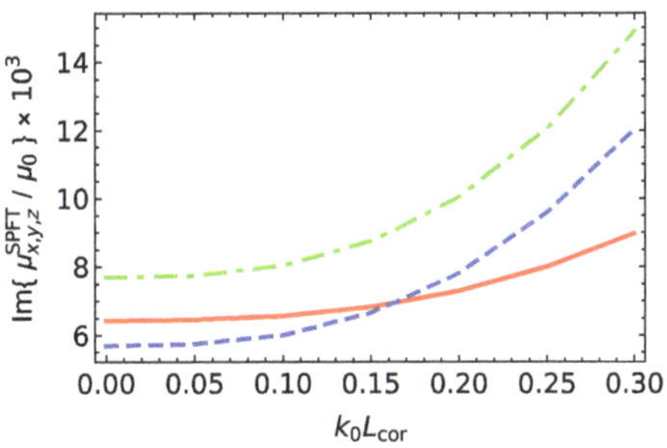

B.6 Figure 6.7

Here follows the code for the Bruggeman formalism in the case of a Faraday chiral material as the HCM, as used for figure 6.7.

```
(* Inputs : Bruggeman for Faraday chiral material *)
I33 = IdentityMatrix[3]; I66 = IdentityMatrix[6];
ϵa = (5 + 0.04I)I33; ξa = (0.2 + 0.003I) I I33; ζa = −ξa; μa = (2 + 0.02I)I33;
ϵb = (1.5 + 0.01I)I33; ξb = 0 I33; ζb = 0 I33;
μb = {{3.5 + 0.02I, (2 + 0.01I)I, 0}, {−(2 + 0.01I)I, 3.5 + 0.02I, 0}, {0, 0, 1 + 0.005I}};
U = 5;
Ua = DiagonalMatrix[{1, 1, U}] / ∛U ;
Ub = DiagonalMatrix[{1, 1, U}] / ∛U ;
Ka = {{ϵa[[1, 1]], ϵa[[1, 2]], ϵa[[1, 3]], ξa[[1, 1]], ξa[[1, 2]], ξa[[1, 3]]},
{ϵa[[2, 1]], ϵa[[2, 2]], ϵa[[2, 3]], ξa[[2, 1]], ξa[[2, 2]], ξa[[2, 3]]},
{ϵa[[3, 1]], ϵa[[3, 2]], ϵa[[3, 3]], ξa[[3, 1]], ξa[[3, 2]], ξa[[3, 3]]},
{ζa[[1, 1]], ζa[[1, 2]], ζa[[1, 3]], μa[[1, 1]], μa[[1, 2]], μa[[1, 3]]},
{ζa[[2, 1]], ζa[[2, 2]], ζa[[2, 3]], μa[[2, 1]], μa[[2, 2]], μa[[2, 3]]},
{ζa[[3, 1]], ζa[[3, 2]], ζa[[3, 3]], μa[[3, 1]], μa[[3, 2]], μa[[3, 3]]}};
Kb = {{ϵb[[1, 1]], ϵb[[1, 2]], ϵb[[1, 3]], ξb[[1, 1]], ξb[[1, 2]], ξb[[1, 3]]},
{ϵb[[2, 1]], ϵb[[2, 2]], ϵb[[2, 3]], ξb[[2, 1]], ξb[[2, 2]], ξb[[2, 3]]},
{ϵb[[3, 1]], ϵb[[3, 2]], ϵb[[3, 3]], ξb[[3, 1]], ξb[[3, 2]], ξb[[3, 3]]},
{ζb[[1, 1]], ζb[[1, 2]], ζb[[1, 3]], μb[[1, 1]], μb[[1, 2]], μb[[1, 3]]},
{ζb[[2, 1]], ζb[[2, 2]], ζb[[2, 3]], μb[[2, 1]], μb[[2, 2]], μb[[2, 3]]},
{ζb[[3, 1]], ζb[[3, 2]], ζb[[3, 3]], μb[[3, 1]], μb[[3, 2]], μb[[3, 3]]}};

(* Initialization of arrays for plotting *)
fmaxcount = 21;
plotReϵBr = Array[0&, {fmaxcount, 2}];
plotReϵBrg = Array[0&, {fmaxcount, 2}];
plotReϵBrz = Array[0&, {fmaxcount, 2}];
plotIϵBr = Array[0&, {fmaxcount, 2}];
plotIϵBrg = Array[0&, {fmaxcount, 2}];
plotIϵBrz = Array[0&, {fmaxcount, 2}];
plotReϵBr[[1]] = {0, Re[ϵb[[1, 1]]]};
plotReϵBrg[[1]] = {0, Re[ϵb[[1, 2]]/(−I)] × 10^5};
plotReϵBrz[[1]] = {0, Re[ϵb[[3, 3]]]};
plotIϵBr[[1]] = {0, Im[ϵb[[1, 1]]] × 10^2};
plotIϵBrg[[1]] = {0, Im[ϵb[[1, 2]]/(−I)] × 10^6};
plotIϵBrz[[1]] = {0, Im[ϵb[[3, 3]]] × 10^2};
plotReϵBr[[fmaxcount]] = {1, Re[ϵa[[1, 1]]]};
plotReϵBrg[[fmaxcount]] = {1, Re[ϵa[[1, 2]]/(−I)] × 10^5};
```

```
plotIϵBrz[[fmaxcount]] = {1, Im[ϵa[[3, 3]]] × 10^2};
plotRξBr = Array[0&, {fmaxcount, 2}];
plotRξBrg = Array[0&, {fmaxcount, 2}];
plotRξBrz = Array[0&, {fmaxcount, 2}];
plotIξBr = Array[0&, {fmaxcount, 2}];
plotIξBrg = Array[0&, {fmaxcount, 2}];
plotIξBrz = Array[0&, {fmaxcount, 2}];
plotRξBr[[1]] = {0, Re[ξb[[1, 1]]/I] × 10};
plotRξBrg[[1]] = {0, Re[ξb[[1, 2]]] × 10^2};
plotRξBrz[[1]] = {0, Re[ξb[[3, 3]]/I] × 10};
plotIξBr[[1]] = {0, Im[ξb[[1, 1]]/I] × 10^3};
plotIξBrg[[1]] = {0, Im[ξb[[1, 2]]] × 10^3};
plotIξBrz[[1]] = {0, Im[ξb[[3, 3]]/I] × 10^3};
plotRξBr[[fmaxcount]] = {1, Re[ξa[[1, 1]]/I] × 10};
plotRξBrg[[fmaxcount]] = {1, Re[ξa[[1, 2]]] × 10^2};
plotRξBrz[[fmaxcount]] = {1, Re[ξa[[3, 3]]/I] × 10};
plotIξBr[[fmaxcount]] = {1, Im[ξa[[1, 1]]/I] × 10^3};
plotIξBrg[[fmaxcount]] = {1, Im[ξa[[1, 2]]] × 10^3};
plotIξBrz[[fmaxcount]] = {1, Im[ξa[[3, 3]]/I] × 10^3};
plotRμBr = Array[0&, {fmaxcount, 2}];
plotRμBrg = Array[0&, {fmaxcount, 2}];
plotRμBrz = Array[0&, {fmaxcount, 2}];
plotIμBr = Array[0&, {fmaxcount, 2}];
plotIμBrg = Array[0&, {fmaxcount, 2}];
plotIμBrz = Array[0&, {fmaxcount, 2}];
plotRμBr[[1]] = {0, Re[μb[[1, 1]]]};
plotRμBrg[[1]] = {0, Re[μb[[1, 2]]/(−I)]};
plotRμBrz[[1]] = {0, Re[μb[[3, 3]]]};
plotIμBr[[1]] = {0, Im[μb[[1, 1]]] × 10^2};
plotIμBrg[[1]] = {0, Im[μb[[1, 2]]/(−I)] × 10^2};
plotIμBrz[[1]] = {0, Im[μb[[3, 3]]] × 10^2};
plotRμBr[[fmaxcount]] = {1, Re[μa[[1, 1]]]};
```

```
plotRμBrg[[fmaxcount]] = {1, Re[μa[[1, 2]]/(−I)]};
plotRμBrz[[fmaxcount]] = {1, Re[μa[[3, 3]]]};
plotIμBr[[fmaxcount]] = {1, Im[μa[[1, 1]]] × 10^2};
plotIμBrg[[fmaxcount]] = {1, Im[μa[[1, 2]]/(−I)] × 10^2};
plotIμBrz[[fmaxcount]] = {1, Im[μa[[3, 3]]] × 10^2};

(* Depolarization dyadic *)
depol[K_, U_]:=Module[{ϵ, ξ, ζ, μ, iU, q, nϵ, nξ, nζ, nμ, b, θ, ϕ, dϵ, dϵz, dξ, dξz, dμ, dμz,
eed, emd, med, mmd},
ϵ = {{K[[1, 1]], K[[1, 2]], K[[1, 3]]}, {K[[2, 1]], K[[2, 2]], K[[2, 3]]}, {K[[3, 1]], K[[3, 2]], K[[3, 3]]}};
ξ = {{K[[1, 4]], K[[1, 5]], K[[1, 6]]}, {K[[2, 4]], K[[2, 5]], K[[2, 6]]}, {K[[3, 4]], K[[3, 5]], K[[3, 6]]}};
ζ = {{K[[4, 1]], K[[4, 2]], K[[4, 3]]}, {K[[5, 1]], K[[5, 2]], K[[5, 3]]}, {K[[6, 1]], K[[6, 2]], K[[6, 3]]}};
μ = {{K[[4, 4]], K[[4, 5]], K[[4, 6]]}, {K[[5, 4]], K[[5, 5]], K[[5, 6]]}, {K[[6, 4]], K[[6, 5]], K[[6, 6]]}};
iU = Inverse[U];
q = {Sin[θ] Cos[ϕ], Sin[θ] Sin[ϕ], Cos[θ]};
nϵ = q.(iU.ϵ.iU).q;
nξ = q.(iU.ξ.iU).q;
nζ = q.(iU.ζ.iU).q;
nμ = q.(iU.μ.iU).q;
b = (nϵ nμ) − (nξ nζ);
dϵ = NIntegrate[(nμ q[[1]] q[[1]] Sin[θ])/b, {θ, 0, Pi}, {ϕ, 0, 2 Pi}];
dϵz = NIntegrate[(nμ q[[3]] q[[3]] Sin[θ])/b, {θ, 0, Pi}, {ϕ, 0, 2 Pi}];
eed = iU.{{dϵ, 0, 0}, {0, dϵ, 0}, {0, 0, dϵz}}.iU/(4 Pi);
dξ = −NIntegrate[(nξ q[[1]] q[[1]] Sin[θ])/b, {θ, 0, Pi}, {ϕ, 0, 2 Pi}];
dξz = −NIntegrate[(nξ q[[3]] q[[3]] Sin[θ])/b, {θ, 0, Pi}, {ϕ, 0, 2 Pi}];
emd = iU.{{dξ, 0, 0}, {0, dξ, 0}, {0, 0, dξz}}.iU/(4 Pi);
med = −emd;
dμ = NIntegrate[(nϵ q[[1]] q[[1]] Sin[θ])/b, {θ, 0, Pi}, {ϕ, 0, 2 Pi}];
dμz = NIntegrate[(nϵ q[[3]] q[[3]] Sin[θ])/b, {θ, 0, Pi}, {ϕ, 0, 2 Pi}];
mmd = iU.{{dμ, 0, 0}, {0, dμ, 0}, {0, 0, dμz}}.iU/(4 Pi);
{{eed[[1, 1]], eed[[1, 2]], eed[[1, 3]], emd[[1, 1]], emd[[1, 2]], emd[[1, 3]]},
{eed[[2, 1]], eed[[2, 2]], eed[[2, 3]], emd[[2, 1]], emd[[2, 2]], emd[[2, 3]]},
```

```
{eed[[3, 1]], eed[[3, 2]], eed[[3, 3]], emd[[3, 1]], emd[[3, 2]], emd[[3, 3]]},
{med[[1, 1]], med[[1, 2]], med[[1, 3]], mmd[[1, 1]], mmd[[1, 2]], mmd[[1, 3]]},
{med[[2, 1]], med[[2, 2]], med[[2, 3]], mmd[[2, 1]], mmd[[2, 2]], mmd[[2, 3]]},
{med[[3, 1]], med[[3, 2]], med[[3, 3]], mmd[[3, 1]], mmd[[3, 2]], mmd[[3, 3]]}}]

(* Initial values for iteration *)
KBr = {{fb μb[[1, 1]], fb μb[[1, 2]], fb μb[[1, 3]], .1 fb μb[[1, 1]], .1 fb μb[[1, 2]], .1 fb μb[[1, 3]]},
{fb μb[[2, 1]], fb μb[[2, 2]], fb μb[[2, 3]], .1 fb μb[[2, 1]], .1 fb μb[[2, 2]], .1 fb μb[[2, 3]]},
{fb μb[[3, 1]], fb μb[[3, 2]], fb μb[[3, 3]], .1 fb μb[[3, 1]], .1 fb μb[[3, 2]], .1 fb μb[[3, 3]]},
{−.1 fb μb[[1, 1]], −.1 fb μb[[1, 2]], −.1 fb μb[[1, 3]], .5 fb μb[[1, 1]], .5 fbμb[[1, 2]], .5 fb μb[[1, 3]]},
{−.1 fb μb[[2, 1]], −.1 fb μb[[2, 2]], −.1 fb μb[[2, 3]], .5 fb μb[[2, 1]], .5 fb μb[[2, 2]], .5 fb μb[[2, 3]]},
{−.1 fb μb[[3, 1]], −.1 fb μb[[3, 2]], −.1 fb μb[[3, 3]], .5 fb μb[[3, 1]], .5 fb μb[[3, 2]], .5 fb μb[[3, 3]]}}/.
fb → 0.5;

tKBr[K_]:=Module[{ta, tb},
ta = fa Inverse[I66 + depol[K, Ua].(Ka − K)]; tb = fb Inverse[I66 + depol[K, Ub].(Kb − K)];
(Ka.ta + Kb.tb).Inverse[ta + tb]];
tol = 0.000001;

For[fcount = 1, fcount < fmaxcount − 1, fcount++,
fa = fcount/(fmaxcount − 1); fb = 1 − fa; delta = 1; n = 1;
For[ncount = 1, ncount < 25, ncount++,
If[delta > tol,
newKBr = tKBr[KBr];
dK = newKBr − KBr;
delta = Max[{Abs[dK][[1, 1]]/Abs[KBr][[1, 1]], Abs[dK][[1, 2]]/Abs[KBr][[1, 2]],
Abs[dK][[3, 3]]/Abs[KBr][[3, 3]], Abs[dK][[1, 4]]/Abs[KBr][[1, 4]], Abs[dK][[1, 5]]/Abs[KBr][[1, 5]],
Abs[dK][[3, 6]]/Abs[KBr][[3, 6]], Abs[dK][[4, 4]]/Abs[KBr][[4, 4]], Abs[dK][[4, 5]]/Abs[KBr][[4, 5]],
Abs[dK][[6, 6]]/Abs[KBr][[6, 6]]}];
KBr = newKBr;
n++;]];
Print ["fa = ", N[fa], ", no. of interations = ", n, ", delta = ", delta];
plotRεBr[[fcount + 1]] = {fa, Re[KBr[[1, 1]]]};
```

```
plotRϵBrg[[fcount + 1]] = {fa, Re[KBr[[1, 2]]/(−I)] × 10^5};
plotRϵBrz[[fcount + 1]] = {fa, Re[KBr[[3, 3]]]};
plotIϵBr[[fcount + 1]] = {fa, Im[KBr[[1, 1]]] × 10^2};
plotIϵBrg[[fcount + 1]] = {fa, Im[KBr[[1, 2]]/(−I)] × 10^6};
plotIϵBrz[[fcount + 1]] = {fa, Im[KBr[[3, 3]]] × 10^2};
plotRξBr[[fcount + 1]] = {fa, Re[KBr[[1, 4]]/I] × 10};
plotRξBrg[[fcount + 1]] = {fa, Re[KBr[[1, 5]]] × 10^2};
plotRξBrz[[fcount + 1]] = {fa, Re[KBr[[3, 6]]/I] × 10};
plotIξBr[[fcount + 1]] = {fa, Im[KBr[[1, 4]]/I] × 10^3};
plotIξBrg[[fcount + 1]] = {fa, Im[KBr[[1, 5]]] × 10^3};
plotIξBrz[[fcount + 1]] = {fa, Im[KBr[[3, 6]]/I] × 10^3};
plotRμBr[[fcount + 1]] = {fa, Re[KBr[[4, 4]]]};
plotRμBrg[[fcount + 1]] = {fa, Re[KBr[[4, 5]]/(−I)]};
plotRμBrz[[fcount + 1]] = {fa, Re[KBr[[6, 6]]]};
plotIμBr[[fcount + 1]] = {fa, Im[KBr[[4, 4]]] × 10^2};
plotIμBrg[[fcount + 1]] = {fa, Im[KBr[[4, 5]]/(−I)] × 10^2};
plotIμBrz[[fcount + 1]] = {fa, Im[KBr[[6, 6]]] × 10^2}; ]

pRe = ListPlot[plotRϵBr, Joined → True, PlotStyle → {Red, Thickness[0.008]}];
pReg = ListPlot[plotRϵBrg, Joined → True,
PlotStyle → {Blue, Dashing[{0.02, 0.02}], Thickness[0.008]}];
pRez = ListPlot[plotRϵBrz, Joined → True,
PlotStyle → {Green, Dashing[{0.035, 0.025, 0.005, 0.025}], Thickness[0.008]}];
pIe = ListPlot[plotIϵBr, Joined → True, PlotStyle → {Red, Thickness[0.008]}];
pIeg = ListPlot[plotIϵBrg, Joined → True, PlotStyle → {Blue, Dashing[{0.02, 0.02}], Thickness[0.008]}];
pIez = ListPlot[plotIϵBrz, Joined → True,
PlotStyle → {Green, Dashing[{0.035, 0.025, 0.005, 0.025}], Thickness[0.008]}];

pR = Show[Graphics[pRe], Graphics[pReg], Graphics[pRez],
Frame → True, Axes → False, PlotRange → All,
FrameLabel → {"fa", "Re{ ϵ1^Br/ ϵ0 }", "Re{ ϵ2^Br/ ϵ0 }", "Re{ ϵ3^Br/ ϵ0 } × 10^5"},
BaseStyle → {FontSize → 14, FontFamily → "Myriad"}]
```

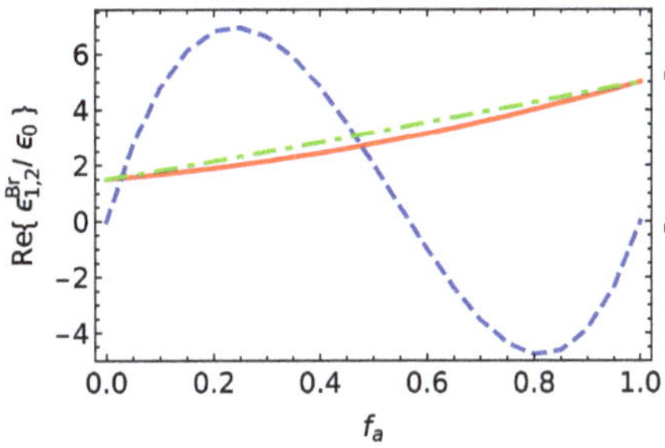


```
pI = Show[Graphics[pIe], Graphics[pIeg], Graphics[pIez],
Frame → True, Axes → False, PlotRange → All,
FrameLabel → {"f_a", "Im{ ϵ_1^Br/ ϵ_0 } × 10^2", "Im{ ϵ_2^Br/ ϵ_0 } × 10^2", "Im{ ϵ_3^Br/ ϵ_0 } × 10^6"},
BaseStyle → {FontSize → 14, FontFamily → "Myriad"}]
```

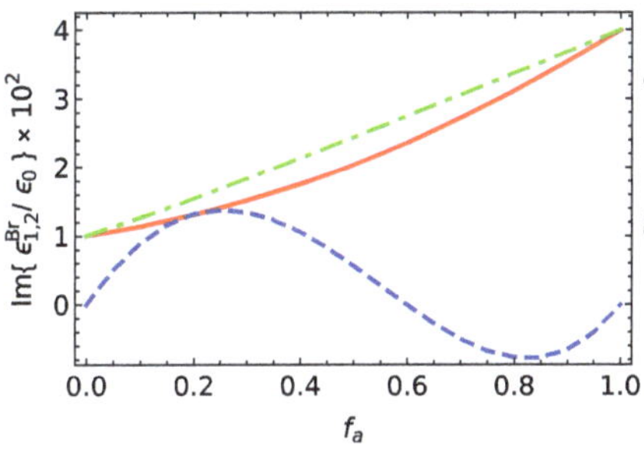


```
pRx = ListPlot[plotRξBr, Joined → True, PlotStyle → {Red, Thickness[0.008]}];
pRxg = ListPlot[plotRξBrg, Joined → True,
PlotStyle → {Blue, Dashing[{0.02, 0.02}], Thickness[0.008]}];
pRxz = ListPlot[plotRξBrz, Joined → True,
PlotStyle → {Green, Dashing[{0.035, 0.025, 0.005, 0.025}], Thickness[0.008]}];
pIx = ListPlot[plotIξBr, Joined → True, PlotStyle → {Red, Thickness[0.008]}];
pIxg = ListPlot[plotIξBrg, Joined → True, PlotStyle → {Blue, Dashing[{0.02, 0.02}], Thickness[0.008]}];
pIxz = ListPlot[plotIξBrz, Joined → True,
PlotStyle → {Green, Dashing[{0.035, 0.025, 0.005, 0.025}], Thickness[0.008]}];

pR = Show[Graphics[pRx], Graphics[pRxg], Graphics[pRxz],
Frame → True, Axes → False, PlotRange → All,
FrameLabel → {"f_a", "Re{ ξ_1^Br/ i√(ϵ_0μ_0) } × 10", "Re{ ξ_2^Br/ i√(ϵ_0μ_0) } × 10",
"Re{ ξ_3^Br/ i√(ϵ_0μ_0) } × 10^2"}, BaseStyle → {FontSize → 14, FontFamily → "Myriad"}]
```

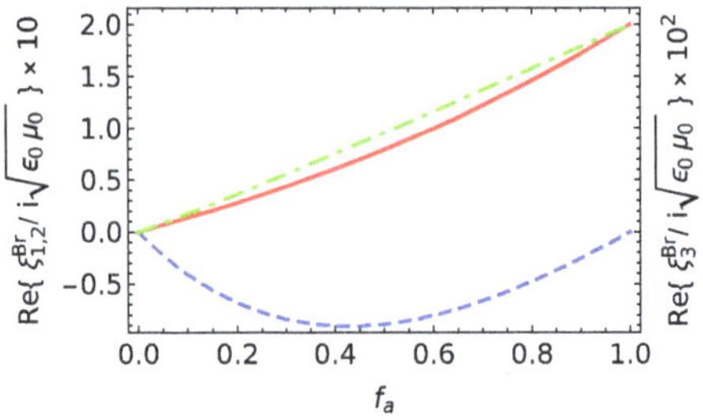

pI = Show[Graphics[pIx], Graphics[pIxg], Graphics[pIxz],

Frame → True, Axes → False, PlotRange → All,

FrameLabel → {"f_a", "Im{ ξ_1^{Br}/ i$\sqrt{\epsilon_0\mu_0}$ } × 10^3", "Im{ ξ_2^{Br}/ i$\sqrt{\epsilon_0\mu_0}$ } × 10^3",

"Im{ ξ_3^{Br}/ i$\sqrt{\epsilon_0\mu_0}$ } × 10^3"} , BaseStyle → {FontSize → 14, FontFamily → "Myriad"}]

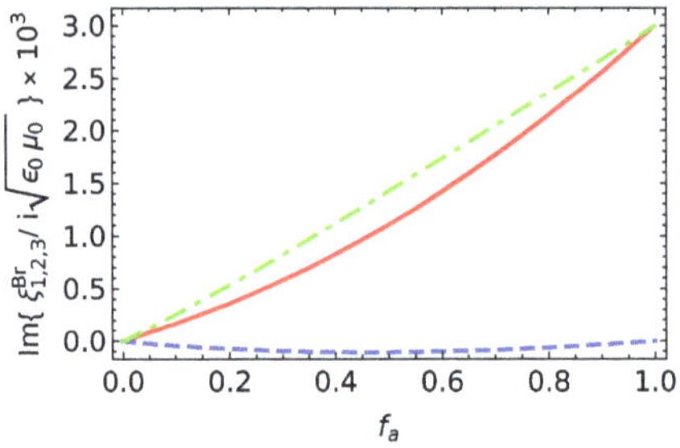

pRm = ListPlot[plotRμBr, Joined → True, PlotStyle → {Red, Thickness[0.008]}];

pRmg = ListPlot[plotRμBrg, Joined → True,

PlotStyle → {Blue, Dashing[{0.02, 0.02}], Thickness[0.008]}];

pRmz = ListPlot[plotRμBrz, Joined → True,

PlotStyle → {Green, Dashing[{0.035, 0.025, 0.005, 0.025}], Thickness[0.008]}];

pIm = ListPlot[plotIμBr, Joined → True, PlotStyle → {Red, Thickness[0.008]}];

pImg = ListPlot[plotIμBrg, Joined → True, PlotStyle → {Blue, Dashing[{0.02, 0.02}], Thickness[0.008]}];

pImz = ListPlot[plotIμBrz, Joined → True,

PlotStyle → {Green, Dashing[{0.035, 0.025, 0.005, 0.025}], Thickness[0.008]}];

pR = Show[Graphics[pRm], Graphics[pRmg], Graphics[pRmz],

Frame → True, Axes → False, PlotRange → All,

FrameLabel → {"f_a", "Re{ μ_1^{Br}/ μ_0 }", "Re{ μ_2^{Br}/ μ_0 }", "Re{ μ_3^{Br}/ μ_0 }"} ,

BaseStyle → {FontSize → 14, FontFamily → "Myriad"}]

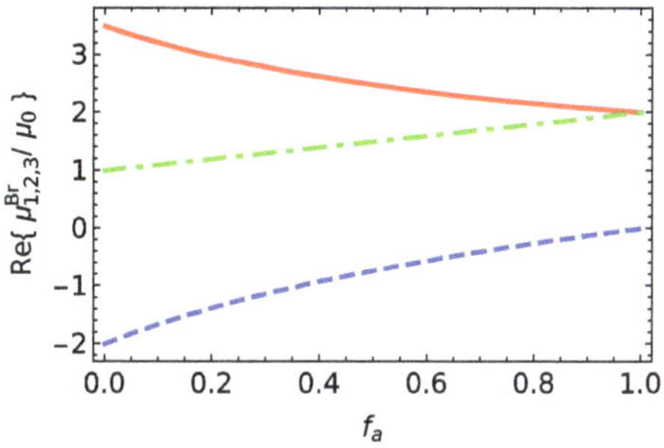


```
pI = Show[Graphics[pIm], Graphics[pImg], Graphics[pImz],
Frame → True, Axes → False, PlotRange → All,
FrameLabel → {"fa", "Im{ μ1^Br/ μ0 } × 10^2", "Im{ μ2^Br/ μ0 } × 10^2", "Im{ μ3^Br/ μ0 } × 10^2"},
BaseStyle → {FontSize → 14, FontFamily → "Myriad"}]
```

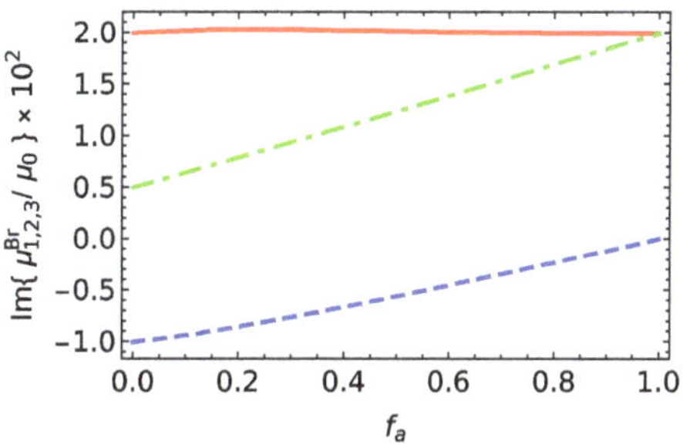

B.7 Figure 6.8

Here follows the code for the Bruggeman formalism in the case of a uniaxial dielectric HCM arising from disk-shaped constituent particles, as used for figure 6.8.

```
(* Bruggeman : uniaxial dielectric HCM from disk-shaped particles *)
ϵ1 = (fa ϵa) + (fb ϵb);
ϵ2 = ϵa ϵb / ((fb ϵa)+(fa ϵb));
fb = 1 − fa;
r = ϵ1/ϵ2;

rp1 = Plot3D[r/.ϵa → 2, {ϵb, .1, 0.5}, {fa, 0, 1}, MaxRecursion → 0, PlotPoints → 30, Mesh → None,
PlotRange → All, AxesLabel → {" ϵb / ϵ0", " fa ", "ϵ⊥^Br/ϵ∥^Br "},
BaseStyle → {FontSize → 14, FontFamily → "Myriad"}]
```

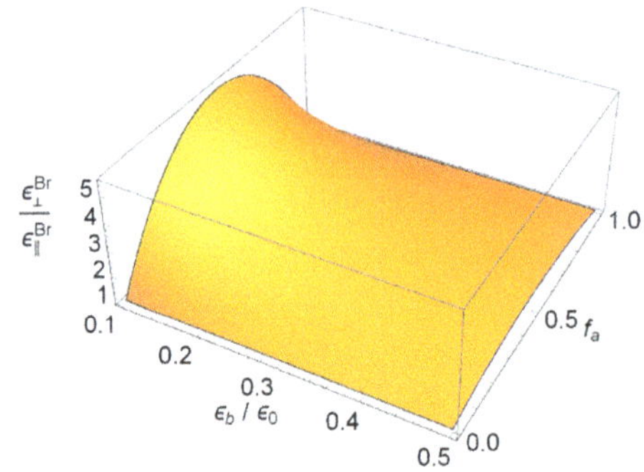


```
rp2 = Plot3D[r/.εa → 2, {εb, 1, 200}, {fa, 0, 1}, MaxRecursion → 0, PlotPoints → 30, Mesh → None,
AxesLabel → {" εb / ε0 ", " fa ", " ε⊥Br/ε∥Br "}, BaseStyle → {FontSize → 14, FontFamily → "Myriad"}]
```

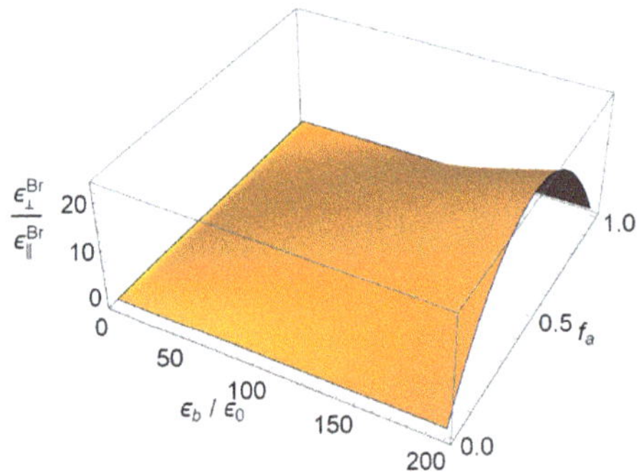

B.8 Figure 6.9

Here follows the code for the Bruggeman formalism in the case of a uniaxial dielectric HCM arising from needle-shaped constituent particles, as used for figure 6.9.

```
(* Bruggeman : uniaxial dielectric HCM from needle-shaped particles *)
ε1 = (((fb−fa) (εb−εa))+√(((fb−fa) (εb−εa))^2+(4 εa εb)))/2;

ε2 = (fa εa) + (fb εb);

fb = 1 − fa;

r = ε2/ε1;


rp1 = Plot3D[r/.εa → 2, {εb, .1, 0.5}, {fa, 0, 1}, MaxRecursion → 0, PlotPoints → 30, Mesh → None,
PlotRange → All, AxesLabel → {" εb / ε0 ", " fa ", " ε∥Br/ε⊥Br "},
BaseStyle → {FontSize → 14, FontFamily → "Myriad"}]
```

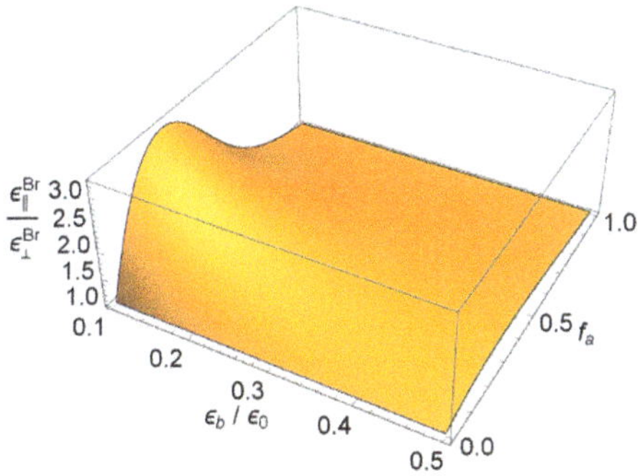

rp2 = Plot3D[r/.ϵa → 2, {ϵb, 1, 200}, {fa, 0, 1}, MaxRecursion → 0, PlotPoints → 30, Mesh → None, AxesLabel → {" ϵ_b / ϵ_0", " f_a ", " $\frac{\epsilon_{\parallel}^{Br}}{\epsilon_{\perp}^{Br}}$ "}, BaseStyle → {FontSize → 14, FontFamily → "Myriad"}]

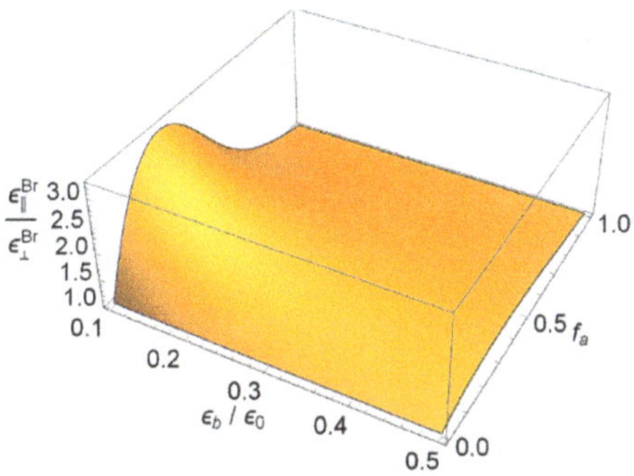

B.9 Figure 6.11

Here follow the codes for the Bruggeman formalism, and the Hashin–Shtrikman and Wiener bounds, in the case of an isotropic dielectric HCM, as used for figure 6.11. The codes used for figures 6.12, 6.13, and 6.14 are straightforward modifications of these codes. Also, straightforward modifications of the Bruggeman code were used for figures 6.15, 6.16, 6.17, 6.18, and 6.19.

```
(* Bruggeman, Hashin-Shtrikman & Wiener for isotropic dielectric HCM *)
α = 2;
β = (ϵb ((3 f) − 2)) + (ϵa (1 − (3 f)));
γ = −ϵa ϵb;
disc = (β β) − (4 α γ);
ϵp = (−β + Sqrt[disc])/(2 α);
ϵm = (−β − Sqrt[disc])/(2 α);
ϵHSL = ϵb + ((3 f ϵb (ϵa − ϵb))/(ϵa + ϵb + ϵb − (f (ϵa − ϵb))));
ϵHSU = ϵa + ((3 (1 − f) ϵa (ϵb − ϵa))/(ϵb + ϵa + ϵa − ((1 − f) (ϵb − ϵa))));
ϵWU = (f ϵa) + ((1 − f) ϵb);
ϵWL = 1/((f/ϵa) + ((1 − f)/ϵb));
ϵb = 6; ϵa = 1.5;

prbr = Plot[If[Im[Sqrt[disc]] ≥ 0, Re[(−β + Sqrt[disc])/(2 α)], Re[(−β − Sqrt[disc])/(2 α)]],
{f, 0, 1}, PlotStyle → {Thickness[0.01], RGBColor[1, 0, 0]}];
prHSU = Plot[Re[ϵHSU], {f, 0, 1},
PlotStyle → {Dashing[{0.035, 0.025, 0.005, 0.025}], Thickness[0.01], RGBColor[0, 1, 0]}];
prHSL = Plot[Re[ϵHSL], {f, 0, 1},
PlotStyle → {Dashing[{0.035, 0.025, 0.005, 0.025}], Thickness[0.005], RGBColor[0, 1, 0]}];
prWU = Plot[Re[ϵWU], {f, 0, 1},
PlotStyle → {Dashing[{0.02, 0.02}], Thickness[0.01], RGBColor[0, 0, 1]}];
prWL = Plot[Re[ϵWL], {f, 0, 1},
PlotStyle → {Dashing[{0.02, 0.02}], Thickness[0.005], RGBColor[0, 0, 1]}];
p2lϵa = Text [Style ["ϵa=6 ϵ0", FontSize → 12, FontFamily → "Myriad"] , {0.85, 5.5}] ;
p2lϵb = Text [Style ["ϵb=1.5 ϵ0", FontSize → 12, FontFamily → "Myriad"] , {0.85, 5}] ;

pr = Show[Graphics[prHSU], Graphics[prHSL], Graphics[prWU], Graphics[prWL], Graphics[prbr],
Graphics[p2lϵa], Graphics[p2lϵb],
Frame → True, Axes → False, FrameLabel → {"fa", "ϵBr / ϵ0", "", ""} ,
BaseStyle → {FontSize → 14, FontFamily → "Myriad"}, DisplayFunction → $DisplayFunction]
```

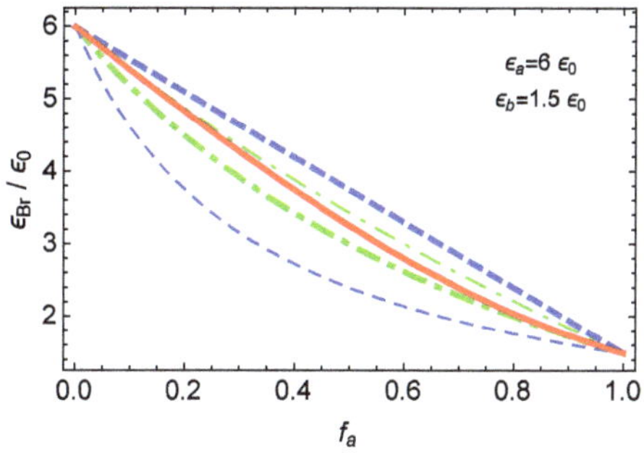

B.10 Figures 6.24 and 6.25

Here follow the codes for the zeroth- and second-order SPFT, in the case of a weakly nonlinear isotropic dielectric HCM, as used for figures 6.24 and 6.25.

```
(* Zeroth- and second-order SPFT for weakly nonlinear isotropic dielectric HCM *)


(* 2nd-order SPFT linear component *)
SPFTL[epsal_, epsb_, vola_, Lcor_]:=Module[{volb, epsBrl, a, b, c},
volb = 1 - vola;

epsBrl = mBruggemanl[epsal, epsb, vola];
a = (Lcor epsBrl)^2;
b = 3 + (2 I Lcor Sqrt[epsBrl]);
c = vola ((epsal-epsBrl)/(epsal+(2 epsBrl)))^2 + volb ((epsb-epsBrl)/(epsb+(2 epsBrl)))^2;
epsBrl + (a b c)]


(* 2nd-order SPFT nonlinear component *)
SPFTNL[epsal_, epsanl_, epsb_, vola_, Lcor_]:=Module[{volb, epsBrl, epsBrnl, gff, a, b, c, d, e},
volb = 1 - vola;
epsBrl = mBruggemanl[epsal, epsb, vola];
epsBrnl = mBruggemannl[epsal, epsanl, epsb, vola];
gff = (Abs[(3 epsBrl)/(epsal+(2 epsBrl))])^2;
a = 2 epsBrl (3 + (2 I Lcor Sqrt[epsBrl]));
b = vola (((epsal-epsBrl)/(epsal+(2 epsBrl)))^2) (((gffepsanl)-epsBrnl)/(epsal-epsBrl) - ((gff epsanl)+2 epsBrnl)/(epsal+(2 epsBrl)));
c = volb (((epsb-epsBrl)/(epsb+(2 epsBrl)))^2) ((-epsBrnl)/(epsb-epsBrl) - (2 epsBrnl)/(epsb+(2 epsBrl)));
```

$d = \text{epsBrnl}\left(6 + \left(5\, I\, \text{Lcor}\sqrt{\text{epsBrl}}\right)\right);$

$e = \left(\text{vola}\left(\left(\frac{\text{epsal}-\text{epsBrl}}{\text{epsal}+(2\,\text{epsBrl})}\right){}^\wedge 2\right)\right) + \left(\text{volb}\left(\left(\frac{\text{epsb}-\text{epsBrl}}{\text{epsb}+(2\,\text{epsBrl})}\right){}^\wedge 2\right)\right);$

$\text{epsBrnl} + (\text{Lcor}{}^\wedge 2\text{epsBrl}((a\,(b+c)) + (d\,e)))]$

(* 0th-order SPFT linear component *)

mBruggemanl[epsal_, epsb_, vola_]:=Module[{volb, epsBrl, a, b, c},

volb = 1 − vola;

$a = 2;$

$b = (\text{epsal}\,(\text{volb} - 2\text{vola})) + (\text{epsb}\,(\text{vola} - 2\text{volb}));$

$c = -\text{epsal}\,\text{epsb};$

$\text{epsBrl} = \frac{-b+\sqrt{b^\wedge 2-(4ac)}}{2a}]$

(* 0th-order SPFT nonlinear component *)

mBruggemannl[epsal_, epsanl_, epsb_, vola_]:=Module[{volb, epsBrl, epsBrnl, gff, deps, a, b, c},

volb = 1 − vola;

epsBrl = mBruggemanl[epsal, epsb, vola];

$\text{gff} = \left(\text{Abs}\left[\frac{3\,\text{epsBrl}}{\text{epsal}+(2\,\text{epsBrl})}\right]\right){}^\wedge 2;$

$\text{deps} = \frac{\text{volaepsBrl(epsb+2 epsBrl) (epsb+2 epsBrl)}}{(\text{vola epsal((epsb+(2 epsBrl))}{}^\wedge 2))+(\text{volb epsb((epsal+(2 epsBrl))}{}^\wedge 2))};$

epsBrnl = gff epsanl deps]

p1 = Plot3D[Re[mBruggemanl[2, eb, f]], {f, 0, 1}, {eb, 3, 15}, MaxRecursion → 0, PlotPoints → 30, Mesh → None, AxesLabel → $\left\{"f_a\ ", "\ \frac{\epsilon_b}{\epsilon_0}", "\frac{\epsilon^L_{\text{Br}}}{\epsilon_0}\ "\right\}$, BaseStyle → {FontSize → 14, FontFamily → "Myriad"}]

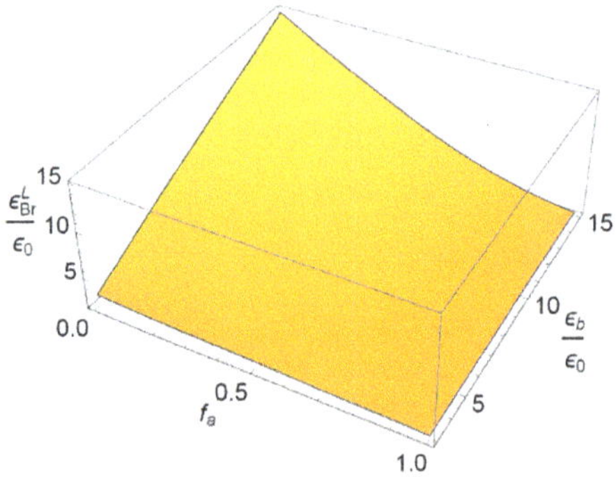

p2 = Plot3D[Re[mBruggemannl[2, 1, eb, f]], {f, 0, 1}, {eb, 3, 15}, MaxRecursion → 0, PlotPoints → 30, Mesh → None, AxesLabel → {" f_a ", " $\frac{\epsilon_b}{\epsilon_0}$ ", " $\frac{\epsilon_{Br}^{NL}}{\epsilon_a^{NL}}$ "}, BaseStyle → {FontSize → 14, FontFamily → "Myriad"}]

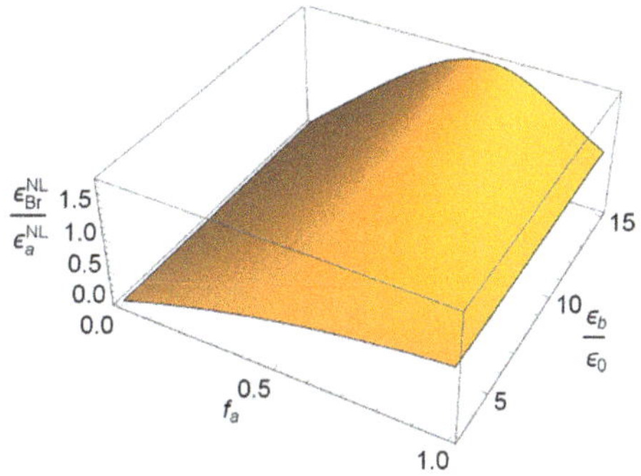

p3 = Plot3D[Re[SPFTL[2, 9, f, L] − mBruggemanl[2, 9, f]], {f, 0, 1}, {L, 0, 0.3}, PlotRange → Full, MaxRecursion → 0, PlotPoints → 30, Mesh → None, AxesLabel → {" f_a ", " $k_0 L_{cor}$", "Re{$\frac{\epsilon_\Delta^L}{\epsilon_0}$} "}, BaseStyle → {FontSize → 14, FontFamily → "Myriad"}]

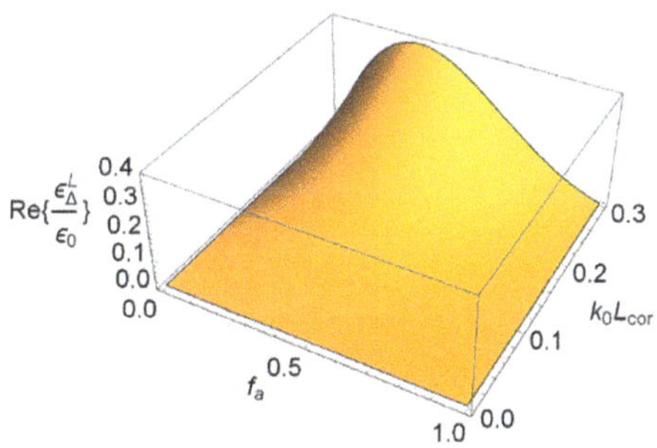

p4 = Plot3D[Im[SPFTL[2, 9, f, L] − mBruggemanl[2, 9, f]], {f, 0, 1}, {L, 0, 0.3}, PlotRange → Full, MaxRecursion → 0, PlotPoints → 30, Mesh → None, AxesLabel → {" f_a ", " $k_0 L_{cor}$", "Im{$\frac{\epsilon_\Delta^L}{\epsilon_0}$} "}, BaseStyle → {FontSize → 14, FontFamily → "Myriad"}]

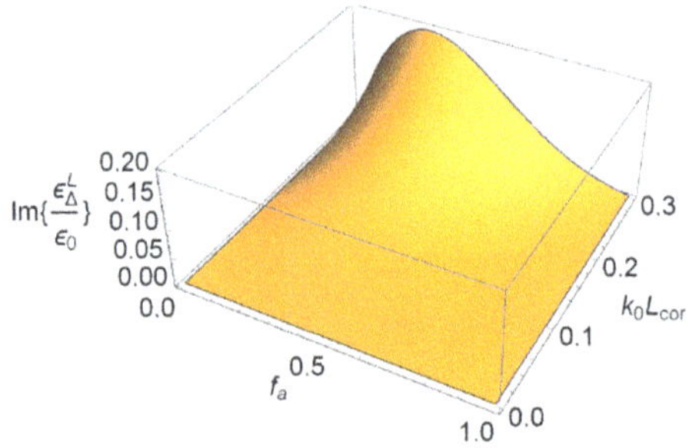


```
p5 = Plot3D[Re[SPFTNL[2, 1, 9, f, L] − mBruggemannl[2, 1, 9, f]],
{f, 0, 1}, {L, 0, 0.3}, PlotRange → {−0.24, 0.015}, MaxRecursion → 0, PlotPoints → 30,
Mesh → None, AxesLabel → {"f_a ", " k_0 L_cor", "Re{ϵ_Δ^NL/ϵ_a^NL} "},
BaseStyle → {FontSize → 14, FontFamily → "Myriad"}]
```

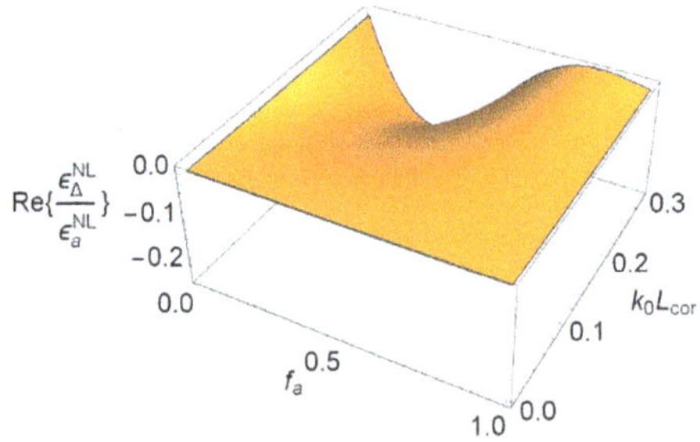


```
p6 = Plot3D[Im[SPFTNL[2, 1, 9, f, L] − mBruggemannl[2, 1, 9, f]],
{f, 0, 1}, {L, 0, 0.3}, PlotRange → Full, MaxRecursion → 0, PlotPoints → 30, Mesh → None,
AxesLabel → {"f_a ", " k_0 L_cor", "Im{ϵ_Δ^NL/ϵ_a^NL} "}, BaseStyle → {FontSize → 14, FontFamily → "Myriad"}]
```

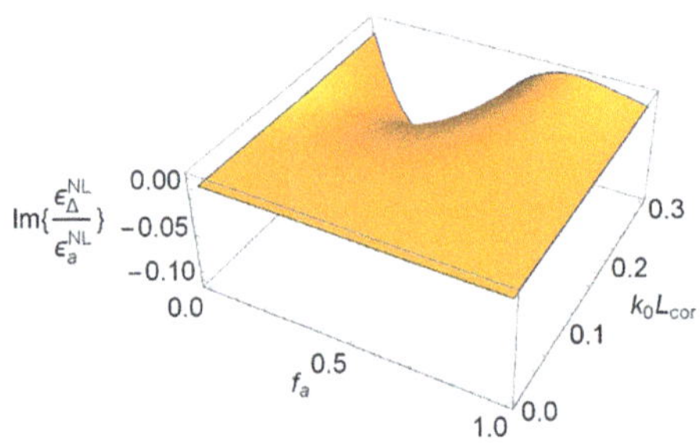

www.ingramcontent.com/pod-product-compliance
Ingram Content Group UK Ltd.
Pitfield, Milton Keynes, MK11 3LW, UK
UKHW051853100926
8556IPUK00005B/609

9 780750 334211